Lecture Notes in Computer Science 9475

Commenced Publication in 1973
Founding and Former Series Editors:
Gerhard Goos, Juris Hartmanis, and Jan van Leeuwen

More information about this series at http://www.springer.com/series/7412

George Bebis · Richard Boyle
Bahram Parvin · Darko Koracin
Ioannis Pavlidis · Rogerio Feris
Tim McGraw · Mark Elendt
Regis Kopper · Eric Ragan
Zhao Ye · Gunther Weber (Eds.)

Advances in Visual Computing

11th International Symposium, ISVC 2015
Las Vegas, NV, USA, December 14–16, 2015
Proceedings, Part II

 Springer

Editors

George Bebis
University of Nevada
Reno, NV, USA

Richard Boyle
NASA Ames Research Center
Moffett Field, CA, USA

Bahram Parvin
Lawrence Berkeley National Laboratory
Berkeley, CA, USA

Darko Koracin
Desert Research Institute
Reno, NV, USA

Ioannis Pavlidis
University of Houston
Houston, TX, USA

Rogerio Feris
IBM T.J. Watson Research Center
Yorktown Heights, NY, USA

Tim McGraw
Purdue University
West Lafayette, IN, USA

Mark Elendt
Side Effects Software
Santa Monica, CA, USA

Regis Kopper
The DiVE
Durham, NC, USA

Eric Ragan
Texas A&M University
College Station, TX, USA

Zhao Ye
Kent State University
Kent, OH, USA

Gunther Weber
Lawrence Berkeley National Laboratory
Berkeley, CA, USA

ISSN 0302-9743   ISSN 1611-3349   (electronic)
Lecture Notes in Computer Science
ISBN 978-3-319-27862-9   ISBN 978-3-319-27863-6   (eBook)
DOI 10.1007/978-3-319-27863-6

Library of Congress Control Number: 2015957779

LNCS Sublibrary: SL6 – Image Processing, Computer Vision, Pattern Recognition, and Graphics

Printed on acid-free paper

This Springer imprint is published by SpringerNature
The registered company is Springer International Publishing AG Switzerland

Preface

It is with great pleasure that we welcome you to the proceedings of the 11th International Symposium on Visual Computing (ISVC 2015), which was held in Las Vegas, Nevada, USA. ISVC provides a common umbrella for the four main areas of visual computing including vision, graphics, visualization, and virtual reality. The goal is to provide a forum for researchers, scientists, engineers, and practitioners throughout the world to present their latest research findings, ideas, developments, and applications in the broader area of visual computing.

This year, the program consisted of 16 oral sessions, one poster session, eight special tracks, and six keynote presentations. The response to the call for papers was very good; we received over 260 submissions for the main symposium from which we accepted 80 papers for oral presentation and 35 papers for poster presentation. Special track papers were solicited separately through the Organizing and Program Committees of each track. A total of 43 papers were accepted for oral presentation in the special tracks.

All papers were reviewed with an emphasis on the potential to contribute to the state of the art in the field. Selection criteria included accuracy and originality of ideas, clarity and significance of results, and presentation quality. The review process was quite rigorous, involving two to three independent blind reviews followed by several days of discussion. During the discussion period we tried to correct anomalies and errors that might have existed in the initial reviews. Despite our efforts, we recognize that some papers worthy of inclusion may have not been included in the program. We offer our sincere apologies to authors whose contributions might have been overlooked.

We wish to thank everybody who submitted their work to ISVC 2015 for review. It was because of their contributions that we succeeded in having a technical program of high scientific quality. In particular, we would like to thank the ISVC 2015 area chairs, the organizing institutions (UNR, DRI, LBNL, and NASA Ames), the industrial sponsors (BAE Systems, Intel, Ford, Hewlett Packard, Mitsubishi Electric Research Labs, Toyota, General Electric), the international Program Committee, the special track organizers and their Program Committees, the keynote speakers, the reviewers, and especially the authors who contributed their work to the symposium. In particular, we would like to express our appreciation to MERL and Drs. Jay Thornton and Mike Jones for sponsoring the "best" paper award this year.

We sincerely hope that ISVC 2015 offered participants opportunities for professional growth.

October 2015

George Bebis
ISVC'15 Steering Committee and Area Chairs

Organization

Steering Committee

Bebis George	University of Nevada, Reno, USA
Boyle Richard	NASA Ames Research Center, USA
Parvin Bahram	Lawrence Berkeley National Laboratory, USA
Koracin Darko	Desert Research Institute, USA

Area Chairs

Computer Vision

Pavlidis Ioannis	University of Houston, USA
Feris Rogerio	IBM, USA

Computer Graphics

McGraw Tim	Purdue University, USA
Elendt Mark	Side Effects Software Inc., USA

Virtual Reality

Kopper Regis	Duke University, USA
Ragan Eric	Texas A&M University, USA

Visualization

Ye Zhao	Kent State University, USA
Weber Gunther	Lawrence Berkeley National Laboratory, USA

Publicity

Erol Ali	Eksperta Software, Turkey

Local Arrangements

Morris Brendan	University of Nevada, Las Vegas, USA

Special Tracks

Wang Junxian	Microsoft, USA

Keynote Speakers

Ravi Ramamoorthi	University of California, San Diego, USA
Benjamin Kimia	Brown University, USA
Claudio Silva	New York University, USA
Oncel Tuzel	Mitsubishi Electric Research Laboratories, USA
Evan Suma	University of Southern California, USA
Luc Vincent	Google, USA

International Program Committee

(Area 1) Computer Vision

Abidi Besma	University of Tennessee at Knoxville, USA
Abou-Nasr Mahmoud	Ford Motor Company, USA
Aboutajdine Driss	National Center for Scientific and Technical Research, Morocco
Aggarwal J.K.	University of Texas, Austin, USA
Albu Branzan Alexandra	University of Victoria, Canada
Amayeh Gholamreza	Foveon, USA
Ambardekar Amol	Microsoft, USA
Angelopoulou Elli	University of Erlangen-Nuremberg, Germany
Agouris Peggy	George Mason University, USA
Argyros Antonis	University of Crete, Greece
Asari Vijayan	University of Dayton, USA
Athitsos Vassilis	University of Texas at Arlington, USA
Basu Anup	University of Alberta, Canada
Bekris Kostas	Rutgers University, USA
Bhatia Sanjiv	University of Missouri – St. Louis, USA
Bimber Oliver	Johannes Kepler University Linz, Austria
Bourbakis Nikolaos	Wright State University, USA
Brimkov Valentin	State University of New York, USA
Cavallaro Andrea	Queen Mary, University of London, UK
Charalampidis Dimitrios	University of New Orleans, USA
Chellappa Rama	University of Maryland, USA
Chen Yang	HRL Laboratories, USA
Cheng Hui	Sarnoff Corporation, USA
Cheng Shinko	HRL Labs, USA
Cui Jinshi	Peking University, China
Dagher Issam	University of Balamand, Lebanon
Darbon Jerome	CNRS-Ecole Normale Superieure de Cachan, France
Demirdjian David	Vecna Robotics, USA
Diamantas Sotirios	Ecole Nationale Superieure de Mecanique et des Microtechniques, France
Duan Ye	University of Missouri – Columbia, USA

Doulamis Anastasios	Technical University of Crete, Greece
Dowdall Jonathan	Google, USA
El-Ansari Mohamed	Ibn Zohr University, Morocco
El-Gammal Ahmed	University of New Jersey, USA
Eng How Lung	Institute for Infocomm Research, Singapore
Erol Ali	ASELSAN, Turkey
Fan Guoliang	Oklahoma State University, USA
Fan Jialue	Northwestern University, USA
Ferri Francesc	Universitat de Valencia, Spain
Ferzli Rony	Intel, USA
Ferryman James	University of Reading, UK
Foresti GianLuca	University of Udine, Italy
Fowlkes Charless	University of California, Irvine, USA
Fukui Kazuhiro	The University of Tsukuba, Japan
Galata Aphrodite	The University of Manchester, UK
Georgescu Bogdan	Siemens, USA
Goh Wooi-Boon	Nanyang Technological University, Singapore
Ghouzali Sanna	King Saud University, Saudi Arabia
Guerra-Filho Gutemberg	Intel, USA
Guevara Angel Miguel	University of Porto, Portugal
Gustafson David	Kansas State University, USA
Hammoud Riad	BAE Systems, USA
Harville Michael	Hewlett Packard Labs, USA
He Xiangjian	University of Technology, Sydney, Australia
Heikkil Janne	University of Oulu, Finland
Hongbin Zha	Peking University, China
Hou Zujun	Institute for Infocomm Research, Singapore
Hua Gang	IBM T.J. Watson Research Center, USA and Stevens Institute, USA
Huang Yongzhen	Chinese Academy of Sciences, China
Imiya Atsushi	Chiba University, Japan
Kamberov George	Stevens Institute of Technology, USA
Kampel Martin	Vienna University of Technology, Austria
Kamberova Gerda	Hofstra University, USA
Kakadiaris Ioannis	University of Houston, USA
Kettebekov Sanzhar	Keane Inc., USA
Kimia Benjamin	Brown University, USA
Kisacanin Branislav	Texas Instruments, USA
Klette Reinhard	Auckland University of Technology, New Zealand
Kokkinos Iasonas	Ecole Centrale de Paris, France
Kollias Stefanos	National Technical University of Athens, Greece
Komodakis Nikos	Ecole Centrale de Paris, France
Kosmopoulos Dimitrios	Technical Educational Institute of Crete, Greece
Kozintsev Igor	Intel, USA
Kuno Yoshinori	Saitama University, Japan
Kim Kyungnam	HRL Laboratories, USA

Latecki Longin Jan	Temple University, USA
Lee D.J.	Brigham Young University, USA
Levine Martin	McGill University, Canada
Li Baoxin	Arizona State University, USA
Li Chunming	Vanderbilt University, USA
Li Xiaowei	Google Inc., USA
Lim Ser N.	GE Research, USA
Lisin Dima	VidoeIQ, USA
Lee Hwee Kuan	Bioinformatics Institute, A*STAR, Singapore
Lee Seong-Whan	Korea University, Korea
Li Shuo	GE Healthcare, Canada
Lourakis Manolis	ICS-FORTH, Greece
Loss Leandro	Lawrence Berkeley National Laboratory, USA
Luo Gang	Harvard University, USA
Ma Yunqian	Honyewell Labs, USA
Maeder Anthony	University of Western Sydney, Australia
Makrogiannis Sokratis	Delaware State University, USA
Maltoni Davide	University of Bologna, Italy
Maroulis Dimitris	National University of Athens, Greece
Maybank Steve	Birkbeck College, UK
Medioni Gerard	University of Southern California, USA
Melenchn Javier	Universitat Oberta de Catalunya, Spain
Metaxas Dimitris	Rutgers University, USA
Ming Wei	Konica Minolta Laboratory, USA
Mirmehdi Majid	Bristol University, UK
Morris Brendan	University of Nevada, Las Vegas, USA
Mueller Klaus	Stony Brook University, USA
Muhammad Ghulam	King Saud University, Saudi Arabia
Mulligan Jeff	NASA Ames Research Center, USA
Murray Don	Point Grey Research, Canada
Nait-Charif Hammadi	Bournemouth University, UK
Nefian Ara	NASA Ames Research Center, USA
Nguyen Quang Vinh	University of Western Sydney, Australia
Nicolescu Mircea	University of Nevada, Reno, USA
Nixon Mark	University of Southampton, UK
Nolle Lars	The Nottingham Trent University, UK
Ntalianis Klimis	National Technical University of Athens, Greece
Or Siu Hang	The Chinese University of Hong Kong, Hong Kong, SAR China
Papadourakis George	Technological Education Institute, Greece
Papanikolopoulos Nikolaos	University of Minnesota, USA
Pati Peeta Basa	CoreLogic, India
Patras Ioannis	Queen Mary University, London, UK
Pavlidis Ioannis	University of Houston, USA
Petrakis Euripides	Technical University of Crete, Greece
Peyronnet Sylvain	LRI, University of Paris-Sud, France

Pinhanez Claudio	IBM Research, Brazil
Piccardi Massimo	University of Technology, Australia
Pietikainen Matti	LRDE/University of Oulu, Finland
Pitas Ioannis	Aristotle University of Thessaloniki, Greece
Porikli Fatih	Australian National University, Australia
Prabhakar Salil	DigitalPersona Inc., USA
Prokhorov Danil	Toyota Research Institute, USA
Qian Gang	Arizona State University, USA
Raftopoulos Kostas	National Technical University of Athens, Greece
Regazzoni Carlo	University of Genoa, Italy
Regentova Emma	University of Nevada, Las Vegas, USA
Remagnino Paolo	Kingston University, UK
Ribeiro Eraldo	Florida Institute of Technology, USA
Robles-Kelly Antonio	National ICT Australia (NICTA), Australia
Ross Arun	Michigan State University, USA
Rziza Mohammed	Agdal Mohammed-V University, Morocco
Samal Ashok	University of Nebraska, USA
Samir Tamer	Allegion, USA
Sandberg Kristian	Computational Solutions, USA
Sarti Augusto	DEI Politecnico di Milano, Italy
Savakis Andreas	Rochester Institute of Technology, USA
Schaefer Gerald	Loughborough University, UK
Scalzo Fabien	University of California at Los Angeles, USA
Scharcanski Jacob	UFRGS, Brazil
Shah Mubarak	University of Central Florida, USA
Shi Pengcheng	Rochester Institute of Technology, USA
Shimada Nobutaka	Ritsumeikan University, Japan
Singh Rahul	San Francisco State University, USA
Skodras Athanassios	University of Patras, Greece
Skurikhin Alexei	Los Alamos National Laboratory, USA
Souvenir Richard	University of North Carolina – Charlotte, USA
Su Chung-Yen	National Taiwan Normal University, Taiwan (R.O.C.)
Sugihara Kokichi	University of Tokyo, Japan
Sun Zehang	Apple, USA
Syeda-Mahmood Tanveer	IBM Almaden, USA
Tan Kar Han	Hewlett Packard, USA
Tavakkoli Alireza	University of Houston – Victoria, USA
Tavares Joao	Universidade do Porto, Portugal
Teoh Eam Khwang	Nanyang Technological University, Singapore
Thiran Jean-Philippe	Swiss Federal Institute of Technology Lausanne (EPFL), Switzerland
Tistarelli Massimo	University of Sassari, Italy
Tong Yan	University of South Carolina, USA

Tsui T.J.	Chinese University of Hong Kong, Hong Kong, SAR China
Trucco Emanuele	University of Dundee, UK
Tubaro Stefano	Politecnico di Milano, Italy
Uhl Andreas	Salzburg University, Austria
Velastin Sergio	Kingston University London, UK
Veropoulos Kostantinos	GE Healthcare, Greece
Verri Alessandro	Università di Genova, Italy
Wang Junxian	Microsoft, USA
Wang Song	University of South Carolina, USA
Wang Yunhong	Beihang University, China
Webster Michael	University of Nevada, Reno, USA
Wolff Larry	Equinox Corporation, USA
Wong Kenneth	The University of Hong Kong, Hong Kong, SAR China
Xiang Tao	Queen Mary, University of London, UK
Xu Meihe	University of California at Los Angeles, USA
Yang Ming-Hsuan	University of California at Merced, USA
Yang Ruigang	University of Kentucky, USA
Yin Lijun	SUNY at Binghampton, USA
Yu Ting	GE Global Research, USA
Yu Zeyun	University of Wisconsin-Milwaukee, USA
Yuan Chunrong	University of Tübingen, Germany
Zabulis Xenophon	ICS-FORTH, Greece
Zervakis Michalis	Technical University of Crete, Greece
Zhang Jian	Wake Forest University, USA
Zheng Yuanjie	University of Pennsylvania, USA
Zhang Yan	Delphi Corporation, USA
Ziou Djemel	University of Sherbrooke, Canada

(Area 2) Computer Graphics

Abd Rahni Mt. Piah	Universiti Sains Malaysia, Malaysia
Abram Greg	Texas Advanced Computing Center, USA
Adamo-Villani Nicoletta	Purdue University, USA
Agu Emmanuel	Worcester Polytechnic Institute, USA
Andres Eric	Laboratory XLIM-SIC, University of Poitiers, France
Artusi Alessandro	GiLab, Universitat de Girona, Spain
Baciu George	Hong Kong Poly, Hong Kong, SAR China
Balcisoy Selim Saffet	Sabanci University, Turkey
Barneva Reneta	State University of New York, USA
Belyaev Alexander	Heriot-Watt University, UK
Benes Bedrich	Purdue University, USA
Berberich Eric	Max Planck Institute, Germany
Bilalis Nicholas	Technical University of Crete, Greece
Bimber Oliver	Johannes Kepler University Linz, Austria

Bouatouch Kadi	University of Rennes I, IRISA, France
Brimkov Valentin	State University of New York, USA
Brown Ross	Queensland University of Technology, Australia
Bruckner Stefan	Vienna University of Technology, Austria
Callahan Steven	University of Utah, USA
Capin Tolga	Bilkent University, Turkey
Chaudhuri Parag	Indian Institute of Technology Bombay, India
Chen Min	University of Oxford, UK
Cheng Irene	University of Alberta, Canada
Chiang Yi-Jen	New York University, USA
Choi Min-Hyung	University of Colorado at Denver, USA
Comba Joao	Universidade Federal do Rio Grande do Sul, Brazil
Cremer Jim	University of Iowa, USA
Culbertson Bruce	HP Labs, USA
Dana Kristin	Rutgers University, USA
Debattista Kurt	University of Warwick, UK
Deng Zhigang	University of Houston, USA
Dick Christian	Technical University of Munich, Germany
Dingliana John	Trinity College, Ireland
El-Sana Jihad	Ben-Gurion University of The Negev, Israel
Entezari Alireza	University of Florida, USA
Fabian Nathan	Sandia National Laboratories, USA
De Floriani Leila	University of Genoa, Italy
Fu Hongbo	City University of Hong Kong, Hong Kong, SAR China
Fuhrmann Anton	VRVis Research Center, Austria
Gaither Kelly	University of Texas at Austin, USA
Gao Chunyu	Epson Research and Development, USA
Geist Robert	Clemson University, USA
Gelb Dan	Hewlett Packard Labs, USA
Gotz David	University of North Carolina at Chapel Hill, USA
Gooch Amy	University of Victoria, Canada
Gu David	Stony Brook University, USA
Guerra-Filho Gutemberg	Intel, USA
Habib Zulfiqar	COMSATS Institute of Information Technology, Lahore, Pakistan
Hadwiger Markus	KAUST, Saudi Arabia
Haller Michael	Upper Austria University of Applied Sciences, Austria
Hamza-Lup Felix	Armstrong Atlantic State University, USA
Han JungHyun	Korea University, Korea
Hand Randall	Lockheed Martin Corporation, USA
Hao Xuejun	Columbia University and NYSPI, USA
Hernandez Jose Tiberio	Universidad de los Andes, Colombia
Hou Tingbo	Google Inc., USA
Huang Jian	University of Tennessee at Knoxville, USA
Huang Mao Lin	University of Technology, Australia

Rudomin Isaac	Barcelona Supercomputing Center, Spain
Rushmeier Holly	Yale University, USA
Sander Pedro	The Hong Kong University of Science and Technology, Hong Kong, SAR China
Sapidis Nickolas	University of Western Macedonia, Greece
Sarfraz Muhammad	Kuwait University, Kuwait
Scateni Riccardo	University of Cagliari, Italy
Sequin Carlo	University of California – Berkeley, USA
Shead Timothy	Sandia National Laboratories, USA
Sourin Alexei	Nanyang Technological University, Singapore
Stamminger Marc	REVES/Inria, France
Su Wen-Poh	Griffith University, Australia
Szumilas Lech	Research Institute for Automation and Measurements, Poland
Tan Kar Han	Hewlett Packard, USA
Tarini Marco	University of Insubria (Varese), Italy
Teschner Matthias	University of Freiburg, Germany
Tong Yiying	Michigan State University, USA
Torchelsen Rafael Piccin	Universidade Federal da Fronteira Sul, Brazil
Umlauf Georg	HTWG Constance, Germany
Vanegas Carlos	University of California at Berkeley, USA
Wald Ingo	University of Utah, USA
Walter Marcelo	UFRGS, Brazil
Wimmer Michael	Technical University of Vienna, Austria
Wylie Brian	Sandia National Laboratory, USA
Wyman Chris	University of Calgary, Canada
Wyvill Brian	University of Iowa, USA
Yang Qing-Xiong	University of Illinois at Urbana, Champaign, USA
Yang Ruigang	University of Kentucky, USA
Ye Duan	University of Missouri – Columbia, USA
Yi Beifang	Salem State University, USA
Yin Lijun	Binghamton University, USA
Yoo Terry	National Institutes of Health, USA
Yuan Xiaoru	Peking University, China
Zhang Jian Jun	Bournemouth University, UK
Zeng Jianmin	Nanyang Technological University, Singapore
Zara Jiri	Czech Technical University in Prague, Czech Republic
Zeng Wei	Florida Institute of Technology, USA
Zordan Victor	University of California at Riverside, USA

(Area 3) Virtual Reality

Alcaniz Mariano	Technical University of Valencia, Spain
Arns Laura	Purdue University, USA
Bacim Felipe	Virginia Tech, USA
Balcisoy Selim	Sabanci University, Turkey

Behringer Reinhold	Leeds Metropolitan University, UK
Benes Bedrich	Purdue University, USA
Bilalis Nicholas	Technical University of Crete, Greece
Billinghurst Mark	HIT Lab, New Zealand
Blach Roland	Fraunhofer Institute for Industrial Engineering, Germany
Blom Kristopher	University of Barcelona, Spain
Bogdanovych Anton	University of Western Sydney, Australia
Brady Rachael	Duke University, USA
Brega Jose Remo Ferreira	Universidade Estadual Paulista, Brazil
Brown Ross	Queensland University of Technology, Australia
Bues Matthias	Fraunhofer IAO in Stuttgart, Germany
Capin Tolga	Bilkent University, Turkey
Chen Jian	Brown University, USA
Cooper Matthew	University of Linkoping, Sweden
Coquillart Sabine	Inria, France
Craig Alan	NCSA University of Illinois at Urbana-Champaign, USA
Cremer Jim	University of Iowa, USA
Edmunds Timothy	University of British Columbia, Canada
Egges Arjan	Universiteit Utrecht, The Netherlands
Encarnaio L. Miguel	ACT Inc., USA
Figueroa Pablo	Universidad de los Andes, Colombia
Friedman Doron	IDC, Israel
Fuhrmann Anton	VRVis Research Center, Austria
Gregory Michelle	Pacific Northwest National Lab, USA
Gupta Satyandra K.	University of Maryland, USA
Haller Michael	FH Hagenberg, Austria
Hamza-Lup Felix	Armstrong Atlantic State University, USA
Herbelin Bruno	EPFL, Switzerland
Hinkenjann Andre	Bonn-Rhein-Sieg University of Applied Sciences, Germany
Hollerer Tobias	University of California at Santa Barbara, USA
Huang Jian	University of Tennessee at Knoxville, USA
Huang Zhiyong	Institute for Infocomm Research (I2R), Singapore
Julier Simon J.	University College London, UK
Johnsen Kyle	University of Georgia, USA
Jones Adam	Clemson University, USA
Kiyokawa Kiyoshi	Osaka University, Japan
Klosowski James	AT&T Labs, USA
Kohli Luv	InnerOptic, USA
Kopper Regis	Duke University, USA
Kozintsev Igor	Samsung, USA
Kuhlen Torsten	RWTH Aachen University, Germany
Laha Bireswar	Stony Brook University, USA
Lee Cha	University of California, Santa Barbara, USA

Liere Robert van	CWI, The Netherlands
Livingston A. Mark	Naval Research Laboratory, USA
Luo Xun	Qualcomm Research, USA
Malzbender Tom	Hewlett Packard Labs, USA
MacDonald Brendan	National Institute for Occupational Safety and Health, USA
Molineros Jose	Teledyne Scientific and Imaging, USA
Muller Stefan	University of Koblenz, Germany
Owen Charles	Michigan State University, USA
Paelke Volker	University of Ostwestfalen-Lippe, Germany
Peli Eli	Harvard University, USA
Pettifer Steve	The University of Manchester, UK
Pronost Nicolas	Utrecht University, The Netherlands
Pugmire Dave	Los Alamos National Lab, USA
Qian Gang	Arizona State University, USA
Raffin Bruno	Inria, France
Ragan Eric	Oak Ridge National Laboratory, USA
Rodello Ildeberto	University of Sao Paulo, Brazil
Roth Thorsten	Bonn-Rhein-Sieg University of Applied Sciences, Germany
Sandor Christian	Nara Institute of Science and Technology, Japan
Sapidis Nickolas	University of Western Macedonia, Greece
Schulze Jurgen	University of California – San Diego, USA
Sherman Bill	Indiana University, USA
Singh Gurjot	Virginia Tech, USA
Slavik Pavel	Czech Technical University in Prague, Czech Republic
Sourin Alexei	Nanyang Technological University, Singapore
Steinicke Frank	University of Würzburg, Germany
Suma Evan	University of Southern California, USA
Stamminger Marc	REVES/Inria, France
Srikanth Manohar	Indian Institute of Science, India
Wald Ingo	University of Utah, USA
Whitted Turner	TWI Research, UK
Wong Kin Hong	The Chinese University of Hong Kong, Hong Kong, SAR China
Yu Ka Chun	Denver Museum of Nature and Science, USA
Yuan Chunrong	University of Tübingen, Germany
Zachmann Gabriel	Clausthal University, Germany
Zara Jiri	Czech Technical University in Prague, Czech Republic
Zhang Hui	Indiana University, USA
Zhao Ye	Kent State University, USA

(Area 4) Visualization

AAndrienko Gennady	Fraunhofer Institute IAIS, Germany
Avila Lisa	Kitware, USA

Apperley Mark	University of Waikato, New Zealand
Balizs Csibfalvi	Budapest University of Technology and Economics, Hungary
Brady Rachael	Duke University, USA
Benes Bedrich	Purdue University, USA
Bilalis Nicholas	Technical University of Crete, Greece
Bonneau Georges-Pierre	Grenoble University, France
Bruckner Stefan	Vienna University of Technology, Austria
Brown Ross	Queensland University of Technology, Australia
Bihler Katja	VRVis Research Center, Austria
Burch Michael	University of Stuttgart, Germany
Callahan Steven	University of Utah, USA
Chen Jian	Brown University, USA
Chen Min	University of Oxford, UK
Chevalier Fanny	Inria, France
Chiang Yi-Jen	New York University, USA
Cooper Matthew	University of Linkoping, Sweden
Chourasia Amit	University of California – San Diego, USA
Crossno Patricia	Sandia National Laboratories, USA
Daniels Joel	University of Utah, USA
Dick Christian	Technical University of Munich, Germany
Duan Ye	University of Missouri-Columbia, USA
Dwyer Tim	Monash University, Australia
Entezari Alireza	University of Florida, USA
Ertl Thomas	University of Stuttgart, Germany
De Floriani Leila	University of Maryland, USA
Geist Robert	Clemson University, USA
Gotz David	University of North Carolina at Chapel Hill, USA
Grinstein Georges	University of Massachusetts Lowell, USA
Goebel Randy	University of Alberta, Canada
Gregory Michelle	Pacific Northwest National Lab, USA
Hadwiger Helmut Markus	KAUST, Saudi Arabia
Hagen Hans	Technical University of Kaiserslautern, Germany
Hamza-Lup Felix	Armstrong Atlantic State University, USA
Healey Christopher	North Carolina State University at Raleigh, USA
Hochheiser Harry	University of Pittsburgh, USA
Hollerer Tobias	University of California at Santa Barbara, USA
Hong Lichan	University of Sydney, Australia
Hong Seokhee	Palo Alto Research Center, USA
Hotz Ingrid	Zuse Institute Berlin, Germany
Huang Zhiyong	Institute for Infocomm Research (I2R), Singapore
Jiang Ming	Lawrence Livermore National Laboratory, USA
Joshi Alark	Yale University, USA
Julier Simon J.	University College London, UK
Koch Steffen	University of Stuttgart, Germany
Laramee Robert	Swansea University, UK

Lewis R. Robert	Washington State University, USA
Liere Robert van	CWI, The Netherlands
Lim Ik Soo	Bangor University, UK
Linsen Lars	Jacobs University, Germany
Liu Zhanping	Kentucky State University, USA
Lohmann Steffen	University of Stuttgart, Germany
Maeder Anthony	University of Western Sydney, Australia
Malpica Jose	Alcala University, Spain
Masutani Yoshitaka	The Hiroshima City University, Japan
Matkovic Kresimir	VRVis Research Center, Austria
McCaffrey James	Microsoft Research/Volt VTE, USA
Melancon Guy	CNRS UMR 5800 LaBRI and Inria Bordeaux Sud-Ouest, France
Miksch Silvia	Vienna University of Technology, Austria
Monroe Laura	Los Alamos National Labs, USA
Morie Jacki	University of Southern California, USA
Moreland Kenneth	Sandia National Laboratories, USA
Mudur Sudhir	Concordia University, Canada
Museth Ken	Linköping University, Sweden
Paelke Volker	University of Ostwestfalen-Lippe, Germany
Papka Michael	Argonne National Laboratory, USA
Peikert Ronald	Swiss Federal Institute of Technology Zurich, Switzerland
Pettifer Steve	The University of Manchester, UK
Pugmire Dave	Los Alamos National Lab, USA
Rabin Robert	University of Wisconsin at Madison, USA
Raffin Bruno	Inria, France
Razdan Anshuman	Arizona State University, USA
Reina Guido	University of Stuttgart, Germany
Rhyne Theresa-Marie	North Carolina State University, USA
Rosenbaum Rene	University of California at Davis, USA
Sadana Samik	Georgia Tech, USA
Sadlo Filip	University of Stuttgart, Germany
Scheuermann Gerik	University of Leipzig, Germany
Shead Timothy	Sandia National Laboratories, USA
Sips Mike	Stanford University, USA
Slavik Pavel	Czech Technical University in Prague, Czech Republic
Sourin XavierAlexei	Nanyang Technological University, Singapore
Thakur Sidharth	Renaissance Computing Institute (RENCI), USA
Theisel Holger	University of Magdeburg, Germany
Thiele Olaf	University of Mannheim, Germany
Xavier Tricoche	Purdue University, USA
Umlauf Georg	HTWG Constance, Germany
Viegas Fernanda	IBM, USA
Wald Ingo	University of Utah, USA
Wan Ming	Boeing Phantom Works, USA

Weinkauf Tino	Max-Planck-Institut für Informatik, Germany
Weiskopf Daniel	University of Stuttgart, Germany
Wischgoll Thomas	Wright State University, USA
Wongsuphasawat Krist	Twitter Inc., USA
Wylie Brian	Sandia National Laboratory, USA
Wu Yin	Indiana University, USA
Xu Wei	Brookhaven National Lab, USA
Yeasin Mohammed	Memphis University, USA
Yuan Xiaoru	Peking University, China
Zachmann Gabriel	Clausthal University, Germany
Zhang Hui	Indiana University, USA
Zhao Jian	University of Toronto, USA
Zhao Ye	Kent State University, USA
Zheng Ziyi	Stony Brook University, USA
Zhukov Leonid	Caltech, USA

Special Tracks

1. Computational Bioimaging Organizers

Tavares João Manuel R.S.	University of Porto, Portugal
Natal Jorge Renato	University of Porto, Portugal

2. 3D Surface Reconstruction, Mapping, and Visualization Organizers

Nefian Ara	Carnegie Mellon University/NASA Ames Research Center, USA
Edwards Laurence	NASA Ames Research Center, USA
Huertas Andres	NASA Jet Propulsion Lab, USA

3. Observing Humans Organizers

Savakis Andreas	Rochester Institute of Technology, USA
Argyros Antonis	University of Crete, Greece
Asari Vijay	University of Dayton, USA

4. Advancing Autonomy for Aerial Robotics Organizers

Alexis Kostas	ETH Zurich, Switzerland
Chli Margarita	University of Edinburgh, UK
Achtelik Markus	ETH Zurich, Switzerland
Kottas Dimitrios	University of Minnesota, USA
Bebis George	University of Nevada, Reno, USA

5. Spectral Imaging Processing and Analysis for Environmental, Engineering and Industrial Applications Organizers

Doulamis Anastasios (Tasos)	National Technical University of Athens, Greece
Loupos Konstantinos	Institute of Communications and Computer Systems, Greece

6. Unconstrained Biometrics: Challenges and Applications Organizers

Proena Hugo	University of Beira Interior, Portugal
Ross Arun	Michigan State University, USA

7. Intelligent Transportation Systems Organizers

Ambardekar	Amol, Microsoft, USA
Morris Brendan	University of Nevada, Las Vegas, USA

8. Visual Perception and Robotic Systems Organizers

La Hung	University of Nevada, Reno, USA
Sheng Weihua	Oklahoma State University, USA
Fan Guoliang	Oklahoma State University, USA
Kuno Yoshinori	Saitama University, Japan
Ha Quang	University of Technology Sydney, Australia
Tran Anthony (Tri)	Nanyang Technological University, Singapore
Dinh Kien	Rutgers University, USA

Organizing Institutions and Sponsors

Contents – Part II

Applications

Pattern Recognition

Recognition

Virtual Reality

Contents – Part I

Motion and Tracking

Segmentation

Recognition

Visualization

ST: 3D Mapping, Modeling and Surface Reconstruction

ST: Advancing Autonomy for Aerial Robotics

Medical Imaging

Virtual Reality

ST: Observing Humans

ST: Spectral Imaging Processing

ST: Intelligent Transportation Systems

Visualization

ST: Visual Perception and Robotic Systems

Applications

Hybrid Example-Based Single Image Super-Resolution

Yang Xian[1]([✉]), Xiaodong Yang[2], and Yingli Tian[1,3]

[1] The Graduate Center, The City University of New York, New York, USA
yxian@gradcenter.cuny.edu
[2] NVIDIA Research, Santa Clara, USA
xiaodongy@nvidia.com
[3] The City College, The City University of New York, New York, USA
ytian@ccny.cuny.edu

Abstract. Image super-resolution aims to recover a visually pleasing high resolution image from one or multiple low resolution images. It plays an essential role in a variety of real-world applications. In this paper, we propose a novel hybrid example-based single image super-resolution approach which integrates learning from both external and internal exemplars. Given an input image, a proxy image with the same resolution as the target high-resolution image is first generated from a set of externally-learnt regression models. We then perform a coarse-to-fine gradient-level self-refinement on the proxy image guided by the input image. Finally, the refined high-resolution gradients are fed into a uniform energy function to recover the final output. Extensive experiments demonstrate that our framework outperforms the recent state-of-the-art single image super-resolution approaches both quantitatively and qualitatively.

1 Introduction

Image super-resolution (SR) aims to estimate a fine-resolution image from one or multiple coarse-resolution images. It is essential in numerous applications such as desktop publishing, video surveillance, remote sensing, etc. Due to limitations of low-grade physical devices and sensors, sometimes we only have low-resolution (LR) images at disposal. Under these circumstances, we reply on SR techniques for the enhancement of image resolution.

Broadly speaking, image SR tasks can be divided into two categories: multi-image SR [1–4] and single-image SR. With only one input LR image available, single image SR is a numerically ill-posed problem. The high-resolution (HR) solution is not unique since it is a many-to-one mapping. Therefore, to alleviate the inherent ambiguity, single image SR generally relies on additional assumptions or priors to finalize a visually pleasing output.

Single image SR methods can be further classified as interpolation based, reconstruction based and example learning based. Interpolation-based approaches exploit pre-defined mathematical formula to predict intermediate pixel values

© Springer International Publishing Switzerland 2015
G. Bebis et al. (Eds.): ISVC 2015, Part II, LNCS 9475, pp. 3–15, 2015.
DOI: 10.1007/978-3-319-27863-6_1

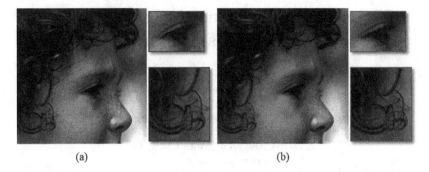

(a)   (b)

Fig. 1. SR result of image 'face' (×2). (a) HR image generated by the proposed SR method. (b) Ground truth image. Our SR approach reconstructs natural and realistic edges and textures close to the original ground truth image. This figure is better viewed on screen with HR display.

given a LR image. Commonly used filters, e.g., bilinear and bicubic, are simple and efficient and thus are widely used in commercial software. These methods hinge on the weak smoothness assumption and the generated HR images suffer from visual artifacts such as aliasing, jaggies, and blurring. To generate results with sharper edges, more sophisticated interpolation-based methods [5,6] were proposed.

Reconstruction-based image upscaling methods tend to enforce some statistical priors during the estimation of the target HR image. This group of SR approaches is often referred to as "edge-directed SR" [7] due to their emphasis on restoring sharp edges. Aly and Dubois [8] incorporated a total-variation regularizer to suppress oscillatory along the edges. In [9], Shan *et al.* built a feedback-control loop that enforces the output image to be consistent with the input image. Gradient profiles [10,11] are popularly utilized to describe the edge statistics due to its heavy-tailed distribution [12]. In [10], Fattal proposed a framework which generates the gradient field of the target image based on a statistical edge dependency relating certain edge features of two different resolutions. Sun *et al.* [11] performed a gradient field transformation to constrain the HR image gradients given the LR image based on a parametric gradient profile model. Reconstruction-based SR approaches achieve satisfying results in constructing clean edges but are less effective in hallucinating rich texture regions. A uniform parametric model for SR task is challenging since it is difficult to describe the diverse characteristics of natural images with a limited number of parameters.

Example-based SR explores the relationship between HR and their corresponding LR exemplars. The learning can be performed either via an external dataset [13–21], within the input image [22–24], or combined [25]. We refer to them as external, internal, and hybrid example-based learning. External example-based SR is brought up by Freeman *et al.* [13,14]. Based on Freeman's framework, many external example-based approaches are proposed to improve the

SR performance and the computational speed. Coupled LR/HR dictionaries are popular representations for the raw patch exemplars or patch-related features. Yang *et al.* [15,16] learnt a compact dictionary based on sparse signal representation which allows it to adaptively choose the most relevant reconstruction neighbors. Timofte *et al.* [19] combined the benefits of neighbor embedding and sparse coding to improve the execution speed. In [20], Zhu *et al.* proposed a deformable-patch based method by viewing a patch as a flexible deformation flow. This leads to a more "expressive" representation without increasing the size of the dictionary. Yang and Yang [21] divided the input feature space of LR source images into subspaces and trained a regression model for each individual subspace.

Internal example-based SR is based on the fact that small patches in a natural image tend to appear repeatedly within the image itself and across different scales. Glasner *et al.* [22] performed the nearest-neighbor search based on a patch pool formed with patches collected through a pyramid structure with the input image at different resolutions. In [23], Freedman and Fattal proposed a real time multi-step algorithm which allows a local search instead of a global search. Huang *et al.* [24] extended internal example-based SR by allowing geometric variations to fully utilize the internal similarity. For hybrid example-based learning, Yang *et al.* [25] combined learning from self-examples and an external dataset into a regression model based on in-place examples.

Effectiveness of internal example-based SR has been demonstrated in [26]. The difficulty in estimating missing high frequency details increases as the scaling factors get larger due to the increment of LR/HR ambiguity. External example-based learning breaks the limitation by introducing new information from a natural image dataset. However, the performance of external example-based SR depends on the similarity between the training dataset and the testing images. Due to diversification in natural images, the lack of relevance between certain testing images and a universal training dataset still exists. Keeping increasing the size of the training dataset provides a limited solution but still leaves the key problem untouched.

In this paper, we propose a novel hybrid example-based single image SR method which incorporates learning image-level statistics from an external dataset and gradient-level self-awareness with internal statistics. The proposed SR scheme consists of three steps: a proxy HR image is constructed through a set of pre-built regression models learnt from external exemplars; the gradients of the proxy image are then fed into a pyramid self-awareness framework guided by the input LR gradients; finally, the refined HR gradients and the input image are integrated into a uniform cost function to recover the final HR image. Figure 1 illustrates the comparison of the generated HR image 'face' and the ground truth (GT) image. Our SR result is very close to the original GT image. Edge details including eye and face contours are natural and realistic. Hair textures are well reconstructed with minimal visual artifacts. The contributions of the proposed SR framework are fourfold:

- A novel single image SR scheme is proposed to benefit from both external and internal statistics. Our framework hinges on image-level hallucination from externally learnt regression models as well as gradient level pyramid self-awareness for edges and textures refinement. A uniform energy function is utilized to restore the final HR image in a manner consistent with the input image.
- In the training of a set of regression models from an external dataset with exemplar patches, we model the input LR feature space with Gaussian Mixture Model (GMM) to ensure the effective and targeted learning.
- To obtain quality edges and textures, a novel gradient-level self-refinement pyramid is proposed to recover the high-frequency details lost during the reconstruction process of proxy HR image.
- The proposed framework is effective and outperforms the recent state-of-the-art single image SR algorithms quantitatively and qualitatively.

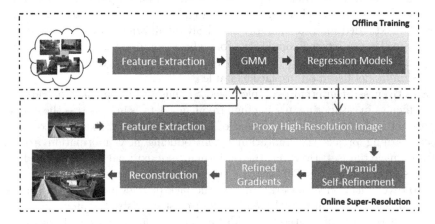

Fig. 2. Flowchart of the proposed SR method. Given a LR image, a proxy HR image is constructed through the regression models trained via an external dataset. The input feature space is modeled with GMM to ensure a targeted learning. With the proxy image, its gradients are refined using gradients of the input image. The refined HR gradients are then integrated into the reconstruction framework to recover the final output image.

2   Hybrid Example-Based Super-Resolution

External example learning-based SR usually relies on learning priors or models from a natural image dataset which leads to a stable SR performance. Different from internal example-based approaches, learning externally is normally performed off-line and is less time consuming when upscaling a testing image. However, natural images vary dramatically especially for edges and textured regions. Given a natural image, certain patches occur rarely in the universal training dataset and this results in a less effective SR performance for those patterns. On the other hand, internal patch redundancy has been validated to be effective both in "expressiveness" (how similar between a small patch and its most similar patches found internally or externally) and "predictive power" (how well the found similar patches can be used in image restoration tasks given a prediction model) [26]. In order to combine the benefits of external and internal example-based learning, we propose a hybrid learning based SR framework. Figure 2 illustrates the schematic pipeline of our approach. The system consists of three steps to upsample an image: proxy image recovery from external statistics, gradient-level self-awareness from internal statistics, and final image reconstruction.

Given a LR image, a proxy HR image is first generated with a group of pre-built regression models. The regression models are trained on an external natural image dataset. To ensure a targeted learning, the input feature space is modeled with GMM where an individual regression model is trained for each Gaussian component. The generated proxy HR image is robust with stable SR performance since the regression models are trained through a large number of natural images in a divide-and-conquer manner. However, certain LR patches in the input image may appear rarely within the training dataset and thus lead to an inaccurate HR prediction, i.e., over-smoothed with missing high-frequency details. Therefore, after obtaining the proxy image, a gradient-level coarse-to-fine self-refinement is performed guided by gradients of the input LR image. Motivated by reconstruction-based SR approaches, we adopt a gradient-level refinement to better preserve the intensity changes. This process aims to replace the high-variance gradient patches in the proxy image with more accurate representations to recover more visually plausible outputs. Finally, the targeted HR image is restored through minimizing a uniform cost function with the refined gradients. The detailed three steps are presented in the following subsections.

2.1 Proxy Image Recovery

Given an input image L, we first recover a proxy HR image from a set of externally-trained regression models. A large set of LR/HR exemplar patch pairs with magnification factor s are collected from a dataset consisted of more than 6,000 images. All images within the dataset are considered HR images and the corresponding LR images are generated with a blur and downsampling process. To better preserve the structure information, for a LR/HR patch pair $\{P_l, P_h\}$, we normalize both patches by extracting the mean value of P_l. After normalization and vectorization, the input LR and HR features are represented as $X \in \mathbb{R}^{l \times M}$ and $Y \in \mathbb{R}^{r \times M}$ respectively where l and r denote the corresponding feature dimensions and M indicates the number of samples.

To ensure a targeted learning, we first model the input LR feature space where later multiple regression models are trained. The most straightforward model to describe the feature space is the normal distribution. However, a single normal distribution is insufficient to capture the complex nature of the features. We therefore employ GMM to represent the feature distribution. GMM is a generative model which has the capacity to model any given probability distribution function when the number of Gaussian components is large enough. Given a GMM with K components, the probability of a feature $\boldsymbol{x}_i$ is

$$p(\boldsymbol{x}_i|\theta) = \sum_{k=1}^{K} w_k \mathcal{N}\left(\boldsymbol{x}_i; \boldsymbol{\mu}_k, \boldsymbol{\sigma}_k\right), \tag{1}$$

where w_k is the prior mode probability which satisfies the constraint $\sum_{k=1}^{K} w_k = 1$, and $\mathcal{N}\left(\boldsymbol{x}_i; \boldsymbol{\mu}_k, \boldsymbol{\sigma}_k\right)$ indicates the kth normal distribution with mean $\boldsymbol{\mu}_k$ and variance $\boldsymbol{\sigma}_k$:

$$\mathcal{N}\left(\boldsymbol{x}_i; \boldsymbol{\mu}_k, \boldsymbol{\sigma}_k\right) = \frac{\exp\left(-\frac{1}{2}\left(\boldsymbol{x}_i - \boldsymbol{\mu}_k\right)^T \left(\boldsymbol{\sigma}_k\right)^{-1} \left(\boldsymbol{x}_i - \boldsymbol{\mu}_k\right)\right)}{(2\pi)^{l/2} |\boldsymbol{\sigma}_k|^{1/2}}, \tag{2}$$

where $\boldsymbol{x}_i \in \mathbb{R}^l$, $\boldsymbol{\mu}_k \in \mathbb{R}^l$, and $\boldsymbol{\sigma}_k \in \mathbb{R}^{l \times l}$. By using the Expectation-Maximization (EM) algorithm to optimize the Maximum Likelihood (ML) from a large number of features, we can estimate the GMM parameters $\theta = \{w_k, \boldsymbol{\mu}_k, \boldsymbol{\sigma}_k, k = 1, \ldots, K\}$. We employ $200,000$ randomly sampled features to learn the parameters θ in our experiment. Though Eq. (1) supports the full covariance matrix, a diagonal matrix in practice is sufficient to model most distributions. Moreover, the GMM with diagonal matrices is more computationally efficient and stable compared to the one with full matrices.

GMM is based on a well-defined statistical model and is computationally tractable. We then assign each LR feature $\boldsymbol{x}_i \in \boldsymbol{X}$ to corresponding Gaussian component with the highest probability. Suppose there are M_k patches associated with the kth Gaussian component and $\boldsymbol{X}_k \in \mathbb{R}^{l \times M_k}$, $\boldsymbol{Y}_k \in \mathbb{R}^{r \times M_k}$ represent the corresponding LR/HR features, a linear regression model is then trained with the regression coefficient A_k learnt through:

$$A_k^* = \underset{A_k}{\operatorname{argmin}}\{|\boldsymbol{Y}_k - A_k \hat{\boldsymbol{X}}_k|^2\}, \tag{3}$$

where $\hat{\boldsymbol{X}}_k = [\boldsymbol{X}_k^T \ \boldsymbol{1}]^T$. During testing phase, given a LR image, we first extract all features by performing normalization and vectorization for every LR patch. Then each feature is assigned to a Gaussian component according to the posterior where the corresponding regression model is applied to obtain the HR patch. We use simple averaging to blend overlapping pixels to generate the proxy HR image.

2.2   Gradient-Level Self-Awareness

Internal patch redundancy has been demonstrated powerful for image restoration tasks [26] and serves as the theoretical foundation for internal example-based SR methods [22–24]. With good performance for SR under relatively small magnification factors, the limitations of internal example-based SR approaches lie on the heavy computational costs to execute on-line exhaustive pair-wise patch comparisons and degraded performance with the increment of the scaling factor.

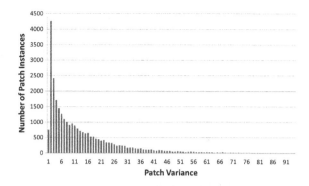

Fig. 3. Variance distributions of 20,000 patches with size 7×7 extracted from BSDS200 [27]. Given a patch, the larger the variance is, the less frequent it tends to appear within the dataset.

Fig. 4. (a) Illustration for gradient-level coarse-to-fine self-awareness procedure. Gradients of the proxy image $H^p_{\{x,y\}}$ are downsampled to $M^p_{\{x,y\}}$ which are refined with the gradients $L_{\{x,y\}}$ of the input image. Gradients with darker frames represent the corresponding refined results of the ones with lighter frames. Afterwards, $H^p_{\{x,y\}}$ is refined with $M_{\{x,y\}}$. Please refer to text for details. (b) Difference map between $H_{\{x\}}$ and $H^p_{\{x\}}$. (c) Difference map between $H_{\{y\}}$ and $H^p_{\{y\}}$

In this step, advantage of self-similarity is absorbed to refine the proxy image generated previously without going through exhaustive patch matching.

The self refinement process aims at recovering the missing high-frequency details for patches which are not frequently seen in the external training dataset. We first verify that patches with higher variances tend to appear less frequently within a natural image dataset. The experiment is performed by randomly extracting 20,000 patches of size 7×7 within the Berkeley Segmentation Database (BSDS200) [27]. As observed from Fig. 3, the number of patch instances decreases quickly as the variance increases.

To refine H^p with L, patches from H^p of size $a \times a$ with variance larger than a pre-set threshold θ are firstly extracted. We utilize self-similarity to recover the missing high-frequency details of those high variance patches which are not frequently seen in the external dataset. For each high variance patch, its k most similar patches with the same size are searched and extracted within L where the similarity of two patches is measured in their mean square error (MSE).

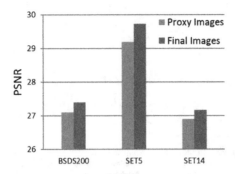

Fig. 5. Average PSNR (dB) comparisons of the proxy images (marked in blue) and the corresponding final output (marked in red) recovered from self-refinement and reconstruction in datasets BSDS200 [27], SET5 [28], and SET14 [29] (scaling factor ×4). There is an obvious boost in PSNR after performing self-awareness and reconstruction for all three datasets. (Color figure online)

Afterwards, the original patch is replaced with the weighted sum of the found k patches in a softmax way.

Motivated by reconstruction-based SR, in our proposed scheme, a gradient-level self-refinement is adopted to better preserve edges and textures information. Moreover, it is validated in [22] that average patch recurrence across scales decays as the resolution difference increases. Therefore, if the magnification factor s is larger than s_0 ($s_0 = 3$ in our experimental setting), the proposed self-refinement is executed in a coarse-to-fine scheme.

Figure 4 illustrates the self-awareness process. After obtaining the proxy image H^p, its gradients in horizontal and vertical (denoted as x and y) directions are computed and refined with the corresponding gradients of the input image L. In later context, for ease of interpretation, we denote the gradients of an image I in x and y directions as $I_{\{x,y\}}$. The refinement is performed separately for each gradient direction. Take direction x as an example, if s is larger than s_0, we first downsample H_x^p by factor $\sqrt{s}$ to obtain M_x^p. After that, high variance patches in M_x^p are refined with L_x to obtain finer-version gradient M_x. Then the final HR gradient H_x is computed by utilizing M_x to refine H_x^p. In the above process, if scaling factor $\sqrt{s}$ is still larger than s_0, we further decompose $\sqrt{s}$ in a similar manner before proceed.

Gradient patches are mostly flat with small variances. Therefore, only a small portion of the patches are refined with missing high frequency details. To ensure a more effective refinement, all the patches are normalized to have zero means and unit standard variances before searching. The combined patch is then readjusted according to the original mean and variance of the input patch. After the self-awareness, proxy HR patches which are over-smoothed during the upsampling process are refined with restored high frequency details.

2.3   Final Image Reconstruction

The final step is to reconstruct the targeted output image from the self-refined HR gradients through the following cost function:

$$H^* = \underset{H}{\mathrm{argmin}}\{|\nabla H - \nabla H_r|^2 + \alpha|(H * G) \downarrow_s -L|^2\}, \tag{4}$$

where ∇H_r represents the refined $H_{\{x,y\}}$ after the pyramid gradient-level self-awareness step. G stands for a Gaussian kernel with standard variance σ varies for different scaling factors s: $\sigma = \{0.8, 1.2, 1.6\}$ for $s = \{2, 3, 4\}$. α is the weighting factor.

Constraints in both the gradient-level and the image-level are integrated into a uniform cost function. The first term states the constraint imposed by the self-refined HR gradients. The second constraint ensures the fidelity between the output and the input images. The energy function can be optimized through the gradient descent algorithm.

We demonstrate the effectiveness of our proposed gradient-level self-refinement and the feasibility of reconstructing images based on gradients experimentally on datasets BSDS200 [27] (200 images), SET5 [28] (5 images), and SET14 [29] (14 images). All the images are downsampled by a factor of 4. Figure 5 presents the average Peak Signal-to-Noise Ratio (PSNR) comparison of the proxy HR images and the final outputs after self-awareness and reconstruction within each dataset. After refining the ambiguous patches, all the images experience an obvious boost in the SR performance measured in PSNR.

3   Experimental Results

In this section, the proposed hybrid example-based SR method is evaluated with multiple natural images on SET5 [28], SET14 [29], and BSDS200 [27]. We also compare our results with recent state-of-the-art single image SR algorithms both quantitatively and qualitatively.

Table 1. Comparison of the proposed approach with recent state-of-the-art methods in SET5 [28], SET14 [29], and BSDS200 [27] in terms of average PSNR (dB). Our results outperform other methods in all three datasets.

Dataset	Bicubic	Shan [9]	ScSR [15]	Zeyde [29]	ANR [19]	Yang [21]	Ours
SET5 [28]	28.4202	28.2488	28.1507	29.6990	29.6959	29.2540	**29.7199**
SET14 [29]	26.1911	26.0938	26.1244	27.1149	27.0918	26.9004	**27.1559**
BSDS200 [27]	26.6339	26.5103	26.5232	27.3286	27.3255	27.1749	**27.3568**

Parameter Selection: Same as many existing SR methods, for color images, we only perform the proposed SR algorithm on the luminance channel in YUV color space while the other two color channels are upsampled with bicubic interpolation.

The training dataset used for regression model learning is the same as in [21] with 6, 152 natural images. We extract all patches with size 7×7 from LR images. Corners of each patch are removed and thus the LR feature dimension is 45. Only the central $3s \times 3s$ pixels in the corresponding HR patch are used to formulate the HR feature where s indicates the magnification factor. We randomly select 200, 000 LR/HR features to train the GMM model with 512 components. To better model the feature space, we filter out the smooth patches before selection. With the trained GMM, each feature is assigned to a Gaussian component with the highest probability. We learn a linear regression model for each Gaussian component using maximum 1, 000 LR/HR features within this component.

In the pyramid gradient-level self-awareness, the maximum magnification factor s_0 between each level is 3. If the scaling factor s is larger than 3, we adopt a coarse-to-fine scheme with a factor of $\sqrt{s}$ per-step. Patch size a in the self-refinement is 7 and the pre-set threshold θ used to differentiate the smooth patches and the high-variance ones is 5. The number k of similar patches captured during the searching is 5. We set α in Eq. (4) to be $4/7$.

Quantitative Analysis: Our proposed approach is evaluated on a variety of natural images and we compare the generated SR results with recent state-of-the-art methods [9,15,19,21,29] quantitatively measured in PSNR. We use the source code [19,21,29] and executable file [9] provided by the authors or a third-party implementation [15] to generate the corresponding HR images based on the same LR input images. To be more specific, given a GT image, the LR image is obtained by performing the bicubic downsampling. These LR images are saved in PNG format and serve as the uniform input for all SR approaches. We then follow the code or executable file provided by the authors to perform the image upsampling. SR methods [15,19,29] generate results with borders shaved. To perform a fair comparison, we crop the borders for all the HR results generated by different SR approaches utilizing the same scheme before the PSNR calculation over the luma channel.

The evaluation is performed on three datasets, i.e., SET5 [28], SET14 [29], and BSDS200 [27] at magnification factor 4. As illustrated in Table 1, for different input images, our proposed approach is robust and outperforms the other methods measured in average PSNR over all three datasets.

Qualitative Analysis: Figure 6 presents a set of SR results on image 'Lenna' with an upscaling factor of 4. Reconstructing the hat contours with minimal visual artifacts is difficult for most of the SR approaches listed. Our method successfully generates clear contours consistent with the GT image.

Figure 7 provides another set of results with a scaling factor of 4 on image 'snow'. It is challenging to reconstruct the rattan textures as illustrated by the zoom-in regions. Results generated by bicubic interpolation and [16] are over-smoothed. Deformed patterns exist within [9] (irregular squared patterns) and [21] (discontinuities). [19] fails to recover several edges and [30] over-sharps the edges. Our recovered HR image reveals natural patterns.

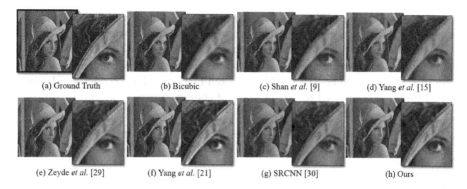

(a) Ground Truth   (b) Bicubic   (c) Shan *et al.* [9]   (d) Yang *et al.* [15]

(e) Zeyde *et al.* [29]   (f) Yang *et al.* [21]   (g) SRCNN [30]   (h) Ours

Fig. 6. SR of image 'Lenna' (×4). Zoom-ins clearly indicates that our proposed approach reconstructs the hat contours with minimal artifacts while other methods suffer from blurring, jaggies, and aliasing artifacts. This figure is better viewed on screen with HR display.

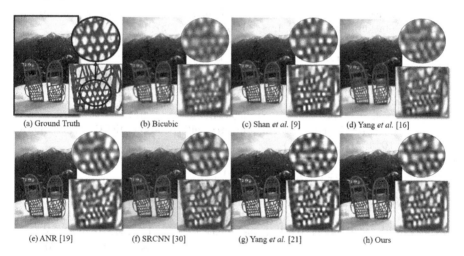

(a) Ground Truth   (b) Bicubic   (c) Shan *et al.* [9]   (d) Yang *et al.* [16]

(e) ANR [19]   (f) SRCNN [30]   (g) Yang *et al.* [21]   (h) Ours

Fig. 7. SR of image 'snow' (×4). It is a challenging task to reconstruct the rattan textures. Results generated by bicubic interpolation and [16] over-smooth the textures. Deformed patterns exist in [9] (squared textures) and [21] (discontinuities). [19] fails to recover several edges as shown in the circled zoom-in and [30] oversharps the edges. Our result best reconstructs the details. This figure is better viewed on screen with HR display.

4   Conclusion

In this paper, we have proposed a novel hybrid example-based single image super-resolution approach which integrates both external and internal statistics. Given an input LR image, a proxy HR image is firstly generated with pre-built regression models learnt from a large external dataset. Then its gradients in horizontal and vertical directions are refined guided by the corresponding gradients of the input image. Finally, the refined HR gradients and the input LR image are fed into our proposed energy function to recover the final output.

As demonstrated by the extensive experimental results, the proposed approach is robust with satisfying super-resolution performance measured quantitatively in PSNR. The generated HR images tend to have sharper edges and more natural textures compared to recent state-of-the-art super-resolution methods.

Acknowledgments. This work was supported in part by ONR grant N000141310450 and NSF grants EFRI-1137172, IIP-1343402.

References

1. Boulanger, J., Kervrann, C., Bouthemy, P.: Space-time adaptation for patch-based image sequence restoration. IEEE Trans. Pattern Anal. Mach. Intell. **29**, 1096–1102 (2007)
2. Farsiu, S., Robinson, M.D., Elad, M., Milanfar, P.: Fast and robust multiframe super resolution. IEEE Trans. Image Process. **13**, 1327–1344 (2004)
3. Protter, M., Elad, M., Tekeda, H., Milanfar, P.: Generalizing the non-local-means to super-resolution reconstruction. IEEE Trans. Image Process. **18**, 36–51 (2009)
4. Shi, B., Zhao, H., Ben-Ezra, M., Yeung, S.-K., Fernandez-Cull, C., Shepard, R.H., Barsi, C., Raskar, R.: Sub-pixel layout for super-resolution with images in the octic group. In: Fleet, D., Pajdla, T., Schiele, B., Tuytelaars, T. (eds.) ECCV 2014, Part I. LNCS, vol. 8689, pp. 250–264. Springer, Heidelberg (2014)
5. Li, X., Orchard, M.T.: New edge-directed interpolation. IEEE Trans. Image Process. **10**, 1521–1527 (2001)
6. Su, D., Willis, P.: Image interpolation by pixel-level data-dependent triangulation. Comput. Graph. Forum **23**, 189–201 (2004)
7. Tai, Y., Liu, S., Brown, M., Lin, S.: Super resolution using edge prior and single image detail synthesis. In: CVPR (2010)
8. Aly, H.A., Dubois, E.: Image up-sampling using total-variation regularization with a new observation model. IEEE Trans. Image Process. **14**, 1647–1659 (2005)
9. Shan, Q., Li, Z., Jia, J., Tang, C.K.: Fast image/video upsampling. In: ACM SIGGRAPH Asia (2008)
10. Fattal, R.: Image upsampling via imposed edge statistics. In: ACM SIGGRAPH (2007)
11. Sun, J., Sun, J., Xu, Z., Shum, H.Y.: Image super-resolution using gradient profile prior. In: CVPR (2008)
12. Huang, J., Mumford, D.: Statistics of natural images and models. In: CVPR (1999)
13. Freeman, W.T., Pasztor, E.C., Carmichael, O.T.: Learning low-level vision. Int. J. Comput. Vis. **40**, 25–47 (2000)
14. Freeman, W.T., Jones, T.R., Pasztor, E.C.: Example-based super-resolution. Comput. Graph. Appl. **22**, 56–65 (2002)
15. Yang, J., Wright, J., Huang, T., Ma, Y.: Image super-resolution as sparse representation of raw image patches. In: CVPR (2008)
16. Yang, J., Wright, J., Huang, T.S., Ma, Y.: Image super-resolution via sparse representation. IEEE Trans. Image Process. **19**, 2861–2873 (2010)
17. HaCohen, Y., Fattal, R., Lischinski, D.: Image upsampling via texture hallucination. In: ICCP (2010)

18. Sun, J., Zhu, J., Tappen, M.F.: Context-constrained hallucination for image super-resolution. In: CVPR (2010)
19. Timofte, R., Smet, V.D., Gool, L.V.: Anchored neighborhood regression for fast example-based super-resolution. In: ICCV (2013)
20. Zhu, Y., Zhang, Y., Yuille, A.L.: Single image super-resolution using deformable patches. In: CVPR (2014)
21. Yang, C.Y., Yang, M.H.: Fast direct super-resolution by simple functions. In: ICCV (2013)
22. Glasner, D., Bagon, S., Irani, M.: Super-resolution from a single image. In: ICCV (2009)
23. Freedman, G., Fattal, R.: Image and video upscaling from local self-examples. ACM Trans. Graph. **28**, 1–10 (2010)
24. Huang, J.B., Singh, A., Ahuja, N.: Single image super-resolution from transformed self-exemplars. In: CVPR (2015)
25. Yang, J., Lin, Z., Cohen, S.: Fast image super-resolution based on in-place example regression. In: CVPR (2013)
26. Zontak, M., Irani, M.: Internal statistics of a single natural image. In: CVPR (2011)
27. Martin, D., Fowlkes, C., Tal, D., Malik, J.: A database of human segmented natural images and its application to evaluating segmentation algorithms and measuring ecological statistics. In: ICCV (2001)
28. Bevilacqua, M., Roumy, A., Guillemot, C., Morel, M.A.: Low-complexity single-image super-resolution based on nonnegative neighbor embedding. In: BMVC (2012)
29. Zeyde, R., Elad, M., Protter, M.: On single image scale-up using sparse-representations. Curves Surf. **6920**, 711–730 (2010)
30. Dong, C., Loy, C.C., He, K., Tang, X.: Learning a deep convolutional network for image super-resolution. In: Fleet, D., Pajdla, T., Schiele, B., Tuytelaars, T. (eds.) ECCV 2014, Part IV. LNCS, vol. 8692, pp. 184–199. Springer, Heidelberg (2014)

Automated Habit Detection System: A Feasibility Study

Hiroki Misawa, Takashi Obara, and Hitoshi Iyatomi[✉]

Applied Informatics, Graduate School of Science and Engineering,
Hosei University, Tokyo, Japan
iyatomi@hosei.ac.jp

Abstract. In this paper, we propose an automated habit detection system. We define a "habit" in this study as some motion that is significantly different from our common behaviors. The behaviors of two subjects during conversation are tracked by the Kinect sensor and their skeletal and facial conformations are detected. The proposed system detects the motions considered as habits by analyzing them using a principal component analysis (PCA) and wavelet multi-resolution analysis (MRA). In our experiments, we prepare a total of 108 movies containing 5 min of conversation. Of these, 100 movies are used to build the average motion model (AMM), and the remainder are used for the evaluation. The accuracy of habit detection in the proposed system is shown to have a precision of 84.0 % and a recall of 81.8 %.

1 Introduction

Although we primarily communicate in words, it is well known that non-verbal communication such as expressions, gestures, and unconscious motions have a significant influence on our communication [1]. Habits are behaviors that we often perform unconsciously or have little awareness. Some habits might make people displeased and, in some cases, could be the cause of a loss of opportunity. Thus, we consider the objective recognition of our habits to be meaningful not only for better communication, but also for a wide range of general purposes.

Many studies on motion analysis have considered a wide range of objectives. The methodologies can be divided into two categories from the perspective of the usage of sensory devices: (1) subjects wear sensory devices and their motions are estimated based on obtained signals, or (2) subjects wear no special devices and their motions are directly estimated from video recordings with image processing techniques. In the former, acceleration sensors are commonly used [2] because of their excellent practical applicability in detecting gradient, motion, and fluctuation. These sensors provide meaningful information, although they are sometimes unavailable because of limitations in terms of cost, weight, and geometry. In the latter category, commercially available video cameras have been widely used. Bobick and Davis [3] identified the behavior of subjects by extracting the transformation areas of the silhouette from movies and generating the binary

© Springer International Publishing Switzerland 2015
G. Bebis et al. (Eds.): ISVC 2015, Part II, LNCS 9475, pp. 16–23, 2015.
DOI: 10.1007/978-3-319-27863-6_2

motion energy image and motion history image. Schuldt et al. [4] also identified subjects' motion by means of a bag-of-features consisting of a histogram
of local features and a support vector machine (SVM) as a classifier. Infrared
cameras have also been used for motion analysis, either alone or in conjunction
with visible light cameras, because of their tolerance for variations in lighting
conditions. In each of these cases, some depth estimation of the target is necessary when there is a need for 3D analysis. In many cases, depth estimation is
achieved by using a stereopsis system with multiple cameras. If highly accurate
analysis is required, marker detection is commonly used, although this method
needs dedicated equipment and/or facilities.

The Kinect sensor was released in 2010 by Microsoft Corp. as a peripheral
device for their Xbox gaming platform. As the Kinect is relatively cheap, but
can track 3D motion with a considerable measure of credibility, it has been
used in many studies [5–7]. Xia et al. [5] recognized human behavior with their
HOJ3D (Histograms of 3D joint locations) method. This method generates 3D
histograms of human posture that are analyzed by a linear discriminant analysis and a vector quantization. Evangelidis et al. [6] performed motion recognition with Fisher vectors based on the location of articulations obtained from
3D skeletal information and the SVM classifier. Miranda et al. [7] proposed a
gesture recognition system. Their method detected characteristic motion from
the observed 3D skeletal information with the SVM, and classified the detected
motion using a trained decision tree. However, to the best of our knowledge,
no systematic research on automated habit detection or habit analysis has been
conducted. We believe this is because the variety of habits is quite broad, making
it difficult to apply conventional methodologies.

In this study, we propose an automated habit detection system that utilizes the Kinect sensor. Considering the further applicability of this system, we
analyze the behavior of the subjects during conversation.

2 Habit Detection System

First, we define a habit as some motion that is significantly different from our
common behaviors. The proposed habit detection system tracks the behavior
of two subjects in conversation with the Kinect sensors, and detects distinctive
behavior as a habit. The schematics of the proposed habit detection system
are shown in Fig. 1. The proposed system has two operation phases: a training
phase and an evaluation phase. In the training phase, we record a large number of
conversations with the Kinect sensor and form the average motion model (AMM)
for each body part as a reference. In the evaluation phase, the behavior of the
subject is compared with the AMM, and significant differences are identified
as habit. The details will be explained in later sections. The proposed system
calculates the velocity of the subjects' body parts. The time-series of the velocity
of each body part is analyzed using a wavelet multi-resolution analysis (MRA).

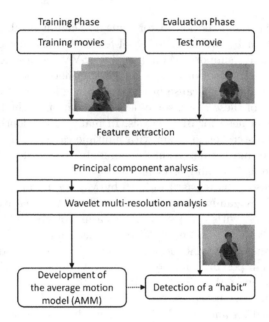

Fig. 1. Schematics of the proposed habit detection system.

2.1   Recording Environment and Material

In this study, we recorded video clips as follows: (1) Two subjects sit face-to-face, (2) the Kinect sensor is set up in front of each subject to track his/her upper body, and (3) their conversation is recorded with the Kinect sensor for at least 300 s. The recording environment is shown in Fig. 2. The Kinects are approximately 1 m from their respective subjects, and the distance between the two subjects is approximately 2.5 m. We prepared several topics for conversation (e.g., school life, hobbies, friends), and the subjects selected one of these topics prior to the recording.

In this experiment, we recorded three movies of each of 36 male subjects (24.3 ± 1.3 years old), i.e., a total of 108 movies, and selected an arbitrary 300 s from each movie for processing. We used the Kinect Studio (SDK 1.5) to record the conversations at 30 fps and detect the 3D motion information. The RGB and depth sensor resolutions of the Kinect were 640×480 and 320×240 pixels, respectively.

2.2   Detection of Tracking Points

In this study, we used skeletal information of the upper body and the facial components of the subjects. For the former, we detected a total of 10 joints from the upper body while the subjects were seated (Fig. 3). For the latter, we used five points (forehead, left eye, right eye, nose, and mouth; see Fig. 4) out of 121 detected facial feature points. Accordingly, a total of 15 three-dimensional

Fig. 2. Experimental environment.

feature points (i.e., 45 features) were extracted from each image frame for each subject. We calculated the velocity of each point by investigating the difference in point locations between successive frames. Accordingly, the 3D velocity vector for each body part p ($p = 1, 2, \cdots, 15$) at time t is expressed as:

$$\mathbf{v}^{\mathbf{p}}(\mathbf{t}) = [v_x^p(t), v_y^p(t), v_z^p(t)]^T. \tag{1}$$

Because the motion of the body has physical and geometrical limitations, we conducted a principal component analysis (PCA) for each velocity vector. Table 1 summarizes the contribution of the first primary component in each body part. According to these results, we can confirm that the first primary component makes a significant contribution to many body parts. Therefore, we decided to approximate the obtained 3D velocity $\mathbf{v}^{\mathbf{p}}(\mathbf{t})$ by the one-dimensional velocity value $v_{1st}^p(t)$ in the direction of the first eigenvector. In the experiment, we analyzed these 15-dimensional time-series data

$$\mathbf{V}(\mathbf{t}) = [v_{1st}^1(t), v_{1st}^2(t), \cdots, v_{1st}^{15}(t)]^T \tag{2}$$

by means of MRA with Daubechies' wavelet (N = 2). According to the results of the preliminary experiment, we focused on wavelet coefficients with a frequency of 1.25 Hz for each body part.

2.3   Definition of "habit" and Its Detection

In the training phase, we formed the AMM of each subject's body parts by averaging the wavelet coefficients of the training dataset (i.e., 100 movies). In the evaluation phase, wavelet coefficients were calculated for each subject's body motions. If the difference between the evaluation target and the AMM was greater than twice the standard deviation (SD) of the AMM, the system considered this motion to be uncommon, and therefore identified it as a habit.

   To conduct a quantitative evaluation of the proposed system, the gold standard is required. As there is no objective definition of a habit in terms of physical motions, the authors manually selected several motions considered as habits from

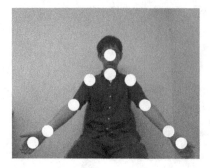

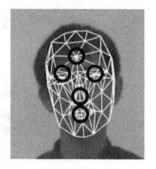

Fig. 3. Ten tracking points in upper body.

Fig. 4. Five tracking points on the face.

Table 1. Contribution ratio of the PCA.

Body part	Contribution ratio (%)	Body part	Contribution ratio (%)
Head	82.6	ShoulderCenter	79.6
ShoulderLeft	81.1	ShoulderRight	79.7
ElbowLeft	64.7	ElbowRight	65.3
WristLeft	45.4	WristRight	45.9
HandLeft	47.6	HandRight	50.6
Forehead	88.1	Nose	88.2
EyeLeft	89.1	EyeRight	89.4
Mouth	87.9		

the evaluation dataset, and determined them as the gold standard. We used the precision and recall as performance criteria. These were calculated as follows:

$$\text{precision} = \frac{\text{\# of correctly detected habit}}{\text{\# of detected habit}}. \tag{3}$$

$$\text{recall} = \frac{\text{\# of correctly detected habit}}{\text{\# of habit (gold standard)}}. \tag{4}$$

If a habit was detected by the system within 2 s of the gold standard, we considered the detection to be appropriate.

3   Results

In this experiment, we used a total of 100 movies to build the AMM, with the reminder used for evaluation. We determined 77 motions as habits for the gold standard. We illustrate our results with an example. Figure 5 is a sample diagram summarizing the habit detection results of our system and the gold standard. Figures 6 and 7 are examples of true positives, Fig. 8 is a false positive, and Fig. 9 is a false negative. In our system, the body part considered to have a habitual

Detected habit by the proposed system

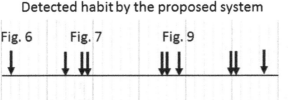

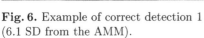

Fig. 5. Comparison of habit detection.

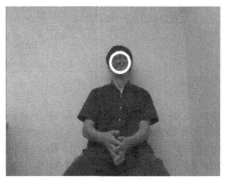

Fig. 6. Example of correct detection 1 (6.1 SD from the AMM).

Fig. 7. Example of correct detection 2 (2.4 SD from the AMM).

motion is highlighted by a circle on the detected body part. In Figs. 6 and 7, the subject scratches his nose with his right hand (at 10 s, 6.1 SD from the AMM), and leans backward (at 90 s, 2.4 SD). These motions were identified as being habits. Recall that the system detects those motions that differ from the AMM by more than 2 SD to be habits. In Fig. 8, we show an example in which the system falsely detected a commonly seen motion as a habit (at 60 s, 2.1 SD). On the contrary, Fig. 9 shows an example in which the system did not detect the habit of falling forward (at 190 s, 0.6 SD). The confusion matrix of our results is shown in Table 2. In summary, the habit detection performance of our system achieved a precision of 84.0 % and a recall of 81.8 %.

4   Discussion

In our experiment, the proposed system achieved good habit detection performance. It can be considered that the proposed methodology has the potential

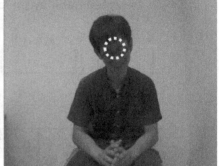

Fig. 8. Example of false positive (2.1 SD from the AMM).

Fig. 9. Example of false negative (0.6 SD from the AMM).

Table 2. Confusion matrix in detection of habits.

Proposed system	Gold standard		
		habit	non-habit
habit		63	12
non-habit		14	

to detect human habits. However, we recognize that several issues need to be addressed to make our system practical. In this experiment, we detected only uncommon motions as habits, focused only on the speed of the motion, and employed a subjective gold standard. We will investigate these issues and develop an improved system in the near future.

5 Conclusion

In this study, we proposed a prototype automated habit detection system to objectively recognize unconscious habits. We used a total of 108 video clips of subjects in conversation, and achieved habit detection accuracy with a precision of 84.0 % and a recall of 81.8 %. Future work will focus on improving the accuracy and reliability of our methodology.

References

1. Merabian, A.: Communication without words. Psychol. Today **2**, 53–55 (1968)
2. Sagawa, K., Abo, S., Tsukamoto, T., Kondo, I.: Forearm trajectory measurement during pitching motion using an elbow-mounted sensor. J. Adv. Mech. Des. Syst. Manuf. **3**, 299–311 (2009)
3. Bobick, A.F., Davis, J.W.: The recognition of human movement using temporal templates. IEEE Trans. Pattern Anal. Mach. Intell. **23**, 257–267 (2001)

4. Schuldt, C., Laptev, I., Caputo, B.: Recognizing human actions: a local SVM approach. In: Proceedings of the 17th International Conference on Pattern Recognition (ICPR), vol. 3, pp. 32–36 (2004)
5. Xia, L., Chen, C.C., Aggarwal, J.K.: View invariant human action recognition using histograms of 3D joints. In: 2012 IEEE Computer Society Conference on Computer Vision and Pattern Recognition Workshops (CVPRW), pp. 20–27 (2012)
6. Evangelidis, G., Singh, G., Horaud, R.: Skeletal quads: human action recognition using joint quadruples. In: 2014 22nd International Conference on Pattern Recognition (ICPR), pp. 4513–4518 (2014)
7. Miranda, L., Vieira, T., Martinez, D., Lewiner, T., Vieira, A.W., Campos, M.F.M.: Online gesture recognition from pose kernel learning and decision forests. Pattern Recogn. Lett. **39**, 65–73 (2014)

Conductor Tutoring Using the Microsoft Kinect

Andrea Salgian[1]($\boxtimes$), Leighanne Hsu[1], Nathaniel Milkosky[1],
and David Vickerman[2]

[1] Department of Computer Science, The College of New Jersey,
Ewing, NJ 08628, USA
{salgian,hsul1,milkosn1}@tcnj.edu
[2] Department of Music, The College of New Jersey, Ewing, NJ 08628, USA
vickermd@tcnj.edu

Abstract. In this paper we present a system that uses the Microsoft Kinect to provide beginner conducting students real time feedback about their performance. Using upper body joint coordinates we detect common mistakes such as swaying, rocking, excessive hinge movement, and mirroring. We compute instant velocities to determine tempo and classify articulation as legato or staccato. Our experiments show that the system performs perfectly when detecting erroneous movements, correctly classifies articulation type most of the time, and can correctly determine tempo by counting the number of beats per minute. The system was well received by conducting students and their instructor, as it allows them to practice by themselves, without an orchestra.

1 Introduction

Musicians spend long hours practicing alone with their instrument. The music they make provides immediate feedback about their performance: if it doesn't sound good, they need to improve. A conductor's instrument however is the orchestra. Without it, the novice conductor is constrained to practice alone, often times in front of a mirror, but without any other feedback. In this paper we describe a system that fills this gap: using a Microsoft Kinect we track and analyze the conductor's gestures and provide immediate feedback. Our system is small and simple enough that a student can set it up and use it in her room.

Conducting has recently become the focus of several computer vision and human computer interaction projects, perhaps because conductors are the only musicians who freely move their hands to create sound and whose gestures are not constrained by a rigid instrument. Many systems were aimed at allowing a musician, or even the general public, to conduct a virtual instrument or an entire virtual orchestra. Examples include Behringer's visual tracking system that can replace sliders and knobs common on electronic instruments with conducting-like gestures performed in front of the camera [1]. Wilson and Bobick's system allowed the general public to "conduct" by waving their hands in the air and controlling the playback speed of a MIDI-based orchestral score [2]. Murphy et al. [3] created computer vision techniques to track a conductor's baton, and analyzed the relationship between the gestures and sound.

© Springer International Publishing Switzerland 2015
G. Bebis et al. (Eds.): ISVC 2015, Part II, LNCS 9475, pp. 24–34, 2015.
DOI: 10.1007/978-3-319-27863-6_3

Pure vision-based systems have had difficulty tracking the conducting baton (or hand) in a variety of backgrounds, so their functionality was limited. Better performance was obtained by systems that used batons equipped with sensors and/or emitters. Most notably, the Digital Baton system implemented by Marrin and Paradiso [4], had an input device that contained pressure and acceleration sensors, and the tip of the baton held an infrared LED which was tracked by a camera with a position-sensitive photodiode. Nakra et al. built the Virtual Maestro [5], which used accelerometer data provided by a Wii remote held by the user.

More recently, researchers turned their attention to conducting real orchestras, as well as pedagogical applications. Salgian et al. used video footage of a professional conductor conducting a professional orchestra to analyze conducting technique [6]. Peng and Gerhard [7] used the infrared camera in a Wii remote to capture conducting gestures with the purpose of comparing them to textbook diagrams. Finally, Ivanova et al. recently published the preliminary design of a conducting practice room that provides an immersive rehearsal experience for beginner conductors [8].

The system we describe in this paper utilizes the Microsoft Kinect to track the hand gestures of a student conductor and provides real time feedback about leaning, swaying, mirroring, as well as conducted tempo and articulation. The system was tested on several students with good results.

2   System Overview

Our system uses the Microsoft Kinect and the Microsoft Kinect SDK to track and analyze the hand gestures of the user. Since the system is targeted towards students learning to conduct an orchestra, it is programmed to detect the main mistakes made by beginner conductors. More specifically, the Conducting Tutor detects the presence of swaying left to right, rocking back and forth, and mirroring, classifies the conducting style as staccato or legato, and displays the tempo.

Conductors traditionally practice in front of a mirror. Our system provides the "mirror" by showing the live video of the user on the screen, and it augments it with information displayed as color and text on the right side of the screen (see Fig. 1).

Ask an outsider to show you what conducting looks like, and they will most likely perform mirroring gestures with their arms. However, if we take a closer look at a conductor, we notice that the two arms move differently. The conductor's right hand generally keeps a steady beat, while the left hand is used for signals such as dynamics (indicating louder or softer music) or cues (alerting musicians that they are expected to perform). Mirroring of the hands means that the hands are making symmetric conducting gestures (Fig. 2 shows an example). Proper technique dictates that some mirroring should occur. The conductor's left hand may mirror the right hand for emphasis or leading into cues, for instance. However, students often continue mirroring with the left hand when it is unnecessary. This often desensitizes the ensemble to the left hand's movements, which

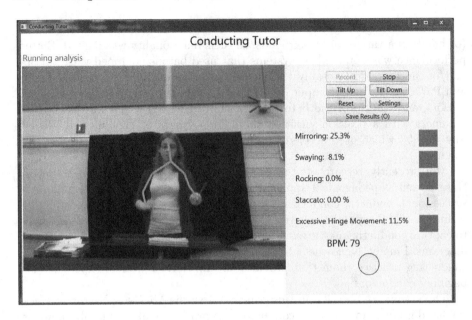

Fig. 1. Screen capture of the Conducting Tutor (Color figure online)

results in a failure to notice any special signals given by this hand. The conducting tutor system continuously displays the percentage of time that mirroring occurred, and alerts the user of mirroring by changing the color of the top square on the right from green to red.

Swaying is simply side-to-side movement of the entire body. Students who are nervous or otherwise uncomfortable will often shift their weight between their feet while conducting, which leads to swaying.

Rocking and excessive hinge movement are often caused by exaggerated movements to instruct the ensemble to bring out more sound or expression. Hinges are defined simply as joints used for conducting gestures. Certain joints, such as the elbow, are supposed remain fairly stationary. Students often fail to understand or realize that subtle changes are generally sufficient in changing the feedback they receive from the ensemble. As a result, they exaggerate their gestures excessively or rock forward in an attempt to bring out more sound. Exaggerated gesturing is also common in students with a marching band or drum major background, who need to do so to ensure visibility across the field. For the same reason, excessive hinge movement frequently occurs with mirroring. It can also be common with faster pieces, as conductors are tempted to conduct these with more gusto. However, exaggerated gestures are generally seen as poor technique in non-marching band situations, where visibility is not an issue. Exaggeration can muddle the message conveyed to the ensemble.

Our system measures the amount of time swaying, rocking, and excessive hinge movement occurs, and displays it on the screen as a percentage. If any one of these mistakes is currently occurring, the indicator color changes from green to red.

Fig. 2. Trace of mirroring motions (Color figure online)

Should this percentage go above a preset threshold, the color of the text box changes from green to red.

Probably the most important information shared through conducting is the tempo, the speed at which the music should be played. The conductor's right hand conveys the beat, and the number of beats per minute defines the tempo. Beginner conductors often misjudge the tempo at which they are conducting, resulting in an incorrect performance. Our system estimates the number of beats per minute (bpm) conducted by the user, and displays it on the screen.

Another important musical characteristic conveyed by the conductor is articulation, the continuity (or lack thereof) between notes and sounds played by the musicians. The two most common articulations are legato and staccato. Legato indicates that musical notes should be smooth and connected, while staccato indicates short notes that are sharply separated. While exact movements between conductors and between pieces for a given conductor will vary greatly, legato conducting tends to be smoother, and staccato conducting tends to be much more pointed, potentially with a more defined beat. Students will occasionally conduct these erroneously. They may fail to change from one to another, or they may start in one style and transition into the other. Either way, such movements are erroneous and therefore undesired. Our system detects the articulation currently being conducted and displays it on the screen as an L on yellow background for legato, and an S on blue background for staccato (see Fig. 1). The percentage of time spent conducting staccato is also displayed.

3   Methodology

The Microsoft Kinect is a motion sensing device that enables gesture recognition by detecting the skeleton of a human figure and tracking its joints. Tracking can

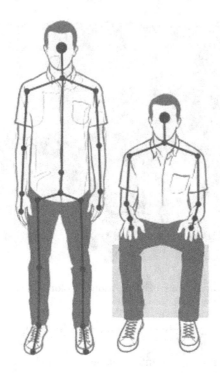

Fig. 3. Joints tracked by the Microsoft Kinect in standing (left) and seated (right) mode [9]

happen in one of two modes: standing mode and seated mode (shown in Fig. 3). Standing mode tracks twenty joints, while seated mode tracks only the ten joints in the upper half of the body (shoulders, elbows, wrists, arms and head). Often, conductors have a music stand in front of them, which obstructs the view of their lower body. Since this can lead to unpredictable system behavior, we opted to use the seated mode instead of the default standing mode.

While some aspects of conducting technique, such as tempo, can be clearly defined in a textbook, others are best demonstrated by example, and conductors have a hard time expressing them in words. Therefore we started by asking a conducting educator and an advanced conducting student to demonstrate correct conducting of a number of musical pieces with varying tempo and articulation, as well as isolated incorrect techniques in a continuous fashion (e.g. continual swaying, continually excessive hinge movement). We recorded these performances and analyzed the motion of the skeletal points as tracked by the Microsoft Kinect.

3.1   Detection of Swaying, Rocking, and Excessive Hinge Movement

Graphing coordinates for relevant joints can easily show where and by how much normal and erroneous movement differ. We picked the "center shoulder" joint, located at the base of the neck, to detect swaying. In seated mode, the spine and other lower joints are not tracked, so we can only identify swaying based on the amount of change we see in the x-coordinates of this joint. The base of the neck does not move in relation to the rest of the body, and it is centralized and infrequently obstructed during conducting.

Figure 4 (left) shows a time series plot of the x-coordinate of this joint for correct and incorrect techniques. As we can see, with correct conducting technique, the conductor does not sway or move from their central location at all. The slight noise is easily accounted for with a small threshold. On the other hand, there is significant movement when the conductor is swaying. By identifying a threshold and a general number of frames it takes to exceed this threshold when the conductor moves, we were able to identify swaying movements using only the x-coordinates.

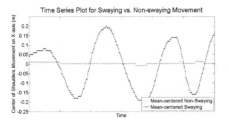 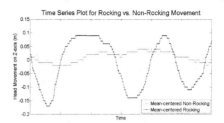

Fig. 4. Left: mean-centered plot of center shoulder coordinate during swaying and non-swaying motions. Right: mean-centered plot of head coordinate during rocking and non-rocking motions

Detection of rocking forward and backward works similarly. Figure 4 (right) shows the z coordinates of the head over time. This plot is very similar to that of swaying, except that there is slightly more motion even when the correct technique is used. We could have used the center shoulder joint again here, but the difference between rocking and non-rocking conducting movements in this joint is less distinct than that of the head. Since the data is very similar to that of swaying, we can process rocking very similarly to swaying by using a larger threshold value.

Excessive hinge movement primarily involves swinging of the elbows, which, like the conductor's body, should remain relatively stationary. Figure 5 shows a scatterplot of the mean-centered x and y coordinates of the elbows for correct and incorrect motion. While we have identified two areas of concern for hinge movement (bad elbow and bad shoulder movement), both resulted in fairly similar elbow movements. Samples from these, graphed in green and blue respectively, bear very similar patterns. The red dots, which indicate correct technique, show much less movement.

It is evident that the movement using a correct technique falls in a smaller radius than the movement using a poor technique. Since incorrect motion happens along all three dimensions (the z dimension is not pictured in our plot

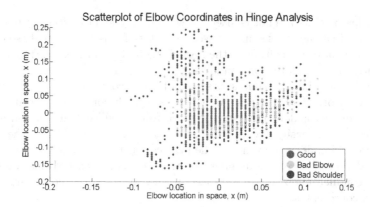

Fig. 5. Scatterplot of mean-centered elbow coordinates in two dimensions (Color figure online)

because of readability), our algorithm calculates the Euclidean distance between consecutive elbow locations over a certain number of frames. A distance larger than a threshold is classified as incorrect conducting technique.

3.2 Mirroring Detection

Mirroring is detected somewhat similarly to swaying and rocking. This time we need to analyze the relative position of two joints that are not stationary, the two hands, over time, by looking at their x, y, and z coordinates. Figure 6 shows two plots: on the left, the x-coordinates of the hands, mirrored over the center of the body using the center shoulder joint; on the right, the y-coordinates of both hands. In both plots, the right hand, which remains on a steady beat, is shown in red. The left hand when mirroring constantly is shown in pink. It is clear that when mirroring, the locations of the two hands is similar, though not perfectly symmetrical.

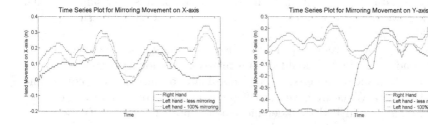

Fig. 6. Plot of hand coordinates on x-axis (left) and y-axis (right) during mirroring and non-mirroring movements (Color figure online)

When mirroring, the left hand mimics the right hand very closely in both dimensions. The blue line in both graphs shows the proper technique. While similar to the right hand in some places, usually the left hand is making some other motion and is not conducting the beat like the right hand. Near the end of the time shown, the conductor is mirroring for about one beat. However, instead of mirroring the entire time, the conductor also cues with the left hand for entrances or cutoffs.

Unlike the previous two errors, mirroring of the hands is not exclusive to incorrect technique. As previously mentioned, some mirroring is permitted, although conductors may disagree as how much does still constitute correct technique. Mirroring may also be dependent on the piece being conducted. Since conducting technique is so situational, the user can adjust the threshold at which mirroring is considered incorrect.

3.3 Tempo Calculation

The tempo of a musical piece is the speed at which it is played. Tempo is measured in beats per minute, where the beat is the basic unit of time, often defined as the rhythm of the music. The beat of the music is indicated by the conductor's right hand. The hand traces a shape in the air depending on the time signature, indicating each beat with a change in the direction of motion.

Our algorithm uses the right hand coordinates provided by the Kinect tracker to compute the instantaneous velocity and looks at neighboring instances to detect changes in direction. Each change in direction is counted as a beat. The timing of the ten most recent beats is averaged to extrapolate the beat per minute (bpm) count, which is then displayed on the screen.

3.4 Recognizing Articulation

Articulation in conducting refers to conveying the continuity between the notes played by the musicians. Legato indicates smooth and connected notes, and is conducted with long and fluid gestures, while staccato indicates short, sharply separated notes, and is conducted with short and sharp movements.

Of all the conducting characteristics analyzed by our system, articulation was the hardest to quantify. The difference between conducting legato and staccato seems obvious, yet one has difficulty describing it in words. The sharpness of staccato movements seemed to suggest that the difference would lie in the velocity or acceleration of the right hand. We used the hand coordinates obtained by the Kinect to compute instantaneous and average velocities and accelerations, and we noticed that the acceleration was the same for staccato and legato pieces. If the tempo was the same, the average velocity was also the same regardless of the articulation.

We have found that due to the tempo patterns, the right hand velocity magnitude varies constantly, peaking right before the hand changes direction. The difference between staccato and legato lies in the height of these peaks, staccato gestures speed up and slow down significantly more than legato gestures. This can be seen in Fig. 7, which shows how the velocity magnitude varies over time. The same musical piece was conducted in staccato, with the velocity shown in

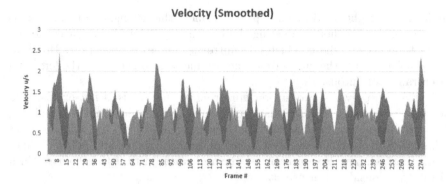

Fig. 7. Smoothed magnitude of velocity of right hand conducting a staccato piece (blue) and a legato piece (orange). Staccato is characterized by peaks of higher magnitude (Color figure online)

blue, and in legato, with the velocity shown in orange. The velocity magnitude had to be smoothed due to occasional tracking errors by the Kinect.

Given this finding, our algorithm computes the peak height in velocity magnitude and uses a threshold to distinguish between legato and staccato. Since conducting styles vary widely, this threshold can be adjusted by the user.

4   Experiments and Results

The system was tested on ten students who conducted the introductory part (called *Theme*) of Edward Elgar's *Enigma Variations* as part of their assessment in the Conducting I course at our institution. The piece is characterized by varying tempo, making conducting (and assessment) more difficult. The articulation also changes, starting out as legato and switching to staccato in the middle.

To make sure testing was not biased, students could not see the computer screen, and received feedback only from the Conducting instructor. The instructor could not see the computer screen either, thus his decisions were independent of the program output. We recorded the program output, as well as the instructor's feedback, and compared the two.

The students conducted anywhere between 40 and 100 seconds (until they were stopped by the instructor), always starting from the beginning of the piece.

We found that the mirroring, as well as swaying, rocking, and excessive hinge movement was always correctly detected. Articulation was correctly classified 70 % of the time.

Tempo measurement performance varies because of the Kinect framerate.

We also tested the system allowing students to see the instant feedback on the screen. Both the students and their instructor found the system very helpful and easy to use.

5   Conclusions and Future Work

We presented the *Conducting Tutor*: a system that uses the Microsoft Kinect to provide real-time feedback about conducting performance.

Our method uses the trajectory of upper body joint coordinates to detect swaying, rocking, excessive hinge movement, and mirroring, common mistakes made by novice conductors. We perform beat per minute calculation to determine tempo by looking at abrupt changes in the direction of right hand motion. Finally we classify articulation as legato or staccato by looking at how the magnitude of right hand velocity changes over time.

The system provides the user with a mirror image and overlaid upper body skeleton as tracked by the Kinect, together with instantaneous analysis of the current gesture and overall performance statistics.

We tested the system on several students conducting a difficult musical piece containing changes in tempo and articulation, and we have found that incorrect gestures are always detected. Articulation is correctly classified 70 % of the time, while tempo calculation still has room for improvement.

The system was very well received by users, as it fills an important need of feedback for novice conductors that practice by themselves without an orchestra or an instructor. Future work includes refinement of the tempo calculation method, extensive rigorous testing, and the addition of more conducting technique elements.

References

1. Behringer, R.: Conducting digitally stored music by computer vision tracking. In: First International Conference on Automated Production of Cross Media Content for Multi-Channel Distribution (AXMEDIS05), Florence, Italy (2005)
2. Wilson, A., Bobick, A.: Realtime online adaptive gesture recognition. In: International Workshop on Recognition, Analysis, and Tracking of Faces and Gestures in Real-Time Systems, Corfu, Greece
3. Murphy, D., Andersen, T.H., Jensen, K.: Conducting audio files via computer vision. In: Camurri, A., Volpe, G. (eds.) GW 2003. LNCS (LNAI), vol. 2915, pp. 529–540. Springer, Heidelberg (2004)
4. Marrin, T., Paradiso, J.: The digital baton: a versatile performance instrument. In: International Computer Music Conference, Thessaloniki, Greece, pp. 313–316 (1997)
5. Nakra, T.M., Ivanov, Y., Smaragdis, P., Ault, C.: The ubs virtual maestro: an interactive conducting system. In: New Interfaces for Musical Expression (NIME), Pittsburgh, PA (2009)
6. Salgian, A., Pfirrmann, M., Nakra, T.M.: Follow the beat? Understanding conducting gestures from video. In: Bebis, G., Boyle, R., Parvin, B., Koracin, D., Paragios, N., Tanveer, S.-M., Ju, T., Liu, Z., Coquillart, S., Cruz-Neira, C., Müller, T., Malzbender, T. (eds.) ISVC 2007, Part I. LNCS, vol. 4841, pp. 414–423. Springer, Heidelberg (2007)
7. Peng, L., Gerhard, D.: A wii-based gestural interface for computer-based conducting systems. In: New Interfaces for Musical Expression (NIME), Pittsburgh, PA (2009)

8. Ivanova, E., Wang, L., Fu, Y., Gadzala, J.: Maestro: a practice system to track, record, and observe for novice orchestral conductors. In: CHI 2014 Extended Abstracts on Human Factors in Computing Systems
9. Library, M.: Kinect for windows sdk. https://msdn.microsoft.com/en-us/library/hh855347.aspx. Accessed 20 August 2015

Lens Distortion Rectification Using Triangulation Based Interpolation

Burak Benligiray$^{(\boxtimes)}$ and Cihan Topal

Department of Electrical and Electronics Engineering,
Anadolu University, Eskişehir, Turkey
burakbenligiray@anadolu.edu.tr

Abstract. Nonlinear lens distortion rectification is a common first step in image processing applications where the assumption of a linear camera model is essential. For rectifying the lens distortion, forward distortion model needs to be known. However, many self-calibration methods estimate the inverse distortion model. In the literature, the inverse of the estimated model is approximated for image rectification, which introduces additional error to the system. We propose a novel distortion rectification method that uses the inverse distortion model directly. The method starts by mapping the distorted pixels to the rectified image using the inverse distortion model. The resulting set of points with subpixel locations are triangulated. The pixel values of the rectified image are linearly interpolated based on this triangulation. The method is applicable to all camera calibration methods that estimate the inverse distortion model and performs well across a large range of parameters.

1 Introduction

Many computer vision and image processing methods assume a general pinhole camera model where linearity is preserved. More specific applications, such as reconstruction or stereovision, benefit from precise estimation of camera parameters [1]. Pinhole camera model is not directly applicable when the camera lens causes a nonlinear distortion in the image. This distortion is commonly modeled using the Brown-Conrady polynomial model [2] or Fitzgibbon's division model [3]. These distortion models either define the mapping from distorted coordinates to rectified coordinates, or vice versa. Parameters for either of these versions of the methods are estimated with distortion calibration methods. However, a problem arises if the application requires the inverse mapping. Since the mapping function is not analytically invertible, the inverse is required to be approximated [4].

The conventional method of distortion rectification of an image requires the mapping from rectified coordinates to distorted coordinates. We will name the parameters that define this mapping as *(forward) distortion parameters*, as this is the most common objective of distortion calibration. The parameters that define the mapping from distorted coordinates to rectified coordinates will be referred to as *inverse distortion parameters*. In calibration cases where a set of

© Springer International Publishing Switzerland 2015
G. Bebis et al. (Eds.): ISVC 2015, Part II, LNCS 9475, pp. 35–44, 2015.
DOI: 10.1007/978-3-319-27863-6_4

distorted and rectified point correspondences can be generated using the estimated intrinsic and extrinsic camera parameters, parameters of the distortion model can be estimated in both ways. However, in self-calibration of lens distortion, the options are limited. The common method is to rectify detected features with different inverse distortion parameters iteratively, until the resulting rectified features are believed to belong to a correctly rectified image [5–7]. The general assumption of line straightness methods is that the correctly rectified image contains more linear features compared to suboptimally rectificated images. The method of quantifying the linearity of the features in an image is one of the main differences among line straightness methods. When applying such methods, the estimated inverse distortion parameters are not used for rectifying the image directly. Instead, the distortion parameters are needed to be approximated using the estimated inverse parameters. Similar to this problem, points from the distorted image cannot be mapped to the rectifed image knowing only the distortion parameters. This mapping is defined by the inverse distortion model.

The distortion parameters can be directly inverted if only a single distortion parameter is used [5,8,9]. If multiple parameters are used, the inverse of the known distortion function is approximated. Doing so will introduce additional error to the system. We propose a novel method for rectifying the lens distortion using the inverse parameters directly, thus skipping the approximation step. The inverse distortion model maps pixels of the distorted image to subpixel locations on the rectified image. Since not all pixels of the rectified image are covered by this mapping, blank regions appear on the rectified image. In the proposed method, the rectified point set is triangulated, such that a pixel inside this point set's convex hull will be surrounded by a triangle whose vertices are three mapped points. These three mapped points are used to linearly interpolate the value at the pixel. By doing this for every pixel, the image can be rectified with no blank regions.

Other methods approximate the solution with some error [4,10–14] and interpolate to rectify the distortion, meanwhile our method does the inversion implicitly in a single interpolation operation. Furthermore, our method can work with more complex approximations of lens distortion without needing any modification or additional processing time.

2   Lens Distortion Model

For self-calibration of lens distortion, polynomial [4–6,9–11,15–18] and division [3,7,8,19,20] radial lens distortion models are used. Division model is generally approximated with a single parameter, exception being [20]. Methods that use polynomial model use either one [5,6,9,16] or two parameters [4,10,11,15,17,18]. Out of these methods, only Ahmed and Farag's estimates a parameter of tangential lens distortion [16]. Many methods estimate a distortion center different from the principal point [5,7,8,16,18,20], which can be said to model some aspects of tangential distortion, along with the ratio of vertical and horizontal focal lengths [5]. We will use the polynomial distortion model with only two radial distortion parameters for brevity. The proposed method for distortion rectification can be easily extended to division model, tangential distortion modeling and additional distortion parameters.

Let us assume $(\tilde{x}_d, \tilde{y}_d)$ are distorted screen coordinates and $(\tilde{x}_u, \tilde{y}_u)$ are rectified screen coordinates. These coordinates can be normalized by subtracting the distortion center, (c_x, c_y), thus yielding (x_d, y_d) and (x_u, y_u), recpectively. The distances of the normalized points from distortion centers give the radii, r_d and r_u. According to this notation, the radial distortion can be modeled in two ways:

$$r_d = r_u(1 + \kappa_1 r_u^2 + \kappa_2 r_u^4) \tag{1}$$

$$r_u = r_d(1 + \kappa_1' r_d^2 + \kappa_2' r_d^4) \tag{2}$$

Equation 1 will be referred to as the distortion model, as it models the mapping from non-distorted points to distorted points. Equation 2 will be referred to as the inverse distortion model, as it does the opposite. The effect of these distortion models on x and y coordinates are as follows:

$$x_d = x_u(1 + \kappa_1 r_u^2 + \kappa_2 r_u^4) y_d = y_u(1 + \kappa_1 r_u^2 + \kappa_2 r_u^4) \tag{3}$$

$$x_u = x_d(1 + \kappa_1' r_d^2 + \kappa_2' r_d^4) y_u = y_d(1 + \kappa_1' r_d^2 + \kappa_2' r_d^4) \tag{4}$$

κ_1 and κ_2 in Eq. 1 are the distortion parameters and κ_1' and κ_2' in Eq. 2 are the inverse distortion parameters. In a majority of applications, this distinction is not made and the more convenient model is used directly. When using a line straightness based method, only the distorted coordinates of the detected features are known. Generating rectified coordinates using a set of parameters uses the inverse distortion model. Thus, inverse distortion parameters are found when such methods are used.

3   Distortion Rectification

Conventional distortion rectification uses the forward distortion model, which is the model that generates distorted coordinates, using the distortion parameters and rectified coordinates (Eq. 3). Assume we have an empty image that will store the result of the rectification. Obviously, each pixel of this image needs to be filled. Distortion parameters (κ_i) and rectified coordinates (x_u, y_u) are known, thus the distorted coordinates (x_d, y_d) can be found. The empty pixel in the rectified image will correspond to a subpixel location in the distorted image. The value of the pixel is commonly estimated with bilinear interpolation, which will use the four surrounding pixels' values. Doing this operation for each pixel in the empty image will result in a complete rectified image. Now let us discuss the cases in which only the inverse distortion parameters are known.

3.1 Inversion by Newton-Raphson Method

Newton-Raphson method is used to find roots of a function by iteration. Using Eq. 2:

$$f(r_d) = r_d + \kappa_1' r_d^3 + \kappa_2' r_d^5 - r_u \tag{5}$$

$$f'(r_d) = 1 + 3\kappa_1' r_d^2 + 5\kappa_2' r_d^4 \tag{6}$$

This function's derivative exists for possible roots (Eq. 6), thus the method is applicable. The method approximates to the solution by iteration in the following manner:

$$r_{dn+1} = r_{dn} - \frac{f(r_{dn})}{f'(r_{dn})} \tag{7}$$

A good starting point for r_d can be r_u, as they will be close when the distortion is weak, and there is no better choice for cases in which the distortion is strong. After estimating r_d, we can find (x_d, y_d) using Eq. 4. By mapping pixel coordinates of rectified image to subpixel locations in the distortion image using this method, conventional distortion rectification can be done [10]. The downside of this method is that this iteration must be done for each pixel and convergence will take longer for stronger distortions.

3.2 Taylor Expansion Based Inverse Model

Taylor expansion [12] and implicit rational polynomials [11,13,14] are utilized in some studies, yet they are argued to be unstable by Mallon and Whelan. Instead, Taylor expansion method is refined to be [4]:

$$p_u = p_d - p_d \left(\frac{\alpha_1 r_d^2 + \alpha_2 r_d^4 + \alpha_3 r_d^6 + \alpha_4 r_d^8}{1 + 4\alpha_5 r_d^2 + 6\alpha_6 r_d^4} \right) \tag{8}$$

In this form, this model estimates the inverse model, knowing the forward model. However, it can also be used to esimate the forward model, knowing the inverse distortion model, which is the problem we have defined.

4 Proposed Method: Rectification Using Inverse Distortion Model

Previously mentioned methods aim to aproximate the distortion model from the estimated inverse distortion model, which will introduce additional error to the system, as will be shown in our experimental results. Instead, directly using the inverse distortion model would have been preferable. To do so, an approach similar to conventional distortion rectification can be followed. Let us start with an empty rectified image. For each pixel in the distorted image, the corresponding pixel in the empty image can be filled using the inverse distortion model. The problem with this approach is that not all pixels in the rectified image are guaranteed to be filled, as addressed in other studies [9,10]. Certain

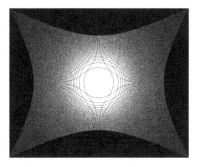

Fig. 1. Rectification of a distorted fully white image using the inverse distortion function with nearest neighbor interpolation

patterns of voids will appear at the rectified image (see Fig. 1). Prescott and McLean vote on the surrounding four pixels of the mapped coordinates instead of using nearest neighbor, yet doing so still does not guarantee the elimination of all voids in the image [18].

The problem can simply be defined as interpolating a nonlinearly scattered sparse image. This is not a very common issue in image processing, but the solution can be derived from a similar problem. Image sequence superresolution aims to create a high resolution image from multiple low resolution images. After fusing the information from lower resolution images, the result is a pseudo-random scattering of points that do not fit the grid of the output image. Lertrattanapanich and Bose propose using Delaunay triangulation for solving this 2D interpolation problem [21]. This approach is applicable to our case.

The pixels from the distorted image are mapped to the rectified image plane using Eq. 4, which will produce a set of points in subpixel locations. Delaunay triangulation [22] is applied to this set of points. Each pixel of the rectified image will be surrounded by a triangle. Using the values of the surrounding triangle's vertices, the value at the pixel can be estimated. This estimation can be done linearly as proposed by Dyn et al. [23], which resembles Phong's illumination model [24]. In this case, each pixel will be voted on by three points, instead of four as is the case in bilinear interpolation done in conventional distortion rectification (see Fig. 2).

The effect of using a different interpolation method is rather unpredictable. The used triangulation method does not guarantee finding the closest three points to the pixel whose value will be interpolated. Especially where points are sparser, suboptimal interpolations can be expected. In such cases, any type of linear interpolation will fail to estimate the values correctly. On the other hand, not approximating the inverse model is a certain advantage.

5   Experimental Results

The proposed method directly uses the inverse distortion parameters to rectify the lens distortion. Other methods approximate the distortion parameters using

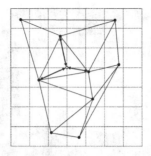

Fig. 2. Triangulation based 2D interpolation. Corners of the grid represent the pixels of the rectified image. Filled circles are distorted pixels mapped to the rectified image. The empty circle represents a single pixel in the rectified image. The vertices of the surrounding triangle will vote on the value of this pixel, weighted by their Euclidian distances

the inverse parameters and apply conventional distortion rectification. Since our method's output is a rectified image, doing experiments with randomly generated point data and measuring the geometric error is not possible. Instead, we will apply artificial lens distortion to an image. Then, the distortion will be rectified using alternative methods and differences from the original image will be evaluated.

(a) $\kappa_1' = 1 \times 10^{-11} pix^{-2}$
$\kappa_2' = 2 \times 10^{-12} pix^{-4}$

(b) $\kappa_1' = 1 \times 10^{-13} pix^{-2}$
$\kappa_2' = 2 \times 10^{-14} pix^{-4}$

Fig. 3. Examples of artificial distortions applied to an image with 1920×1080 resolution. The distortion center is assumed to be the center of the image and the ratio of horizontal and vertical focal lengths is unity

The composition of scenes or the actual lens distortion model for the test images are not critical, as the images will not be used for calibration. To avoid any bias, we used the means of results produced from 10 uncompressed photographs taken in indoor and outdoor environments. We have discussed in Sect. 3 that forward distortion parameters are used to apply conventional distortion rectification. In a similar manner, the inverse distortion parameters can be used to apply distortion. The images are distorted using sets of inverse distortion parameters diverse enough to cover both subtle and substantial distortions. κ_1' is given values between $1 \times 10^{-11} pix^{-2}$ and $1 \times 10^{-13} pix^{-2}$, κ_2' is set to be one fifth of κ_1'. See Fig. 3 for the minimum and maximum effects of the applied distortion. Since the signs of parameters do not have a significant effect on the performance of the methods, we omitted the negative range of parameters.

After distorting the images, the images are rectified using different methods and differences of their results from the original image are quantified using RMSE and PSNR. The existing methods use bilinear interpolation after approximating the distortion model. For fair comparison between the methods, we also used bilinear interpolation after triangulation, although more complex interpolation methods such as bicubic are available. The resulting image is aligned with the input image for all methods. The interpolation by triangulation method had 2–3 pixel wide artifacts along the borders (see Fig. 4), hence these parts are cropped.

Fig. 4. Upper border of an image rectified with the proposed method. Triangular patterns appear at the borders of the image due to the interpolation method. For the methods that use bilinear interpolation, edges of the images appear as straight lines

RMSE and PSNR results for different levels of artificial distortion are given in Fig. 5. The proposed method performs stably well across all parameters. The custom inverse model is approximated by using all normalized pixels on the distorted image in a nonlinear optimization scheme. Rate of convergence for Newton-Raphson method to a single zero is quadratic. Starting point is chosen as the undistorted point and the method iteratively converges to its distorted correspondence. These points are expected to be close to each other, especially where distortion is less pronounced. Therefore, the convergence is fairly quick, to the point that applying a single iteration of Newton-Raphson may yield acceptable results. In the experimental results, we chose to show the single iteration case and the case where full convergence is met. For the distortion coefficients chosen in our experiments, full convergence was met after 5 iterations. One interesting result of Newton-Raphson method was that it created artifacts on the image along the void patterns shown in Fig. 1. The sampling points are farther away along the patterns, hence the interpolation may have failed to approximate the nonlinearity. This is not the weakness of Newton-Raphson, as given enough iterations, it should be able to approximate the inverse of the distortion function. The difference in results between the proposed method and Newton-Raphson originates from the interpolation performance where data points are sparse.

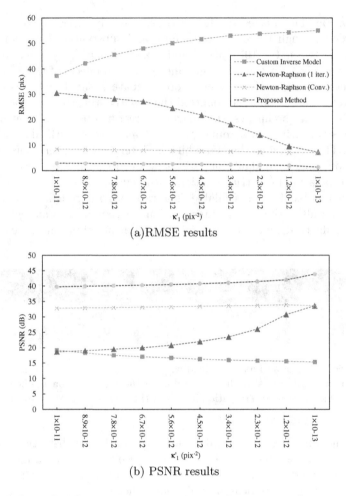

(a)RMSE results

(b) PSNR results

Fig. 5. Results with different inverse distortion coefficients. For all steps, κ'_2 is set to be $\kappa'_1/5$

The rectification map is a look up table that lists contributing pixel coordinates from the distorted image for each pixel in the rectified image, along with their respective weights. This map is used to rectify multiple images with the same distortion characteristics in an efficient manner. The average running times of different methods for building a rectification map are presented in Table 1. The proposed method gives improved results in a comparable amount of time. The running time of the proposed method can be improved by processing the rectified points in smaller partitions. Doing the triangulation operation in small batches will be faster than triangulating the entire image. Obviously, this may change the final result of the triangulation. The imposed difference may either be insignificant, or negative. After the rectification map is built for a distortion model, rectification of images with each method will take the same amount of processsing power and memory.

Table 1. Average running times for building the distortion rectification map using a 1920×1080 image

Method	Average running times (s)
Custom Inverse Model	76.7
Newton-Raphson (Single Iteration)	26.8
Newton-Raphson (Complete Convergence)	32.3
Proposed	33.4

6 Conclusion

In this paper, a method of image rectification using directly the inverse distortion model is introduced. Instead of approximating the forward distortion model, the pixels of the distorted image are mapped to the rectified image. The resulting point set is triangulated using Delaunay's algorithm. The pixel values of the rectified image are interpolated according to this triangulation.

The method can be applied to any distortion model, as it only uses the results of the mapping function. Linear interpolation is chosen as it is commonly used in conventional distortion rectification. However, interpolation accuracy can be improved by using a nonlinear approach. Delaunay triangulation is used as it is a well-studied and optimized algorithm. Memory and processing time tradeoffs can be explored using different implementations. Furthermore, the effect of using different triangulation algorithms may be significant.

References

1. Hartley, R., Zisserman, A.: Multiple View Geometry in Computer Vision. Cambridge University Press, New York (2003)
2. Brown, D.C.: Decentering distortion of lenses. Photometric Eng. **32**, 444–462 (1966)
3. Fitzgibbon, A.W.: Simultaneous linear estimation of multiple view geometry and lens distortion. In: Computer Vision and Pattern Recognition, IEEE (2001)
4. Mallon, J., Whelan, P.F.: Precise radial un-distortion of images. In: Proceedings of the 17th International Conference on Pattern Recognition ICPR 2004, vol. 1, pp. 18–21. IEEE (2004)
5. Devernay, F., Faugeras, O.: Straight lines have to be straight. Machine Vision and Applications **13**, 14–24 (2001)
6. Brauer-Burchardt, C., Voss, K.: A new algorithm to correct fish-eye-and strong wide-angle-lens-distortion from single images. In: International Conference on Image Processing, vol. 1, pp. 225–228. IEEE (2001)

7. Wang, A., Qiu, T., Shao, L.: A simple method of radial distortion correction with centre of distortion estimation. J. Math. Imaging Vision **35**, 165–172 (2009)

8. Bukhari, F., Dailey, M.N.: Automatic radial distortion estimation from a single image. J. Math. Imaging Vision **45**, 31–45 (2013)

9. Cucchiara, R., Grana, C., Prati, A., Vezzani, R.: A hough transform-based method for radial lens distortion correction. In: 2003 Proceedings of 12th International Conference on Image Analysis and Processing, pp. 182–187. IEEE (2003)

10. Gonzalez-Aguilera, D., Gomez-Lahoz, J., Rodríguez-Gonzálvez, P.: An automatic approach for radial lens distortion correction from a single image. IEEE Sens. J **11**, 956–965 (2011)

11. Alvarez, L., Gómez, L., Sendra, J.R.: An algebraic approach to lens distortion by line rectification. J. Math. Imaging Vision **35**, 36–50 (2009)

12. Heikkila, J.: Geometric camera calibration using circular control points. IEEE Trans. Patt. Anal. Machine Intell. **22**, 1066–1077 (2000)

13. Heikkila, J., Silvén, O.: A four-step camera calibration procedure with implicit image correction. In: Proceedings of 1997 IEEE Computer Society Conference on Computer Vision and Pattern Recognition, pp. 1106–1112. IEEE (1997)

14. Wei, G.Q., De Ma, S.: Implicit and explicit camera calibration: theory and experiments. IEEE Trans. Patt. Anal. Mach. Intell. **16**, 469–480 (1994)

15. Grammatikopoulos, L., Karras, G., Petsa, E.: An automatic approach for camera calibration from vanishing points. ISPRS J. Photogrammetry Remote Sens. **62**, 64–76 (2007)

16. Ahmed, M., Farag, A.: Nonmetric calibration of camera lens distortion: differential methods and robust estimation. IEEE Transactions on Image Processing **14**, 1215–1230 (2005)

17. Thormählen, T., Broszio, H., Wassermann, I.: Robust line-based calibration of lens distortion from a single view. In: Proceedings of MIRAGE, pp. 105–112 (2003)

18. Prescott, B., McLean, G.: Line-based correction of radial lens distortion. Graph. Models Image Process. **59**, 39–47 (1997)

19. Strand, R., Hayman, E.: Correcting radial distortion by circle fitting. In: BMVC (2005)

20. Brauer-Burchardt, C., Voss, K.: Automatic correction of weak radial lens distortion in single views of urban scenes using vanishing points. In: International Conference on Image Processing, vol. 3, pp.865–868. IEEE (2002)

21. Lertrattanapanich, S., Bose, N.K.: High resolution image formation from low resolution frames using Delaunay triangulation. IEEE Transactions on Image Processing **11**, 1427–1441 (2002)

22. Lee, D.T., Schachter, B.J.: Two algorithms for constructing a delaunay triangulation. Int. J. Comput. Inf. Sci. **9**, 219–242 (1980)

23. Dyn, N., Levin, D., Rippa, S.: Data dependent triangulations for piecewise linear interpolation. IMA J. Numer. Anal. **10**, 137–154 (1990)

24. Phong, B.T.: Illumination for computer generated pictures. Commun. ACM **18**, 311–317 (1975)

A Computer Vision System for Automatic Classification of Most Consumed Brazilian Beans

S.A. Araújo$^{(\boxtimes)}$, W.A.L. Alves, P.A. Belan, and K.P. Anselmo

Informatics and Knowledge Management Graduate Program,
Universidade Nove de Julho, São Paulo, Brazil
saraujo@uninove.br

Abstract. In this work we propose a computer vision system (CVS) for automatic classification of beans. It is able to classify the beans most consumed in Brazil, according to their skin colors and is composed by three main steps: (i) image acquisition and pre-processing, (ii) segmentation of grains and (iii) classification of grains. In the conducted experiments, we used an apparatus controlled by a PC that includes a conveyor belt, an image acquisition chamber and a camera, to simulate an industrial line of production. The results obtained in the experiments indicate that proposed system could be used to support the visual quality inspection of Brazilian beans.

Keywords: Beans · Granulometry · Computer vision system

1 Introduction

Beans are a legume rich in protein and energy and compose, with rice, the basic diet of Brazilian people. As most food products, their visual properties comprise an important criterion for the choice of the consumers. The quality control of this product in Brazil follows a set of standards and procedures of the Brazilian Ministry of Agriculture, Livestock and Supply BMALS [1].

Basically, the visual quality inspection of beans in Brazil is conducted manually as follows: a sample of at least 250 g is extracted of a batch of beans. After separating the foreign matter and impurities using a circular sieve with holes of 5 mm of diameter and performing a visual inspection of sample, is determined the group, class and type, based on the operating procedures established by BMALS. The group is related to the botanical species. The class is determined according to the mixture of beans, taking into account their skin colors (e.g. black, white, colors or mixed), independent of the group. The type refers to defects found in the sample, such as broken, grains of other plant species, foreign matter and some defects as moldy, burned, crushed, damaged by insects (chopped), sprouted, wrinkled, stained, discolored and damaged by various causes [1].

Although we have a range of works in the literature dealing with analysis and classification of seeds and grains, only the papers of [2–6], address the development of automated systems for automatic classification of beans.

© Springer International Publishing Switzerland 2015
G. Bebis et al. (Eds.): ISVC 2015, Part II, LNCS 9475, pp. 45–53, 2015.
DOI: 10.1007/978-3-319-27863-6_5

Kiliç et al. [2] developed a CVS for classification of beans, based on their colors. For this purpose, known techniques for segmentation, edge detection and color features quantification, besides mathematical morphology operators were implemented in MATLAB software. For the classification task, the authors used a Multilayer Perceptron Artificial Neural Network. In the conducted experiments, the hit rate obtained was 90.6 %.

Aguilera et al. [3] proposed an application based on computer vision to classify seeds, beans, and grains. However, in this paper only experiments using different types of rice and lentils grains provided by a food company were conducted.

Venora et al. [4,5] used techniques of analysis and image processing, and statistical classifiers to identify varieties of beans grown in two regions of Italy, based on features like color, shape and size of the grains. In the experiments, they obtained the hit rates of 99.56 % and 98.49 %, considering, respectively, six and fifteen landraces of beans.

Laurent et al. [6] evaluated the relationship between changes in the skin color of beans during storage with the phenomenon known as *hard-to-cook beans*. In the experiments, the authors used color histograms and statistical analysis and the results showed the existence of the investigated relationship.

The researches of [2–6] demonstrate the need and importance of automatic processes focused on visual inspection of beans. However, it is possible to observe in these investigations a severe limitation: the spacing required between the grains in the samples to be analyzed, in order to facilitate the segmentation process. This occurs because it is still a great challenge for a CVS to identify the edges of each object when they are very close or glued to each other. The problem is that this limitation hinders the applicability of such systems in industrial processes, considering the need for additional equipment to automatically make the arrangement of grains/seeds in each sample to be analyzed. Furthermore, these researches do not include beans most consumed in Brazil.

This limitation in the segmentation process was overcome by a CVS presented by Araújo et al. [7], which proposed a robust correlation-based granulometry module for segmentation of grains and conducted several experiments considering Brazilian beans. However, the drawback of the Araújo's CVS is the time consumption.

In this paper we present a CVS for automatic classification of Brazilian beans aiming to improve the time consumption of the CVS proposed in [7]. The developed system is composed by the modules of pre-processing and classification of Araujo's CVS [7] and a new module based on Watershed Transform (WT) for segmentation of grains.

2   Materials and Experimental Setup

2.1   Materials

For conducting the experiments, we used the same apparatus developed in [7]. It consists of a low cost prototype composed by a conveyor belt with 1,500 mm of length and 250 mm of width and an image acquisition chamber containing the lamp for illumination and a camera Microsoft® LifeCam HD-5000. The velocity of the conveyor, the camera and the lighting of lamps are totally controlled by the PC that is running the system. The system proposed in this work was developed in C/C++ using libraries for image processing and computer vision ProEikon[1] and OpenCV[2].

2.2 Experimental Setup

The experiments were conducted aiming to evaluate the performance of developed CVS in the process of beans classification, based on their skin colors, so that it was possible to determine the amount of mixed grains present in each sample. The idea is that only after evaluating a pre-determined set of samples, the class information is determined, based on the tolerance limits of the mixtures for each class of beans.

We take into account most consumed beans for Brazilian people [8], that is, carioca (70% of market share), black (20% of market share) and mulatto (included in the 10% of all other varieties), illustrated in Fig. 1.

(a) carioca beans   (b) mulatto beans   (c) black beans

Fig. 1. Samples of beans most consumed in Brazil [7].

We evaluated 100 samples of beans, each one containing 100 grains. These samples, presented in Table 1, were divided into 10 subsets, according to the percentage of grains black, mulatto and carioca. The difference among the 10 samples in each subset was only the grains positions.

The results obtained by our system were compared with results achieved by Araújo's CVS [7]. Summarizing, the analysis of each sample was done according to the following steps:

– A sample of beans was placed in the conveyor belt via input box;
– The conveyor belt transported the sample to image acquisition chamber;
– The proposed system stopped the conveyor belt, acquired the image from the sample, processed it and computed the percentage of the mixture in the sample.

---

[1] Available at http://www.lps.usp.br/~hae/software/.
[2] Available at http://opencv.org/

Table 1. Subsets of samples used in the experiments.

Subset	Total of samples	Number of carioca grains in each sample	Number of black grains in each sample	Number of mulatto grains in each sample
1	10	100	0	0
2	10	95	0	5
3	10	95	5	0
4	10	90	5	5
5	10	85	5	10
6	10	85	10	5
7	10	80	10	10
8	10	85	0	15
9	10	85	15	0
10	10	70	15	15
Total	100	870	65	65

3 Proposed CVS

The proposed system was integrated into the apparatus described in Subsect. 2.1, and is able to analyze the sample of beans placed on the conveyor belt. It is composed by three main steps briefly presented in next subsections.

3.1 Image Acquisition and Pre-processing

First, consider an image as a function f from a set of pixels $\mathcal{D} \subseteq \mathbb{Z}^2$ to set of values $\mathbb{K} = \{0, 1, ..., 2^{bits} - 1\}$, i.e., $f : \mathcal{D} \to \mathbb{K}$. Then, a rgb image f_{rgb} of the sample with 800×600 pixels with 24-bits of depth per pixel is acquired and after that is pre-processed.

In the first pre-processing step the rgb image f_{rgb} is mapped to a grayscale image f_{gray} as follows: each pixel $p \in \mathcal{D}$ is mapped to black if the pixel color $f_{rgb}(p)$ is a typical color of beans (foreground), and to white if the pixel color $f_{rgb}(p)$ is a typical background color (see Fig. 2b). Then, as described in [7], when the procedure cannot decide whether the pixel color is foreground or background, it maps the unknown pixel to different shades of gray according to similarities to known foreground or background colors. Formally, let B and F be two sets of colors constructed from training images containing colors of beans and their background, respectively. The pre-processing filtering is given by

$$\forall p \in \mathcal{D}, f_{gray}(p) = \frac{dist(\text{ColorF}, f_{rgb}(p))}{dist(\text{ColorF}, f_{rgb}(p)) + dist(\text{ColorB}, f_{rgb}(p))}, \qquad (1)$$

where $dist(\text{ColorA}, \text{ColorB})$ is the Euclidian distance between ColorA and ColorB,

$$\text{ColorF} = \arg \min\{dist(\text{Color}, f_{rgb}(p)) : \text{Color} \in F\} \qquad (2)$$

and

$$\text{ColorB} = \arg\min\{dist(\text{Color}, f_{rgb}(p)) : \text{Color} \in B\}. \tag{3}$$

Differently from [7], in this work we used a look-up table to compute this filter very efficiently. So, instead to conduct a training for each image, the filtering is done directly based on the look-up table previously created. The modifications in this step allowed a significant reduction in processing time.

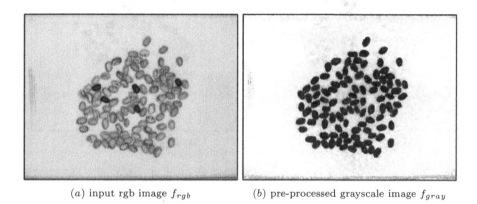

(a) input rgb image f_{rgb} (b) pre-processed grayscale image f_{gray}

Fig. 2. Acquired and pre-processed images.

3.2 Segmentation of Grains

In this step, the Watershed Transform (WT) [9] is applied to segment the grains. WT is as well-known region-based segmentation approach and the intuitive idea behind this method comes from of geography: suppose a landscape immersed in a lake, with holes pierced in local minima. Basins will fill up with water starting at these local minima, and, at points where water coming from different basins would meet, dams are built. When the water level has reached the highest peak in the landscape, the process is stopped. As a result, the landscape is partitioned into regions or basins separated by dams, called watershed lines or simply watersheds.

Segmentation by WT normally suffers from the problem of over-segmentation [9,10], especially if the image is corrupted with different kinds of noise during its acquisition. To overcome this problem, besides the pre-processing performed to obtain the grayscale image f_{gray}, another sequence of pre-processing and post-processing is performed, as follows:

1. First, a binary image f_{ugf} is obtained, from the grayscale image f_{gray} (Fig. 2b), by means of the ultimate grain filter (UGF) operator [11,12] with shape informations based on maximally stable extremal regions [13].
2. In the sequence, we compute the distance transform [9] in binary image f_{ugf} obtained in the step 1 and thus we can extract a set $M \subset \mathcal{D}$ of local maximum to start the flooding process.

3. Finally, we compute the WT using the marked set M obtained in the step 2. Then, it is performed a post-processing through of merge regions in order to improve the result of segmentation, as illustrated in Fig. 3.

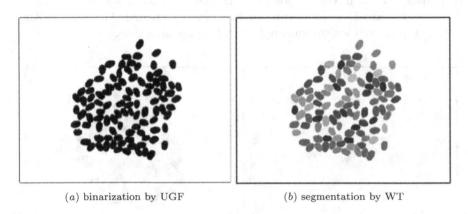

(a) binarization by UGF   (b) segmentation by WT

Fig. 3. Result of segmentation using Watershed technique.

3.3 Classification of Grains

In this last step, the system maps out each grain to one of the three most consumed beans in Brazil, using the technique based on k-means and k-NN algorithms developed in Araújo's CVS [7]. The result of this final step of processing is illustrated in Fig. 4.

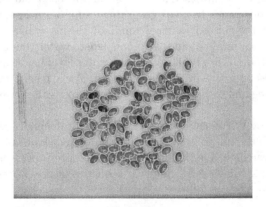

Fig. 4. Final result of the proposed system.

As can be seen in Fig. 4, a Carioca grain (marked as 94) was incorrectly classified because there are brown streaks occupying a considerable area of the bean, making it very similar to a Mulatto bean.

4   Experimental Results

The performance of proposed system regarding the segmentation and clasifica-
tion steps is summarized in Table 2. The subset is placed on first column, while
the following columns present, respectively, the number of true positives (TP),
the number of false negatives (FN), the number false positives (FP), the hit rate
of segmentation of grains and the success rate of classification.

Table 2. Results of the proposed system.

| Subset | Detection/Segmentation of grains | | | | Classification |
	TP	FN	FP	Hit rate (%)	Success rate (%)
1	994	6	0	99.4	100.0
2	995	4	1	99.7	99.9
3	996	3	1	99.7	99.8
4	972	28	0	97.2	100.0
5	994	6	0	99.4	99.9
6	985	15	0	98.5	100.0
7	986	14	0	98.6	100.0
8	993	7	0	99.3	100.0
9	985	15	0	98.5	100.0
10	984	16	0	98.4	99.9
Total	9884	114	2	98.87	99.95

As can be seen in Table 2, the proposed system was able to detect/segment
correctly 98.87 % of grains in all analyzed samples. This rate is less than that
one presented in Araújo el al. [7] for the same set of images. On the other hand,
the proposed system is about 5 times faster than the CVS proposed in [7], as
demonstrated in Table 3.

Table 3. Comparison of computational time.

Step of processing	Time spend (s) (Araújo's CVS [7])	Time spend (Proposed CVS)
Image acquisition and pre-processing	6 s	0.6 s
Segmentation of grains	11 s	2 s
Classification of grains	1 s	1 s
Total	18 s	3.6 s

With respect to classification, the proposed CVS classify correctly 9979 out
of 9984 grains detected (99.95 %), following the high rates showed in [7].

5   Conclusions

In this work we presented a CVS for classification of beans most consumed in Brazil, according to their skin colors. The rates obtained in the experiments (98.87 % for correct detections of beans and 99.95 % for correct classifications) shows good performance of proposed system, indicating that it could be used to determine the class of Brazilian beans. However, it remains improvements in the segmentation process to make it more robust. In future works we intend to incorporate some heuristics to this end.

Acknowledgments. The authors would like to thank UNINOVE and FAPESP São Paulo Research Foundation (Process 2014/09194-5) by financial support.

References

1. BMALS - Brazilian Ministry of Agriculture, Livestock and Supply. Law n.6.305, Decree n.93.563, of 11/11/86, normative instruction n.12 April 2015. http://sistemasweb.agricultura.gov.br/sislegis
2. Kiliç, K., Boyacl, I.H., Koksel, H., Kusmenoglu, I.: A classification system for beans using computer vision system and artificial neural networks. J. Food Eng. **78**, 897–904 (2007)
3. Aguilera, J., Cipriano, A., Erana, M., Lillo, I., Mery, D., Soto, A.: Computer vision for quality control in latin American food industry, a case study. In: International Conference on Computer Vision (ICCV2007): Workshop on Computer Vision Applications for Developing Countries (2007)
4. Venora, G., Grillo, O., Ravalli, C., Cremonini, R.: Tuscany beans landraces, on-line identification from seeds inspection by image analysis and linear discriminant analysis. Agrochimica **51**, 254–268 (2007)
5. Venora, G., Grillo, O., Ravalli, C., Cremonini, R.: Identification of Italian landraces of bean (phaseolus vulgaris l.) using an image analysis system. Sci. Hortic. **121**, 410–418 (2009)
6. Laurent, B., Ousman, B., Dzudie, T., Carl, M.M., Emmanuel, T.: Digital camera images processing of hard-to-cook beans. J. Eng. Technol. Res. **2**, 177–188 (2010)
7. Araújo, S.A., Pessota, J.H., Kim, H.Y.: Beans quality inspection using correlation-based granulometry. Eng. Appl. Artif. Intell. **40**, 84–94 (2015)
8. Souza, T.L.P.O., Pereira, H.S., Faria, L.C., Wendland, A., Costa, J.G.C., Abreu, A.F.B., Dias, J.L.C., Magaldi, M.C.S., Sousa, N.P., Peloso, M.J.D., Melo, L.C. (2013) Common bean cultivars from Embrapa and partners available for 2013. Technical Statement, 211, 16. http://ainfo.cnptia.embrapa.br/digital/bitstream/item/97404/1/comunicadotecnico-211.pdf. Accessed on December 2013
9. Soille, P.: Morphological Image Analysis: Principles and Applications. Springer Science & Business Media, Heidelberg (2013)
10. Najman, L., Talbot, H.: Mathematical Morphology: From Theory to Applications. ISTE-Wiley, Hoboken (2010). ISBN: 9781848212152 (p. 520)
11. Alves, W.A.L., Morimitsu, A., Hashimoto, R.F.: Scale-space representation based on levelings through hierarchies of level sets. In: Proceedings of the 12th International Symposium on Mathematical Morphology and its Applications to Image and Signal Processing ISMM 2015 (2015)

12. Alves, W., Hashimoto, R.: Ultimate grain filter. In: 2014 IEEE International Conference on Image Processing (ICIP), pp. 2953–2957 (2014)
13. Matas, J., Chum, O., Urban, M., Pajdla, T.: Robust wide-baseline stereo from maximally stable extremal regions. Image and Vision Comput. **22**, 761–767 (2004). British Machine Vision Computing (2002)

3D Computer Vision

Stereo-Matching in the Context
of Vision-Augmented Vehicles

Waqar Khan[1]([✉]) and Reinhard Klette[2]

[1] School of Business and Information Technology, Petone Campus,
Wellington Institute of Technology, Wellington, New Zealand
wkha011@aucklanduni.ac.nz
[2] EEE Department, School of Engineering, Auckland University of Technology,
Auckland, New Zealand

Abstract. Stereo matching accuracy is determined by comparing results
with ground truth. However, the kind of detail remains unspecified in
regions where a stereo matcher is more accurate. By identifying feature
points we are identifying regions where the data cost used can easily
represent features. We suggest to use feature matchers for identifying
sparse matches of high confidence, and to use those for guiding a belief-
propagation mechanism.

Extensive experiments, also including a semi-global stereo matcher,
illustrate achieved performance. We also test on data just recently made
available for a developing country, which comes with particular challenges
not seen before. Since KITTI ground truth is sparse, for most of identified
feature points ground truth is actually missing. By using our novel stereo
matching method (called FlinBPM) we derive our own ground truth and
compare it with results obtained by other matching approaches including
our novel stereo matching method (called WlinBPM).

Based on this we were able to identify circumstances in which a census
transform fails to define an appropriate data cost measure. There is not a
single all-time winner in the set of considered stereo matchers, but there
are specific benefits when applying one of the discussed stereo matching strategies. This might point towards a need of adaptive solutions for
vision-augmented vehicles.

1 Introduction

In the context of vision-augmented vehicles, object detection and tracking are
the most important aspects to build an effective collision warning or avoidance
system. A human brain can easily detect objects via different cues like stereo
vision, motion, object recognition, trained models based on previously seen scenarios, and so forth. The object-detection performance of a computer program
depends on implemented algorithms, limited memory resources, and its processing power, which is still relatively low compared to a human brain. There is also
a power consumption limit for computers to be implemented in a vehicle.

Especially due to more serious memory limitations at that time, Franke
et al. [4] decided in 2000 still for the use of sparse features only when computing

© Springer International Publishing Switzerland 2015
G. Bebis et al. (Eds.): ISVC 2015, Part II, LNCS 9475, pp. 57–69, 2015.
DOI: 10.1007/978-3-319-27863-6_6

stereo correspondence via correlation. They relied on a ground-plane estimation to detect vertical objects. Due to analysing two different views of the same scene, objects can have self-occlusions. In these regions, a sparse feature matcher can often produces mismatches.

Recent computing developments supported the design and application of dense stereo matchers. A rectified pair of *base* and *match image* define the input. The goal is to assign a *disparity value* to each pixel in the *base image*. For example, a semi-global correspondence approach [7] or a belief-propagation-based approach [2] can generate in general better (compared to the earlier sparse stereo matchers) dense disparity values even at sparse feature-matching positions, due to consulting neighbourhoods of pixels for confirming disparity calculations.

Thus, sparse stereo matching appeared to be of no interest any-more. In this paper we will return to sparse feature matching. We use selected sparse feature matching results for guiding a dense stereo matcher. We demonstrate results not only on common data sets such as KITTI 2012 [5] and KITTI 2015 [8], but also on a just published dataset CCSAD 2015 [1] which contains challenging scenarios as appearing in a developing country.

For embedding a sparse feature detector into dense stereo matching, we decided for *belief-propagation* (BP) matching. However, the proposed method in Sect. 2.3 can easily improve any dense stereo matcher.

Due to disparity mismatches, there are two general approaches to perform object tracking, either by computing the centroid of an object (after matching the silhouette) or via a set of feature points [11]. Both have their specific limitations.

If using the centroid, the nearest point of an object is ignored which can be crucial for vision-augmented vehicle systems.

If using feature points, the disparity mismatches are handled through a filter. The accuracy of this filter depends on its training parameters as well as the accuracy of the matched disparity. To improve the matching accuracy, a "good" descriptor is needed; the validity of invariance properties is experimentally evaluated in [11] for popular feature detectors. The performance of a descriptor also changes with the kind of intensity details available in recorded images.

In this paper, the objective is to make sure that feature points being detected (i.e. passing the "filter") also have quality-optimised disparity values.

Due to computational limitations, it turns out that we cannot have a very strong descriptor for every disparity level. We use the census transform [14] as data term for evaluating pixel correspondences, which is known to support improved stereo-matching results for outdoor scenes [6]. After feature-point matching, confidence in the detected correspondence is adjusted based on how close the feature point is to the corresponding pixel in the match image at a particular disparity level. This analysis uses the census transform.

The paper is structured as follows. Section 2 describes the used and proposed stereo algorithms. Section 3 reports about our experiments and our performance evaluation. Section 4 concludes.

2   Stereo Matching Algorithms

Each recorded image is of size $N_{rows} \times N_{cols}$, and pixel locations define a domain Ω. Dense stereo matching aims at calculating a labelling f on Ω, with a value f_p for each $p \in \Omega$. Values are in a range between 0 to $d_{\max}$.

2.1   Applied Basics

For the purpose of stereo matching, we consider a pair of left L and right R camera images, taken (as a result of calibration) by exactly identical (with respect to all the relevant parameters) cameras. The pair of L and R describes a canonical stereo configuration [11], and subsequent image pairs are taken time-synchronised.

For a pixel location $p = (x_L, y_L) \in \Omega$, a *disparity offset* f_p is in the row direction, between the *base image* L and the *match image* R, defining a corresponding pixel location $s = (x_R, y_R)$, with $y_L = y_R$ (specifying the epipolar line) and $f_p = x_L - x_R$.

Markov Random Field Model. Stereo matchers which follow the *Markov random field* (MRF) model consider two kinds of cost values when deciding for a corresponding pixel location: a data-cost C_d and a smoothness cost C_s. Together, they define the *energy* (or *error*) $E(f)$ of a considered labelling f. C_d is the matching cost between corresponding candidate pixels in base and match image, and C_s basically a penalty for having non-identical disparities at adjacent pixels.

For a given image pair L and R, it is time efficient to calculate at first data costs at each $p \in \Omega$ for each potential disparity value $d \in \{0, \ldots, d_{\max}\}$, and to use smoothness costs C_s while communicating (in iterations) consequences for assigning particular disparity at adjacent pixel locations.

For example, stereo matchers such as *semi-global matching* (SGM) [7], *iterative SGM* (iSGM) [6], *belief-propagation matching* (BPM) [2], or a special variant (linBPM) defined by truncated linear cost, use of the census as data-cost function, and other properties [9], are all examples of strategies which follow the MRF model.

Census Data-Cost. Assume a window W of a size $w \times h$ around a candidate pixel location $p = (x, y)$ in an image I. Every pixel intensity $I(x_n, y_n)$, with $(x_n, y_n) \in W$, participates in computing a binary bit for the data signature $\rho_I(x_n, y_n)$ for the candidate pixel location at $p = (x, y)$, as follows:

$$\rho_{I,p}(x_n, y_n) = \begin{cases} 0 & \text{if } I(x, y) > I(x_n, y_n) \\ 1 & \text{otherwise} \end{cases} \tag{1}$$

where I stands for L or R, and data signatures $\rho_{L,p}$ and $\rho_{R,s}$ are then compared for potentially corresponding pixel locations p and s by using the Hamming distance, which is the resulting census data-cost C_d.

Semi-global Matching. Basic SGM is provided by `OpenCV`. iSGM also incorporates multi-resolution processing and different weights for propagation rays (scan-lines), as known from standard SGM [7]. We use iSGM with census data-cost for comparing with various BPM algorithms.

After computing the census data-cost C_d for each pixel, correspondence costs are propagated (*dynamic programming* (DP)) along rays, from p_0 at the image border via p_i to $p_m = p$, the considered pixel location in the base image. At each stage $i \geq 0$ of a dynamic process, the energy is updated along a considered scan-line as follows:

$$E(p_i, f_{p_i}) = C_d(p_i, f_{p_i}) + C_s(p_i, p_{i-1}) - \min\{0 \leq d \leq d_{\max} : E(p_{i-1}, d)\} \quad (2)$$

where the smoothness cost C_s is defined by a three-level cost increase: stay with $E(p_{i-1}, f_{p_{i-1}})$ if $f_{p_i} = f_{p_{i-1}}$, increase by a small value if $|f_{p_i} - f_{p_{i-1}}| = 1$, or otherwise increase by a large value, which is inversely scaled by the intensity difference between $L(p_i)$ and $L(p_{i-1})$.

2.2   Belief-Propagation Matching

Like in the case of SGM, BPM too starts off by computing data costs for all disparities in $0 \leq d \leq d_{max}$. When BPM was popularized (particularly due the availability of the source code to the public), it had a C_d defined by absolute value difference only [2]. Later, various modifications were made in C_d to match the challenges in the real-world sequences, like the census transform [9].

Unlike SGM, BPM does not have a fixed path along scan-lines to communicate between adjacent pixels. Instead, it uses pixel-neighbourhoods to propagate the message (also known as belief) among adjacent pixels. The message in the current iteration t is derived from the belief in the previous iteration $t - 1$. Figure 1 illustrates the use of 4-adjacency for propagating a message from p to q, where q received before messages from r_1, r_2, and r_3, excluding q such that p can deliver an unbiased message to q at iteration t.

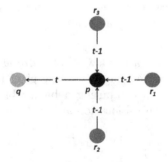

Fig. 1. Belief propagation: pq message passing at time t.

An optimised labelling function f assigns labels (i.e. disparities) by minimizing the energy $E(f)$, derived from applied measures C_d and C_s. In iteration t, the message $m^t_{p\to q}(f_q)$ is sent from p to adjacent pixel location q for the case of having disparity f_q at q:

$$m^t_{p\to q}(f_q) = \min_{0\le d\le d_{max}} \left\{ C_d(p,d) + C_s(f_q,d) + \sum_{r\in A(p),r\ne q} m^{t-1}_{r\to p}(d) \right\} \qquad (3)$$

where $A(p)$ are the 4-adjacent pixel locations of p in Ω.

Linear Belief Propagation. Besides using a linear truncated smoothness cost and a census data-cost function, the message passing scheme in linBPM [9] was derived from a tree-based message passing approach in [12]. In each iteration t, a message is transported between adjacent pixels in two passes, an *inward pass* and an *outward pass*. After completing the inward pass, the outward pass is performed. [12] designed the strategy for FPGA implementation, whereas linBPM converted the strategy into a sequential algorithm.

Hierarchical LinBPM. No matter how messages are transferred, either loopy [2] or tree-based, a global minimum of the energy

$$\sum_{p\in\Omega} \left(C_d(f_p) + \sum_{q\in A(p)} C_s(f_p,f_q) \right) \qquad (4)$$

is the ultimate goal for a labelling function f calculated based on communications between all pairs of pixels. Hence, a significantly large number of iterations are needed in BPM. However, due to time limits, other alternates can be used like using the hierarchical coarse-to-fine refinement like in the case of loopy BPM [2]. We use five levels of coarse-to-fine refinement for our *hierarchical linBPM* (HlinBPM).

2.3   Hierarchical Feature LinBPM

We are now specifying our new proposal for combining sparse features with dense stereo matching. For the purpose of object tracking, sparse feature points are often used to track the temporal position of features. To determine their real-world position, disparity maps from matchers like linBPM stereo and iSGM stereo are needed [10]. Similarly, the same feature points can be used to perform a sparse stereo correspondence which can be used as priors when defining a cost function C_d. This specifies our *hierarchical feature-based LinBPM* method, denoted by FlinBPM.

Outlier Removal. Depending on the used feature detector (note: feature = keypoint + descriptor), the number of detected keypoints varies, as well as the used descriptor. Let g'_L and g'_R be the initial sets of feature points matched in the base and the match image using simple k-nearest neighbour matching, with $k = 2$. A ratio test is performed between the distances of two of the best matches to remove outliers. Let g_L be the set of matched features in the base image after outlier removal. Furthermore, let h_L and h_R be the remaining matched features in base and match image after removing more outliers by a *random sample consensus* (RANSAC) approach [3].

Both steps of outlier removal guarantee "reasonably" that, among the detected features, the matched ones are very likely correct. Of course, this still does not guarantee that matched features are actually correct. In fact, sometimes a matched feature is even not on the same epipolar line (i.e. row).

Let (x_r, y_r) be the query feature point in the base image at location (x_L, y_L). Let (x_m, y_m) be the matched feature in the match image. Since $d = x_L - x_R$, so, by identifying how close x_R is to x_m, we can identify which disparity levels have C_d cost defined by feature matching, and which by the census cost as used in the applied dense stereo matcher. To identify whether a matched feature can be considered or not for disparity level d, a window of size $(2m_w, 2m_h)$ around (x_m, y_m) is taken into consideration. A feature is considered if

$$\{ x_m + d - m_w \leq x_r \leq x_m + d + m_w \} \quad \wedge \quad |y_m - y_r| \leq m_h \qquad (5)$$

If the condition is satisfied then $C_d(x_L, y_L) = C_d(x_r, y_r) = \lambda \frac{|x_r - x_m|}{m_w}$, where $\lambda > 0$ is a data-cost scaling factor. In our experiments, we used $[m_w, m_h] = [1, 1]$.

Hierarchical Weighted Linear BP. Instead of assigning the distance between matched features as data cost, the *hierarchical weighted linear BPM* (WlinBPM) assigns confidence to the already computed census cost.

A matched feature point is considered for disparity d if the pixel-level distance

$$D_p = \sqrt{(|x_r - x_m| - d)^2 + (y_r - y_m)^2} \qquad (6)$$

is less than the maximum-allowed distance

$$D_{max} = \left| \sqrt{m_w^2 + m_h^2} \right| + \alpha \qquad (7)$$

where $\alpha > 0$. If the condition is satisfied then *confidence* is assigned based on the distance between $(x_r - d, y_r)$ and (x_m, y_m):

$$C_{d,new}(x_L, y_L) = \frac{\lambda \cdot C_d(x_r, y_r) \cdot D_p}{D_{max}} \qquad (8)$$

$C_d(x_r, y_r)$ represents the census-based cost at the discrete pixel (x_L, y_L), and $C_{d,new}(x_L, y_L)$ is the new cost function due to combined use of feature matching and census transform.

3   Experiments

In our experiments, we evaluated various feature-point detectors using the *difference of Gaussian* (DoG) scale space and *features from an accelerated segment test* (FAST) for keypoint detection. Regarding the descriptors, we tested as potential descriptors the *scale-invariant feature transform* (SIFT), also its variant in

Table 1. Win-counts for FAST and DoG detectors using FlinBPM and WlinBPM on KITTI 2012 training dataset with $\tau = 1$

		Descriptors				
Detector	stereo matcher	SIFT	SURF	FREAK	BRIEF	BRISK
FAST	FlinBPM	38	31	43	38	44
	WlinBPM	26	38	44	26	60
DoG	FlinBPM	45	21	30	48	50
	WlinBPM	47	25	49	35	38
Total wins by a matcher	FlinBPM	83	52	73	86	94
	WlinBPM	73	63	93	61	98
Total wins by a descriptor		156	115	166	147	192

the form of SURF, the *binary robust independent elementary feature* (BRIEF) descriptor, the *fast retina keypoint* (FREAK) descriptor, and the *binary robust invariant scalable keypoints* (BRISK) descriptor.

We found the newly published data on [1] of particular interest due to new challenges not discussed before elsewhere. We also test on common test data published on KITTI [5].

For each image, *accuracy* β is determined by computing the percentage of pixels with a disparity error greater than τ pixels. In our experiments, we only test the matchers with at least $\tau = 1$ pixel. We also use β to identify which matcher is the best one for each image. This gives us the win-count.

3.1 KITTI 2012 Dataset

The KITTI training dataset consists of 194 images with sparse ground truth (with only around 25 % pixels with ground truth greater than 0). In our analysis we consider occlusions. Similarly, the KITTI test dataset consists of 195 images. However, the ground truth is unavailable publicly. Instead, the matcher is evaluated by KITTI.

Best Feature Point Detector Descriptor Combination. For the matchers FlinBPM and WlinBPM, Table 1 shows the win-counts for $\tau = 1$ for both detectors on the KITTI training dataset. To identify the best descriptors, we compute the sum of win-counts for each descriptor. This showed us that the BRISK descriptor performed best due to a total win-count of 192.

The FREAK descriptor is the second best with a total win-count of 166. Furthermore, by considering the win-counts for each stereo matcher (i.e. FlinBPM, WlinBPM), the win-count of BRISK $(94, 98)$ was still better than for FREAK $(73, 93)$. Based, on this analysis we prefer BRISK only to validate the performance of FlinBPM and WlinBPM, compared to the other stereo matchers (i.e. linBPM, iSGM, and HlinBPM in this paper).

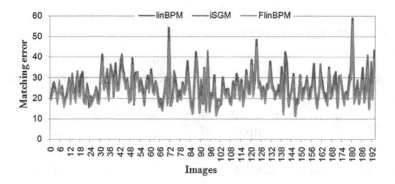

Fig. 2. linBPM vs iSGM vs FlinBPM with error > 1 pixel on the KITTI 2012 training dataset. Curves are very much correlated

Although the DoG detector is equally good as the FAST detector on the considered dataset, we prefer FAST over the FAST detector. The reason for that is the average number of feature points matched by FAST: $g_L = (935.9)$ is much larger than the number for DoG $g_L = (295.7)$ on the whole dataset. Since the idea is to improve the belief around a corresponding pixel, so after the outlier removal, the matcher can benefit more often due to a larger number of feature points with FAST.

LinBPM Vs. iSGM Vs. HlinBPM Vs. FlinBPM Vs. WlinBPM. We use the FAST detector and the BRISK descriptor for the feature-based matchers FlinBPM and WlinBPM. We compare them with linBPM, iSGM and HlinBPM on the KITTI training dataset. Figure 2 highlights the accuracy of linBPM, iSGM, and FlinBPM.

Note that, due to sparse ground truth, the number of pixels being evaluated are around 25 % of the pixels in the whole image. This means that for the remaining 75 % pixels, the matching performance cannot be evaluated (at least on this dataset). Therefore, the difference in HlinBPM, FlinBPM and WlinBPM is very marginal. So, instead of showing their comparison as curves, we have to use the win-count.

As there are in total five matchers, we compare them as pairs. Starting with the well known ones like iSGM and linBPM, linBPM is winning more often due to a win-count of 129 compared to 65 for $\tau = 1$; whereas, for $\tau = 2$, the win-count difference between iSGM and linBPM reduces significantly (win-count of iSGM increases to 94). Since linBPM is winning more often, so we use it for our next comparison with HlinBPM. See Table 2 for comparison data in detail.

The Table 2 shows that HlinBPM, due to coarse-to-fine refinement, brings a significant improvement compared to linBPM. When comparing HlinBPM and FlinBPM, FlinBPM is winning more often with a win-count of 114 compared to 80, for $\tau = 1$. The results remain similar for $\tau = 2$. This highlights that the

Table 2. Win-count for stereo matcher pairs on the KITTI training dataset with $\tau = 1$ and $\tau = 2$

	linBPM	HlinBPM	HlinBPM	FlinBPM	FlinBPM	WlinBPM
$\tau = 1$	21	173	80	114	117	77
$\tau = 2$	19	175	71	123	125	69

introduction of sparse feature-based correspondences as priors into the matching process decreases indeed the overall matching error.

The comparison between FlinBPM and WlinBPM shows that, on this dataset, FlinBPM is better for both $\tau = 1$ and $\tau = 2$. The results are surprising because WlinBPM uses the sparse feature correspondences to adjust the census-based confidence; so, it considers both census and BRISK matching. Whereas, FlinBPM is assigning confidence applying BRISK-based matching only. If FlinBPM can produce better results without considering the census transform then this means that WlinBPM is "misguided" by the census transform.

Figure 2 illustrates the comparison between linBPM, iSGM, and FlinBPM. The comparison is based on accuracy β defined by $\tau = 1$ pixel. Although, the error is "quite correlated", however, linBPM and iSGM have higher peaks than FlinBPM which signifies more error percentage for these matchers compared to FlinBPM. Note that difference in β values between FlinBPM and WlinBPM is very minor, and to avoid redundancy it is absent in Fig. 2.

LinBPM Vs. iSGM Vs. WlinBPM. We evaluated WlinBPM with the FREAK descriptor on KITTI test dataset[1]. Based on our experiments, we identified the weakness of the census transform at closer distances. And, the coarse-to-fine refinement assists our WlinBPM matcher in these regions.

3.2 KITTI 2015 Dataset

The KITTI 2015 dataset has two categories for training and testing. Both categories consist of colour image pairs, each of size $N_{rows} \times N_{cols} = 375 \times 1242$ pixels. The training dataset consists of $(200 + 200)$ image pairs captured at two different time frames ($t = 10$ and $t = 11$). In our experiments on the training dataset we use only the $t = 10$ image pairs. Also, instead of using the colour images, we use their grey-scale intensity channel only. The training dataset comes with ground truth images (with up to 25 % of pixels with ground truth values greater than 0).

As in the case of the KITTI 2012 training dataset, also on the KITTI 2015 training dataset, the performance of HlinBPM still remains better than linBPM due to a win-count of 164 compared to a win-count of 36. However, when comparing HlinBPM and FlinBPM, HlinBPM is winning more often due the win-count

---

[1] On 28th August, 2015 our stereo matcher WlinBPM ranked 46th at [5] compared to linBPM and iSGM which ranked 57th and 32nd, respectively.

of 112 compared to 88. One of the reasons could be that, due to sparse ground truth, the pixels being evaluated do not contain many of the detected feature points. When replacing the use of the BRISK descriptor by the FREAK descriptor (which, potentially, can produce different matches in the outlier removal step), the win-count remained unchanged.

To understand this further, we also analyse the performance of WlinBPM compared to HlinBPM for $\tau = 1$, first for the BRISK descriptor and then also with the FREAK descriptor. See Table 3.

Table 3. Win-count for stereo matcher pairs. KITTI 2015 training dataset with $\tau = 1$

	HlinBPM	FlinBPM	HlinBPM	WlinBPM
BRISK descriptor	112	88	104	96
FREAK descriptor	112	88	99	101

For the BRISK descriptor, the win-count of HlinBPM decreases (from 112 to 104), or, in other words, WlinBPM offers some improvement due to the census transform in this feature-based matcher. Whereas, with the FREAK descriptor, the win-count of WlinBPM exceeds the win-count of HlinBPM. This highlights that, depending on the intensity details around the detected feature points, a weak descriptor like the census transform used in HlinBPM can outperform strong descriptors (like BRISK or FREAK) used in FlinBPM. However, when confidence for feature-based matching is mixed with census-transform confidence (as in WlinBPM), then the performance of the matcher becomes dependent on the descriptor.

The FREAK descriptor is performing better here with WlinBPM. Similarly, when WlinBPM is compared with HlinBPM on the KITTI 2012 training dataset, the win-count of WlinBPM with the FREAK descriptor also improves to 121 compared to 117 with the BRISK descriptor. Furthermore, on the same dataset, when WlinBPM is again compared with FlinBPM but now using the FREAK descriptor, then the win-count of WlinBPM improves to 85 instead of 77.

This concludes that WlinBPM with the FREAK descriptor is a good choice, whereas for FlinBPM the BRISK descriptor is the better choice.

3.3 CCSAD Dataset

This dataset consists of various challenging sequences. Here we report about experiments with the session 3 sequence named *Colonial Town Streets*. It consists of 1,998 pairs of images. The hazards in this sequence are vehicles and pedestrians, either static or moving. The road is cobble stones, hence plenty of texture there. However, it not a smooth road surface; it has occasional humps. The lighting is inconsistent as the ego-vehicle moves through a windy road.

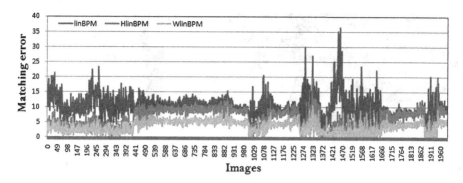

Fig. 3. linBPM vs HlinBPM vs WlinBPM with error > 0.5 pixel on the CCSAD dataset. The error changes depending on how close the hazards are

In our experiments, we evaluate linBPM, HlinBPM, and WlinBPM on the whole sequence of rectified 8-bit images, with $d_{max} = 239$.[2]

There is no ground truth available for a direct evaluation. Instead, we use results provided by FlinBPM with the BRISK descriptor, being our best matcher. Instead of comparing based on the whole disparity image, we only compare the accuracy β on h_L FAST features. We use $\tau = 0.5$ pixel. This allows us to analyse the effect of features on the output. We compare FlinBPM with WlinBPM (which also uses the BRISK descriptor), along with HlinBPM and linBPM (both only using the census transform). Figure 3 illustrates their comparison.

Let k be the image number. For CCSAD, k is from 0 to 1,997. linBPM produces very large errors for the pedestrians in frames $0 \le k \le 75$. The error reduces when a larger object, like a vehicle, becomes visible in the scene at $76 \le k \le 110$. Due to the inward and outward passes in linBPM, the belief is propagated easily along the scan-line, however the vertical communication is less efficient. Therefore, streaks are visible for pedestrians, which are quite similar to the streaking effect known from single-scanline DP stereo matching.

This streaking effect becomes more prominent when pedestrians walk across, in front of the ego-vehicle, for $628 \le k \le 795$. During those frames, the disparity maps from HlinBPM are much more meaningful, without the streaks (see Fig. 4). This shows that the coarse-to-fine refinement allows us that belief is also communicated very well in the vertical directions of the image.

Furthermore, for the whole sequence illustrated by Fig. 3, the performance of HlinBPM and WlinBPM is relatively poor when the objects on the road are very close to the ego-vehicle. This signifies that the census transform fails to extract the appropriate C_d for larger (closer) objects. Whereas, FlinBPM relies only on its properties to compute better disparity maps.

---

[2] Authors acknowledge the support by Mick Jays, Ian Armstrong and Jeff Echano for allowing overnight access to multiple computers for this sequence.

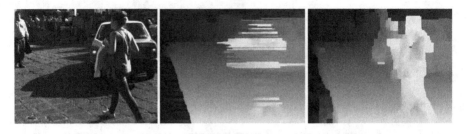

Fig. 4. *Left to right*: Base image, linBPM, HlinBPM. Disparity maps for frame $k = 647$ of the CCSAD dataset

However, when the road in front is clear, then the performance improves for HlinBPM and WlinBPM. The reduced size of objects can be one of the reasons for the improvement.

4   Conclusions

We successfully improved the stereo matching by incorporating sparse feature-point correspondences into the matching process. Like the performance of census transform, the performance of a sparse feature detector and descriptor depends on the intensity details in an image. Furthermore, not all descriptors fail in describing the same features. In our experiments we found that the census transform fails more often for larger (or closer) objects. Therefore, two variants are proposed here. WlinBPM improves the census transform cost by adjusting the confidence. Whereas, FlinBPM replaces the census transform cost with feature based cost. The later was found to be performing better with FAST detector and BRISK descriptor.

References

1. CCSAD dataset. CIMAT, Guanajuato, http://camaron.cimat.mx/Personal/jbhayet/ccsad-dataset (2015)
2. Felzenszwalb, P.F., Huttenlocher, D.P.: Efficient belief propagation for early vision. Int. J. Comput. Vision **70**, 41–54 (2006)
3. Fischler, M.A., Bolles, C.R.: Random sample consensus: a paradigm for model fitting with applications to image analysis and automated cartography. J. Comm. ACM **24**, 381–395 (1981)
4. Franke, U., Joos, A.: Real-time stereo vision for urban traffic scene understanding. In: Proceedings of IEEE Symposium on Intelligent Vehicles, pp. 273–278 (2000)
5. Geiger, A., Lenz, P., Urtasun, R.: Are we ready for autonomous driving?. The KITTI Vision benchmark suite. In: Proceedings of IEEE International Conference on Computer Vision Pattern Recognition (2012)
6. Hermann, S., Klette, R.: Iterative semi-global matching for robust driver assistance systems. In: Lee, K.M., Matsushita, Y., Rehg, J.M., Hu, Z. (eds.) ACCV 2012, Part III. LNCS, vol. 7726, pp. 465–478. Springer, Heidelberg (2013)

7. Hirschmüller, H.: Accurate and efficient stereo processing by semi-global matching and mutual information. In: Proceedings of IEEE International Conference on Computer Vision Pattern Recognition, vol. 2, pp. 807–814 (2005)
8. Menze, M., Geiger, A.: Object scene flow for autonomous vehicles. In: Proceedings IEEE Conference on Computer Vision Pattern Recognition (2015)
9. Khan, W., Suaste, V., Caudillo, D., Klette, R.: Belief propagation stereo matching compared to iSGM on binocular or trinocular video data. In: Proceedings of IEEE Symposium on Intelligent Vehicles (2013)
10. Khan, W., Klette, R.: Stereo accuracy for collision avoidance for varying collision trajectories. In: Proceedings of IEEE Symposium on Intelligent Vehicles (2013)
11. Klette, R.: Concise Computer Vision. Springer, London (2014)
12. Park, S., Jeong, H.: A fast and parallel belief computation structure for stereo matching. In: Proceedings of IASTED European Conference on Internet Multimedia Systems Applications, pp. 284–289 (2007)
13. Scharstein, D., Szeliski, R.: A taxonomy and evaluation of dense two-frame stereo correspondence algorithms. Int. J. Computer Vision **47**, 7–42 (2002)
14. Stein, F.J.: Efficient computation of optical flow using the census transform. In: Rasmussen, C.E., Bülthoff, H.H., Schölkopf, B., Giese, M.A. (eds.) DAGM 2004. LNCS, vol. 3175, pp. 79–86. Springer, Heidelberg (2004)

A Real-Time Depth Estimation Approach for a Focused Plenoptic Camera

Ross Vasko[1($\boxtimes$)], Niclas Zeller[2,3], Franz Quint[2], and Uwe Stilla[3]

[1] The Ohio State University, Columbus, USA
vasko.38@osu.edu
[2] Karlsruhe University of Applied Sciences, Karlsruhe, Germany
{niclas.zeller,franz.quint}@hs-karlsruhe.de
[3] Technische Universität München, Munich, Germany
stilla@tum.de

Abstract. This paper presents an algorithm for real-time depth estimation with a focused plenoptic camera. The described algorithm is based on pixel-wise stereo-observations in the raw image recorded by the plenoptic camera which are combined in a probabilistic depth map. Additionally, we provide efficient methods for outlier removal based on a Naive Bayes classifier as well as depth refinement using a bilateral filter. We achieve a real-time performance for our algorithm by an optimized parallel implementation.

1 Introduction

The plenoptic camera, first described more than a century ago [1,2], provides a way for photographers to capture more information from a scene, compared to a traditional camera, by placing a micro lens array (MLA) in front of the camera sensor. The MLA allows for light-field information to be captured as a 4D function and the recorded image retains information about the structure of the scene itself [3,4]. In a single image captured from a plenoptic camera, information is recorded from several different viewpoints. By finding correspondences between these view points, depths for objects in the scene are able to be estimated.

In contrast with binocular stereo systems, the plenoptic camera is able to estimate depth information from a single image. This fact combined with the small size of a plenoptic camera makes it suitable for a compact visual odometry system which is able to measure metric scale.

In this paper, we present the adaptations made to parallelize a probabilistic depth estimation algorithm [5] and a method of depth map refinement for plenoptic cameras to allow for their use in real-time visual odometry systems.

1.1 Related Work

Due to the complexity of the light-field depth estimation problem, many different types of approaches to estimate a depth map are possible. Each of these

© Springer International Publishing Switzerland 2015
G. Bebis et al. (Eds.): ISVC 2015, Part II, LNCS 9475, pp. 70–80, 2015.
DOI: 10.1007/978-3-319-27863-6_7

approaches will be better suited towards different applications depending on their run-times and accuracies.

Some methods seek a globally optimal solution and find a dense depth map for the image such as [6,7].

Other approaches will calculate a sparse depth map using only local constraints by considering only the textured regions of the image such as [8,9].

In visual odometry and SLAM systems, we are specifically interested in having depth map estimation algorithms with a low complexity to allow for real-time systems. The algorithm presented in [5] uses a probabilistic approach to generate a sparse depth map from a single image captured by a plenoptic camera. This algorithm's low complexity makes it feasible to complete depth calculations in real-time and move closer to a real-time visual odometry system based on a plenoptic camera. We aim towards a depth estimation algorithm that will enable a visual odometry system for plenoptic cameras.

1.2   Outline of Work

In this paper we provide an approach for adapting an existing algorithm to be able to perform in real-time. In Sect. 2 we briefly introduce the focused plentoptic camera and in Sect. 3 we describe the already existing algorithm that we proposed in previous work [5]. Section 4 continues on to describe the post-processing of the depth map to increase its reliability in visual odometry systems. Section 5 details the adapation of this algorithm to allow it to run in real-time on a GPU. Lastly, Sect. 6 shows the results of our depth estimation algorithm along with the effects of the post-processing.

2   The Focused Plenoptic Camera

Rather than only capturing light intensities like a traditional camera, the plenoptic camera is able to record the direction of the light as well. The camera model used in this research is a focused plenoptic camera constructed by Raytrix. A diagram of this camera model is shown in Fig. 1. Additionally, the Raytrix camera features micro lenses of three different focal lengths arranged in a hexagonal grid to increase the depth of field. A subsection of an example raw image captured from the Raytrix camera is shown in Fig. 2.

Figure 1 also shows how one is able to estimate the virtual depth for some image point. Given a correspondence between pixels on the image sensor, the distance b in Fig. 1 can be estimated by triangulation [12]:

$$b = \frac{d \cdot B}{p_x} \tag{1}$$

where d is the distance between the corresponding micro lens centers, B is the constant, but not precisely known, distance between the image sensor and the MLA, and p_x is the disparity between the corresponding pixels under different micro lenses. A full derivation of this formula can be found in [5]. Note that

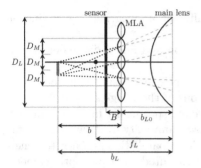

Fig. 1. Diagram of the inside of a focused plenoptic camera [10,11]. The image sensor is positioned in front of the virtual image, which is in the distance b behind the sensor.

Fig. 2. Section of a raw image that was captured by a Raytrix camera. The image shows the edge of a chessboard with a portion of the background.

because we do not have a recorded value for B, we will calculate a virtual depth [9], v, relative to B instead. Notice that as the disparity between pixels increases, the virtual depth will decrease.

$$v = \frac{b}{B} = \frac{d}{p_x} \qquad (2)$$

3   A Probabilistic Virtual Depth Map Estimation Algorithm

Once an image is captured from a plenoptic camera, we are able to seek a correspondence between pixels in the raw image (Fig. 2) to estimate the virtual depths for pixels in textured regions. We are able to complete this correspondence quickly by only searching regions of the image that a match could possibly be for a pixel. To address the sensor noise, we use a probabilistic approach and calculate a variance for each depth estimate as well. This section is provided as a brief overview for how the depth estimations are performed for each pixel to allow for greater insight in later sections. A full description of this algorithm can be found in [5].

The MLA of the Raytrix camera has the micro lenses arranged in a hexagonal pattern. On the MLA, we define a graph of baselines from the center of one micro lens to the center of all micro lenses to the right of it. The search for the matching pixels is done across these baselines because it is guaranteed that corresponding pixels will be on the same baseline. It is only necessary to search to the right of a microlens because it is only required that a disparity between pixel pairs is found once. Figure 3 shows the micro lens array along with several of the defined baselines. Each baseline is defined by a vector $\mathbf{e}_p$ where $\|\mathbf{e}_p\| = 1$ pixel and a distance in pixels, d. Since the micro image can be considered to be rectified, $\mathbf{e}_p$ also defines the direction of the epipolar line for each pixel under the microlens.

To begin the matching process for pixel $\mathbf{x}_R = (x_R, y_R)^T$ with intensity $I(\mathbf{x}_R)$, the baseline $\mathbf{e}_p$ with the shortest distance that has not yet been considered for the current pixel is chosen. It is then checked that the gradient of $\mathbf{x}_R$ meets the following condition, where T_H is a predefined threshold.

$$|\mathbf{g}_I(\mathbf{x}_R)^T \mathbf{e}_p| \geq T_H \qquad (3)$$

This is done to verify that $\mathbf{x}_R$ has sufficient contrast in the direction of the epipolar line. The search for a corresponding pixel is then performed along the line

$$\mathbf{x}_R^s(p_x) = \mathbf{x}_{R0}^s + p_x \cdot \mathbf{e}_p \qquad (4)$$

where $\mathbf{x}_{R0}^s$ is defined by

$$\mathbf{x}_{R0}^s = \mathbf{x}_R + d \cdot \mathbf{e}_p \qquad (5)$$

The disparity that best matches two pixels is the p_x that minimizes the sum of squared intensity error between two 5×1 pixel patches below

$$e_{ISS}(p_x) = \sum_{k=-2}^{2} [I(\mathbf{x}_R + k\mathbf{e}_p) - I(\mathbf{x}_R^s(p_x) + k\mathbf{e}_p)]^2 \qquad (6)$$

An inverse virtual depth, $z(\mathbf{x}_R)$ is then calculated from the disparity that minimizes $e_{ISS}(p_x)$ for the current pixel. Additionally, a variance σ_z^2 can be defined for the inverse virtual depth observation as given in [5]. We record and work with the inverse virtual depths because they can be considered to be Gaussian distributed [5]. We continue to advance and consider the next shortest unprocessed baseline for the current pixel until none of the baselines at a single distance were able to find pixel correspondences under a certain error threshold. The multiple pixel correspondences are incorporated into a probabilistic estimate, similar to the update step in a Kalman Filter.

If a correspondence was able to be found for pixel $\mathbf{x}_R$, we consider this pixel to be valid and assign it an inverse virtual depth estimation based on all observed disparities. If a correspondence was not found, the pixel is marked invalid. Once the correspondence is attempted for each pixel, we project the raw inverse virtual depth map to a 3D space [5].

4   Virtual Depth Map Post-processing

With the probabilistic approach, there is expected to be missing and incorrect depth values in the virtual depth map once the algorithm completes. For this depth map to be used later in visual odometry applications, it is necessary that we fill in missing depth values in regions where we can confidently assign a depth value as well as remove and smooth incorrect estimates in our depth map.

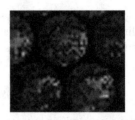

Fig. 3. The hexagonal microlens grid of the camera used with various baselines that can be searched for matches.

Fig. 4. A heatmap of the time spent on correspondence calculations for each pixel. The lighter the pixel color is, the longer is spent performing the correspondence calculations.

4.1 Removing Noise and Estimating Missing Depth Values

Due to only searching in image regions with a high gradient, we receive a sparse depth map with depth information only for the textured regions of the image. But from noise and imperfections in the image, there are incorrect depth values recorded for untextured regions as well as missing values in textured regions. We seek to remove depth values in unreliable or unstable regions and fill in data where we can confidently estimate a value.

First, we are able to remove depth pixels with a variance σ_z^2 that do not meet a previously defined threshold,

$$\sigma_z^2(\mathbf{x}_v) < T(z) = \beta \cdot z(\mathbf{x}_v)^3 \qquad (7)$$

This threshold removes inaccurate depth estimates with an unexpectedly high variance for their virtual depth. A full explanation of this threshold can be found in [5].

Additionally, depth values are filled in or removed based on how much depth information is in some nearby region. If the depth calculations for a region were indeed correct for a pixel, we expect that there are other pixels with valid depth calculations in a nearby window. Due to the perspective projection, which is performed by each of the micro lenses, objects with a high virtual depth occur smaller in the micro images than objects with a low virtual depth. Additionally, virtual image points with a high virtual depth are observed by more micro lenses, as can be seen by Fig. 1. Consequently, regions with high virtual depths that are projected back from the micro images consist of more points which are spread over a larger region than regions that have small virtual depths.

Thus, we define the following relationship in Eq. (8), which defines the neighborhood for a pixel,

$$m = c \cdot v_n \qquad (8)$$

where m is the width of a window around pixel p, v_n is the average virtual depth in some fixed neighborhood around p, and c is a fixed constant. Within this $m \times m$ window, we expect to find other pixels with valid recorded depth values in a high density if p is indeed a valid depth estimation. Also, if p is the result of noise in this image, the density of valid pixels in the $m \times m$ neighborhood should be low.

Once m is calculated, we measure the density of valid pixels with recorded depth information and then a Naive Bayes classifier is used to make an estimate on whether or not the pixel can be assumed to be a part of some textured region with valid depth information:

$$P(V_k|\rho) \propto P(V_k)P(\rho|V_k) \qquad (9)$$

where V_k is a label representing the validity of the current pixel and ρ represents the density information of neighboring pixels. The most probable label for the pixel's validity is then selected given the density information in the $m \times m$ window.

4.2 Depth Map Refinement

In the final depth map refinement step, we wish to smooth the depth estimates to reduce noise. To accomplish this, we use a bilateral filter [13] to remove the noise while preserving the edges. The filtered inverse virtual depth after an iteration of the bilateral filter is defined by:

$$Z^{filtered}(\mathbf{x}) = \sum_{\mathbf{p} \in N_{\mathbf{x}}} \frac{w_i(|Z(\mathbf{x}) - Z(\mathbf{p})|)w_d(|\mathbf{x} - \mathbf{p}|)w_v(V_{\mathbf{p}})Z(\mathbf{p})}{\sum_{\mathbf{p} \in N_{\mathbf{x}}} w_i(|Z(\mathbf{x}) - Z(\mathbf{p})|)w_d(|\mathbf{x} - \mathbf{p}|)w_v(V_{\mathbf{p}})} \qquad (10)$$

where $\mathbf{x}$ and $\mathbf{p}$ are pixel coordinates, $N_{\mathbf{x}}$ is a window around $\mathbf{x}$, $Z(\mathbf{x})$ is an inverse virtual depth, and V is a variance. The weighting functions were chosen to be Geman-McClure functions. In general, $w_a(x)$ is defined to be:

$$w_a(x) = \frac{x^2}{x^2 + \sigma_a^2} \qquad (11)$$

where σ_a^2 is a predefined variance for each of the weighting functions. The weighting functions are designed so that the inverse virtual depth value for each pixel is only smoothed with other inverse virtual depth values that are reliable and similar in depth, to prevent the smoothing of edges. The weighting functions include w_i for the inverse virtual depth difference, w_d for the Euclidean distance difference, and w_v for the variance.

5 Implementation on a CUDA Device

As previously described, we attempt to find a correspondence between pixels in neighboring micro images by searching through various baselines. For a single pixel in the image, the correspondence search consists of several independent operations along the baselines. These independent calculations are able to be completed much more quickly with the use of parallelization. This algorithm was adapted to run on a CUDA capable GPU so that instead of finding the correspondences for each pixel sequentially, threads work in parallel to find correspondences for their assigned pixel.

To eliminate any communication between threads during the correspondence search, a single thread completes all of the correspondence calculations required for one pixel. The time to complete the correspondence calculations for a single pixel depends both on the contrast in the pixel's region as well as the virtual depth of the pixel. If a pixel does not have much contrast in its region the correspondence calculations will finish quickly, as correspondences will not be able to be found due to a low gradient. Also, as the virtual depth of a pixel increases, the correspondence calculations will take longer because the pixel will appear in many micro images and many correspondences will have to be calculated, based on the relationship shown in Fig. 1. Figure 4 shows a heatmap of the time taken for the correspondence calculations for each pixel in a micro image. It can be seen by Fig. 4 that it is possible for the correspondence calculations of pixels to all take significantly different times.

The varying computation times will cause warp divergence, as each thread in a warp will be spending different amounts of time on its correspondence calculations. This could lead to a performance loss because the threads that were able to finish their correspondence calculations will not be doing anything while the remaining threads in the warp finish their computations. To address this, a list of pixels to be processed is built and threads process the next available pixel to minimize the effect of correspondence calculations taking different amounts of time for different pixels. This allows for a single thread to move on to a new pixel to process instead of being required to wait for all of the warp's threads to complete their correspondence calculations.

To gain a further decrease in run-time, it is possible to skip calculating the correspondences for some of the pixels in the image. Skipping the correspondence calculations for pixels in the image decreases the number of recorded depths in the final depth map, but this decrease in density is able to be rectified by the post-processing methods described.

To make the Naive Bayes classifier computations suitable for real-time and interactive applications, a Summed Area Table (SAT) [14] is used to count the number of valid pixels in the previously described $m \times m$ neighborhoods. Rather than requiring a number of memory accesses dependent on m, an SAT allows for these computations to be done in constant time, requiring only four memory accesses for each neighborhood to be computed.

We find that the CUDA implementation allows this algorithm to be suitable for real-time systems. Evaluations for the algorithm and the effects of skipping pixels are given in the Sect. 6.

6   Results

In this section we present the performance of the parallel implementation of our algorithm along with the effects of the post-processing by displaying the results generated from test images. We wish to show that the modifications

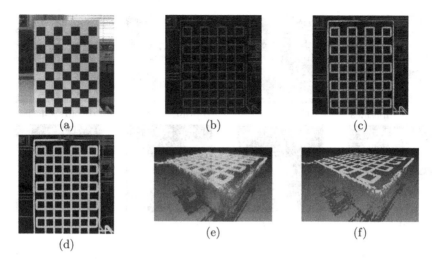

Fig. 5. (a) shows the totally focused image of the chessboard, (b) shows the results from the commercial software [9] with similar settings, (c) shows the original depth map generated by our algorithm, (d) shows the depth map with pixel validities adjusted, (e) shows a point cloud of the unfiltered depth information, (f) shows a point cloud of the filtered depth data. The output depth map has a resolution of 1024×1024.

made to this algorithm allow for it to be used with plenoptic cameras in real-time visual odometry systems. All results were computed on a NVIDIA GeForce GTX TITAN GPU. In the sample images, we compare our algorithm with software from Raytrix [9]. The Raytrix software [9] has many parameters that affect the results of the depth map. We chose the settings by finding parameters that produced results with similar pixel densities. We have chosen to not include the timings for the commercial software [9], because the timings were very sensitive to changes in the settings. Nevertheless, we found our timings to always be comparable or faster than the timings for the commercial software.

First, a walkthrough of the algorithm will be shown in detail for a sample image of a chessboard and then the results from additional images will be shown. Once the depth estimates above the previously discussed variance threshold are removed, the validity of pixels are readjusted using the Naive Bayes classifier. This resulting image is then refined using the bilateral filter. Results from this process are displayed in Fig. 5. For this chessboard image, the depth estimation takes 25 ms, the pixel validity adjustment takes approximately 3 ms, and the filtering takes approximately 4 ms. A longer period of time can be spent on the filtering to further smooth the image if required for the application. Additionally, Fig. 6 displays several filtered depth maps generated along with the total time taken to perform the depth estimations and then filter the image. In the displayed images, the black pixels are in regions of weak texture and are marked invalid with no recorded depth information.

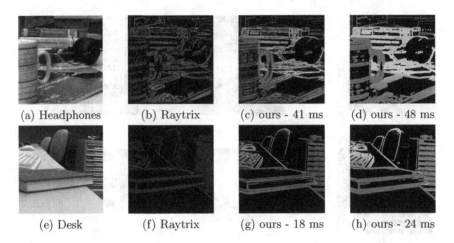

Fig. 6. (a and e) show the totally focused intensity images, (b and f) show depth maps from the commericial software, (c and g) show our original depth maps for the scene, (d and h) show the final post-processed depth maps from our algorithm. The time under the images is the total time taken to produce the depth map. Both output depth maps have resolutions of 1024×1024.

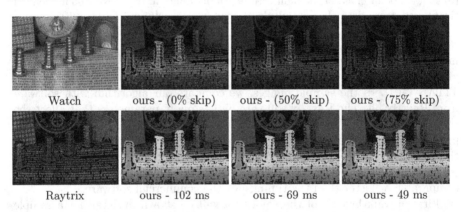

Fig. 7. The effects of pixel skipping on the Watch image. Column 1 contains an image of the original scene along with Rayrix's depth map. Columns 2–4 contain original and post-processed images from our algorithm with a percentage of pixels skipped and the computation time. The output depth maps have resolutions of 2004×1332.

The resulting depth maps for skipping pixels, as well as post-processing these images, are shown in Fig. 7. It can be seen by this figure that by skipping pixels we are able to save a significant amount of time in our depth map calculations. Although the quality of the original depth map is decreased significantly, through post-processing, we are able to restore the depth map.

It can be seen from the results that as the complexity of the scene increases, the run-time of the algorithm also increases. The algorithm seeks pixel correspondences, so if many pixel matches are found, the algorithm will take longer. Additionally, scenes with close objects take longer to calculate a depth map for because points near the camera appear in many micro lenses and many correspondences are calculated. But, the many parameters available in the algorithm make it possible to decrease the computation time and keep reasonable results, as needed for the application.

7 Conclusion

In this paper we described the modifications made to a previous algorithm to make it suitable for use in a real-time visual odometry system as well as post-processing methods to improve the image quality. We gave a method for removing and adding depth estimations using a Naive Bayes classifier. Also, a refinement method using a bilateral filter was described. We then detailed the process of parallelizing and optimizing the existing algorithm.

Our results section shows that we are able to compute and post-process many depth maps per second. The computation times can be decreased further through the refinement of parameters, if it is found that an application does not require as precise depth information. From our timings and results, we find that our algorithm is suitable for real-time visual odometry systems.

References

1. Ives, F.E.: Parallax stereogram and process of making same. US Patent 725,567. Google Patents (1903). http://www.google.com/patents/US725567
2. Lippmann, G.: Epreuves reversibles. photographies integrales. Comptes Rendus De l'Academie Des Sciences De Paris, vol. 146, pp. 446–451 (1908)
3. Adelson, E.H., Wang, J.Y.A.: Single lens stereo with a plenoptic camera. IEEE Trans. Pattern Anal. Mach. Intell. **14**, 99–106 (1992)
4. Gortler, S.J., Grzeszczuk, R., Szeliski, R., Cohen, M.F.: The lumigraph. In: Proceedings of the 23rd Annual Conference on Computer Graphics and Interactive Techniques, SIGGRAPH, pp. 43–54. ACM, New York, NY, USA (1996)
5. Zeller, N., Quint, F., Stilla, U.: Establishing a probabilistic depth map from focused plenoptic cameras. In: Proceedings of International Conference on 3D Vision (3DV), pp. 91–99 (2015)
6. Wanner, S., Goldluecke, B.: Globally consistent depth labeling of 4D lightfields. In: Proceedings of IEEE Conference on Computer Vision and Pattern Recognition (CVPR) (2012)
7. Wanner, S., Goldluecke, B.: Variation light field analysis fo disparity estimation and super-resolution. IEEE Trans. Pattern Anal. Mach. Intell. **36**, 606–619 (2014)
8. Bishop, T.E., Favaro, P.: Full-resolution depth map estimation from an aliased plenoptic light field. In: Kimmel, R., Klette, R., Sugimoto, A. (eds.) ACCV 2010, Part II. LNCS, vol. 6493, pp. 186–200. Springer, Heidelberg (2011)
9. Perwaß, C., Wietzke, L.: Single lens 3D-camera with extended depth-of-field. In: Proceedings of SPIE 8291, Human Vision and Electronic Imaging XVII, Burlingame, California, USA (2012)

10. Lumsdaine, A., Georgiev, T.: Full resolution lightfield rendering. Technical report, Adobe Systems, Inc. (2008)
11. Lumsdaine, A., Georgiev, T.: The focused plenoptic camera. In: Proceedings of IEEE International Conference on Computational Photography (ICCP), San Francisco, CA, pp. 1–8 (2009)
12. Zeller, N., Quint, F., Stilla, U.: Calibration and accuracy analysis of a focused plenoptic camera. ISPRS Ann. Photogrammetry Remote Sens. Spatial Inf. Sci. **II–3**, 207–212 (2014)
13. Paris, S., Kornprobst, P., Tumblin, J., Durand, F.: Bilateral Filtering: Theory and Applications. Now Publishers Inc, Hanover (2009)
14. Crow, F.C.: Summed-area tables for texture mapping. ACM SIGGRAPH Comput. Graph. **18**, 207–212 (1984)

Range Image Processing for Real Time Hospital-Room Monitoring

Alessandro Mecocci, Francesco Micheli[(✉)], and Claudia Zoppetti

Department of Information Engineering and Mathematics,
University of Siena, Siena, Italy
{alessandro.mecocci,francesco.micheli,claudia.zoppetti}@unisi.it

Abstract. In this paper we describe a robust and movable real-time system, based on range data and 2D image processing, to monitor hospital-rooms and to provide useful information that can be used to give early warnings in case of dangerous situations. The system auto-configures itself in real-time, no initial supervised setup is necessary, so is easy to displace it from room to room, according to the effective hospital needs. Night-and-day operations are granted even in presence of severe occlusions, by exploiting the 3D data given by a Kinect[©] sensor. High performance is obtained by a hierarchical approach that first detects the rough geometry of the scene. Thereafter, the system detects the other entities, like beds and people. The current implementation has been preliminarily tested at "Le Scotte" polyclinic hospital in Siena, and allows a 24 h coverage of up to three beds by a single Kinect[©] in a typical room.

1 Introduction

Falls are the most common adverse events in hospitals, and may generate substantial consequences. It is known that 30 % of falls result in injury, disability, or death. Injury falls led to as much as a 61 % increase in patient care costs. In 2013 about 2.5 million non-fatal falls among older adults were treated in emergency departments, and more than 29 % of these patients were hospitalized [1–3]. This is why hospital-room monitoring systems designed to automatically give early warnings about falls are important. Some previous works have been proposed about fall detection, but the majority of the available systems restrict their attention only to moving entities, without considering the information that can be obtained by analyzing the fixed part of the scene. In general, even when using a sensor that gives 3D range data, the available approaches apply a background removal step, to focus only on the regions that are changing. This is the case, for example, in [4–6] that use background subtraction to focus only on the dynamic part of the scene, while in [7] the Kinect[©] sensor is placed on the ceiling and depth is only used to segment the humans based on their height. Similarly, simple standard techniques are used in [8–10] to completely disregard the context. Other approaches concentrate on the human skeleton [11,12] obtained by available software libraries (NITE middleware [13], Microsoft Skeleton [14]). Unfortunately, skeleton extraction is weak in presence of occlusions, and imposes

© Springer International Publishing Switzerland 2015
G. Bebis et al. (Eds.): ISVC 2015, Part II, LNCS 9475, pp. 81–92, 2015.
DOI: 10.1007/978-3-319-27863-6_8

a frontal body pose with respect to the sensors, at least for the skeleton initialization. Moreover, in a real hospital environment, a lot of "visual clutter" (continuously changing in time and location within the room) is generally present: chairs for visitors, medical devices, drip rods, and lockers for the comfort of patients. Even the position of the beds can slightly change from time to time. This is why previous approaches are not trustworthy for real applications. To solve the "visual clutter" problem, we take advantage of a full characterization of the room environment obtained by analyzing 3D range data given by a Kinect$^{©}$ sensor. Instead of disregarding the context/background, the proposed system uses it to improve the overall performances. Robust results are obtained through a hierarchic approach that starts by locating the floor plane and then obtains a reliable description of the overall room structure as seen by the sensor. As a first step, the 3D point cloud given by Kinect$^{©}$ is segmented to locate the floor plane by using a novel approach which is stable even when the floor is slightly visible. In this way, the difficulties experienced by algorithms such as those of the OpenNI library are overcome [15]. Once the floor plane has been extracted, eventual walls are searched for, by imposing constraints on the perpendicularity of the walls to the floor. Once the floor and the walls have been recovered, this knowledge is used to analyse the rest of the scene in real-time, by projecting the 3D space onto suitable planes and then applying fast 2D processing algorithms.

The paper is organized as follows: Sect. 2 describes the floor and walls retrieval procedure. Section 3 explains how beds are located inside a hospital room. Section 4 describes how people can be detected over the beds. Section 5 shows the experimental framework and results, then concluding remarks and future works are reported in Sect. 6.

2   Floor-Plane and Walls Detection

In general, previous algorithms for floor retrieval, like those proposed in [15–17], work well only if a relatively large flat portion of the floor is visible, which is rarely the case in a real environment. Recently, more advanced techniques have been proposed in [18,19], based on the use of RGB image segmentation mixed with iterative or not iterative schemes to refine the plane-fitting to the point cloud. Unfortunately, in the 24 h hospital-room monitoring scenario, the RGB information is highly unreliable or unavailable, in particular during the night, so such approaches are unsuitable for this specific application. To solve the problem, we have developed a novel algorithm that detects the floor even in presence of a very cluttered room and does not use RGB information to obtain the initial hypothesis about the possible plane location. In the following, a mathematical description of the algorithm is presented.

Define $C = \{P_i = (P_{i_x}, P_{i_y}, P_{i_z}) \mid i = 1, \ldots, N\}$ as a set of N 3D points P_i provided by the Kinect$^{©}$. Each point $(P_i \in C)$ can be associated to a normal vector $\hat{n}_i = (n_{i_x}, n_{i_y}, n_{i_z})$, according to [20]. Assuming that the sensor points down towards the scene, we select a subset $T \subset C$ (out of all available normals), defined as $T = \{(P_i \in C) \mid (n_{iy} < 0) \wedge (n_{iz} \geq 0) \wedge (n_{i_x} \leq \epsilon)\}$. The first two conditions state that only the normal vectors with negative y and positive z components (with respect to the reference system centered on the Kinect$^©$) are selected (see the blue arrows with open blue dot in Fig. 1(a)). The second constraint is not stringent, and basically imposes that the sensor points toward the floor (ϵ should be chosen so that the roll angle of the sensor is less than $45°$, otherwise the floor could be confused with a wall). If we consider the points in T, it has been experimentally verified that those belonging to the floor are greater in number, so that they can be identified by computing the most probable orientation of the normals. This is why the z-component of the normals has been quantized, with a bin-width $\Delta_{n_z} \in [0,1]$. A histogram with $N_{bin} = \left[\frac{1}{\Delta_{n_z}}\right] \cdot 2 + 1$ bins is accumulated and its mode ($ModeBin$) is computed. The points of the point cloud whose normals show a $z - component$ belonging to the mode, are aggregated into a new set $T_{mode} = \{(P_i \in T) \wedge (n_{iz} \in ModeBin)\}$. It is evident that the set T_{mode} comprises the points that lie on the floor or on a plane parallel to the floor. The *mean* of the normals belonging to T_{mode} is computed to obtain a first estimation of the floor normal vector $\hat{n}'_{floor} = \frac{1}{\#T_{mode}} \sum_{P_i \in T_{mode}} \hat{n}_i$, where the $\#$ unary operator gives the cardinality of its argument. To improve the estimate of $\hat{n}'_{floor}$, a further property of the points lying on the floor plane $P_{floor,i} \in T_{mode}$ is used: such points lie at the greatest possible distance from the sensor in the direction $\hat{n}'_{floor}$ (see Fig. 1(a)). The distance of a generic point $P_i \in T_{mode}$ from the Kinect$^©$ in the direction $\hat{n}'_{floor}$ is computed as a scalar product: $d_i = \left\langle P_i, \hat{n}'_{floor} \right\rangle$. Again, a histogram of the distances d_i is accumulated. By selecting the peak d_{far} of the histogram corresponding to the higher values of the distance (Fig. 1(a)), the subset of the point cloud associated to the floor-plane can be retrieved as:

$$C_{floor} = \left\{ P_i \in T_{mode} \wedge \left| \left\langle P_i, \hat{n}'_{floor} \right\rangle - d_{far} \right| < \epsilon_f \right\} \qquad (1)$$

where ϵ_f is a threshold constant (in our experiments we used a value of 10). The set C_{floor} represents the input to a RANSAC procedure that gives the refined estimate $\hat{n}_{floor}$ of $\hat{n}'_{floor}$. For convenience, $\hat{n}_{floor}$ points toward the sensor (see Fig. 1(b)-left).

Once the floor plane (the normal) has been obtained, a new reference system can be defined (Fig. 1(b)-right). The origin O' is located by projecting the 3D Kinect$^©$ position O_s on the floor-plane; the axis Z' (equal to $\hat{n}_{floor}$) is orthogonal to the floor-plane; the axis Y' is given by the normalized projection of the Z-vector on the floor-plane; the axis X' is given by the cross product of Y' and Z'. The new coordinate system is useful to speed up the analysis by building 2D synthetic images, called height map (Fig. 1(c)), obtained by suitably projecting the 3D point cloud.

Even if the detection of the walls is, in general, more complicated due to the presence of a lot of occluding entities, the knowledge of the floor plane and the orthogonality constraint between floor and walls, allows to robustly detect the walls in a way similar to the one used to find the floor itself. A maximum number of 3 walls are searched for.

At the very end, a characterization of the room conformation is provided, consisting of: the floor description (its orientation and position), the number of walls actually seen by the sensor, their orientations and positions. From an implementation point of view, a specific processing thread is devoted to the room conformation analysis, and runs in real-time in the background. So, in case the sensor is displaced, the floor segmentation and the detection of the walls is automatically updated and the new conformation of the room is obtained.

3   Real-Time Bed Detection

To ensure full automatic analysis, the system must be capable of detecting the objects of interest in the scene. In the case of a hospital-room, the position and the number of beds to be monitored, are important. In a real scenario, beds can be occluded (partially or severely) by other objects or people. Furthermore, they can posses a not planar surface. For these reasons, it is hard to use iterative algorithms similar to that proposed in [21] which searches for planar surface from raw data. Instead, our experiments show that approaches based on statistical properties and on template matching, give good and robust results while making the procedure adaptive also in relation to "caterpillar shaped" (not-flat) beds. To find the beds location, their height above the floor is preliminarily estimated. This is achieved by using the knowledge about the room conformation, gained as explained in Sect. 2. In particular, each point of the cloud is projected on the Z'-axis and the probability density function $PDF(h_i)$ of these projections is computed, where $h_i = \langle P_i, Z' \rangle$. When beds are present, the PDF generally presents two modes: one associated to the *floor-plane* (around zero height), while the second is associated to the *beds-plane* (around bed height). Even if the beds surfaces are not planar, the second mode is still present although with bigger variance. We take h_i^* (the second mode) as bed height. We use it to adaptively set up the Kernel K shown in Fig. 3(a). In particular, the value of each pixel belonging to K_A (the rectangle marked with A) is set to h_i^*, while the other white pixels are set to zero. By doing so we ensure that both the template matches the bed shape (that is rectangular), and the matching refers to points

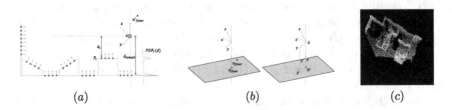

(a)   (b)   (c)

Fig. 1. (*a*) Geometric representation of the method used to find the floor plane. (*b*) Reference system centered on the floor-plane. (*c*) Example of height map, where colors indicate the height along Z' axis of the floor reference system. (Color figure online)

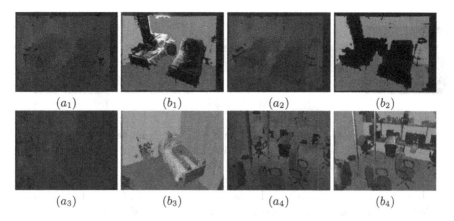

(a_1) (b_1) (a_2) (b_2)

(a_3) (b_3) (a_4) (b_4)

Fig. 2. Example of room segmentation. Each image in gray level represents the depth acquired with the Kinect$^©$. The adjacent colored image represents the correspondent output of floor/walls retrieval method: floor is indicated in blue, the first wall is highlighted in green and the second wall in pink. Images from (a_1) to (b_2) show a real hospital room scenario with varying light conditions, until night in (a_2)-(b_2). Images (a_3)-(b_3) show a room scenario with only one bed. Images (a_4)-(b_4) show our lab, where many elements of furniture are present. (Color figure online)

that are at the correct height. Instead, black pixels are set to *don't care* value and the corresponding pixels are discharged during the matching step. The kernel $K(h_i^*)$ is then applied on the height map H_{map} through a 2D correlation. A bed is detected if the correlation exceeds a predetermined threshold (126 in our experiments). To compensate for possible rotations, the highest value of the correlation function is found by rotating the H_{map} and evaluating the maximum kernel response among the different orientations. Then it is easy to obtain the location mask $B_{image} = Mask\{H_{map_{\vec{\theta}_{opt}}}, K\}$ relative to each bed[1]. The $Mask\{A, B\}$ operator simply selects those pixels of region A that correspond to values different from *don't care* in region B. There is a $B_{image_k}, k = 1, 2, 3$, for each bed present in the room.

4   Real-Time People Segmentation

A further step is to determine if a bed is empty or not: this information can be used to indirectly infer possible dangerous situations like a patient fall down (for example, if a bed becomes void, and at the same time there is no human standing nearby the bed, possibly there has been a fall down). The problem of deciding whether a bed is *full* or *empty* is hard in a real scenario: a first difficulty arises from the bed conformation that can be not flat, so is difficult to state if

---

[1] $\vec{\theta}_{opt} \triangleq \underset{\theta_i \in \Theta}{\arg\max} \left\{ H_{map_{\vec{\theta}_i}} \otimes K \right\}$, where $H_{map_{\vec{\theta}_i}}$ is the original image rotated and translated by $\vec{\theta}_i$ and Θ is the range of all admissible rotations and translations.

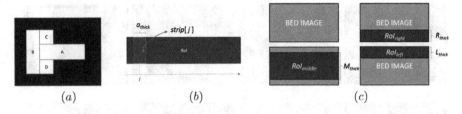

Fig. 3. (*a*) Customized kernel K for bed detection. (*b*) Example of *strip* applied to a generic RoI. (*c*) RoIs used for bottom and top profile definition.

the change in height is due to a human body or not. In fact, a typical bed is divided into movable subparts that can assume many different conformations. Secondly, there can be cushions, covers or other objects left over the bed surface. Besides, the patient can lie undercover or not, and can assume a huge variety of poses. To solve the previous problems, the bed is preliminarily divided into three horizontal Region of Interest: RoI_{left}, RoI_{right} and RoI_{middle}, partially superimposed (see Fig. 3(c)). Generally, if a patient is on the bed his/her body covers only one of the two RoIs RoI_{left} or RoI_{right} at a time, while the middle RoI (RoI_{middle}) is almost always covered. So, if we analyze the upper RoI_{left} and the lower RoI_{right} at the same time, it is possible to construct a *bottom* profile, that follows the bed surface, by scanning the two previous RoIs and comparing the outcomes. The bottom profile is denoted by $\psi_B[j]$, where $j = 1, \ldots, L$ and L is the number of pixel used to sub-sample the bed length. The scanning is performed by using a sliding transverse window, a *strip*, of fixed width a_{thick} and height equal to that of the RoI under analysis[2]. For example, if we consider the $strip[j]$ and the RoI_{left} (Fig. 3(c)), the output is the $P_{left}^{\alpha\%}$ value retrieved by the following formula:

$$\sum_{m=0}^{P_{left}^{\alpha\%}} hist_{RoI_{left}}[m] \leqslant \alpha_{left}^{\%} \tag{2}$$

In the formula, $hist[\cdot]$ is the normalized histogram obtained by accumulating the values $RoI_{left}[i, k]$ of the points $[i, k]$ inside the $strip[j]$ and outputting the value $P_{left}^{\alpha\%}$ corresponding to the $\alpha_{left}^{\%}$ percentile. The value for the RoI_{right} is obtained in a similar way. At this point the value of the *bottom* profile is computed as:

$$\psi_B[j] = min(P_{left}^{\alpha_{left}^{\%}}(j), P_{right}^{\alpha_{right}^{\%}}(j)) \tag{3}$$

In our experiment the 50 % percentile has been used for both RoIs. By construction, the *bottom* profile is a vector with a number of element corresponding to the length of the bed under analysis. Moreover, for each longitudinal position j along the bed, it gives an estimate of the height of the part of the bed not covered by the patient's body (the patient's body cannot cover all the three

---

[2] a_{thick} is expressed in pixel and in our set up it is fixed a value equivalent to 0.05 m.

RoIs at the same time). In this way, even if the bed profile is not flat, it can be accurately approximated by the *bottom* profile. As a second experimental observation the patient, while lying over the bed, tends to occupy the central part of it (the RoI_{middle}). This is why a second type of profile, called the *top* profile, is introduced. The *top* profile is obtained by applying the sliding *strip* window to RoI_{middle}. In this case, the reported percentile is high[3]. It is evident that, if a patient is present, the values obtained by the *top* profile will be very different from those obtained by the *bottom* profile. Summarizing: it is highly probable that in the left or right parts of a bed, the true profile of the bed is visible (no patient's body is there), while in the central zone of the bed it is more probable to find the patient's body (if the bed is not empty). On the other end, the use of robust (percentile) estimators make the whole procedure less influenced by eventual outlier measurements due to noise or to eventual processing errors (e.g. points belonging to the floor near the border of a bed and erroneously attributed to the bed due to errors during the bed location phase). Once the *bottom* profile for the current bed has been computed, it is used as an adaptive threshold to obtain the binary image[4] I_k^{binary} that comprises those regions that depart from the bed surface and are possible indicators of the patient's body position. In general, the binary image contains some blobs that are subsequently regularized by morphological closing. At this point, a feature vector comprising the area between the *top* and *bottom* profiles and the ratio between the global blobs area and the bed area, is used to feed a Support Vector Machine (SVM) to decide if a patient is present or not. In Fig. 4(b)–(d) the results of people segmentation are shown for two different situations. The binary images are used to provide input data to the SVM. As we can see from Fig. 4(c) the system is also able to detect standing people. This is done by using an height map similarly to [22].

5   Experimental Framework and Result

In this section we first present some numerical results of our procedure for floor detection, then some results about people segmentation, and finally an evaluation of a real scenario. To show the robustness of the floor detection algorithm, we have set up two experiments. The first one evaluates the reliability of the normal estimation by comparing the direction of the computed floor-normal-vector with its actual direction. To obtain the actual normal direction, we have measured the sensor tilt using a software tool that shows a viewfinder at the center of the depth image. Then a not reflecting[5] slim object is put on the floor in the position pointed by the viewfinder. By hand measurements and triangulation, the real tilt is evaluated. During these experiments the portion of floor inside the sensor field-of-view has been quite large (flat floor area bigger than 50 % of the whole view

---

[3] Fixed at 90 % in our experiments. Therefore, $\psi_T[j] = P_{middle}^{90\%}(j)$.

[4] $I_k^{binary}(i,j) = B_{image_k}(i,j) > \psi_B[j]$.

[5] In this way it is easy to see the exact position in the depth view since the object appears like a black stain.

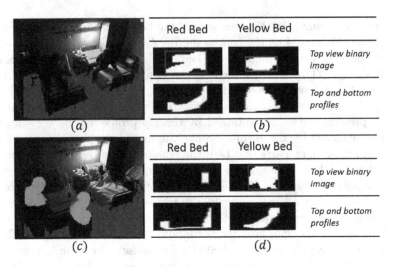

Fig. 4. (a)-(c) Hospital room with detected bed. (b)-(d) At the top, the I_1^{binary} and I_2^{binary} obtained by thresholding the height map by using the bottom profile. At the bottom, the image representing the side view, that is the region defined by the two profiles (*top* and *bottom*).

area). The case of cluttered floor has been evaluated by a further test. In Table 1 we report a comparison between hand made and automatic measurements.

To evaluate the performance in case of a cluttered floor, we have kept the sensor at a fixed position and we have done different measurements with an increasing amount of stuff lying on the floor so to reduce the proportion of visible flat area. For each measure we have registered: (1) the percentage of the floor directly visible with respect to the whole visible area, (2) the tilt, (3) the height, both as estimated by the system. Table 2 reports the obtained results.

Tables 1 and 2 show that the procedure is very robust even in presence of a very cluttered environment. Moreover many other tests have been done, in particular: the sensor has been placed in different rooms, in different light conditions, with different materials, different furnitures, and in different positions. In all tests the floor parameters have been correctly recovered with errors substantially similar to those reported in the two previous tables. As regards the people and bed segmentation, we have considered three possible decisions: people lying, empty bed and people sitting. By manual inspecting the system during its work, it turns out that the real time classification process is very confident and robust, errors are very unlikely and temporary in nature. To quantify the performance level, we have selected some parts from the hospital recordings. We have chosen parts where different bed and patients conditions are present: empty bed, sloped or not-flat beds, uncovered patient, covered patient, sitting patient, partially occluded bed, sleeping patient. The whole set of selected recordings have been assembled into a 3 h video that has been manually annotated. The assembled video has been sampled with a period of 10 s and a total of 1080 frames has been collected. These frames are used as test set and the SVM result are reported as confusion matrix in Table 3. It shows that the classification has a total accuracy of 97.31 %. The more critical situations are when standing people occlude the bed (when they stand up between the sensor and the bed) or when the patient is undercover. In this last case the patient mass is occluded by the sheet and an empty bed is occasionally detected for very short time intervals (usually 3–8 s). However, these kind of temporary occlusions and incorrect classifications, can be limited by temporal filtering.

Table 1. Results of the floor detection algorithm in *free floor condition*. The estimated angle is obtained by averaging the measures for one second (about 25 measures). The sensor has been positioned at 1.9 m above the floor.

Real tilt angle	Estimated height	Estimated angle	Angle error
15°	1.90 m	14.77°	0.23°
20°	1.88 m	19.38°	0.62°
25°	1.89 m	24.42°	0.58°
30°	1.88 m	29.65°	0.35°
35°	1.87 m	34.31°	0.69°

Table 2. Results of the floor detection algorithm vs floor clutter level. The estimated angle is obtained by averaging the measures for one second (about 25 measures). The real tilt angle is set to 25° and 35°. The sensor is located at 1.9 m above the floor.

Real tilt angle	Estimated height	Floor percentage in scene	Estimated angle	Error
25°	1.89 m	54 %	24.71°	0.29°
25°	1.90 m	42 %	24.23°	0.77°
25°	1.90 m	22 %	24.35°	0.65°
35°	1.87 m	51 %	34.42°	0.58°
35°	1.88 m	40 %	34.24°	0.76°
35°	1.87 m	21 %	34.32°	0.68°

As regards real scenario experiments, we have equipped a room of the polyclinic hospital 'Le Scotte' in Siena with two our mobile systems. Each system consists of a Kinect© sensor plugged to a DualCore i5 compact PC@2.5GHz with 2 Gb RAM and can process about 25 frames per second. We have not put any constraints on: light conditions, number of bed(s), number of people present in the room, specific orientation of the sensor. We have done all-night-long benchmarking in cooperation with the polyclinic hospital personnel for a 1

Table 3. Confusion matrix of the people segmentation algorithm. It shows a total accuracy of 97.31 %

Ground Truth	Classified as LYING	Classified as EMPTY	Classified as SITTING	Number of frames
LYING	577	15	2	594
EMPTY	2	321	1	324
SITTING	6	3	153	162

week duration. During the tests, the sensors have also been displaced from their original location, to verify their self awareness and self calibration abilities. The outcomes we have had, are very promising: the floor-plane detection was accurate and stable in all different situations (walking people, furniture movements, light changing, etc.); the system has detected the correct room conformation during the whole test time. The robustness and adaptability of the procedure has been confirmed and the errors remained more or less equal to that previously reported in Tables 1 and 2. In Fig. 5 it is possible to see that the beds are correctly detected and located within the room, thanks to the walls detection combined with the adaptive bed analysis. Each bed is indexed so that a further bed by bed analysis, can be easily done. The processing speed is very high: the initial floor/walls and the bed(s) location phase takes only few seconds (2–4 s, no more). If a new pose is given to the sensor, the new location of the floor, walls, and beds is reliably recovered again by the dedicated processing thread (again in 2–4 s).

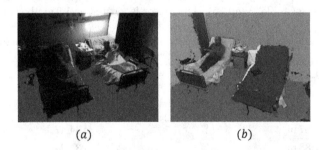

(a)   (b)

Fig. 5. Segmented depth image from Kinect© sensor. The procedure explained in 3 lead to detect the bed 1 (in red in (a) and in blue in (b)) and the bed 2 (in red in (a) and in cyan in (b)) (Color figure online).

6   Conclusion and Future Works

A novel, automatic, and robust method for recovering the geometric conformation of an hospital room has been proposed in this work. Even if originally developed for hospital rooms, the system is capable of working in general contexts like elderly-people-at-home. The proposed approach becomes autonomously aware of the context by locating the important entities in the scene like: beds, walls, floor, and people present in the room. Moreover it is capable of detecting if a bed is free or busy even if the patient is under the covers. Currently, we are working on improving the fall detection process and on deepening the analysis of patient's dangerous behavior. Regarding fall detection our system will do a double parallel analysis. This is due to the fact that falls can happen in different way and from different sides of the bed. Sometimes people may fall from a side directly visible from the sensor, while other times they may fall from an invisible side. To manage as many situations as possible, the first analysis tracks over time the segmented blobs (see paragraph 4) and uses the speed, direction and acceleration, to detect falls. For partially visible falls, both the patient's body and the pavement cannot be reliably seen, so indirect inference must be used based on temporal sequences of higher level events related to the state of people and beds. For example, if the bed changes state from busy to empty and no one is standing nearby the bed, an alarm can be raised. Furthermore, features tracking and saliency will be investigated to determine the patient level of agitation.

References

1. Wolf, M., Alexander, B., Rivara, F.: The cost and frequency of hospitalization for fall related injuries in older adults. Am. J. Public Health **82**, 1020–1023 (1992)
2. Hennessy, A., Bell, A., Talbot-Stern, J.: Characteristics and outcomes of older patients presenting to the emergency department after a fall: a retrospective analysis. Med. J. Aust. **173**, 176 (2000)
3. Finkelstein, E., Miller, T., Stevens, J., Corso, P.: The costs of fatal and nonfatal falls among older adults. Inj. Prev. **12**, 290 (2006)
4. Skubic, M., Stone, E.: Silhouette classification using pixel and voxel features for improved elder monitoring in dynamic environments. In: Workshop on Smart Environments to Enhance Health Care, Seattle, USA (2011)
5. Li, M., Popescu, M., Stone, E., Skubic, M., Banerjee, T., Rantz, M., Scott, S.: Monitoring hospital rooms for safety using depth images. In: AI for Gerontechnology, Arlington, Virginia (2012)
6. Fosty, B., Konig, A., Romdhane, R., Thonnat, M., Crispim Junior, C., Bathrinarayana, V., Bremond, F.: Evaluation of a monitoring system for event recognition of older people. In: International Conference on Advanced Video and Signal-Based Surveillance, pp. 4321–4326 (2013)
7. Gasperini, S., Cippitelli, E.: A depth-based fall detecion system using a kinect sensor. Sensor **14**, 2756–2775 (2014)
8. Brumitt, B., Toyama, K., Krumm, J., Meyers, B.: Wallflower: principles and practice of background maintenance. In: Proceedings of the Seventh IEEE International Conference on Computer Vision, vol. 1, pp. 255–261 (1999)
9. Grimson, W., Stauffer, C.: Learning patterns of activity using real time tracking. IEEE Trans. Pattern Anal. Mach. Intell. **22**, 747–757 (2000)
10. Ganesan, D., Williams, A., Hanson, A.: Aging in place: fall detection and localization in a distributed smart camera network. In: Proceedings of the 15th International Conference on Multimedia, pp. 892–901 (2007)

11. Kawatsu, C., Li, J., Chung, C.J.: Development of a fall detection system with microsoft kinect. In: Kim, J.-H., Matson, E., Myung, H., Xu, P. (eds.) Robot Intelligence Technology and Applications. AISC, vol. 208, pp. 623–630. Springer, Heidelberg (2013)
12. Zhang, C., Tian, Y.: Rgb-d camera-based daily living activity recognition. J. Comput. Vis. Image Process. **2**, 12 (2012)
13. Nite. (http://www.openni.org/files/nite/)
14. Microsoft. (http://www.microsoft.com/en-us/kinectforwindows/)
15. OpenNI. (http://www.openni.ru/index.html)
16. Rusu, R., Cousins, S.: 3d is here: point cloud library. In: ICRA (2011)
17. Lingemann, K., Borrmann, D., Elseberg, J., Nuchter, A.: The 3D hough transform for plane detection in point clouds - a review and a new accumulator design. 3D Res. **2**, 1–13 (2011)
18. Guan, L., Yu, T., Tu, P., Lim, S.: Simultaneous image segmentation and 3D plane fitting for rgb-d sensors. an iterative framework. In: IEEE Computer Vision and Pattern Recognition Workshop (CVPRW) (2012)
19. Cowley, A., Taylor, C.: Segmentation and analysis of rgb-d data. In: RSS 2011 Workshop on RGB-D Cameras (2011)
20. Holz, D., Holzer, S., Rusu, R.B., Behnke, S.: Real-time plane segmentation using RGB-D cameras. In: Röfer, T., Mayer, N.M., Savage, J., Saranlı, U. (eds.) RoboCup 2011. LNCS, vol. 7416, pp. 306–317. Springer, Heidelberg (2012)
21. Triebel, R., Burgard, W.: Using hierarchical EM to extract planes from 3d range scans. In: Proceedings of the IEEE International Conference on Robotics & Automation (ICRA) (2005)
22. Harville, M.: Stereo person tracking with adaptive plan-view statistical templates. In: ECCV Workshop on Statistical Methods in Video Processing (SMVP) (2002)

Real–Time 3-D Surface Reconstruction
from Multiple Cameras

Yongchun Liu[1(✉)], Huajun Gong[1], and Zhaoxing Zhang[2]

[1] College of Automation Engineering, Nanjing University of Aeronautics and Astronautics,
No. 29 Yudao Street, Nanjing, China
liuyc2015@nuaa.edu.cn
[2] Key Laboratory of Complex System and Intelligence Science, CAS, Institute of Automation,
Chinese Academy of Sciences, No. 95 Zhongguancun East Road, Beijing, China

Abstract. Recently, by means of the cheap GPUs and appropriate parallel algorithms, it is possible to perform real-time 3-D reconstruction. In this paper, a real-time 3-D surface reconstruction system has been set up to achieve dense geometry reconstruction from multiple cameras. Pose of the cameras are accurately estimated with the help of a self-calibration system. The depth map of the recorded scene is computed by means of a dense multi-view stereo algorithm. Matching cost aggregation and global optimization method are used to obtain the accurate depth values. We merge our works into the Meshlab, where the depth information is used for generating the surface model. High-quality results are finally presented to prove the feasibility of our system and reconstruction algorithms.

Keywords: Reconstruction · Parallel algorithms · Depth map · Real-time · 3-D

1 Introduction

3-D reconstruction is an inverse process from the 2-D imaging, including four main steps such as intrinsic camera parameters calibration, camera poses evaluation, dense stereo matching and reconstruction. The ability to reconstruct real scene in real-time promises new possibilities for augmented reality applications.

In past several years, a lot of research works make contribution to the development of 3-D reconstruction. The nearly real-time structure from motion (SFM), allowing to recover sparse 3-D points and camera poses from image sequence, has produced many compelling results [1, 2]. With the emergence of the simultaneous localization and mapping (SLAM) technology, many systems [3, 4] have achieved real-time tracking and scene reconstruction based on a single camera. However, these systems only provided sparsely mapped 3-D points. Later, online systems of dense reconstruction with embracing active depth sensors have been developed [5, 6]. Although active sensors have the advantage of providing accurate depth information, there are certain scenarios where traditional passive camera are preferred because of power consumption. With the development of GPU hardware, it's capable to combine depth map computation and fusion with camera tracking to enable sophisticated dense surface reconstruction in real-time [7, 8].

© Springer International Publishing Switzerland 2015
G. Bebis et al. (Eds.): ISVC 2015, Part II, LNCS 9475, pp. 93–101, 2015.
DOI: 10.1007/978-3-319-27863-6_9

Inspired by the MonoFusion system [9] presented by Pradeep, we develop a 3-D surface reconstruction system which permits recovering dense structure of scene surface in real-time via multiple cameras. Either Parallel Tracking and Mapping (PTAM) or SFM can be used for camera-pose evaluation. We utilize a fast and efficient multi-stereo matching method to generate the depth map of reference image, in which matching cost aggregation is added to ensure the accurate depth values. We also use global optimization to make depth map smooth and with little noise. Finally, depth information is used to generate surface model in Meshlab, which provides various 3-D image process library modules including mesh generation, cleaning and re-meshing. This paper details the algorithmic components that make up our system. Some experiment results have been shown to demonstrate the feasibility of our method with high quality reconstructions.

Our study contains two main contributions:

1. We introduce a matching cost aggregation method for multiple stereo matching to calculate more accurate depth map.
2. We set up a surface reconstruction system from multiple cameras which allows dense reconstruction in real-time.

2   System Overview

The system in this paper works on eight images of a scene captured by our sparse camera array (SCA), the resolution of which is 640*360, finally generate a dense 3-D reconstruction of the scene. The live processing of the image data can be visualized as three computation blocks shown in Fig. 1. The three blocks are described in the following sections.

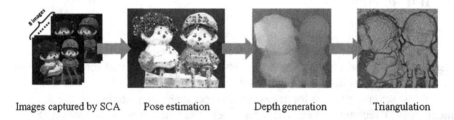

Images captured by SCA   Pose estimation   Depth generation   Triangulation

Fig. 1. System overview. The 6DOF camera pose is estimated from the images captured by SCA in advance. The pose information is used to construct the depth map of reference image using our depth generation method. The surface model is reconstructed in real-time from the accurate depth map

3   Pose Estimation

The essence of 3-D reconstruction is a process of back-projecting 2-D image pixels to 3-D space points. This can be expressed as

$$x = sPX = sKCX = sK[R\,t]X \tag{1}$$

with a scaling factor s. The projective matrix $P \in R^{3\times4}$ contains internal matrix K and external matrix C consisting of the rotation R and translation t with 6DOF.

For the purpose of a precise pose estimation, we use a algorithm presented by ZHANG [10] to conduct the calibration of K, from which the external matrix C can be obtained by a minimum number of 3 points correspondences. Generally, a solution of n-point problem [11] can be selected to solve this problem. Especially, for the case $n \geq 5$, a random sample and consensus (RANSAC) algorithm can be added to improve robustness.

In our system, we can use either SFM [12] or PTAM [4] methods to do the pose estimation for the reason that both the two methods perform bundle adjustment [12] to globally optimize camera positions.

4   Depth Generation

In this section, we present a depth map generation method mainly based on planesweep [13]. A method similar to semi-global matching [14] is introduced to get the most possible correspondence. To achieve accurate depth value, a global optimization step was implemented. Firstly, we start our method from planesweep algorithm.

4.1   Planesweep

A plane sweeps through the volume of space along a line perpendicular to the plane. Without loss of generality, it is assumed that the plane is swept along the Z axis of the scene, which is parallel to the image plane of the reference view at different depth values (see Fig. 2).

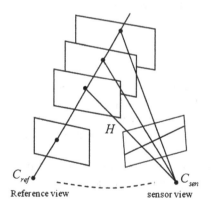

Fig. 2. Illustration of the planesweep

Each of the sensor views is mapped onto an arbitrary depth plane with different homography induced by the plane. In the case of constant lighting brightness, if the plane passes through the surface of the object, the pixel values of the reference image and the sensor images mapped onto the plane match (under the assumption of lambertian surfaces). So it is feasible to compute the matching cost of correspondence pixels to determine their depth, and assigning each pixel the depth where the similarity is maximal. As long as the plane sweeps through the whole volume, a depth map of the reference image can be computed.

Firstly, we explain some symbols with their corresponding meaning. K means internal matrix of a camera. $C_{ref} = [R_{ref}|t_{ref}]$ and $C_{sen} = [R_{sen}|t_{sen}]$ are external matrix of reference view and sensor view respectively. $\pi(d) = (n^T, d)^T$ is a plane defined in reference frame, where $n = (0, 0, 1)^T$ is a unit vector along Z axis, d is the depth to the reference image. Considering the generic case, the homography induced by the plane $\pi(d)$ can be expressed as

$$H = K(R_{rel} - t_{rel}n^T/d)K^{-1} \tag{2}$$

where $R_{rel} = R_{sen}R_{ref}^{-1}$ and $t_{rel} = t_{sen} - R_{rel}t_{ref}$ are the relative rotation and translation between two different positions [15].

Each sensor view will be mapped onto the planes with the corresponding homography. Matching process can be performed by non-parametric local transforms (NPLT) [16] in the pixel-wise manner as the following figure presents:

And matching cost is computed via the Hamming distance, i.e. the number of bits that differ in the two bit strings (Fig. 3).

$$
\begin{array}{ccc}
86 & 72 & 58 \\
49 & p_{ref} & 77 \\
82 & 59 & 92
\end{array}
\longrightarrow 10000101
\qquad
\begin{array}{ccc}
86 & 72 & 58 \\
49 & p_{sen} & 77 \\
82 & 59 & 70
\end{array}
\longrightarrow 10000100
$$

Fig. 3. NPLT of two corresponding pixels result from planesweep

4.2 Cost Aggregation

We introduce a cost aggregation procedure to increase the accuracy of stereo matching, similar to the semi-global matching [14]. Starting from the first pixel row(column) towards the last row(column) and repeating the two processes in a reverse order, we perform the aggregation in four directions through all the pixels in the reference image as the Fig. 4 shows:

Furthermore, a sub-pixel interpolation (SPI) strategy is used by interpolating the cost values of sub-pixels between two pixels in the same row or column. The minimal cost of direction r is

$$COST_r(x, d) = COST(x, d) + \min(COST_r(x_{sw} - r, d - 1) + P_1,$$
$$COST_r(x_{sw} - r, d), COST_r(x_{sw} - r, d + 1) + P_1, \min_i COST_r(x_{sw} - r, i) + P_2) \tag{3}$$

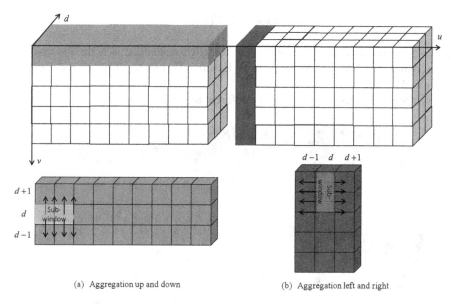

(a) Aggregation up and down   (b) Aggregation left and right

Fig. 4. Cost aggregation

where $x = (u, v)^T$ is the image pixel coordinates and x_{sw} means sub-pixels which are in the sub-window. P_1 is a small constant penalty for the neighbouring pixels, for which the depth value change a little bit. P_2 is a larger constant penalty for all larger depth changes. All *COST* values are computed from the corresponding Hamming distance. Thus, the aggregated cost of all four directions is

$$COST_{agg}(x, d) = \sum_r COST_r(x, d). \tag{4}$$

4.3 Global Optimization

The initial depth value is extracted by a simple winner-takes-all (WTA) scheme. The depth value for each pixel of the reference image can be selected if the aggregated matching cost is minimal. We use a global optimization [17] to improve depth map quality. The energy functional consisting of data term and regularization term is given by

$$E = \int_\Omega \{g(x)\,|\nabla d(x)| + \lambda COST_{agg}(x, d(x))\}dx \tag{5}$$

where λ is the weight of data term, and $g(x)$ is the weight of regularization term. The data term provides robustness against outliers, whereas the total variation regulariser allows discontinuities of edges while simultaneously smoothing homogenous regions. Because depth discontinuities usually appear at the edges of the reference image, the weight of regulariser $g(x)$ is computed by the gradient magnitude of the initial depth map as

$$g(x) = \exp(-\alpha \left| \nabla I'(x) \right|) \qquad (6)$$

reducing the regularization strength where the edge magnitude is high, and ensuring global optimization process respects edges. Equation (5) can be solved by the approach mentioned in [18].

We show some results generated by different methods in Fig. 5.

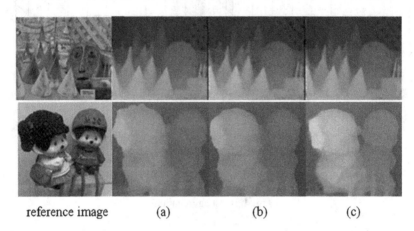

reference image   (a)   (b)   (c)

Fig. 5. Depth maps generated by different methods. (a) Planesweep + Global optimization, (b) Planesweep + Cost aggregation, (c) Planesweep + Cost aggregation + Global optimization. The "Cone" consists of 8 captured images from the Middlebury standard image database [19] and the "Twins" is from two figurines captured 8 perspectives by our SCA. The external parameters of each image are computed by VisualSFM software [20]. All of the results are obtained based on planesweep. Specially, compare (c) to (a) and (b), we can found the quality of the depth map has been improved with little noise and clear edges

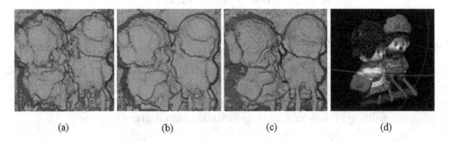

(a)   (b)   (c)   (d)

Fig. 6. Reconstruction results comparison. (a) Planesweep + Global optimization, (b) Planesweep + Cost aggregation, (c) Planesweep + Cost aggregation + Global optimization, (d) Reconstruction result with texture

5   Triangulation and Results

We merge our works into the Meshlab developed by Visual Computing Lab of ISTI-CNR [21]. Meshlab provides various 3-D image process library modules including mesh generation, cleaning and re-meshing that helps us generate the surface model out of our disparity information. We write plug-in under the Meshlab framework to accomplish the real-time reconstruction.

We perform the reconstruction comparisons with the corresponding depth map shown in Fig. 5 above. The reconstruction results is in following figure. It could be generally concluded that our method (Planesweep + Cost aggregation + Global optimization) has the edge in terms of restraining the noise to smooth the surface while keeping as much details as possible in the results (Fig. 6).

We discuss the effect two major factors in the expression (5) and (6) impose on the reconstruction result respectively as the Fig. 7 shows:

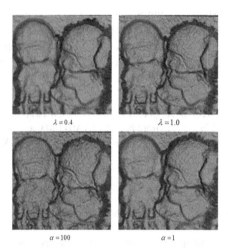

Fig. 7. Reconstruction results with different parameters in the Global optimization process

Note that the quality of the depth map significantly affects the final reconstruction result. Changing the value of data term weight λ and parameter α in global optimization process will make the quality of the depth map different. In theory, within a certain range, a bigger λ means the effect of smoothing decreasing, further getting a more accurate depth map to make reconstruction result better. We demonstrate this in Fig. 7. When setting λ to 1.0, the edge of reconstructed objects becomes more clear, and shows more obvious surface details than the condition of $\lambda = 0.4$. The value of α is inversely proportional to that of the regularization item. Therefore the reconstruction object is smoother as we accelerate the regularization item by decreasing α. So we can try various λ and α to get a desired reconstruction result.

The time consumption of our major 3-D reconstruction steps is listed in the following table. All computation was performed on a commodity system consisting of an nVidia GeForce Titan Black GPU (Table 1).

Table 1. The time consumption of our major 3-D reconstruction steps

Reconstruction Step	Time (s)
Planesweep	0.020
Cost aggregation (four directions)	0.008
Global optimization	0.019

6   Conclusion

We have presented a system offering a significant advance in real-time dense reconstruction via multiple cameras. Our system allows to reconstruct dense structure of scene surface on-the-fly once the multiple images of a scene are captured. The proposed method is very suitable to obtaining a accurate 3-D surface with GPU-accelerated implementation.

Acknowledgments. This work is supported by the National High Technology Research and Development Program of China under Grant No. 2012AA011903, and by Postgraduate Research Innovation Projects of Jiangsu Province of China under Grant No. CXLX 13_158.

References

1. Dellaert, F., et al.: Structure from motion without correspondence. Proc. Comput. Vis. Pattern Recogn. **2**, 557–564 (2000)
2. Pollefeys, M., et al.: Detailed real-time urban 3D reconstruction from video. Int. J. Comp. Vision **78**(2), 143–167 (2008)
3. Durrant-Whyte, H., Bailey, T.: Simultaneous localisation and mapping (slam): Part I the essential algorithms. IEEE Rob. Autom. Mag. **13**(2), 99–110 (2006)
4. Klein, G., Murray, D.W.: Parallel tracking and mapping for small AR workspaces. In: Proceedings the International Symposium on Mixed and Augmented Reality (ISMAR), pp. 225–234. IEEE, Nara, Japan (2007)
5. Izadi, S., et al.: KinectFusion: real-time 3D reconstruction and interaction using a moving depth camera. In: Proceedings ACM Symposium on User Interface Software and Technology, pp. 559–568. ACM, New York (2011)
6. Stuckler, J., Behnke, S.: Robust real-time registration of RGB-D images using multi-resolution surfel representations. In: Proceedings of the German Conference on Robotics (ROBOTIK), pp. 1–4. VDE, Munich, Germany (2012)
7. Newcombe, R.A., Davison, A.J.: Live dense reconstruction with a single moving camera. In: Proceedings of the IEEE Conference on Computer Vision and Pattern Recognition (CVPR), pp. 1498–1505. IEEE, San Francisco (2010)
8. Stuehmer, H., Gumhold, S., Cremers D.: Real-time dense geometry from a handheld camera. In: Proceedings of the DAGM Symposium on Pattern Recognition, pp. 11–20. Springer, Heidelberg (2010)

9. Pradeep, V., et al.: MonoFusion: Real-time 3D reconstruction of small scenes with a single web camera. In: Proceedings of the International Symposium on Mixed and Augmented Reality (ISMAR), pp. 83–88. IEEE, Adelaide, SA (2013)
10. Zhengyou, Z.: A flexible new technique for camera calibration. IEEE Trans. Pattern Anal. Mach. Intell. **22**(11), 1330–1334 (2000)
11. Fischler, M.A., Bolles, R.C.: Random sample consensus: a paradigm for model fitting with applications to image analysis and automated cartography. Commun. ACM **24**(6), 381–395 (1981)
12. Changchang, W.: Towards linear-time incremental structure from motion. In: Proceedings of the 2013 International Conference on 3D Vision, pp. 127–134 (2013)
13. Cornells, N., Van Gool, L.: Real-time connectivity constrained depth map computation using programmable graphics hardware. In: Proceedings of the IEEE Computer Society Conference on Computer Vision and Pattern Recognition, vol. 1, pp. 1099–1104 (2005)
14. Hirschmüller, H.: Stereo processing by semiglobal matching and mutual information. IEEE Trans. Pattern Anal. Mach. Intell. **30**(2), 341–382 (2008)
15. Hartley, R., Zisserman, A.: Multiple view geometry in computer vision. Cambridge University Press, New York (2003)
16. Zabin, R., Woodfill, J.: Non-parametric local transforms for computing visual correspondence. In: Proceedings of European Conference on Computer Vision, pp. 151–158 (1994)
17. Richard, A., et al.: DTAM: Dense tracking and mapping in real-time. In: Proceedings of the 2011 International Conference on Computer Vision (ICCV), pp. 2320–2327. IEEE Computer Society, Washington, DC, USA (2011)
18. Chambolle, A., Pock, T.: A first-order primal dual algorithm for convex problems with applications to imaging. J. Math. Imaging Vis. **40**(1), 120–145 (2011)
19. http://vision.middlebury.edu/stereo/
20. http://ccwu.me/vsfm/
21. http://meshlab.sourceforge.net/

Stereo Correspondence Evaluation Methods: A Systematic Review

Camilo Vargas[1(✉)], Ivan Cabezas[2], and John W. Branch[1]

[1] Departamento de Ciencias de la Computación Y de la Decisión,
Facultad de Minas, Universidad Nacional de Colombia., Bogotá, Colombia
{cjvargas,jwbranch}@unal.edu.co
[2] Laboratorio de Investigación para el Desarrollo de la Ingeniería de Software,
Universidad de San Buenaventura, Avenida 10 de Mayo, La Umbría,
Vía a Pance, Cali, Colombia
imcabezas@usbcali.edu.co

Abstract. The stereo correspondence problem has received significant attention in literature during approximately three decades. During that period of time, the development on stereo matching algorithms has been quite considerable. In contrast, the proposals on evaluation methods for stereo matching algorithms are not so many. This is not trivial issue, since an objective assessment of algorithms is required not only to measure improvements on the area, but also to properly identify where the gaps really are, and consequently, guiding the research. In this paper, a systematic review on evaluation methods for stereo matching algorithms is presented. The contributions are not only on the found results, but also on how it is explained and presented: aiming to be useful for the researching community on visual computing, in which such systematic review process is not yet broadly adopted.

1 Introduction

Matching corresponding points is a fundamental problem in computer vision. It is commonly discussed and termed in literature as the stereo correspondence problem. Stereo correspondence has several application fields, such as autonomous navigation [1], pedestrian detection [2] and agriculture [3]. A plethora of algorithms have been proposed for tackling the stereo correspondence problem. Stereo correspondence algorithms take as input a rectified image pair, and compute a disparity map as output. The estimation of an accurate disparity map still remains a challenging task, mainly due to the presence of occluded pixels, and textureless regions, among other factors inherent to the problem [4].

Stereo correspondence algorithms can be broadly classified as local or global, according to the used optimization strategy. On the one hand, local algorithms are based on sliding windows, where the disparity computation at a given point depends–mainly– on intensity values. On the other hand, global algorithms are based on energy functions, which make explicit smoothness and continuity assumptions [5].

Whilst there are plenty of stereo correspondence algorithms, there are only few methods available for analyzing the quality of a disparity map. Nevertheless,

© Springer International Publishing Switzerland 2015
G. Bebis et al. (Eds.): ISVC 2015, Part II, LNCS 9475, pp. 102–111, 2015.
DOI: 10.1007/978-3-319-27863-6_10

an assessment on the progress of stereo correspondence can only be achieved if quantitative and objective performance results are reported for proposed algorithms.

Stereo correspondence evaluation methods are classified in [6] as ground-truth based methods or methods performed in the absence of ground-truth. Ground-truth based methods rely on independent measurements by active sensors [7, 8]. Disparity maps resulting from stereo correspondence algorithms are compared against ground-truth information using metrics such as Bad Matched Pixels (BMP) [5], Bad Matched Pixels Relative Errors (BMPRE) [9], SZE [10], among others. Evaluation methods in the absence of ground-truth estimate disparity maps quality by computing errors on predicted views [1] or using confidence metrics [11]. In practice, many researchers on the area might be relying blindly on a single evaluation method, ignoring which their strengths and weaknesses really are.

In this paper a systematic review on stereo correspondence evaluation methods is presented. A systematic review a means of identifying, evaluating and interpreting all available information relevant to a specific field [12]. This research method has well-defined steps to be performed according to a protocol [13]. It begins with the definition of research questions, keywords and search string, followed by selecting and searching in information sources. Obtained results are validated and filtered using control papers and inclusion/exclusion criteria, respectively. At this point found papers should be reviewed in detail in order to be able to classify them and produce a visual summary.This review technique, which has a growing acceptance and use in other research fields, is not yet commonly used by the visual computing community.

The remainder of this paper is structured as follows. The conducted systematic review process is presented in Sect. 2. Findings are summarized in Sect. 3. Finally, concluding remarks are stated in Sect. 4.

2 Systematic Review

The developed protocol was defined as follows.

Define Research Question. Research questions are related to the concerns that should be answered during the review. This review considers the following research question: (i) which are the evaluation methods and evaluation frameworks for assessing the quality of disparity maps obtained from stereo correspondence algorithms.

Define Keywords and Search String. Keywords allow determining the search string that is going to be used on the web search engines. Used keywords were classified in two groups: stereo correspondence related terms (i.e. stereo matching, stereo correspondence, stereo algorithm, stereo vision and disparity map) and quality related terms (i.e. evaluation, measure, quality, metric, assessment and performance).

The following search string was used: *(("stereo" AND "matching") OR ("stereo" AND "correspondence") OR ("stereo" AND "algorithm") OR ("stereo" AND "vision") OR ("disparity" AND "map") OR ("Stereoscopic" AND "image")) AND (("evaluation") OR ("measure") OR ("quality") OR ("metric") OR ("assessment") OR ("performance"))*

Select and Search in Information Sources. Information sources are selected according to the defined research question. Among multiple information sources, bibliographic databases have high reliability. The Scopus database was chosen since it integrates important digital libraries addressing visual computing topics. A total of 5937 papers were obtained as searching results.

Validate and Filter Obtained Results. The use of control papers allows to quickly verifying the coherence between search string and obtained results. It requires of some background on the addressed topic. Inclusion criteria allows to consider specific works based on such background, whilst exclusion criteria filters out obtained results not closely related to research questions. [1, 9, 14, 15] were defined as control papers. The selected inclusion/exclusion criteria were: (i) the study should approach a method, strategy, metric or dataset for assessing quality of disparity maps or stereo correspondence algorithms. (ii) Stereo image evaluation for comfort measuring or 3DTV applications that does not include assessment of disparity maps will be excluded. (iii) The study publication date should be equal or greater than 2005. [5, 7, 16–18], were included based on their relevance. In this way, the protocol indicates that the review should be focused in a total of 44.

Summarize Results. Categorizations of results allow constructing a visual summary, indicating trends. Figure 1 shows the quantity of published papers per year. Figure 2 shows the quantity of published papers by trend, classified as papers proposing or applying strategies without ground-truth (confidence metrics and prediction error approaches), stereo datasets and ground-truth methods.

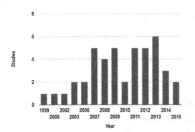

Fig. 1. Quantity of published studies per year.

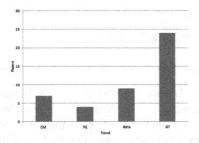

Fig. 2. Quantity of published studies by trend, confidence measures, prediction error, datasets, ground-truth methods, respectively.

3 Findings on the Area

3.1 Evaluation Methods Using Ground-Truth

The Middlebury dataset and evaluation method is presented in [5]. The dataset introduced on this work is available at Middlebury's website, including several stereo images and ground-truth data. This method measures the estimated disparity map quality using the BMP and RMS metrics against ground-truth data. Different error criteria are associated to image segments: all, the entire image; nonocc, non-occluded pixels; disc, areas near depth discontinuities and occluded regions; and textureless, areas of low texture.

In [17] an evaluation is performed using two separate approaches: a comparison against ground-truth data and a prediction error approach. This work defines an error as an estimation disagreeing from the ground-truth disparity value in more than 1 pixel. An error criterion is used in order to measure the algorithms performance under different situations such as occluded pixels or low texture regions.

In [19] is pointed out that it might be possible to quantify quality of recorded stereo images with respect to some measures, which may be used for indicating domain of relevant scenarios when performing evaluations for some particular test data. The aim of the work is to judge the complexity of a specific stereo dataset and its qualitative relation to other datasets.

Robustness to radiometric changes between views is required in stereo correspondence algorithms for real world applications. In [15, 16] images under noise and radiometric changes are used to assess the performance of cost functions and stereo correspondence algorithms.

In [20] a quality assessment of stereo correspondences based on histogram differences is proposed. The improvement of this study is based on the idea of assessing when an object is missed from a non-dense disparity map. The proposed method divides the image in small sub-regions where disparity histograms are calculated. For each region the histogram distances are calculated using the earth mover's distance.

[21] Proposes a method to create arbitrary stereo ground-truth datasets with reliable per-pixel error bars with. It is based on previously measured point clouds and arbitrary calibrated cameras and is therefore versatile for indoor and outdoor applications.

An evaluation method for parameter setting is proposed in [22]. It considers two error types: the error rate and the sparsity rate, for accuracy and completeness measuring respectively. These error definitions are based on four principles: orthogonality, symmetry, completeness and algorithm independence.

A cluster ranking evaluation method is proposed in [23]. The proposed method consists on using a statistical inference technique (ANOVA) to rank the accuracy of disparity estimation algorithms combining ranks from rom multiple stereo pairs. In this study the BMP measure is used, only, according to the nonocc error criterion.

The R-SSIM measure is proposed in [24]. The R-SSIM is a modification of the Multi-scale Structural Similarity index. The obtained results by using R-SSIM measure are statistically correlated to obtained results from BMP measure.

In [25] the SSIM and PSNR measures are compared for disparity maps with added salt and pepper noise. The authors conclude that obtained PSNR values are closer to the scores assigned by subjective evaluation.

[9] Proposes a quality metric for disparity map using ground-truth data. The proposed BMPRE metric offers a clear and concise interpretation of a disparity estimation error considering both the error magnitude and the inverse relation between depth and disparity.

Several evaluation methods oriented to specific contexts are proposed or applied in the stereo correspondence field; these studies are summarized in Table 1.

Table 1. Stereo correspondence evaluation methods oriented or applied to specific contexts.

Application/Context	Reference
Autonomous vehicle applications	[14] (2011), [26] (2006), [27] (2007), [28] (2009), [29] (2009), [30] (2012), [31] (2013)
Face reconstruction	[32] (2006)
Real time oriented evaluation	[33] (2006), [34] (2010)
Agriculture applications	[3] (2007)
Pedestrian detection	[2] (2011), [35] (2007), [36] (2008)
Silicon retina stereo cameras	[37] (2013)
Remote sensors	[38] (2014)

3.2   Evaluation Methods Without Ground-Truth

Evaluation methods that do not use ground-truth data can be classified as prediction error approaches and confidence measure approaches [14, 39]. The prediction error approach proposed in [18] suggest the estimation of a novel view of the scene. The predicted view is compared to a reference view obtained from a third camera in a known position. However, error scores reflect not only the accuracy of the disparity estimation algorithm, but also the accuracy of the selected rendering algorithm [5, 20]. Confidence metrics are used to measure the reliability of the estimated disparity value for each pixel [14]. Several stereo correspondence algorithms use confidence metrics as part of their estimation processes in order to refine the resulting disparity maps.

Prediction Error Approaches. In [1] three stereo correspondence algorithms are evaluated using the prediction error approach for an autonomous navigation context. The evaluation is performed using synthetic data from [40]. The reference and predicted images are compared using root mean squared (RMS) and normalized cross correlation (NCC).

More sophisticated metrics can be applied to compare the predicted and reference images. In [41] the authors use a prediction error approach applied to the view interpolation problem. This study uses structural similarity index (SSIM) and peak signal to noise ratio (PSNR) measures to calculate a quality metric.

Recently, in [42] an evaluation method for stereo video sequences is proposed. Matching error and temporal instability metrics are used to estimate the disparity map quality. The matching error measure is a prediction error based approach, but particularly in this work the evaluation does not require a third view. Instead, the evaluation method predicts the right view using the left view and the estimated disparity map. The temporal error is measured using motion estimations, where disparity maps with high temporal instability will lead to a higher temporal error.

Confidence Measures. Confidence metrics are commonly used as a supporting step on stereo correspondence algorithms and can also be used as a quality metric. A quantitative and qualitative comparison for confidence metrics is presented on [43].

Confidence metrics are expected to be high for correct disparities and low for errors. The evaluation is performed by comparing the confidence measures against the disparity maps errors through ROC curves. Finally the study shows a detailed performance analysis where advantages and disadvantages for each metric are discussed.

Two indexes for measuring smoothness of a noised disparity map–the disparity gradient and the disparity acceleration– are proposed in [44]. Disparity maps from the Middlebury repository [5] were artificially corrupted with noise in order to test the indexes.

A classifier using confidence measures as input features is proposed in [45]. The proposed stereo correspondence algorithm is supported by an AdaBoost approach, classifying the estimated disparities as either 'correct' or 'incorrect'. The feature vector used on the classifier includes confidence metrics such as average color components, texture, color variation, disparity variation, among others.

In the same way, [46] proposes the use of confidence metric as features for a random decision forest classifier. This study is developed using stereo images and ground-truth from the KITTI dataset [30]. The confidence metrics used as features include Entropy of disparity costs, peak ratio measure, consistency between left and right disparity, horizontal gradient, among others. This work shows that a classifier using confidence measures can be an appropriate approach to increase accuracy in stereo error detection.

A quality metric for depth maps using unsupervised no reference segmentation quality metrics is proposed in [47]. The quality metric is calculated by checking consistency between a segmented depth map and one input image of the same scene. The results show some correlation degree between the proposed quality metric and the prediction error approach where a MSE metric is used to compare the predicted and reference views.

A correlation assessment between the prediction error method and 2D image metrics is performed in [39]. The assessment is done for stereo video sequences under absence of ground-truth [48]. The three proposed data measures deal with image homogeneity, standard deviation of the Sobel image and similarity between stereoscopic images.

4 Concluding Remarks

This paper presents a taxonomy of the different state-of-art methods for assessing stereo correspondence algorithms. The introduced taxonomy is shown in Fig. 3. As it is shown here, the quantity of disparity map evaluation methods is relatively low. Moreover, difficulties associated to obtain ground-truth data and the fact that the assessment of stereo correspondence algorithms may vary among datasets can be considered as research opportunity.

Middlebury method is one of the most used ground-truth based evaluation methods. In Middlebury's method BMP and RMS are used as metrics [5, 7, 8]. Several metrics including SSIM, PSNR [25], R-SSIM [24], BMPRE [9], SZE [10], disparity gradient and disparity acceleration [44] have been also proposed in order to estimate the disparity quality. Nevertheless, there is a lack of consistency on the evaluation results achieved by considering different error measures [6].

Robustness to noise and radiometric changes are addressed in [15, 16]. These approaches consider radiometric changes artificially generated on the Middlebury datasets and artificial noise on a synthetic dataset, respectively.

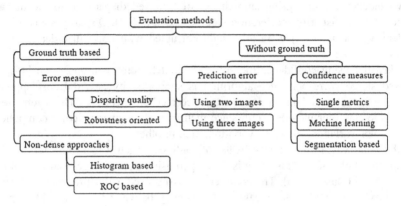

Fig. 3. Disparity map evaluation methods taxonomy.

Ground-truth based proposals also include histogram [20] and ROC [22] based evaluations, where the sparsity of estimated disparity maps its handled explicitly. The histograms approach is focused on disparity distribution and outliers. The ROC approach is focused on studying a wide range of parameter settings for a single algorithm based on the defined error and sparsity rates.

Regarding to evaluation methods in the absence of ground-truth data, prediction error is proposed in [18]. This approach requires the use of a third camera and therefore the modification of the standard stereo acquisition system. In [42] the prediction error method is performed using the two standard stereo images, removing the additional work at the acquisition stage.

According to [43], confidence metrics are grouped as matching cost metrics, local properties of the cost curve, entire cost curve metrics, consistency between the left and right disparity maps and distinctiveness based confidence measures. Confidence measures can be used as input features for classifiers as is presented in [45, 46]. The confidence measure proposed in [47] is calculated by checking consistency between a disparity based segmentation against a color based segmentation of a view of the stereo image pair used as input. This approach is limited by the assumption of smooth disparity changes over color based segments.

Datasets for stereo correspondence algorithms evaluation include the Middlebury [5], KITTI [30] and the enpeda Image Sequence Analysis Test Site (EISATS) [40], where several stereo images with their respective ground-truth are available. Additionally, methods to create and compare datasets are discussed in [19, 21] respectively.

Although the progress on the stereo correspondence problem can be qualitatively inferred, for instance, by the application of different optimization strategies, or by the approaches proposed on different aspects of the disparity estimation process, an objective and quantitative assessment is required not only to determine if a particular algorithm can be considered as superior to other or others-within a particular context-, but also, in order to properly provide feedback to the researcher or practitioner. In this sort of ideas, this paper may result interesting to the reader for two main reasons: by the particular findings on the stated question, and highlighting how a systematic review can be used on visual computing research.

References

1. Morales, S., Klette, R.: A third eye for performance evaluation in stereo sequence analysis. In: Jiang, X., Petkov, N. (eds.) CAIP 2009. LNCS, vol. 5702, pp. 1078–1086. Springer, Heidelberg (2009)
2. Keller, C.G., Enzweiler, M., Gavrila, D.M.: A new benchmark for stereo-based pedestrian detection. In: Intelligent Vehicles Symposium (IV), pp. 691–696 IEEE (2011)
3. Nielsen, M., Andersen, H.J., Slaughter, D.C., Granum, E.: Ground truth evaluation of computer vision based 3D reconstruction of synthesized and real plant images. Precis. Agric. **8**, 49–62 (2007)
4. Wang, Z.-F., Zheng, Z.-G.: A region based stereo matching algorithm using cooperative optimization. In: IEEE Conference on Computer Vision and Pattern Recognition, CVPR 2008, pp. 1–8. IEEE (2008)
5. Scharstein, D., Szeliski, R.: A taxonomy and evaluation of dense two-frame stereo correspondence algorithms. Int. J. Comput. Vis. **47**, 7–42 (2002)
6. Cabezas, I.: Evaluation of disparity maps, Doctoral thesis. Universidad del Valle (2013)
7. Scharstein, D., Szeliski, R.: High-accuracy stereo depth maps using structured light. In: IEEE Computer Society Conference on Computer Vision and Pattern Recognition, Proceedings. 2003, pp. I-195. IEEE (2003)
8. Scharstein, D., Hirschmüller, H., Kitajima, Y., Krathwohl, G., Nešić, N., Wang, X., Westling, P.: High-resolution stereo datasets with subpixel-accurate ground truth. In: Jiang, X., Hornegger, J., Koch, R. (eds.) GCPR 2014. LNCS, vol. 8753, pp. 31–42. Springer, Heidelberg (2014)
9. Cabezas, I., Padilla, V., Trujillo, M.: BMPRE: An error measure for evaluating disparity maps. In: ICSP, pp. 1051–1055 (2012)
10. Cabezas, I., Padilla, V., Trujillo, M.: A measure for accuracy disparity maps evaluation. In: San Martin, C., Kim, S.-W. (eds.) CIARP 2011. LNCS, vol. 7042, pp. 223–231. Springer, Heidelberg (2011)
11. Haeusler, R., Klette, R.: Evaluation of stereo confidence measures on synthetic and recorded image data. In: International Conference on Informatics, Electronics & Vision (ICIEV), pp. 963–968. IEEE (2012)
12. Kitchenham, B.: Procedures for performing systematic reviews. Keele University, Keele, UK, pp. 1–26 (2004)
13. Petersen, K., Feldt, R., Mujtaba, S., Mattsson, M.: Systematic mapping studies in software engineering. In: EASE, pp. 68–77. British Computer Society, Italy (2008)
14. Morales, S., Klette, R.: Ground truth evaluation of stereo algorithms for real world applications. In: Koch, R., Huang, F. (eds.) ACCV 2010 Workshops, Part II. LNCS, vol. 6469, pp. 152–162. Springer, Heidelberg (2011)
15. Hirschmuller, H., Scharstein, D.: Evaluation of stereo matching costs on images with radiometric differences. IEEE Trans. Pattern Analy. Mach. Intell. **31**, 1582–1599 (2009)

16. Leclercq, P., Morris, J.: Robustness to noise of stereo matching. In: Proceedings of 12th International Conference on Image Analysis and Processing, pp. 606–611 IEEE (2003)
17. Szeliski, R., Zabih, R.: An experimental comparison of stereo algorithms. In: Triggs, B., Zisserman, A., Szeliski, R. (eds.) ICCV-WS 1999. LNCS, vol. 1883, pp. 1–19. Springer, Heidelberg (2000)
18. Szeliski, R.: Prediction error as a quality metric for motion and stereo. In: The Proceedings of the Seventh IEEE International Conference on Computer Vision, pp. 781–788 IEEE (1999)
19. Haeusler, R., Klette, R.: Benchmarking stereo data (not the matching algorithms). In: Goesele, M., Roth, S., Kuijper, A., Schiele, B., Schindler, K. (eds.) Pattern Recognition. LNCS, vol. 6376, pp. 383–392. Springer, Heidelberg (2010)
20. Sellent, A., Wingbermühle, J.: Quality assessment of non-dense image correspondences. In: Fusiello, A., Murino, V., Cucchiara, R. (eds.) ECCV 2012 Ws/Demos, Part II. LNCS, vol. 7584, pp. 114–123. Springer, Heidelberg (2012)
21. Kondermann, D., Nair, R., Meister, S., Mischler, W., Güssefeld, B., Honauer, K., Hofmann, S., Brenner, C., Jähne, B.: Stereo ground truth with error bars. In: Cremers, D., Reid, I., Saito, H., Yang, M.-H. (eds.) ACCV 2014. LNCS, vol. 9007, pp. 595–610. Springer, Heidelberg (2015)
22. Kostlivá, J., Čech, J.: Others: Feasibility boundary in dense and semi-dense stereo matching. In: IEEE Conference on Computer Vision and Pattern Recognition, CVPR 2007, pp. 1–8 IEEE (2007)
23. Neilson, D., Yang, Y.-H.: Evaluation of constructable match cost measures for stereo correspondence using cluster ranking. In: IEEE Conference on Computer Vision and Pattern Recognition, CVPR, pp. 1–8. IEEE (2008)
24. Malpica, W., Bovik, A.C.: Range image quality assessment by structural similarity. In: Furht, B. (ed.) Encyclopedia of Multimedia, pp. 757–762. Springer, New York (2008)
25. Shen, Y., Lu, C., Xu, P., Xu, L.: Objective quality assessment of noised stereoscopic images. In: ICMTMA, pp. 745–747. IEEE (2011)
26. van der Mark, W., Gavrila, D.M.: Real-time dense stereo for intelligent vehicles. IEEE Trans. Intell. Transp. Syst. 7, 38–50 (2006)
27. Leibe, B., Cornelis, N., Cornelis, K., Van Gool, L.: Dynamic 3D scene analysis from a moving vehicle. In: IEEE Conference on Computer Vision and Pattern Recognition, CVPR 2007, pp. 1–8. IEEE (2007)
28. Steingrube, P., Gehrig, S.K., Franke, U.: Performance evaluation of stereo algorithms for automotive applications. In: Fritz, M., Schiele, B., Piater, J.H. (eds.) ICVS 2009. LNCS, vol. 5815, pp. 285–294. Springer, Heidelberg (2009)
29. Morales, S., Vaudrey, T., Klette, R.: Robustness evaluation of stereo algorithms on long stereo sequences. In: Intelligent Vehicles Symposium, pp. 347–352. IEEE (2009)
30. Geiger, A., Lenz, P., Urtasun, R.: Are we ready for autonomous driving? the kitti vision benchmark suite. In: IEEE Conference on Computer Vision and Pattern Recognition (CVPR), pp. 3354–3361. IEEE (2012)
31. Hamilton, O.K., Breckon, T.P., Bai, X., Kamata, S.: A foreground object based quantitative assessment of dense stereo approaches for use in automotive environments. In: 20th IEEE International Conference on Image Processing (ICIP), pp. 418–422 IEEE (2013)
32. Woodward, A., Leclercq, P., Delmas, P., Gimel'farb, G.: Generation of an accurate facial ground truth for stereo algorithm evaluation. In: Wojciechowski, K., Smolka, B., Palus, H., Kozera, R.S., Skarbek, W., Noakes, L. (eds.) CVG 2006. Computational Imaging and Vision, pp. 534–539. Springer, Heidelberg (2006)

33. Gong, M., Yang, R., Wang, L., Gong, M.: A performance study on different cost aggregation approaches used in real-time stereo matching. Int. J. Comput. Vis. **75**, 283–296 (2007)
34. Tombari, F., Mattoccia, S., Di Stefano, L.: Stereo for robots: quantitative evaluation of efficient and low-memory dense stereo algorithms. In: 11th International Conference on Control Automation Robotics & Vision (ICARCV), pp. 1231–1238. IEEE (2010)
35. Kelly, P.: Pedestrian detection and tracking using stereo vision techniques. Dublin City University (2007)
36. Kelly, P., O'Connor, N.E., Smeaton, A.F.: A framework for evaluating stereo-based pedestrian detection techniques. IEEE Trans. Circ. Syst. Video Technol. **18**, 1163–1167 (2008)
37. Kogler, J., Eibensteiner, F., Humenberger, M., Gelautz, M., Scharinger, J.: Ground truth evaluation for event-based silicon retina stereo data. In: CVPRW, pp. 649–656. IEEE (2013)
38. Aguilar, M.A., del Mar Saldana, M., Aguilar, F.J.: Generation and quality assessment of stereo-extracted DSM from GeoEye-1 and WorldView-2 imagery. IEEE Trans. Geosci. Remote Sens. **52**, 1259–1271 (2014)
39. Shin, B.-S., Caudillo, D., Klette, R.: Evaluation of two stereo matchers on long real-world video sequences. Pattern Recogn. **48**, 1113–1124 (2015)
40. Wedel, A., Rabe, C., Vaudrey, T., Brox, T., Franke, U., Cremers, D.: Efficient dense scene flow from sparse or dense stereo data. In: Forsyth, D., Torr, P., Zisserman, A. (eds.) ECCV 2008, Part I. LNCS, vol. 5302, pp. 739–751. Springer, Heidelberg (2008)
41. Fuhr, G., Fickel, G.P., Dal'Aqua, L.P., Jung, C.R., Malzbender, T., Samadani, R.: An evaluation of stereo matching methods for view interpolation. In: 20th IEEE International Conference on Image Processing (ICIP), pp. 403–407. IEEE (2013)
42. Vandewalle, P., Varekamp, C.: Disparity map quality for image-based rendering based on multiple metrics. In: International Conference on 3D Imaging (IC3D), pp. 1–5. IEEE (2014)
43. Hu, X., Mordohai, P.: A quantitative evaluation of confidence measures for stereo vision. IEEE Trans. Pattern Analy. Mach. Intell. **34**, 2121–2133 (2012)
44. Zhang, Z., Hou, C., Shen, L., Yang, J.: An objective evaluation for disparity map based on the disparity gradient and disparity acceleration. In: ITCS, pp. 452–455. IEEE (2009)
45. Varekamp, C., Hinnen, K., Simons, W.: Detection and correction of disparity estimation errors via supervised learning. In: 2013 International Conference on 3D Imaging (IC3D), pp. 1–7. IEEE (2013)
46. Haeusler, R., Nair, R., Kondermann, D.: Ensemble learning for confidence measures in stereo vision. In: IEEE Conference on Computer Vision and Pattern Recognition (CVPR), pp. 305–312. IEEE (2013)
47. Milani, S., Ferrario, D., Tubaro, S.: No-reference quality metric for depth maps. In: 20th IEEE International Conference on Image Processing (ICIP), pp. 408–412. IEEE (2013)
48. Hermann, S., Morales, S., Klette, R.: Half-resolution semi-global stereo matching. In: Intelligent Vehicles Symposium (IV), pp. 201–206. IEEE (2011)

Computer Graphics

Guided High-Quality Rendering

Thorsten Roth[1,2]($\boxtimes$), Martin Weier[1,3], Jens Maiero[1,2], André Hinkenjann[1],
and Yongmin Li[2]

[1] Institute of Visual Computing, Sankt Augustin, Germany
thorsten.roth@h-brs.de
[2] Brunel University London, Uxbridge, UK
[3] Saarland University Computer Graphics Lab, Saarbrücken, Germany

Abstract. We present a system which allows for guiding the image quality in global illumination (GI) methods by user-specified regions of interest (ROIs). This is done with either a tracked interaction device or a mouse-based method, making it possible to create a visualization with varying convergence rates throughout one image towards a GI solution. To achieve this, we introduce a scheduling approach based on Sparse Matrix Compression (SMC) for efficient generation and distribution of rendering tasks on the GPU that allows for altering the sampling density over the image plane. Moreover, we present a prototypical approach for filtering the newly, possibly sparse samples to a final image. Finally, we show how large-scale display systems can benefit from rendering with ROIs.

1 Introduction

Photorealistic image synthesis largely depends on the complex, physically-based process of GI. The underlying rendering equation [1] is often solved by employing Monte Carlo methods, which in turn depend on random variables. Basically, these variables are used to create rays, which are traced through the scene and reflected at their intersection points with the scene geometry according to material and illumination properties. Computing GI with MC methods poses two strongly related problems due to its random nature: Noise and convergence behaviour. At first, MC approaches converge quickly, but with an $\mathcal{O}(1/\sqrt{n})$ asymptotic error, this convergence slows down quickly.

As a consequence, cutting the noise for n samples in half requires the computation of $4n$ samples. This is a serious issue when physically-based GI is used in performance-critical applications in which quality is crucial for judging certain properties of the rendered image, which is the case in areas like design review or architectural visualization. Also, this statement can certainly be extended to games. However, determining the number of samples required to achieve a specific quality without performing the actual rendering process is generally not possible. Modern large display setups like 4k projections allow for detailed visualizations, while modern PCs allow for the parallel implementation of the aforementioned rendering algorithms on GPUs. While a single GPU can handle a 4k setup memory-wise, even when a large number of buffers is required

© Springer International Publishing Switzerland 2015
G. Bebis et al. (Eds.): ISVC 2015, Part II, LNCS 9475, pp. 115–125, 2015.
DOI: 10.1007/978-3-319-27863-6_11

(e.g., for post-processing), its performance is still not high enough to perform real-time MC-based GI at such resolutions. We present an approach for coping with today's situation, where designers would like to have a quick, high-quality visualization of products at high resolutions, while modern GPUs are not yet fast enough to compute high-resolution, physically-based GI in an acceptable time frame. Our basic assumption is that an image can be divided into areas with different significance for the user. Consequently, the focus will mostly be put on regions with higher significance, making the convergence of outer areas less important as they are not perceived in a detailed way by the human visual system. Nevertheless, outer areas are still important in order to provide a visual context. Our main contributions are:

1. A general framework for adaptive rendering based on a user-centered context
2. Two interaction metaphors for choosing the relevant image regions
3. A scheduling method based on *Sparse Matrix Compression (SMC)* for efficiently rendering an image with a varying sample density on the GPU
4. A basic filtering approach providing an improved visual context

The GI rendering method we employ throughout this paper is Path Tracing. However, the presented methods are applicable to all rendering methods dealing with similar issues. All benchmarks and other results are based on a resolution of 3840×2160 pixels, unless stated otherwise. Our work focuses mainly on large projection systems. Lacking a 4k projection system in our lab, we use our large, high-resolution display wall HORNET, consisting of 7×5 displays with a 1080p resolution each, to mimic such a system.

2   Related Work

Over the last decades several visualization algorithms that make use of focus and context techniques and ROIs have been published. Most of them try to support the user in understanding complex data in 2D and 3D, e.g., the work in [2] for reading documents or [3] for graphs and hierarchies. Another focus and context technique was introduced in [4] to display giga-pixel images on a large, high resolution display wall. For Giga-pixel images, dealing with a system's limited bandwidth is the major challenge. To overcome this issue, ROIs that can be specified by the user to guide which data to load, defined by a specific position and resolution are employed. Focus and context techniques for 3D data sets are often used in combination with direct volume rendering as the sampling rate for individual rays can be altered at run time. Volume rendering with focus and context can be categorized into (a) distortion lenses, which use approaches similar to the Document Lens, altering the sampling with lenses like ROIs [5], (b) cutaway illustrations, which allow exploring the data using cut planes or x-ray vision, as proposed in [6] and (c) multi-resolution focus and context approaches like the system in [7], where a multi-resolution technique based on eye-tracking is introduced, altering the resolution and sampling rates of volume rendering based on the actual viewing direction. These systems use ROIs specified by the user's gaze in combination with a linear falloff of the visual acuity in the visual peripheral (foveated rendering). They either use eye-tracking or only the head-tracked view-direction. User-centered approaches like [8] use the latter method to generate constant frame rates in virtual environments, combining different rendering approaches. Content-based models of visual attention are used in [9]. This approach also includes a user-centered GI method for prerendered animations. Another approach that utilizes foveated rendering was introduced in [10], employing raster graphics to create three layers with different sampling rates that are blended to a final image. In addition, they use different geometric levels of detail employing tesselation of parametric surfaces to further reduce the computational workload. Geometric simplification based on ROIs and gaze has been proposed in [11–13]. The system proposed in [11] uses ray tracing in combination with gaze-based geometric simplification with multi-resolution meshes. Weier et al. [13] present a simplification based on a hybrid voxel representation. Another method that is based on ray tracing without geometric simplification was introduced in [14]. This system uses eye tracking in combination with VR Headsets. Their main contribution consists of thread creation methods for ray tracing with OpenCL on modern graphics hardware. In contrast to the existing methods our approach concentrates on (a) large projection-based systems and (b) enhancing rendering performance and image quality with global illumination. Our ROI-based rendering technique enables us to render the ROIs at a high quality and the periphery based on a level-of-detail approach, dynamically altering the sampling densities.

3 Framework and Implementation

The basic idea is to provide a framework for user-centered, adaptive high-quality rendering. Therefore, we suggest a number of building blocks taking care of the various tasks that have to be carried out throughout the rendering process. These are described in Sect. 3.1. As a proof of concept and a basic guide, we provide exemplary implementations of each building block and plug them together into a complete system in Sects. 3.2 to 3.6.

3.1 Building Blocks

Figure 1 shows an overview of the building blocks we suggest. The interaction block is responsible for getting information on how to construct the ROI and also for computing it, which is done in a per-pixel manner. In order to do this, it is necessary to provide a method for determining the actual importance of a pixel. This is done by a user-defined distance measure taking into account the image resolution, the focused pixel coordinates and a set of user-defined parameters like the user's position relative to the display system. Based on this information, the measure then has to determine the distance between each pixel and the ROI. Section 3.2 provides an example for such a distance measure.

The ROI resulting from the distance computation can be represented with per-pixel values stored as a distance-guide image (DGI); alternatively, it would

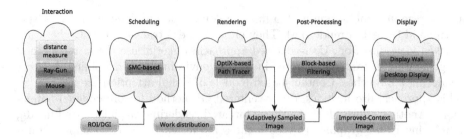

Fig. 1. Building blocks of our suggested framework, exemplary implementations and data flow; the clouds represent the building blocks and contain exemplary implementations represented by green boxes. The orange nodes illustrate the data provided by each of the blocks.

also be possible to do the actual computation of these values on-the-fly in the scheduling block without an intermediate storage. However, as the ROI is only modified on-demand in our current implementation, it makes sense to store the current DGI instead of recomputing it in each iteration. The DGI is assumed to have the same resolution as the rendered image and stores values between 0 and 1. The scheduling block takes this actual distance information and applies a suitable scheduling algorithm. The scheduling depends on the actual representation of distance values as well as the targeted processor architecture employed for rendering. The generated task distribution is then used to efficiently render an image accounting for the user-defined ROI in the rendering block with a suitable rendering method. As the resulting image is potentially sampled sparsely (see Sect. 3.3), there might be a high variance due to low sample numbers or even unsampled pixels outside the ROI. These areas tend to be perceptually disturbing. In order to increase the visual context's quality for the user, a post-processing block is suggested. Here, image parameters such as brightness can be locally adapted to yield a more homogenous appearance and actual filtering algorithms can be applied. The resulting image is then forwarded to the display-block, the abstract representation of the actual display system, such as a simple desktop display, a projection system or a high-resolution multi-display system with an underlying management middleware. The following subsections provide further information on how the individual components can be implemented by giving a few examples.

3.2   Interaction and Distance Measure

In this section we present the two interaction methods we implemented for selecting a region of interest. Besides the mentioned methods we also experimented with head-tracking, which seemed to be a natural approach to select a region for convergence. However, this forces the user to stare at the same region until the desired image quality is achieved, which can be anywhere from seconds to hours depending on the scene and the lighting configuration. This is a huge drawback and renders this approach practically useless in the presented context. Obviously, the same argument is also valid for the additional use of eye-tracking.

Ray-Gun-based Tracking. Our first interaction method is based on a "Ray-Gun" approach. We use a tracked interaction device (FlyStick) for pointing at the display system and "shooting" a ray through the center of the ROI on demand. The FlyStick's analog stick can also be used to modify the horizontal and vertical fields of view. The distance measure is defined by the projection of an elliptical shape onto the display system, oriented orthogonally to the user's pointing direction. To determine the field of view's (FOV) projection on the display system, the view center and four corner points are used, defining the horizontal and vertical size of the ROI. We determine the pixel coordinates of these points by intersecting rays with a virtual model of the display system. As we make no fixed assumption about the shape of the display system but use an exact geometric model, the projected shape might not be an elliptic or oval shape. In our case, the display wall is curved with a 10° angle between each display column and the resulting DGI looks similar to the one shown in Fig. 2. For each pixel in the image, the importance is now determined by computing the minimum distance to any point inside the projected elliptic shape, normalized to $[0, 1]$. All pixels inside the projected shape get assigned the distance value 0, which means that they are sampled in each iteration of the Path Tracing process. A hybrid approach between bisection and Newton's method can be used to find the closest point on the projected shape and thus the shortest distance. For a detailed explanation see [13].

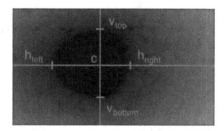

Fig. 2. The Detail-guide Image (DGI) used to guide the rendering process

Mouse-based. An ROI can also be defined in a mouse-based manner by simply defining its center with a click. In a projection system, this is more of a fallback method when no tracking is available. However, it can be useful in a high-resolution desktop environment. Due to the user's lower distance to the display system and the higher pixel density, the aforementioned ROI generation method is also employed in this case. Although, as we assume that there is no tracking system in a desktop environment, the direction for projecting the field of view is perpendicular to the display system's surface, which is considered to be plane. This projection is performed in an orthogonal way, as no distance between the user and the display system is known. Instead, the size of the ROI can be defined in a graphical user-interface when using the mouse.

3.3 Scheduling

Based on the generated DGI, a method has to be developed to compute the sampling densities used during the rendering process in a way so that they are inverse to the DGI's distances. In order to do that, we interpret the distance values $d(x, y) \in [0, 1]$ as inverse probability values for sampling the individual pixels, i.e., $p(x, y) = 1 - d(x, y)$. We then make a binary sampling decision for each pixel individually based on these probabilities and store this decision in a binary image with the same resolution as the rendered image (1 for pixels to be sampled, 0 otherwise). Note due to the probabilistic approach, there is a variation in the number of samples computed between the individual iterations.

This step is necessary because simply computing the samples for pixels with a positive sampling decision would lead to a high thread divergence, thus resulting in idle threads on the GPU. Instead, the binary image computed in the aforementioned step is processed with cuSparse's *Sparse Matrix Compression (SMC)*, which effectively yields the pixel coordinates of non-zero values, i.e., the pixels that have to be sampled. Also, this provides the total number of non-zero values, which means that the Path Tracing process can now be launched with an appropriate number of threads. In turn, the pixel area they have to sample can be directly determined by looking it up in the coordinate array.

Figure 3a shows an unfiltered rendering resulting from SMC-based scheduling. One major issue that arises from rendering an image with these sparsely sampled areas is the potentially high variance in brightness which the human visual system can perceive still well outside the area of sharp vision. Section 3.5 suggests a basic filtering method to overcome this issue, also providing an image brightness similar to the full-resolution image early in the rendering process.

3.4 Rendering

Rendering is performed using our path tracer *Spark*, which is based on NVIDIA's OptiX framework. The rendering process is done completely on the GPU.

3.5 Filtering

As shown in Fig. 3a, the varying noise and sampling density outside the ROI can be visually disturbing. We suggest a Gaussian-based filter with varying kernel size in order to improve the user's visual context. However, the implementation of this filter is still prototypical and does not deliver sufficient performance for a continuous use in an interactive environment. Thus, it serves only as an example for how a filtering approach for an improved visual context could work in general. Note that due to the probabilistic sampling decision made for each pixel, there may always be pixels whose area has not been sampled at all so far. Thus, during the filtering process, only pixels that have been sampled at least once are accounted for, which means that there is no negative influence of unsampled pixels on the image brightness. We use a Gaussian filter kernel with $\sigma = k_i/3$, with k_i being the kernel radius for pixel i. The maximum kernel size k_{max} is set by the user, so that the actual kernel size $k_i = w_i \cdot k_{max} \cdot d_i$ for a pixel i can be computed, based on the distance value d_i contained in the DGI. Also, $w_i = \max\{0, 1 - (s_i/c)\}$ is used to linearly blend between the values from the original image and the filtered results, so that the original, unblurred image becomes more and more visible with the number of samples approaching c. Here, s_i is the current number of samples rendered for the specific pixel and c is a blending constant. This blending constant effectively leads to a slowly decreasing kernel size with the rendering process going on. This also means that image regions are kept at their respective number of samples regardless of whether the ROI is changed.

Instead of performing the blending based on a fixed number of samples, which does not allow for a reliable estimation of the actual noise contained in the image, pixel variance could be employed. Figure 3b shows an example of Gaussian filtering for unconverged, non-focused areas.

(a) An unfiltered image with at most 128 samples per pixel

(b) Result of the Gaussian filtering approach applied to the same image

Fig. 3. Unfiltered vs. filtered image: The perceived difference in brightness is clearly visible.

3.6   Display

As mentioned above, we employ our large, high-resolution display consisting of 7×5 1080p displays to mimic a 4k projection system. Figure 4 shows an example of the rendering results displayed on this system with and without filtering. In order to bring the 4k rendering to the display wall, the DVI output of a regular PC is connected to a video grabber card, which streams the desktop to the SAGE framework, which is in turn responsible for displaying it.

4   Benchmarks

For the presented system, we performed various measurements. First, we are interested in how the decreased ray coherence and density influence rendering efficiency. Thus, we measured the rendering time per one million samples for various fields of view, with larger fields of view corresponding to an increased coherence because of the overall higher sample density. This should in turn lead to a lower thread divergence on the GPU. Also, the pixel density itself is presented. All benchmark results regarding rendering efficiency represent the average of 100 runs.

(a) max. 1 sample per pixel, no filtering   (b) max. 8 samples per pixel, no filtering   (c) max. 64 samples per pixel, no filtering

(d) max. 1 sample per pixel, Gaussian-based filtering   (e) max. 8 samples per pixel, Gaussian-based filtering   (f) max. 64 samples per pixel, Gaussian-based filtering

Fig. 4. Comparison between unfiltered and filtered images displayed on our large, high-resolution display system. Filtering leads to a good approximation of the overall image brightness.

4.1 Setup

With our current implementation we mainly target 4k projection systems. As we do not have a large-scale 4k projection system available, we mimic it using our large, tiled display wall by streaming and upscaling a PC's 4k video output to the SAGE framework. For rendering, we use a Xeon E5-2637 with 16GiB RAM, equipped with a GeForce GTX 980 graphics adapter running Linux. As test scenes we use a simple Cornell Box and the more complex city model *Urban Sprawl 2 by Stonemason* (around 350k triangles), both with environmental illumination only. All benchmarks have been performed with the viewpoint set to an exemplary position 1 m in front of the display wall, looking towards its center. The rendering resolution is 3840×2160 pixels.

4.2 Results

DGI generation using a separate CUDA kernel at a resolution of 3840×2160 pixels remained almost constant for all fields of view, taking 2–3 milliseconds. Figure 5a shows the relation between the number of samples computed per frame and a varying FOV. Clearly, an increased FOV results in a greater number of samples that need to be rendered per iteration. The influence of this number on rendering performance is shown in Fig. 5c and d, which show rendering time in milliseconds per one million samples, thus denoting rendering efficiency. It becomes clear that a greater FOV with a consequently higher number of samples usually results in an efficiency increase, as overheads can be reduced and

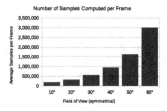
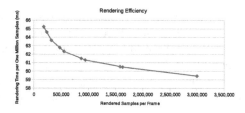

(a) samples computed per frame
for FOVs from 10 to 60 degrees

(b) Rendering efficiency

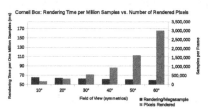
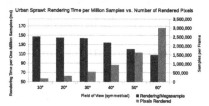

(c) Number of computed samples vs. rendering time per one million samples for the Cornell Box scene

(d) Number of computed samples vs. rendering time per one million samples for the Urban Sprawl scene

Fig. 5. Benchmarks: Computed samples for various FOVs, rendering efficiency

scheduling efficiency increased, where the latter is strongly related to increased ray coherence and sample density. Nonetheless, even with a reduced rendering efficiency it has to be kept in mind that fewer samples have to be rendered per iteration for smaller fields of view. This means that the focused areas usually still converge much faster than with the original efficiency achieved for the full resolution, as shown below. The general behaviour of rendering efficiency shown in Fig. 5b is similar for the Urban Sprawl scene.

We also compared our results to plain Path Tracing without any guiding methods, corresponding to the maximum achieveable rendering efficiency. The pure rendering time for a full 3840×2160 image with one sample per pixel is 460 milliseconds for Cornell Box and 750 milliseconds for Urban Sprawl. Thus, the rendering times per megasample result to 55.46 and 90.42 milliseconds, respectively. For us however, the rendering time per iteration is very important, as this yields some information on how far the image has converged in the focused area after a certain amount of time. Looking at the numbers for several fields of view, we can see that for a small FOV the convergence rate in the focused area is enormous when compared to standard Path Tracing; a 10° FOV with SMC-based scheduling yields a rendering time of 12.29 ms vs. 460 ms for full-resolution rendering, which is a factor of 37.43, corresponding to approximately 84 % reduced noise according to MC methods' probabilistic error of $\mathcal{O}(1/\sqrt{n})$. For a 60° FOV, SMC-based scheduling yields a rendering time of 188 ms vs. 460 ms of standard Path Tracing, which is still a factor of 2.44. The numbers for Urban Sprawl are similar.

Note that with the rendering process progressing, the error difference between areas converging at different rates decreases. With a sampling probability of $p_0 = 1$ for pixels inside the ROI and $p_1 \in [0, 1]$ for some pixel outside the ROI, the number of drawn samples after k iterations should be $n_0 = k$ for pixels inside the ROI and $n_1 \approx p_1 k = p_1 n_0$ for the outside pixel. With the MC asymptotic error of $\mathcal{O}(1/\sqrt{n})$ the error difference between pixels inside and outside the ROI is in

$$\mathcal{O}\left(\frac{1}{\sqrt{n_1}} - \frac{1}{\sqrt{n_0}}\right) = \mathcal{O}\left(\frac{1}{\sqrt{p_1 n_0}} - \frac{1}{\sqrt{n_0}}\right) = \mathcal{O}\left(\frac{1 - \sqrt{p_1}}{\sqrt{p_1 n_0}}\right).$$

With p_1 being constant for the specific pixel, this difference approaches 0 with an increasing number of samples.

5   Conclusion and Future Work

We have shown how techniques using ROIs guiding the Path Tracing process can be employed to increase performance and local quality in global illumination rendering with a focus on, but not limited to large, 4k projection and desktop systems. Moreover, we presented an approach to schedule the rendering tasks guided by an ROI on GPUs and presented a prototypical filtering method to combine the computed samples to a final image. To interpret our results correctly, it is important to note that our work is based on the assumption that there actually is an important area in the image which is desired to be rendered at a high quality, while the quality of the remaining image is not as important. Though, the context of the non-focused area is preserved. We achieved our goal of improving performance for such cases, as shown in the benchmark section, where we also compared our runtime results with standard full-resolution Path Tracing.

Future research should include scheduling methods for increased ray coherence, which may result in a further performance benefit. Sparsely sampled as well as focused areas could be processed with denoising methods such as Ray Histogram Fusion [15], which has recently also been optimized for GPUs [16]. It has to be analyzed whether denoising should be performed at full resolution or at area-specific fractions of the original resolution.

Also, performance and perceptual aspects of such a reconstruction/filtering step have to be taken into account. Even though the basics of gaze-based rendering are well-understood, a user study has to be performed on how the noise introduced by the stochastic nature of Path Tracing affects the perceived quality of rendered images using ROIs based on the user's gaze. Depending on their performance, reconstruction methods could also be implemented to be executed once every few seconds or itereatively for static viewpoints. Generally, the presented work aims for the next step to be the realization of such a system with the full 72 megapixel resolution of our lage display wall. As this rendering should take place on a rendering cluster with several dozens of GPUs, a hybrid scheduling approach for adaptive resolutions is one of the major challenges for achieving this goal.

References

1. Kajiya, J.T.: The rendering equation. ACM SIGGRAPH Comput. Graph. **20**, 143–150 (1986)
2. Robertson, G.G., Mackinlay, J.D.: The document lens. In: Proceedings of the 6th Annual ACM Symposium on User Interface Software and Technology, UIST 1993. ACM, New York, NY, USA, pp. 101–108 (1993)
3. Lamping, J., Rao, R.: Laying out and visualizing large trees using a hyperbolic space. In: Proceedings of the 7th Annual ACM Symposium on User Interface Software and Technology, UIST 1994. ACM, New York, NY, USA, pp. 13–14 (1994)
4. Papadopoulos, C., Kaufman, A.E.: Acuity-driven gigapixel visualization. IEEE Trans. Vis. Comput. Graph. **19**, 2886–2895 (2013)
5. LaMar, E., Hamann, B., Joy, K.I.: A magnification lens for interactive volume visualization. In: Proceedings of the Ninth Pacific Conference on Computer Graphics and Applications, pp. 223–232 (2001)
6. Krüger, J., Schneider, J., Westermann, R.: Clearview: an interactive context preserving hotspot visualization technique. IEEE Trans. Vis. Comput. Graph. **12**, 941–948 (2006)
7. Levoy, M., Whitaker, R.: Gaze-directed volume rendering. In: Proceedings of the 1990 Symposium on Interactive 3D Graphics, I3D 1990. ACM, New York, NY, USA, pp. 217–223 (1990)
8. Funkhouser, T.A., Squin, C.H.: Adaptive display algorithm for interactive frame rates during visualization of complex virtual environments. In: Proceedings of the 20th Annual conference on Computer Graphics and Interactive Techniques, pp. 247–254, ACM (1993)
9. Yee, H., Pattanaik, S., Greenberg, D.P.: Spatiotemporal sensitivity and visual attention for efficient rendering of dynamic environments. ACM Trans. Graph. **20**, 39–65 (2001)
10. Guenter, B., Finch, M., Drucker, S., Tan, D., Snyder, J.: Foveated 3d graphics. ACM Trans. Graph. (TOG) **31**, 164 (2012)
11. Murphy, H., Duchowski, A.T.: Gaze-contingent level of detail rendering. Euro-Graphics 2001 (2001)
12. Weaver, K.A.: Design and evaluation of a perceptually adaptive rendering system for immersive virtual reality environments. Master's thesis, Iowa State University (2007)
13. Weier, M., Maiero, J., Roth, T., Hinkenjann, A., Slusallek, P.: Enhancing rendering performance with view-direction-based rendering techniques for large, high resolution multi-display systems. In: 11. Workshop Virtuelle Realität und Augmented Reality der GI-Fachgruppe VR/AR (2014)
14. Fujita, M., Harada, T.: Foveated real-time ray tracing for virtual reality headset (2014)
15. Delbracio, M., Mus, P., Buades, A., Chauvier, J., Phelps, N., Morel, J.M.: Boosting Monte Carlo rendering by ray histogram fusion. ACM Trans. Graph. **33**, 8:1–8:15 (2014)
16. Szeracki, S., Roth, T., Hinkenjann, A., Li, Y.: Boosting histogram-based denoising methods with gpu optimizations. In: Workshop Virtuelle Realität und Augmented Reality der GI-Fachgruppe VR/AR (2015)

User-Assisted Inverse Procedural Facade Modeling and Compressed Image Rendering

Huilong Zhuo[1], Shengchuan Zhou[2], Bedrich Benes[1(✉)], and David Whittinghill[1]

[1] Purdue University, West Lafayette, USA
bbenes@purdue.edu
[2] Qingdao Geotechnical Investigation and Surveying Research Institute, Qingdao, China

Abstract. We take advantage of human intuition by encoding facades into a procedural representation. Our user-assisted inverse procedural modeling approach allows users to exploit repetitions and symmetries of facades to create a split grammar representation of the input. Terminal symbols correspond to repeating elements such as windows, window panes, and doors and their distributions are encoded as the production rules. Our participants achieved a compression factor that averaged 57 % (min = 12 %, max = 99 %) while taking on average 7 min (min = 1, max = 25) to compress an image. The compressed facades do not suffer from occlusion problems present in the input, such as trees or cars. Our second contribution is a novel rendering algorithm that directly displays the compressed facades in their procedural form by interpreting the procedural rules during texture lookup. This algorithm provides considerable memory savings while achieving comparable rendering performance.

1 Introduction

The most common facade representation is by texture images which occupy a large amount of memory, often even more than the actual model geometry. Although images can be compressed, few techniques exist for their compressed rendering. Facades contain repeated, structures, making them suitable for procedural representation [1–3]. The goal of inverse procedural modeling (IPM) is to find a procedural representation of an input model [4] and it can be thought of as image compression. Existing IPM methods require a preprocessed input in which the scene includes higher semantic information [5,6]. This hinders the applicability of IPM in practice and is exacerbated by the large variability of details in facade images.

We present a user-assisted approach for inverse procedural facade modeling. Our key observation is that human visual perception provides global insight into structures and can detect subtle details at varying scales while simultaneously judging their importance. Humans are good at detecting repetitions and, due to previous experiential knowledge, can account for missing data or incoherencies. Humans may not achieve an (algorithmically) optimal facade compression such

© Springer International Publishing Switzerland 2015
G. Bebis et al. (Eds.): ISVC 2015, Part II, LNCS 9475, pp. 126–136, 2015.
DOI: 10.1007/978-3-319-27863-6_12

as [6], but they work directly with unstructured images, they can complete missing information, and they can generate representations that will be perceived as realistic.

We encoded the input image by indicating symmetries and repetitions during an interactive session. The procedural rules are immediately available and the system also shows the resulting compression of the facade as well as the original. The output of this process is a set of repeating elements (terminal symbols of the grammar), and a set of procedural rules that define the distribution of the terminals.

Our second contribution is an approach for direct *compressed* image rendering. We store only the terminal symbols of the grammar while pixel color is calculated using a direct lookup through the generated procedural rules. This presents considerable savings of the GPU memory while maintaining comparable efficiency of rendering. While individual image rendering is slower by an average of 23 % depending on the rules' complexity. In an application testing the rendering of multiple houses the compressed facade rendering is slower an average 17 %.

Figure 1 shows an input facade that includes missing parts, large variability, and obstacles. This facade was encoded procedurally in less than two minutes. In the upper row the user aimed for a precise representation of the input leading to the compression factor 49 % and a high number of 44 procedural rules. The lower row shows the image with compression of 85 % and only 30 rules. The higher compression is achieved by sacrificing various unique details. These choices derive directly from aesthetic objectives; an automated system would be poorly suited to achieve acceptable results. Moreover, the user implicitly accounts for missing details.

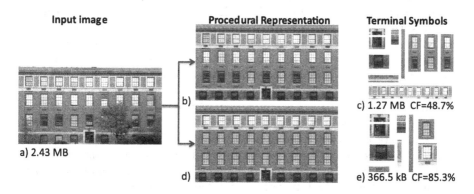

Fig. 1. The user encoded a facade (a) procedurally with attention paid to conveying the varying details (b) resulting in 44 procedural rules, 49 % compression, and 10 terminal symbols (c). The same facade was encoded again with less precision (d) resulting in only 30 procedural rules, 85 % compression, and 9 terminal symbols (e). The user removed the occluding trees and each encoding took under two minutes. The original image was rendered in 1.32 ms and the direct compressed facade rendering was 1.73 ms for (b) and 1.54 ms for (d).

2   Related Work

Procedural representation has been addressed in Computer Graphics by work ranging from methods for pure procedural generation [2,7–10] to inverse approaches that attempt to find procedural representation of real buildings [11–13] or even vegetation [14].

Recently, attention has been focused on procedural *facade* representation. Mueller et al. presented what is likely the first method for procedural representation of facades in Computer Graphics in [15]. Their approach builds on shape grammars and allows for an automatic extrusion of facade features into a 3D representation. This approach was later extended to interactive facade editing in [7]. Various approaches aim at automatic generation of a grammar representation based upon an input facade. They can be categorized into techniques that use images as input, or those that use Laser Interferometry Detection and Ranging (LiDAR) data sets. In the first class of algorithms that work with images, there are a number of techniques that attempt to find symmetries within the image's structures. Musialski et al. introduced an automatic approach for facade repair that exploits the facade symmetry in [16]. Teboul et al. addressed segmentation in [17], where they use a combination of grammars, supervised classification, and random walks to synthesize facades and recently a 3D reconstruction was presented in [18].

Grammar generation of facades from input images has been addressed only recently. Boulch et al. use bottom-up automated parsing to discover attributes from a predefined grammar [19], and similarly Martinović et al. learns attributed context-free grammar using Bayesian networks in [20]. Zhang et al. use an automated approach for automatic grammar generation by using symmetry detection and evaluate different groups and layers in [3]. A similar approach for automatic grammar generation from segmented facades was presented in [6] in which the objectives were to generate minimal grammar and a new facade synthesis. We are also acquiring an inverse procedural representation, but rather than new facade synthesis, we focus on compressed facade rendering. Also, our approach uses human intuition to complete missing parts and irregularities. Moreover, previous approaches require some kind of pre-segmentation and user identification of terminal symbols. This is a common requirement for the majority of inverse procedural approaches such as [4].

LiDAR data have been addressed by a number of previous studies as well. A technique called adaptive recursive facade splitting has been used to reconstruct facades from LiDAR data in [21]. Vanegas et al. used a generalized rewriting rule to describe Manhattan-world reconstructed buildings [22]. Ceylan et al. introduced an image-based 3D building reconstruction framework that focused upon facades in [23]. Wan and Sharf used grammars to aid in 3D reconstruction of faces from scanned urban facades [24]. Kerber et al. used symmetry detection with a feature descriptor for large scale urban scenes in [25]. Li et al. used a combined approach in which images and LiDAR data were used to create depth layers to complete 3D facades in [26].

Close to our approach is the interactive system for facade modeling [27]. Our approach exploits symmetries and user intuition in order to create facade representations. The work [27] differs from ours in that it does not attempt to perform direct facade rendering, nor does it show the actual procedural representation of the input.

3   Split Grammar

The objective of our work is to describe an input image as a context-free attributed split grammar [2,9,15]. The split grammar is a quadruple

$$G = \langle \omega, N, T, R \rangle, \tag{1}$$

where $\omega \in N$ is the starting symbol of the grammar (the axiom) that usually corresponds to the entire facade. The non-terminal symbols $N = \{N_1, N_2, \ldots, N_{|N|}\}$ represent transformations and intermediate steps and can be thought of as groups or structures. The terminal symbols $T = \{T_1, T_2, \ldots, T_{|T|}\}$ are the final sub-images that compose the input and R is a non-empty set of production rules in the form

$$N(\mathbf{p}) \rightarrow op_1(\mathbf{p})\{x_1(\mathbf{p})\}op_2(\mathbf{p})\{x_2(\mathbf{p})\}\ldots. \tag{2}$$

The rules denote rewriting of the non-terminal symbol $N(\mathbf{p})$ from the left-hand side with a sequence of nonterminal, and terminal symbols $x \in N \cup T$. Each symbol on the right-hand side can be modified by an operation op. The operation op can be one of the $splitX$, $splitY$, $subdivX$, $subdivY$, $flipX$, $flipY$, and $rotateX$.

Every (terminal and non-terminal) symbol is associated with a (rectangular) shape. Terminal symbols represent a certain part of the input image, whereas non-terminals correspond to areas that are composed of other non-terminals and/or terminals. We define the *compression* of the input image as

$$cf = \frac{1}{\sum scale(T_i)}, \tag{3}$$

where $scale(T_i)$ measures the relative size of the terminal symbol in the input image. Though we could also add the size of the grammar to the compression, its size is usually significantly smaller than the size of the images and can safely be ignored.

4   Interactive Application

We have developed an application that allows the user to compress a facade during an interactive session and represent it as a split grammar, Eq. (1). The application has three main windows (see Fig. 2). The first is the editing window that shows the input facade and allows all user interaction. The second window displays the facade that is the result of the procedural representation. The image in the second window is always immediately interpreted from the rules. The grammar (non-terminals, terminals, rules) and grammar statistics (compression factor, number of rules) are displayed on the third window. This window is observed by the user when she attempts different procedural representations.

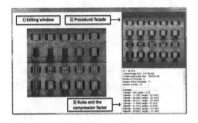

Fig. 2. The user represents the original facade (1) as a procedural representation. The application shows the result in the second window, while the third window (3) shows the procedural rules and the compression factor (Color figure online).

The application implements rules from Sect. 3, namely terminal and non-terminal node selection, facade horizontal $splitX$ and vertical $splitY$, symbol $flip$, symbol $rotation$, and regular distribution of symbols via the $subdivide$ command.

The user interface shows selected symbols with dark blue color, terminal symbols in lighter blue, and non-terminal symbols with a pattern. In this way the user has visual control over the amount of encoded parts of the input image and can therefore make a better estimate of the total area of terminal symbols that will need to be stored.

The interactive session starts with the complete facade that is represented as the axiom ω. The user selects a symbol and applies an operation that also creates a procedural rule. This process is repeated until the user decides that further modeling is not necessary at which point the procedural model is saved. The result of the session is the grammar G from Eq. (1) where each terminal symbol is stored as a separate image. The user can create rules that may not be optimal, however we do not perform any rule optimization to avoid biasing the results.

5   Rendering

We present a new approach for direct procedural facade rendering. The terminal symbols and the rules are stored on the GPU and the (u, v) texture coordinates are evaluated directly from the grammar. Depending on the encoding, the compressed image may be significantly smaller than the input image. However, as rule interpretation is required, there is an overhead imposed by the texture color look-up.

There are two possible ways to implement this function: (1) the procedural rules could be interpreted once and the expanded form stored as a long array of terminal symbols and accompanying symbol operations, (2) the rules can be interpreted for every pixel. We have implemented both approaches. Although the first approach would save the overhead of rule evaluation, we observed performance to be significantly slower (2.97 ms vs 1.54 ms for the texture from Fig. 1d)) on the current GPU architectures (Shader Model 5.0, OpenGL 4.3) that are sensitive to branching and loops. For the fully expanded rules, the GPU must traverse the array to check if the searched pixel belongs inside the displayed region of terminals. This results in extensive branching. The looping and branching operations cannot be unrolled by the compiler and the expanded rule includes various repeating *if* statements that present a significant penalty for rule runtime execution performance. The second approach, where the split rules are stored on the GPU and interpreted for each pixel, was more efficient on current GPUs, but it can be different for future GPU architectures or rendering systems.

Let us recall (see Sect. 3) that the terminal symbols (images) are denoted by $T = \{T_1, T_2, \ldots, T_{|T|}\}$, non-terminals by $N = \{N_1, N_2, \ldots, N_{|N|}\}$ and the split rules have form $N(\mathbf{p}) \to op_1(\mathbf{p})\{x_1(\mathbf{p})\}op_2(\mathbf{p})\{x_2(\mathbf{p})\} \ldots$. Each rule is encoded in a data structure that stores the operation, parameters, and the set of terminal and non-terminal symbols on the right-hand side. Each rule has an ID and the rules are stored in a table that is indexed by the left hand side and has the right hand side as the table elements. This table is stored on the GPU.

The rule interpretation begins with the axiom ω and the texture coordinates (u, v) for which the color is queried. The rule with ω on the left-hand side is found in the table, the right-hand side is scanned, and each operation (*split*, *subdivide*, *flip*, etc.) is evaluated. As the operations lead to a smaller part of the ω we check if the searched (u, v) coordinates fall inside the evaluated operation. If yes, the symbol that was on the right-hand side is located and further interpreted. If the new symbol is a non-terminal, the interpretation continues with interpreting the right-hand side of the corresponding rule. The scanning stops if a non-terminal symbol is found. The (transformed) coordinates of the (u, v) are used to look up the color.

The algorithm has theoretical complexity $\mathcal{O}(|R|)$, where $|R|$ is the number of rules. This case would correspond to the ω being replaced by the non-terminals covering the same area and eventually ending with an image that is the same size as the input. In a practical implementation the rules split the input image into smaller parts, so the behavior is $\approx \mathcal{O}(log_2|R|)$. This, of course, depends on the strategy for the creation of the procedural rules that is discussed in Sect. 6.1.

6   Implementation and Results

The system is implemented in C++ and uses an OpenGL for rendering, shaders were implemented in GLSL 5.0, the system has been tested on Intel i7 CPU clocked at 3.2 GHz with 16 GB of memory and a NVIDIA K5000 Quadro GPU.

6.1   User Testing

We tested our approach on ten participants aged 20–30 years. All held under-graduate degree in computer science or a related discipline, and all had some knowledge of procedural modeling. The objective was to interactively encode eleven facades (images available on the project page at hpcg.purdue.edu) while maintaining visual similarity with the input and maximizing the compression factor. We selected test facades such that they (1) provided variability of struc-tures, (2) allowed for comparison with previous work [3,28], and (3) included facades with occlusion problems. The participants were first introduced to the goal of the task. Next, they completed a short training tutorial that led them step-by-step to compress a simple schematic facade and explained every action and result in detail. There was no time limit on the duration of the training and all participants finished the training in an average of 11 minutes.

The overall compression was 57 % with the standard deviation equal to 10 %. The maximum compression factor was 99 % and the minimum 12 %. Higher compression was achieved by visually degrading the input and vice versa. The image can be better compressed by replacing similar windows with some variation of a single instance. We demonstrate this in Fig. 4 where the first column shows the original image, the middle column is the figure that corresponds to the highest compression factor, and the last column shows images that have the lowest compression factor. Note the results for the facade #8 are quite subtle, yet the compression went from 65 % to 22 %. The compression of 99 % of the facade 10 is an extreme example and washes out the visual differences.

The results also show a degree of variability among individual subjects. The second column of Fig. 4 shows the average compression per participant for the entire set of images (min = 22 %, max = 69 %).

The participants spent in average of 7 min compressing each image (min = 1, max = 25). The average number of production rules was 52 (min = 4, max = 146).

Rule Selection. We analyzed the selected rules and discussed the strategy for image compression with the participants in a group discussion. All participants independently tested two different strategies. First they would attempted to isolate the most frequently repeating large elements, such as windows, and convert them into terminal symbols. That is also a common strategy in automated systems [3, 6]. Surprisingly this did not lead to the best compression strategy and most participants quickly discarded this approach as inefficient.

The common strategy to which all participants independently converged was to find the largest symmetrical parts of the image and divide it. This strategy was then repeated until small elements, such as windows, were found. The terminal symbols were then analyzed and some of them were merged to replace windows with small visual changes.

Occlusion. Users use their intuition to quickly account for occluded parts of missing structures as can be seen in Figs. 1 and 3. Interestingly, users can create similar facades by discarding obstacles as can be seen in Fig. 3, where the right side of the facade was replaced with the left one that was completed by non-occluded windows.

Comparison to Previous Work. Images 1 and 2 from our testing set (hpcg.purdue.edu) correspond to the images used by [28] (Figs. 9 and 11) and Images 4, 6, 7, and 8 are from [3] (Fig. 9). While both methods focused on an automatic detection of structural similarities in our work we have completely

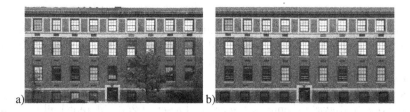

Fig. 3. An image with occluded parts (1) can be quickly resolved by user intuition (b).

offloaded the detection of similarities to human. The compression factor of the previous work was not their main aim but a side effect. In order to find the compression factor we have manually calculated the number of terminal symbols and the compression factor of each facade from [3,28]. Table 1 shows the comparison between A: [28], B: [3], and C: our approach. The compression factors reported by our method are the average numbers from the user study.

Our user-assisted approach leads to better results for all cases. The average CFs are A:32 %, B:17 %, and C:54 %. The number of terminal symbols has the major effect on the size of the final representation and the averages were A:7, B:18, and C:13. It seems the main reason for a higher compression factor of the user-assisted approach is in the human tolerance to variability in the input images. It would likely be difficult to tune an automatic system to be as tolerant to the variance in input images as are humans. Yet, humans produce facades that are visually plausible.

Table 1. Comparison of the compression factor and the number of terminal symbols among A: [28], B: [3], and C: our approach.

Image	CF [%]			# of T		
	A	B	C	A	B	C
1	31		76	3		8
2	32		71	4		6
4		25	63		10	17
6		17	39		7	11
7		8	24		14	23
8		18	48		5	21

6.2 Compressed Image Rendering

Compressed facade rendering comes at an additional cost for extra computation. We have used the result from the user study to demonstrate the performance hit as a function of the number of rules. The rules are different and having the same number of rules can vary significantly in their composition. To account for this we have sorted the results according to the number of rules. We have then calculated the average over intervals of 5 rules, displayed the normalized rendering time. For the worst case of 110 procedural rules the rendering was two times slower than direct image rendering. For the average case of 50 rules the performance was 23 % slower (1.234 ms vs. 1.525 ms).

As rendering an individual image is a rare case, we have also created an example of a real-world scenario. We created a set of 250 different buildings and rendered them with and without compressed facades from the user-generated examples. The average compression factor for the textures from the entire scene was 81 %. The loading time was 31 s for uncompressed textures and 6 s for compressed ones; the rendering time was 1.39 ms per frame for uncompressed and 1.62 ms per frame for compressed (16 % performance hit).

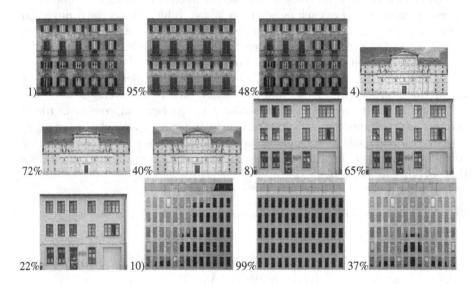

Fig. 4. Example results of the testing. The first column shows the original image, the middle column image with the highest compression factor, and the rightmost column shows the image with the lowest compression factor. The high compression factor is compensated for by losing variety of detail amongst structures (e.g., reflections in windows).

7   Conclusions

We have presented a user-assisted approach for inverse procedural modeling of facade images and their compressed rendering. Our approach treats the facade as a set of pixels and visual consistency is judged by the user. Users quickly account for occluded sections and use their own intuition to compress global symmetries and identify visual differences between similar elements. Images are compressed in an interactive session that takes approximately two minutes per facade depending upon its complexity. During our testing users achieved average compression of 57 %. The original image is represented as a set of sub-images and directly rendered by the GPU from the procedural rules. The overhead for the image rendering depends upon the complexity of the procedural representation and was on average 23 % for an image composed from 50 procedural rules. In a real-world scenario of a small city rendering, the performance hit was 16 %. An interesting observation is that subjects used different strategies for splitting, such as dividing the elements that are usually considered atomic (i.e. windows or doors).

One obvious limitation of our method is the need for manual editing. Though it would have been possible to directly render images compressed by some existing automatic lossless method, users nonetheless achieved a good lossy compression factor while preserving the quality of the output, since they were allowed to apply their own individual judgment as to what could be discarded. Another potential limitation of our study was the selection of our test group. It would have been interesting to see how a more heterogeneous group of users might have performed.

Many possible avenues for future work exist. Recently, a number of automatic methods for facade encoding have appeared. One is an approach [28] that allows an accounting for varying facade visual properties which are then encoded as a parameter. It would be interesting to implement their approach using direct GPU image rendering. Lastly, a potentially promising caching strategy derived from this work could be implemented in which similar pixels are identified and then processed by grammatical rules similar to those created using our approach.

References

1. Bao, F., Schwarz, M., Wonka, P.: Procedural facade variations from a single layout. ACM Trans. Graph. **32**, 8:1–8:13 (2013)
2. Müller, P., Wonka, P., Haegler, S., Ulmer, A., Van Gool, L.: Procedural modeling of buildings. ACM Trans. Graph. **25**, 614–623 (2006)
3. Zhang, H., Xu, K., Jiang, W., Lin, J., Cohen-Or, D., Chen, B.: Layered analysis of irregular facades via symmetry maximization. ACM Trans. Graph. **32**, 121:1–121:13 (2013)
4. Stava, O., Benes, B., Mech, R., Aliaga, D.G., Kristof, P.: Inverse procedural modeling by automatic generation of L-systems. Comput. Graph. Forum **29**, 665–674 (2010)
5. Haegler, S., Wonka, P., Arisona, S.M., Gool, L.V., Müller, P.: Grammar-based encoding of facades. In: Proceedings of the EGSR, Eurographics Association, pp. 1479–1487 (2010)
6. Wu, F., Yan, D.M., Dong, W., Zhang, X., Wonka, P.: Inverse procedural modeling of facade layouts. Technical report, arXiv:1308.0419 [cs.GR] (2013)
7. Lipp, M., Wonka, P., Wimmer, M.: Interactive visual editing of grammars for procedural architecture. ACM Trans. Graph. **27**, 102:1–102:10 (2008)
8. Parish, Y.I.H., Müller, P.: Procedural modeling of cities. In: Proceedings SIG-GRAPH, pp. 301–308. ACM Press (2001)

9. Wonka, P., Wimmer, M., Sillion, F., Ribarsky, W.: Instant architecture. ACM Trans. Graph. **22**, 669–677 (2003)
10. Smelik, R.M., Tutenel, T., Bidarra, R., Benes, B.: A survey on procedural modelling for virtual worlds. Comput. Graph. Forum **33**, 31–50 (2014)
11. Aliaga, D.G., Rosen, P.A., Bekins, D.R.: Style grammars for interactive visualization of architecture. IEEE TVCG **13**, 786–797 (2007)
12. Hohmann, B., Krispel, U., Havemann, S., Fellner, D.: Cityfit - high-quality urban reconstruction by fitting shape grammars to image and derived textured point clouds. In: Proceedings of the International Workshop 3D-ARCH 2009 (2009)
13. Vanegas, C.A., Garcia-Dorado, I., Aliaga, D.G., Benes, B., Waddell, P.: Inverse design of urban procedural models. ACM Trans. Graph. **31**, 168:1–168:11 (2012)
14. Stava, O., Pirk, S., Kratt, J., Chen, B., Mch, R., Deussen, O., Benes, B.: Inverse procedural modelling of trees. Comput. Graph. Forum **33**, 118–131 (2014)
15. Müller, P., Zeng, G., Wonka, P., Van Gool, L.: Image-based procedural modeling of facades. ACM Trans. Graph. **26**, 85 (2007)
16. Musialski, P., Wonka, P., Recheis, M., Maierhofer, S., Purgathofer, W.: Symmetry-based facade repair. In: Vision, Modeling, and Visualization Workshop 2009 (2009)
17. Teboul, O., Simon, L., Koutsourakis, P., Paragios, N.: Segmentation of building facades using procedural shape priors. In: Proceedings of CVPR, pp. 3105–3112 (2010)
18. Demir, I., Aliaga, D.G., Benes, B.: Coupled segmentation and similarity detection for architectural models. ACM Trans. Graph. **34**, 104:1–104:11 (2015)
19. Boulch, A., Houllier, S., Marlet, R., Tournaire, O.: Semantizing complex 3D scenes using constrained attribute grammars. Comput. Graph. Forum **32**, 33–42 (2013)
20. Martinovic, A., Van Gool, L.: Bayesian grammar learning for inverse procedural modeling. In: Proceedings of CVPR, pp. 201–208 (2013)
21. Shen, C.H., Huang, S.S., Fu, H., Hu, S.M.: Adaptive partitioning of urban facades. ACM Trans. Graph. **30**, 184:1–184:10 (2011)
22. Vanegas, C.A., Aliaga, D.G., Beneš, B.: Building reconstruction using manhattan-world grammars. In: Proceedings of CVPR, pp. 358–365 (2010)
23. Ceylan, D., Mitra, N.J., Li, H., Weise, T., Pauly, M.: Factored facade acquisition using symmetric line arrangements. Comp. Graph. Forum **31**, 671–680 (2012)
24. Wan, G., Sharf, A.: Applications of geometry processing: grammar-based 3D facade segmentation and reconstruction. Comput. Graph. **36**, 216–223 (2012)
25. Kerber, J., Bokeloh, M., Wand, M., Seidel, H.P.: Scalable symmetry detection for urban scenes. Comput. Graph. Forum **32**, 3–15 (2013)
26. Li, Y., Zheng, Q., Sharf, A., Cohen-Or, D., Chen, B., Mitra, N.J.: 2D–3D fusion for layer decomposition of urban facades. In: Proceedings of ICCV, pp. 882–889 (2011)
27. Musialski, P., Wimmer, M., Wonka, P.: Interactive coherence-based facade modeling. Comp. Graph. Forum **31**, 661–670 (2012)
28. AlHalawani, S., Yang, Y.L., Liu, H., Mitra, N.J.: Interactive facades analysis and synthesis of semi-regular facades. Comput. Graph. Forum **32**, 215–224 (2013)

Facial Fattening and Slimming Simulation Based on Skull Structure

Masahiro Fujisaki[1(✉)] and Shigeo Morishima[2]

[1] Waseda University, Tokyo, Japan
fujisaki.1000@gmail.com
[2] Waseda Research Institute for Science and Engineering,
3-4-1 Okubo Shinjuku-Ku, Tokyo 169-8555, Japan

Abstract. In this paper, we propose a novel facial fattening and slimming deformation method in 2D images that preserves the individuality of the input face by estimating the skull structure from a frontal face image and prevents unnatural deformation (e.g. penetration into the skull). Our method is composed of skull estimation, optimizing fattening and slimming rules appropriate to the estimated skull, mesh deformation to generate fattening and slimming face, and generation background image adapted to the generated face contour. Finally, we verify our method by comparison with other rules, precision of skull estimation, subjective experiment, and execution time.

1 Introduction

Accurate simulations of facial fattening or slimming are required in many areas like beautification, health, and entertainment. For example, fattening and slimming simulation can be used to motivate dieters. Before performing cosmetic surgery, such simulations can be used to inform the surgeon of patient's expectations. In criminal investigations, the results can be useful for investigator to know about the face feature variation when a wanted person's weight has changed.

We propose a novel facial fattening and slimming simulation method in 2D images that includes a facial restoration [1–3]. A Facial restoration reconstructs the facial surface shape by adding statistical skin depth data to skull bone areas. Thus, to generate an accurate simulation, it is inevitable to consider the precise shape of skulls.

Ideally, facial fattening and slimming simulation should have data about the target person's facial characteristics, i.e., fatness or thinness. However, it is almost impossible in practice to get such data. Consequently, previous works compared a target person's face with the other faces statistically and estimated optimum parameters for fattening and slimming deformation [4–9]. In the proposed method, we estimate the feature in inner facial structure by predicting the target (input) person's skull structure with considering correlation between facial feature points (FPs) and skull FPs. Moreover, optimum fattening and slimming rule is calculated from various face database whose skull feature is very close to the target person's. We refer to "various face database whose skull feature is fit to match the target (input) person's" as *facial fattening variations*. Our facial fattening variations are standardized relative to the skull

© Springer International Publishing Switzerland 2015
G. Bebis et al. (Eds.): ISVC 2015, Part II, LNCS 9475, pp. 137–149, 2015.
DOI: 10.1007/978-3-319-27863-6_13

structure; therefore, the surface differences are only differences in the volume of fat. Consequently, an individual facial fattening and slimming rule can be calculated using only the fattening and slimming components extracted by applying principal component analysis (PCA) to these variations.

In the proposed method, the input is only a frontal face image. Other requirements are the user defined fattening parameter, some frontal face images of the other people, MRI images of the other people and 661 vertices from an average face model as preparations. The output is a fattened or slimmed face image.

In this paper, due to the assumed situations, we do not consider drastic fattening or slimming changes that cause the facial depth and the facial parts to change. Hence, the facial fattening and slimming simulation is represented by changes in the facial contour shape. The advantages of the proposed method are summarized as follows:

1. Reflects individuality based on the target person's estimated skull structure
2. Prevents unnatural deformation of the facial surface, i.e., penetration into the skull
3. Requires only a single frontal face image as input
4. Versatile, fast and fully automatic

2   Related Work

There are great many studies about human fattening and slimming simulation. Allen et al. simulated 3D global-shape fattening and slimming for an arbitrary person [4, 5]. They searched corresponding points from 3D point data of human body, applied PCA to the corresponding points, and manipulated their parameters.

The simulations proposed in these previous studies realized natural deformation for global fattening of the whole body. However, unlike bodies, human faces have detailed characteristics and intricate structures. The methods proposed in these previous studies are not sufficient to represent faces because faces consist of detailed parts (eyes, nose, mouth and so on). In addition, we focus on facial characteristics and examine the venial face difference when human beings look at other human beings. Facial simulation must be realistic and detailed, because if the results differ from the ground truth, observers feel uncomfortable.

Blanz et al. focused on facial fattening and slimming [9]. They used 3D morphable model generated by applying PCA to 3D facial models. They represented facial fattening and slimming by fitting the 3D morphable model to each part of the face and manipulating parameters. However, their method has two problems. First, they fattened all faces in the same way by considering only the difference in facial surfaces. However, human face fattening depends on the individual skull structure. The results of their method are not realistic because facial fattening and slimming deformation should be unique for each person. Second, without considering skull structure, there is a possibility of deforming facial contour unnaturally. For example, the facial surface can penetrate into the skull, which is also not realistic. Therefore, they could not simulate facial fattening and slimming realistically.

We focus on these problems. Blanz et al. fattened faces by manipulating the parameters of principal components generated by applying PCA to the facial surface

Table 1. Database properties

	Ave.	Min.	Max.
BMI	20.7	18.3	25.0
Age	32.8	22	47
BMI distribution			
18–20	20–22	22–24	24–26
11 people	8 people	3 people	2 people
Age distribution			
20 s	30 s		40 s
9 people	10 people		5 people

data. However, facial shape is determined by both skull structure and skin thickness [1–3]. Therefore, their method could not distinguish differences in the facial surface by the difference in the skull structure from the skin thickness. A principal component considered as a fattening parameter in their method included differences of skull structure. For this reason, as long as the skull structure is not considered, principal components that include only the fattening parameter cannot be extracted.

We proposed a facial fattening simulation method that improves the quality of fattening and slimming deformation by considering the skull structure. The proposed method maintains the individuality of the input face image.

3   Database Construction

In this section, we describe the process to construct a database that consists of frontal face images and MRI images as a pre-process.

We generated frontal face images of 24 participants and the corresponding MRI images for the database. The database properties are shown in Table 1. All participants are women. Because there are different fat features between female and male, as a first step, we construct a fattening rule for women who have a higher priority in the dieting market. In addition, the reason we use MRI images to obtain skull structure is that MRI scans have no risk of radiation exposure unlike CT scans and X-ray exams.

3.1   Photography Conditions

The photography conditions of the frontal face images and MRI images in the database are described. The frontal face images were taken using a Canon EOS Kiss X5. The photography conditions are as follows: focal length, 53–55 [mm]; aperture value, f/9; exposure time, 1/50 [s]; and ISO speed, ISO-800. Expressionless front images of participants' face were taken. These images were normalized (Sect. 3.2) to eliminate the differences of facial FPs coordinate (including the roll angle of face) without the effect of fattening; however, pitch and yaw angles were not normalized. Thus, we made

(a)Facial FPs   (b)MRI image   (c)Skull contour FPs

Fig. 1. An example of database

these images align as frontally as possible. In addition, MRI images were taken using a Siemens MAGNETOM Trio. The MRI conditions are as follows: magnet length, 198 [cm]; gantry inner diameter, 60 [cm]; and static magnetic flux density, 3.0 [T]. Expressionless MRI images of the participants' heads were taken.

3.2   Facial FPs Normalization

86 facial FPs were detected from the frontal face images via a face alignment technique OKAOVision [10] (based on [11]). The 86 facial FPs are shown in Fig. 1(a).

In addition, we normalized 24 frontal face images in the database. We normalized images relative to facial position, angle and scale. For facial position, we arranged the median point of the 86 facial FPs on the origin. Next, for facial angle, we rotated the images until the inner corners of both eyes were even. At last, facial scale is normalized by the distance between the inner corners of both eyes. Thus, the frontal face images were normalized. We then calculated the average facial FPs of the 24 participants.

3.3   Skull Contour FPs Normalization

23 skull contour FPs were extracted from the MRI images. First, we elucidated skull shapes in the MRI images by deciding transparency from the luminance value and its derivative value using a volume rendering technique [12] because skull structure in the MRI image is very obscure. We manually extracted 23 skull contour FPs elucidated from the MRI images. One MRI image and the skull contour FPs on an elucidated MRI image are shown in Fig. 1(b) and (c). In addition, to align the skull contour FPs with the facial FPs, seven FPs were extracted from the face contour in the elucidated MRI image. We aligned these seven FPs with the facial contour FPs manually. We then calculated the average aligned skull contour FPs of the 24 participants.

4   Facial Fattening and Slimming Simulation

In this section, we describe the real-time process to simulate facial fattening and slimming using a frontal face image. An overview of the proposed method is shown in Fig. 2. The proposed method requires only the input image and fattening parameter k (discussed below), which represents the degree of fattening. All processes are executed

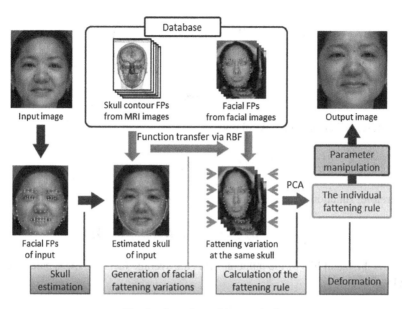

Fig. 2. Overview of our method

automatically. Facial fattening and slimming simulation process includes skull estimation, generation of facial fattening variations, calculation of the facial fattening rule and deformation.

4.1 Skull Estimation

The skull structure is estimated from the input frontal face image. 86 facial FPs are detected from the frontal face image using OKAOVision. Detected FPs are normalized as well as the database's normalization process. The differences between the normalized FPs and the average facial FPs in the database are calculated. Next, the average skull contour FPs in the database are warped to match the input facial FPs via a radial basis function (RBF) corresponding to the difference [13]. These warped FPs are the estimated skull contour FPs of the input image. The movement of the reference FPs is transferred to FPs $\mathbf{y} = (y_1, \cdots, y_N)$ via RBF $\phi(\mathbf{x})$. Function f is expressed using weight $\mathbf{w}$ as follows:

$$f(\mathbf{x}) = \sum_i^N \omega_i \varphi(\mathbf{x}, \mathbf{x}_i), \tag{1}$$

where $\mathbf{w} = (\omega_1, \cdots, \omega_N)$, $\phi(\mathbf{x}) = \sqrt{\mathbf{x}^2 + \beta^2}$ (multi quadric), and N is the number of FPs. $\mathbf{x}$ is two-dimensional (i.e., x and y coordinates of each FPs). Here, $\mathbf{w}$ is calculated for the j th data (x_j, y_j) as follows:

$$y_j = \sum_i^N \omega_i \varphi(\mathbf{x}_j, \mathbf{x}_i), \tag{2}$$

For $j = 1, \cdots, N$:

$$\mathbf{\Phi} = \begin{pmatrix} \varphi(x_1, x_1) & \cdots & \varphi(x_1, x_N) \\ \vdots & \ddots & \vdots \\ \varphi(x_N, x_1) & \cdots & \varphi(x_N, x_N) \end{pmatrix}, \tag{3}$$

$$\mathbf{y} = \mathbf{\Phi}\mathbf{w}, \tag{4}$$

Therefore, $\mathbf{w}$ is calculated as follows:

$$\mathbf{w} = \mathbf{\Phi}^{-1}\mathbf{y} \tag{5}$$

Therefore, function f can be expressed as follows:

$$f(x) = f(\mathbf{A} - \mathbf{B}) = \sum_i^N \omega_i \varphi(\mathbf{A} - \mathbf{B}, (\mathbf{A} - \mathbf{B})_i), \tag{6}$$

where $\mathbf{A}$ is the facial FPs of the input and $\mathbf{B}$ is the average facial FPs.$\mathbf{w}$ is calculated by Eqs. (1)–(5).

$$\mathbf{D} = f(\mathbf{C}). \tag{7}$$

Here, $\mathbf{C}$ is the average skull contour FPs and $\mathbf{D}$ is estimated skull contour FPs of the input. Thus, $\mathbf{D}$ is calculated. We refer to these processes as *RBF interpolation*. The method to estimate skull structure uses examples from the literature [14, 15]. An example of the estimated skull contour FPs is shown in Fig. 3.

4.2   Generation of Facial Fattening Variations

Facial fattening variations for the input image are generated from the estimated skull contour FPs and the facial FPs in the database.

The proposed method uses the concept of facial restoration [1–3]. If images in the database have the same skull structure as input image and the degree of fattening or slimming is different, the difference of the facial contour shape depends only on the difference of the degree of the fat. Thus, a precise fattening rule can be formulated. However, it is nearly impossible to obtain such a database. To solve this problem, we artificially generated faces in the database with the same skull structure and different degrees of fat that match the skull structures of the input skull structure. This is owing to the fact that the degree of fat of each participant differs. Thus, facial fattening variations are generated.

In addition to the Eqs. (1)-(7), the difference between the estimated skull con tour FPs of the input ($\mathbf{A}$) and the skull contour FPs of a participant in the database ($\mathbf{B}$) is calculated first. Next, the facial FPs of this participant ($\mathbf{C}$) are warped to match $\mathbf{A}$ via RBF interpolation corresponding to this difference. This process is applied to all images in the database. Thus, the facial fattening variations ($\mathbf{D}$) of the input can be generated while maintaining the original thickness of the facial fat. An example of facial fattening variations is shown in Fig. 4.

Input: source skull FPs(pink), source
facial FPs(green), target skull FPs(red)
Output: target facial FPs
(facial fattening variation)

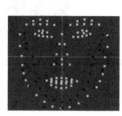

Fig. 3. Estimated skull(light blue) and facial
FPs(red) (Color figure online)

Fig. 4. Facial FPs deformed to estimated
skull (Color figure online)

4.3 Calculation of the Facial Fattening Rule

To extract the differences in the degree of the fat from the generated facial fattening
variations (Sect. 4.2), PCA was applied to facial fattening variations for the 24 participants. Facial fattening variations were warped to match the estimated skull of the
input, the most sufficient factor of the difference for each facial contour FPs coordinates
should be the difference in the degree of fat by fattening or slimming. Therefore, the
upper principal components should be facial fattening rule.

When each principal component is added to the facial FPs after applying PCA, it
was observed that S_1 uniformly spreads facial contour FPs and S_2 deforms facial
contour FPs from an inverted triangle to a square (where S_1 is i th principal component). In particular, S_2 lowers the FPs on the cheeks. We interpreted S_1 as the change of
facial size and S_2 as the change of facial sag by fattening. The changes of facial FPs by
S_1 and S_2 are shown in Fig. 5(a) and (b). We interpreted all principal components
except S_1 and S_2 as the component that changes the angle of the face or noisy components. The facial fattening rule is defined as follows:

$$S = \alpha_1 S_1 + \alpha_2 S_2 \tag{8}$$

where α_1 is ith contributing rate. Here, the directions of S_1 and S_2 were redefined such
that each positive direction of α_1 and α_2 became the fattening. The change of facial FPs
by S is shown in Fig. 5(c). Table 2 shows the average contributing rate of each
principal component when PCA was applied to the 24 participants in the database. As
shown in Table 2, the sum of S_1 and S_2 is greater than 80 %. Note that the contributing
rates of all principal components except S_1 and S_2 are negligible.

(a)Change of S_1 (b)Change of S_2 (c)Change of S

Source facial FPs(white), Source + each component(pink), Source − each component (light blue)

Fig. 5. Change of each component (Color figure online)

Table 2. The average of contribution ratio [%]

S_1	S_2	S_3	S_4	S_5
50.11	30.02	7.61	5.00	3.55

4.4   Deformation and Synthesizing Background Image

The facial fattening rule was added to the facial contour FPs of the input face. To maintain the individuality of the input face, the facial FPs coordinates of the input image are used relative to eyes, nose, mouth and brow. Facial FPs after fattening or slimming deformation X_{dst} is expressed as follows:

$$X_{dst} = X_{src} + kS \tag{9}$$

where S is the facial fattening rule, k is the fattening parameter and X_{src} is the facial FPs of the input image. The degree of fattening of the output image can be changed arbitrarily by manipulating the value of k.

Here, if the value of k becomes too small, the facial contour becomes an unnatural shape such as penetrating into the skull. To prevent such deformation, the movement of each facial contour FP is constrained when the distances between the facial contour FPs and the skull contour FPs are less than a certain threshold. Thus, X_{dst} is determined by the value of k and the mentioned constraint rule. Next, the difference between X_{dst} (**A**) and 86 FPs of 661 vertices of an average face model corresponding to the facial FPs of OKAOVision (**B**) is calculated. Then, 661 vertices of an average face model (**C**) are warped to match **A** via RBF interpolation corresponding to this difference. Finally, the texture of the input facial image is mapped to the warped **C** (**D**). This mapped texture is the result of fattening. **C** and a texture mapped to **D** are shown in Fig. 6. As shown in Fig. 6, this mapped texture does not have a background. The background of the input image is warped via RBF interpolation and synthesized. First, the differences between the facial contour FPs of the result of fattening (**A**) and the facial contour FPs of the input (**B**) are calculated. Next, the background of the input (**C**) is warped to match **A** via RBF interpolation corresponding to this difference. This is the background of the result of fattening (**D**). Finally, **D** and the result of fattening are synthesized. Thus, the result of fattening simulation is output. The results of two inputs simulated by the proposed method are shown in Fig. 7. The images at the center are the

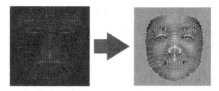

Fig. 6. 661 vertices deformed FPs for an average face model

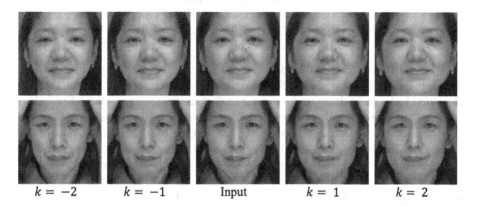

$k = -2$ $k = -1$ Input $k = 1$ $k = 2$

Fig. 7. Results of the proposed method

inputs. The images on the left show slimming and those on the right show fattening (each value for parameter k is shown). As can be seen, the proposed method realizes natural fattening and slimming deformation while maintaining the individuality of the input face. In addition, for slimming deformation, especially for the jaw, unnatural deformation penetrating into the skull is prevented using an estimated skull. Facial contour FPs are stopped by the constraint of the estimated skull structure between $k = -1$ and $k = -2$.

5   Discussions

5.1   Comparison with Other Rules and the Precision of Skull Estimation

To evaluate our results, we take two images that have different degrees of fattening to an extent of the same person. The error in facial contour FPs coordinates between the ground truth image and the simulated image is calculated. The original fattening rule (the proposed method) is compared with other fattening rules. The evaluation process is as follows:

1. Fattening and slimming images are taken of the same person.
2. Fattening simulation is performed using the individual fattening rule with slimming (fattening) images as the input such that the results are the same as the fattening (slimming) images relative to the width of the faces.

Table 3. Comparison of individual error and other rules

	Individual RMSE	Other RMSE (Ave. of 10)	Max.	Min.	The rank of the individual
Slimming	1.181	1.235	1.351	1.114	3/11
Fattening	1.042	1.100	1.180	1.039	2/11

3. Fattening simulation is performed using other fattening rules.
4. 86 facial FPs are detected from each result via OKAOVision. Facial FPs are then normalized. The root mean square error (RMSE) is calculated with the ground truth image's facial contour FPs coordinates.

The results were compared to determine which rule is closer to the ground truth. 10 people were selected at random for the other rule. The results are shown in Table 3. The individual rule is closer to the ground truth than the average of the others rules. However, the closest result of the other rule is closer to the ground truth than the individual rule. For slimming, the best result was 0.067 closer than the individual rule. These facial FPs were normalized by the distance between the inner corners of both eyes. The average distance between the inner corners of both eyes of all humans is approximately 3 [cm]. Therefore, we consider that this error negligible. In addition, when the other rule was closer, the other rule's facial FPs coordinates were very close to the input. In the proposed method, the closer both faces are to each other, the closer both rules will be to each other. Consequently, even though the other rule is closer to the ground truth than the individual rule, it is likely that applying the individual rule is more sensible and effective than applying the common rule to all the inputs.

In addition, the precision of skull estimation was calculated. The RMSE between the ground truth and the estimation result was calculated after normalizing the skull contour FPs. The average of 24 people was 0.175147. This error is also considered negligible due to the scale.

5.2   Subjective Experiment with Facegen

The results of the proposed method were compared with Facegen [16]. Each result is shown in Fig. 8. For, fairness, the background in image used for the proposed method was deleted. The subjects compared the input image with the proposed meth od or Facegen at the same time. A subjective experiment was conducted with 40 subjects using 25 inputs × 2 (fattening and slimming) × 2(the proposed method and Facegen) = 100 slides. Note that no restriction was placed on observation time. The average observation time for one slide was approximately 15 [s]. The questions average were as follows.

Q.1   *(compared to the input) Can you identify if fattening or slimming has been applied?* {1:Slimming, 2:No change, 3:Fattening}

Q.2   *(compared to the input) Was individuality maintained?* {1:Different, 2:Looking like the other, 3:Neither, 4:Looking like the original, 5:Same as original}

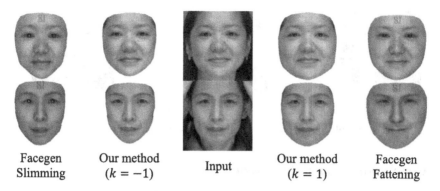

| Facegen Slimming | Our method ($k = -1$) | Input | Our method ($k = 1$) | Facegen Fattening |

Fig. 8. Comparison of the proposed method and Facegen [16]

Table 4. Results of subjective evaluation

	Q.1		Q.2				
	Accuracy rate(%)	Ave. (out of 5)	Distribution of each point (%)				
			1	2	3	4	5
Proposed	90.1	4.26	0.25	5.00	12.1	33.5	49.2
Facegen	81.6	2.08	35.1	40.0	10.1	12.1	2.75

For Q.1, the accuracy rate is of detecting fattening or slimming was calculated. For Q.2, individuality was evaluated in 5 stages. The results are shown in Table 4. The proposed method obtained higher scores than Facegen for both Q.1 and Q.2.

5.3 Execution Time

We created a system that can output the results of the proposed method from a frontal face image to measure execution time. The measurement conditions and system specifications are as follows: input image size: 600×600 [pixel], CPU: Core i7-3770 (clock frequency 3.40 [GHz]), 8 [GB] main memory, GPU: AMD Radeon HD 7770. Parallel processing was not used. The average time taken 5 measurements was 4.175 [s]. We consider this time to be sufficiently short for practical application.

5.4 Limitation

Since our results are represented by changes in the facial contour shape, we cannot generate drastic fattening or slimming changes. Over-fattening or over-slimming results are shown in Fig. 9. As can be seen, for over-fattening deformation, shading due to fattening does not appear. In addition, the deformation of contour is stopped at the estimated skull by the constraint. Therefore, unnatural edges occasionally appear on contour for over-slimming deformation.

$k = -4$ Input $k = 4$

Fig. 9. Results of the proposed method (over-fattening or over-slimming)

6   Conclusions and Future Work

We have proposed a facial fattening and slimming method in 2D images. The proposed method realizes fattening deformation that maintains the individuality of the input person's face by calculating an individual facial fattening rule based on an estimated skull structure. In addition, we avoid unnatural deformation (e.g., penetrating into the skull) by considering the skull structure. The proposed method is versatile and can be applied to many facial fattening simulations, because the inputs of the proposed method are only a frontal face image and the fattening parameter k. In addition, the proposed method is fully automatic.

In this paper, we did not consider drastic fattening or slimming changes that cause facial depth changes and facial parts to change because of the assumed application. We considered only fattening from the center of the face to the facial contour, i.e., the change of the facial contour shape when observed from the front. In future, we plan to realize fattening and slimming deformation of facial parts and facial depth while maintaining the individuality of the input. In addition, we will attempt to add shading to the jaw, cheeks and so on. Moreover, we plan to apply the proposed method to 3D models to realize 3D facial fattening and slimming simulation. Methods for skull estimation [14, 15], will help realize 3D facial fattening and slimming simulation.

References

1. Pei, Y., Zha, H., Yuan, Z.: The craniofacial reconstruction from the local structural diversity of skulls. Comput. Graph. Forum **27**(7), 1711–1718 (2008)
2. Jiang, L., Ma, X., Lin, Y., Yu, L., Ye, Q.: Craniofacial Reconstruction Based on MLS Deformation. WSEAS Trans. Comput. **9**(7), 758–767 (2010)
3. Claes, P., Vandermeulen, D., Greef, S.D., Willems, G., Suetens, P.: Craniofacial reconstruction using a combined statistical model of face shape and soft tissue depths. Forensic Sci. Int. Suppl. **159**, S147–S158 (2006)
4. Allen, B., Curless, B., Popović, Z.: The space of human body shapes: reconstruction and parameterization from range scans. ACM Trans. Graph. Proc. ACM SIGGRAPH 2003 **22**(3), 587–594 (2003)

5. Allen, B., Curless, B., Popović, Z., Hertsmann, A.: Learning a correlated model of identity and pose-dependent body shape variation for real-time synthesis. In: Proceedings of the 2006 ACM SIGGRAPH/Eurographics Symposium on Computer Animation, pp. 147–156 (2006)
6. Baek, S.Y., Lee, K.: Parametric human body shape modeling framework for human-centered product design. Comput. Aided Des. **44**(1), 56–57 (2012)
7. Zhou, S., Fu, H., Liu, L. Cohen-Or, D., Han, X.: Parametric reshaping of human bodies in images. ACM Trans. Graphics Proc. ACM SIGGRAPH 2010, **29**(4) Article no. 126 (2010)
8. Jain, A., Thormählen, T., Seidel, H. P., Theobalt, C.: MovieReshape: tracking and reshaping of humans in videos. In: Proceedings of ACM SIGGRAPH Asia 2010, 29(6), Article no. 148 (2010)
9. Blanz, V., Vetter, T.: A morphable model for the synthesis of 3D faces. In: Proceedings of the 26th Annual Conference on Computer Graphics and Interactive Techniques, pp. 187–194 (1999)
10. OKAOVision. http://plus-sensing.omron.co.jp/technology/
11. Irie, A., Takagiwa, M., Moriyama, K., Yamashita, T.: Improvements to facial contour detection by hierarchical fitting and regression. In: First Asian Conference on Pattern Recognition, pp. 273–277 (2011)
12. Kniss, J., Kindlmann, G., Hansen, C.: Multidimensional transfer functions for interactive volume rendering. IEEE Trans. Visual. Comp. Graphics **8**(3), 270–285 (2002)
13. Noh, J.Y., Fidaleo, D., Neumann, U.: Animated deformations with radial basis functions. In: Proceedings of the ACM Symposium on Virtual Reality Software and Technolog, pp. 166–174 (2000)
14. Beeler, T., Brandley, D.: Rigid stabilization of facial expressions. ACM Trans. Graphics Proc. ACM SIGGRAPH 2014, **33**(4), Article no. 44 (2014)
15. Ali-Hamadi, D., Liu, T., Gilles, B., Kavan, L., Faure, F.: Anatomy transfer. ACM Trans. Graphics, **32**(6), Article no. 188 (2013)
16. Facegen. http://www.facegen.com/

Many-Lights Real Time Global Illumination Using Sparse Voxel Octree

Che Sun and Emmanuel Agu[⊠]

Computer Science Department, Worcester Polytechnic Institute, Worcester, USA
emmanuel@cs.wpi.edu

Abstract. The many-lights real time Global Illumination (GI) algorithm is promising but requires many shadow maps to be generated for Virtual Point Light (VPL) visibility tests, which reduces its efficiency. Prior solutions restrict either the number or accuracy of shadow map updates, which may lower the accuracy of indirect illumination or prevent the rendering of fully dynamic scenes. In this paper, we propose a hybrid real-time GI algorithm that utilizes an efficient Sparse Voxel Octree (SVO) ray marching algorithm for visibility tests instead of the shadow map generation step of the many-lights algorithm. Our technique achieves high rendering fidelity at about 50 FPS, is highly scalable and can support thousands of VPLs generated on the fly.

1 Introduction

Global illumination (GI) simulates the propagation of light through a 3D volume and its interaction with surfaces, dramatically increasing the fidelity of computer generated images. While offline GI algorithms such as ray tracing and radiosity can generate physically accurate images, their rendering speeds are too slow for real-time applications.

A new wave of GI algorithms targeting real-time applications such as video games have recently emerged. One class of these approaches is based on the many-lights method that is derived from Kellers instant radiosity technique [1]. The many-light method transforms solving the lighting transport equation into the calculation of direct illumination from many virtual light sources. This algorithm is hardware-friendly and can easily be implemented on modern GPUs. One shortcoming of the many-lights algorithm is that it requires many shadow maps to be generated for Virtual Point Light (VPL) visibility tests, which reduces the its efficiency. While imperfect shadow maps [2] can alleviate this issue, they are noticeably inaccurate for indirect shadows.

Another class of real-time GI methods discretize the original scene into voxels, which has several advantages: First, voxels are geometry-independent and many efficient scene voxelization methods have previously been proposed. Secondly, ray-geometry intersection and visibility tests on voxel data structures are very fast. Thirdly, high quality anti-aliasing techniques can also be implemented using voxel data [3].

© Springer International Publishing Switzerland 2015
G. Bebis et al. (Eds.): ISVC 2015, Part II, LNCS 9475, pp. 150–159, 2015.
DOI: 10.1007/978-3-319-27863-6_14

Fig. 1. Sample results using our technique: Dynamic Cornell box scene with dragon and running elephant rendered using two spot lights at 50 fps on an NVIDIA GeForce Titan X GPU. Indirect illumination is generated by 512 one-bounce virtual point lights

Inspired by discussions in Ritschel et al. [4], in this paper we introduce a hybrid real-time GI algorithm that combines the advantages of the many-lights method with those of voxelization. We use an efficient Sparse Voxel Octree (SVO) ray marching algorithm for visibility tests in place of the expensive shadow maps generation step of the many-light method. Our technique achieves high fidelity rendering quality at real time speeds (about 50FPS, shown in Fig. 1). Moreover, our technique is highly scalable and can support thousands of VPLs generated on the fly. Our main contributions are:

- An alternative real-time many-lights GI method that can support thousands of VPLs generated dynamically.
- Sparse Voxel Octree (SVO)-based VPL visibility tests at interactive rates without using computationally expensive shadow maps.

2   Related Work

Instant Radiosity Methods: Instant radiosity [1] uses many VPLs to approximate indirect illumination and several is the basis of many proposed real-time GI methods. To achieve high performance, some earlier methods only gather near-field VPLs, ignoring visibility tests [5]. Others such as Laine et al. [6] restrict movement in parts of the scenes in order to reuse VPL shadow maps over multiple frames. Ritschel et al. [2] efficiently generate inaccurate VPL shadow maps by using a point-based scene representation. Their method is very fast for one-bounce indirect illumination, but consumes lots of GPU memory bandwidth for second and third bounce VPLs. Knecht [7] exploits a temporal coherence technique to alleviate a flickering issue that occurs in real-time instant radiosity when there is an insufficient number of VPLs.

The distribution of VPLs within the 3D scene also affects the rendering quality of instant radiosity methods. Poor one-bounce VPL sampling leads to artifacts and unshaded scene surfaces [8]. To construct robust light transport paths, Tokuyoshi et al. [9] changed the VPL sampling strategy to bidirectional path tracing. Their algorithm uses global ray-bundles [10], which are groups of parallel rays sampling the scene geometry. Their technique is implemented via a

GPU-based per fragment concurrent link list [11]. However, global ray-bundles introduce additional burdens, since every time a VPL is used, scene polygons must be rasterized again in order to create a corresponding global ray-bundle. Consequently, while their technique produces photorealistic images for complex scenes, it does not run at high frame rates.

Voxel-based Methods: Voxel-based methods discretize the scene into 3D grid cells (voxels). An outstanding feature of scene voxelization is that it generates an approximation of scene geometric information extremely fast (usually less than several milliseconds on current commodity GPUs for a scene with over a hundred thousand triangles). Using this discretized scene representation, an iterative diffusion process can be performed to transfer light energy between neighboring grid cells [12]. Thiedemann et al. [13] introduces a regular grid-based voxelization technique that supports fast near-field VPL visibility tests. However, voxelization using regular grids consumes a lot of GPU memory. To solve this problem and its artifacts related to instant radiosity methods, Crassin [3] develops a fast dynamic GPU SVO generation method and uses it for high quality GI rendering: they utilize an injection and pre-filtering process to filter data related to indirect illumination in a bottom-up fashion. The resulting hierarchical octree structure is then used by their voxel cone tracing technique to produce high quality indirect illumination that supports both diffuse and high glossy indirect illumination.

3   Our Algorithm

Our lighting system is a deferred renderer, which supports illumination by multiple scene lights. First, we voxelize the scene and create an SVO to represent geometric occlusion. Similar to other deferred rendering systems, we then create a geometry buffer (G-buffer) [14] which stores position, normal and materials information for direct and indirect illumination. The G-buffer is then split for VPL interleaved sampling [15]. For each scene light we render the scene from the lights view and create a reflective shadow map [5] into which we store position, normal and flux information that is needed by VPL importance sampling [16].

Direct illumination is calculated using standard shadow maps and light contributions of all scenes lights are accumulated into a direct illumination buffer. Indirect illumination is calculated by accumulating contributions of VPLs into an indirect illumination buffer. To accelerate indirect illumination, we adopt a technique similar to Segovia [17]: each pixel being shaded is only assigned a subset of the VPLs using a pre-generated interleaved sampling pattern. A visibility test is performed by shooting shadow rays from the pixel toward the VPL subset. Here, instead of performing shadow map lookup and comparison, we perform SVO ray-marching to query if the shadow ray is occluded. To utilize cache coherence of modern GPUs, we group pixels that use the same VPL subset together by splitting the G-buffer into 4 × 4 tiles. On completion, the indirect illumination buffer is merged, filtered and combined with the direct illumination buffer to create the final image. Figure 2 shows our steps.

Voxel Data Representation: We use SVO as a spatial data structure for fast VPL visibility tests. Inspired by Crassin [3], we first build the SVO non-recursively and store it in GPU memory. Later the SVO structure is read by multiple GPU threads concurrently during VPL visibility tests. To exploit GPU thread-group caches, groups of eight tree nodes are placed together.

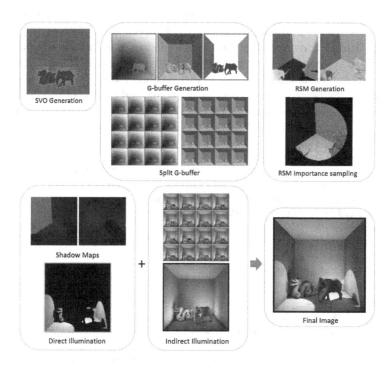

Fig. 2. Lighting pipeline steps. Top left: Scene voxelization and SVO generation. Top middle: G-buffer generation. World position, normal and diffuse materials rendered into floating-point textures. Position and normal buffers are split for calculating indirect illumination. Top right: RSM generated from scene lights view. First-bounce VPLs are created by sampling the RSM. Bottom left: Direct illumination using scene lights and corresponding shadow maps. Bottom middle: Indirect illumination calculated with split G-buffer and SVO ray-marching. Merging and filtering are performed to produce the final indirect illumination result. Bottom right: Adding direct and indirect illumination together. A simple HDR tone mapping is applied to produce the final image.

3.1   VPL Visibility Tests Using Sparse Voxel Octree

We use SVO to perform shadow ray marching between each surface point being shaded and the VPLs. To exploit data coherence, surface positions accessing the same VPL subset are grouped together. We also assume that the surface material only has a diffuse property for indirect illumination. This assumption leads to incoherent tree node access patterns due to the nature of diffuse reflection. Thus, packet traversal [18] cannot be applied easily as grouping individual shadow rays which have a uniform distribution over a hemisphere does not benefit much from fetching the same nodes during ray marching. However, dividing shadow rays into packets based on VPLs that reside in the same solid angle is a viable option to improve node access coherence, which we will explore in future work. Currently we only implement a per-ray traversal method. In Sect. 4, we will show that the algorithm has a good performance for diffuse indirect illumination.

The kd-restart traversal algorithm is an efficient method for ray marching thousands of rays concurrently on GPUs [19,20]. Consequently, we implement an SVO-restart traversal algorithm for VPL visibility tests. We also selected a kd-tree since it splits spatial regions in two as compared to the octree which splits the space into eight parts, making packet traversal more efficient on kd-trees. Our ray marching algorithm in pseudocode is shown in Fig. 3.

```
1  Transform world space ray start and end positions to SVO space
2  Offset SVO space ray start and end positions by epsilon
3  rayDirSVO = rayendpositionraystartposition
4  sceneMaxT = length(rayDirSVO)
5  normalize(rayDirSVO)
6  hit = 0
7  minT = maxT = 0.0
8  while( maxT < sceneMaxT )
9      curNode = root
10     mint = maxT
11     maxT = sceneMaxT
12     rayentryposition = raystartposition + raydirection * mint
13     while( !isLeafNode(curNode) )
14         Figure out which child node of curNode the ray entry position is in
15         Update curNode to be the child node
16     if( Flaged(curNode) )
17         hit = 1 and break
18     Fetch AABB for curNode
19     Compute ray-AABB intersection positions and possibly return hit = 0
20     Update maxT = t1 + SVO_RAY_T_EPSILON
21 return hit
```

Fig. 3. Pseudocode for our ray marching algorithm

The ray marching algorithm takes world space ray end positions and SVO root as input. It first transforms the ray end positions to SVO space. In line 2 the ray segment is shrunk slightly to avoid intersection issues because both the shading position and VPL position are on geometry surfaces that have been flagged as SVO leaf nodes. From line 3 to 7 we initialize some variables used by the subsequent while loop. Here, sceneMaxT is the distance between the shrink ray end positions. It is used as the ray marching termination condition when the ray goes through the leaf nodes. The hit variable indicates whether or not the ray hits a leaf node which has geometric information in it. Since we use SVO ray marching primarily for VPL visibility tests, whenever the ray hits geometry, the algorithm terminates and returns immediately. If we were to implement glossy reflection in future, this SVO-restart method could be extended easily to fetch hit position geometric data. Lines 8 to 21 lists the main while loop. The idea is similar to that of kd-restart algorithm: marching the ray through the leaf nodes iteratively and checking if shadow ray occlusion occurs. The differences here are (1) how we determine through which subdivided region we should keep marching the ray and (2) how we clip the ray each time it intersects a subdivided region. In our implementation we use the octree nodes AABB, which is easy and efficient to be implemented on a GPU. The implementation on line 19 is based on the ray-AABB intersection detection algorithm described in PBRT. In line 20 we update maxT using the result t1 from line 19, which is the maximum t value of the intersection position with the current nodes AABB. To make the algorithm numerically robust, we offset maxT using a user specified epsilon value. Otherwise the ray may stop marching forward in the next iteration due to inaccuracies caused by a floating point representation. Finally in line 21, the result is returned indicating whether the ray passed the VPL visibility test.

3.2   Rendering

Since we replace the time-consuming VPL shadow map generation with a scene voxelization pass, the GPUs main burden changes from performing hundreds or thousands of scene drawing and rasterization operations to a one-time scene voxelization and subsequent shadow ray tests for indirect illumination. One important advantage of our method is that implementing the quasi-random walk [1] for n-bounce VPL distribution is much more straightforward than shadow maps based real-time instant radiosity method. The reason is simple: since we already have a voxelized scene representation, against which shooting rays is extremely efficient. We just needed to extend our VPL sampling pass by adding second-bounce VPL sampling from first-bounce VPLs sampled with reflective shadow maps. In complex scenes such as the Crytek Sponza, second and third-bounce VPLs are useful for illuminating the scenes complex geometry.

In our system, standard deferred shading with scene light shadow maps is applied to produce a direct illumination buffer. For indirect illumination, we accumulate all the VPLs that have influence on a shading fragment and store the results in an indirect illumination buffer. Since we have split the G-buffer into 4×4 tiles with an interleaved sampling pattern, all the fragments inside the same tile use exactly the same subset of VPLs. In this way, data access to VPLs and SVO nodes exhibits good locality, which in turn improves the performance of shadow ray marching significantly. Similar to Dachsbacher et al. [5], given a VPLs flux Φ_{VPL}, world space position p_{VPL}, world space normal n_{VPL}, shading surface points position p and normal n, the indirect illumination at a surface position due to a VPL is formulated as follows:

$$E_{VPL}(p, n, p_{VPL}, n_{VPL}) = \Phi_{VPL}G(1 - V) \tag{1}$$

$$G = \frac{max(0, cos\theta_i)max(0, cos\theta_o)}{max(B, |p - p_{VPL}|^2)} \tag{2}$$

$$cos\theta_i = dot(n, -\frac{p - p_{VPL}}{|p - p_{VPL}|}) \tag{3}$$

$$cos\theta_o = dot(n_{VPL}, \frac{p - p_{VPL}}{|p - p_{VPL}|}) \tag{4}$$

We fetch the VPL related information from our VPL buffer, which has been generated by the VPL sampling stage of the system. V is the visibility term between the VPL and shading surface point. To evaluate V, we have to perform a shadow ray test using our SVO structure. Note that for shadow maps based instant radiosity method, evaluating V is just a simple shadow map lookup and comparison. In our case, since the evaluation of V is time-consuming, we could avoid a fruitless V term evaluation by first checking the luminance of the VPL and the G term. If either the luminance of the VPL is too small or the G term equals zero, we then skip accumulating this VPL to the indirect illumination buffer. B is the bouncing singularity constant factor used to clamp the distance between the VPL and the surface point. Since our technique currently only handles diffuse lighting, once the indirect illumination is complete, the buffer is merged and filtered for final combination with the direct illumination buffer.

4   Results

Our real-time global illumination method was rendered on an Intel i7 3930k CPU with an NVIDIA GeForce Titan X GPU. We tested using two scenes: A Cornell box with complex models (dragon and running elephant) and Crytek Sponza. All scenes are fully dynamic and no preprocessing methods are required.

The time spent on critical stages of our system are summarized in Table 1. The Cornell box scene is rendered at 768×768 (128^3 SVO) and Crytek Sponza scene is rendered at 1280×720 (256^3 SVO). The G-buffer, direct illumination buffer and indirect illumination buffer all have the same resolution as the back-buffer. For interleaved sampling of VPLs, the G-buffer is split into 4×4 tiles. Therefore, the VPL set is divided into 16 subsets, each of which is assigned to a G-buffer tile. The number of VPLs is varied and their influence on rendering speed and image quality is observed. Figure 4 shows the Cornell box scene rendered with 512 and 2048 VPLs respectively. Note that the visual difference between two settings is very small due to the accurate VPL visibility tests, which makes the integration of incoming radiance from VPLs converge very quickly. We also tested the Cornell box scene using a SVO resolution of 256^3, but the visual improvement is negligible. For Crytek Sponza scene, one-bounce indirect illumination is not enough to illuminate all the regions of the scene. Artifacts may appear in regions where insufficient number of VPLs are accumulated. Figure 5 shows the rendering result of the sponza scene using 1024 and 2048 first-bounce VPLs.

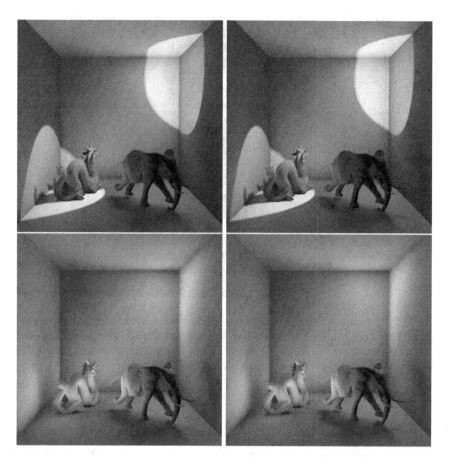

Fig. 4. Fully dynamic Cornell box scene with dragon and running elephant (150k triangles) rendered using two spot lights. SVO grid dimension is 128^3. Top left: Final image (512 first-bounce VPLs, 50 fps). Top right: Final image (2048 first-bounce VPLs, 15 fps). Bottom left: Indirect illumination (512 first-bounce VPLs, 50 fps). Bottom right: Indirect illumination (2048 first-bounce VPLs, 15 fps)

Table 1. Detailed timings for the Cornell box (128^3 SVO, 768 × 768 resolution, no MSAA) and Crytek Sponza scene (256^3 SVO, 1280 × 720 resolution, no MSAA) measured in milliseconds for critical stages of our lighting system. Three VPL number settings are used: 512, 1024 and 2048.

	Cornell box with dragon and running elephant	Crytek Sponza
Scene Voxelization and SVO Generation	0.57 / 0.57 / 0.58	4.65 / 4.65 / 4.65
RSM rendering	0.46 / 0.46 / 0.46	0.61 / 0.62 / 0.62
VPL Generation	0.02 / 0.02 / 0.02	0.02 / 0.02 / 0.02
Direct Illumination	0.17 / 0.17 / 0.18	0.31 / 0.31 / 0.31
Indirect Illumination	16.0 / 33.0 / 69.0	19.5 / 36.2 / 73.0

Fig. 5. Fully dynamic Crytek Sponza scene (180k triangles) rendered using one spot light and 1024 first-bounce VPLs. SVO grid dimension is 256^3. Top left: Final image from camera position 1 (19 FPS). Top right: Final image from camera position 2 (20 FPS). Bottom: Final image from camera position 3 (23 FPS).

5   Conclusion and Future Work

We have presented an alternate method to perform VPL visibility tests for instant radiosity-based real-time global illumination. While our method has a fixed voxelization cost and is not as efficient as fast shadow maps for computing one-bounce indirect illumination. However, for second- and third-bounce VPL sampling our technique is promising especially for complex scenes while maintaining real-time frame rates. Our method is automatic, supports fully dynamic scenes, and requires no scene preprocessing.

Similar to other real-time instant radiosity techniques, our method inherits limitations from instant radiosity. For instance, highly glossy surfaces cannot be reconstructed without having adequate VPLs evenly distributed in the entire scene. Meanwhile, applying interleaved sampling to scene surfaces not directly illuminated exhibits noticeable artifacts if a shading surface point fetches VPLs, most of which are blocked due to poor VPL distribution. To solve these issues, as mentioned before, our next step is implementing a robust VPL distribution and gathering method. We believe this can be done by taking advantage of the SVO data structure, using it as a scene geometric data representation for n-bounce VPL generation and rendering-time VPL acquisition. To support large scale scenes, a cascaded SVO grid management scheme could be implemented by following the idea of cascaded Light Propagation Volumes (LPV) [12] as well.

References

1. Keller, A.: Instant radiosity. In: Proceedings of the ACM SIGGRAPH, pp. 49–56 (1997)
2. Ritschel, T., Grosch, T., Kim, M.H., Seidel, H.P., Dachsbacher, C., Kautz, J.: Imperfect shadow maps for efficient computation of indirect illumination. ACM Trans. Graph. (TOG) **27**, 129 (2008)
3. Crassin, C., Neyret, F., Sainz, M., Green, S., Eisemann, E.: Interactive indirect illumination using voxel cone tracing. CG Forum **30**, 1921–1930 (2011)
4. Ritschel, T., Dachsbacher, C., Grosch, T., Kautz, J.: The state of the art in interactive global illumination. CG Forum **31**, 160–188 (2012)
5. Dachsbacher, C., Stamminger, M.: Reflective shadow maps. In: Proceedings of the ACM Symposium on Interactive 3D graphics and games, pp. 203–231 (2005)
6. Laine, S., Saransaari, H., Kontkanen, J., Lehtinen, J., Aila, T.: Incremental instant radiosity for real-time indirect illumination. In: Proceedings of the Eurographics Conference on Rendering Techniques, pp. 277–286 (2007)
7. Knecht, M.: Real-time global illumination using temporal coherence (2009)
8. Segovia, B., Iehl, J.C., Péroche, B.: Metropolis instant radiosity. CG Forum **26**, 425–434 (2007)
9. Tokuyoshi, Y., Ogaki, S.: Real-time bidirectional path tracing via rasterization. In: Proceedings of the ACM Symposium on Interactive 3D Graphics and Games, pp. 183–190 (2012)
10. Sbert, M., i Sàndez, X.P.: The Use of global random directions to compute radiosity: global Montecarlo techniques (1996)
11. Yang, J.C., Hensley, J., Grün, H., Thibieroz, N.: Real-time concurrent linked list construction on the GPU. CG Forum **29**, 1297–1304 (2010)
12. Kaplanyan, A., Dachsbacher, C.: Cascaded light propagation volumes for real-time indirect illumination. In: Proceedings of the ACM SIGGRAPH Symposium on Interactive 3D Graphics and Games, pp. 99–107 (2010)
13. Thiedemann, S., Henrich, N., Grosch, T., Müller, S.: Voxel-based global illumination. In: ACM Symposium on Interactive 3D Graphics and Games, pp. 103–110 (2011)
14. Saito, T., Takahashi, T.: Comprehensible rendering of 3-D shapes. ACM SIGGRAPH Comput. Graph. **24**, 197–206 (1990)
15. Keller, A., Heidrich, W.: Interleaved sampling. Springer (2001)
16. Clarberg, P., Jarosz, W., Akenine-Möller, T., Jensen, H.W.: Wavelet importance sampling: efficiently evaluating products of complex functions. ACM Trans. Graph. (TOG) **24**, 1166–1175 (2005)
17. Segovia, B., Iehl, J.C., Mitanchey, R., Péroche, B.: Non-interleaved deferred shading of interleaved sample patterns. In: Proceedings of the ACM SIGGRAPH/Eurographics Symposium on Graphics hardware: Vienna, Austria, vol. 3, pp. 53–60 (2006)
18. Aila, T., Laine, S.: Understanding the efficiency of ray traversal on GPUs. In: Proceedings of the ACM Conf High Performance graphics, vol. 2009, pp. 145–149 (2009)
19. Foley, T., Sugerman, J.: KD-tree acceleration structures for a GPU raytracer. In: Proceedings of the ACM SIGGRAPH/EUROGRAPHICS Conference on Graphics hardware, pp. 15–22 (2005)
20. Horn, D.R., Sugerman, J., Houston, M., Hanrahan, P.: Interactive Kd tree GPU raytracing. In: Proceedings of the ACM Symposium on Interactive 3D graphics and games, pp. 167–174 (2007)

WebPhysics: A Parallel Rigid Body Simulation Framework for Web Applications

Robert (Bo) Li[⊠], Tasneem Brutch, Guodong Rong, Yi Shen, and Chang Shu

Samsung Research America, Mountain View, USA
robert.li@samsung.com

Abstract. Due to the ubiquity of web browser engines and the advent of modern web standards (like HTML5), software industry tends to use web application as an alternative to traditional native application. Web app development commonly uses script language (like JavaScript, CSS), the low performance language which significantly hinders real-time execution of physics simulation. We design a new framework to achieve real-time physics simulation engine. The key novelty lies at: we choose native implementation for computing intensive functions in physics simulation, and bind native implementation with JavaScript APIs, then we only expose JavaScript APIs through web browser engine to developers but still calling native implementation. Based on this model, we build WebPhysics: the first 2D simulation engine targeting on real-time web applications, which is seamlessly compatible to both de-facto standard simulation engine (Box2D) and browser engine (Webkit). We also explore and implement a parallel rigid body simulation (Box2DOCL) in the context of web app framework to obtain further performance improvement. Our experiments show significant performance improvement in simulation time.

1 Introduction

Web application development is a promising trend in the smart mobile era. A web application (web app) can run directly in a web browser. Normally a web app is created in a browser-supported programming language, including the combination of JavaScript, HTML and CSS. Recently, the whole software industry is trying to use web app development as a viable and pervasive alternative to native application development like Objective-C and Java. The major advantage is its cross-platform ability. Currently web browsers are ubiquitous on any operating system, and all mainstream browsers support modern web standards like HTML5. A web app developer can now develop software with any complex function under modern web standards and users can run the developed web app directly on the universal browsers instead of various operating systems to solve the cross-platform development issue. Particularly, the cross-platform ability can solve the challenge that there is no universal or dominant operating system/native SDK for mobile devices, fragmented by IOS/Android and their countless versions in history. The native app developers are forced to double the

© Springer International Publishing Switzerland 2015
G. Bebis et al. (Eds.): ISVC 2015, Part II, LNCS 9475, pp. 160–169, 2015.
DOI: 10.1007/978-3-319-27863-6_15

precious resource on developing the same software in two completely different development environments. By using web app, developers can now completely eliminate this unnecessary work by targeting on following web standard only, and rely on standard-support browsers on all operating systems to execute the app with the same result.

Unfortunately, computer graphics-related app is far behind this trend of web app. For example, although gaming app has already become the largest category in every mobile app store, game developers rarely consider web app development despite the blissful cross-platform advantage. Our investigation with game developers identifies the following major concerns. The first major concern is *"Performance"*. The low performance of JavaScript implementation significantly blocks real-time physics simulation. (2) The second one is *"Portability"*. Many existing algorithms, libraries and toolkits for physics simulation, as de-facto standard like Box2D, are implemented as the native source code like C++ thus web app cannot directly utilize them through JavaScript.

The effort of designing new physics simulation algorithm can not solve the intrinsic language and framework deficiencies in web app development. The **main idea** of our methodology is JavaScript binding in web browser. The key novel architecture of this model lies at: We first abstract computing intensive functions in physics simulation and implement them in native code. Then we bind these native functions with JavaScript APIs. Finally we only expose JavaScript APIs through web browser engine to developers for development. When the developed app is rendered by the web browser, the browser still calls native implementation of simulation functions. The native implementation and execution provide significant acceleration; Since we expose the Javascript API through browsers to frond-end game developers, they can consider all the bindings as transparent to them and still follow the web and JavaScript standard to focus on game design.

To our knowledge, we first leverage this model and create the rigid body physics simulation toolkit for web app development. Based on this model we design the whole framework pipeline and investigate each step of the pipeline to achieve performance optimization. We identify collision detection and constraint solver as the major computationally intensive function, and bind them with JavaScript APIs. Further, we design a native parallel engine to accelerate collision detection and constraint solver. Unlike native implementation, the parallel simulation algorithms are carefully designed to achieve optimized parallelization in the context of web app based framework. Our experiments show a visible difference in speedup compared to entire JavaScript implementation.

In addition, our contributions to the web app development also include: (1) We implement JavaScript API binding through WebKit-based browser engine, the standard web browser engine in both IOS and Android. (2) Our native implementation integrates the de-facto standard simulation engine Box2D. We also extend native functions in Box2D to OpenCL accelerated native functions (Box2DOCL).

2   Related Work

We recommend the reader the latest state-of-the-art report [1] for obtaining a thorough knowledge of the related research work. Parallelization is our focus since multi-core systems and massively parallel GPUs are common on every device. Some examples of parallel rigid body solvers include [2,3]. Unlike these pure-algorithm design research, in this paper we also focus on designing widely compatible hardware acceleration system to embed parallel computer power. Also, we have to design the parallel acceleration algorithm that is optimized in the context of a web app instead of a native environment.

We choose OpenCL to implement our parallelization for web application. The introduction of high-level languages for general purpose GPUs like OpenCL and CUDA architecture makes programming much easier for developers. Currently high level language is more widely utilized for algorithm implementation like OpenCL in [4] and CUDA in [5]. CUDA's power is limited by the fact that it can only work on NVIDIA GPU. By comparison, OpenCL is widely accepted for almost every GPU vendor on both PC and mobile devices.

In our paper we choose Box2D [6] because currently it works as the de-facto standard of 2D rigid body simulation. To facilitate JavaScript development, there are some JS implementations such as Box2DWeb [7]. Box2DWeb provides JavaScript implementation and interface to Box2D. Our ambitious goal in this paper, in contrast, is to provide JS implementation while still preserving near-native performance.

Meanwhile, a generic approach to achieve high performance web app is by browser acceleration like [8]. A common method, for example, is offloading the computation. We can use a server-side proxy browser to render an application and send compressed computing results back to mobile devices. This generic approach is insufficient to gaming development because it brings severe server-client latency. Another non-technical difficulty of this method is that: any redesign of the overall browser requires adoption widely by the community and browser users. In contrast, our approach of web application acceleration seeks to use bindings between browser to native implementation to achieve acceleration without redesigning/violating any browser standard.

3   Methodology Overview and Binding

The whole framework includes the following steps. Figure 1 (Left) shows the differences between our method and conventional method like Box2DWeb [7].

1. Binding Analysis. As the first step, game developers analyze and make the clear boundary between front-end JavaScript game design tasks and back-end rigid body simulation function implementation tasks. Then they abstract the rigid body simulation framework into high-level functions and consolidate the name of each function as a JavaScript-style interface.

2. Native Implementation. Each native function is implemented as an object in C++ based WebKit Browser's rendering engine WebCore (Fig. 1 Left, orange box in WebCore). Since each function is implemented natively by C++, we can easily port existing open source native toolkit Box2D [6] and achieve the portability-objective. We design parallel algorithms here by utilizing standard native parallel library like OpenCL to get better performance (Fig. 1 Left, red box).

3. JavaScript Binding. The interfaces are bound and exposed to external build-in JavaScript engine for WebKit (JavaScriptCore). Now JavaScript can have access to the implemented native functions as objects in WebCore. When game designers finish the HTML game design and run the web app, the JavaScript will directly call the native functions to achieve near-native performance.

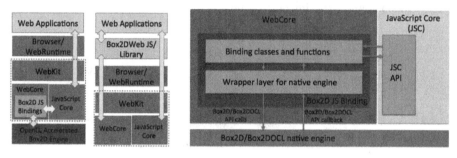

Fig. 1. Left: Comparison between WebPhysics & Box2DWeb approaches. Right: Web-Physics binding framework(Color figure online).

3.1   JavaScript Binding

Our WebPhysics implementation enables new JavaScript physics engine APIs, through an expanded web application runtime (WebKit). This enables efficient access from HTML5 application classes to the native accelerated physics engine APIs. Our goal is to provide efficient, maintainable and comprehensive implementation, with 100 % Box2D feature/API support. Figure 1 (Right) illustrates the implementation of WebPhysics JavaScript bindings: (1) In the Webkit Web-Core, we first insert the wrapping layer of Box2D engine to achieve Box2D API calls and callbacks. (2) Then we build an interface layer with all the binding classes and functions that we plan to expose outside of WebCore. (3) The interface establishes the communication between WebCore and JavaScript-Core through internal WebCore interfaces/function tables and internal JavaScriptCore interfaces.

4   Box2DOCL: Parallel Rigid Body Simulation

Besides building a practical open source parallel toolkit Box2DOCL, (An OpenCL implementation), we also investigate and test where to best introduce parallelization in the context of web applications, and then choose the best parallel algorithms which would provide most performance improvement to web applications. We first analyze some of the commonly used rigid body simulation and parallelization of physics methods (e.g., [1,3,9,10]). We benchmark different components of the pipeline to identify computationally intensive and parallelization friendly components. Our experiments show that on the average, 64 % of the total time is spent in the Collision Detection stage (21 % in Broad Phase, and 43 % in Narrow Phase), and 35 % of time is spent in Constraint Solver stage. Thus we abstract the whole pipeline into two major functions, and prioritize the parallelization effort on them. The detailed pipeline of Box2DOCL is shown in Fig. 2. Here we provide detailed features related to our Box2DOCL architectural design, tradeoffs and experimental probes.

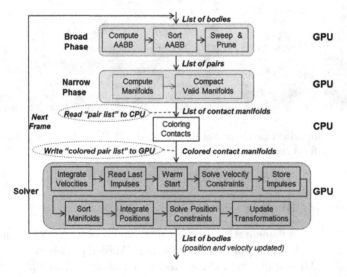

Fig. 2. Pipeline of Box2DOCL. The Broad Phase, Narrow Phase and Solver stages are parallelized using OpenCL, and executed on the GPU (or multi-core CPU).

4.1   Broad Phase

Our choice of parallelization for Broad Phase is highly related to the choice of rigid body bounding volume. Object bounding volumes are used to encapsulate one or more complex objects, which are tested for overlap before geometry intersection test is performed. Commonly used bounding volumes include Sphere, Axis-Aligned Bounding Box (AABB), Oriented Bounding Box (OBB), and Discrete Oriented Polytope (DOP). Between AABB, OBB and DOP, both the complexity of performing overlap test and the resulting accuracy increase from left to right across the list from AABB to DOP. Following a tradeoff analysis between complexity and accuracy, our practical experiment make us proceed with implementation of the Axis Aligned Bounding Box (AABB) algorithm.

The sequential Box2D Broad Phase implements the Bounding Volume Hierarchy (BVH) algorithm. However, a naive parallel implementation of BVH may result in low parallelism for the BVH data structure. Box2DOCL parallelization follows the method like Sweep-and-Prune (SaP) algorithm [11]. Specially in Box2DOCL, all AABBs are sorted along either x- or y- axis, using parallel sort (in Box2DOCL we use bitonic sort [12]). Compared with [11], to simplify the computation of the best suited sweep direction, we compute variance between minimum x and y values of all AABBs and select the axis with the maximum variance.

4.2 Narrow Phase

For each collision pair generated using AABB from Broad Phase, we check it in Narrow Phase to determine if two shapes intersect with each other ("compute manifolds"). We parallelize the Separating Axis Theorem (SAT) algorithm used by sequential Box2D [13].

The Separating Axis Theorem (SAT) states that two convex shapes are disjoint if and only if there exists a line (named separating axis), where the projections of the two shapes on to the line are disjoint. In Box2DOCL, we also have the practical problem about parallelizing projection tests for different types of collisions. There are two ways to handle different types in OpenCL: (1) We use separate OpenCL kernels for different types of collisions. These kernels are launched one by one. Each kernel handles only one type of collisions. (2) Another alternative approach is to use a single OpenCL kernel for all types of collision pairs (e.g., we use a switch statement to cover all types of collisions in different branches). The second method results in varying performance depending upon the underlying hardware architecture. This dependency lies at: some GPU architectures choose to execute all branches first and then discard results from irrelevant branches, resulting in unnecessary computations. Therefore, we use the first method to accelerate launching of one kernel for each type.

At the end of Narrow Phase, each collision query is either marked as valid (containing one or two contact points) or invalid (empty, with zero point). To enhance efficiency of parallelism in the next step Constraint Solver, we use parallel scan [14] to pack all valid contacts compactly. This packing operation enables selective execution of solver kernels only for valid pairs and thus greatly improves solver performance.

4.3 Constraint Solver

Constraint Solver receives all contact pairs obtained from the collision detection, and updates every body's velocity and position. When parallelizing Box2DOCL,

Table 1. Performance comparison of Box2DWeb (JS), Binding+Box2D, and Binding+Box2DOCL on Mac.

Brownian Motion			Simulation Time (ms)			Speedup over JS	
No. of Particles	Col	Row	JS	Binding + Box2D	Binding + OCL	Binding + Box2D	Binding + OCL
2500	50	50	82	10.8	18.7	7.59	4.39
3025	55	55	112	13.7	20.2	8.18	5.54
3600	60	60	142	18.6	21.4	7.63	6.64
4225	65	65	182	23.9	22.1	7.62	8.24
4900	70	70	220	29.7	22.9	7.41	9.61
5625	75	75	270	36.6	24.7	7.38	10.93
6400	80	80	350	45.4	26.9	7.71	13.01
7225	85	85	420	54	28.7	7.78	14.63
8100	90	90	510	63.8	30.2	7.99	16.89
9025	95	95	640	76.3	33.4	8.39	19.16
10000	100	100	710	88.8	36.2	8.00	19.61
11025	105	105	830	107.6	38.4	7.71	21.61
12100	110	110	1000	124.8	42.4	8.01	23.58

we have to address the challenge that not all contact constraints could be solved in parallel. For example, two contacts sharing one identical body cannot be solved in parallel. We use a coloring scheme that assigns colors to contacts. Contacts with the same color do not share the same body. These contacts with the same color are then computed in parallel. A greedy algorithm is used for coloring. For a given color associated with a contact, we mark two bodies involved as "locked", and skip all other contacts involving these two bodies. When no more contacts can be "colored" with the current color, we change to another color, we loop over all the uncolored contacts with the new color. This procedure is continued until all contacts are colored. Besides using coloring scheme to handle contact parallelization, we also explore other approaches and evaluate the relative performance. We have the following observations based on our exploration of different approaches:

1. We implement the space division algorithm in [4], but find it to be slower than the greedy algorithm, especially when applied to complex body movement scenes. In a complex scene a slight resolution change of space division can significantly change the time cost.
2. We test end-to-end time for both the sequential and OpenCL parallelized versions of the greedy algorithm on CPU and GPU respectively, and find that the sequential version on CPU is faster than OpenCL parallelized version tested on GPU.

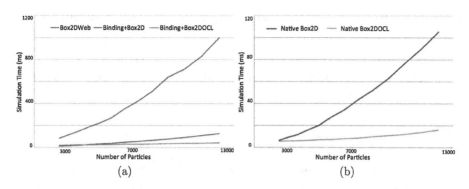

Fig. 3. (a) Performance comparison of Box2DWeb, Binding+Box2D, and Binding+Box2DOCL on Mac using the Brownian Motion demo. (b) Performance comparison of native Box2D and native Box2DOCL on Windows using the Brownian Motion demo.

3. We also explore a fully parallelized approach [5]. When each body has more than two contacts associated with it, we virtually split this body's mass by the total number of assigned contacts. We find that the cost of preparing mass splitting negatively impacts the end to end acceleration from parallelization.

5   Experimental Results

We develop a group of examples for performance evaluation. Here we use one benchmark demo application "Brownian Motion" for performance testing and comparison. The demo has a boundary box and numerous small dynamic bodies (particles) moving freely inside the boundary. Our experiments focus on performance on different platforms. First, we test our Box2DOCL on a MacBook Pro laptop with Intel Core i7 2.6 GHz CPU and 16 GB DDR3 RAM. We run our OpenCL parallelized Box2DOCL on multi-core CPU and compare the performance of binding plus sequential Box2D and a pure JS physics engine Box2DWeb. Table 1 lists the performance of pure JavaScript implementation Box2DWeb (JS), Binding+Box2D, and Binding+Box2DOCL on the Mac laptop. Compared with JavaScript (JS) implementation of Box2D (Box2DWeb), WebPhysics bindings with sequential Box2D native physics can achieve around 8× speedup, while WebPhysics bindings with OpenCL parallelized Box2D (Box2DOCL) can achieve > 20× speedup. Figure 3(a) uses curves corresponding to numbers in Table 1 to show the significant speedup. We also test the Brownian Motion demo on a Windows desktop with Intel Core i7 3.5 GHz CPU, 8 GB DDR3 RAM, and AMD Radeon HD 7770 GPU. The performance numbers are listed in Table 2 as well as in Fig. 3(b).

To better understand performance of Box2DOCL in a complex scene, we use a demo application ("MixDemo") with a large number of dynamic bodies formed by all types of joints, including distance, prismatic, gear, pulley, rope and wheel joints. Table 3 provides the performance results on a Windows system with

Table 2. Performance comparison of native Box2D and native Box2DOCL on Windows.

Bownian Motion			Time (ms)		Speedup
No. Bodies	Col	Row	Box2D Native	Box2DOCL Native	Box2DOCL vs. Box2D
2500	50	50	6.34	5.52	1.15
3025	55	55	9.03	5.73	1.58
3600	60	60	11.51	5.93	1.94
4225	65	65	15.83	6.28	2.52
4900	70	70	20.16	6.9	2.92
5625	75	75	27.63	7.23	3.82
6400	80	80	34.15	7.97	4.29
7225	85	85	43.19	8.51	5.08
8100	90	90	51.72	9.76	5.3
9025	95	95	62.32	10.77	5.79
10000	100	100	75.81	11.89	6.37
11025	105	105	89.34	13.72	6.51
12100	110	110	105.17	16.01	6.57

Table 3. Performance comparison of native Box2D and native Box2DOCL on Windows using MixDemo with different types of bodies and joints.

Mix Demo		Simulation Time (ms)		Speedup
No. of Bodies	No. of Joints	Box2D Native	Box2DOCL Native	Box2DOCL vs. Box2D
4,022	865	13.02	9.69	1.34
6,329	1,326	23.02	11.88	1.94
8,642	1,790	32.38	13.92	2.33
10,949	2,251	47.35	16.31	2.9
13,262	2,715	61.87	18.39	3.36
15,569	3,176	73.76	20.33	3.63
17,882	3,640	90.8	21.79	4.17
20,189	4,101	100.74	23.42	4.3
22,502	4,565	115.49	25.07	4.61
24,809	5,026	128.35	27.01	4.75

sequential Box2D and Box2DOCL. We can obtain approximately 5× speedup using the MixDemo.

6   Conclusion

To enable high performance gaming use cases, especially across mobile platforms, we implement JavaScript bindings for the Box2D physics engine and enable seamless access to an OpenCL accelerated Box2DOCL physics engine. Our hybrid approach provides portable acceleration of over 20× speedup compared to a JavaScript implementation, obtained from WebPhysics JavaScript bindings implemented in a WebKit-based browser and OpenCL parallelized Box2DOCL Physics Engine.

References

1. Bender, J., Erleben, K., Trinkle, J.: Interactive simulation of rigid body dynamics in computer graphics. Comput. Graph. Forum **33**, 246–270 (2014)
2. Tasora, A., Negrut, D., Anitescu, M.: A GPU-based implementation of a cone convex complementarity approach for simulating rigid body dynamics with frictional contact. In: ASME 2008 International Mechanical Engineering Congress and Exposition, pp. 107–118 (2008)
3. Harada, T.: Parallelizing the physics pipeline: physics simulations on the GPU. In: Game Developers Conference (2009)
4. Harada, T.: A parallel constraint solver for a rigid body simulation. In: SIGGRAPH Asia 2011 Sketches, pp. 22:1–22:2. ACM (2011)
5. Tonge, R., Benevolenski, F., Voroshilov, A.: Mass splitting for jitter- free parallel rigid body simulation. ACM Trans. Graph. **31**, 105:1–105:8 (2012)
6. Box2D: A 2d physics engine for games. http://box2d.org
7. Box2DWeb: Box2d port to javascript. http://code.google.com/p/box2dweb/
8. Jones, C., Liu, R., Meyerovich, L., Asanovic, K., Bodik, R.: Parallelizing the web browser. In: USENIX HotPar, First NSENIX Workshop on Hot Topics on Parallelism, USENIX, pp. 124–125 (2009)
9. Tsuda, J.: Practical rigid body physics for games. In: ACM SIGGRAPH ASIA 2009 Courses. SIGGRAPH ASIA 2009, pp. 14:1–14:83 (2009)
10. Coumans, E.: Accelerating game physics. In: Game Developer Conference (2013)
11. Liu, F., Harada, T., Lee, Y., Kim, Y.J.: Real-time collision culling of a million bodies on graphics processing units. ACM Trans. Graph. **29**, 154:1–154:8 (2010)
12. Kipfer, P., Westermann, R.: Improved GPU sorting. In: GPU Gems 2. Addison-Wesley (2005)
13. Gottschalk, S.: Separating axis theorem. Technical report TR96-024, Department of Computer Science, UNC Chapel Hill (1996)
14. Harris, M., Sengupta, S., Owens, J.: Parallel prefix sum (scan) with CUDA. In: GPU Gems 3. Addison-Wesley (2007)

Segmentation

Segregation

A Markov Random Field and Active Contour Image Segmentation Model for Animal Spots Patterns

Alexander Gómez[1]([✉]), German Díez[1], Jhony Giraldo[1,2,3], Augusto Salazar[1,2], and Juan M. Daza[3]

[1] Grupo de Investigación SISTEMIC, Facultad de Ingeniería,
Universidad de Antioquia UdeA, Calle 70 No. 52–21, Medellín, Colombia
{alexander.gomezv,german.diezv,heriberto.giraldo}@udea.edu.co
[2] Grupo de Investigación AEyCC, Facultad de Ingenierías,
Instituto Tecnológico Metropolitano ITM, Carrera 21 No. 54–10, Medellín, Colombia
[3] Grupo Herpetológico de Antioquia, Instituto de Biología,
Universidad de Antioquia, Calle 67 No. 53–108, Bloque 7–121, A.A. 1226,
Medellín, Colombia

Abstract. Non-intrusive biometrics of animals using images allows to analyze phenotypic populations and individuals with patterns like stripes and spots without affecting the studied subjects. However, non-intrusive biometrics demand a well trained subject or the development of computer vision algorithms that ease the identification task. In this work, an analysis of classic segmentation approaches that require a supervised tuning of their parameters such as threshold, adaptive threshold, histogram equalization, and saturation correction is presented. In contrast, a general unsupervised algorithm using Markov Random Fields (MRF) for segmentation of spots patterns is proposed. Active contours are used to boost results using MRF output as seeds. As study subject the *Diploglossus millepunctatus* lizard is used. The proposed method achieved a maximum efficiency of 91.11 %.

1   Introduction

Animal biometrics has increased in recent years, identifying indivi0dual animals and recognizing them at different places and time is an important requirement in many biological tasks like calculating animal population density, survival, emigration, examination of a particular behavior and planning conservation measures [1]. Commonly applied strategies can be categorized in two classes: intrusive and non-intrusive. Intrusive approaches include marking animals, which involves capture and risk the animal to injury, modify its behavior and even changes survival possibilities [2]; also marking strategies are not suitable in large populations or for long time. Non-intrusive approaches include the identification of genetic markers in excrement [3] and photographic mark recapture (PMR) [4]. The PMR method is based on visual identification using phenotypic features like spots, stripes or morphology. Those features must be stable over time, unique,

© Springer International Publishing Switzerland 2015
G. Bebis et al. (Eds.): ISVC 2015, Part II, LNCS 9475, pp. 173–184, 2015.
DOI: 10.1007/978-3-319-27863-6_16

and photographed under different conditions. This method is a two photo comparison of one target and hundreds of possible subjects to test similarity between patterns. For this reason large animal populations impede the identification by a human observer, because subjectivity, skill or experience of the expert would affect the objectivity of the study [1]. Automatic biometric identification (possible in PMR) is a time-saving alternative that provides clearness to the identification process. Previous semi-automatic approximations include shapes for marine mammals [5], elephants, and some lizard species; stripes for zebras [3] and tigers; also spots for cheetahs, giraffes [4], marine turtles and polar bears. There are two possible scenarios for a computer vision perspective. First, photos taken in the wild as photo trap framework; this media is commonly cluttered, with low contrast, containing trees, shrubs, other subjects, and the target in multiple poses [6]. Second, the subject is photographed under controlled conditions and position. Additionally to the scenario, both cases present problems in natural appearance of skin, brightness, 3D shape, contamination produced by sand or environmental components, and scars. Due to the spot concept that is related to a contrast change between two or more regions, the purpose of this paper is to find a general algorithm for segmentation, whatever kind of spots set on animals that deal with the previous declared problems. A general algorithm for spots patterns give the opportunity to identify a great variety of animals, e.g., the above mentioned salamanders and whale sharks. Our study subjects are the endangered lizards *Diploglossus millepunctatus* from Malpelo Island (Colombia) [7]. These reptiles present an unique spot pattern per subject and currently are studied using mark based methods. Due to the structured scales comprising the lizard's skin it has 3D variations influencing the illumination. The spots have an non-uniform color distribution and can be blurred or highly defined, or occluded by residuals from food, garbage or excrement. All possible variations in the spot segmentation problem for animal biometrics are present in this scenario. In general, biometrics approaches analyze a region of interest (ROI), which is manually selected and then a segmentation is done giving seeds (also manually selected) to an adaptive shape algorithm as deformable shapes or active contours. We propose an automatic method for spot segmentation that avoids user initialization or seeds. Our results show that simple cost functions with MRF framework can perform powerful and effective segmentation of these patterns in multiple illumination variations and under noisy conditions. Related work is mentioned in Sect. 2. In Sect. 3 the methods used in the model are explained. Section 4 describes the experiments used to test the model, and Sect. 5 shows and discusses the results. Finally, in Sect. 6 conclusions and future work are presented.

2   Related Work

The *Diploglossus millepunctatus* spots have no the same intensity values throughout the whole subject. This issue is most critical when high amounts of light irradiate the lizard and mask the spots in the illuminated regions. This issue can be modeled with a Markov Random Field (MRF) that can deal with uncertainty of pixel intensities that belong to a spot in a determinate region based on multiple soft criteria like local intensity, neighborhood relations and a broad number of patterns.

MRFs have been proven to be a suitable image model to resolve computer vision tasks like image segmentation. Boikov and Jolly [8] showed that with some seeds set by the user any object can be segmented using hard constraints and histograms for object and background. In [9] the histograms of user seeds were replaced by Gaussian Mixture Models (GMM), one for background and one for foreground, and also a border matting algorithm was developed to fix transparency on segmented object edges. Another approach is [10] where a shape model was imposed through Layered Pictorial Structures to MRF, which favored specific trained shapes (cows) but need user initialization and a training stage. The method in [11] did not need user interaction or training, it is based on color values from CIE-L*u*v* color space and texture features from Gabor filtered images as data term, with a GMM parameterized automatically with EM algorithm. However, estimating the number of classes highly depends on the image appearance. In [12] the authors propose a multi-region segmentation method based on geometric interactions between objects that were previously segmented with user interaction or automatic framework. Previous segmentation algorithms showed excellent performance but all of them need seeds or depend on image conditions.

Our approach uses monogrid model-based segmentation, does not need user interaction and targets a specific object (spots) in challenging scenarios, without previous training, using an appearance model based on RGB color space, gray-level image and smoothness constraints. Segmentation has been proven in hard light contamination conditions, noisy and blurry images with 3 types of models and 2 inference algorithms.

3   Methods

Here the proposed method is described, as shown in Fig. 1, where the method is subdivided in its main processes. The preprocessing step highlights characteristics and helps to enhance the models score. The MRF model block extracts parameters from the input image to feed the mathematical model, and the inference solves the maximum a posteriori probability problem of the MRF model and gives a mask with spots.

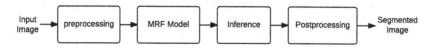

Fig. 1. Scheme of the proposed method.

3.1   Preprocessing

Non-uniform illumination and non-constant color of *Diploglossus millepunctatus* spots are essential objectives for preprocessing steps, since there is no threshold that can separate spots from foreground, a low value in binarization lets pass all the spots, but also large amounts of light (Fig. 2(b)). Moreover a high threshold (Fig. 2(c)) lets only pass the desired pattern, but misses low-intensity spots. There is no prior knowledge about the optimal threshold value on every image. A common solution is Otsu's method (Fig. 2(d)), which assumes binarization like bi-class clustering problem and selects a threshold value that minimizes intra-class variation.

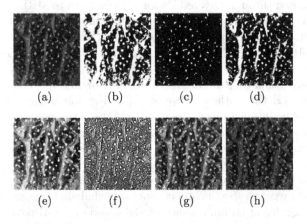

(a)   (b)   (c)   (d)

(e)   (f)   (g)   (h)

Fig. 2. (a) Raw image. (b) Low threshold value. (c) High threshold value. (d) Otsus threshold method. (e) histogram equalization. (f) adaptive threshold. (g) CLAHE (h) Preprocessed image. A single method only cannot isolate the spots without a significant loss of spots or a noise introduction.

Histogram preprocessing techniques to enhance contrast include histogram equalization and contrast correction. Histogram equalization is a global method that sparse the histogram of an image; however this approximation does not produce good results (Fig. 2(e)), because it makes the spots closest to brightness regions intensities. Contrast correction is a point operation that enhances contrast multiplying intensities of a pixel by a fixed value between 1 and 3 and casting it to a value between 0 and 255, this causes a significant contrast enhancement in dark regions, but gaps among spots and higher intensity regions remain unchanged. Global techniques as histogram equalization or point operations like contrast correction are strategies that use global statistics of an image or just modified pixel values with a constant; they do not observe local variation on contrast and assume equal distribution of intensities in an image.

Local operations like adaptive thresholding (AT) and contrast adaptive histogram equalization (CLAHE) observe a local window in each pixel and calculate the optimum threshold value or intensity to a split histogram. Local algorithms depend on window selections and since dark and bright regions, size and distributions, are aleatory, windowing size has to vary throughout the image. Results using AT and CLAHE are exhibited in Fig. 2(f) and (g), both reflect bad choices of correction values, caused by the fixed size of the observed window.

The proposed method equalizes light and keeps the color values constant to exploit spot color information, this reasoning is done using color spaces that convert RGB color space to representations independent of brightness, HSV, L*a*b* and HSI color spaces are representations that deal with this problem. Due to the equalization of aleatory light distribution a CLAHE was applied in the brightness channel on 3 spaces, thus the L*a*b* space shows a more uniform distribution. In order to separate spots from light regions, a saturation correction was implemented. This process highlights Red and Green channels; hence spots were turned brighter than the light regions (see Fig. 2(h)).

Figure 3 shows the proposed preprocessing method applied to the input images. First, local correction (CLAHE) in the Luminance channel of L*a*b* space is applied, followed by a point operation (saturation correction) in HSI color space, and finally the image is transformed to RGB space.

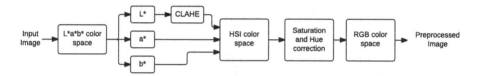

Fig. 3. Schema of the proposed preprocessing method

3.2 MRF Model for Segmentation

Image segmentation ideology assumes that a scene consists of a finite number of regions with characteristics, which change slowly and could be identified with the image constitutive elements. The segmented image is a simplification of the source image where every identified region has a label that classifies them into an image feasible class.

Several prominent traditional segmentation algorithms have been based on probabilistic graphical methods, in which there are sets of observed random variables, hidden random variables and observations over some random variables. Probabilistic approaches try to calculate the probability of a pixel or number of pixels belonging to a certain feasible image class. These classes are discrete random variables, taking values in $L = \{1, 2, \ldots, L\}$, with L as the maximum number of feasible classes in the image. The set of these labels is a random field, called the label process [11]. In this approach, the random variables are related through energy functions that determine whether the pixel belongs to a determined class. The inference process is based on both, the individual values of the pixel or group of pixels and the neighborhood relations. These relations are computed using cliques [13]. This graph topology allows the interaction of

each pixel or group of pixels only with their closer neighborhood, which is called
a first order Markov blanket.

One form to define an energy function E is to define it in terms of the
disagreement between the observed data or E_{data} and the measurement of
the extent to which E is not piecewise smooth or E_{smooth}, such as $E = E_{data} + E_{smooth}$. The selection of the energy functions is a difficult task because
different elections in the E_{smooth} and E_{data} produce different results in the final
segmented image. E_{data} term could consider different factors as the interaction
with the user, the shape or the different characteristics of the target object [14].
In non-supervised object segmentation the introduction of previously known
information about the target object to the energy function as foreground specific intensity range of values or background-foreground contrast information
could even improve the inference process.

In this work, three different energy functions were tested in order to properly
represent the task to solve in the segmentation process. In those energy functions
I_{Gp} represents the gray-scale intensity value of the pixel p. Table 1 shows selected
energy functions.

Table 1. Energy functions

Function	E_{data}^1	E_{data}^0	E_{smooth}
1	$\sum_{p \in P} 250 - I_{Gp}$	$\sum_{p \in P} I_{Gp}$	Potts
2	$\sum_{p \in P} 250 - I_{Gp}$	$\sum_{p \in P} I_{Gp} * I_{Lp}$	Potts
3	$\sum_{p \in P} 250 - I_{Gp}$	$\sum_{p \in P} I_{Gp} * I_c$	Potts

Function 1 approaches the problem based on a priori knowledge of the spot
structure, i.e., high values in grayscale values and lower values of background.
Since spots and the background vary in the image, some possible spots regions
will have more probability in light regions than dark ones. Also, the gap energy
between spots and background regions is higher in dark regions. *Function 2*
uses a grayscale image discretized to 2 levels. Assuming that spots must become
to higher values and background to lower ones, this knowledge is incorporated
through the binary variable I_{Lp}, which takes value 1 when I_{Gp} is 255 or 0 otherwise. Finally, *Function 3* uses previous information about the spot color histogram, using a binary weight I_c with value 1 when red and green channels in
the input image have an intensity lower than the blue one, and 0 otherwise.

A common smooth term energy function is the Potts model, which is the
simplest discontinuity preserving model, where discontinuities between any pair
of labels are penalized equally and can be reduced to the multi-way minimization
problem [8], which is known to be a NP-complete problem where NP means
non-deterministic polynomial time.

In order to solve the inference task, this work uses two algorithms: Graph Cuts (GC) using a push relabel approach and Loopy Belief Propagation (LBP); each algorithm has a different approximation to the solution, and thus, the final results differ, too.

3.3   Postprocessing

Active Contours are energy-minimizing spline curves, guided by external constraint forces and influenced by image forces that pull it towards features such as lines and edges [15]. The internal forces in a spline curve impose smoothness constraints. The image forces moves the *snake* toward salient image features like edges and lines. Finally, external constraints put the *snake* in a local minimum. The energy of the *snake* can be written as a sum of E_{int} and E_{image}. E_{int} term controls elasticity and stiffness, and E_{image} uses image features to reduce the energy of the expression, e.g. edges and gradients. This energy is minimized by iterative algorithms. Because Active Contours require a seed to begin MRF segmentation, results are used to initialize the algorithm and to enhance the performance. Active contours have a parameter called contraction bias that is part of E_{int} and the best value was found through experimentation and set to -0.4.

4   Experimental Framework

4.1   Dataset

The database used in this work was provided by the Exact and Natural Science Department of the University of Antioquia and consists of images from 19 individuals taken under controlled conditions. The database is limited, because the ground truth must be obtained by an expert segmenting manually each image, which is quite time-consuming. As is usual in visual expert identification, two homologous regions are selected in order to perform an identification between individuals as shown in Fig. 4. Five images from each individual were taken, 3 frontal images and 2 lateral images, and with this images the expert obtains the ground truth.

Fig. 4. ROIs from the *Diploglossus millepunctatus* lizard.

4.2   Experiments

Based on the methods previously explained, a set of experiments was planned, in order to find the combination that provides the best segmentation. Table 2 lists the experiments performed where the energy functions described in Table 1 are proved to find which one fits more to the problem. Preprocessing and post-processing stages were used to enhance the functions performance and reduce the final error, respectively.

Table 2. Description of the experiments

Test	Preprocessing	Energy	Inference	Postprocessing
Exp1	None	*Function 1*	LBP/GC	None
Exp2	None	*Function 2*	LBP/GC	None
Exp3	None	*Function 3*	LBP/GC	None
Exp4	None	*Function 1*	LBP/GC	Active Contours
Exp5	None	*Function 2*	LBP/GC	Active Contours
Exp6	None	*Function 3*	LBP/GC	Active Contours
Exp7	Proposed	*Function 1*	LBP/GC	None
Exp8	Proposed	*Function 2*	LBP/GC	None
Exp9	Proposed	*Function 3*	LBP/GC	None
Exp10	Proposed	*Function 1*	LBP/GC	Active Contours
Exp11	Proposed	*Function 2*	LBP/GC	Active Contours
Exp12	Proposed	*Function 3*	LBP/GC	Active Contours

Based on the ground truth, two different formal metrics, confusion matrix and Hoover metrics, were implemented in order to measure the segmentation performance. The confusion matrix compares the ground truth with the machine segmented image and weighs the percentage of pixels matched and mismatched based on the total number of pixels. Hoover metrics [16] consider five types of regions in the ground truth and machine segmented image comparison, either classified as correctly detected, over-segmented, under-segmented, missed and noise, and then plots the number of areas in each class weighted by total amount of areas based on a threshold (tolerance %) term that is the free term in which the graphics are based.

5   Results

Table 3 shows the performance of the model with each cost function and inference algorithm, and the performance of each model after Active Contours were applied. The values correspond to the mean of efficiency in each condition, where efficiency is calculated as sum of true positive and true negative terms from the confusion matrix.

Table 3. Efficiency of the segmentation for each of the experiments performed

Test	LBP	GC	Test	LBP	GC
Exp1	71.86	71.92	Exp7	84.52	84.64
Exp2	66.75	66.56	Exp8	66.75	66.75
Exp3	78.57	78.62	Exp9	82.48	82.54
Exp4	78.52	84.91	Exp10	88.97	88.38
Exp5	54.58	54.58	Exp11	64.64	64.64
Exp6	91.11	90.23	Exp12	90.21	90.2

The results show that a cost function built with intensity differences *Function 1* performs bad per-pixel segmentation when the image has low contrast between foreground and background. However preprocessing enhances this performance significantly pushing the efficiency from 71.86 % to 84.52 % raising the contrast gap and reducing false positives. *Function 2* showed the worst results due to insufficient seed provision. The color-based cost function *Function 3* shows the best results in raw images, owing to the color nature of lizard spots; in preprocessed images it reaches similar results to *Function 1*. Using active contours, with MRF as seeds, enhanced the results up to 91.11 % in raw images and 90.21 % in preprocessed images, because active contours correct under-segmented instances augmenting regions from the seed.

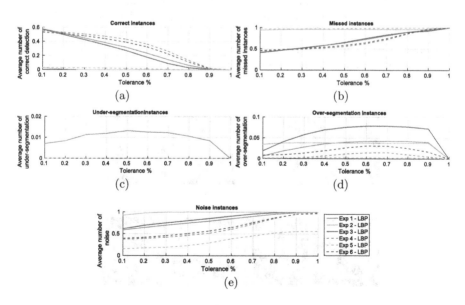

Fig. 5. Average of the Hoover metrics. (a) Correct instances. (b) Missed instances. (c) Under-segmented instances. (d) Over-segmented instances. (e) Noise instances

Since confusion matrices do not expose segmentation quality and just give an idea of correct identified pixels, Hoover metrics [16] gives a wide intuition of the method performance and are presented in Fig. 5 just LPB inference is exposed due to inference algorithms produce slightly differences in the graphics.

Hoover metrics shows that *Function 1* produces the better region performance in all metrics proving that color information is not determinant for good segmentation. The Fig. 5(a) and (b) exhibit that the intensity based function has problems delimiting spots. This probably is caused by the short gap between spots and background in light regions. Figure 5(d) shows how non-uniform color distribution inside spots causes over-segmentation in color based on the energy *Function 3*.

Figure 5(c) and (d) demonstrate that the models does not suffer meaningful under-over segmentation problems, giving less than 2 % and 10 %, respectively. The Active Contours enhance over-segmented images merging regions that are inside a spot in the input image and adjust to the original spot of the input image in under-segmented images. Noise regions (see Fig. 5(e)) are reduced with Active Contours, because the input image does not have a valid region's contour to adapt. In contrast, missed regions increment when a spot in the input image does not have a visual region to adapt.

To give a wide insight into the algorithm performance Fig. 6 compares the ground truth and the machine segmented image; Fig. 6(a)–(d) are input images. The Fig. 6(e)–(h) are the output of the MRF with Active Contours postprocessing. Pixels the model identifies as spots are green, but do not appear as a spot in the ground truth. Red pixels are ground truth spot pixels the model did not catch and yellow regions mean zones where the model accords with the ground truth. The proposed model can solve the spot segmentation task under inner image variant illumination conditions, nevertheless, it is common that the model ignores spots that only span few pixels and also dark spots with intensity

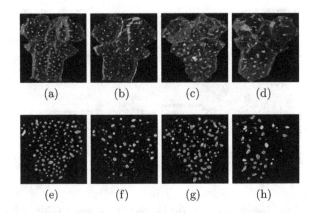

(a)   (b)   (c)   (d)

(e)   (f)   (g)   (h)

Fig. 6. Qualitative segmentation results. Raw images (upper row). Output images (lower row)

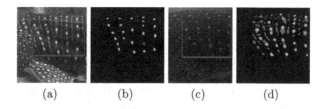

(a)	(b)	(c)	(d)

Fig. 7. Result on whale shark database. (a) Input image 1. (b) Segmented image 1. (c) Input image 2. (d) Segmented image 2

similar to the background. Large spots with narrow parts are usually divided into two parts with the narrow part as the break point, as common in energy segmentation approaches.

6 Conclusions

In this paper a segmentation model for spots on animals based on Markov Random Fields and Active Contours is proposed and tested on *Diploglossus millepunctatus* lizard images. Extensive experiments using energy functions based on pixel intensities, quantization, and color information as cost functions were carried out. Also two inference methods, loopy belief propagation and Graph cuts were tested. A preprocessing approximation dealing with color spaces, global and local enhancing, and segmentation methods was performed. The best performance was achieved with an intensity build data term function that reached 84.52 % using proposed preprocessing stage. Using Active Contours as postprocessing boosts the results up to 91.11 %. The model shows promising performance to automatize segmentation processes in photographic mark recapture and to reduce processing time and subjectivity.

In future work, the cost functions will have extra terms that include considerations of shape through a pictorial structures concept [10]. Color constrains will be modeled through GMM framework training and specifically modeled to *Diploglossus millepunctatus* spots. The work will be extended to other animals and species. For instance, Fig. 7 shows the cost function based on intensity applied to some samples from a whale shark dataset [17] to extract spot patterns. In this dataset, just one region of the whale shark is needed for identification (marked inside a red rectangle as in Fig. 7(a) and (c)). Figure 7(b) and (d) show qualitative segmentation that follows the same notation from Fig. 6. These results show promising performance assuming that the model segments all spots in the image without any additional tuning procedure.

Acknowledgment. The authors express thanks to projects 512C-2013 Ruta N and P13124 from ITM for their support. Also, Fundación Malpelo and Parques Nacionales de Colombia provided funding and research permits for collecting data at the Santuario de Fauna y Flora Isla de Malpelo.

References

1. Kühl, H.S., Burghardt, T.: Animal biometrics: quantifying and detecting phenotypic appearance. Trends Ecol. Evol. **28**(7), 432–441 (2013)
2. Kelly, M.J.: Computer-aided photograph matching in studies using individual identification: an example from serengeti cheetahs. J. Mammal. **82**(2), 440–449 (2001)
3. Lahiri, M., Tantipathananandh, C., Warungu, R., Rubenstein, D.I., Berger-Wolf, T.Y.: Biometric animal databases from field photographs: identification of individual zebra in the wild. In: Proceedings of the 1st ACM International Conference on Multimedia Retrieval, page 6. ACM (2011)
4. Bolger, D.T., Morrison, T.A., Vance, B., Lee, D., Farid, H.: A computer-assisted system for photographic mark-recapture analysis. Methods Ecol. Evol. **3**(5), 813–822 (2012)
5. Gope, C., Kehtarnavaz, N., Hillman, G., Würsig, B.: An affine invariant curve matching method for photo-identification of marine mammals. Pattern Recogn. **38**(1), 125–132 (2005)
6. Ardovini, A., Cinque, L., Sangineto, E.: Identifying elephant photos by multi-curve matching. Pattern Recogn. **41**(6), 1867–1877 (2008)
7. López-Victoria, M.: The lizards of malpelo (colombia): some topics on their ecology and threats. Caldasia **28**(1), 129–134 (2006)
8. Boykov, Y.Y., Jolly, M.-P.: Interactive graph cuts for optimal boundary & region segmentation of objects in nd images. In: Proceedings Eighth IEEE International Conference on Computer Vision, 2001. ICCV 2001, vol. 1, pp. 105–112. IEEE (2001)
9. Rother, C., Kolmogorov, V., Blake, A.: Grabcut: Interactive foreground extraction using iterated graph cuts. ACM Trans. Graph. (TOG) **23**(3), 309–314 (2004)
10. Prema Kumar, M., Ton, P.-H.-S., Zisserman, A.: Obj cut. In: IEEE Computer Society Conference on Computer Vision and Pattern Recognition, 2005. CVPR 2005, vol. 1, pp. 18–25. IEEE (2005)
11. Kato, Z., Pong, T.-C.: A markov random field image segmentation model for color textured images. Image Vis. Comput. **24**(10), 1103–1114 (2006)
12. Delong, A., Boykov, Y.: Globally optimal segmentation of multi-region objects. In: 2009 IEEE 12th International Conference on Computer Vision, pp. 285–292. IEEE (2009)
13. Koller, D., Friedman, N.: Probabilistic Graphical Models: Principles and Techniques. MIT press, Cambridge (2009)
14. Lézoray, O., Grady, L.: Image processing and analysis with graphs: theory and practice. CRC Press, Boca Raton (2012)
15. Kass, M., Witkin, A., Terzopoulos, D.: Snakes: active contour models. Int. J. Comput. Vis. **1**(4), 321–331 (1988)
16. Hoover, A., Jean-Baptiste, G., Jiang, X., Flynn, P.J., Bunke, H., Goldgof, D.B., Bowyer, K., Eggert, D.W., Fitzgibbon, A., Fisher, R.B.: An experimental comparison of range image segmentation algorithms. IEEE Trans. Pattern Anal. Mach. Intell. **18**(7), 673–689 (1996)
17. Holmberg, J., Norman, B., Arzoumanian, Z.: Estimating population size, structure, and residency time for whale sharks rhincodon typus through collaborative photo-identification. Endangered Species Res. **7**, 39–53 (2009)

Segmentation of Building Facade Towers

Gayane Shalunts[(⊠)]

SAIL LABS Technology GmbH, Vienna, Austria
Gayane.Shalunts@sail-labs.com

Abstract. Architectural styles are phases of development that classify architecture in the sense of historic periods, regions and cultural influences. The article presents the first approach, performing automatic segmentation of building facade towers in the framework of an image-based architectural style classification system. The observed buildings, featuring towers, belong to Romanesque, Gothic and Baroque architectural styles. The method is a pipeline unifying bilateral symmetry detection, graph-based segmentation approaches and image analysis and processing technique. It employs the specific visual features of the outstanding architectural element tower - vertical bilateral symmetry, raising out of the main building and solidity. The approach is robust to high perspective distortions. It comprises two branches, targeting facades with single and double towers correspondingly. The performance evaluation on a vast number of images reports extremely high segmentation precision.

1 Introduction

Geographical, geological, climate, religion, social, political and historical factors influenced the formation of architectural styles [1]. Facade images from online image databases usually do not have labels of architectural styles. To know the style of an observed building, one can search for the building name. Whereas this is a solution for famous buildings with well-known names, it is not applicable to buildings lacking names. In this case facade visual information holds the only clue to its style. Here an alternative to an architectural style visual classification tool could be visual search engines, like Google visual search[1] or landmark recognition engines, like [2]. Nevertheless observations show that whereas those engines work well for famous landmarks, succeeding in mining those buildings and all information about them, they fail for ordinary buildings, not even delivering facades belonging to the same architectural style. An automatic tool for image-based classification of architectural styles will solve this problem, not depending on building popularity or its importance as a cultural heritage object.

The research related to architectural style classification of facades emerged in computer vision community recently. The approach introduced in [3] is a four-stage method, incorporating steps of scene classification, image rectification, facade splitting and style classification [3]. The authors in [4] adopt Deformable Part-based Models to capture the morphological characteristics of architectural

---

[1] Google Visual Search Engine http://images.google.com.

© Springer International Publishing Switzerland 2015
G. Bebis et al. (Eds.): ISVC 2015, Part II, LNCS 9475, pp. 185–194, 2015.
DOI: 10.1007/978-3-319-27863-6_17

components and propose Multinomial Latent Regression that introduces the probabilistic analysis and tackles the multi-class problem in latent variable models. [5] performs classification using mining of word pairs and semantic patterns. The authors in [6] propose a hierarchical sparse coding algorithm to model blocklets, representing basic architectural components. [7] segments 2D facades into balconies, walls, windows and doors, using shape grammars for facade parsing.

The current work, being the first approach addressing the automatic segmentation of facade towers, is a unit in a novel computer vision system, called STYLE, the objective of which is to classify the architectural style of a facade given its image. Architectural elements, such as windows, towers, domes, columns, are the component parts of buildings. The algorithm of the proposed architectural style classification system comprises three major steps: (1) Semantic segmentation of facade architectural elements is performed. (2) Each segmented element is passed to an appropriate module for architectural style classification. (3) Architectural style voting of the classified elements determines the architectural style of the whole facade.

The method permits to cluster building image databases by historic periods. Such a semantic categorization confines the search of building image databases to certain category portions and may be useful in building recognition [2,8], Content Based Image Retrieval (CBIR) [9], 3D reconstruction and 3D city-modeling [10]. Inverse procedural modeling aims to reconstruct a detailed procedural model of a building from a set of images [3]. The search space of building images is narrowed down by implicitly assuming an architectural style [3]. Architectural style classification system may also find an interesting application in virtual and real tourism. The publications within the scope of the STYLE project have addressed the tasks of classification of windows [11], classification of architectural elements called tracery, pediment and balustrade [12], as well as the classification [13] and segmentation [14] of domes.

Among numerous styles there are such that have influenced large regions and dominated a few centuries. The presented article classifies towers of such influential European styles: **Romanesque** (8th - 12th centuries), **Gothic** (12th - 16th centuries), **Baroque** (17th - mid 18th centuries). Tower is a building or part

   a) Romanesque   b) Gothic   c) Baroque

Fig. 1. Sample facades of Romanesque, Gothic and Baroque styles.

of a building that is exceptionally high in proportion to its width and length [15]. Towers are featured on buildings of religious and secular great significance, as a display of power. The tower segmentation module addresses buildings with single and double towers. Samples of Romanesque, Gothic and Baroque facades featuring towers are shown in Fig. 1a, b and c correspondingly.

The tower segmentation approach is built on the first method of segmentation of the architectural element dome, proposed by [14] in the context of the current project. It is likewise a multi-step pipeline of bilateral symmetry detection, graph-based segmentation approaches, image analysis and processing technics. The tower segmentation approach is a modification and extension of the dome segmentation method [14], taking into account the architectural specificities and visual distinguishing features of towers and developed into two branches, segmenting the single and double tower facades respectively. The experimental image dataset, containing a wide variety of buildings featuring towers, is an additional contribution of this paper.

The article is organized as follows. Section 2 explains the methodology of the tower segmentation. The experiments and results of are detailed in Sect. 3. The conclusions are drawn in Sect. 4.

2   The Methodology

Segmentation of towers is a high-level semantic segmentation by an architectural element. Color-based segmentation approaches are not applicable, since color is not a distinctive feature and a single tower may contain multiple color segments. Shape analysis is also not suitable for segmentation, as tower shapes cannot be modeled due to unlimited variety.

The tower segmentation approach is based on the dome segmentation method, proposed by [14], modifying and extending it taking into account the architectural specificities of towers. Such a choice is justified by the following common visual characteristics of towers and domes: (1) Both elements display vertical bilateral symmetry. (2) Both elements are situated high above the building and have the sky and clouds as a background. The algorithm in [14] is a pipeline processing the image through several stages, delivering the dome segment at the final step. The tower segmentation approach has two branches, aiming to handle buildings, featuring a single tower and double towers respectively. The detailed illustration of the algorithm follows below on sample facade images with single and double towers.

The branch of single tower segmentation is clarified on the example of Vienna City Hall (Fig. 2a), of double tower segmentation – Vienna Church of Mariahilf (Fig. 3a). At the first step the image bilateral symmetry axes are detected, using the method proposed by [16]. The matches of symmetric points are found by modern feature-based methods, such as [17], from which bilateral symmetry axes [16]. The method is independent of the feature detector and descriptor used, requiring only robust, rotation-invariant matching and an orientation measure for each feature [16]. The bilateral symmetry axes and the supporting symmetric

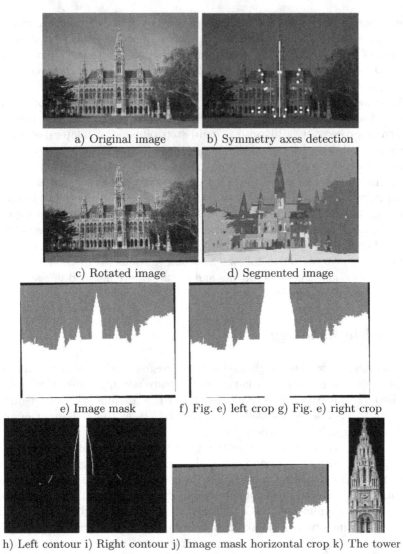

a) Original image   b) Symmetry axes detection

c) Rotated image   d) Segmented image

e) Image mask   f) Fig. e) left crop g) Fig. e) right crop

h) Left contour i) Right contour j) Image mask horizontal crop k) The tower

Fig. 2. Algorithm steps for segmentation of a single tower.

points of images in Figs. 2a and 3a are displayed in Figs. 2b and 3b respectively. In case multiple symmetry axes are found, the axis with the strongest symmetry magnitude (supported by the biggest number of symmetry points) is chosen. Here the false positive symmetry descriptors are ignored, as they never succeed in building the strongest symmetry axis. There is priori knowledge that the dominant symmetry axis is vertical for historic facades, so the author proceeds with rectifying the images by rotation, to make the strongest symmetry axis vertical (Figs. 2c and 3c). Images, whose strongest symmetry axis is vertical,

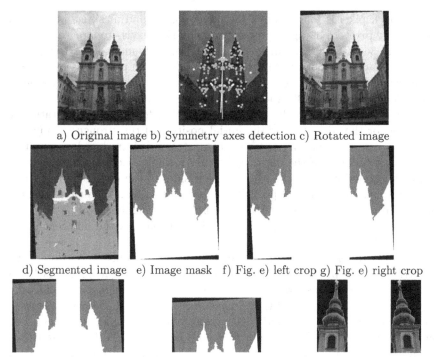

a) Original image b) Symmetry axes detection c) Rotated image

d) Segmented image   e) Image mask   f) Fig. e) left crop g) Fig. e) right crop

h) Left crop i) Right crop j) Image mask horizontal crop k) Left tower l) Right tower

Fig. 3. Algorithm steps for segmentation of double towers.

skip this step. Bilateral symmetry detection [16] is performed once more on the rotated image to find the position (column) of the strongest symmetry axis.

The purpose of the next step is the segmentation of the original image background and foreground. Since towers like domes are situated high above the buildings, the sky and clouds form their background. The original image is segmented using the methodology introduced by [18] (Figs. 2d and 3d). The pairwise comparison of neighboring vertices, i.e. partitions is used to check for similarities [18,19]. In [18] a definition of a pairwise group comparison function $Comp(\cdot, \cdot)$ is given, judging if there is evidence for a boundary between two image segments or not. $Comp(\cdot, \cdot)$ contains a scale parameter k, where bigger k prefers larger segmented regions. The function measures the difference along the boundary of two components relative to a measure of differences of components' internal differences. This definition tries to encapsulate the intuitive notion of contrast. Images are preprocessed by Gaussian blurring with σ before the segmentation, as well as postprocessed by merging small regions with the biggest neighboring one [18]. Afterwards the segmented image is also rotated to make the strongest symmetry axis vertical. As the images are already justified so that the sky is on the top of it, the segment (color) of the sky is found by tracing down the strongest symmetry axis starting from the first upper row until the first non-black pixel is found. Then the image foreground mask is obtained by setting all sky pixels to background color and non-sky pixels - to foreground color (Figs. 2e and 3e). Having the image foreground mask found, the next step is to analyze its shape in order to crop the tower(s). Now the purpose is to find the bottom row of the tower(s). The image masks are cut from the main symmetry axis into left (Figs. 2f and 3f) and right (Figs. 2g and 3g) parts.

To this point the author followed the algorithm for dome segmentation in [14], except for the correction of the initial position of the strongest symmetry axis. Here a new step specific for tower segmentation is introduced: a condition is analyzed to find out, if the building features a single or double towers. If the local neighborhood both on the right and left sides of the main symmetry axis has higher situated foreground pixels than that on the main symmetry axis, the building features double towers. More precisely, the neighborhood on the right and left sides from the main symmetry axis is defined as the adjacent 1/8th parts of the right and left images.

The single tower segmentation branch proceeds as follows. As domes are symmetric in all 2D projections, for optimization only the right image mask was analyzed in [14]. Whereas towers are more sensitive to camera viewpoint, that is why analysis of both left and right image masks is needed. Domes and towers possess a common visual property: they raise out of the main building. This means that the facade left and right contours, formed by the foreground pixel adjacent to a background pixel in each row (Fig. 2h and i), have leaps on the rows, where the tower meets the main building. The observed leaps are found by tracing down row by row the facade left and right contours in Fig. 2h and i until the condition in Eq. 1 is satisfied:

$$Leap_k/(Row_k - Row_1) > \text{LeapThreshold} \;\&\&\; Leap_k > \text{minLeapThreshold} \quad (1)$$

where $Leap_k$ is the column difference of contour pixels on kth and $(k-1)$th sequential rows: $Leap_k = Col_k - Col_{k-1}$.

$Row_k - Row_1$ is a normalization factor and is the difference of the kth and first rows of the contour. minLeapThreshold excludes too small leaps between two subsequent rows and is set to 18 pixels for images with resolution lower than 1 million pixels and to 26 pixels otherwise. LeapThreshold was found empirically and is equal to 0.24. As the final row, containing the leap, is chosen the one of right and left leap rows, which lies higher. After the image mask is cropped from the leap row found (Fig. 2j) to discard the image part below, which does not include the segment of interest. In order to obtain the final tower segment, the blob, through which the main symmetry axis passes, is picked up. All the other blobs formed by clouds, trees and any other objects present in the image are discarded. Multiplication of the blob mask with the same segment of the original image delivers the segmented tower (Fig. 2k).

The branch, addressing the segmentation of double towers, is also a novel introduction of this article and performs the following processing. Image left and right masks (Fig. 3f and g) are cut horizontally from the lowest background pixel. After the blobs which are not connected with the strongest symmetry axis are removed (Fig. 3h and i). Note that the strongest symmetry axis is located on the last column of the left mask (Fig. 2f) and on the first one – of the right (Fig. 2g). The removal of the disconnected blobs aims to eliminate the objects (trees, clouds, etc.), which can be situated higher than the towers. Afterwards the columns of the global foreground minimums[2] are located on the left and right image masks. Then the global background maximum rows are found between the strongest symmetry axis and the columns mentioned. The image is cut horizontally from the row situated higher (Fig. 2j). And finally the left (Fig. 2k) and right (Fig. 2l) towers are detected by picking up the blobs, containing the highest foreground pixels on the left and right masks.

To address the images, in which the 1st strongest symmetry axis is not placed on the tower (single tower case) or between the towers (double tower case) due to too high perspective distortions or other symmetric objects in the image, an additional feature is introduced, called solidity. Solidity is a region property, specifying the proportion of the pixels in the convex hull that are also in the region and is computed by Eq. 2. If tower solidity and bounding box resolution pass the thresholds in Eq. 3, segmentation is considered successful.

$$\text{Solidity} = \text{Area}/\text{ConvexArea} \qquad (2)$$

$$\text{Solidity} > 0.72 \ \&\& \ \text{TowerBoundingBox} > 5000 \qquad (3)$$

The thresholds in Eqs. 1 and 3 were found empirically, exercising multiple images. Setting a threshold for tower bounding box resolution excludes too small blobs. In case the condition in Eq. 3 is not met, the whole segmentation algorithm pipeline is rerun by taking the 2nd strongest symmetry axis in the initial step. The tower segment search is limited by the 2nd strongest symmetry axis (if such exists), since the database lacks images, for which further search would lead to segmentation success (Sect. 3). Note for comparison, that the dome segmentation approach in [14] judges the segmentation success by feature roundness, instead of solidity and seeks for the dome segment in the vicinity of up to the 5th strongest symmetry axis, instead of the 2nd, as in the current method.

3   Experimental Setup, Evaluation and Results

An extensive study was conducted to evaluate the tower segmentation approach. One of the challenges for testing the approach is the lack of image databases labelled by architectural styles and featuring towers. Thus the author compiled such an dataset from own and Flickr[3] image databases. The author's own share of the images is freely available to the scientific community[4]. The database includes 325 images of 35 cathedrals, churches, basilicas and city halls located in Austria,

---

[2] Note that the rows in an image count down from the upper row, so the minimum row is the highest and the maximum - the lowest.

[3] http://www.flickr.com.

[4] https://www.flickr.com/photos/lady_photographer/sets/72157636149550844/.

Table 1. The segmentation rate.

	Image seg. rate (%)	Tower seg. rate (%)
1st symmetry axis	303 (93.23 %)	539 (91.20 %)
2nd symmetry axis	5 (1.54 %)	5 (0.85 %)
1 tower of 2	18 (5.54 %)	18 (3.05 %)
1 tower as if 1 of 2	6 (1.85 %)	6 (1.02 %)
Success	308 (**94.77 %**)	544 (**92.05 %**)
Fail	17 (5.23 %)	47 (7.95 %)
Total	325	591

Germany, Sweden, Czech Republic, Hungary, France, Spain, Luxemburg, England, Belgium, Switzerland and China. Famous landmarks, like Notre Dame Cathedral in Paris, city halls of Vienna and Brussels are among the buildings. Original and revived Romanesque, Gothic and Baroque styles are represented in the database. The approach is capable to handle both day and night images due to being color-independent. It also achieves success on buildings in complex scenes and exposed to high perspective distortions. The only limitations of the method are: (1) the segmentation of occluded towers is not supported, (2) the rare cases, when the tower background is formed by cityscape, not the sky, as a result of shooting from a level higher than the ground (building roofs, helicopter, etc.) are also not handled, (3) the images, in which the tower(s) touch the top of the image, are also excluded, since on such images the background segmentation by [18] algorithm fails to deliver a single background (sky) segment.

The allowed tilt of the towers is $(-90$ to $90)$ degrees related to the vertical axis. This should not be considered a limitation, as the search showed that building images taken upside down or tilted more than 90 degrees related to the vertical axis rarely meet. The values by default for both parameters of graph-based segmentation algorithm [18] σ and k are 1000. The goal of obtaining coarse segmentation of sky and non-sky segments is the ground for the chosen big values. For images taken by night illumination, foggy weather condition or having low resolution the values of σ and k should be tuned down to obtain the non-sky segment with the precise tower edge. Whereas for images with strong cloud edges in the tower vicinity the values of σ and k should be tuned up to blur the cloud edges. Only the rare cases, when cloud-sky edges are stronger than cloud-tower edges fail. Here the cloud gets attached to the tower(s) in the foreground segment, leading to failure in the final step of solidity thresholding. Clouds not touching the tower do not affect the segmentation output, as they are discarded segments. For the current database the empirically found values of σ and k are in the range from 200 to 5000.

Interesting cases to study are images, where facades are exposed to high perspective distortions. Here the challenge is in the very first step of symmetry detection. The experiments revealed 2 curious ways of behavior of symmetry detection, determining the final segmentation output. (1) On images with double towers, the strongest symmetry axis is located on one of the towers, instead of passing through them. As a consequence the algorithm follows the single tower branch, ending up with the delivery of one of double towers. (2) On facade images with single towers, the strongest symmetry axis is located on another vertical axis, but not on the one passing through the tower. This makes the segmentation algorithm follow the double tower branch. However the segmentation output is successful, since the objects that could be falsely segmented as the second tower fail to pass the solidity and bounding box size thresholds at the final step.

Table 1 summarizes the results of the tower segmentation approach performance. The second column in Table 1 presents the segmentation rate in terms of images. Here the segmentation is assumed successful in the case, when one of double towers is segmented (bullet 1) above), since the single tower will pass to further classification and voting. 308 out of 325 images achieved successful segmentation, indicating an average **94.77 %** segmentation rate in terms of images. On 303 out of 308 successfully segmented images the position of tower(s) was found by the 1st strongest symmetry axis and only on 5 images - by the 2nd strongest symmetry axis. On 18 images 1 of double towers was segmented. 6 images passed successful segmentation, though the single tower was segmented as if it were 1 of the double towers. The segmentation was unsuccessful only on 17 images (5.23 %). 15 images owe the failure to the symmetry detector and the remaining 2 images failed, because the cloud-sky edges were stronger than the cloud-tower edges. The third column in Table 1 evaluates the segmentation rate in terms of towers. 47 out of 591 towers failed the segmentation, 18 of them being one of the towers on a double tower facade. The average segmentation rate in terms of towers yielded **92.05 %**.

4   Conclusion

The article presented the first method in computer vision literature for automatic segmentation of Romanesque, Gothic and Baroque towers. The algorithm was a pipeline uniting bilateral symmetry detection, graph-based segmentation approaches and image analysis and processing technics. The performance evaluation on a self-collected image database, including a vast variety of buildings, reported very high average segmentation rate. The work was carried out in the scope of a computer vision system, targeting the architectural style classification problem of building facades and proposing a solution as successive phases of segmentation, classification and voting of architectural elements.

References

1. Fletcher, B.: A history of architecture on the comparative method, 5th edn. Scribner's, Batsford, London (1920)
2. Zheng, Y.T., Zhao, M., Song, Y., Adam, H., Buddemeier, U., Bissacco, A., Brucher, F., Chua, T.S., Neven, H.: Tour the world: building a web-scale landmark recognition engine. In: Proceedings of the 20th CVPR, Miami, Florida, USA, pp. 1085–1092. IEEE (2009)

3. Mathias, M., Martinovic, A., Weissenberg, J., Haegler, S., Gool, L.V.: Automatic architectural style recognition. In: Proceedings of the 4th International Workshop on 3D Virtual Reconstruction and Visualization of Complex Architectures. International Society for Photogrammetry and Remote Sensing, Trento, Italy, pp. 280–289 (2011)

4. Xu, Z., Tao, D., Zhang, Y., Wu, J., Tsoi, A.C.: Architectural style classification using multinomial latent logistic regression. In: Fleet, D., Pajdla, T., Schiele, B., Tuytelaars, T. (eds.) ECCV 2014, Part I. LNCS, vol. 8689, pp. 600–615. Springer, Heidelberg (2014)

5. Goel, A., Juneja, M., Jawahar, C.V.: Are buildings only instances?: exploration in architectural style categories. In: Proceedings of the ICVGIP, Mumbai, India, pp. 1–8 (2012)

6. Zhang, L., Song, M., Liu, X., Sun, L., Chen, C., Bu, J.: Recognizing architecture styles by hierarchical sparse coding of blocklets. Inf. Sci. **254**, 141–154 (2014)

7. Teboul, O., Kokkinos, I., Simon, L., Koutsourakis, P., Paragios, N.: Parsing facades with shape grammars and reinforcement learning. IEEE Trans. Pattern Anal. Mach. Intell. **35**, 1744–1756 (2013)

8. Zhang, W., Kosecka, J.: Hierarchical building recognition. Image Vis. Comput. **25**(5), 704–716 (2004)

9. Li, Y., Crandall, D., Huttenlocher, D.: Landmark classification in large-scale image collections. In: Proceedings of 12th ICCV, Kyoto, Japan, pp. 1957–1964. IEEE (2009)

10. Cornelis, N., Leibe, B., Cornelis, K., Gool, L.V.: 3d urban scene modeling integrating recognition and reconstruction. IJCV **78**, 121–141 (2008)

11. Shalunts, G., Haxhimusa, Y., Sablatnig, R.: Architectural style classification of building facade windows. In: Bebis, G., et al. (eds.) ISVC 2011, Part II. LNCS, vol. 6939, pp. 280–289. Springer, Heidelberg (2011)

12. Shalunts, G., Haxhimusa, Y., Sablatnig, R.: Classification of gothic and baroque architectural elements. In: Proceedings of the 19th IWSSIP. LNCS, Vienna, Austria, pp. 330–333. IEEE (2012)

13. Shalunts, G., Haxhimusa, Y., Sablatnig, R.: Architectural style classification of domes. In: Bebis, G., et al. (eds.) ISVC 2012, Part II. LNCS, vol. 7432, pp. 420–429. Springer, Heidelberg (2012)

14. Shalunts, G., Haxhimusa, Y., Sablatnig, R.: Segmentation of building facade domes. In: Alvarez, L., Mejail, M., Gomez, L., Jacobo, J. (eds.) CIARP 2012. LNCS, vol. 7441, pp. 324–331. Springer, Heidelberg (2012)

15. Illustrated architecture dictionary: Tower (2015). http://www.buffaloah.com/a/DCTNRY/t/tower.html, Accessed August 4, 2015

16. Loy, G., Eklundh, J.-O.: Detecting symmetry and symmetric constellations of features. In: Leonardis, A., Bischof, H., Pinz, A. (eds.) ECCV 2006. LNCS, vol. 3952, pp. 508–521. Springer, Heidelberg (2006)

17. Lowe, D.G.: Distinctive image features from scale-invariant keypoints. IJCV **60**(2), 91–110 (2004)

18. Felzenszwalb, P.F., Huttenlocher, D.P.: Efficient graph-based image segmentation. IJCV **59**(2), 167–181 (2004)

19. Laurent, G., Men, H.L., Jean-Pierre, C.: The hierachy of the cocoons of a graph and its application to image segmentation. PRL **24**, 1059–1066 (2003)

Effective Information and Contrast Based Saliency Detection

Aditi Kapoor[1]([✉]), K.K. Biswas[2], and M. Hanmandlu[3]

[1] Amar Nath and Shashi Khosla School of Information Technology,
Indian Institute of Technology Delhi Hauz Khas, New Delhi 110016, India
`aditi@cse.iitd.ernet.in`
[2] Department of Computer Science and Engineering, Indian Institute of Technology
Delhi Hauz Khas, New Delhi 110016, India
`kkb@cse.iitd.ernet.in`
[3] Department of Electrical Engineering,
Indian Institute of Technology Delhi Hauz Khas, New Delhi 110016, India
`mhmandlu@gmail.com`

Abstract. Human attention tends to get focused on the most prominent objects in a scene which are different from the background. These are termed as salient objects. The human brain perceives an object of salient type based on its difference with the surroundings in terms of color and texture. There have been many color based approaches in the past for salient object detection. In this paper, we augment information set features with color features and detect the final single salient object using a set of color, size and location based features. The information set features result from representing the uncertainty in the color and illumination components. To locate the salient parts of the image, we make use of the entropy to find the uncertainties in the color and luminance components of the image. Extensive comparisons with the state-of-the-art methods in terms of precision, recall and F-Measure are made on two different publicly available datasets to prove the effectiveness of this approach.

1 Introduction

Detection of a salient object or regions in images has received a lot of attention in the recent years. Saliency detection for identifying the most prominent object in an image or video in computer vision is attempted by mimicking the human visual perception. Everything focused on our visual field is not given the same importance. The perceptual visual mechanism in humans is responsible for editing the incoming information [1]. Salient object detection facilitates solving many diverse computer vision problems that include automatic cropping [2], image summarization [3], non-photorealistic rendering [4] and even aesthetic quality and interestingness of image [5].

One of the earliest computational models for saliency based spatial attention is presented in [6]. Here the center surround is used to find the difference between different scales. Saliency maps are created using color, intensity and orientation

© Springer International Publishing Switzerland 2015
G. Bebis et al. (Eds.): ISVC 2015, Part II, LNCS 9475, pp. 195–204, 2015.
DOI: 10.1007/978-3-319-27863-6_18

features. Liu et al. [7] propose a supervised approach for salient object detection. Their features describe multi-scale contrast, center-surround histogram and color spatial distribution. These features are combined using the Conditional Random Field (CRF) method in a probabilistic framework for labeling and segmenting the structured data. Achanta et al. in [8] detect salient regions in an image using a contrast determination filter that operates at various scales and generates saliency maps. The K-means algorithm is used on the overall map to segment the whole objects. The saliency maps of [9] are constructed using the maximum symmetric surround method. This method is similar to the center surround of [7] but the surround is varied based on the nearness of a pixel to the center.

Saliency can be determined by quantifying the self-information of local image patch [10] computed using a neural network. Given an input image, its local neighborhood characteristics and the collected basis functions are computed and these are then input to the neural network to get the saliency map based on the information maximization in that image.

Cheng et al. [4] propose a histogram contrast based method which computes the global contrast while maintaining the spatial coherence. Here the emphasis is laid on color contrast. Margolin et al. [11] use the concept of distinctness color and pattern for detecting the salient objects. They use PCA to represent a set of patches of an image and determine the distinctness. The pattern distinctness rests on the scatter of the various patches whereas the color distinctness is determined using Simple Linear Iterative Clustering (SLIC) that generates uniform superpixels.

In this paper, we present a novel approach for salient object detection using information set based features along with color features and a combination of size and location based features. We want to extract the "information content" in the windows by means of representation of uncertainty in the fuzzy set formed from the intensity values and the membership function fitted to them. The membership function gives the degree of association of each intensity value to fuzzy set but it doesn't give the overall uncertainty because summing up all the membership function values will only the average degree of association. We are concerned with the uncertainty associated with the information source values which could be attributes, features, property values etc. using the entropy and to integrate it with color, size and location based features. The rest of the paper is organized as follows. Section 2 presents the information set based approach. Section 3 details out the analysis of results while Sect. 4 concludes the paper.

2   Proposed Approach

An approach is presented here for extracting information set based features from the predefined image windows. We judge whether the uncertainty of the information source values in a window corresponds to saliency or background. We augment the information set based features and the color clustering based features using thresholding and the connected component analysis.

2.1   Information Set Based Features

Since CIELab colorspace closely resembles the human visual perceptivity, we convert the images into this colorspace. Features are extracted corresponding to each of the 3 components, namely, L, 'a' and 'b' to represent both color and intensity (Luminance). To extract the local features, an image is divided into nxn sized windows and from each window three features are extracted corresponding to the three components. In order to extract the neighbourhood information effectively, we take n to be 10.

Effective Information Feature. The color component value at each pixel within a window along with its respective membership function value form a pair that constitutes an element of the fuzzy set. The membership function gives the degree of association of each color component to the set but it doesn't provide the overall uncertainty in the fuzzy set. We want to quantify the uncertainty present in each component separately. This uncertainty is sought to be quantified by an entropy function. Here the values of the color component are termed as the information source values as in [12]. Our aim is to find the uncertainty associated with the information source values using an appropriate entropy function.

To this end consider the Hanman-Anirban entropy [12] function with $a = 0, b = 0$ and $c = 1/f$, $d = -I_{av}/f$ and $p_{ij} = I_{ij}$:

$$H = \sum_{i=1}^{n} \sum_{j=1}^{n} p_{ij} e^{-(cp_{ij}-d)/f} \tag{1}$$

where I_{ij} is the Information source value for ij^{th} pixel and I_{av} is the average of all I_{ij} in a chosen nxn window given as $I_{av} = 1/n^2 \sum_{i=1}^{n} \sum_{j=1}^{n} I_{ij}$.

The fuzzifier f is chosen as suggested in [12]:

$$f = \sum_{i=1}^{n} \sum_{j=1}^{n} (I_{av} - I_{ij})^4 / \sum_{i=1}^{n} \sum_{j=1}^{n} (I_{av} - I_{ij})^2 \tag{2}$$

This substitution converts the exponential gain function into an exponential membership function. As a result the entropy function takes the form:

$$H = \frac{1}{n^2} \sum_{i=1}^{n} \sum_{j=1}^{n} I_{ij} \mu_{ij} \tag{3}$$

where $\mu_{ij} = e^{-(|I_{ij}-I_{av}|)/f}$.

Notice that the information source values I_{ij} and its membership function value as the pair (I_{ij}, μ_{ij}) now becomes a product denoted by $h_{ij} = (I_{ij}\mu_{ij})$. The n^2 term is included for normalization. It is possible to obtain different forms of information sets by choosing the type of the membership function, μ_{ij}. In this paper, we derive the effective information feature [12] by using the information values $\mu_{ij} I_{ij}$ in a nxn window as follows:

$$EI_{kL} = \frac{\sum_{i=1}^{n} \sum_{j=1}^{n} \mu_{ij} I_{ij}}{\sum_{i=1}^{n} \sum_{j=1}^{n} \mu_{ij}} \tag{4}$$

Here EI_{kL} is the effective information for the kth window of the 'L' component. Similarly the effective information features EI_{ka} and EI_{kb} for 'a' and 'b' color components of the image are computed. The collection of individual features EI_{kL} constitutes the effective information feature vector EI_L for the entire image. Each pixel in a window is assigned the common effective information (EI_{kL}). This way the size of the original image is kept the same.

Maximum Symmetric Surround. Similarly, the effective information features EI_a and EI_b for 'a' and 'b' components of the image are computed. A single contrast map is produced by combining all three components using symmetric surround saliency as outlined in [9]. This saliency value at the given pixel, $EI_{ssL}(x, y)$ in the feature image for L component is given by:

$$EI_{ssL}(x, y) = |EI_{\mu L}(x, y) - EI_L(x, y)| \tag{5}$$

where $EI_L(x, y)$ is the value of the pixel at position (x,y) and $EI_{\mu L}(x, y)$ is the average effective information in the neighbourhood of this pixel and is defined as:

$$EI_{\mu L}(x, y) = 1/A \sum_{i=x-w_0}^{x+w_0} \sum_{j=y-h_0}^{y+h_0} EI_L(i, j) \tag{6}$$

$EI_{\mu L}(x, y)$ is the average L value of the sub-image whose center pixel is at position (x, y) with offsets w_0, h_0, and area A. These are computed from:

$$w_0 = min(x, W - x), h_0 = min(y, H - y), A = (2w_0 + 1)(2h_0 + 1) \tag{7}$$

where W and H represent the width and height of the image.

The sub-images obtained in (5) and (6) are called the maximum possible symmetric surround regions for a given pixel at the center. Consequently, the closer a pixel is to the edges, the narrower is its surround. The advantage of narrowing the bandwidth near the borders is to make the background less highlighted thus giving less prominence as described in [9]. In a similar manner we compute EI_{ssa} and EI_{ssb} for the 'a' and 'b' components of the LAB space. We combine these to obtain a single saliency map EI_c as follows:

$$EI_c(x, y) = \sqrt{EI_{ssL}^2(x, y) + EI_{ssa}^2(x, y) + EI_{ssb}^2(x, y)} \tag{8}$$

We then normalize this saliency map image in the range [0 255] as follows:

$$EI_{norm} = 255 * (EI_c - EI_{min})/EI_{max} - EI_{min} \tag{9}$$

where EI_{max} and EI_{min} denote the maximum and minimum values of EI_c. EI_{norm} is used to construct a contrast map of the image. This contrast map segregates the image into various components based on the information. Next we need to convert this map into a saliency cut.

Saliency Cut Computation. The contrast image EI_{norm} is a grayscale image possessing both local and global contrasts. Thus it basically segments out the salient and background regions. However it doesn't indicate which region is salient. To segregate the salient region from the background, we first quantize the contrast image into *black*, *white* and *gray* segments.

Next we need to segregate the background color from the possible salient colors. This is necessary because an object is considered salient if it is uniquely distinguishable from its surroundings. To segregate the background color, a one window-wide strip is taken along the boundary of the image and each pixel is marked black, white or gray. We count the number of pixels of each type and the maximum count helps us to declare the pixels as background. We use the labels of the maximum number of boundary windows as belonging to the background and the rest as belonging to salient levels. This results in a binary saliency cut image, S_{EI} for the input image.

2.2   Color Based Clustering

In addition to obtaining the information based features, we perform color clustering for the segmentation of the background and creation of saliency cuts. As in Sect. 2.1, we divide the image into nxn windows (with n = 10). The mean values for L, 'a' and 'b' (μ_{Lk}, μ_{ak}, μ_{bk}) are calculated for each window.

As mentioned above, we need to segregate the background portion to locate the possible salient portions. This is done by considering a one window-wide strip is taken along the boundary. The μ values so extracted form the feature vector for the boundary. This vector is input to the fuzzy c-means clustering which is set to return four clusters (C_1, C_2, C_3 & C_4). We compute four clusters corresponding to the four sides of an image, as we want to label a color as background color only if it occupies at least 25 % of all the boundary windows. Using these centers, we compute for each cluster, the difference between the average value of each window of the image and each dominant boundary color. The difference is taken as the sum of the distances between the average 'a', 'b', 'L' values of an window and the corresponding elements of cluster centres, such as C_1, as shown below:

$$D_{1k} = |\mu_{ak} - C_{1a}| + |\mu_{bk} - C_{1b}| + |\mu_{Lk} - C_{1L}| \tag{10}$$

This is repeated for other clusters leading to D_{2k}, D_{3k} and D_{4k}. Then we compute the minimum of the four distances:

$$D_k = min(D_{1k}, D_{2k}, D_{3k}, D_{4k}) \tag{11}$$

This D_k is the minimum distances of the kth block from all centers. From the values of the minimum distances, we compute D_{max} as

$$D_{max} = max(D_1, D_2, ..., D_k, ...) \tag{12}$$

A grade based on D_{max} is used to create a D_{map} saliency map by comparing each D_k with D_{max}. The closer a particular D_k is to D_{max}, the more salient D_k is considered to be. Saliency map values vary from white for most salient to black for least salient.

2.3   Combining Color Clustered and Information Based Images

Salient object segmentation is done by combining the color clustered and information set based features. For this, we take a threshold for the color clustered image and another threshold based on the background information for the salient part. For the color cluster, the threshold is a function of D_{max} which results in a binary saliency cut image (S_C) from the saliency map D_{map}. We combine this with the saliency cut (S_{EI}) obtained from the contrast images that are created using the effective information features.

A single saliency cut (S_{cut}) image is generated by combining two binary saliency cuts. S_{cut} consists of one or more multiple salient objects out of which the most salient object needs to be selected. For this purpose, we consider the size and location features. The connected component analysis is carried out on the final image to include the most salient component based on size and location. In this stage, if a particular object is too near the boundary, it will be less likely to be salient as compared to an object near the center. Also, it has been observed that either the largest or the second largest object is salient. However, if there are many components of notable size then all of them are considered as salient. The connected component analysis operates with the aim of removing the noisy components and aggregating the salient components. If there are more than one salient objects present in the saliency cut, the connected component analysis selects the most salient object based on size and location.

3   Analysis of Results

An empirical analysis is done on two popular saliency databases: MSRA [7] and Berkley-300 [13]. For the quantitative comparison, the precision, recall and f-measure rates of various models are utilized. For qualitative analysis, some visual comparisons are made available.

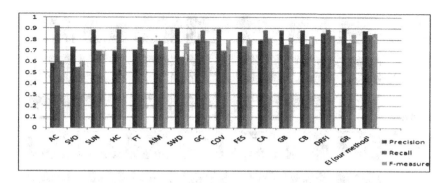

Fig. 1. Comparison with other approaches in terms of Precision, Recall and F-measure for MSR dataset

3.1 Evaluation on MSRA Dataset

In our study, we have randomly taken 5000 images from MSRA dataset for testing as in [7]. The dataset also provides separately the ground truth for each image. Once the salient object is identified by our approach from the given image, a bounding rectangle R_{our} is drawn. The overlap of this rectangle R_{our} is compared with the ground truth rectangle R_{gt} to compute the average precision, recall and F-Measure as defined in [7]. We have compared these values with those obtained by 15 other methods for which the saliency cuts (binary maps) are viewable in [14]. These are: AC [8], SVO [15], SUN [16], HC [4], FT [17], AIM [10], SWD [18], GC [19], COV [20], CA [3], FES [21], CB [22], GB [23], DRFI [24] and GR [25]. We compare the results by enclosing the salient objects in the smallest bounding rectangle. The binary map is then bound by a rectangle. Our method performs either close to or outperforms these state-of-the-art algorithms as can be seen in Fig. 1. Out of the three measures, F-measure is considered the best since its results are dependent on both precision and recall as shown in [7]. As depicted in Fig. 1, our results are best in terms of F-measure. Visual comparison with other results is shown in Fig. 2 where our results provide the most compact rectangle with the maximum intersection with the ground truth in most cases.

3.2 Evaluation on Berkley-300 Database

The Berkley-300 is a more challenging dataset that contains 300 images with more complex background or multiple objects of different sizes and positions. The groundtruth masks are provided in [13]. We compare our approach with 12 other approaches: FT [17], LC [26], HC [4], SR [27], RC [4], CA [3], SC [28], GS_GD, GS_SP [13], GMMS [29], CRF [7] and EQ [30] in terms of precision, recall and F-measure as shown in Fig. 3. For FT, HC, LC, RC, SR, SC, CA and GMMS we use the results given in [29]. For GS_GD and GS_SP, we use the results as given in [13]. For EQ, we use the saliency maps provided in [30]. As discussed

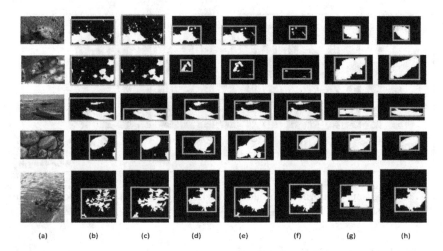

Fig. 2. Comparison with other approaches (a) Original (b) FT [17] (c) HC [4] (d) FES [21] (e) DRFI [24] (f) GR [25] (g) Our Result (h) Ground Truth for MSR dataset

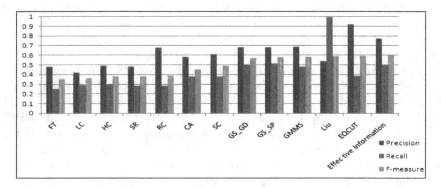

Fig. 3. Comparison with other approaches in terms of Precision, Recall and F-measure for Berkley dataset

in the previous section, F-measure is considered the best for comparison. Our method outperforms all methods in terms of F-measure as can be seen in Fig. 3.

4   Conclusions

A novel approach is presented for the detection of salient objects based on the information set and color features. Each of the LAB components in a window is converted into the effective information type in a two-step process by first converting it into entropy value and then modifying it into the effective information feature. Color components are combined using the symmetric surround to generate both the contrast maps and the subsequent saliency cuts. Similar saliency cuts are formed with the help of fuzzy color clustering. Both saliency cuts are combined and the salient object is selected using the connected component analysis. Extensive comparisons with the state-of-the-art methods in terms of precision, recall and F-Measure are made on a publicly available dataset to prove the effectiveness of this approach. The contributions of this paper include: (i) Use of information set concept in formulating the effective information features, (ii) Generation of contrast maps and the subsequent saliency cuts, including another saliency cuts based on fuzzy clustering, and (iii) Detection of a salient object from the combination of both types of saliency cuts using the connected component analysis.

References

1. Ma, Y.F., Zhang, H.J.: Contrast-based image attention analysis by using fuzzy growing. In: Proceedings of the Eleventh ACM International Conference on Multimedia, New York, NY, USA, pp. 374–381. ACM (2003)
2. Chen, L.Q., Xie, X., Fan, X., Ma, W.Y., Zhang, H.J., Zhou, H.Q.: A visual attention model for adapting images on small displays. Multimedia Syst. **9**, 353–364 (2003)
3. Goferman, S., Zelnik-Manor, L., Tal, A.: Context-aware saliency detection. IEEE Trans. Pattern Anal. Mach. Intell. **34**, 1915–1926 (2012)
4. Cheng, M.M., Mitra, N.J., Huang, X., Torr, P.H.S., Hu, S.M.: Global contrast based salient region detection. IEEE TPAMI **37**, 569–582 (2015)
5. Dhar, S., Ordonez, V., Berg, T.: High level describable attributes for predicting aesthetics and interestingness. In: 2011 IEEE Conference on CVPR, pp. 1657–1664 (2011)
6. Itti, L., Koch, C., Niebur, E.: A model of saliency-based visual attention for rapid scene analysis. IEEE Trans. Pattern Anal. Mach. Intell. **20**, 1254–1259 (1998)
7. Liu, T., Yuan, Z., Sun, J., Wang, J., Zheng, N., Tang, X., Shum, H.Y.: Learning to detect a salient object. IEEE Trans. Pattern Anal. Mach. Intell. **33**, 353–367 (2011)
8. Achanta, R., Estrada, F.J., Wils, P., Süsstrunk, S.: Salient region detection and segmentation. In: Gasteratos, A., Vincze, M., Tsotsos, J.K. (eds.) ICVS 2008. LNCS, vol. 5008, pp. 66–75. Springer, Heidelberg (2008)
9. Achanta, R., Susstrunk, S.: Saliency detection using maximum symmetric surround. In: 2010 17th IEEE International Conference on Image Processing (ICIP), pp. 2653–2656 (2010)
10. Bruce, N.D., Tsotsos, J.K.: Saliency, attention, and visual search: An information theoretic approach. J. Vis. **9**, 1–24 (2009)
11. Margolin, R., Tal, A., Zelnik-Manor, L.: What makes a patch distinct? In: 2013 IEEE Conference on Computer Vision and Pattern Recognition (CVPR), pp. 1139–1146 (2013)
12. Mamta, H.M.: Robust ear based authentication using local principal independent components. Expert Syst. Appl. **40**, 6478–6490 (2013)
13. Wei, Y., Wen, F., Zhu, W., Sun, J.: Geodesic saliency using background priors. In: Fitzgibbon, A., Lazebnik, S., Perona, P., Sato, Y., Schmid, C. (eds.) ECCV 2012, Part III. LNCS, vol. 7574, pp. 29–42. Springer, Heidelberg (2012)
14. Borji, A., Cheng, M.M., Jiang, H., Li, J.: Salient object detection: A benchmark. ArXiv e-prints (2015)

15. Chang, K.Y., Liu, T.L., Chen, H.T., Lai, S.H.: Fusing generic objectness and visual saliency for salient object detection. In: 2011 IEEE International Conference on Computer Vision (ICCV), pp. 914–921 (2011)
16. Zhang, L., Tong, M.H., Marks, T.K., Shan, H., Cottrell, G.W.: Sun: A bayesian framework for saliency using natural statistics. J. Vis. 8(7), 32 (2008)
17. Achanta, R., Hemami, S., Estrada, F., Ssstrunk, S.: Frequency-tuned salient region detection. In: IEEE International Conference on Computer Vision and Pattern Recognition (CVPR), pp. 1597–1604 (2009)
18. Duan, L., Wu, C., Miao, J., Qing, L., Fu, Y.: Visual saliency detection by spatially weighted dissimilarity. In: IEEE Conference on Computer Vision and Pattern Recognition (CVPR), pp. 473–480 (2011)
19. Cheng, M.M., Warrell, J., Lin, W.Y., Zheng, S., Vineet, V., Crook, N.: Efficient salient region detection with soft image abstraction. In: IEEE ICCV, pp. 1529–1536 (2013)
20. Erdem, E., Erdem, A.: Visual saliency estimation by nonlinearly integrating features using region covariances. J. Vis. 13, 1–20 (2013)
21. Rezazadegan Tavakoli, H., Rahtu, E., Heikkilä, J.: Fast and efficient saliency detection using sparse sampling and kernel density estimation. In: Heyden, A., Kahl, F. (eds.) SCIA 2011. LNCS, vol. 6688, pp. 666–675. Springer, Heidelberg (2011)
22. Jiang, H., Wang, J., Yuan, Z., Liu, T., Zheng, N., Li, S.: Automatic salient object segmentation based on context and shape prior. In: BMVC. vol. 6 (2011)
23. Harel, J., Koch, C., Perona, P.: Graph based visual saliency. In: NIPS, pp. 545–552 (2007)
24. Jiang, H., Wang, J., Yuan, Z., Wu, Y., Zheng, N., Li, S.: Salient object detection: A discriminative regional feature integration approach. In: 2013 IEEE Conference on (CVPR), pp. 2083–2090. IEEE (2013)
25. Yang, C., Zhang, L., Lu, H.: Graph-regularized saliency detection with convex-hull-based center prior. Sign. Process. Lett. IEEE 20, 637–640 (2013)
26. Zhai, Y., Shah, M.: Visual attention detection in video sequences using spatiotemporal cues. In: Proceedings of the 14th Annual ACM International Conference on Multimedia, pp. 815–824. ACM (2006)
27. Hou, X., Zhang, L.: Saliency detection: A spectral residual approach. In: CVPR 2007, pp. 1–8. IEEE (2007)
28. Fu, K., Gong, C., Yang, J., Zhou, Y.: Salient object detection via color contrast and color distribution. In: Lee, K.M., Matsushita, Y., Rehg, J.M., Hu, Z. (eds.) ACCV 2012, Part I. LNCS, vol. 7724, pp. 111–122. Springer, Heidelberg (2013)
29. Zhou, L., Fu, K., Li, Y., Qiao, Y., He, X., Yang, J.: Bayesian salient object detection based on saliency driven clustering. Sign. Process. Image Commun. 29, 434–447 (2014)
30. Aytekin, C., Kiranyaz, S., Gabbouj, M.: Automatic object segmentation by quantum cuts. In: Pattern Recognition (ICPR), pp. 112–117. IEEE (2014)

Edge Based Segmentation of Left and Right Ventricles Using Two Distance Regularized Level Sets

Yu Liu[1], Yue Zhao[1], Shuxu Guo[1], Shaoxiang Zhang[2], and Chunming Li[3]($\boxtimes$)

[1] College of Electronic Science and Engineering, Jilin University,
Changchun, People's Republic of China
[2] Institute of Digital Medicine, Third Military Medical University (TMMU),
Chongqing, People's Republic of China
[3] School of Electronic Engineering, University of Electronic Science and Technology
of China (UESTC), Chengdu, People's Republic of China
chunming.li@uestc.edu.cn

Abstract. In this paper, we present a new approach for segmentation of left and right ventricles from cardiac MR images. A two-level-set formulation is proposed which is the extension of distance regularized level set evolution (DRLSE) model in [1], with the 0-level set and k-level set representing the endocardium and epicardium, respectively. The extraction of endocardium and epicardium is obtained as a result of the interactive curve evolution of the 0 and k level sets derived from the proposed variational level set formulation. The initialization of the proposed two-level-set DRLSE model is generated by performing the original DRLSE from roughly located endocardium. Experimental results have demonstrated the effectiveness of the proposed two-level-set DRLSE model.

1 Introduction

Currently, cardiovascular disease is becoming one of the leading causes to death. Cine-magnetic resonance imaging (cine-MRI) is an important modality which contains accurate information for non-invasively diagnosing and treating cardiac disease. To quantify cardiac function through ventricle volumes, masses, and cavity ejection function (EF), segmentation of left (LV) and right (RV) ventricles from cine-MRI is crucial for physicians and radiologists [2–4]. Manual segmentation of LV and RV which requires delineation slice by slice is a time consuming and error-prone task. Although various automatic segmentation techniques have been proposed to segment the cardiac structure, the LV and RV segmentation is still an open problem because of the poor image contrast across the desired ventricle boundaries [5,6].

Active contour models and level set methods have been widely implemented to segment different biological structures from medical images [7–10]. Several desirable advantages exist for active contour models over other classical image segmentation methods, such as thresholding, edge detection and region grow.

© Springer International Publishing Switzerland 2015
G. Bebis et al. (Eds.): ISVC 2015, Part II, LNCS 9475, pp. 205–212, 2015.
DOI: 10.1007/978-3-319-27863-6_19

Firstly, active contour models have capable to achieve sub-pixel accuracy of object boundaries. Second, these models can be easily formulated in a principled energy minimization framework and facilitate incorporation of various prior knowledge, such as shape or intensity distribution, for robust image segmentation [11,12]. Third, active contour models can provide smooth and closed contours as segmentation outcomes, which are essential for segmenting most of the biological structures and can be readily used for further applications, such as shape analysis and recognition [13].

In order to segment the inner and outer contours of both ventricles from cine-MRI, we propose a two-level-set approach, which based on the strengths from DRLSE in [1]. In this new method, endocardial and epicardial contours are mathematically represented by two specified level contours of a level set function. Biventricular segmentation is expressed as an optimization problem of the level set function such that both level set contours best capture the biological structures of epicardium and endocardium.

This paper is further structured as follows: Sect. 2 describes the details of proposed algorithm, Sect. 3 proposes a two-step approach for segmentation of left and right ventricles and Sect. 4 presents the implementation details as well as segmentation results, which is followed by the concluding remarks in Sect. 5.

2   Two-Level-Set Approach

2.1   Anatomical Knowledge for Left and Right Ventricle Segmentation

First, we take account of the anatomy of the both ventricles in the formulation of proposed model. Endocardium is the innermost contour of the ventricle, which is a smooth membrane of endothelial cells that lines the cavities of the heart and the valves [14]. Myocardium is a thick layer of cardiac muscle which is responsible for the contraction and relaxation of the ventricles and atria, and this layer is composed almost completely of cardiomyocytes [3]. The outside of the myocardium is covered with a thin layer called the epicardium, which consists mostly of connective tissue and fat [3,14].

Based on several informative observations of the cardiac anatomy, the desired segmentation outputs should fulfill the following two criteria: firstly the endocardial and epicardial contours should be smooth contours and secondly the interval between endocardial and epicardial contours should vary smoothly. In particular, the first criterion will be introduced to smooth the epicardial and endocardial contours individually, while the second criterion will be introduced to provide an interaction between the two contours such that the distance between them is gradually varying.

2.2   Edge Based Segmentation of LV and RV Using Two Distance Regularized Level Sets

We formulate the segmentation of both ventricles as an problem of seeking an optimal level set function such that its 0-level and k-level contours best fit the epicardial and endocardial contours respectively. What is more, according to the anatomical properties of endocardium and epicardium, as discussed in Sect. 2.1, the two-level-set function should satisfy the following two properties: first, the 0-level and k-level contours are smooth; and second, the distance between the 0-level and k-level contours is smoothly changing. For the above considerations, we propose a variational framework with an energy functional in the following form:

$$\mathcal{F}(\phi) = \mathcal{L}(\phi) + \mathcal{A}(\phi) + \mathcal{R}_p(\phi) \tag{1}$$

where $\mathcal{L}(\phi)$ is an energy functional defined from edge-based image information, such that it is minimized when the 0-level and k-level contours of the function ϕ are on the endocardium and epicardium; $\mathcal{A}(\phi)$ is weighted area term and introduced to speed up the motion of the 0-level and k-level contour in the level set evolution process, which is necessary when the initial contour is placed far away from the desired object boundaries. Energy $\mathcal{R}_p(\phi)$ is the double-well potential defined in [1],

$$\mathcal{R}_p(\phi) = \mu \int p(|\nabla\phi|)d\mathbf{x} \tag{2}$$

In this paper, we let the level set function ϕ take positive values inside the 0-level contour C_0 and negative values outside C_0.

The energy $\mathcal{L}(\phi)$ is defined by,

$$\mathcal{L}(\phi) = \int g[\lambda_1\delta(\phi) + \lambda_2\delta(\phi - k)]\,|\nabla\phi|\,d\mathbf{x} \tag{3}$$

where g is an edge indicator function, which is defined as,

$$g \triangleq \frac{1}{1 + |\nabla G_\sigma * I|}. \tag{4}$$

The above defined energy $\mathcal{L}(\phi)$ computes the line integral of the function g along the 0-level and k-level contours. Obviously, this energy is minimized when the 0-level and k-level contours of the level set function ϕ are located on desired object boundaries, where the function g takes smaller values than other non-edge locations. The weighted area term is defined by,

$$\mathcal{A}(\phi) = \alpha_0 \int gH(-\phi)\mathbf{x} + \alpha_k \int gH(-\phi + k)d\mathbf{x} \tag{5}$$

The minimization of the energy $\mathcal{A}_g$ is achieved by shrinking and expanding the 0-level and k-level contours, depending on the sign of α_0 and α_k of the banded region between the 0-level and k-level contour when they arrive at object boundaries where take larger values.

2.3   Energy Minimization

With the energy terms including $\mathcal{L}(\phi)$, $\mathcal{A}(\phi)$ and $\mathcal{R}_p(\phi)$ defined above, we propose to minimize the following energy functional:

$$\mathcal{F}(\phi) = \mathcal{L}(\phi) + \mathcal{A}(\phi) + \mathcal{R}_p(\phi). \tag{6}$$

This energy functional can be minimized by alternately minimizing $\mathcal{F}$ with respect to each of its variables. The energy minimization process starts with an initialization of the level set function ϕ and the smooth function α. We minimise the function $\mathcal{F}$ with respect to ϕ applying gradient flow method and get,

$$
\begin{aligned}
\frac{\partial \phi}{\partial t} &= -\frac{\partial \mathcal{E}}{\partial \phi} \\
&= div\left(g\frac{\nabla\phi}{|\nabla\phi|}\right)[\lambda_1\delta(\phi) + \lambda_2\delta(\phi - k)] \\
&\quad + \partial g[\delta(\phi - k) - \delta(\phi)] + \mu div(d_p|\nabla\phi|\nabla\phi)
\end{aligned}
\tag{7}
$$

where d_p is the double-well potential defined in [1].

3   Application of Proposed Method to Segmentation of Left and Right Ventricles

In this section, we describe a two-step approach for segmentation of LV and RV. In the first step, we use the DRLSE to perform a preliminary segmentation of LV and RV to roughly locate the endocardial contours of the LV and RV. The level set function obtained in the first step is used as the initial level set function of the proposed level set evolution in the second step, with the 0-level and k-level contours representing the initial endocardial and epicardial contours, respectively. The final endocardial and epicardial contours of LV and RV is then obtained as the result of the level set evolution in the proposed model. The details of this two-step approach are described below.

3.1   Roughly Locate Endocardial Contours of LV and RV Using DRLSE

In the first step, we use distance regularized level set evolution (DRLSE) model to obtain a preliminary segmentation of left and right ventricles, which is then applied to define the initial level set function for the distance regularized two-level-set model described in Sect. 2. We put two square blocks inside LV and RV as the initialization of DRLSE. Then, the final zero level contour in DRLSE is evolved to capture the inner contours of both LV and RV.

3.2   Extraction of Both Endocardial and Epicardial Contours Using Proposed Model

The second step of our method aims to accurately capture both the endocardial contours as well as the epicardial contours of both LV and RV at the same time, based on the previously initialized results. With the above two-level-set representation of endocardial and epicardial contours, we implement the mathematical model discussed in Sect. 2 to accurately segment left and right ventricles from cine-MRI based on previously initialized contours from DRLSE method.

4   Experimental Results

In this section, we will demonstrate the implementation details as well as the segmentation results of the proposed two-step approach with the application to segment ventricles from cardiac cine-MRI data, where the data are obtained from MICCAI 2012 right ventricle segmentation challenge.

4.1   Parameters Selection

In first step, we set the parameters for the DRLSE model with $\Delta t = 1$, $\mu = 0.2$, and $\lambda = 10$. In the second step, we numerically solve the level set evolution equation of proposed model presented in Eq. 1 by following a standard finite difference scheme proposed in [1]. The details of the selected parameters for the proposed model are disclosed in the following. The time step Δt used in the approximation of temporal derivative is set to $\Delta t = 0.1$ in our implementation. For the datasets used in this chapter, we set the other parameters $\alpha_0 = -3$, $\alpha_k = -3$, $\rho = 3$, $\mu = 1$, $\omega = 0.5$, $\lambda_1 = 0.002$, $\lambda_2 = 0.05$, and $\nu_1 = 0.001 \times 255 \times 255$, $\nu_2 = 0.001 \times 255 \times 255$. The choice of the level $k = 140$.

4.2   Segmentation Results

Results for MICCAI 2012 Right Ventricle Segmentation Challenge. Our two-step approach has been tested on the dataset of MICCAI 2012 right ventricle segmentation challenge (http://www.litislab.eu/rvsc), and this two-step approach has already showed the promising capability over other existing segmentation algorithms.

Figure 1 shows the illustrative segmentation results of both left and right ventricles from a selected patient cine-MRI using proposed method after applying DRLSE initialization, where each of these figures corresponds to one of the middle steps within iteration (iterates from left to right and from top to bottom). It is easy to find out the output of the proposed segmentation model (as shown in the bottom right figure in Fig. 1) has capable to accurately capture both the endocardial and epicardial contours, even in presence of the intensity inhomogeneity and image noises. Figure 2 shows the results and ground truth of the subject named P15 in the training set. We choose two slices at end-diastole and end-systole and we can reach promising results through our proposed method.

Table 1 shows evaluation results for MICCAI 2012 right ventricle segmentation challenge of our method with others. We use Dice metric (DSC) and Hausdorff distance (HD) assess the RV segmentation results. DSC is a measure of the overlap of two regions and is defined by

$$DSC = \frac{2 \times |A \cap B|}{|A| + |B|} \tag{8}$$

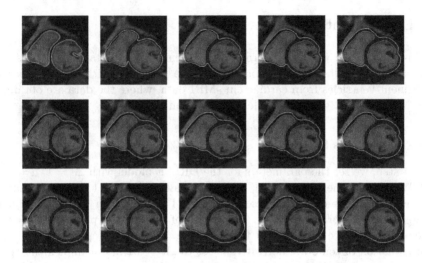

Fig. 1. Process of implementing Distance Regularized Two-Level-Set model to segment both LV and RV from a selected patient's cine-MRI. Note, the iteration starts from left to right and from top to bottom.

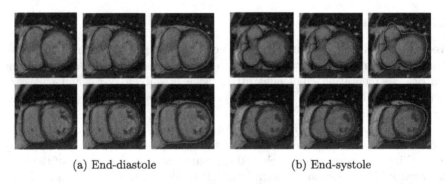

(a) End-diastole   (b) End-systole

Fig. 2. Results of our method (right column) and ground truth (middle column) for the images (left column) at end-diastole shown in (a) and end-systole in (b). Each row shows one of two slices in the images from one case.

where $| * |$ is the area of region $*$. Hausdorff distance provides a symmetric distance measure of the maximal discrepancy between two labeled contours, which is defined as,

$$HD(A, B) = max \left(\max_{a \in A}(\min_{b \in B} d(a, b)), \max_{b \in B}(\min_{a \in A} d(a, b)) \right) \quad (9)$$

where $d(a, b)$ donates Euclidean distance. From Table 1, our proposed method shows promising results comparing to the other three methods.

Table 1. Right ventricle segmentation results of different methods for 2012 RV MICCAI test1 data sets

		[15]		[16]		[17]		Our method	
		ED	ES	ED	ES	ED	ES	ED	ES
En	DM	0.83(0.17)	0.72(0.27)	0.86(0.11)	0.69(0.25)	0.88(0.11)	0.77(0.18)	0.90(0.15)	0.82(0.23)
	HD	9.77(7.88)	11.41(10.49)	7.70(3.74)	11.16(5.53)	7.69(6.03)	10.71(7.69)	7.51(5.47)	10.50(8.03)
Ep	DM	0.86(0.13)	0.77(0.23)	0.88(0.08)	0.77(0.17)	0.90(0.08)	0.82(0,13)	0.89(0.21)	0.83(1.57)
	HD	10.23(7.22)	11.81(9.46)	7.93(3.72)	11.72(5.44)	8.02(5.96)	11.52(7.70)	9.36(8.19)	12.58(9.03)

5   Conclusions

In this paper, we present the development of a new segmentation framework for left and right ventricles from cardiac MR short-axis images. This framework contains two key steps: firstly we apply the DRLSE method as initialization of endocardial contour and then we use the two-level-set model to accurately capture both endocardial and epicardial contours for LV and RV, simultaneously. Moreover, this proposed approach based on the strengths from the distance regularized level set evolution (DRLSE) method in [1]. Experimental results have demonstrated the effectiveness of this proposed two-step level set approach for segmenting cardiac left and right ventricles from cine-MRI.

References

1. Li, C., Xu, C., Gui, C., Fox, M.D.: Distance regularized level set evolution and its application to image segmentation. IEEE Trans. Image Process. **19**, 3243–3254 (2010)
2. Paragios, N.: A variational approach for the segmentation of the left ventricle in cardiac image analysis. Int. J. Comput. Vis. **50**, 345–362 (2002)
3. Petitjean, C., Dacher, J.N.: A review of segmentation methods in short axis cardiac MR images. Med. Image Anal. **15**, 169–184 (2011)
4. Kurkure, U., Pednekar, A., Muthupillai, R., Flamm, S.D., Kakadiaris, I.A.: Localization and segmentation of left ventricle in cardiac cine-MR images. IEEE Trans. Biomed. Eng. **56**, 1360–1370 (2009)
5. Qian, X., Lin, Y., Zhao, Y., Wang, J., Liu, J., Zhuang, X.: Segmentation of myocardium from cardiac mr images using a novel dynamic programming based segmentation method. Med. Phys. **42**, 1424–1435 (2015)
6. Frangi, A.F., Niessen, W.J., Viergever, M.A.: Three-dimensional modelling for functional analysis of cardiac images: a review. IEEE Trans. Med. Image **20**, 2–5 (2001)
7. Li, C., Kao, C., Gore, J.C., Ding, Z.: Minimization of region-scalable fitting energy for image segmentation. IEEE Trans. Image Process. **17**, 1940–1949 (2008)
8. Chan, T., Vese, L.: Active contours without edges. IEEE Trans. Image Process. **10**, 266–277 (2001)
9. Qian, X., Wang, J., Guo, S., Li, Q.: An active contour model for medical image segmentation with application to brain ct image. Med. Phys. **40**, 8 (2012)
10. Kass, M., Witkin, A., Terzopoulos, D.: Snakes: Active contour models. Int. J. Comput. Vis. **1**, 321–331 (1988)

11. Chen, Y., Tagare, H.D., Thiruvenkadam, S., Huang, F., Wilson, D., Gopinath, K.S., Richard Briggs, W., Geiser, E.A.: Using prior shapes in geometric active contours in a variational framework. IJCV **50**, 315–328 (2002)
12. Leventon, M., Grimson, W., Faugeras, O.: Statistical shape influence in geodesic active contours. In: 2000 Proceedings of the IEEE Conference on Computer Vision and Pattern Recognition, vol. 1, pp. 316–323 (2000)
13. Wu, J., Brigham, K.G., Simon, M.A., Brigham, J.C.: An implementation of independent component analysis for 3d statistical shape analysis. Biomed. Sign. Process. Control **13**, 345–356 (2014)
14. Jolly, M.: Automatic segmentation of the left ventricle in cardiac MR and CT images. Int. J. Comput. Vis. **70**, 151–163 (2006)
15. Zuluaga, M.A., Cardoso, M.J., Ourselin, S.: Automatic right ventricle segmentation using multi-label fusion in cardiac mri. In: Workshop on RV Segmentation Challenge in Cardiac MRI Medical Image Computing and Computer-Assisted Intervention (2012)
16. Bai, W., Shi, W., O'Regan, D., Tong, T., Wang, H., Jamil-Copley, S., Peters, N., Rueckert, D.: A probabilistic patch-based label fusion model for multi-atlas segmentation with registration refinement: Application to cardiac mr images. IEEE Trans. Med. Imaging **32**, 1302–1315 (2013)
17. Ringenberg, J., Deo, M., Devabhaktuni, V., Berenfeld, O., Boyers, P., Gold, J.: Fast, accurate, and fully automatic segmentation of the right ventricle in short-axis cardiac MRI. Comput. Med. Imaging Graph. **38**, 190–201 (2014)

Automatic Crater Detection Using Convex Grouping and Convolutional Neural Networks

Ebrahim Emami[1], George Bebis[1(✉)], Ara Nefian[2], and Terry Fong[2]

[1] Department of Computer Science and Engineering, University of Nevada, Reno, USA
ebrahim@nevada.unr.edu, bebis@cse.unr.edu
[2] Intelligent Robotics Group (IRG), NASA Ames Research Center, Mountain View, USA
{ara.nefian,terry.fong}@nasa.gov

Abstract. Craters are some the most important landmarks on the surface of many planets which can be used for autonomous safe landing and spacecraft and rover navigation. Manual detection of craters is laborious and impractical, and many approaches have been proposed in the field to automate this task. However, none of these methods have yet become a standard tool for crater detection due to the challenging nature of this problem. In this paper, we propose a new crater detection algorithm (CDA) which employs a multi-scale candidate region detection step based on convexity cues and candidate region verification based on machine learning. Using an extensive dataset, our method has achieved a 92 % detection rate with an 85 % precision rate.

1 Introduction

Craters are topographic features on planetary surfaces that result from impacts with meteoroids. They are found on all hard-surface bodies in the solar system, but are most abundant on planets like the Moon or Mars where they can accumulate due to slow surface erosion rates. Craters are important landmarks for autonomous spacecraft and rover navigation and control, which have become key technologies in deep space exploration [2]. Craters can be used for high-precision spacecraft landing missions, and accurate identification of potential hazards [3]. Crater surveys also contain important information about planetary surfaces; for example, crater counting can be used for establishing relative chronology of planetary surfaces.

Currently, all crater databases have been gathered manually via visual inspection of images. However, they are not comprehensive as they mostly contain large craters only. On the other hand, advances in gathering planetary data by space probes has resulted in high resolution images that can show smaller craters on planets like Mars, and Moon. Clearly, manual crater detection is not be appropriate for generating comprehensive catalogues of craters and this task can only be achieved by automating the process of crater surveying [1, 4].

There have been numerous efforts to develop CDAs using image processing and machine learning techniques. However, most of the previous approaches are not capable of achieving high performance in real world crater detection applications [1]. Variations in illumination and surface properties as well as variations in shape and size make automatic

© Springer International Publishing Switzerland 2015
G. Bebis et al. (Eds.): ISVC 2015, Part II, LNCS 9475, pp. 213–224, 2015.
DOI: 10.1007/978-3-319-27863-6_20

crater detection a very challenging task. Specifically, crater dimensions in an image might differ by orders of magnitude. Crater shapes may also vary depending on their interior morphologies (central peaks, peak rings, central pits, and wall terraces), level of degrada- tion, and degree of overlap with other craters [3, 4]. These challenges make it hard to design a robust and usable CDA which maintains high accuracy.

In this paper, we propose a new CDA to better deal with these issues. The proposed method consists of two main phases. In the first phase, candidate crater regions are extracted using convexity, an important perceptual organization cue. In the second phase, candidate regions are the classified as crater or non-crater regions using machine learning techniques. In contrast to other CDAs, the proposed method does not rely on strong assumptions about crater shape and properties. Instead of assuming strong circular or elliptical shapes or bright and shadow regions with specific shapes, sizes, and orientations, we only assume that crater regions have a nearly convex shape which is a much weaker assumption. Moreover, we allow crater boundaries to contain gaps which is often the case due to imperfect edge detection results. Using this multi-scale scheme based on convexity, we are able to detect almost all craters while rejecting many non- crater regions. The candidate regions are then verified using a Convolutional Neural Network (CNN).

The rest of the paper is organized as follows: Sect. 2 discussed related work in crater detection. Section 3 presents the proposed approach in detail. Section 4 presents our experimental results and comparisons. Finally, Sect. 5 presents our conclusions and directions for future research.

2   Background

There has been extensive research on crater detection over the past years. Kamarudin et al. [5] have reviewed several methods on crater detection; they claim that the most accepted method of crater detection is based on edge detection and the Hough Transform, while there exist other techniques based on detection of a bright to dark shading pattern inside the crater due to lighting orientation. Salamuniccar and Loncaric [6] have proposed a framework for evaluating crater detection algorithms, however, the evalua- tion is not from a machine vision point of view. We classify previous CDAs into two categories: unsupervised and supervised.

Unsupervised methods mainly employ basic image processing and pattern recogni- tion techniques such as thresholding, circle detection, and ellipse detection [7, 8]. Specifically, Troglio et al. [8] perform crater detection by extracting elliptical regions using watershed segmentation and the Generalized Hough transform. Smirnov performs crater detection by detecting shadow regions; he assumes very specific geometric prop- erties for craters and limits the geometry of shadow regions to three main shapes which are detected using thresholding, pixel clustering and circle fitting. In [9], Kim et al. propose a CDA based on edge detection and ellipse fitting followed by template matching to discriminate between crater and non-crater regions.

Supervised methods use machine learning techniques to learn how to distinguish between crater and non-crater regions. These methods rely on a large number of labeled data for training. Meng et al. [10] perform candidate crater region selection using the Kanade–Lucas–Tomasi (KLT) detector, while MatLSSVM is employed for verifying the candidate crater regions. They claim that their method detects 88 % of craters on their dataset which consists of 160 preprocessed image patches from Google Mars. In [11], Martins et al. use the popular Adaboost algorithm [12] for crater detection using 3216 Haar-like features.

Wetzler et al. [13] have employed several supervised machine learning approaches for small size crater detection. According to their experiments, Support Vector Machines (SVMs) outperform Feed-Forward Neural Networks, AdaBoost (with feed-forward neural networks as base learners) and Continuously Scalable Template Models (CSTM) for crater detection. The classifiers were trained on normalized size image blocks and applied on test images using a sliding window approach. To detect different size craters, image pyramids were utilized. In a similar approach, Palafox et al. [14] evaluated the performance of SVMs and CNNs for the detection of craters and volcanic rootless cones. Although quantitative results were not presented in that work, CNNs were reported to perform better than SVMs in classifying randomly extracted patches from HiRISE images.

In general, unsupervised approaches are fast and more appropriate for the detection of relatively large craters; however, their performance degrades when dealing with smaller craters or more challenging terrains. Therefore, these techniques cannot be used as a general purpose crater detection tool [10, 13]. Supervised methods, on the other hand, are more robust but usually slower and their performance depends on the quality and number of training data.

It should be mentioned that although the majority of methods in the literature employ image data for crater detection, other types of data, such as Digital Elevation Map (DEM) data, have also been used for crater detection [15, 16]. Our interest in this study is on using image data for crater detection.

3   Proposed Method

In this paper, we propose a supervised CDA which consists of two main phases: (i) multi-scale candidate crater region detection, and (ii) candidate crater region verification. In the first phase, we extract candidate crater regions by applying multi-scale edge detection and convex grouping. Candidate crater regions are then classified into crater and non-crater regions using machine learning techniques. A set of discriminative features that can accurately separate craters from non-craters should be chosen in this phase. For training, we use a representative set of crater and non-crater training examples. After verification, a post-processing step is applied to combine detections corresponding to the same crater regions.

3.1   Candidate Crater Region Extraction

The primary goal in this step is to detect all true crater regions in order to avoid searching the whole image and speed-up the verification step. Our method is based on a perceptual organization approach which is a bottom up process that clusters image features into higher level organizations, each likely to come from a single object. Many different cues have been proposed in the literature for extracting perceptually salient structures including continuity, parallelism, and proximity; here, we employ convexity [17]. It has been demonstrated that groups of edges forming convex polygons rarely occur at random and are very likely to have resulted from the same convex object. Our method consists of the following steps: (1) multi-scale edge detection, (2) extraction of convex groups, (3) combination of convex groups, and (4) expansion of candidate crater regions.

Multi-scale Edge Detection. The first step of our method is extracting crater edges using the Canny edge detector. However, using a single scale to extract crater edges would be insufficient since craters typically appear at different sizes in an image.

To deal with this issue, we perform edge detection at multiple scales by varying the scale parameter of the Canny edge detector. Figure 1 shows an example using three different scales. As it can be observed, larger craters are more prominent at higher scales while smaller craters are more prominent at lower scales.

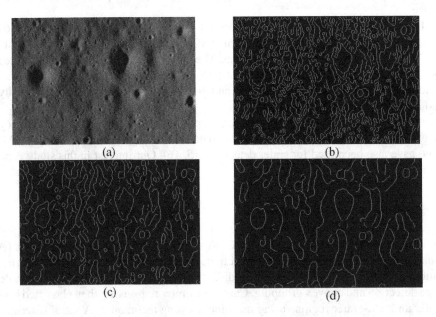

Fig. 1. Multi-scale Canny edge detection: (a) input image, (b) Canny edges at scale 3 using a threshold of 25 (c) Canny edges at scale 5 using a threshold of 25, and (d) Canny edges at scale 9 using a threshold of 25.

Extraction of Convex Groups. Many methods for crater detection in the literature assume that craters have a circular or elliptical shape. By observing many craters in our data set, however, we have concluded that this is not always the case. Other methods assume that craters consist of a pair of dark and bright regions with specific (relative) sizes, orientations and distances from each other. However, this assumption can be violated depending on the position of the sun.

In this paper, we make a weaker assumption about the shape of craters; specifically, we assume that the shape of craters is nearly convex and we use an efficient convex grouping algorithm [17] to extract candidate crater regions. This algorithm is simple, efficient, and robust to noise, occlusion (i.e., gaps), and clutter. Initially, the image is processed to find line segments by performing edge detection followed by line approximation. Here, we use the split-and-merge algorithm [18] which approximates curves with lines, such that the curve points are no more than a fixed threshold from the line segments. Figure 2 shows an example.

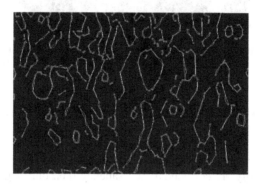

Fig. 2. Line fitting results on a sample edge image. The line segments are shown in green and their end points are shown in red (Color figure online).

Each line segment is characterized by its length, orientation and direction (by distinguishing one endpoint as the first endpoint). One way to define convexity is by considering the sum of absolute values of the angles turned as we traverse the line segments of the group. In the case of convex groups, the sum should be 360 degrees. Alternatively, a group of line segments is convex if for each -directed- line segment of the group, all other line segments are on the same side as its normal (it points to the right of the line segment when we traverse it from the first endpoint to the second).

Since the number of convex groups in an image can be very large, the algorithm considers only finding the most *salient* convex groups. A group is considered to be salient, if the sum of gap lengths between line segments is smaller than some fixed proportion of the sum of line lengths in the group. Let us assume that a convex group S_n contains line segments line segments $(l_1, l_2 \ldots l_n)$. We define L_i to be the length of line segment l_i and G_i to be the length of the gap between l_i and l_{i+1}, then the sum of line lengths $L_{1,n}$ and the sum of gap lengths $G_{1,n}$ are defined as follows:

$$L_{1,n} = \sum_{i=1}^{n} L_i \text{ and } G_{1,n} = \sum_{i=1}^{n} G_i \tag{1}$$

Then, S_n is called a salient convex group if:

$$\frac{L_{1,n}}{L_{1,n} + G_{1,n}} > k \tag{2}$$

where k is a fixed threshold. Figure 3 shows an example.

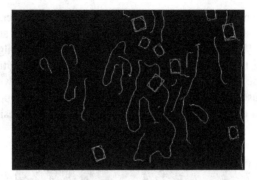

Fig. 3. Convex grouping results on a sample image using k=0.85. The detected convex groups are represented with their bounding boxes.

Initially, the algorithm considers every line segment as defining a new convex group. Each group is then grown by adding more segments to it using backtracking. To avoid considering every possible case, several constraints are imposed based on distance (i.e., only segments within a certain distance are considered), convexity (i.e., only segments that preserve convexity are considered) and saliency (i.e., only segments that do not degrade saliency are considered). To facilitate efficient implementation of these constraints, information about the line segments is precomputed and pre-stored in appropriate data structures. To increase system's robustness, several heuristics were used (e.g., the convexity criterion was relaxed to accept not perfectly convex groups). It should be mentioned that the salience criterion is rotation and scale independent. Finding the m most salient groups in an image containing n segments has $O(n^2 log(n) + mn)$ complexity [17].

Combination of Convex Groups. Since the same crater might be detected multiple times at different scales or even within the same scale (i.e., by using slightly different line segments each time), it is desirable to combine multiple detections to reduce verification cost but also to improve the extraction of candidate crater regions. We apply the following two steps in order to combine multiple detections: (1) cluster convex groups based on the overlap of their bounding boxes and (2) represent each cluster by the average of bounding boxes. Specifically, two convex groups are clustered together if their intersection to union ratio is above a threshold:

$$\frac{Area\,(b1 \cap b2)}{Area\,(b1 \cup b2)} > T \tag{3}$$

where b_1 and b_2 are their bounding boxes. A low threshold results in less candidate crater regions while a high threshold results in more accurate candidate crater regions.

Expansion of Candidate Crater Regions. The convex groups resulting from the previous step might not be perfectly localized around craters. This can affect verification performance especially since the training data comes from manually extracted craters which are typically well localized using a square window. To address this issue, candidate regions are expanded to become square in size (i.e., by making the shorter side of tits bounding box equal to its longer side). In particular, each region is expanded in three different ways: (i) from its left side, (ii) from its right side, and (iii) both from its left and right sides; Fig. 4 shows an example. It should be mentioned that although this step increases the number of candidate crater regions, our experimental results show that expanding the candidate regions improves accuracy considerably.

Fig. 4. Expanding a candidate region (green box) in three different ways (red boxes) (Color figure online).

3.2 Candidate Crater Region Verification

Once the candidate crater regions have been detected, they need to be verified in order to reject non-crater regions. We have opted for a machine learning approach since crater appearance might vary a lot. We have experimented with different features and classifiers including raw pixels, Haar features, Histograms of Oriented Gradient (HOG) features, SVMs, and CNNs. The best performance was obtained using raw features and CNNs which has been often the case in many classification applications. Next, we provide a brief overview CNNs.

Classification using CNNs. Deep learning systems have recently achieved state-of-the-art performance on many classification tasks. CNNs are among the most prominent deep learning techniques [19]. CNNs are feedforward neural networks with a special architecture inspired from the human visual system. They consist of alternating convolution and sub-sampling/pooling layers and work directly on 2D data (maps). The convolution layers compose feature maps by convolving kernels over feature maps in layers below them while the sub-sampling layers, down-sample the feature maps by a constant factor. The activation of a single map j in convolution layer l is given by:

$$a_j^l = f\left(b_j^l + \sum_{i \in M_j^l} a_i^{l-1} * k_{ij}^l\right) \qquad (4)$$

where f is a non-linear function (e.g., *tanh*), and b is a scalar bias. M_j^l is a vector of indices of feature maps in layer *l-1* which feature map j in layer l should sum over, " * " is the 2D convolution operator and k is the kernel used on feature map i in layer *l-1*. For a single feature map j in sub-sampling layer l:

$$a_j^l = down\left(a_j^{l-1}, N^l\right) \tag{5}$$

where *down* means down-sampling by a factor N.

To discriminate between C classes, a fully connected output layer with C neurons is added. The output layer takes as input the concatenated feature maps of the layer below it, denoted by the feature vector fv:

$$o = f\left(b^o + W^o fv\right) \tag{6}$$

where b^o is a bias vector and W^o is a weight matrix which can be determined using the back-propagation learning algorithm.

Combination of Verified Regions. As mentioned in Sect. 3.1 we combine candidate crater regions by thresholding the ratio of their area of overlap over the union of their areas. Using a fairly high threshold in that step but also adding extra regions by expanding the candidate regions allow us to improve verification performance, however, we might end up with verifying the same crater multiple times. To eliminate multiple detections, we apply the same methodology described in Sect. 3.1 to the verified regions, however, using a lower threshold this time.

4   Experimental Results and Comparisons

4.1   Data Set

Our data set consists of 448 images, each having a size of 600×400, obtained from the Lunar Reconnaissance Orbiter (LRO) [21]. Craters with a size between 20×20 and 200×200 have been partially labeled by NASA scientists in this dataset. We have used 428 for training and 20 images for testing. A total of 1830 craters (i.e., ground truth) exist in the training images. To increase the variability of crater appearance in the training set, we generate more training samples by randomly shifting the original ones. Specifically, we generate 3 samples for each ground truth crater by slightly changing its position and size. The new samples are still well localized and have more 90 % overlap with the ground truth. Combining these samples with the original ground truth craters make up our training set of 7320 samples. All training samples are then normalized to size 24×24.

It should be mentioned that the original partially labeled images are not suitable for testing and fully labeled images are needed for this purpose. Therefore, we have manually labeled all craters larger than 20×20 in the 20 test images. We use 7320 non-crater training examples which are chosen randomly; to reduce the number of false positives, we use bootstrapping [20] to augment the non-crater training samples.

4.2   Performance Evaluation Measures

Standard recall and precision rates are used to evaluate the performance of our CDA. These measures are defined as follows:

$$Recall = \frac{TP}{TP + FN}, \quad Precison = \frac{TP}{TP + FN} \tag{7}$$

where TP, FN, and FP are the number of true positives, false negatives, and false positives respectively. A verified region is a TP if it has more than 40 % overlap with a ground truth crater; otherwise, it is a false positive. The overlap between a candidate region and a ground truth crater is calculated using Eq. 3. It should be mentioned that our algorithm is designed to detect craters bigger than 20 × 20, but it is common that smaller craters are also detected. These craters are not considered as true or false detections in our performance evaluation.

4.3 Performance Analysis of Candidate Crater Region Detection

To evaluate the performance of the proposed candidate region detection method discussed in Sect. 3.1, we have applied it on all 448 images which include 2480 labeled craters.

Table 1. Statistical performance analysis of the proposed candidate crater region detection.

Total Number of ground truth craters	2480
Number of detected ground truth craters	2464 (99.4 %)
Average number of detections per ground truth crater	18.22 (std: 12.27)
Average overlap between true detections and the coresponding ground truth crater	52.70 (std: 11.70)
Average overlap between the best candidate crater regions and the corresponding ground truth crater	75.70 (std: 10.69)
Average number of candidate crater regions per image	7889 (std: 1675)

For edge detection, we used the Canny edge detector at scales 3, 5, 7, and 9. A low threshold of 25 was used to keep most of the detected edges (the high threshold was twice the low threshold). Convex grouping was then applied using a gap tolerance parameter k = 0.51. As it can be inferred from our parameter selection, our main goal is detecting all true craters. To combine the detected convex groups, we used a 70 % overlap threshold. While this threshold allows for combining many convex groups, it still allows multiple detections of the same crater region which lead to better verification performance as discussed in Sect. 3.1.

Table 1 shows the performance of our candidate crater region detection step along with some useful statistics; as it can be seen, we can detect almost all ground truth craters (99.4 %). On average, 18 candidate crater regions are detected for each ground truth crater. The detected regions corresponding to a ground truth crater have more than 50 % average overlap with it which is higher than our desired 40 % overlap. More interestingly, the best candidate regions (i.e., the regions with highest overlap) have more 75 % average overlap with their ground truth craters.

4.4 Performance Analysis of Candidate Crater Region Verification

The performance of the complete crater detection algorithm has been evaluated on the 20 test images. There is a total of 251 ground truth craters in these images which were all detected in the candidate crater region detection step. These regions along with other detections were passed to the verification step. We have performed several different experiments using the CNN and SVM classifiers. The CNN classifier is trained using raw pixel intensities since it extracts its own features. The SVM classifiers was trained using raw pixel intensities, Haar features, and HOG features. Table 2 shows our experimental results without using bootstrapping.

Table 2. Experimental results using different clssifiers and features witout bootstrapping.

Type of classifier	Recall (%)	Precision (%)
SVM using raw pixel intensities	84	24.11
SVM using Haar features	82.21	27.9
SVM using HoG features	93.67	27.31
CNNs using raw pixel intensities	94.46	58.9

As it can be observed from Table 2, the CNN classifier outperforms the SVM classifier both in terms of recall and precision. However, both classifiers have low precision which is mostly due to the lack of challenging non-crater samples in the training set. We have tried to improve the quality of the training set using bootstrapping. Table 3 shows our verification results for the CNN classifier using bootstrapping. By performing two iterations of bootstrapping, we have added around 3000 false positive samples to the training set. This has increased the precision of the CNN classifier from 58.29 % to 85.66 % while its recall rate has slightly dropped from 94.46 % to 92.09 %. Figure 5 shows the verified crater regions for a sample test image.

Table 3. CNN classifier`s performance improvement on test set using bootstapping

Bootstrapping round	Number of samples added to the data set	Recall	Precision
#0	–	94.46	58.29
#1	1920	91.30	74.51
#2	1140	92.09	85.66

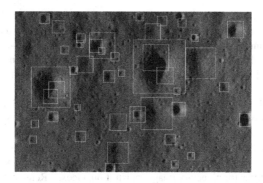

Fig. 5. Verified regions (blue boxes) for a sample test image (Color figure online).

It should be mentioned that most of the false detections are regions which look very similar to eroded craters in our training set. Since it is not clear whether these regions are true craters, we have considered them as false positives. Many of the false negatives are also craters which have very low contrast, lack of edges, and overlap with other surface features. Since these properties are not abundant in our training set, they could not be learnt effectively.

5 Conclusions

In this paper, we proposed CDA based on perceptual organization and machine learning. Using a multi-scale candidate crater region detection method, we were able to include almost all ground truth crater regions in our candidate list. Using a CNN classifier, we were able to verify 92 % of ground truth craters with an 85 % precision rate. The proposed approach can be enhanced in several ways. First of all, using a more informative training set with more diverse crater and non-crater samples would improve performance. Second, combining the responses of several classifiers trained on different features is also expected to improve verification performance. Finally, fusing crater detection results from images and DEMs is expected to improve overall performance.

Acknowledgements. This material is based upon work supported by NASA EPSCoR under cooperative agreement No. NNX11AM09A.

References

1. Bandeiraa, L., Ding, W., Tomasz, F.: Detection of sub-kilometer craters in high resolution planetary images using shape and texture features. Adv. Space Res. **49**(1), 64–74 (2012)
2. Yu, Z., Zhu, S., Cui, P.: Sequence detection of planetary surface craters from DEM data. In: World Congress on Intelligent Control and Automation (2012)
3. Maoyin, A., Pan, W.: Crater Detection algorithm with part PHOG features for safe landing. In: International Conference on Systems and Informatics, pp. 103–106 (2012)

4. Salamunićcara, G., Lončaricb, S., Mazarico, E.: LU60645GT and MA132843GT catalogues of lunar and martian impact craters developed using a crater shape-based interpolation crater detection algorithm for topography data. Planet. Space Sci. **60**(1), 236–247 (2012)
5. Kamarudin, N., Ghani, N., Mustapha, M., Ismail, A., Daud, N.: An overview of crater analyses, tests and various methods of crater detection algorithm. Front. Environ. Eng. **1**(1), 1–7 (2012)
6. Salamunićcara, G., Lončarić, S.: Open framework for objective evaluation of crater detection algorithms with first test-field subsystem based on MOLA data. Adv. Space Res. **42**(1), 6–19 (2008)
7. Smirnov, A.: Exploratory Study of Automated Crater Detection (2012)
8. Troglio, G., Le Moigne, J., Benediktsson, A., Moser, G., Serpico, S.: Automatic extraction of ellipsoidal features for planetary image registration. Geosci. Remote Sens. Lett. **9**(1), 95–99 (2012)
9. Kim, J., Muller, J.: Impact Crater Detection on Optical Images and DEMS, International Society for Photogrammetry and Remote Sensing, Working Group IV/9: Extraterrestrial Mapping Workshop, Advances in Planetary Mapping (2003)
10. Ding, M., Caob, Y., Wub, Q.: Novel approach of crater detection by crater candidate region selection and matrix-pattern-oriented least squares support vector machine. Chin. J. Aeronaut. **26**(2), 385–389 (2013)
11. Martins, R., Pina, P., Marques, J., Silveira, M., Silveira, M.: Crater detection by a boosting approach. Geosci. Remote Sens. Lett. **6**(1), 127–131 (2009)
12. Viola, P., Jones, M.: Robust real-time face detection. Int. J. Comput. Vis. **57**(2), 137–154 (2004)
13. Radu, V.: Application. In: Radu, V. (ed.) Stochastic Modeling of Thermal Fatigue Crack Growth. ACM, vol. 1, pp. 178–184. Springer, Heidelberg (2015)
14. Palafox, L., Alvarez, A., Hamilton, C.: Automated Detection of impact craters and volcanic rootless cones in mars satellite imagery using convolutional neural networks and support vector machines. In: 46th Lunar and Planetary Science Conference (2015)
15. Salamuniccar, G., Loncaric, S.: Method for crater detection from martian digital topography data using gradient value/orientation, morphometry, vote analysis, slip tuning, and calibration. IEEE Trans. Geosci. Remote Sens. **48**(5), 2317–2329 (2010)
16. Xie, Y., Tang, G., Yan, S., Hui, L.: Crater detection using the morphological characteristics of Chang'E-1 digital elevation models. Geosci. Remote Sens. Lett. IEEE **10**(4), 885–889 (2013)
17. Jacobs, D.: Robust and efficient detection of convex groups. IEEE Trans. Patern Anal. Mach. Intell. **18**(1), 23–37 (1996)
18. Pavlidis, T., Horowitz, S.: Segmentation of plane curves. IEEE Trans. Comput. **C-23**(8), 860–870 (1974)
19. Bengio, Y., Courville, A., Vincent, P.: Representation learning: A review and new perspectives. IEEE Trans. Pattern Anal. Mach. Intell. **35**(8), 1798–1828 (2013)
20. Sung, K., Poggio, T.: Example-based learning for view-based human face detection. IEEE Trans. Pattern Anal. Mach. Intell. **20**(1), 39–51 (1998)
21. http://www.nasa.gov/mission_pages/LRO/main/index.html

ST: Biometrics

Segmentation of Saimaa Ringed Seals for Identification Purposes

Artem Zhelezniakov[1,3], Tuomas Eerola[1]([✉]), Meeri Koivuniemi[2],
Miina Auttila[2,4], Riikka Levänen[2], Marja Niemi[2], Mervi Kunnasranta[2],
and Heikki Kälviäinen[1,5]

[1] Machine Vision and Pattern Recognition Laboratory,
School of Engineering Science, Lappeenranta University of Technology,
Lappeenranta, Finland
a.zheleznyakof@yandex.ru, {tuomas.eerola,heikki.kalviainen}@lut.fi
[2] Department of Biology, University of Eastern Finland, Joensuu, Finland
{meeri.koivuniemi,riikka.levanen,marja.niemi,mervi.kunnasranta}@uef.fi
[3] Department of Computer Technologies and Control Systems,
ITMO University, Saint Petersburg, Russia
[4] Parks and Wildlife Finland, Metsähallitus, Savonlinna, Finland
Miina.Auttila@metsa.fi
[5] School of Information Technology, Monash University Malaysia,
Bandar Sunway, Selangor, Malaysia
heikki.kalviainen@monash.edu

Abstract. Wildlife photo-identification is a commonly used technique
to identify and track individuals of wild animal populations over time.
It has various applications in behavior and population demography studies. Nowadays, mostly due to large and labor-intensive image data sets,
automated photo-identification is an emerging research topic. In this
paper, the first steps towards automatic individual identification of the
critically endangered Saimaa ringed seal (*Phoca hispida saimensis*) are
taken. Ringed seals have a distinctive permanent pelage pattern that is
unique to each individual making the image-based identification possible.
We propose a superpixel classification based method for the segmentation
of ringed seal in images to eliminate the background and to simplify the
identification. The proposed segmentation method is shown to achieve a
high segmentation accuracy with challenging image data. Furthermore,
we show that using the obtained segmented images promising identification results can be obtained even with a simple texture feature based
approach. The proposed method uses general texture classification techniques and can be applied also to other animal species with a unique fur
or skin pattern.

1 Introduction

Wildlife photo-identification (Photo-ID) provides a tool to study and monitor
animal populations over time based on captured images of individuals. It has
various applications in studying key aspects of the populations such as survival,

© Springer International Publishing Switzerland 2015
G. Bebis et al. (Eds.): ISVC 2015, Part II, LNCS 9475, pp. 227–236, 2015.
DOI: 10.1007/978-3-319-27863-6_21

dispersal, site fidelity, reproduction, health, population size and density. Due to its non-invasive nature the photo-ID is a great alternative to more destructive techniques, such as tagging that requires catching the animal and may cause stress to the animal, as well as, change its behavior or increase mortality. The identification of individuals is based on distinctive permanent characteristics, such as fur patterns, pigmentations, scars, or shape. Traditionally, the identification has been performed manually by researchers. However, due to the rapid increase in the amount of image data, there is a demand for automated methods. Computer vision techniques provide an attractive tool to replace the laborious and time-consuming manual work.

This work focuses on the Saimaa ringed seal (*Phoca hispida saimensis*) which is a critically endangered subspecies of ringed seal only found in Lake Saimaa, Finland. At present, around 300 seals inhabit the lake, and some 60 pups are born annually. This small and fragmented population is threatened by various anthropogenic factors, especially by-catch and climate change [1]. The long-term and accurate assessment of the population is needed for conservation purposes and camera trapping has recently launched as a new monitoring tool for the Saimaa ringed seal [2,3]. Camera trapping is especially effective method for Saimaa ringed seal monitoring due to high site fidelity of the seals. The pelage of the Saimaa ringed seal has a distinctive patterning of dark spots surrounded by light gray rings (see Fig. 1). These patterns are unique to each seal enabling the identification of individuals over their whole lifetime. Photo-ID data of the Saimaa ringed seal is mostly collected by game cameras during the molting season in spring. The camera trapping produces large number of seal images, which have been identified manually until these days.

Fig. 1. Saimaa ringed seal

In this work, a novel supervised segmentation method for the ringed seal is proposed. The method starts with unsupervised segmentation to produce a set of superpixels and proceeds to the superpixel classification to obtain two segments: the seal and background. Furthermore, a simple identification method exploiting texture features is proposed. The methods are evaluated with challenging image data consisting wide range of different images, such as images captured using camera traps and various camera types. The obtained results indicate a high segmentation success rate. Also, the preliminary identification results can be considered promising.

2   Related Work

Several approaches for automatic image-based animal identification can be found in the literature. Methods have been developed, for example, for polar bears [4], cattle [5], newts [6], giraffes [7], salamanders [8], snakes [9], insects [10], and marine mammals [11]. All of these methods use image processing and pattern recognition techniques to identify individuals. Most of the studies consider the identification of a certain animal species or species groups. For example, in [7], the effectiveness of wild-ID [7] software in identifying individual Thornicroft's giraffes from a dataset of 552 images was studied. The approach uses a Scale Invariant Feature Transform (SIFT) algorithm [12] to extract and match distinctive image features regardless of scale and orientation.

In [6], the suitability of biometric techniques for the identification of the great crested newt was investigated. Distinctive belly patterns were used to compare images of newts with an image database. Two different methods were used for the comparison: (1) the correlation coefficient of the gray-scale pixel intensities, and (2) the Hamming distance between the binary image segments.

Most of the current methods are developed for one species only and are not generalizable to other animals. However, there have been also research efforts towards creating an unified approach applicable for identification purposes for several animal species. For example, in [13], HotSpotter method to identify individual animals in a labeled database was presented. This algorithm is not species specific and has been applied to Grevy's and plains zebras, giraffes, leopards, and lionfish. HotSpotter uses viewpoint invariant descriptors and a scoring mechanism that emphasizes the most distinctiveness keypoints and descriptors. In [14], a species recognition algorithm based on sparse coding spatial pyramid matching (ScSPM) was proposed. It was shown that the proposed object recognition techniques can be successfully used to identify animals on sequences of images captured using camera traps in nature.

3   Methodology

The proposed method for the identification of Saimaa ringed seals has two main parts: the segmentation and identification (Fig. 2). First, the segmentation is applied to detect the seal and to eliminate the background that could complicate the identification process. Saimaa ringed seals tend to use same sites or areas inter-annually for molting and haul out, and the images are captured using static camera traps. herefore, the same seal is often captured with the same background increasing the risk that a supervised identification algorithm learns to "identify" the background instead of the actual seal if the full image or bounding box around the seal is used. This may further leads to a system that is not able to identify the seal in a new environment making the separation of the animal from the background (segmentation) important. After the segmentation, the seal is identified based on texture features computed from the pelage pattern.

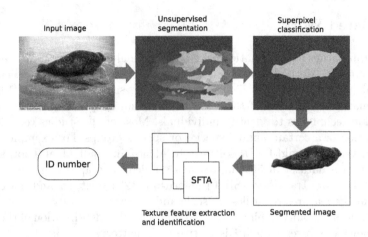

Fig. 2. The proposed method.

3.1   Segmentation

Automatic segmentation of animals is often difficult due to the camouflage colors of animals, i.e. the coloration and patterns are similar to the visual background of the animal. This makes it difficult to define a single criterion to distinguish the animal from the background. The proposed method to segment ringed seals start with unsupervised segmentation step to divided image into segments or superpixels. After this, the superpixels are classified into two different classes: seal or background. Similar approach have been used, for example, to segment the optical disc in retinal images [15]. Ideally, superpixels should be completely inside or outside the object, i.e. there should not be such superpixels that contain both seal and background pixels. Furthermore, superpixels should be large enough to make the classification possible. Generally, the larger the superpixels are the better as long as they fulfill the first criterion.

In this work, the method proposed in [16,17] is used for unsupervised segmentation. The method has shown to produce the state-of-the-art performance in the Berkeley Segmentation Dataset [18]. It combines Globalized Probability of Boundary (gPb) detector, Oriented Watershed Transform (OWT), and Ultrametric Contour Map (UCM) to produce a weighted contour image that can be thresholded to produce superpixels. The size of the superpixels can be controlled by varying the threshold value. By selecting a proper value, both criteria mentioned above can be fulfilled.

After the unsupervised segmentation step, texture features are computed from the superpixels and each superpixel is classified into the seal or background class in a supervised manner. In this work, blur invariant Local Phase Quantization (LPQ) features [19] and support vector machine (SVM) classifier were used for superpixel classification. The LPQ descriptor uses locally computed phase information and decorrelated low-frequency coefficients to create a code word histogram that can be used for the texture classification. Since the

low-frequency phase components are ideally blur invariant, and the phase information is invariant to uniform illumination changes, the LPQ texture descriptor
is a good choice for processing camera trap images which often are out of focus
and suffer from large illumination changes between day and night time. After
the superpixel classification step, all the superpixels that are classified to the
seal class are combined to form the seal segment for identification purposes.

3.2   Identification

The identification of Saimaa ringed seals is based on the unique pelage pattern of the individuals. The ring pattern forms a texture making the texture
classification techniques an apparent choice for the identification.

In this work, the identification of seal is performed using segmentation-based
fractal texture analysis (SFTA) features [20] computed from the seal segment.
SFTA features are calculated by decomposing the segmented image into a set
of binary images and by computing fractal dimension from the regions of the
each binary image. The identification is formulated as a classification problem
and solved using a naive Bayesian classifier. Given the vector of SFTA features
$\mathbf{x}$, posterior probabilities

$$p(C_k|\mathbf{x}) = \frac{p(C_k)p(\mathbf{x}|C_k)}{p(\mathbf{x})} \qquad (1)$$

are computed for each class C_k representing earlier identified individual in a
database. Class priors $p(C_k)$ are calculated by assuming equiprobable classes.
Posterior probabilities are used to rank classes, i.e., individual animals in the
database, from most likely to less likely. Finally, a set of most likely matches can
be presented to an expert for the final decision making.

4   Experiments

4.1   Data

To evaluate the method, a unique photo-ID database of Saimaa ringed seal
images collected by University of Eastern Finland was used. The database contained total of 785 images of 131 individual seals. Most of the images contains
one individual Saimaa ringed seal, and only few images contains two or more
individuals. Example images from the database are shown in Fig. 3.

The database contained images of individual seals from the right and left sides
because both flanks of the ringed seal have different pelage patterns. In addition,
the database contained images from belly and back sides of the seals. However,
most of the individual seals in the database had only one or few images making
the training of the identification method practically impossible. Therefore, the
seals with less than 5 images were omitted from the experiments. Also, images
with multiple seals or too low quality for identification purposes were screened
out from the data set. The final data set contained 40 ringed seal individuals
and total of 363 images with reasonable quality.

Fig. 3. Examples from the Saimaa ringed seals database.

The seals were identified from the images by experts to form the ground truth for identification. The segmentation ground truth was constructed by manually drawing the contour of the animals.

4.2   Segmentation

The gPb-OWT-UCM segmentation algorithm [17] selected for the unsupervised segmentation allows to choose the threshold value that effects on the size of the superpixels obtained as an output. The success of the ringed seal segmentation depends highly on the selected threshold value. To select the value, two criteria should be considered: (1) superpixels that contains both seal and background pixels should be minimized, and (2) the size of the superpixels should be as large as possible without violating the first criteria to make the superpixels classification as robust as possible.

The evaluation of the UCM thresholds was carried out using 121 images. A superpixel was considered as bad superpixel if the percentage of background pixels inside the superpixel was more than 0.1 and less than 0.9. The number of bad superpixels was computed for each image using the ground truth information. The results of the experiment are presented in Fig. 4. Based on the results a threshold value of 0.3 was selected for the further experiments.

In order to train and test the superpixel classification methods, the unsupervised segmentation was performed to all images, and resulting superpixels were manually labeled into two classes: seal and background. The images were divided into training and test set randomly. The training set contained 343 images and the test set 20 images. LPQ features were compared to SFTA and Local Binary Pattern Histogram Fourier features (LBP-HF) [21]. SVM classifier was compared to Naive Bayes and k-NN ($k = 9$) classifiers. The segmentation accuracy was measured as percentage of images that were successfully segmented. Image was considered as successfully segmented if the percentage of pixels that were correctly classified was higher than 95 % (Table 1).

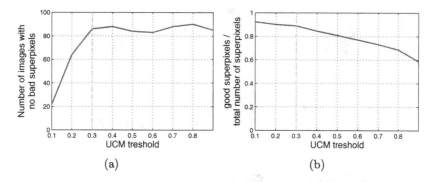

(a)   (b)

Fig. 4. The effect of UCM threshold: (a) Number of images with no bad superpixels; (b) Portion of good superpixels over all images.

Table 1. Segmentation accuracy with different texture features and classifiers.

Features/classifier	k-NN ($k = 9$)	Naive Bayes	SVM
SFTA	0.24	0.52	0.44
LBP-HF	0.32	0.00	0.28
LPQ	0.60	0.00	0.81

The results shows that the LPQ texture features and SVM classifier out-perform the other feature-classifier combinations. Examples of the segmentation results with the LPQ features and the SVM classifier are shown in Fig. 5.

4.3   Identification

The identification was studied using a set of 40 seals. One image of each seal was randomly selected to test set and rest of the images (323) were used as training set. The identification algorithms ranked the seals in the training set from most likely to least like match to the seal in each test image. The performance of different texture features and ranking methods for identification were measured using Cumulative Match Score (CMS) histogram commonly used in the face recognition research [22]. It measures how well the identification system ranks the identities in the database with respect to input image. The Nth bin in the CMS histograms tells the percentage of test images where the correct individual seal was in the set of N best matches proposed by the identification algorithm.

The proposed SFTA texture features for identification were compared with the LBP-HF and LPQ features. For each set of features, the raw feature set was compared with sets of features reduced using Principal Component Analysis (PCA), and the feature set producing the best results was used. The posterior probability provided by the Bayesian classifier was compared to class scores provided by k-NN and SVM classifiers. The results are shown in Fig. 6.

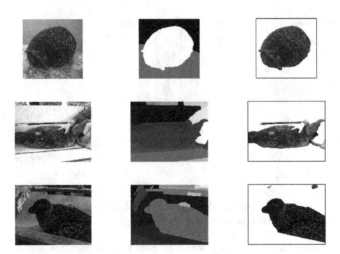

Fig. 5. Segmentation results (from the left to the right): the input image, unsupervised segmentation (superpixels), the segmentation result.

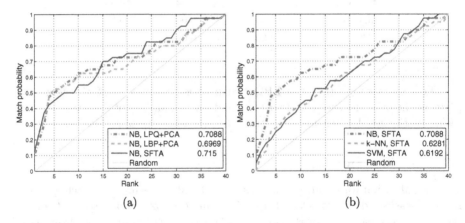

Fig. 6. Cumulative match score histograms (CMSH) with the areas under CMSH curve for identification methods: (a) Comparison of feature extractors with naive Bayesian classifier; (b) Comparison of classifiers with SFTA features.

The SFTA features have the highest area under CMS histogram value. However, there are no significant differences in the performance between different texture features. Posterior probabilities provided by Naive Bayesian classifier, on the other hand, outperforms all the other ranking methods. The correct seal was in the set of 15 best matches in 70 % of the cases. Although, the identification performance is not yet good enough for most practical applications, the fact that the results were obtained using rather simple texture classification methods with no preprocessing besides the segmentation suggests that the automatic identification when further developed has potential to become a useful tool for studying the Saimaa ringed seal.

5   Conclusions

In this paper, a segmentation method for Saimaa ringed seals using unsupervised segmentation and texture based superpixel classification was proposed. Different texture features and classification techniques were compared and the proposed combination of LPQ features and SVM classifier was shown to produce the best segmentation results. Furthermore, a simple texture based approach for the ringed seal identification was evaluated. The best identification results were obtained using SFTA features and posterior probabilities provided by a Bayesian classifier. Although, the identification performance is not yet good enough for most practical applications, the results of this pilot study can be considered promising.

Acknowledgements. The authors would like to thank the Wildlife Photo-ID Network funded by the Finnish Cultural Foundation.

References

1. Kovacs, K.M., Aguilar, A., Aurioles, D., Burkanov, V., Campagna, C., Gales, N., Gelatt, T., Goldsworthy, S.D., Goodman, S.J., Hofmeyr, G.J.G., Härkönen, T., Lowry, L., Lydersen, C., Schipper, J., Sipilä, T., Southwell, C., Stuart, S., Thompson, D., Trillmich, F.: Global threats to pinnipeds. Mar. Mammal Sci. **28**, 414–436 (2012)
2. Auttila, M., Niemi, M., Skrzypczak, T., Viljanen, M., Kunnasranta, M.: Estimating and mitigating perinatal mortality in the endangered saimaa ringed seal (phoca hispida saimensis) in a changing climate. Annal. Zool. Fenn. **51**, 526–534 (2014)
3. Koivuniemi, M., Auttila, M., Niemi, M., Levänen, R., Kunnasranta, M.: Photo-ID as a tool for studying and monitoring the critically endangered saimaa ringed seal. (2015) manuscript under review
4. Anderson, C.J.: Individual identification of polar bears by whisker spot patterns. Ph.D. thesis, University of Central Florida, Orlando, Florida (2007)
5. Tharwat, A., Gaber, T., Hassanien, A., Hassanien, H.A., Tolba, M.F.: Cattle identification using muzzle print images based on texture features approach. In: Proceedings of the Fifth International Conference on Innovations in Bio-Inspired Computing and Applications, pp. 217–227 (2014)
6. Hoque, S., Azhar, M., Deravi, F.: ZOOMETRICS-biometric identification of wildlife using natural body marks. Int. J. Bio-Sci. Bio-Technol. **3**, 45–53 (2011)
7. Halloran, K.M., Murdoch, J.D., Becker, M.S.: Applying computer-aided photo-identification to messy datasets: a case study of Thornicroft's giraffe (Giraffa camelopardalis thornicrofti). Afr. J. Ecol. **53**, 147–155 (2014)
8. Bendik, N.F., Morrison, T.A., Gluesenkamp, A.G., Sanders, M.S., O'Donnell, L.J.: Computer-assisted photo identification outperforms visible implant elastomers in an endangered salamander, Eurycea tonkawae. PLoS One **8**, e59424 (2013)
9. Albu, A.B., Wiebe, G., Govindarajulu, P., Engelstoft, C., Ovatska, K.: Towards automatic modelbased identification of individual sharp-tailed snakes from natural body markings. In: Proceedings of ICPR Workshop on Animal and Insect Behaviour, Tampa, FL, USA (2008)

10. Yılmaz Kaya, L.K., Tekin, R.: A computer vision system for the automatic iden-
    tification of butterfly species via gabor-filter-based texture features and extreme
    learning machine: GF+ ELM. TEM J. **2**, 13–20 (2013)
11. Adams, J.D., Speakman, T., Zolman, E., Schwacke, L.H.: Automating image
    matching, cataloging, and analysis for photo-identification research. Aquat. Mam-
    mals **32**, 374 (2006)
12. Lowe, D.G.: Distinctive image features from scale-invariant keypoints. Int. J. Com-
    put. Vis. **60**, 91–110 (2004)
13. Crall, J., Stewart, C., Berger-Wolf, T., Rubenstein, D., Sundaresan, S.: Hotspotter
    - patterned species instance recognition. In: IEEE Workshop on Applications of
    Computer Vision (WACV), pp. 230–237 (2013)
14. Yu, X., Wang, J., Kays, R., Jansen, P., Wang, T., Huang, T.: Automated identifi-
    cation of animal species in camera trap images. EURASIP J. Image Video Process.
    2013, 52 (2013)
15. Cheng, J., Liu, J., Xu, Y., Yin, F., Wong, D.W.K., Tan, N.M., Tao, D., Cheng,
    C.Y., Aung, T., Wong, T.Y.: Superpixel classification based optic disc and optic
    cup segmentation for glaucoma screening. IEEE Trans. Med. Imaging **32**, 1019–
    1032 (2013)
16. Arbelaez, P., Maire, M., Fowlkes, C., Malik, J.: Contour detection and hierarchical
    image segmentation. IEEE Trans. Pattern Anal. Mach. Intell. **33**, 898–916 (2011)
17. Arbelaez, P., Maire, M., Fowlkes, C., Malik, J.: From contours to regions: An
    empirical evaluation. IEEE Conf. Comput. Vis. Pattern Recogn. **2009**, 2294–2301
    (2009)
18. Martin, D., Fowlkes, C., Tal, D., Malik, J.: A database of human segmented natural
    images and its application to evaluating segmentation algorithms and measuring
    ecological statistics. In: Proceedings of the 8th International Conference on Com-
    puter Vision vol. 2, pp. 416–423 (2001)
19. Ojansivu, V., Heikkilä, J.: Blur insensitive texture classification using local phase
    quantization. In: Elmoataz, A., Lezoray, O., Nouboud, F., Mammass, D. (eds.)
    ICISP 2008 2008. LNCS, vol. 5099, pp. 236–243. Springer, Heidelberg (2008)
20. Costa, A., Humpire-Mamani, G., Traina, A.: An efficient algorithm for fractal
    analysis of textures. In: 25th Conference on Graphics, Patterns and Images vol.
    2012, pp. 39–46 (2012)
21. Ahonen, T., Matas, J., He, C., Pietikäinen, M.: Rotation invariant image descrip-
    tion with local binary pattern histogram fourier features. In: Salberg, A.-B., Hard-
    eberg, J.Y., Jenssen, R. (eds.) SCIA 2009. LNCS, vol. 5575, pp. 61–70. Springer,
    Heidelberg (2009)
22. Phillips, P.J., Moon, H., Rauss, P.J., Rizvi, S.: The feret evaluation methodology
    for face recognition algorithms. IEEE Trans. Pattern Anal. Mach. Intell. **22**, 1090–
    1104 (2000)

Fingerprint Matching with Optical Coherence Tomography

Yaseen Moolla$^{(\boxtimes)}$, Ann Singh, Ebrahim Saith, and Sharat Akhoury

Council for Scientific and Industrial Research, Pretoria, South Africa
ymoolla@csir.co.za

Abstract. Fingerprint recognition is an important security technique with a steadily growing usage for the identification and verification of individuals. However, current fingerprint acquisition systems have certain disadvantages, which include the requirements of physical contact with the acquisition device, and the presence of undesirable artefacts, such as scars, on the surface of the fingerprint. This paper evaluates the accuracy of a complete framework for the capturing of undamaged, undistorted fingerprints from below the skins surface using optical coherence tomography hardware, the extraction and conversion of the subsurface data into a usable fingerprint and the matching of such fingerprints. The ability of the framework to integrate with existing fingerprint recognition systems and its ability to operate as an independent stand-alone system are both evaluated.

1 Introduction

1.1 Fingerprint Matching

Fingerprint matching is a biometric technique that is often used for the identification or verification of individuals. It is used for access control to physical areas, such as at the entrances to buildings; for access to virtual areas, such as unlocking virtual devices; in forensics and the criminal justice system; and for border control. It is also use in authentication systems, such as verification during the transfer of electronic funds [1].

Fingerprint recognition requires the extraction and matching of minutia details from the fingerprint. The most common minutiae used in fingerprint recognition are ridge endings and bifurcations. A ridge ending is a point where a ridge stops. A bifurcation is a point where a ridge splits into two separate ridges. Figure 2(a) shows examples of ridge endings and bifurcations. These are collected from the surface of a users finger. A user presses their finger against the platen of a surface scanner and the fingerprint is scanned.

However, the fingerprints on the surface of the skin originate below the surface of the skin, in the papillary layer. The fingerprint patterns are formed at the papillary junction, which is the junction of the dermal layer and the papillary layer [1, 2]. Through the use of optical coherence tomography (OCT), the subsurface information of the fingerprint can be extracted and used. Several advantages of OCT over conventional surface fingerprint scanners are discussed below.

© Springer International Publishing Switzerland 2015
G. Bebis et al. (Eds.): ISVC 2015, Part II, LNCS 9475, pp. 237–247, 2015.
DOI: 10.1007/978-3-319-27863-6_22

1.2 Optical Coherence Tomography

OCT is a non-invasive, non-contact, optical imaging technique that is able to yield sub-surface morphology (2D or 3D) of scattering samples in situ and in real time. OCT is often described as the optical analogue to ultrasound. However the back scattered light cannot be measured electronically due to the high speed of light. Therefore OCT uses the technique of low coherence interferometry, first demonstrated by Huang in 1991 [3]. Since then it has been applied extensively in biomedical applications especially ophthalmology, dermatology and cardiology [4–7] with some applications in material science and in artwork and has in the last few years been gaining momentum in the field of biometrics [8].

The advantage of OCT over other biometric techniques lies in its inherent ability to deliver contactless, subsurface 3D fingerprints with high resolution. The 3D capability allows one to differentiate between a real and a fake fingerprint to counter identity theft spoofing attacks [8] and also allows for the detection of the inner subsurface fingerprint which is a replica of the outer fingerprint. By using OCT technology, the fingerprint patterns can be extracted from the papillary junction (PJ). The fingerprints found at the papillary junction are unaffected by scarring, cuts and wrinkles that are present on the surface of the skin. This invariance allows PJ fingerprints to have a greater clarity and greater reliability than surface fingerprints. This is also an advantage for people such as miners and people who are subject to extensive manual labour or old age whose outer fingerprints are less readable by conventional surface scanners due to abrasions, scars and wrinkles.

Due to the contactless nature of OCT technology, fingerprints will not experience the distortion that occurs when fingers are pressed against a surface, as occurs with conventional surface contact scanners [9]; and it may also be used in environments in which contact with surfaces is detrimental, such as at access points to sterile operating theatres or laboratories.

Furthermore, while compliance with FBI standards requires a resolution of 500 dots per inch (dpi) for fingerprint scanners, there is a growing trend towards higher resolutions or 1,000 dpi. Greater resolutions allow for accurate detection of incipient ridges and sweat pores, which allows for great accuracy in fingerprint recognition [1]. OCT technology is capable of detecting sweat pores on the surface of fingerprints, as well as the subsurface eccrine glands from which the pores originate [10]. Doppler OCT scanners are also capable of resolutions greater than 10,000 dpi, which are fine enough and with non-invasive optical penetration deep enough to detect the blood capillaries below the fingerprint [11]. This may be used for liveness detection, and as a further biometric recognition modality.

While techniques for the extraction of fingerprints have been described [12], little has been done to evaluate the ability to match these extracted fingerprints against both legacy surface fingerprints, as well as other extracted OCT fingerprints. The remainder of this paper describes a framework for the detection and extraction of information from the papillary junction, mapping of the extracted

3D fingerprint to a 2D image, and extraction and matching of fingerprint minutiae. Subsurface OCT fingerprints obtained in this manner are compared against surface fingerprints and other subsurface fingerprints to determine the reliability of this technique.

2 Methodology

2.1 Fingerprint Extraction

The process for the capturing and extraction and matching of OCT fingerprints is shown in Fig. 1. Fingerprints are captured using a Swept Source Optical Coherence Tomography (SSOCT) device (Model OCM1300SS) - Thorlabs Inc., USA. The system consists of a laser source that scans in over a wavelength band (frequency band). The centre wavelength is 1325 nm, scanning 110 nm around the central wavelength with a spectral width of 3 dB spectral bandwidth. The system resolution is 15-20 um with a probing depth of 3 mm in air. The laser output power is 10 mW with a coherence length of 6 mm and an axial scan speed of 16 kHz. It can image a physical volume of $15 \times 15 \times {\sim} 3 mm^3$ with a representation of $512 \times 512 \times 512$ pixels (867 pixels per inch) in approximately 22 seconds; and a volume of $13 \times 13 \times {\sim} 3\ mm^3$ with a representation of $256 \times 256 \times 512$ pixels (500 pixels per inch) in approximately 6 seconds. The papillary junction (PJ) is extracted, using a technique similar to Akhoury and Darlow [12], in which the stratum corneum is first detected. Using the stratum corneum as a reference point, the papillary junction is then detected and extracted, to form a three dimensional structure.

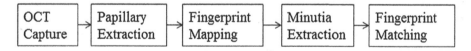

Fig. 1. An overview of the process for fingerprint matching using OCT technology

Figure 2(b) shows a slice of a finger from an OCT image. The papillary junction is outlined with a solid line. This junction has a wave-like form due to the existence of ridges and valleys. This is the micro-curvature of the finger. A broken line represents the greater overall curvature of the finger, which is due to its approximately cylindrical nature. This is the macro-curvature of the finger. For accurate mapping of the 3D structure into 2D, the macro-curvature needs to be eliminated, while the micro-curvature information of ridges and valleys needs to be maintained. This conversion is performed, by projecting the 3D structure onto a 2D plane, as described in Akhoury and Darlow [12].

Since the OCT images are degraded by speckle noise, it is natural to anticipate the effects of such artefacts to influence the location of the algorithmically

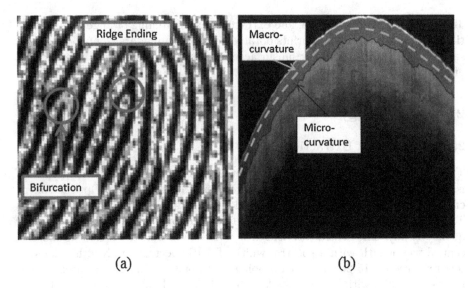

(a)   (b)

Fig. 2. *(a)* A sample fingerprint with a ridge ending and a bifurcation highlighted. *(b)* A sample slice of an OCT scan from a tip-on view. The *thick white line* represents the skins surface. Data below the *thick white line* is subsurface information collected using an OCT scanner. The *green dashed line* depicts the macro-curvature of a finger and the *solid red line* depicts the micro-curvature created by the fingerprint pattern (Color figure online).

determined papillary junction contours. Hence, prior to any further processing - errors in the detected contours must be accounted for. This is done by applying the non-linear median filter. The resultant surface is a smoothed 3D representation of the ridges and valleys of the fingerprint from the internal fingerprint. The ridge-valley fingerprint structure is situated along the varying depth dimension of the finger. As we are only concerned with the peak-crest sinusoidal-like pattern of the ridge-valley structure, the effects of the varying depth parameter can be eliminated by differentiation. In image processing, this is more formally known as edge detection. As in Akhoury and Darlow [10], we also apply the phase congruency approach [13]. The resultant derivatives are then exemplified by a sigmoid function in an attempt to amplify the low powered sinusoidal pattern. A second stage non-linear median filter is then applied to smooth the contours and remove anomalous saturated peaks. The resultant signal is then enhancement using contrast-limited adaptive histogram equalization (CLAHE) [14] and contrast stretching.

Once the fingerprint is successfully converted into a 2D format, fingerprint minutiae may be extracted. Once the minutiae are extracted from two fingerprints, these fingerprints may be matched. The extraction and matching of minutiae is thoroughly covered in literature and there are many commercial solutions available [1,15].

2.2   Parameter Optimization

The framework for the mapping from a three dimensional (3D) structure to a two dimensional (2D) image, contains a very large set of parameters. In order to determine the best values for each of these parameters, search optimization techniques may be used. The chosen technique must find an optimal combination of values for each parameter such that the ideal 2D image of the fingerprint is acquired. An example of common search optimization technique in such cases is the genetic algorithm, which mimics evolution theory. It is a stochastic approach used to search a wide space of unknown gradient, and reduces to an optimal solution while overcoming the risks that local minima present to search techniques [16,17].

Key to designing an effective genetic algorithm is determining the objective function, which evaluates the overall score for each combination of parameter values. In this case, the objective function was implemented to optimize the match score between the mapped 2D fingerprint and a conventional contact surface fingerprint.

3   Results and Discussion

Two sets of tests were performed. In the first set of tests, the PJ fingerprints, extracted from below the surface of the skin using OCT technology, were compared against surface fingerprints, which are collected using a surface fingerprint scanner. This is used to evaluate the accuracy of matching between OCT and contact surface fingerprints for the purposes of backwards compatibility with existing fingerprint recognition systems. These comparisons were performed against 22 fingers of various individuals.

In the second set of tests, two OCT samples were collected from each of 5 fingers. The samples were compared against each other to evaluate the potential accuracy of a system in which fingerprint recognition is performed using contactless PJ fingerprints only.

The performance of matching before and after parameter optimization is also evaluated. Due to the newness and novelty of the hardware technology and extraction technique, there are currently no publicly available databases of OCT fingerprints. Data sets were collected by the research team, conforming to the South African Protection of Personal Information Act [18].

3.1   Papillary Junction Compared to Surface Fingerprints

Subsurface scans of 22 fingers were collected from various individuals using the hardware described in Sect. 2.1. In the first 12 samples, an area of $15 \times 15\,mm^2$ was captured at 867dpi and subsequently scaled down to 500 dpi. In samples 13 to 22, an area of $13 \times 13\,mm^2$ was captured at 500 dpi. Surface scans of the fingers, using contact fingerprint scanners, were also obtained for matching comparisons.

The internal subsurface fingerprints were extracted and mapped to a 2D plane, using parameters which were optimized as described in Sect. 2.2. The minutia extraction and matching was performed using the SecuGen Software Development Kit [19]. Figure 3 shows a comparison of a subsurface fingerprint mapped to 2D using random unoptimized parameters in Fig. 3(a), and optimized parameters in Fig. 3(b). In Fig. 3(a), data is lost during the mapping process, due to using sub-optimal parameters. With optimized parameters, in Fig. 3(b), the ridges and valleys are more clearly defined and sufficient information is retained to accurately detect ridge ending and bifurcation minutia points. In Fig. 3(c), a surface scan of the same fingerprint, taken with a conventional surface scanner, is shown. The detected minutia points are marked with circles. There is a high similarity between the optimized subsurface fingerprint and the surface fingerprint. Additionally, the subsurface fingerprint is free of the scarring that is visible on the surface fingerprint.

Table 1 shows a comparison of acceptance and rejection rates for both the optimized and unoptimized parameters when matched against surface fingerprints. 22 true comparisons were performed, by comparing the collected subsurface OCT fingerprints with the corresponding surface scanner fingerprints. 462 false comparisons were performed, by comparing each of the 22 subsurface OCT fingerprints with each of the other 21 non-corresponding surface scanner fingerprints. While no false matches were accepted, several true matches were rejected. However, the parameter optimization reduced the false rejection rate and improved true acceptance. Figure 4 shows a comparison of the match scores before and after the optimized parameters were determined. Optimizing the parameters improved the accuracy of matching in all cases. It is seen that match scores in some cases were still below the threshold which determines whether a match is accepted or not. This undesirable outcome may be caused by several factors: errors in capturing, and errors in mapping from 3D to 2D. Descriptions of the source of the errors and proposed solutions to mitigate them are provided below.

There are two known errors which may occur during the capturing process. The first of these is caused by excessive movement of the finger during capturing, since it is difficult to keep a finger still during a lengthy capturing time. The second is due to the maximum scanning depth of the OCT hardware. The scanning depth is set at a pre-determined value prior to the initiation of the capturing sequence. However, due to the irregular cylindrical nature of a finger in three-dimensional space, parts of the finger may be outside of the depth range of the scanner. This leads to a reduction in the total scanned area. A reduced area leads to a lower number of detected minutia points, which in turn reduces the ability to accurately match two fingerprints. This problem is called the depth dependency roll-off. Both of these known errors may be countered by the use of better hardware.

The current drive in OCT technology is to develop systems that can scan faster so as to remove motion artefacts and that is able to penetrate deeper into the sample. The SSOCT system offers the fastest scan times which are dependent on the sweeping speed of the laser used. The system used for this study had a scan speed of 16 kHz. The very same system can be implemented with higher scan rates from newly available or soon to be available sources [20,21]. One such source is source with a 200 kHz scan speed theoretically resulting in a 12.5× decrease in acquisition time. Such a laser, which is available from Thorlabs Inc., USA, also has an imaging depth of at least 12 mm. Currently, faster speeds are limited by the available detection electronics so there is a possibility of even faster and more efficient systems in the not so distant future. The goal is to reduce capturing time to a speed of less than 1 second, which is compatible with current fingerprint capturing standards [1]. A faster acquisition time and greater scanning depth would eliminate the errors which occur during the capturing process and thus improve the matchability of the fingerprints.

Improvements in the mapping of 3D to 2D may also have an impact on the matching. Currently, the 3D structure is projected directly onto a plane. However, direct projection may impact the perceived distance between ridges and the thickness of ridges. Additionally, due to the elastic nature of the skin, distortions occur in touch-based fingerprint matching. Ridges contract at the first point of contact with a surface, and expand further out. The touchless OCT fingerprint acquisition system is free from such distortions. Techniques similar to those employed by Zhao et al. [9], which take into consideration the macro-curvature of the finger and model the distortions of the skin may allow for better matching and better backwards compatibility of touchless subsurface fingerprints with the conventional touch-based surface fingerprint scanners.

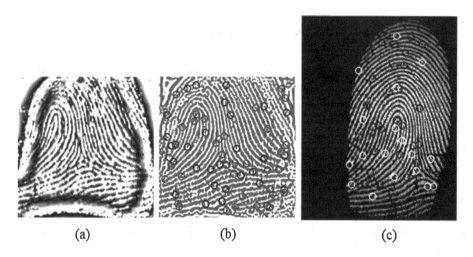

(a)   (b)   (c)

Fig. 3. A comparison of subsurface OCT scanner derived and conventional surface scanner fingerprints. *(a)* shows an OCT subsurface fingerprint before parameter optimization; *(b)* shows an OCT subsurface fingerprint after parameter optimization; and *(c)* shows a surface fingerprint from a conventional scanner. The *blue circles* signify detected ridge bifurcations, and the *green circles* signify detected ridge endings (Color figure online)

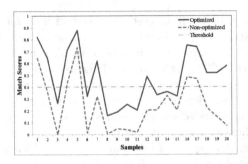

Fig. 4. A comparison of scores for the matching of OCT derived subsurface fingerprints using random unoptimized parameters and optimized parameters against corresponding conventionally scanned surface fingerprints

Table 1. A comparison of acceptance and rejection rates for the matching of optimized and unoptimized OCT fingerprints against conventional surface scanned fingerprints

	Unoptimized		Optimized	
	Acceptance \%	Rejection \%	Acceptance \%	Rejection \%
False	0	77	0	41
True	23	100	59	100

3.2 Papillary Junction Compared Against Other Papillary Junction Fingerprints

For these tests, 10 samples were collected, with 2 each per finger, with a physical coverage of $13 \times 13\,mm^2$ at 500 dpi. The subsurface fingerprints from the papillary junction were extracted from each and compared against each other. Figure 5 shows a visual comparison of two fingerprints captured in this manner. The two captured fingerprints are almost identical in nature which results in a very high match score. Since both fingerprints were captured using a contactless system, neither fingerprint suffers from the adverse effects of distortion which plagues touch-based fingerprint scanners. The quality of surface scanned fingerprints is also affected by the dryness of the users finger and the pressure applied. If a finger is too dry or if too little pressure is applied, the ridges will be discontinuous. With too much pressure, the width of detected ridges increases and causes valleys to become indistinct. The contactless OCT scanner is not affected by these factors.

Table 2 shows a comparison of acceptance and rejection rates for both the optimized and unoptimized parameters when subsurface fingerprints were matched against other subsurface fingerprints. 5 true comparisons were performed, by comparing two collected subsurface OCT fingerprints with each other for each finger. 80 false comparisons were performed, by comparing the two OCT fingerprints of each finger with each of the other 8 non-corresponding

OCT subsurface fingerprints. While no false matches were accepted, several true matches were rejected. However, as with the surface comparisons, the parameter optimization reduced the false rejection rate and improved true acceptance. Figure 6 shows a comparison of matching for each finger. As before, parameter optimization improved the results in all cases. The two lowest scores (for samples 2 and 5) were due to depth dependence roll-off. Hardware solutions to counteract this issue were discussed in the previous section. Due to the large capturing area; lack of distortion, which occurs in contact-based systems; elimination of surface artefacts; and the potential for higher resolution data capture, the matching together of PJ fingerprints, captured with OCT technology, has the potential to provide more accurate fingerprint recognition than comparisons with surface fingerprints.

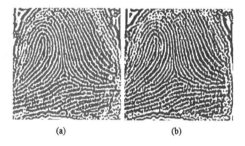

(a)   (b)

Fig. 5. Comparison of two separate OCT scans of the same finger, showing a very high match

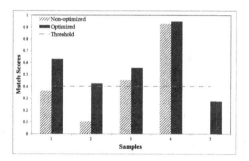

Fig. 6. A comparison of scores for the matching of OCT derived subsurface fingerprints using random unoptimized parameters and optimized parameters against other OCT derived prints from the same fingers

These preliminary studies show promise for a subsurface fingerprint matching system. Further statistical analysis of the parameter optimization and the match scores of the proposed system will be performed once a larger dataset is collected.

Table 2. A comparison of acceptance and rejection rates for the matching of optimized and unoptimized OCT fingerprints against other OCT derived fingerrints

	Unoptimized		Optimized	
	Acceptance \%	Rejection \%	Acceptance \%	Rejection \%
False	0	60	0	20
True	40	100	80	100

4   Conclusion

It has been shown that the subsurface papillary junction fingerprint captured using OCT technology can be matched with surface fingerprints with a high degree of reliability. Further, this can be reproduced by the ability to match subsurface fingerprints with each other. The proposed framework has several advantages over current fingerprint acquisition systems, which include the elimination of surface artefacts, the avoidance of distortion which is introduced by surface contact acquisition devices, and the ability to capture greater details at higher resolutions. Improvements to the current hardware system have also been proposed to increase the quality of the capturing process and subsequently improve the reliability of matching.

Acknowledgements. Acknowledgments of support go to Dr. Christiaan van der Walt and Luke Darlow of the Council for Scientific and Industrial Research's Modelling and Digital Science unit, and the CSIR National Laser Centre.

References

1. Maltoni, D., Maio, D., Jain, A.K., Prabhakar, S.: Handbook of fingerprint recognition. Springer Science & Business Media (2009)
2. Babler, W.J.: Embryologic development of epidermal ridges and their configurations. Birth Defects Original Art. Ser. **27**, 95–112 (1991)
3. Huang, D., Swanson, E.A., Lin, C.P., Schuman, J.S., Stinson, W.G., Chang, W., Hee, M.R., Flotte, T., Gregory, K., Puliafito, C.A., et al.: Optical coherence tomography. Science **254**, 1178–1181 (1991)
4. Fercher, A.F., Hitzenberger, C.K., Kamp, G., El-Zaiat, S.Y.: Measurement of intraocular distances by backscattering spectral interferometry. Opt. Commun. **117**, 43–48 (1995)
5. Hamdan, R., Gonzalez, R.G., Ghostine, S., Caussin, C.: Optical coherence tomography: from physical principles to clinical applications. Arch. Cardiovasc. Dis. **105**, 529–534 (2012)
6. Romania, C.P.: Optical coherence tomography in oncological imaging. Rom. Rep. Phy. **66**, 75–86 (2014)
7. Ju, M.J., Lee, S.J., Min, E.J., Kim, Y., Kim, H.Y., Lee, B.H.: Evaluating and identifying pearls and their nuclei by using optical coherence tomography. Opt. Express **18**, 13468–13477 (2010)

8. Chang, S., Larin, K.V., Mao, Y., Flueraru, C., Almuhtadi, W.: Fingerprint spoof detection using near infrared optical analysis, pp. 57–84. Croatia, State of the art in biometrics. Intechopen (2011)
9. Zhao, Q., Jain, A., Abramovich, G.: 3D to 2D fingerprints: Unrolling and distortion correction. In: 2011 International Joint Conference on Biometrics (IJCB), pp. 1–8. IEEE (2011)
10. Liu, M., Buma, T.: Biometric mapping of fingertip eccrine glands with optical coherence tomography. Photonics Technol. Lett. IEEE **22**, 1677–1679 (2010)
11. Liu, G., Chen, Z.: Capturing the vital vascular fingerprint with optical coherence tomography. Appl. Opt. **52**, 5473–5477 (2013)
12. Akhoury, S.S., Darlow, L.N.: Extracting subsurface fingerprints using optical coherence tomography. In: 2015 Third International Conference on Digital Information, Networking, and Wireless Communications (DINWC), pp. 184–187. IEEE (2015)
13. Kovesi, P.: Phase congruency detects corners and edges. In: The Australian Pattern Recognition Society Conference: DICTA 2003 (2003)
14. Zuiderveld, K.: Contrast limited adaptive histogram equalization. In: Graphics gems IV, Academic Press Professional, Inc., pp. 474–485 (1994)
15. ISO: Biometric sample quality - Part 4: Finger image. ISO 29794-4:2010, International Organization for Standardization, Geneva, Switzerland (2010)
16. Davis, L.: Handbook of genetic algorithms (1991)
17. Spall, J.C.: Introduction to Stochastic Search and Optimization: Estimation, Simulation, and Control, vol. 65. John Wiley & Sons, New York (2005)
18. South African Department of Justice,: Protection of Personal Information Act (2013). http://www.justice.gov.za/legislation/acts/2013-004.pdf Accessed 2015-08-27
19. SecuGen Corporation,: Secugen software development kits (2015). http://secugen.com/products/sdk.htm Accessed 2015-07-14
20. Drexler, W., Liu, M., Kumar, A., Kamali, T., Unterhuber, A., Leitgeb, R.A.: Optical coherence tomography today: speed, contrast, and multimodality. J. Biomed. Opt. **19**, 071412–071412 (2014)
21. Breithaupt, R., Sousedik, C., Meissner, S.: Full fingerprint scanner using optical coherence tomography. In: 2015 International Workshop on Biometrics and Forensics (IWBF), pp. 1–6. IEEE (2015)

Improve Non-graph Matching Feature-Based Face Recognition Performance by Using a Multi-stage Matching Strategy

Xianming Chen[1]([⊠]), Wenyin Zhang[2], Chaoyang Zhang[1], and Zhaoxian Zhou[1]

[1] School of Computing, University of Southern Mississippi,
Hattiesburg, MS 39406, USA
xianming.chen@eagles.usm.edu
[2] School of Informatics, Linyi University, Linyi 276000, Shandong, China

Abstract. In this paper, a multi-stage matching strategy that determines the recognition result step by step is employed to improve the recognition performance of a non-graph matching feature-based face recognition. As the gallery size increases, correct correspondence of feature points between the probe image and training images becomes more and more difficult so that the recognition accuracy degrades gradually. To deal with the recognition degradation problem, we propose a multi-stage matching strategy for the non-graph matching feature-based method. Instead of finding the best match, each step picks out one half of the best matching candidates and removes the other half. The behavior of picking and removing repeats until the number of the remaining candidates is small enough to decide the final result. The experimental result shows that with the multi-stage matching strategy, the recognition performance is remarkably improved. Moreover, the improvement level also increases with the gallery size.

1 Introduction

As an indispensable part of pattern recognition, face recognition has been gaining growing attention in the past three decades. It can be apparently proven by its increasingly prevalent applications in our daily life, such as facial capture of social media, access control system, anti-terrorism system and so on. Although some other methods of pattern recognition, such as iris recognition, fingerprint recognition, are more reliable in terms of recognition accuracy, the noninvasive feature of face recognition has its own incomparable advantage, especially in the security system where requires unaware cooperation from the target participants. Various face recognition methods have been proposed to achieve higher accuracy, less computation cost, and better performance reliability. Technically, those methods can be categorized into holistic methods and feature-based methods [1]. The "eigenfaces" of Principal Component Analysis (PCA) [2], and the "fisherfaces" of Linear Discriminant Analysis (LDA) [3] are the typical representations of holistic methods, while the "AAM search" of Active Appearance Models (AAM) [4] and the "jets" of Elastic Bunch Graph Match (EBGM) [5]

© Springer International Publishing Switzerland 2015
G. Bebis et al. (Eds.): ISVC 2015, Part II, LNCS 9475, pp. 248–257, 2015.
DOI: 10.1007/978-3-319-27863-6_23

represent the feature-based methods. In our work, a non-graph matching feature-based method [6] that applies Gabor Wavelets Transform (GWT) to extract features is studied.

After extracting features, it is also critical to employ a matching algorithm to find out the recognition result. As a holistic method, PCA calculates the distances of "eigenfaces" and chooses the smallest one [7]. EBGM calculates the similarities of "jets" and finds out the one that has the maximum value [5]. Other identification methods, such as "Face-ARG Matching" [8], "Probabilistic Models" [9], are also effective strategies. Although the supported theories of these methods differ from each other, they share a similar identification process, which is to find a function to express the similarity and choose the optimal one. For the studied non-graph matching feature-based methods, the recognition result depends on the correspondence of feature points between the probe image and the training images. However, as the gallery size increases, correct correspondence of feature points becomes more and more difficult so that the recognition accuracy degrades gradually and becomes eventually unacceptable [10]. The degradation problem appears more evident when one probe image corresponds to only one training image [11,12].

In this paper, we aim to improve the recognition performance of the non-graph matching feature-based method by enhancing the matching process. Other than deciding recognition result with only one round of matching procedure, a multi-stage matching strategy that determines recognition result step by step is employed. Each step picks out one half of the best candidates as the temporary result and removes the other half. The step-by-step selection behavior repeats until the number of remaining images is small enough to output the final recognition. Two multi-stage matching algorithms, binary elimination and divide and conquer, are introduced to the matching procedure of the tested non-graph matching method, from the perspectives of global and local respectively. For binary elimination, each step of matching procedure picks out half of the involved training images by ranking the similarities and removes the others. A new step of matching procedure restarts with the remaining training images, until the number of the involved training images is no greater than a small certain number (we set it with 15). Finally, the recognition result comes out from the remaining images. For divide and conquer, the training images are divided into groups by every two. And then, the training image that has the greater similarity of each group is picked out as a candidate for the next step of matching procedure. The behavior of dividing and selecting repeats, until the number of remaining images is no greater than a small certain number (15 the same with that of binary elimination). Eventually, the recognition result is generated from the remaining images. Binary elimination searches the recognition result globally, while divide and conquer seeks it locally, yet they share the common strategy of multi-stage matching. The experimental results of both show that recognition accuracy of the tested non-graph matching feature-based method has been improved remarkably. Moreover, it seems that the improved level also increases with the number of images in the gallery.

2   Non-graph Matching Feature-Based Face Recognition

2.1   Implementation of Non-graph Matching Method

In our work, the studied non-graph matching feature-based method [6] is implemented as next:

1. Gabor Wavelets Transform (GWT) calculating.
   For each image, calculate the GWT response value of each pixel with the 40 Gabor filters.
2. Feature points searching.
   Divide images into blocks with size of $w \times w$ and find a feature point for each block. A feature point located at (x_0, y_0) in a window W_0 should satisfy

$$R_j(x_0, y_0) = \max\left(R_j(x, y)\right),$$

and

$$R_j(x_0, y_0) > \frac{1}{N_1 N_2} \sum_{x=1}^{N_1} \sum_{y=1}^{N_2} R_j(x, y),$$

where $(x, y) \in W_0, j = 0, \cdots, 40$, and N_1, N_2 is the width and height of the image.

3. Feature points generating.
   Generate the bunch graphs by combining the 40 GWT coefficients and 2 coordinates of those feature points:

$$v_k = \{x_k, y_k, R_j\left(x_k, y_k\right), j = 1, 2, \cdots, 40\}.$$

4. Feature points matching.
   For each feature point of each probe image, measure the similarities with those feature points of all training images within a certain distance by:

$$S_i(k, j) = \frac{\sum_l |v_{i,k}(l)||v_{t,j}(l)|}{\sqrt{\sum_l |v_{i,k}(l)|^2 |v_{t,j}(l)|^2}},$$

where $S_i(k, j)$ represents the similarity of the jth feature vector of the probe image, $(v_{i,j})$, to the kth feature vector of the jth reference training image, $(v_{i,k})$, where l is the number of vector elements.

5. Feature points corresponding.
   For each feature point of each probe image, find out the training image that has the maximum similarity and mark it with one score.

6. Overall similarity calculating.
   For each probe image, calculate the overall similarities of training images with the formula:

$$OS_i = \frac{C_i}{N_i}.$$

where C_i is the score of the ith training image in the scoreboard, and N_i is the amount of feature points that the ith training image has.

7. Recognition result output.
   For each probe image, find out the training image that has the greatest overall similarity.

3   The Recognition Degradation of Feature-Based Method

3.1   Dispersion of Feature Points Correspondence

The larger the gallery size is, the more challenging the face recognition becomes, particularly for the non-graph matching feature-based method. The matching procedure of it is to correspond the feature points of the training images and the probe image. As the number of training images increases, for one single feature point from a probe image, it becomes more possible to correspond to a wrong training image.

Table 1 illustrates a case that a false recognition happens due to gallery size increasing. A probe image indexed 10 has about 382 feature points to be corresponded. Three training images that have most corresponding feature points are listed in Table 1. The training image indexed 10 succeeds in matching 53 and 45 feature points with gallery sizes of 100 and 200 respectively, and achieves correct recognition. However, when the gallery size goes to 300, it just matches 37 feature points, which is less than that of another training image indexed 51, and a false recognition occurs. Actually, it can be seen that the matching points of the three most similar candidates are all reducing as the gallery size increases, but the reduced speed may varies with each other. And this is the so-called dispersion of feature points correspondence, which causes the recognition degradation problem.

Table 1. Correspondence Dispersion of a probe image indexed 10

Gallery size=100		Gallery size=200		Gallery size=300	
Trian Index	Matched points	Trian Index	Matched points	Trian Index	Matched points
10	53	10	45	51	39
51	45	51	42	10	37
89	23	89	18	89	15
Correct recognition		Correct recognition		False recognition	

3.2   Accuracy Degradation

We tested the recognition accuracy of the non-graph matching feature-based method presented in Sect. 2.1 by varying the gallery size from 100 to 1000 with an increment of 100. One probe image only references to one training image and the other experimental parameters will be described in Sect. 5. The experimental accuracies of the non-graph matching method along with different gallery sizes are collected and plotted in Fig. 1, from which we can see that the accuracy decreases steadily with the increasing number of image. With a size of 1000 images, the recognition accuracy drops below 60 %, which is unacceptable for face recognition.

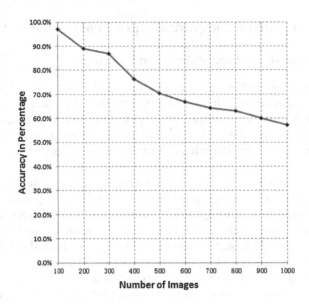

Fig. 1. The accuracy degradation of the non-graph feature-based matching method

4   Multi-stage Matching Algorithms

Two multi-stage matching algorithms, binary elimination and divide and conquer, are proposed to deal with the recognition degradation problem from the
perspectives of global and local, respectively.

4.1   The Core of Multi-stage Matching Strategy

To deal with the recognition degradation problem brought by the dispersion of
feature points correspondence, a multi-stage matching strategy that determines
the recognition result step by step is introduced to the matching process of the
non-graph matching feature-based method. Instead of finding the best match,
each step picks out one half of the best matching candidates and removes the
others. The behavior of picking and removing repeats until the number of the
remaining candidates is small enough to output the final result.

In authors' previous works, it has been found in practice that when the
number of the training images is no greater than 15, the disturbance from other
training images becomes trivial and thus the multi-stage searching process should
stop [13].

4.2   Binary Elimination

Recognizing a probe image from a bunch of training images can be considered
as deciding the championship in a sports tournament, which has preliminary,
intermediary, and final. When matching a probe image with gallery, training

images are ranked by the similarity with the probe image, and the first half of them are picked as the candidates for the next step, while the others are eliminated. With binary convergence rate, the picking and elimination process repeats, until the number of remaining training images is no more than a certain number. And finally the recognition result comes from the remaining candidates. In our work, when the number of the remaining training images is no more than 15, an empirical number. Figure 2 detailedly illustrates the binary elimination matching algorithm.

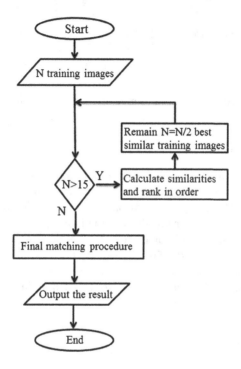

Fig. 2. The non-graph matching feature-based method with binary elimination

4.3   Divide and Conquer

Different with binary elimination that solves from global, divide and conquer algorithm breaks a big problem into several sub-problems of the same type, solves them individually and combines them until they are simple enough to be solved directly. In other words, divide and conquer solves searching from local. In matching procedure, training images are divided into groups by every two. And then, the matching process executes in each sub-group, and the training image that has greater similarity of each group is selected as a candidate for the next step. The divide and conquer behavior repeats, until the number of the remaining candidates is not greater than 15 to output the final recognition result.

Figure 3 demonstrates a case of recognizing a probe image indexed 1 from 100 training images indexed from 1 to 100 with the divide and conquer matching algorithm.

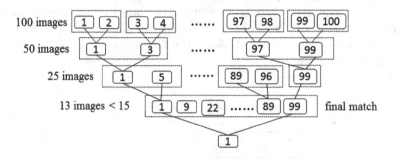

Fig. 3. The demonstration of matching with divide and conquer

4.4   A Typical Case

Table 2 demonstrates a typical case that a corresponding training image does not rank the best in similarity in the first round of matching but is correctly recognized finally with binary elimination matching algorithm. A probe image indexed 15 is to be recognized from 100 training images, one to one corresponding in numbers. In the four rounds of matching, the candidates are eliminated in binary, with the amount of 100, 50, 25 and 13, until no more greater than 16. From Table 2, we can see that the corresponding training image indexed 15 dose not rank the best in the first and second round of matching, but it stands out in the next two rounds.

5   Experiments

5.1   Database and Sample Images

To get a convincing comparison of performances, FERET [14] that can be tested up to 1000 individuals is chosen as the database. 1000 fa individuals and 1000

Table 2. A typical case of multi-stage binary elimination algorithm

Matching Round	Candidates Amount	Probe Index	Top 10 Similarity Ranking Indexes
First	100	15	13 21 15 34 25 55 29 91 37 35
Second	50	15	21 15 13 7 34 25 37 2 55 67
Third	25	15	15 21 13 34 7 67 29 2 4 25
Final	13	15	15 21 37 4 7 34 2 13 41 25

fb individuals with different race, gender, age, expression, illumination, etc. are picked randomly as the training images and probe images, respectively. Several sample images are shown in Fig. 4.

Fig. 4. Sample images

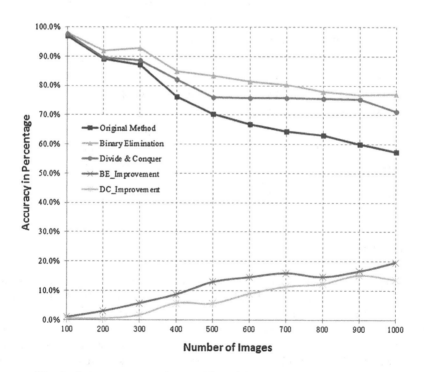

Fig. 5. Accuracy comparisons with multi-stage matching algorithms

5.2   Results

The original non-graph matching feature-based method and that with multi-stage matching algorithms, binary elimination and divide and conquer, are tested for recognition accuracy on galleries of varying sizes ranging from 100 to 1000, with an increment of 100. The recognition accuracies of original non-graph matching method and that with two multi-stage matching algorithms, binary elimination and divide and conquer, are collected in Table 3 and plotted in Fig. 5. It can be obviously seen that the non-graph matching feature-based method with multi-stage matching algorithms out-performances the original method with all gallery sizes. Moreover, the improved level also increases steadily with the size of gallery. For instance, even when the gallery reaches to 1000 images, with binary elimination, the recognition accuracy maintains at a high level of 77.0 %, with a improvement of 19.7 %, which means there are 197 recognition cases rectified from false.

Table 3. Accuracy comparison with multi-stage matching algorithms

Gallery Size	Original Matching	Binary Elimination	Divide and Conquer	Improved By BE	Improved By DC
100	97.0 %	98.0 %	97.5 %	1.0 %	0.5 %
200	89.0 %	92.0 %	89.5 %	3.0 %	0.5 %
300	87.0 %	92.7 %	88.7 %	5.7 %	1.7 %
400	76.3 %	85.0 %	82.0 %	8.8 %	5.8 %
500	70.4 %	83.4 %	76.0 %	13.0 %	5.6 %
600	66.8 %	81.5 %	75.8 %	14.7 %	9.0 %
700	64.3 %	80.3 %	75.7 %	16.0 %	11.4 %
800	63.1 %	77.9 %	75.5 %	14.8 %	12.4 %
900	60.1 %	76.8 %	75.4 %	16.7 %	15.3 %
1000	57.3 %	77.0 %	71.1 %	19.7 %	13.8 %

6   Conclusion

In this paper, a multi-stage matching strategy was used to improve the performance of a non-graph matching feature-based face recognition method, of which recognition accuracy degrades with the gallery size increasing, due to the dispersion of feature points correspondence. Rather than only one matching round to decide the recognition result, the multi-stage matching strategy determines it step by step. Each step picks out one half of the best similar training images as the new candidates for the next step and removes the others. The behavior of picking and removing repeats until the remaining training images are few enough to produce the final result. Two multi-stage matching algorithms, binary elimination and divide and conquer, were introduced to the non-graph matching feature-based method, from the perspective of global and local, respectively. The experimental results showed that with multi-stage matching algorithms, the recognition accuracy was remarkably improved. Moreover, the improved level also increased with the gallery size. For future work, besides feature-based methods, we believe that the multi-stage matching strategy is also promising to other types of face recognition methods.

References

1. Zhao, W., Chellappa, R., Phillips, P.J., Rosenfeld, A.: Face recognition: A literature survey. ACM Comput. Surv. **35**, 399–458 (2003)
2. Turk, M., Pentland, A.: Eigenfaces for recognition. J. Cogn. Neurosci. **3**, 71–86 (1991)
3. Peter, N., Belhumeur, J.P.H., Kriegman, D.J.: Eigenfaces vs. fisherfaces: Recognition using class specific linear projection. IEEE Trans. Pattern Anal. Mach. Intell. **19**, 711–720 (1997)
4. Matthews, I., Baker, S.: Active appearance models revisited. Int. J. Comput. Vis. **60**, 135–164 (2004)
5. Wiskott, L., Fellous, J.M., Kruger, N., von der Malsburg, C.: Face recognition by elastic bunch graph matching. IEEE Trans. Pattern Anal. Mach. Intell. **19**, 775–779 (1997)
6. Kepenekci, B.: Face recognition using gabor wavelet transform. Master's thesis, The Middle East Technical University September 2001
7. Turk, M., Pentland, A.: Face recognition using eigenfaces. In: Proceedings of the IEEE Conference on Computer Vision and Pattern Recognition, Maui, HI (1991)
8. Park, B.G., Lee, S.U.: Face recognition using face-arg matching. IEEE Trans. Pattern Anal. Mach. Intell. **27**, 1982–1988 (2005)
9. Fu, Y., Mohammed, U., Elder, J.H., Li, P.: Probabilistic models for inference about identity. IEEE Trans. Pattern Anal. Mach. Intell. **34**, 144–157 (2012)
10. Lu, J., Plataniotis, K.N.: Boosting face recognition on a large-scale database. In: Proceedings of the 2002 International Conference on Image Processing. vol. 2, Rochester, New York (2002)
11. Wu, J., Zhou, Z.: Face recognition with one training image per person. Pattern Recogn. Lett. **23**, 1711–1719 (2002)
12. Mohammadzade, H., Hatzinakos, D.: Projection into expression subspaces for face recognition from single sample per person. IEEE Trans. Affect. Comput. **4**, 69–82 (2013)
13. Chen, X., Zhang, C., Zhou, Z.: Improve recognition performance by hybridizing principal component analysis (pca) and elastic bunch graph matching (ebgm). In: 2014 IEEE Symposium Series on Computational Intelligence, Orlando, FL (2014)
14. Phillips, P., Wechslerb, H., Huang, J., Raussa, P.J.: The feret database and evaluation procedure for face recognition algorithms. Image Vis. Comput. **16**, 295–306 (1998)

Neighbors Based Discriminative Feature Difference Learning for Kinship Verification

Xiaodong Duan$^{(\boxtimes)}$ and Zheng-Hua Tan

Department of Electronic Systems, Aalborg University, Aalborg, Denmark
{xd,zt}@es.aau.dk

Abstract. In this paper, we present a discriminative feature difference learning method for facial image based kinship verification. To transform feature difference of an image pair to be discriminative for kinship verification, a linear transformation matrix for feature difference between an image pair is inferred from training data. This transformation matrix is obtained through minimizing the difference of L2 norm between the feature difference of each kinship pair and its neighbors from non-kinship pairs. To find the neighbors, a cosine similarity is applied. Our method works on feature difference rather than the commonly used feature concatenation, leading to a low complexity. Furthermore, there is no positive semi-definitive constrain on the transformation matrix while there is in metric learning methods, leading to an easy solution for the transformation matrix. Experimental results on two public databases show that the proposed method combined with a SVM classification method outperforms or is comparable to state-of-the-art kinship verification methods.

1 Introduction

Facial image based kinship verification is an interesting and challenging research topic. Its objective is to determine whether two persons are biologically related based on a pair of their facial images. There are many potential applications for kinship verification, including missing child search, social media information analysis and family photo annotation [1].

Kinship verification through facial image analysis has been a recently focused research topic. To realize this, hand-crafted features are used in paper [2], while a type of local feature called Gabor-based gradient orientation pyramid feature is applied in paper [3]. However, it is not easy to have a hand-crafted feature that can capture all the information needed for kinship verification. On the other hand, local feature captures too much detail of facial images, including pose, illumination, and inter-person variations. Therefore, hand-crafted or local features, when directly used for kinship verification, do not perform well.

In order to improve the performance of local feature based methods, Zhou et al. [4] use a K-means learning method to obtain a spatial pyramid learning-based (SPLE) feature descriptor. A mid-level feature for kinship verification is

This work is supported by the Danish Council for Independent Research | Technology and Production Sciences under grant number: 1335-00162 (iSocioBot).

G. Bebis et al. (Eds.): ISVC 2015, Part II, LNCS 9475, pp. 258–267, 2015.
DOI: 10.1007/978-3-319-27863-6_24

learned from one low-level local feature or several low-level local features in paper [5]. A local feature is subtracted by a linear transformation of the symmetry feature for kinship verification in paper [6]. Besides these feature learning methods, metric learning [7] methods are also widely used for kinship verification. Distance metric learning [8] is used for kinship verification in paper [9]. Lu et al. [1] use metric learning methods called repulsed metric learning (NRML) of a single feature and multi-view NRML (MNRML) of multiple features for kinship verification. In paper [10], multiple distance metrics for multiple features are learned for kinship verification.

In this paper, we present a new feature difference learning method aiming to make feature difference between an image pair to be discriminative for kinship verification. Opposite to the existing feature learning methods, our method works on the feature difference, which makes it of low complexity. Furthermore, unlike metric learning methods, there is no positive semi-definitive constrain on the transformation matrix. This is done through a linear transformation matrix applied on the feature difference of an image pair. The transformation matrix is learned through minimizing the difference of L2 norm of feature difference between kinship and non-kinship pairs neighbors. The neighbors are defined by a cosine similarity.

The rest of this paper is organized as follows. Section 2 details the proposed method. Experimental results and conclusion can be found in Sects. 3 and 4, respectively.

2   Local Feature Learning

Suppose we have one set of kinship verification face images consisting of two subsets,

$$\begin{cases} S = \{(x_i, x_j) \mid x_i \text{ and } x_j \text{ are with kinship}\} \\ D = \{(x_i, x_j) \mid x_i \text{ and } x_j \text{ are without kinship}\}, \end{cases} \tag{1}$$

where we call S as a positive pair subset and D as a negative pair subset. We can compute the difference $\mathbf{d}$ between extracted local features from a pair of images as follows,

$$\mathbf{d}_{ij} = \mathbf{f}_i - \mathbf{f}_j, \tag{2}$$

where $\mathbf{f}$ is the extracted feature from a face image, e.g., SIFT, LBP and HOG, and we call $\mathbf{d}$ the original difference.

To obtain good performance, we would want $\|\mathbf{d}_{ij}\|_2$ to be close to 0 when $ij \in S$, while large when $ij \in D$. This means the feature difference between a positive pair should be less than that from a negative pair. We illustrate an ideal case in Fig. 1(a), where the data are generated by two uniform distributions. Unfortunately, in reality, the L2 norm of $\mathbf{d}_{ij}$ from a positive pair is not always less than that of a negative pair as exemplified in Fig. 1(b). There are two reasons for this. First, a local feature can capture much detail of an image including not only the facial information but also the environmental factors. Second, kinship

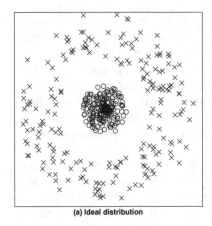 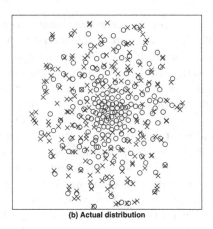

| (a) Ideal distribution | (b) Actual distribution |

Fig. 1. Distributions of the original difference of local features between image pair in ideal and actual situations. ∘ denotes positive pair while × negative one. In the ideal case, the norm and angle of the feature difference are generated using two uniform distributions. For the actual situation, we use t-SNE [11] method to reduce the dimension of **d** to two for 200 negative and 200 positive pairs for visualization

related image pairs share less identical facial information than those from the same person.

To overcome these two problems, we propose to apply a linear transformation through a transformation matrix **A** on original difference **d** as follows,

$$\mathbf{l}_{ij} = \mathbf{A}\mathbf{d}_{ij} = \mathbf{A}(\mathbf{f}_i - \mathbf{f}_j), \tag{3}$$

to make the feature difference between a kinship pair less than that of a non-kinship pair. Then the feature difference between an image pair will be discriminative for kinship verification. Here **A** is learned from training data through minimizing the L2 norm difference between the feature difference of each positive pair and its neighbors from negative pairs. Therefore we call **l** as the learned difference.

If we only use positive pairs, the learning of **A** will not benefit from the information of negative pairs and correlation between positive and negative pairs. When the distribution of negative-pair difference and positive-pair difference are far apart from each other, the feature difference is more discriminative. Therefore, it is important that the L2 norm of the positive-pair difference is less than those of the neighboring negative-pairs. To this end, the L2 norm of positive-pair feature difference is minimized based on those of negative pairs which are close to the positive data point. Consequently, we define the following cost function,

$$J(A) = \sum_{ij \in S} \left(\sum_{mn \in N_{ij}} max \left\{ \|\mathbf{l}_{ij}\|_2 - \|\mathbf{l}_{mn}\|_2, \beta \right\} \right) + \|\mathbf{A}\|_F^2, \tag{4}$$

where

$$N_{ij} = \{pq \mid coss(\mathbf{l}_{pq}, \mathbf{l}_{ij}) \geq \alpha \wedge pq \in D\}, \tag{5}$$

$$coss(\mathbf{l}_{pq}, \mathbf{l}_{ij}) = \frac{\mathbf{l}_{ij}^T \mathbf{l}_{pq}}{\|\mathbf{l}_{ij}\|_2 \|\mathbf{l}_{pq}\|_2}, \tag{6}$$

and $\|\mathbf{A}\|_F^2$ is a regularizer and F is the Frobineus norm.

For each positive-pair difference, its neighbors from negative training dataset are found based on the cosine similarity, which is controlled by α. The L2 norm of positive-pair feature difference should be less by a certain value (defined by β) than those of its negative-pair neighbors. The geometric interpretation of Eqs. (4) to (6) can be found in Fig. 2.

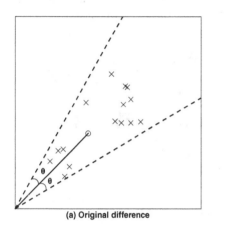
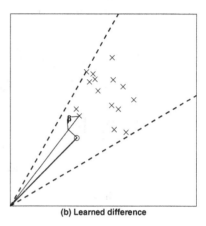

(a) Original difference   (b) Learned difference

Fig. 2. Geometric interpretation of the feature difference learning method. The neighbors of one positive-pair difference are the negative-pairs differences within a certain angle ($\theta = cos^{-1}\alpha$) from the positive-pair difference. Through $\mathbf{A}$ we want the L2 norm of learned difference is less than its negative neighbors by β

As shown in Eq. (7), we can estimate $\mathbf{A}$ by minimizing Eq. (4).

$$\hat{\mathbf{A}} = \arg \min_{\mathbf{A}} J(\mathbf{A}). \tag{7}$$

By taking the derivative of Eq. (4) to $\mathbf{A}$, we can get following equations,

$$\nabla_{\mathbf{A}} J = \sum_{ij \in S} \sum_{mn \in N_{ij}} G_{ijmn} + 2\mathbf{A}, \tag{8}$$

where

$$G_{ijmn} = \begin{cases} 0, & if \; \|\mathbf{l}_{ij}\|_2 - \|\mathbf{l}_{mn}\|_2 \leq \beta, \\ \frac{\mathbf{Ad}_{ij}\mathbf{d}_{ij}^T}{\|\mathbf{l}_{ij}\|} - \frac{\mathbf{Ad}_{mn}\mathbf{d}_{mn}^T}{\|\mathbf{l}_{mn}\|}, & else, \end{cases} \tag{9}$$

and then we can use gradient descent method to minimize $J(\mathbf{A})$.

3   Experiments

3.1   Datasets

The proposed method is evaluated on two public kinship face image databases, namely KinFaceW-I [12] and KinFaceW-II [12]. The face images are collected from the Internet under uncontrolled conditions. There is no restriction in terms of pose, lighting, background, expression, age, ethnicity, and partial occlusion [13]. Both databases consist of four different kin relation subsets: Father-Son (F-S), Father-Daughter (F-D), Mother-Son (M-S), and Mother-Daughter (M-D). Some positive pair samples from these two datasets are shown in Fig. 3.

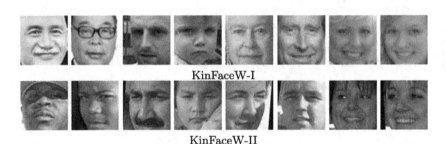

KinFaceW-I

KinFaceW-II

Fig. 3. Positive pair samples from the two databases. From left to right in each respective row, the adjacent pairs are with F-S, F-D, M-S, and M-D relations.

3.2   Experimental Settings

We use five-fold cross validation and follow the image-restricted protocol [13] to evaluate our method, under which only the image pairs and the given kin relation information within the databases are used during the training procedure. The pairs used in our experiment are from pair files within the two databases [13]. In each fold, the number of negative facial image pairs is the same as that of positive pairs as shown in Table 1.

Three popular local features, LBP [14], HOG [15], and SIFT [16], are used in our experiments. We use publicly available LBP and HOG features from the databases. To extract SIFT feature, the images from the two databases are used directly without any preprocessing. Each image is of 64×64 dimensions and is segmented into overlapped cells with the size of 9×9 for calculating SIFT feature. The dimensions of LBP, HOG and SIFT are 3776, 2880 and 18432, respectively. For practical reason, we use principal component analysis (PCA) to reduce the feature dimensions to 100. The feature difference learning method is applied to the PCA processed features. L-BFGS-B [17] method is used to minimize the cost function in Eq. (4). As kinship verification is a binary classification task, we use support vector machine (SVM) with radial basis function (RBF) kernel to learn a classifier based on the learned difference. A SVM classifier is also learned using the original difference as one of the baselines.

Table 1. Number of positive pairs in each fold for the two databases. Each fold consists of the same number of negative pairs as that of positive ones

Fold	KinFaceW-I				KinFaceW-II
	F-S	F-D	M-S	M-D	All subsets
1	27	31	25	23	50
2	27	33	25	23	50
3	27	32	25	23	50
4	27	28	26	23	50
5	26	32	26	24	50

For each five-fold cross validation experiment, samples from four folds are used for PCA, feature difference learning and SVM classifier training while those from the remaining one are used for testing. Libsvm [18] with default settings is used for SVM classification. All the experiments are conducted using the same parameter settings: $\alpha = 0$ and $\beta = -2\sigma$, where σ is the standard deviation of the L2 norm of the positive-pair difference from the training data. Setting α as zero means that θ in Fig. 2 is equal to $\pi/2$. This could help to include many negative-pair differences with very small L2 norm in the neighbors. **A** is initialized as **I**.

3.3   Examples of the Learned Difference

To show the difference between before and after the feature difference learning, we visualize the distributions of original and learned differences of HOG feature from one testing fold of F-S in KinFaceW-II in Fig. 4 using t-SNE method. We can see that, with the learned difference, the positive-pair learned differences are distributed more centrally around the center than the original difference.

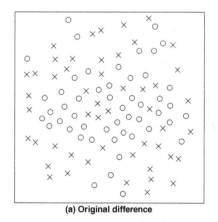

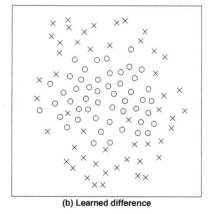

(a) Original difference   (b) Learned difference

Fig. 4. Feature difference distributions before and after feature difference learning

3.4 Experimental Results on KinFaceW-I

The results of the proposed method are listed in Table 2. As can be seen from Table 2, except for LBP for F-D and HOG for F-S and M-S for SIFT, the performances are improved by using the learned difference compared to the original difference. We also compare our result with SILD [19] method, which is also based on single feature following the image-restricted protocol. Our results are very close to those of SILD on KinFace-I.

Table 2. The verification accuracy on KinFaceW-I. Mean stands for the mean accuracy across four different kinships. The results of SILD [19] are from the database website [13]

	LBP			HOG			SIFT	
	SILD [19]	Original	Learned	SILD [19]	Original	Learned	Original	Learned
F-S	78.2	77.6	78.2	80.5	82.4	79.2	74.3	75.2
F-D	69.4	66.8	66.1	72.4	70.2	73.1	68.0	68.3
M-S	66.8	62.4	66.3	69.8	66.8	70.3	65.1	65.1
M-D	70.1	65.7	70.9	77.1	72.1	72.5	70.2	72.4
Mean	71.1	68.1	70.4	74.9	72.9	73.8	69.4	70.3

As shown in Fig. 5, we have compared our method with five types of previously published state-of-the-art methods, ULPGC [20], BIU [20], MNRML [1], DMML [10] and MPDFL [5], which use multiple features and/or image-unrestricted protocol. The mean result of the proposed method based on HOG ranks the second, while ULPGC achieves the best performance. But ULPGC is based on six different local features.

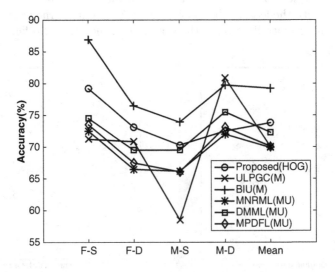

Fig. 5. Comparison to other state-of-the-art methods on KinFaceW-I, where M represents multiple feature and U means image-unrestricted protocol

3.5   Experimental Results on KinFaceW-II

Table 3 shows that, on KinFacwW-II, all learned differences of three local features lead to a performance improvement over original differences. The proposed method also achieves better result than SILD using LBP and HOG. Based on HOG feature, the proposed method obtains the best result with 4.4 percent absolute improvement compared to using the original difference.

Table 3. The verification accuracy on KinFaceW-II. Mean stands for the mean accuracy across four different kinships. The results of SILD [19] are from the database website [13]

	LBP			HOG			SIFT	
	SILD [19]	Original	Learned	SILD [19]	Original	Learned	Original	Learned
F-S	78.2	77.0	77.6	79.6	81.6	86.2	77.6	78.6
F-D	70.0	72.6	74.6	71.6	75.0	77.2	73.0	76.4
M-S	71.2	70.6	74.4	73.2	74.2	78.6	73.2	75.6
M-D	67.8	66.6	70.6	69.6	68.8	75.0	72.0	73.4
Mean	71.8	71.7	74.3	73.5	74.9	79.3	74.0	76.0

As can be seen from Fig. 6, the mean accuracy of the proposed method with HOG is very close to ULPGC and BIU (about one percent difference) and is better than the other three methods that use multiple features and extra training data on KinFaceW-II.

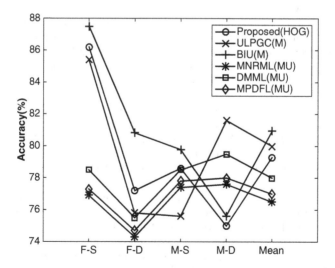

Fig. 6. Comparison to other state-of-the-art methods on KinFaceW-II, where M represents multiple feature and U means image-unrestricted protocol

4   Conclusion

In this paper we presented a novel feature difference learning method for kinship verification. Through minimizing the difference of L2 norm of feature difference between each kinship pair and its neighbors from non-kinship pairs, a transformation matrix is learned. The neighbors are defined by a cosine similarity. There is no positive semi-definitive constrain on the transformation matrix meaning an easy solution for the transformation matrix. Furthermore, our method is based on the feature difference instead of concatenation of the features of an image pair, leading to a low complexity. Experimental results under the image-restricted protocol have shown, on two image-unconstrained public databases, the comparable even superior performance of the proposed method over state-of-the-art methods including fusion methods that combine multiple local features and/or under unrestricted protocol. In the end, the performance of the proposed method may be further improved using multiple features and generating more training data.

References

1. Lu, J., Zhou, X., Tan, Y.P., Shang, Y., Zhou, J.: Neighborhood repulsed metric learning for kinship verification. IEEE Trans. Pattern Anal. Mach. Intell. **36**, 331–345 (2014)
2. Fang, R., Tang, K., Snavely, N., Chen, T.: Towards computational models of kinship verification. In: 2010 17th IEEE International Conference on Image Processing (ICIP), pp. 1577–1580 (2010)
3. Zhou, X., Lu, J., Hu, J., Shang, Y.: Gabor-based gradient orientation pyramid for kinship verification under uncontrolled environments. In: Proceedings of the 20th ACM International Conference on Multimedia. MM 2012, New York, NY, USA, pp. 725–728. ACM (2012)
4. Zhou, X., Hu, J., Lu, J., Shang, Y., Guan, Y.: Kinship verification from facial images under uncontrolled conditions. In: Proceedings of the 19th ACM International Conference on Multimedia, pp. 953–956. ACM (2011)
5. Yan, H., Lu, J., Zhou, X.: Prototype-based discriminative feature learning for kinship verification. IEEE Trans. Cybern. **9**(1), 51–61 (2014)
6. Duan, X., Tan, Z.H.: A feature subtraction method for image based kinship verification under uncontrolled environments. In: IEEE International Conference on Image Processing (ICIP) (2015)
7. Kulis, B.: Metric learning: A survey. Found. Trends Mach. Learn. **5**, 287–364 (2012)
8. Xing, E.P., Jordan, M.I., Russell, S., Ng, A.Y.: Distance metric learning with application to clustering with side-information. In: Advances in Neural Information Processing Systems, pp. 505–512 (2002)
9. Shao, M., Xia, S., Fu, Y.: Genealogical face recognition based on ub kinface database. In: 2011 IEEE Computer Society Conference on Computer Vision and Pattern Recognition Workshops (CVPRW), pp. 60–65 (2011)
10. Yan, H., Lu, J., Deng, W., Zhou, X.: Discriminative multimetric learning for kinship verification. IEEE Trans. Inform. Forensics Secur. **9**, 1169–1178 (2014)
11. Van der Maaten, L., Hinton, G.: Visualizing data using t-sne. J. Mach. Learn. Res. **9**, 85 (2008)

12. Lu, J., Hu, J., Zhou, X., Shang, Y., Tan, Y.P., Wang, G.: Neighborhood repulsed metric learning for kinship verification. In: 2012 IEEE Conference on Computer Vision and Pattern Recognition (CVPR), pp. 2594–2601. IEEE (2012)
13. Lu, J., Hu, J.: Kinship face in the wild. (http://www.kinfacew.com/) Accessed August 20, 2015
14. Ahonen, T., Hadid, A., Pietikäinen, M.: Face recognition with local binary patterns. In: Pajdla, T., Matas, J.G. (eds.) ECCV 2004. LNCS, vol. 3021, pp. 469–481. Springer, Heidelberg (2004)
15. Dalal, N., Triggs, B.: Histograms of oriented gradients for human detection. In: IEEE Computer Society Conference on Computer Vision and Pattern Recognition, 2005. CVPR 2005, vol. 1, pp. 886–893. IEEE (2005)
16. Lowe, D.G.: Distinctive image features from scale-invariant keypoints. Int. J. Comput. Vis. **60**, 91–110 (2004)
17. Zhu, C., Byrd, R.H., Lu, P., Nocedal, J.: Algorithm 778: L-bfgs-b: Fortran subroutines for large-scale bound-constrained optimization. ACM Trans. Math. Softw. (TOMS) **23**, 550–560 (1997)
18. Chang, C.C., Lin, C.J.: LIBSVM: A library for support vector machines. ACM Trans. Intell. Syst. Technol. **2**, 27:1–27:27 (2011). http://www.csie.ntu.edu.tw/cjlin/libsvm
19. Kan, M., Shan, S., Xu, D., Chen, X.: Side-information based linear discriminant analysis for face recognition. In: BMVC, pp. 1–12 (2011)
20. Lu, J., Hu, J., Liong, V., Zhou, X., Bottino, A., Ul Islam, I., Figueiredo Vieira, T., Qin, X., Tan, X., Chen, S., Mahpod, S., Keller, Y., Zheng, L., Idrissi, K., Garcia, C., Duffner, S., Baskurt, A., Castrillon-Santana, M., Lorenzo-Navarro, J.: The fg 2015 kinship verification in the wild evaluation. In: 2015 11th IEEE International Conference and Workshops on Automatic Face and Gesture Recognition (FG), pp. 1–7 (2015)

A Comparative Analysis of Two Approaches to Periocular Recognition in Mobile Scenarios

João C. Monteiro[1]([✉]), Rui Esteves[2], Gil Santos[3], Paulo Torrão Fiadeiro[4],
Joana Lobo[2], and Jaime S. Cardoso[1]

[1] INESC TEC and Faculdade de Engenharia, Universidade do Porto,
Campus da FEUP, Rua Dr. Roberto Frias, 378, 4200-465 Porto, Portugal
jcmonteiro89@gmail.com
[2] Associação Fraunhofer Portugal Research,
Rua Alfredo Allen 455/461, 4200-135 Porto, Portugal
[3] IT - Instituto de Telecomunicacoes, Lisboa, Portugal
[4] Departamento de Física, Unidade de Detecção Remota
Universidade da Beira Interior, Rua Marquês D'Ávila e Bolama,
6201-001 Covilhã, Portugal

Abstract. In recent years, periocular recognition has become a popular alternative to face and iris recognition in less ideal acquisition scenarios. An interesting example of such scenarios is the usage of mobile devices for recognition purposes. With the growing popularity and easy access to such devices, the development of robust biometric recognition algorithms to work under such conditions finds strong motivation. In the present work we assess the performance of extended versions of two state-of-the-art periocular recognition algorithms on the publicly available CSIP database, a recent dataset composed of images acquired under highly unconstrained and multi-sensor mobile scenarios. The achieved results show each algorithm is better fit to tackle different scenarios and applications of the biometric recognition problem.

1   Introduction

Over the past few years face and iris have been on the spotlight of many research works in biometrics. The *face* is a easily acquirable trait with a high degree of uniqueness, while the *iris*, the coloured part of the eye, is composed by a set of irregular textural patterns resulting from its random morphogenesis during embryonic development [1]. These marked advantages, however, fall short when low-quality images are presented to the system. With the increasing popularity and availability of mobile devices capable of performing the whole biometric recognition framework, from data acquisition to final decision, serves as further motivation for research in the field of unconstrained biometrics [2]. Several recent works have tried to explore alternative hypotheses to overcome this challenge, either by developing more robust algorithms or by exploring new traits to allow or aid in the recognition process.

ⓒ Springer International Publishing Switzerland 2015
G. Bebis et al. (Eds.): ISVC 2015, Part II, LNCS 9475, pp. 268–280, 2015.
DOI: 10.1007/978-3-319-27863-6_25

The *periocular* region is one of such unique traits. It is common to describe the periocular region as the region in the immediate vicinity of the eye. Periocular recognition can be motivated as a representation in between face and iris recognition. It has been shown to present increased performance when only degraded facial data or low quality iris images are made available. Even in mobile application scenarios, the periocular region does not require rigid capture or complex imaging systems, thereby making it easy to acquire even by an inexperienced user. Nevertheless, several problems arise when attempting to perform periocular biometrics in mobile environments. The wide variety of camera sensors and lenses used in mobile devices produce discrepancies in working images, as they might be acquired with both color distortions and multiple resolutions. On-the-go acquisition by inexperienced subjects will result in demanding pose, illumination, and expression changes, thereby yielding variable acquisition angles and scales, or rotated images. All these limitations are intrinsic to the nature of mobile devices and must, thus, be handled by the recognition algorithm.

On the present work we aim to compare the performance of two state-of-the-art approaches to periocular recognition - an extension of Monteiro et al. [3] to multiple features and fusion strategies and Santos et al. [2] - when exposed to images acquired on multiple mobile scenarios, using a recently collected multi-sensor periocular database [2]. We evaluate both approaches with regards both to recognition performance as well as the processing time, with real-world applications in mind. Finally, we present some preliminary results on cross-sensor periocular recognition, thorough the analysis of whether or not multiple sensors from varying manufacturers present meaningful interoperability.

2   Related Work

Periocular biometrics is a recent area of research, proposed by the first time in a feasibility study by Park et al. [4]. In this pioneer work, the authors suggested the periocular region as a potential alternative to circumvent the significant challenges posed to iris recognition systems working under unconstrained scenarios. The same authors analysed the effect of degradation on the accuracy of periocular recognition [5]. Padole and Proença [6] also explore the effect of scale, pigmentation and occlusion, as well as gender, and propose an initial region-of-interest detection step to improve recognition accuracy.

Ross et al. [7] explored information fusion based on several feature extraction techniques, to handle the significant variability of input periocular images. Information fusion has become one of the trends in biometric research in recent years and periocular recognition is no exception.

Some works have explored the advantages of the periocular region as an aid to more traditional approaches based on iris. Joshi et al. [8] proposed feature level fusion of wavelet coefficients and LBP features, from the iris and periocular regions respectively, with considerable performance improvement over both singular traits. A more recent work by Tan et al. [9] has also explored the benefits of periocular recognition when highly degraded regions result from the traditional iris segmentation step. The authors have observed discouraging performance when the iris region alone is considered in such scenarios, whereas introducing information from the whole periocular region lead to a significant improvement. A thorough review of the most relevant method in recent years concerning periocular recognition and its main advantages can be found in the work by Santos and Proença [10].

On the present work we chose to perform a comparative analysis of two recent approaches to the issue of periocular recognition, when working with images acquired in unconstrained mobile scenarios. These works, by Monteiro et al. [3] and Santos et al. [2], will be analyzed in further detail in the following section.

3 Recognition Algorithms

This section will detail both previously referred methodologies for periocular recognition. Both approaches will be analysed in a comparative scenario so as to ascertain their main advantages and disadvantages in the mobile acquisition environments that serve as motivation for the present work.

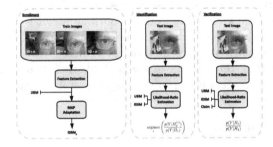

Fig. 1. Flow diagram of the main composing blocks of the methodology proposed in [3].

3.1 Method 1: GMM-UBM

The GMM-UBM algorithm for periocular recognition, first proposed by Monteiro et al. [3], is schematically represented in Fig. 1. During the enrollment, a set of N models describing the unique statistical distribution of biometric features for each individual $n \in \{1, \ldots, N\}$ is trained by maximum *a posteriori* (MAP) adaptation of an Universal Background Model (UBM). The UBM is a representation of the variability that the chosen biometric trait presents in the universe of all individuals. MAP adaptation works as a specialization of the UBM based on each individual's biometric data. The idea of MAP adaptation of the UBM was first proposed by Reynolds [11], for speaker verification. The tuning of the UBM parameters in a maximum *a posteriori* sense, using individual specific biometric data, provides a tight coupling between the individual models and the UBM, resulting in better performance and faster scoring than uncoupled methods, as well as a robust and precise parameter estimation, even when only a small amount of data is available.

The recognition phase is carried out through the projection of the features extracted from an unknown sample onto both the UBM and the individual specific models (IDSM) of interest. A likelihood-ratio between both projections outputs the final recognition score. Depending on the functioning mode of the system - verification or identification - decision is carried out by thresholding or maximum likelihood-ratio respectively. The use of a likelihood-ratio score with an universal reference works as a normalization step, mapping the likelihood values in accord to their global projection. Without such step, finding a global optimal value for the decision threshold would be a far more complex process.

Gaussian Mixture Models (GMM) were chosen to model both the UBM and the individual specific models (IDSM). Regarding feature extraction while the original algorithm used SIFT keypoint descriptors alone, on the present work we present an extension to four distinct descriptors, as detailed later.

3.2   Method 2: Santos et al.

This algorithm, proposed in [2], may be divided into four main blocks: normalization using a device-specific color correction and region-of-interest (ROI) definition; feature encoding using information from both the iris and the periocular region; feature matching and score-level fusion. The flow of information through the aforementioned blocks is schematically depicted in Fig. 2.

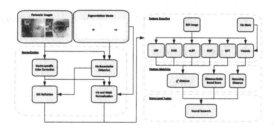

Fig. 2. Flow diagram of the main composing blocks of the methodology proposed in [2].

During the normalization block a device-specific color-correction was applied so as to compensate for possible chromatic distortions observed in real-life scenarios. Another variability source that is commonly observed in data acquired with mobile devices is variable scale. In order to overcome such problem, and making use of a state-of-the-art iris segmentation algorithm [12], the authors propose a segmentation of the iris boundary, to serve as a reference for the periocular region. Using the previously calculated radius of the iris, r_i, the periocular ROI was defined as 35 square patches that formed a 7×5 grid, where each patch had an area equivalent to $1.4r_i^2$.

Periocular data was encoded using a similar feature extraction scenario as the one described for the previous methodology. SIFT, HOG, uLBP (as well as the original LBP) and GIST, were also tested independently and used in a conjugated manner. For the iris region, in addition to these descriptors, a fifth approach was also explored, using the original iriscode algorithm proposed by Daugman [13]. For a single image a total of 11 feature descriptors is, therefore, extracted: 5 for the periocular region and the same 5 plus the iriscode for the iris region.

Matching was carried out by comparing the 11 pairs of feature descriptors extracted from a pair of images, using the matching algorithm specified for each of them (χ^2 for histogram-based algorithms, distance-based score for pairs of SIFT keypoints and Hamming Distance for iriscode), resulting, thus, in 11 individual scores. Performance can then be evaluated either for each descriptor individually or by exploring more complex fusion strategies. In the original algorithm, a multi-layer perceptron artificial neural network was used to achieve such fusion.

3.3   Algorithm Extensions

Besides the comparative analysis between the two methodologies described in the previous sections, we also present some extensions to their original formulations. Such modifications will be present in the following sections.

Pre-processing. An no pre-processing strategy was included in the original formulation of the GMM-UBM algorithm, we aimed to assess if its presence could bring about a significant improvement in performance. We chose the Discrete Cosine Transform (DCT), as proposed by Chen et al. [14] for illumination normalization in face images, as it yielded the best overall performance. This technique is based on the removal of low-frequency coefficients of the DCT, in order to compensate for the variations in lighting conditions, that are known to lie, mainly, on such frequency band. We also tested the device-specific colorcorrection technique proposed in Method 2.

Periocular Segmentation. As the GMM-UBM approach presented no preliminary periocular segmentation in its original form, and in order to achieve an uniform set of conditions for performance comparison, we chose to perform segmentation with the same methodology used for Method 2, as described in Sect. 3.2.

Feature Descriptors. Similar to segmentation, we chose to explore the performance of multiple features using Method 1, as we believed that multiple sources of information might offer the algorithm an increased robustness when dealing with more complex and realistic datasets. Similar to Method 2 we chose to test the GMM-UBM with the LBP, HOG and GIST descriptors, besides the original SIFT formulation.

Fusion Strategies. Fusion scenarios can contribute, in some complex situations, to an overall improvement of system performance. On the present work, two fusion strategies at score level were evaluated: performance-weighted score-level fusion and neural network score-level fusion. Both are novel to Method 1, which did not include any score-level fusion in its original formulation, while the simpler performance-weighted strategy is tested as an alternative to the original version of Method 2.

- **Performance-Weighted Score-Level Fusion:** the fusion score, s_f, is obtained by a weighted sum of the individual scores obtained for each feature individually:

$$s_f = w_{feat_1} \times s_{feat_1} + w_{feat_2} \times s_{feat_2} + ... + w_{feat_N} \times s_{feat_N} \qquad (1)$$

where s_{feat_n} is the individual score obtained for feature $n \in 1...N$ and w_{feat_n} is its corresponding weight in the final score. The weight of each feature is computed in relation to its individual performance:

$$w_{feat_n} = \frac{p_{feat_n}}{\sum_{i=1}^{N} p_{feat_i}} \qquad (2)$$

where p_{feat_n} is the individual performance obtained with a specific metric for feature n and the denominator term is introduced so that $\sum_i w_{feat_i} = 1$.

- **Neural Network Score-Level Fusion:** the final recognition score, s_f, is obtained by a multilayer perceptron artificial neural network (MLP-NN), trained on a small data partition which is not included in the test phase. NN-based methods have been applied widely to classification problems because of their high learning capacities and good generalization [2]. In the present study, two hidden layers NN were trained using back-propagation. The architecture of the NN was as follows: the first hidden layer had the same number of neurons as the number of individual scores derived from the matching stage, i.e. 4 for Method 1 and 11 for Method 2; the second hidden layer had 2 and 6 neurons for Methods 1 and 2 respectively, while the final (output) layer presented a single layer outputting the final s_f value.

Besides these strategies, we also analyzed the effect of fusing information from different channels of the RGB colorspace. Integration of information, in this case, was performed by treating each color channel individually and computing three independent recognition scores for each one: r_R, r_G and r_B. The final score was obtained by simple averaging these three values.

4   Results and Discussion

The present section will serve as a detailed analysis of the comparison carried out between the two algorithms presented in the last section. We start by offering some insight regarding the specific details of the multi-sensor periocular database on which both algorithms were assessed, as well as the experimental setups and performance metrics used for such assessment. We then present and discuss the main results regarding both recognition performance as well as processing time and possible limitations and advantages of each algorithm in real-world scenarios.

4.1 CSIP Database

The CSIP database, created for the assessment of the original version of Method 2 [2], is a recent and publicly available dataset, designed with the main goal of gathering periocular images from a representative group of participants, acquired using a variety of mobile sensors under a set of variable acquisition conditions. Given the heterogeneity of the camera sensors and lens setups of consumer mobile devices, 10 different setups were used during the dataset acquisition stage: four different devices, some of which had both frontal and rear cameras, and LED flash. This variety of sensors confers a strong appeal to the CSIP database regarding its potential use for the assessment of algorithms under a highly heterogeneous set of conditions. A visual example of an image for each subset of the same individual is depicted in Fig. 3. Each participant was imaged using all of the test setups.

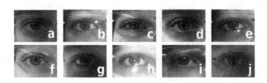

Fig. 3. Examples of images from each subset of the CSIP database. From (a-j) respectively: $AR0$, $AR1$, $BF0$, $BR0$, $BR1$, $CF0$, $CR0$, $CR1$, $DF0$ and $DR0$.

To simulate the variable noise associated with on-the-go recognition, participants were not imaged at a single location, but instead they were enrolled at multiple sites with artificial, natural, and mixed illumination conditions. In total, 50 participants were enrolled, all Caucasian and mostly males (82 %), with ages ranging between 21 and 62 years (mean = 31.18 ± 9.93 years). For each periocular image acquired by the mobile devices, a binary iris segmentation mask was also produced. The masks were obtained automatically using the state-of-the-art iris segmentation approach proposed by Tan et al. [12], which is particularly suitable for uncontrolled acquisition conditions, as demonstrated by its first place ranking at Noisy Iris Challenge Evaluation - Part 1 (NICE.I) (NICE.I) [15].

4.2 Experimental Setup

In order to achieve a fair comparison between both tested algorithms, a uniform experimental setup was defined and adapted to fit the specifics of each method. With that in mind, the set of all images of the CSIP dataset was divided as follows: 50 % of the images per individual and per subset were kept to either train the models in Method 1 or to serve as reference for each identity in Method 2; the remaining 50 %, apart from a small independent set used to train the fusion neural networks, were used to assess the performance of both methodologies. Performance assessment was adapted to fit the nature of the originals algorithms:

- **Method 1:** given an input image I of an unknown source and an associated identity claim S, the score, r_s, associated with this image/claim pair is computed by the likelihood-ratio, $r_s = \frac{proj(desc(I), IDSM_S)}{proj(desc(I), IDSM_{UBM})}$, where $desc(I)$ represents the feature descriptor extracted from image I and $proj(X, GMM)$ represents the projection of feature vector X onto a specific GMM (either the claimed ID's IDSM or the UBM). This process is repeated for every possible ID, so that for image I the assessment block outputs a total of N scores, with N being the total number of individuals enrolled in the database.
- **Method 2:** given an input image I of an unknown source and an associated identity claim S, the score, r_s, associated with this image/claim pair is computed by the averaging of image/image pair similarities, $r_s = \frac{\sum_{k=1}^{n_{ID}} score(I, I_{ID,k})}{n_{ID}}$, where $score(I, I_{ID,k})$ is the comparison score obtained using Method 2 between the unknown image I and the k-th reference image from a specific known ID. The averaging is made in relation to the total number of reference images for the given ID, n_{ID}. This process is repeated for every possible ID, so that for image I the assessment block outputs a total of N scores, with N being the total number of individuals enrolled in the database.

After the N scores are extracted for each image, using both methodologies, performance is assessed for either identification or verification modes. For identification we chose to use the *rank-1 recognition rate* metric, which represents the rate of images for which the highest of the N recognition scores corresponded to the true ID. On the other hand, for verification, we computed the *equal error rate*. This value corresponds to the error rate observed when a specific acceptance threshold is applied to the recognition scores and the resulting false positive and false negative rates are equivalent.

4.3 Performance Comparison

The main results obtained for the setups outlined in the previous sections are summarized in Table 1. All results concern the average performance observed with 10-fold cross-validation for a specific methodology (GMM-UBM or Santos et al. [2]), pre-processing strategy (device specific color correction - DS-CC - or discrete cosine transform - DCT), and single feature (LBP, SIFT, GIST or HOG) or fusion of multiple features (performace-weighted sum-rule - PW-SR - or multilayer perceptron artificial neural network - NN).

Careful observation of the values presented in Table 1 allows for some interesting conclusions to be achieved. Regarding both tested metrics, there is no significantly better algorithm for all the tested subsets. While the UBI algorithm achieves the best average identification performance (Table 1) amongst the whole set of tested subsets, the UBM algorithm still manages to achieve significantly better performances for two of such subsets - $CF0$ and $DR0$ - while managing to achieve values in a very similar range for five other subsets - $AR0$, $AR1$, $BR0$, $BR1$ and $CF0$. In fact, only for the $BF0$, $CR0$ and $DF0$ subsets does the difference in performance between the two algorithms become significant. This non-uniformity in the relative behaviour of performance between the two methodologies might indicate that even though the sources of information

Table 1. Rank-1 recognition rates obtained for each subset of the CSIP database for some variations of both tested methodologies.

	Pre-Proc.	Feat(s)	Trait(s)	CSIP Subset									
				AR0	AR1	BF0	BR0	BR1	CF0	CR0	CR1	DF0	DR0
M.1	DS-CC	PW-SR	P	88.4	97.8	73.5	86.5	93.1	85.0	72.1	92.1	45.5	81.7
	DCT	PW-SR	P	94.9	97.7	75.0	86.5	91.7	91.3	83.7	93.2	55.5	**91.7**
$r_1(\%)$	DCT	RGB PW-SR	P	**97.4**	100	83.8	93.2	95.8	**92.5**	82.6	94.3	50.0	95.0
	DCT	NN	P	94.3	99.4	83.8	89.3	96.8	84.2	79.8	94.2	54.8	82.9
M.2	DS-CC	LBP	P	89.8	100	75.6	81.7	91.8	75.3	79.6	96.8	59.4	72.0
	DS-CC	SIFT	P	81.8	100	73.2	84.2	100	61.8	61.3	96.8	82.8	57.3
	DS-CC	GIST	P	96.6	100	87.8	93.0	98.8	84.3	**92.5**	94.6	75.0	84.0
$r_1(\%)$	DS-CC	HOG	P	68.2	92.6	56.1	64.6	72.9	42.7	55.9	78.5	40.6	54.7
	DS-CC	NN	P+I	95.5	100	**92.7**	**95.3**	95.3	85.4	90.3	97.9	**76.6**	80.0
	DS-CC	PW-SR	P+I	93.2	100	90.2	91.4	98.8	87.6	81.7	**98.9**	73.4	82.7
M.1	DS-CC	PW-SR	P	3.3	3.0	8.7	**4.0**	1.4	6.2	8.0	4.5	17.4	7.5
	DCT	PW-SR	P	2.5	2.9	7.9	4.3	1.7	4.6	**5.1**	4.2	**14.8**	5.0
$r_1(\%)$	DCT	RGB PW-SR	P	**1.7**	**0.5**	7.5	4.4	1.5	**4.5**	5.3	4.0	19.3	**4.1**
	DCT	NN	P	2.1	0.7	**7.3**	5.3	1.3	5.9	9.6	4.0	24.0	11.3
M.2	DS-CC	LBP	P	9.6	4.3	14.7	17.4	10.1	13.5	11.0	5.7	23.9	16.0
	DS-CC	SIFT	P	9.9	0.6	16.5	14.5	**0.1**	16.0	19.2	**2.1**	28.3	18.1
	DS-CC	GIST	P	5.2	3.3	11.1	12.0	6.3	10.3	8.9	5.3	20.2	12.0
$r_1(\%)$	DS-CC	HOG	P	17.1	7.4	24.4	23.7	13.6	21.8	20.8	11.8	28.9	22.5
	DS-CC	NN	P+I	6.3	0.6	9.1	11.7	7.1	10.1	9.7	4.0	19.1	13.1
	DS-CC	PW-SR	P+I	7.5	0.8	10.4	12.2	5.9	9.3	10.5	3.5	20.5	12.6

and the feature descriptors used for its encoding are very similar, the modeling strategies used in both works adapt better for some acquisition scenarios. By visual observation of the images it is readily understandable that the $BF0$ and $DF0$ images are the ones that present lower resolution and overall image quality (Fig. 3). The UBM modeling strategy might, therefore, not be able to aptly train GMMs capable of correctly and in a robust way describe such low quality data. As the UBI data uses direct matching algorithms that are optimized for each feature descriptor, the results in low quality data might be improved. On the other hand, when the quality of input images is a bit higher, as far as images acquired with mobile devices can go, the UBM algorithm either reaches the same ranges of the UBI results, or even exceeds it for moderately more complicated scenarios such as $CF0$ and $DR0$. A different set of observations can, however, be carried out by the analysis of the EER values used to assess performance in verification scenarios (Table 1). Here the UBM algorithm consistently achieves better performance regardless of the tested subset. This variable behavior might indicate that while the UBI algorithm present a higher discriminative power between individuals, the UBM algorithm, probably due to the score normalization effect inherent to the method, is more fit to distinguish between classes (genuine and impostor users) in an identity check application.

A behavior that is easily observable, regardless of the methodology and subset that we choose to focus on, is that images acquired with flash illumination present considerably better results than their non-flash counterparts. This observation was somewhat expected, as flash illumination might serve as a solution to overcome the variable lighting conditions that were referred as a natural limitation of mobile device acquisition in Sect. 1. To the extent of our knowledge no ocular health problems are commonly associated to overexposure to flash illumination in mobile devices, and with the growing technological advances in the manufacturing of such devices, flash illumination might play a crucial role in the implementation of image-based biometrics in mobile environments in the near future. Concerning pre-processing, Table 1 shows that DCT normalization far exceeds the performance obtained with the device-specific color-correction proposed in the original work with the CSIP database. Using a fixed transformation rather than a device-specific approach, that relies on the definition of new transformation matrices for each new device, presents a more robust and reliable alternative as far as the integration of periocular recognition in real-life applications is concerned. Furthermore, it can be seen that the results obtained using information from the three available color channels also results in a non-negligible increase in performance for a variety of subsets. Even though the recognition performance is increased, it must be noted that processing time is increased three times, as the same algorithm must be run in three separate instances. Even though these instances could be ran in parallel, the technological burden for such approach in mobile devices might exceed the current limits.

One more topic to take into consideration regards the alternative fusion strategies that were tested. From Santos et al.s results, it can be observed that the use of neural networks over a simpler performance-weighted approach, results mostly in non-significant variations in the performance. A similar set of conclusions can be drawn for the GMM-UBM results.When comparing this results with the ones obtained for individual features, however, the positive effect of fusion, regardless of its details, is readily discernible. The choice of the fusion strategy should, therefore, be constrained by the specific scenario of application and on how each strategy performs.

On a final note, regarding the processing time of each tested methodology, some considerations can be taken. Given an unknown image I and a single identity check, either by likelihood-ratio or average image similarity, the single-image processing time was computed and averaged for all test images. It was observed that Method 1 using DCT normalization and performance-weighted sum-rule fusion spent an average of 0.018 seconds on this process, whereas Method 2 spent an average of 0.130 seconds for an analogous computation, using device-specific color correction and neural network fusion. This discrepancy is easily explained by the larger amount of features used by Method 2 (11 vs. 4). Real-life applications based on periocular recognition are expected to work as fast as possible, so as to accurately replicate real-time operation. With that in mind, the UBM approach, with its uniform and fast matching algorithm based on GMM projections, seems to present an interesting alternative for further research. Even though the performance obtained for more unconstrained scenarios (for examples $DF0$) is still far from acceptable, future work on more robust representations or future improvements to the intrinsic architecture of the algorithm might help overcome such limitations. The same thought is applicable to the UBI approach, where a more efficient matching strategy might bring about considerable decreases in processing time, with no significant nefarious effect over performance.

4.4 Cross-Sensor Recognition

Some preliminary experiments were also carried out in cross-sensor scenarios. In this alternative to the biometric recognition problem, enrollment and recognition are carried out using data acquired using different sensors. Some of the most interesting results obtained were that, for both methods, only some specific setups showed considerable interoperability. For example, using the GMM-UBM approach, and using the AR1 set for training and the BR1 set for testing, a recognition rate of 86.1 % was achieved, whereas the single setup scenarios yielded 100 % and 95.8 % respectively. This relatively small loss in performance, when compared to other cross-sensor scenarios, might relate to similarities in the hardware of the rear cameras of devices A and B, as well as to the more uniform conditions in lighting, as a result of flash illumination. A similar behavior was observed for a few other pairs of setups - $BR1/CR1$, $AR0/BR0$ and $BR0/DR0$ for example - but, in general, a signifi- cant drop in performance is observed for cross-sensor scenarios, regardless of the tested methodology. Further research is, therefore, needed to achieve stability in performance when enrollment and testing are carried out in highly variable acquisition conditions.

5 Conclusions and Future Work

In the present work we assess the performance of two extended versions of state-of-the-art algorithms for periocular recognition in mobile devices. We extend Monteiro et al. to multiple feature representations and score-level fusion, and adapt Santos et al. to an alternative fusion strategy. The comparative analysis of both approaches shows that, depending on the specific real-world application for which a system is developed, each algorithm presents its advantages and disadvantages. Santos et al. is more fit to identification problems with less restrictions concerning processing time, presenting high performance for a wide variety of noise factors. On the other hand, the GMM-UBM approach presents a faster matching time, with better performance in verification scenarios. Regarding mobile device applications, it would be interesting to explore a joint methodology that managed to keep the fast matching step from Method 1, while achieving good performance in both identification and verification. Further research on both methodologies could lead to interesting improvements. It would also be of relevance to assess how the presented methodologies behave, in their current state, when implemented in existing mobile devices, so as to better understand their current limitations in more realistic scenarios. Another interesting focus of future work would concern on the improvement of performance in the cross-sensor scenario presented in Sect. 4.4, so that enrollment in a single device could serve for recognition purposes in multiple environments.

Acknowledgments. The first author would like to thank Fundação para a Ciência e Tecnologia (FCT) - Portugal the financial support for the PhD grant SFRH/BD/87392/2012. The second and fifth authors would like to acknowledge the financial support obtained from North Portugal Regional Operational Programme (ON.2 - O Novo Norte), Portuguese National Strategic Reference Framework (NSRF) and the European Regional Development Fund (EDRF) from European Union through project ICT4DCC (NORTE-07-0124-FEDER-000042.). The third and fourth authors would like to acknowledge the financial support provided by FCT through the research grant SFRH/BD/80182/2011, and the RSU - Remote Sensing Unit through PEst-OE-FIS/UI0524/2014.

References

1. Bakshi, S., Kumari, S., Raman, R., Sa, P.K.: Evaluation of periocular over face biometric: A case study. Procedia Eng. **38**, 1628–1633 (2012)
2. Santos, G., Grancho, E., Bernardo, M.V., Fiadeiro, P.T.: Fusing iris and periocular information for cross-sensor recognition. Pattern Recogn. Lett. **57**, 52–59 (2015)
3. Monteiro, J.C., Cardoso, J.S.: Periocular recognition under unconstrained settings with universal background models. In: Proceedings of the International Conference on Bio-inspired Systems and Signal Processing (BIOSIGNALS) (2015)
4. Park, U., Ross, A., Jain, A.K.: Periocular biometrics in the visible spectrum: A feasibility study. In: IEEE 3rd International Conference on Biometrics: Theory, Applications, and Systems, pp. 1–6 (2009)
5. Park, U., Jillela, R.R., Ross, A., Jain, A.K.: Periocular biometrics in the visible spectrum. IEEE Trans. Inf. Forensics Secur. **6**, 96–106 (2011)
6. Padole, C.N., Proenca, H.: Periocular recognition: Analysis of performance degradation factors. In: 5th IAPR International Conference on Biometrics, pp. 439–445 (2012)
7. Ross, A., Jillela, R., Smereka, J.M., Boddeti, V.N., Kumar, B.V., Barnard, R., Hu, X., Pauca, P., Plemmons, R.: Matching highly non-ideal ocular images: An information fusion approach. In: 2012 5th IAPR International Conference on Biometrics (ICB), pp. 446–453. IEEE (2012)
8. Joshi, A., Gangwar, A.K., Saquib, Z.: Person recognition based on fusion of iris and periocular biometrics. In: 2012 12th International Conference on Hybrid Intelligent Systems (HIS), pp. 57–62. IEEE (2012)
9. Tan, C.W., Kumar, A.: Towards online iris and periocular recognition under relaxed imaging constraints. Image Process. IEEE Trans. **22**, 3751–3765 (2013)
10. Santos, G., Proença, H.: Periocular biometrics: An emerging technology for unconstrained scenarios. In: 2013 IEEE Workshop on Computational Intelligence in Biometrics and Identity Management (CIBIM), pp. 14–21. IEEE (2013)
11. Reynolds, D., Quatieri, T., Dunn, R.: Speaker verification using adapted gaussian mixture models. Digital Signal Process. **10**, 19–41 (2000)
12. Tan, T., Zhang, X., Sun, Z., Zhang, H.: Noisy iris image matching by using multiple cues. Pattern Recogn. Lett. **33**, 970–977 (2011)

13. Daugman, J.: High confidence visual recognition of persons by a test of statistical independence. IEEE Trans. Pattern Anal. Mach. Intell. **15**, 1148–1161 (1993)
14. Chen, W., Er, M.J., Wu, S.: Illumination compensation and normalization for robust face recognition using discrete cosine transform in logarithm domain. IEEE Trans. Syst. Man Cybern. Part B: Cybern. **36**, 458–466 (2006)
15. Proença, H., Alexandre, L.: The nice.i: noisy iris challenge evaluation - part i. In: First IEEE International Conference on Biometrics: Theory, Applications, and Systems, pp. 1–4. IEEE (2007)

Applications

Visual Perception and Analysis as First Steps Toward Human–Robot Chess Playing

Andreas Schwenk$^{(\boxtimes)}$ and Chunrong Yuan

Faculty of Information, Media and Electrical Engineering,
Technische Hochschule Köln – University of Applied Sciences,
Betzdorferstr. 2, 50679 Cologne, Germany
{andreas.schwenk,chunrong.yuan}@th-koeln.de

Abstract. We propose in this paper a novel visual computing approach for the automatic perception of chess gaming states, where a standard chessboard with original chess pieces is used. Our image analysis algorithm uses only grayscale images captured from a single mobile camera for interactive gaming under natural environmental conditions. On the one hand, we apply computer vision techniques to detect and localize the grid corners, obtaining a 2D representation of the 8×8 chess grid based on the grayscale information of the input image. On the other hand, we exploit computer graphics techniques for the 3D modeling and rendering of the pieces together with the chessboard. Using 2D–3D correspondences, we are able to recover the 3D camera poses and determine game state transitions during the gaming process. Experimental study based on both simulated and real–world scenarios demonstrates the feasibility and effectiveness of our approach.

1 Introduction

Automatons which can play chess exist since a long time. Nowadays, chess computers exist in various forms, ranging from supercomputers to microprocessors. Dedicated hardware has been constructed for chess–playing purpose. Typical examples are industry–like robotic systems with manipulators in the form of robot arms or grippers connected with specially constructed chessboards.

Although it is possible to build a robotic system which can grasp chess pieces and move them around, automatic perception of the chessboard and chess pieces remains a challenging problem. In fact, many chess–playing robotic systems act without "eyes", using often specific electronic boards with magnetic or RFID or other types of digital/analog sensors to extract movements of chess pieces [1,2].

With the drawbacks of stationary industry–like chess automatons, it would be more desirable to be able to play chess with a mobile robot. Actually, most of the current mobile systems are facilitated with optical sensors, which could be used for the purpose of chessboard perception. However, automatic perception and recognition of scenes captured in an everyday environment is a general unsolved problem, making it yet difficult to build robust vision systems capable of autonomous perception of the ordinary chessboard and pieces.

© Springer International Publishing Switzerland 2015
G. Bebis et al. (Eds.): ISVC 2015, Part II, LNCS 9475, pp. 283–292, 2015.
DOI: 10.1007/978-3-319-27863-6_26

Until 2010, published research papers on visual chess perception have investigated only the problem of chessboard detection and segmentation [3,4]. Now some researchers have managed to determine the square occupancy, i.e., to decide whether each of the chess grid is occupied by a chess piece or not [5,6], while the problem of game sate perception remains rarely addressed [7]. Among the available works, several systems use color information obtained by a fixed overhead camera for the extraction of the chessboard area. For chess grid detection and segmentation, some systems use point–based detection method, while others use line–based method. Besides a color camera, the system presented in [7] uses also a Kinect–style visual sensor, with which both color and depth information of the scene can be extracted, making it possible to recognize pieces under non–fixed camera viewpoints. Here, the system is still a stationary industry–like robot arm with limited working space, and recognition relies on offline training, which is time consuming.

In this work, we will present an approach for facilitating mobile agents with visual perception abilities. We believe that the ultimate goal of chess playing with a robot under natural environmental conditions would become a reality, if robust scene perception and accurate grasping and placement of the chess pieces have been solved properly. As first steps toward this goal, we have developed novel methods for the full automatic perception and analysis of the chess game, where a standard chessboard with original chess pieces has been used. In our approach, we need neither color information nor offline training and have achieved robust performance in real time using only a single perspective camera.

The remainder of the paper is organized as follows. Section 2 presents our algorithm for chessboard detection and extraction. The contribution of viewpoint estimation of the dynamic camera is discussed in Sect. 3. Game state recognition is elaborated in Sect. 4, followed by experimental study in Sect. 5. Finally, Sect. 6 summarizes the whole paper.

2   Chessboard Extraction

For the extraction of the chessboard structure, one can either use Hough Transform for line extraction or corner detection methods for grid point extraction. However, images taken in natural environments often contain artifacts and lines and corners may be occluded by the pieces. In order to overcome illumination variations as well as occlusions, we construct a filter kernel w for image normalization and contrast enhancement.

The matrix w is estimated directly from the input image itself. We divide the input image $f(x,y)$ into a set of blocks with equal size. Using randomly 2% of the pixels contained in each block, an average intensity of each block can be estimated. The average intensities from all blocks constitute a matrix w_0. After filtering w_0 with an average filter, we obtain the desired weighting matrix w. An intensity corrected image $g(x,y)$ can be obtained by weighting $f(x,y)$ with matrix w using a bilinear filtering method.

As an example, we show in Fig. 1(a), (b) and (c) respectively the original color image f_c, the converted grayscale image f and the normalized image g.

(a)   (b)

(c)   (d)

Fig. 1. Results of chessboard extraction

As can be seen, g appears more homogeneously and has a better contrast than f. It is also observable that some grid corners are occluded. In order to extract the chessboard and estimate its location in the image, a set of grid corners has to be extracted accurately and yet with a high computational efficiency.

From the enhanced image g, we use a standard corner detection method to extract candidate grid points. In order to eliminate false positives, the corner candidates are processed further to construct candidate lines based on their edge orientations. The candidate lines are then grouped into horizontal and vertical ones and arranged in a sequential manner for possible mapping onto the chessboard. The less the number of corners a line goes through, the less reliable is the line. Line candidates will only be kept, if they go through at least three grid corner points. As a consequence, undesired lines can be filtered out.

It is possible that two or more lines, which have been found with similar parameters, actually should belong to a single line of the chessboard. By a simple averaging of the line parameters, they are integrated into one line. Another possibility is the missing of lines, since not all lines can be constructed due to unfavorable situations like bad illumination, shadows or occlusions. However, by comparing the distances among adjacent lines, we are able to determine the number of missing lines so that all reconstructed lines can be ordered sequentially and mapped properly onto the chessboard.

We can see in Fig. 1(c) the extracted corner candidates together with their edge orientations. Visualized in color orange in Fig. 1(d) are the constructed horizontal and vertical lines. The line shown in color red is an outlier, which has been removed successfully. Since all vertical lines have been detected, they are numbered from 0 to 8. The 6 horizontal lines have also been ordered properly, with the index numbers identified as 0 and then 3 to 7. The missing of lines 1, 2 and 8 are identified, although these three lines are occluded completely by chess pieces.

We then calculate the intersections of these detected horizontal and vertical lines, resulting in a more accurate localization of the grid corners.

3   Camera Viewpoint Estimation

Since we use an approach based on 2D–3D correspondences for camera pose estimation as well as game state recognition, we need to build 3D models of individual chess pieces and make a fast rendering of them for establishing correspondences between the rendered view and the current camera view.

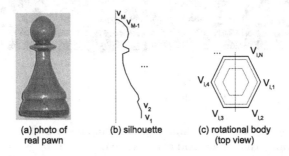

(a) photo of   (b) silhouette   (c) rotational body
real pawn   (top view)

Fig. 2. 3D modeling and rendering of the chess pieces

3.1   3D Modeling and Rendering

The 3D model of a chess piece, which is either symmetric or approximately symmetric, is generated by rotating a discrete set of 2D planar views. Each plane can be indexed by a vertice on the silhouette, as is shown in Fig. 2(b). As depicted in Fig. 2(c), each rotating plane can be approximated by a polygon. In order to accelerate the rendering process, we implement a fast triangle rasterizer as follows:

– Triangle rasterization supported by Bresenham's line algorithm
– Solid–shaded rendering, i.e. same color within a triangle
– Depth ordering instead of depth buffering, via priority fill

During the gaming process, scenes are rendered by projecting 3D models of all existing pieces together with an empty 3D chessboard. For performance optimization during game state recognition, only the vertices of the 3D chess models are rendered. Compared to subsampling, we can keep model granularity of the chess pieces and increase at the same time the rendering speed.

3.2   Parameter Estimation

The projection of a 3D scene onto a 2D image can be calculated using a 3×4 matrix A as $p_{2D} = A\,P_{3D}$ where $P_{3D} = [\mathcal{X}\ \mathcal{Y}\ \mathcal{Z}\ 1]^T$ is a point in 3D space, $p_{2D} = [x\ y\ 1]^T$ the projected one in 2D space, and A consists of both the intrinsic and extrinsic camera parameters.

If enough corresponding point pairs of (P_{3D}, p_{2D}) are known, the projection matrix A can be calculated by solving a linear system of equations. From the localized grid corners using the method illustrated in Sect. 2, we first select six of them. The only criterion for selection is that these points should be distributed evenly on the chessboard. Suppose they are c_0 to c_5. For all of them, the corresponding points C_0 to C_5 on the 3D models are known.

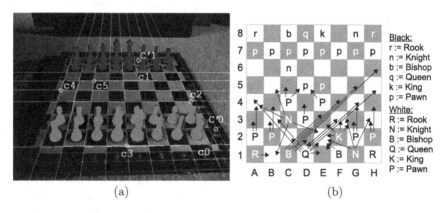

(a)   (b)

Fig. 3. (a): Camera pose estimation. The input image with detected corners and lines plus the rendering of the underlying 3D models. (b): All valid movements of the white pieces from an example game state

Since c_0 to c_5 are all coplanar points, additional points lying outside the chessboard are necessary. We use directly four point pairs (C_i, c_i) with $i = 2$ to 5 and select additional two 3D points C'_0 and C'_1, with both C'_i lying with a fixed vertical distance $= 1$ unit above C_i ($i = 0, 1$) on the 3D model (see Fig. 3(a)). Now the problem becomes solving the corresponding 2D points $c'_i, i = 0, 1$ together with the projection matrix A. With a reasonable initialization, all the unknowns can be jointly estimated using an iterative nonlinear optimization algorithm [8].

4   Game State Recognition

We represent each game state as a matrix

$$
\mathcal{S}_i = \begin{bmatrix} s_i(A,1) & s_i(B,1) & \dots & s_i(H,1) \\ s_i(A,2) & s_i(B,2) & \dots & s_i(H,2) \\ & \dots & & \\ s_i(A,8) & s_i(B,8) & \dots & s_i(H,8) \end{bmatrix}, \tag{1}
$$

where $s_i(a, b)$ represents the individual square occupancy of the chessboard. Each square can be either empty or contain a specific piece. In the very beginning of a chess game, its state is S_0, with all the pieces in their original positions.

During the game, the movement of a chess piece leads to a state transition from S_i to S_{i+1}. Such a state transition may either cause a permutation in the occupancy of two squares (i.e. a piece is moved to an empty field) or can be represented by a permutation followed by the removal of one piece from the field (i.e., a beat occurs). The set of valid moves is determined by the chess gaming rules and hence bounded. The number of possible state transmissions depends largely on the current game state S_i. In average, about 20 or more valid movements can be made from a given state, as is illustrated in Fig. 3(b).

The chess engine we have used is the Toledo Nanochess, developed by the Mexican Oscar Toledo Gutierrez and publicly available at www.nanochess.org. With this integrated into our system, a set of valid state transitions can be determined. Using the estimated camera pose parameters, we render a set of synthesized views corresponding to the set of possible state changes. Game state can hence be inferred by finding among these views the one with the minimal distance to the current camera image.

Figure 4 illustrates the recognition process in detail. Suppose a state transition has been occurred with the pawn moved from C2 to C3. The image captured after state transition is shown in Fig. 4(a). Binarization of it is carried out for the purpose of matching, resulting in an image f_b, as is shown in Fig. 4(b). Figure 4(c) is the rendering of an empty chessboard (noted as R_0), while Fig. 4(d) is the rendered pieces based on a hypothesis of state transition (noted as R_{hyp}). The moved piece whose earlier location is in C2, has also been rendered (see the yellow points Fig. 4(d)). Afterwards, we perform an XOR operation of R_{hyp} with R_0. The result of XOR operation is compared to f_b, with the difference image shown as Fig. 4(e), where the not matched pixels are visualized in color red. By finding the particular hypothesis with the lowest error (see the error curve shown in Fig. 4(f)), the new game state can be determined.

The reason for the XOR operation is to remove the disturbance of the background, particularly in situations where a white piece lies in a white square or a black piece lies in a black square. Such an XOR operation facilitates the search for the minimum and leads to improved robustness in state recognition.

5   Experimental Study

The proposed approach for visual perception and analysis has been implemented in C++ and tested on an Apple iMac with a 3.5 GHz Intel CPU, running as a single–thread application.

For the initial concept validation, the Webots simulation environment has been used, where a physical simulation of a humanoid robot, namely the Nao, is available. Integrated with 3D rendering as well as speech-based interaction mechanism, visual perception and game state recognition can be tested within Webots. Shown in Fig. 5 is a scene with Nao standing before a table playing the

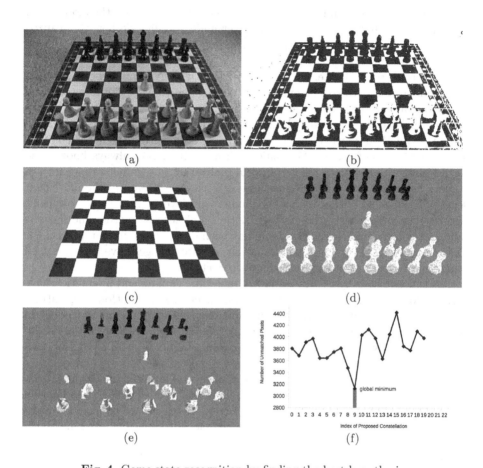

Fig. 4. Game state recognition by finding the best hypothesis

Fig. 5. Nao playing chess in the Webots simulation environment

chess game with a human player (human player not shown, since the modeling of human player is not necessary). For a better understanding on how the realized system works in simulated environment, please refer to the YouTube video linked at https://youtu.be/7E_l-Gqzhhw.

For the purpose of interactive chess playing with Nao in Webots, we have implemented a text-to-speech engine, simulating the conversation between Nao and a human player. Since the grasping of chess pieces by Nao itself is not possible in Webots, Nao asks the human player to move the pieces. And the human player then informs Nao with speech, once the movement has been made (either for Nao or for oneself). Using such a method, Nao always knows who should make the next move. The physical realization of a game state change in Webots has been realized by rendering a new view according to the movement decisions just made (either by Nao or the human player). After each movement, Nao detects immediately the state transition and checks whether the move is valid or perhaps some sort of cheating has been done by the human player. For continuous playing, perception and analysis of state changes have been achieved using the proposed approach introduced in Sects. 2 to 4.

As demonstrated by the linked video, real–time gaming and interaction have been performed successfully in the simulation environment. However, natural environmental conditions like illumination changes are difficult to model in Webots. We have hence carried out another experimental study in real–world, where images have been captured by a Panasonic Lumix DMC-TZ5 digital camera.

In the real–world experiment, sequences of several hundred images have been taken under different lighting conditions, including also very dark illumination under variations of the camera viewing angles. Between two captured frames, either no state transition happens at all, or the pieces have been moved manually. In the longest sequence, a series of 21 moves have been made.

Our system is capable of recognizing both cases, i.e., either with or without state transitions. The state recognition can be performed in real time and is robust against bad conditions such as unfavorable lighting or partial occlusions.

Depicted in Fig. 6 are three chess moves, with the original image shown in (a), the state found with the best hypothesis in (b), and the corresponding matching result in (c). The average time consumption of all major computing steps are listed in Table 1. In average, we achieve a frame rate of 12.7 frames/s for camera pose estimation in case of chess state transition and much higher in case no game action occurs. For chess state recognition, a rate of 26.3 frames/s has been achieved. Overall, the system runs at roughly 10 frames/s.

The proposed visual system needs only to recognize game state transitions. The advantages are that sate recognition is performed without computationally intensive processing steps such as image segmentation and that piece recognition does not need the training of a supervised classifier. Together with the special features of the rendering process, a high recognition speed has been achieved.

The limitation of the current approach is that a known state S_i has been assumed, before S_{i+1} can be recognized. However, this is not a critical issue,

Table 1. Performance measurement

Processing step	Average time
Chess grid extraction	1.0 ms
Viewpoint estimation:	78.6 ms
(a) Rendering	51.0 ms
(b) Nonlinear pose estimation	25.0 ms
(c) Other	2.6 ms
Game state recognition:	38.0 ms
(a) Hypothesis rendering	26.4 ms
(b) Hypothesis verification	10.8 ms
(c) Other	0.8 ms

as a chess game usually starts from the initial sate S_0, with all pieces in their original positions. In the worst case, the game can start again from S_0. Another remedy is to save all the bypassed game states so that the game may return to an earlier state.

Currently, we are making further optimization on the proposed approach. In the future, we would like to have it working on a physical mobile agent like Nao in our lab. Once the visual guided grasping mechanism has been implemented and integrated with the perception system, we would like to make the implementation public within the research community so as to advance further the state of the art in natural human–robot gaming.

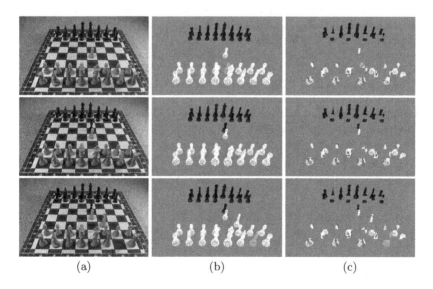

(a)   (b)   (c)

Fig. 6. State recognition during three consecutive movements of chess pieces

6   Conclusion

A novel visual perception system has been demonstrated for the automatic analysis of the chess game under natural environmental conditions. In order to recognize game states, candidate corner points of the chess grids are extracted from the input image. These points are then refined and organized into lines to get a correct 2D grid–representation of the chessboard. By integrating computer vision and computer graphics techniques, we are able to recover the dynamic camera pose and determine game state transitions in real time. Future work consists of embedding the visual perception system into a mobile robot like NAO, facilitating it with robust interaction capabilities and performing extensive user study so as to make further technical improvement and scientific contributions.

References

1. Tong, G., Qu, Y., Cheng, T.: Human-computer interactive gaming system-a chinese chess robot. In: IEEE/RSJ International Conference on Intelligent Robots and Systems, 2009. IROS 2009, pp. 984–987. IEEE (2009)
2. Gracie, S., et al.: Lego chess robot, project report (2005)
3. Tam, K.Y., Lay, J.A., Levy, D.: Automatic grid segmentation of populated chessboard taken at a lower angle view. In: Digital Image Computing: Techniques and Applications (DICTA), 2008, pp. 294–299. IEEE (2008)
4. Neufeld, J.E., Hall, T.S.: Probabilistic location of a populated chessboard using computer vision. In: 2010 53rd IEEE International Midwest Symposium on Circuits and Systems (MWSCAS), pp. 616–619. IEEE (2010)
5. Coens, J.A.: Taking Tekkotsu out of the plane. Ph.D. thesis, Carnegie Mellon University (2010)
6. Wang, V., Green, R.: Chess move tracking using overhead rgb webcam. In: 2013 28th International Conference of Image and Vision Computing New Zealand (IVCNZ), pp. 299–304. IEEE (2013)
7. Matuszek, C., Mayton, B., Aimi, R., Deisenroth, M.P., Bo, L., Chu, R., Kung, M., LeGrand, L., Smith, J.R., Fox, D.: Gambit: An autonomous chess-playing robotic system. In: 2011 IEEE International Conference on Robotics and Automation (ICRA), pp. 4291–4297. IEEE (2011)
8. Hartley, R., Zisserman, A.: Multiple view geometry in computer vision. Cambridge University Press, Canberra (2004)

A Gaussian Mixture Representation
of Gesture Kinematics for On-Line Sign
Language Video Annotation

Fabio Martínez[1,2]([✉]), Antoine Manzanera[2],
Michèle Gouiffès[1], and Annelies Braffort[1]

[1] LIMSI, CNRS, Université Paris-Saclay, Paris, France
[2] U2IS/Robotics-Vision, ENSTA-ParisTech, Université Paris-Saclay, Paris, France
fabio.martinez-carillo@ensta-paristech.fr

Abstract. Sign languages (SLs) are visuo-gestural representations used by deaf communities. Recognition of SLs usually requires manual annotations, which are expert dependent, prone to errors and time consuming. This work introduces a method to support SL annotations based on a motion descriptor that characterizes dynamic gestures in videos. The proposed approach starts by computing local kinematic cues, represented as mixtures of Gaussians which together correspond to gestures with a semantic equivalence in the sign language corpora. At each frame, a spatial pyramid partition allows a fine-to-coarse sub-regional description of motion-cues distribution. Then for each sub-region, a histogram of motion-cues occurrence is built, forming a frame-gesture descriptor which can be used for on-line annotation. The proposed approach is evaluated using a bag-of-features framework, in which every frame-level histogram is mapped to an SVM. Experimental results show competitive results in terms of accuracy and time computation for a signing dataset.

1 Introduction

Sign languages (SLs) are natural languages used to communicate with and among the deaf communities, which, like spoken languages, differ from one country to another. Additionally, SLs are less-resourced languages with very few reference books describing them (grammar rules, etc.), a limited number of dictionaries and corpora, and even less dedicated processing tools.

Annotation software are tools used for linguistic studies, that allow researchers to visualize their data (mainly videos for SLs), annotate them with linguistic inputs, and analyze these inputs [1]. These corpus-based studies allow to create statistically-informed models that are useful for SL description, but also for SL processing. At this moment, such software are limited to only include automatic processing on the secondary data, the annotations, which are textual data. They do not include automatic processing on the primary data, the video. However, there is a growing interest on image and video processing tools, to characterize particular recorded gestures from local and global primitives such as motion, shape, body parts interactions, among others [2,3].

© Springer International Publishing Switzerland 2015
G. Bebis et al. (Eds.): ISVC 2015, Part II, LNCS 9475, pp. 293–303, 2015.
DOI: 10.1007/978-3-319-27863-6_27

This paper introduces a new proposal to support SLs annotations based on a motion descriptor that characterize temporal gestures in video sequences. The proposed approach is developed for French Sign Language corpus annotation. It starts by computing semi-dense trajectories, provided by point tracking in consecutive frames, over a set of gestures recorded in a video. Then, kinematic-cue words, represented as local mixture of Gaussians, are recursively computed at each time and for each trajectory during the video. These features are extremely fast to compute, and the action descriptor is available at each frame, thus allowing prediction on partial video sequences, and then on-line gesture recognition capability.

2   Sign Language

2.1   Main Linguistic Properties

SLs are visuo-gestural representations that follow specific rules induced by use and interaction among corporal articulators and the visual perception. This language promotes the simultaneous use of a number of articulators, the linguistic use of the space in front of the signer so-called 'signing space', and the omnipresence of iconicity at all levels of the language [4]. The main linguistic specificities and challenges are the followings:

- Signs can be broken down into smaller constituents whose linguistic nature, definition and detection are still subject to debate;
- Signs can bear strong modification of their constituents depending on the context, and modelling all possible variations can require too many different training examples to keep the categories consistent;
- Signs can be more or less lexicalised, and the most productive ones are built on the fly and are not indexed in a dictionary, which makes them extremely difficult to be modelled from a classical approach.

SLs characterization must also consider non-manual activity that convey meaningful information. For instance, SL production involves non-manual articulators such as head, face, and torso which are relatively synchronized on different spatial and temporal scales. In fact, the signer uses *the signing space* to support and topologically structure his discourse. This spatial and multi-component property, as well as the importance of the productive signs make the design of SL processing tools a very challenging task.

2.2   The LSF (French Sign Language) Corpora

The corpora used in this study is extracted from the corpus collected during DictaSign, a three-year FP7 ICT project that aimed to improve the state of web-based communication for deaf people [5]. It is composed of nine videos that contain isolated dates, such as *'Lundi 2 novembre 2013'* (Monday, November 2nd, 2013). The lexicon is constituted of the seven days, the twelve months, and a set of numbers. In LSF, a date is composed of four elements following the order: DAY NUMBER MONTH YEAR. The day and month signs are simple gestural units than can present regional variants. These dates are less complex than SL utterance, but they include various issues such as lexicon variability, co-articulation. They also include some spatial constraints, but limited to the image plane. This seems to us good candidates as a first step, to evaluate the performance of our method on motion description with this kind of data.

3 Background on Motion Descriptors

Motion analysis is a fundamental tool to segment potential region of interest, quantify, detect, recognize gestures or describe spatio-temporal interactions. One of the advantages of motion based characterization is the relative independence to appearance, which has potential applications in uncontrolled conditions.

Motion descriptors based on tracked local space-time trajectories from optical flow fields, currently provide the best performance to represent gestures and understand video sequences [6–8]. To recognize human activities, these strategies namely integrates local features along the trajectories to capture shape, appearance and motion information, namely using HOF (Histograms of Optical Flow), MBH (Motion Boundary Histograms) and HOG (Histograms of Oriented Gradients) [7]. This descriptor was also used in [9], but using improved motion trajectories obtained by correcting the camera motion in video sequences. In general, these descriptors are dependent on the appearance and structural image features computed around trajectories, fact that could be critical in language recognition in which shape signs and appearance have high inter subject variability. Additionally, the spatio-temporal volumes are heuristically cut off from a fixed temporal length (for example: 15 frames in [7]) that may be a problem to represent series of gestures of a SL utterance that can vary from one subject to another and also depending of the represented dialog. Besides, the dynamic trajectory information is poorly exploited, $i.e.$, the action descriptors only use the trajectory as information support to compute static frame features (namely spatial features such as image gradients), neglecting relevant kinematic information that is naturally available on the trajectory.

Other works have characterized the dynamic of dense beams of trajectories to describe actions in video sequences. For instance, in [10] a set of k cut trajectories are characterized using first order derivatives in the (x, y) axes, which may be sensitive to motion direction and to scale. In [11] is firstly considered strong sparse coding assumptions to filter out motion trajectories. Then, the remaining trajectories are characterized using *Largest Lyapunov Exponent* and the *correlation dimension*.

Specifically, in the domain of SLs, several works have been focused in the automatic recognition of atomic gestures by characterizing postures, shape regions, global movements among others (see [3] for an overview of the domain). These works include the use of a broad spectrum of methods such as tracking of articulated shapes, colour segmentation to characterize postures, and the

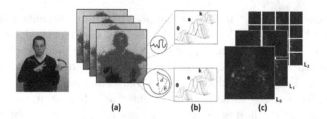

Fig. 1. Pipeline of the proposed approach for SL recognition and annotation support. (a) First, a set of trajectories are computed. (b) For each trajectory, a set of kinematic-cue words are computed using recursive Mixture of Gaussian. (c) A pyramidal partition is applied at each frame to support hierarchical BoW representation, which is in turn used to recognize particular FSL gestures.

static and temporal characterization shape articulators. In terms of annotation support, some projects have tried to integrate video analysis modules into annotation software. For instance, *Ancolin* is a prototype annotation software [12] developed onto a distributed architecture, that includes several external plugins for sign language video processing such as colour skin detection, characterization of head shape and size, and motion history images to code arm movements. This video characterization provides additional useful information to the annotation but remains dependent on accurate segmentation of human silhouettes and is also highly dependent on the user. *SignStream* [1] is another annotation software, currently used for linguistic analysis that includes components for 3D head detection and tracking to estimate head gesture: Currently, this application includes new modules to automatically characterize hand gestures in ASL using a tracking system [2]. Additionally, the *SLMotion* toolkit provides a framework for automatic and semi-automatic analysis, feature extraction and annotation of individual sign language videos. The program includes support for exporting the annotations in ELAN[1] format.

4   The Proposed Method

The proposed strategy recognizes SL gestures by using an on-line spatio-temporal characterization of the signer movements recorded in a video. The Fig. 1 illustrates a pipeline of the proposed approach.

4.1   Computing Semi-dense Trajectories

Point trajectories are useful motion features based on tracking salient points along the video sequence, allowing in most cases a relevant representation of action present in the video. The proposed approach requires a set of trajectories with a suitable trade-off between accuracy and computation time, in order

---

[1] Software widely used for linguistic analysis of video data.

Fig. 2. Spatio-temporal representation of trajectories and MoG kinematic representation. Each row represents a spatio-temporal gesture corresponding to two days. The second and third columns illustrate the dense trajectories and their kinematic descriptors, respectively. The fourth and fifth columns correspond to the semi-dense trajectories and their kinematic descriptors.

to support fast annotation prediction. In this work were considered two different methods to compute motion trajectories (see examples in Fig. 2), described hereunder:

Dense trajectories [7] are extracted from a dense optical flow field estimated at multiple spatial scales and regularized using a median filter. Additionally, a trajectory is considered as outlier and removed if it meets any of the two following conditions: (1) the standard deviation of the velocity along the trajectory is above a given threshold, and (2) it presents sudden displacements, corresponding to vectors whose magnitude is larger than a certain proportion of the overall displacement of the trajectory.

Semi-Dense trajectories [13] are computed from a set of weakly salient points, tracked using a coarse-to-fine prediction and matching approach, allowing a high degree of parallelism and dominant movement estimation. This technique produces high density trajectory beams, robust to large camera accelerations and allowing statistically significant trajectory based representation, with a good trade-off between accuracy and performance.

4.2   Gaussian Mixture Representation of Kinematic Features

Each computed trajectory $\boldsymbol{\Gamma}(t) \in \mathbb{R}^2$ represents a particle traveling in the 2d space (x, y) from time t_1 to t_n. At each time t, the trajectory motion information can be characterized by a collection of kinematic features $\{F_t^i\}_i$, such as the velocity, acceleration, curvature among others, using finite difference approximation. In this work, each computed kinematic feature is modelled as a random variable following a mixture of K Gaussian densities, whose parameters are defined as: $\sum_{k=1}^{K} w_t^k \mathcal{N}(\mu_t^k, \sigma_t^k)$, where (μ_t^k, σ_t^k) are the mean and standard deviation of each Gaussian mode and w_t^k represents the contribution of each mode, with $\sum_{k=1}^{K} w_t^k = 1$.

Algorithm 1. Recursive Mixture of Gaussian estimated locally for each computed trajectory and for each considered kinematic feature.

---

for each time t **do**
  for each trajectory Γ **do**
    if $|\Gamma| > 3$ **then**
      for each kinematic feature i **do**
        calculate feature $F_t = F(i)$
        for each mode k **do**
          if $\left| F_t - \mu_t^k \right| \leq \lambda \sigma_t^k$ **then**
$$\mu_t^k = \mu_{t-1}^k + \alpha(F_t - \mu_{t-1}^k)$$
$$(\sigma_t^k)^2 = \alpha \left(\mu_{t-1}^k - F_t \right)^2 + (1-\alpha)\,(\sigma_{t-1}^k)^2$$
$$\omega_t^k = \omega_{t-1}^k + \alpha$$
          end if
        end for
        normalize ω_t^k **such that** $\displaystyle\sum_{k=1}^{K} \omega_t^k = 1$
        rank the modes in decreasing order of $\frac{\omega_t^k}{\sigma_t^k}$
        keep the first B **modes such that:** $B = \arg\min_{b=1}^{K} \left\{ b; \displaystyle\sum_{k=1}^{b} \omega_t^k > \mathbf{T} \right\}$
      end for
    end if
  end for
end for

The MoG representation is herein implemented as described in Algorithm 1 [14]. This algorithm allows an on-line MoG updating and therefore a kinematic gesture representation is available at each frame. First, the density parameters are initialized, assigning to the mean the first value of each kinematic feature computed, to the standard deviation any fixed value and the weight $\omega_{t,k}$ being the same for each mode k. Then, the distributions that are most likely matched by the current kinematic sample (i.e. when the sample distance to the mode is less than λ times its standard deviations) are updated. The density parameters $\{\mu_t^k, \sigma_t^k\}$ are updated using a on-line cumulative filter with a learning rate parameter $\alpha \in [0,1]$ which takes into account the history of the kinematic measure along the trajectory, with $t \approx 1/\alpha$. Each ω_t^k is also updated according to the matched distribution at each time. After that, the distributions are sorted in decreasing ordered according to $\frac{\omega_t^k}{\sigma_t^k}$. Finally, only the B first distributions of the MoG are considered. If any distribution is initialized (no existing one is matched), then the parameters of the distribution with lowest weight are replaced by the initial values. This recursive representation has the main advantage of computational speed which is essential to on-line annotation tools, the recent history of each kinematic measure being available at each frame.

Kinematic Features F_t**:** In order to keep the computation fast, the kinematic features considered in this work were: the velocity $v(t) = \Gamma'(t)$, depicted by its direction $\theta(t) = \arg v(t)$ and modulus (speed) $s(t) = ||v(t)||$. The curvature was also included; it is related with how rapidly the trajectory is bending to one side, and corresponds to the normal acceleration when the curvilinear speed is constant. The curvature is herein implemented as proposed in [15], using finite difference on consecutive points of Γ as follows: $\kappa(t-1) = \frac{\sqrt{\zeta(\zeta-b)(\zeta-d)(\zeta-e)}}{bde}$, where $\zeta = (b+d+e)/2$, as illustrated in Fig. 1.

Each trajectory is then characterized at time t by the set of kinematic features $F_t = \{\theta(t), s(t), \kappa(t)\}$. The proposed strategy is flexible to include any other local kinematic measure computed along a 2d trajectory. An additional advantage of the proposed strategy is that any kinematic feature can access independently to the recognition or a set of features can be chosen through a learning stage, in order to reach higher execution times or reduce memory requirements, preserving a proper accuracy. Figure 2 shows computed trajectories. The recursive means of computed kinematic features are represented using a RGB color map representation, the blue being the curvature and the red and green being respectively the modulus and the direction of the velocity.

4.3   Spatial Pyramid Representation and Codebooks Learning

In SLs, the signs are visuo-gestural representations, which require a temporal and spatial characterization. In the proposed approach, a regional analysis of the MoG features are carried out by following a fine-to-coarse partition of each frame. This *spatial pyramid* forms a set of partition layers $\{L_i\}_{0 \le i \le N_r}$ (see in Fig. 1-(c)), whose total number of sub-regions is: $s\_r = \sum_{i=0}^{N_r} 4^i = \frac{4^{N_r+1}-1}{3}$.

In the training step, different configurations of the spatial pyramid representation were used to learn the codebooks of kinematic words computed from the MoG recursive representation. Each codebook is made up by a set of MoG kinematic words, formed by the output of a classical k-means algorithm computed using a random selection of 10 % of the MoG features extracted over the whole training video set. All the feature words are computed with the same α, with n the number of kinematic measures estimated at each time on each trajectory and b the number of modes retained from the MoG distribution. Then, each codebook contains k_l representative feature words, each word having a dimension of $3nb$. During a first configuration, a global codebook $\{D_0\}$ was learned from the region of L_0, and then a histogram of motion word occurrences was considered for each sub-region of the spatial pyramid. This histogram is constructed by counting the number of times each one of the k_l kinematic centroid is closest to the computed features, based on the Euclidean distance on $\mathbb{R}^{3n}$. In this case, the total size of the descriptor is the concatenation of histograms computed for each sub-region, with size of $s\_r \times k_{l_0}$. In a second configuration, for each sub-region of the spatial pyramid representation was considered a independent codebook. From the set of codebooks $\{D_l\}_{1 \le l \le \Lambda}$ is then computed histogram of occurrences with variable size according to the size of each regional codebook, resulting a more compact descriptor w.r.t the first version. Finally, the labeling of each potential sign gesture is performed by a Support Vector Machine (SVM) using the standard LIBSVM [16] implementation, using the *one-against-one* multi-class SVM classification with a Radial Basis Function (RBF) kernel.

5   Results

A first exploration over a SL corpus of signatures representing *Dates* was carried out to evaluate the proposed approach in the task of sign recognition to support annotation. The experimental evaluation was performed under a leave-one-out cross validation scheme by using different segments of the videos. The best performance of the proposed approach was obtained with a pyramid of $N_r = 2$ levels and a learning rate of $\alpha = 0.25$ corresponding, to a time depth of 4 frames. The number of estimated modes in each MoG was set to 7, taking into account the 5 dominant modes.

Evaluations over the SL dataset were carried out taking into account different lexical complexities of the signs. First, the *words* recognition related with days and months was performed. Because the approach is based on statistical representations of spatio-temporal gestures, it was only considered gestures with more that 5 samples available into the dataset, corresponding to 4 days and 5 months. In Table 1 is shown the performance obtained by the proposed approach for recognizing these spatio-temporal gestures. In general the proposed approach is able to recognize different atomic gestures that correspond to localized movements. The best performance of the proposed approach was achieved by using dense trajectories with a compact spatial pyramid representation of regional dictionaries. Some mistakes in the recognition may be attributed to regional variations of gestural signs.

Second, the performance of the approach to recognize *dates* was evaluated. Complete dates have more complex lexical structure and they are composed

Table 1. Classification rate of individual gestures corresponding to days and months

Gesture	Spatial Configuration	Trajectories	
		Dense	Semi-dense
Days	Pyr single Dic (L_0)	74.07	66.66
	Pyr mult Dics	81.48	74
	without Sub-regions	70.37	62.96
–			
Months	Pyr single Dic (L_0)	75.05	63.45
	Pyr mult Dics	80	72.21
	without Sub-regions	65	55.3

Table 2. Classification rate using different spatial configurations and trajectories for complete date phrases

Spatial Configuration	Trajectories	
	Dense	Semi-dense
Pyr single Dic (L_0)	75.03	67.13
Pyr mult Dics	77.21	70.3
Without Sub-regions	72.45	62

by the ordered sign information of day, month and year[2]. The proposed approach achieves a recognition rate of 75 % on a total of 7 different dates. In Table 2 is shown the results obtained by using different spatial configurations and the different types of trajectory. The best results is obtained using a spatial pyramid configuration of multiple dictionaries learned by region and the dense trajectories. The both pyramidal representations herein implemented allows a more robust representation than a global space-frame description, i.e., without a sub-regional division. Some mistakes are due to the natural variability between different signers and to the limited number of samples available for each date.

Action Recognition Evaluation: Because the proposed approach is based on the recognition of spatio-temporal patterns, it can be extended to recognize other motion activities. An additional evaluation was herein considered in public action recognition datasets. Two different datasets were considered: the *KTH* (six action classes, contained in a total of 2 391 videos) and the *UT-Interaction* (six different interactions in 120 videos) [17]. Table 3 summarizes the results obtained for the proposed approach with other state-of-the art approaches. It generally achieves competitive results with the great advantage

Table 3. The right table reports the comparison with state-of-the-art methods using the KTH database following the original experimental setup [18]. The figures marked with (*) have been computed using a k-fold validation with $k = 5$ [19]. The left table shows the comparison with state-of-the-art methods using the UT-database using k-fold validation with $k = 10$, as described in [17]

Methods	Accuracy
Proposed approach	**92.23**
Wang et. al. [7]	94.2
Laptev et. al. [20]	91.8
Proposed approach *	**97.0**
Liu et. al. *[19]	93.8

Methods	Accuracy
Proposed approach	**90.3**
Laptev et. al. [6]	87.6
Yu et. al. [21]	83.3
Daysy [22]	71

---

[2] An example of a considered date is: *Vendredi douze septembre mille six cent quatre vingt dix*, which means Friday, September the 12th, 1690.

of being computationally efficient and usable in real-time applications. In contrast, other motion descriptors typically use a lot of features for each trajectory, including appearance information.

The proposed approach achieved memory efficiency, taking in average 0.30 milliseconds for each frame to build the descriptor. The experiments were carried out on a single core i3-3240 CPU @3.40GHz.

6   Conclusions

This work introduced a new motion descriptor that is able to recognize motion gestures related with SLs. The proposed approach allows an on-line support to SL annotation, by combining trajectory beams and Mixture of Gaussian representation of kinematic cues. The motion cues are spatially aggregated at each frame using a pyramid representation. The proposed approach can also be included as a plugin in SL systems and used as part of more sophisticated SL analysis. A more exhaustive evaluation with a larger dataset will be performed in order to increase the statistical samples of each gesture.

Acknowledgements. This research is funded by the RTRA Digiteo project MAPOCA.

References

1. Neidle, C.: Signstream: A database tool for research on visual-gestural language. Sign Lang. Linguist. **4**, 203–214 (2001)
2. Gavrilov, Z., Sclaroff, S., Neidle, C., Dickinson, S.: Detecting reduplication in videos of american sign language. In: 5th Workshop on the Representation and Processing of Sign Language (RPSL) (2014)
3. Cooper, H., Holt, B., Bowden, R.: Sign language recognition (2011)
4. Braffort, A., Filhol, M.: Constraint-based sign language processing. Constraints and Language. Cambridge Scholar Publishing, Cambridge (2014)
5. Matthes, et. al: Elicitation tasks and materials designed for dicta sign's multilingual corpus. In: 4th Workshop on the Representation and Processing of Sign Language (RPSL 2010) of (LREC 2010)
6. Kantorov, V., Laptev, I.: Efficient feature extraction, encoding and classification for action recognition. cvpr (2014)
7. Wang, H., Klaser, A., Schmid, C., Liu, C.L.: Action recognition by dense trajectories. CVPR 2011, Washington, DC, USA, pp. 3169–3176. IEEE Computer Society (2011)
8. Jain, M., Jegou, H., Bouthemy, P.: Better exploiting motion for better action recognition. In: CVPR. CVPR 2013, pp. 2555–2562 (2013)
9. Wang, H., Schmid, C.: Action recognition with improved trajectories. In: ICCV 2013, Sydney, Australia, pp. 3551–3558. IEEE (2013)
10. Matikainen, P., Hebert, M., Sukthankar, R.: Trajectons: Action recognition through the motion analysis of tracked features. In: ICCV Workshops, pp. 514–521. IEEE (2009)

11. Wu, S., Oreifej, O., Shah, M.: Action recognition in videos acquired by a moving camera using motion decomposition of lagrangian particle trajectories. In: ICCV 2011, pp. 1419–1426 (2011)
12. Braffort, A., Choisier, A., Collet, C., Dalle, P., Gianni, F., Lenseigne, B., Segouat, J.: Toward an annotation software for video of sign language, including image processing tools and signing space modelling. In: LREC 2004 (2004)
13. Garrigues, M., Manzanera, A.: Real time semi-dense point tracking. In: Campilho, A., Kamel, M. (eds.) ICIAR 2012, Part I. LNCS, vol. 7324, pp. 245–252. Springer, Heidelberg (2012)
14. Gaber, M.M., Stahl, F., Gomes, J.B.: Background. In: Gaber, M.M., Stahl, F., Gomes, J.B. (eds.) Pocket Data Mining. SBD, vol. 2, pp. 7–22. Springer, Heidelberg (2014)
15. Boutin, M.: Numerically invariant signature curves. Int. J. Comput. Vis. **40**, 235–248 (2000)
16. Chang, C.C., Lin, C.J.: Libsvm: A library for support vector machines. ACM Trans. Intell. Syst. Technol. **2**, 27:1–27:27 (2011)
17. Ryoo, M.S., Aggarwal, J.K.: UT-Interaction Dataset, ICPR contest on Semantic Description of Human Activities (SDHA) (2010)
18. Schuldt, C., Laptev, I., Caputo, B.: Recognizing human actions: A local svm approach. In: ICPR 2004. ICPR 2004, Washington, DC, USA, pp. 32–36 (2004)
19. Liu, J., Luo, J., Shah, M.: Recognizing realistic actions from videos "in the wild". IEEE ICVPR (2009)
20. Laptev, I., Marszałek, M., Schmid, C., Rozenfeld, B.: Learning realistic human actions from movies. In: Conference on Computer Vision and Pattern Recognition (2008)
21. Yu, T.H., Kim, T.K., Cipolla, R.: Real-time action recognition by spatio-temporal semantic and structural forest. BMVA Press **52**(1-52), 12 (2010)
22. Cao, X., Zhang, H., Deng, C., Liu, Q., Liu, H.: Action recognition using 3d daisy descriptor. Mach. Vision Appl. **25**, 159–171 (2014)

Automatic Affect Analysis: From Children to Adults

Rizwan Ahmed Khan[✉], Alexandre Meyer, and Saida Bouakaz

Université Claude Bernard Lyon 1, CNRS, LIRIS, UMR5205, 69622 Lyon, France
{Rizwan-Ahmed.Khan,Alexandre.Meyer,Saida.Bouakaz}@liris.cnrs.fr

Abstract. This article presents novel and robust framework for automatic recognition of facial expressions for children. The proposed framework also achieved results better than state of the art methods for stimuli containing adult faces. The proposed framework extract features only from perceptual salient facial regions as it gets its inspiration from human visual system. In this study we are proposing novel shape descriptor, facial landmark points triangles ratio (LPTR). The framework was first tested on the "Dartmouth database of children's faces" which contains photographs of children between 6 and 16 years of age and achieved promising results. Later we tested proposed framework on Cohn-Kanade (CK+) posed facial expression database (adult faces) and obtained results that exceeds state of the art.

1 Introduction

Facial expressions play crucial role in human interaction and is the most effective form of non-verbal communication. Facial expressions provides clue about emotional state, mindset and intention [1]. Last decade has seen paradigm shift in computing environment, from computer-centered designs to human-centered designs [2]. Thus analysis of user affective states has become utmost important and inevitable. Advancements of many application areas rely on it, i.e. social robots, immersive gaming environment, medical applications, behavior monitoring etc.

There are many challenges in designing framework for affect recognition in general and specifically for children i.e. variability in pose, illumination variation, cluttered faces and the way different children show facial expressions. In this paper we have focused on the problem of automatic recognition of facial expression for children. We have considered six universal facial expressions for this study as these expressions are proved to be consistent across cultures [3]. These six expressions are anger, disgust, fear, happiness, sadness and surprise. There exist different frameworks for automatic facial expression recognition [4–12], but to the best of our knowledge none of the proposed method was tested on stimuli containing children's faces.

Human visual system (HVS) decodes and analyzes facial expressions in real time despite having limited neural resources. As an explanation for such performance, it has been proposed that only some visual inputs are selected by considering "salient regions" [13], where "salient" means most noticeable or most

© Springer International Publishing Switzerland 2015
G. Bebis et al. (Eds.): ISVC 2015, Part II, LNCS 9475, pp. 304–313, 2015.
DOI: 10.1007/978-3-319-27863-6_28

important. Proposed framework gets its inspired from HVS and thus extract features only from perceptual salient facial regions. We have taken results from our psycho-visual experimental study [14], to know which facial regions are salient according to HVS. Proposed framework recognizes children's facial expression by analyzing shape and appearance information of the face as according to Lucey et al. [15] both shape and appearance are important factors for it. Proposed framework extracts shape information using novel shape descriptor, facial landmark points triangles ratio (LPTR) (refer Sect. 4.1). Appearance information is extracted using pyramid local binary pattern (PLBP) features [12].

2   Related Work

Feature selection and regions from where these features are extracted, is one of the most important step for automatic affect analysis. In literature, various methods are employed to extract facial features and these methods can be categorized either as appearance-based methods or geometric feature-based methods.

Appearance-based Methods. One of the widely studied method to extract appearance information is based on Gabor wavelets [4,6]. Littlewort et al. [4] has shown a high recognition accuracy (93.3 % for Cohn-Kanade facial expression database [16]) using Gabor features. Tian [6] has used Gabor wavelets of multiscale and multi-orientation and obtained promising results. Another promising approach to extract appearance information is by using Haar-like features. Yang et al. [17] extracted Haar-like features from the facial image patches (49 subwindows). Their proposed method was tested on Cohn-Kanade facial expression database [16] and it achieved average recognition accuracy of 92.3 % and 80 % in two different scenarios.

Geometric-based Methods. Geometric feature-based methods [8,9,11] extracts shapes and locations of facial components information to form feature vector. For expression recognition, Zhang et al. [8] has measured and tracked the facial motion using Kalman filters. In [9] authors have presented Action Unit (AU) [18] detection scheme by using features calculated from the "Particle filter" tracked fiducial facial points. They trained the system on the MMI-Facial expression database [19] and tested on the Cohn-Kanade database [16] and achieved recognition rate of 84 %. Bai et al. [11] extracted only shape information using Pyramid Histogram of Orientation Gradients (PHOG) and showed "smile" detection accuracy as high as 96.7 % on Cohn-Kanade database [16].

3   Summary of Psycho-Visual Experiment

We conducted psycho-visual experiment using an eye-tracker to determine which facial region(s) are the most important or salient according to HVS for a particular expression. The aim of experiment was to record the eye movement data of human observers in free viewing conditions. During the experiment eye movements of fifteen human observers were recorded using video based eye-tracker

| Happiness | Surprise | Sadness | Anger | Fear | Disgust |

Fig. 1. Six Universal expressions: first row show example images from Cohn-Kanade (CK+) database [20] while second row show images from the Dartmouth database [21]

(EyelinkII system, SR Research), as the subjects watched the collection of 54 videos selected from the CK+ database [20], showing one of the six universal facial expressions [3] (See Fig. 1 for example images). Observers include both male and female aging from 20 to 45 years with normal or corrected to normal vision. All the observers were naïve to the purpose of an experiment. Eye position was tracked at 500 Hz with an average noise less than 0.01°. Head mounted eye-tracker allows flexibility to perform the experiment in free viewing conditions as the system is designed to compensate for small head movements. Conclusions drawn are summarized in Table 1. For detailed discussion refer to our previous publication [14].

Table 1. Facial regions that emerged as salient for six universal expressions

Expression	Salient facial region(s)
Happiness	Mouth region
Surprise	Mouth region
Sadness	Mouth and eye regions. Biased towards mouth region
Disgust	Nose, mouth and eye regions. Wrinkles on the nose region gets little more attention than the other two regions
Fear	Mouth and eye regions
Anger	Mouth, eye and nose regions

4   Affect Analysis Framework

Feature selection along with the region(s) from where these features are going to be extracted is one of the most important step to successfully recognize expressions. As the framework draws its inspiration from the human visual system, it processes only perceptual salient facial region(s) for the feature extraction. The proposed framework creates a novel feature space by extracting pyramid local binary pattern (PLBP) appearance features [12] and facial landmark points triangles ratio (LPTR) shape features (explained in Sect. 4.1) from the perceptually

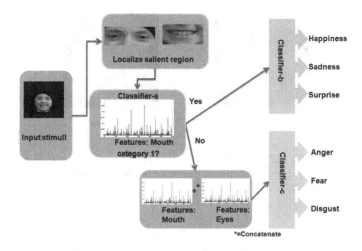

Fig. 2. Schematic overview of the proposed framework

salient facial regions. Schematic overview of the framework is illustrated in Fig. 2. Steps of the proposed framework are as follows:

Step 1: The framework initializes with the localization of the mouth region (using Viola-Jones object detection algorithm [22]) from the input stimuli. Then PLBP and LPTR features are extracted from the localized mouth region. The classification ("Classifier-a" in Fig. 2) is carried out on the basis of extracted features in order to make two groups of facial expressions. First group comprises of those expressions that has one perceptual salient region i.e. happiness, sadness and surprise while the second group is composed of those expressions that have two or more perceptual salient regions i.e. anger, fear and disgust (refer Table 1). Purpose of making two groups of expressions is to reduce feature extraction computational time.

Step 2: If the sequence is classified in the first group, then it is classified either as happiness, sadness or surprise by the "Classifier-b". Classification is carried out on the already extracted features from mouth region which is perceptually salient region for these three expressions.

Step 3: If the input sequence is classified in the second group, then the framework extract texture and shape features from the eyes region and concatenates them with already extracted features from the mouth region (expressions of anger, fear and disgust have more than one salient region). Then, the concatenated feature vector is fed to the classifier ("Classifier-c") for the classification.

4.1 Landmark Points Triangles Ratio (LPTR)

We are proposing novel shape descriptor, landmark points triangles ratio (LPTR). LPTR extracts shape information by calculating area of triangle (normalized) from different facial regions. These triangles are formed from six landmarks points (six points from each facial region) which are obtained automatically

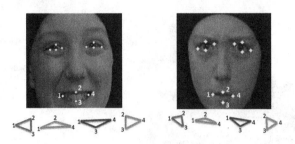

Fig. 3. Toy example for LPTR feature extraction. First row shows stimuli, while second row shows four unique triangles obtained from four landmark points of the mouth region of above placed stimuli.

using method proposed by Silva et al. [23]. Twenty unique triangles are formed from these six facial region landmark points. Then area of obtained unique triangle is calculated and normalized by the sum of area of all the triangles to get the shape descriptor. Thus, for any specific facial region LPTR extracts twenty dimensional feature vector.

$$f(i) = \frac{\delta_i}{\sum_{j=1}^{20} \delta_j} \qquad (1)$$

where δ is the area of triangle.

Look at the toy example (see Fig. 3) to get intuition of how LPTR works. Figure 3 shows images of same kid showing two different expression (stimuli from the Dartmouth database [21]). Four landmark points are obtained on mouth and eyes regions each using method proposed by Silva et al. [23]. In the second row of the referred image, toy example shows four triangles obtained from the mouth region (four landmark points give rise to four unique triangles). It can be observed that the triangle areas are different as the shape of lips is deformed in a different way for two facial expressions. The rational behind LPTR descriptor is that for different facial expressions, different facial regions will be deformed in a unique way. Aim of the LPTR shape descriptor is to capture those unique deformations. Thus, providing discriminative ability for the expression classification. Finally, to achieve invariance for different face sizes LPTR descriptor normalizes area of triangles by the sum of area of all the triangles in that region.

Figure 4 shows LPTR shape descriptor for six universal expressions. Figure shows sixty dimensional (60 D) LPTR feature vector (twenty dimensional feature vector each from mouth, left eye and right eye regions). It can be observed in the referred figure that LPTR features have a discriminative trend for six universal expressions as the shape for extracted LPTR features for different expression differs substantially. This discriminative ability is used in the proposed framework along with discriminative ability of PLBP features to robustly classify children's facial expression.

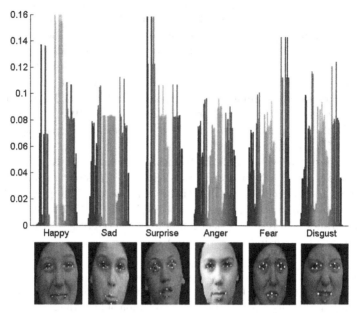

Six universal expressions

Fig. 4. LPTR features for different expressions. First row shows extracted LPTR features for the stimuli shown in the second row. Landmark points shown in the second row are obtained automatically using method proposed by Silva et al. [23]

5   Affect Recognition Experiment

The performance of proposed framework was evaluated for four different state-of-the-art classifiers in order to study its robustness. These four classifiers are: Support vector machine (SVM), C4.5 decision tree (DT), random forest (RF) and Multi-Layer Perceptron (MLP).

The proposed framework was evaluated for two different databases, so that its performance can be measured against children and adult faces. The two databases are: the Dartmouth database of children's faces [21] and Cohn-Kanade (CK+) database (adult faces) [20]. In this study we have focused on the problem of children affect analysis but for the sake of completeness and comparison with the state-of-the-art methods (to the best of our knowledge computer vision scientist have not shown results of their proposed algorithms for children database) we have tested proposed framework on adult faces as well. Results for these two databases are presented in Sects. 5.1 and 5.2 respectively.

5.1   Effectiveness of Framework on Children Faces

To test the effectiveness of the proposed framework on children faces, we conducted experiment on the Dartmouth database [21]. The Dartmouth Database of children Faces [21] contains faces of 40 male and 40 female Caucasian children (Fig. 1 show example images from the database). All faces in the database were assessed by at least 20 raters for facial expression identifiability and intensity (as opposed to CK+ database which is FACS [18] compliant/coded). Expression of happy was most accurately identified while fear was least accurately identified by human raters. Human raters correctly classified 94.3 % of the happy faces while expression of fear was correctly identified in 49.08 % of the images, least identifiable by human raters. On average human raters correctly identified expression in 79.7 % of the images [21].

For the experiment we used all the frames in the database. Region of interest was obtained automatically by using Viola-Jones object detection algorithm [22] and processed to obtain PLBP and LPTR feature vector. The proposed framework achieved average recognition rate of 78.9 %, 74.6 %, 73.1 % and 71.9 % for SVM, MLP, random forest and decision tree respectively (see Table 2). These values were calculated using 10-fold cross validation.

Table 2. Recognition rate (%) for three pyramid levels. Pyramid levels are used only for extracting PLBP feature. LPTR features are extracted from stimuli only in given spatial resolution. Final feature vector is a concatenation of both the extracted features.

	Pyramid Level		
	Level 0	Level 1	Level 2
SVM	60.8	72.5	78.9
MLP	66.4	68.8	74.6
Random Forest	60.1	66.1	73.1
Decision tree	62.2	68.2	71.9

Proposed framework achieved results in the range of 95 % (better than state-of-the-art results) for CK+ database (refer Sect. 5.2) but for Dartmouth database of children faces [21] it achieved result in the range of 75 %. This is due to the fact that database of children faces [21] have actors ranging from 6 to 16 years of ages. Usually children show expressions in a subtle way, thus creating large inter and intra population variations as opposed to adults. But still SVM produced results (78.9 % correct classification) similar to human raters (human raters correctly identified expression in 79.7 % of the images).

Generalization Capabilities. Aim of this experiment is to study how well the proposed framework generalizes on unseen data or new database. According to our knowledge only Valstar et al. [9] have reported such data earlier. In this experiment we trained classifier using the Dartmouth database [21] and tested its performance on frames from NIMH child emotional faces picture set (NIMH-ChEFS) database [24]. NIMH-ChEFS database has 482 frames containing expressions of fear, anger, happy and sad with two gaze conditions: direct and averted gaze. The databases is validated by 20 adult raters.

Table 3. Average recognition accuracy (%): training classifier on Dartmouth database and testing it with NIMH-ChEFS database

	SVM	MLP	Random forest	Decision tree
Training samples	78.9	74.6	73.1	71.9
Test samples	68.4	63.3	61.5	60.1

Results obtained in this configuration are presented in Table 3. This experiment simulates the real life situation when the proposed framework would be employed to recognize facial expressions on the unseen data. Obtained results are encouraging and they can be further improved by training classifiers on more than one databases before using in real life scenario.

5.2 Effectiveness of Framework on Adult Faces

This experiment measures the performance of the proposed framework on the classical database i.e. extended Cohn-Kanade (CK+) database [20]. Most of the methods in literature report their performance on this database, so this experiment could be considered as the benchmark experiment for facial expression recognition framework. For comparison and reporting results, we have used the classification results obtained by the SVM as it is the most cited method for classification in the literature.

Table 4. Comparison with the state-of-the-art methods for adult faces database (extended Cohn-Kanade (CK+) database [20]).

	Sequence Num	Class Num	Performance Measure	Recog Rate (%)
[4]	313	7	leave-one-out	93.3
[25]	374	6	2-fold	95.19
[25]	374	6	10-fold	96.26
[26]	374	6	5-fold	94.5
[6]	375	6	-	93.8
[17]a	352	6	66 % split	92.3
[17]b	352	6	66 % split	80
Ours	**309**	**6**	**10-fold**	**96.89**

Table 4 shows the comparison of the achieved average recognition rate of the proposed framework with the state-of-the-art methods using same database (i.e. Cohn-Kanade database). Results from [17] are presented for the two configurations. "[17]a" shows the result when the method was evaluated for the last three frames from the sequence while "[17]b" presents the reported result for the frames which encompasses the status from onset to apex of the expression. It can be observed from the Table 4 that the proposed framework is comparable to any other state-of-the-art method in terms of expression recognition accuracy. The method discussed in "[17]b" is directly comparable to our method, as we also evaluated the framework on similar frames. In this configuration, our framework is better in terms of average recognition accuracy.

6 Conclusion

In this article we presented robust framework for children affect recognition. This study could be considered as pioneering study for children affect analysis as previously computer vision scientist have generally overlooked the problem of children affect analysis. The proposed framework performed very similar to human raters for Dartmouth database of children's faces and achieved results better than state of the art methods for extended Cohn-Kande database (CK+). Proposed framework is inspired from human visual system (HVS) and selects only salient facial regions (as it happens in HVS) for feature extraction.

Secondly, we proposed to extract facial regions shape information using novel shape descriptor, landmark points triangles ratio (LPTR). Facial regions texture information is extracted using pyramid local binary pattern (PLBP) features. PLBP descriptor is pyramidal-based spatial representation of local binary pattern (LBP) descriptor. Both of these features were concatenated during the learning process. Results better than state of the art methods for CK+ database suggests robustness of the proposed framework.

Acknowledgment. This study is financially supported by BPI France (http://www. bpifrance.fr/) and FUI-KURIO EYE Project.

References

1. Ekman, P.: Telling Lies: Clues to Deceit in the Marketplace, Politics, and Marriage, 3rd edn. W. W. Norton & Company, New York (2001)
2. Pantic, M., Pentland, A., Nijholt, A., Huang, T.: Human computing and machine understanding of human behavior: A survey. In: ACM International Conference on Multimodal Interfaces (2006)
3. Ekman, P.: Universals and cultural differences in facial expressions of emotion. In: Nebraska Symposium on Motivation, pp. 207–283 (1971)
4. Littlewort, G., Bartlett, M.S., Fasel, I., Susskind, J., Movellan, J.: Dynamics of facial expression extracted automatically from video. Image Vis. Comput. **24**, 615–625 (2006)
5. Khan, R., Meyer, A., Konik, H., Bouakaz, S.: Facial expression recognition using entropy and brightness features. In: 11th International Conference on Intelligent Systems Design and Applications (2011)
6. Tian, Y.: Evaluation of face resolution for expression analysis. Comput. Vis. Pattern Recogn. Workshop **68**, 179–201 (2004)
7. Khan, R.A., Meyer, A., Konik, H., Bouakaz, S.: Human vision inspired framework for facial expressions recognition. In: IEEE International Conference on Image Processing (2012)

8. Zhang, Y., Ji, Q.: Active and dynamic information fusion for facial expression understanding from image sequences. IEEE Trans. Pattern Anal. Mach. Intell. **27**, 699–714 (2005)
9. Valstar, M., Patras, I., Pantic, M.: Facial action unit detection using probabilistic actively learned support vector machines on tracked facial point data. In: IEEE Conference on Computer Vision and Pattern Recognition Workshop, pp. 76–84 (2005)
10. Khan, R., Meyer, A., Konik, H., Bouakaz, S.: Pain detection through shape and appearance features. In: 2013 IEEE International Conference on Multimedia and Expo (ICME) (2013)
11. Bai, Y., Guo, L., Jin, L., Huang, Q.: A novel feature extraction method using pyramid histogram of orientation gradients for smile recognition. In: International Conference on Image Processing (2009)
12. Khan, R.A., Meyer, A., Konik, H., Bouakaz, S.: Framework for reliable, real-time facial expression recognition for low resolution images. Pattern Recogn. Lett. **34**, 1159–1168 (2013)
13. Zhaoping, L.: Theoretical understanding of the early visual processes by data compression and data selection. Netw. Comput. Neural Syst. **17**, 301–334 (2006)
14. Khan, R.A., Meyer, A., Konik, H., Bouakaz, S.: Exploring human visual system: study to aid the development of automatic facial expression recognition framework. In: Computer Vision and Pattern Recognition Workshop (2012)
15. Lucey, P., Cohn, J., Matthews, I., Lucey, S., Sridharan, S., Howlett, J., Prkachin, K.: Automatically detecting pain in video through facial action units. IEEE Trans. Syst., Man, Cybern., Part B: Cybern. **41**, 664–674 (2011)
16. Kanade, T., Cohn, J.F., Tian, Y.: Comprehensive database for facial expression analysis. In: Fourth IEEE International Conference on Automatic face and Gesture Recognition (FG 2000), pp. 46–53 (2000)
17. Yang, P., Liu, Q., Metaxas, D.N.: Exploring facial expressions with compositional features. In: Computer Vision Pattern Recognition, pp. 2638–2644 (2010)
18. Ekman, P., Friesen, W.: The facial action coding system: A technique for the measurement of facial movements. Consulting Psychologist, Palo Alto (1978)
19. Pantic, M., Valstar, M.F., Rademaker, R., Maat, L.: Web-based database for facial expression analysis. In: International Conference on Multimedia and Expo (2005)
20. Lucey, P., Cohn, J.F., Kanade, T., Saragih, J., Ambadar, Z., Matthews, I.: The extended cohn-kande dataset (CK+): A complete facial expression dataset for action unit and emotion-specified expression. In: Computer Vision and Pattern Recognition Workshops (2010)
21. Dalrymple, K.A., Gomez, J., Duchaine, B.: The Dartmouth database of children's faces: Acquisition and validation of a new face stimulus set. PLoS ONE **8**, e79131 (2013)
22. Viola, P., Jones, M.: Rapid object detection using a boosted cascade of simple features. In: IEEE Conference on Computer Vision and Pattern Recognition (2001)
23. Silva, C., Schnitman, L., Oliveira, L.: Detection of facial landmarks using local-based information. In: Brazilian Conference on Automation (2012)
24. Egger, H., Pine, D., Nelson, E., Leibenluft, E., Ernst, M., K.E., T., Angold, A.: The NIMH child emotional faces picture set (NIMH-ChEFS): A new set of children's facial emotion stimuli. Int. J. Methods Psychiatr. Res. 20(3), 145–56 (2011)
25. Zhao, G., Pietikäinen, M.: Dynamic texture recognition using local binary patterns with an application to facial expressions. IEEE Trans. Pattern Anal. Mach. Intell. **29**, 915–928 (2007)
26. Kotsia, I., Zafeiriou, S., Pitas, I.: Texture and shape information fusion for facial expression and facial action unit recognition. Pattern Recogn. **41**, 833–851 (2008)

A Study of Hand Motion/Posture Recognition in Two-Camera Views

Jingya Wang and Shahram Payandeh[(✉)]

Experimental Robotics and Imaging Laboratory, School of Engineering Science,
Simon Fraser University, Burnaby, BC, Canada
{jingyaw,payandeh}@sfu.ca

Abstract. This paper presents a vision-based approach for hand gesture recognition which combines both trajectory recognition and hand posture recognition. With two calibrated cameras, the 3D hand motion trajectory can be reconstructed. The reconstructed trajectory is then modeled by dynamic movement primitives (DMP) and a support vector machine (SVM) is trained to recognize five classes of gestures trajectories. Scale-invariant feature transform (SIFT) is used to extract features on segmented hand postures taken from both camera views. Based on various hand appearances captured by the two cameras, the proposed hand posture recognition method has shown a very good success rate. A gesture vector is proposed to combine the recognition result from both trajectory and hand postures. For our experimental set-up, it was shown that it is possible to accomplish a good overall accuracy for gesture recognition.

1 Introduction

Hand gesture is represented by four aspects: hand shape, position, orientation and movement [1]. The same semantic paths are usually made in different scales, speed and shapes due to the individual differences. As a statistical model, Hidden Markov model (HMM) has been found to efficiently model spatiotemporal properties of time series where the same gesture can have different shapes and durations [2]. Other feature extraction methods such as Gaussian mixture model and principle component analysis [3,4] can be used to enhance HMM recognition process. Finite state machine (FSM) which is similar to HMM models hand movement as an ordered sequence of states in a spatiotemporal configuration space [5,6]. Dynamic movement primitives (DMP) has been employed for 2D trajectory recognition which has shown to obtined increased accuracy [6,7]. DMP encodes gesture paths into weight vectors which preserve the topological structures of paths. Advantages of DMP can be stated to be: (a) robust to the spatiotemporal variations in gesture paths, and (b) it is easy to adjust the dimension of the weight vector based on the complexities of gesture paths in order to adapt for different applications.

Hand posture recognition is another key components of gesture recognition. In general, template matching is a popular approach of posture recognition where

G. Bebis et al. (Eds.): ISVC 2015, Part II, LNCS 9475, pp. 314–323, 2015.
DOI: 10.1007/978-3-319-27863-6_29

it is easy to add or remove template classes. To extract features on hand posture, a convex hull on silhouette [8], or fingertip detection using circular mask as a correlation techniques [9] can be employed. However, these recognition approaches based on hand silhouette or contour usually requires a uniform background where the hand can be well segmented. In the case study of this paper, hand posture is made in a clutter and moving background which is then segmented using YCbCr color space. For cases when the hand can be properly segmented or is in partial occlusion, a feature detector that is insensitive to illumination variation and partial occlusion can be utilized. SIFT is robust to affine distortion within a given range. This feature detector can be used in posture recognition since the relative position of the hand changes that causes affine distortion between the input hand postures and posture templates. In our study, SIFT is used as a feature detector for the posture recognition. Combining recognition results of hand trajectory and hand postures, a gesture vector is then defined for gesture recognition. Hand gesture is referred to the trajectory of the moving hand and the hand posture which defines the hand shape and appearance. Two calibrated cameras are employed to record hand gestures. In our study, these cameras are utilized under a steady lighting condition. Due to exposure compensation of the cameras, YCbCr color space is adopted for hand area segmentation.

2   Preliminaries

Hand Segmentation: Hand segmentation is a pre-acquisition for any vision-based hand tracking and posture recognition. Considering the numerous different hand appearances, postures, angels and orientations, color cue is an important tool to separate hand area from the background. However, in general, segmenting hand from cluttered background is very challenging [10]. For example, there exists different skin colors among people which also change under different illumination. [11] studied skin color segmentation on different color representations, color quantization and classification algorithms. YCbCr as a color space, represents color into the chrominance (Cb and Cr) and luminance (Y) which have shown to have a good performance on skin color detection against illumination changes. Here, fixed thresholds in the Cb and Cr channels ($77 \leq Cb \leq 127$ and $133 \leq Cr \leq 173$ [12]) are selected which can obtain a detection rate of up to 93 % with false detection rate of 28 % [11].

Hand Tracking: Since unwanted areas such as face, neck and other skin-color-liked objects can also be segmented, position and area size constraints are adopted to locate the hand area while eliminates noise. Kalman Filter (KF) [13] is applied for tracking hand in each frame, where the hand location is represented by the center of the hand blob.

Trajectory Reconstruction: In our study, two calibrated cameras are employed to record the hand gestures which are made in their overlapping camera view area. Based on the hand tracking result in both camera view, the 3D coordinates of a point in the overlapping camera view can be reconstructed from the pixel values on the projected image plane [14]. For each time instance, the pixel values of both hand centers can be extracted from both camera views used to reconstruct the hand position in the world reference coordinate.

3   Gesture Recognition

Dynamic Movement Primitives for Trajectory Recognition: Dynamic movement primitives (DMP) [7] and its further extensions [15,16] models movements for a given start and end states into a set of differential equations. These differential equations characterize the spatiotemporal evolution of a dynamical system with the given start and end states: $\tau \dot{z} = \alpha_z(\beta_z(g - y) - z) + f$ and $\tau \dot{y} = z$. A second-order linear damped spring model with a non-linear function f (e.g. Gaussian-like forcing function) is added in to capture the complexities of motion patterns made by human: $f(x) = \sum_{i=1}^{N} \phi_i(x)w_i / \sum_{i=1}^{N} \phi_i(x)x(g - y_0)$. y, z and $\dot{z}$ represent the position, velocity and acceleration of the hand motion dynamics. τ is a constant represent trajectory duration and g represents the final hand position of the gesture path. For a suitable selection of parameters α_z and β_z, the forcing term f decays to zero over time which allow the system converge to the goal position, e.g. $(y, z) = (g, 0)$.

The weight vector $\mathbf{w} = [w_1, ..., w_N]^T$ in f preserves the shape information of the trajectories. For instance, if $\mathbf{w}$ is fixed and other parameters such as the goal state g or time constant τ changes, the DMP will generate topological similar trajectories. In other words, similar trajectories would have similar feature vector $\mathbf{w}$ which is called the invariance properties of the DMP model [15]. With such property, trajectories can be classified based on the weight vectors.

Given a trajectory in one dimension $Q = (q_0, ..., q_{t-1})$, let $q_s = (y, z, \dot{z})^T$ denotes the measured state vector at time s. To learn the weight vector for a given trajectory, we can extract its initial and goal states and the its time duration. From the above, the forcing function f can be rewrite as: $f = \alpha_z(\beta_z(g - y) - z) - \tau \dot{z} = \sum_{i=1}^{N} \phi_i(x)w_i / \sum_{i=1}^{N} \phi_i(x)x(g - y_0)$. The weight vector $\mathbf{w}$ with the dimension $D_{\mathbf{w}}$ can be learned for example using the locally weighted regression (LWR) [7].

SIFT on Hand Postures: is a feature detector which is shown to provide robust matching when there are partial occlusions, affine distortions, addition of noises and changes in illumination [17]. This method is also implemented on segmented hand postures using the notion of bag-of-words extracted from the feature points. Each segmented posture is then represented by a histogram vector for classification [18].

Gesture Vector: Based on the previous recognition results, a gesture vector is constructed by two components: posture elements p_{ij} and trajectory element T. T indicates the recognized trajectory class. Depending on the complexity of gestures, each gesture can be separated into i segments to deal with posture variations. p_{ij} represents the occurrence number of the recognized posture class j in segment i. For example, gesture vector can be written as: $\mathbf{v}_g = [p_{11}, p_{12}, p_{13}, ..., p_{ij}, ...p_{mn}, T]$ having i segments and j posture class.

Gestures made at different speed would generate differen number of recognized postures. For example, due to the fixed frame rate of tracking system, faster the hand is moving, the fewer postures can be recognized. As a results, we can normalize the gesture vector by taking into account the total number of recognized postures P_m. The above gesture vector can be rewritten as: $[p_{11}/P_1, p_{12}/P_1, p_{13}/P_1, ..., p_{ij}/P_i, ..., p_{mn}/P_m, T]$.

Multiclass SVM for Recognition: SVM is a supervised learning model which became popular for classification, regression, and novelty detection. An important property of SVM is the determination of the model parameters corresponding to a convex optimization problem, and so any local solution is also a global optimum [19]. In this paper, for each component, the recognition part is accomplished using multi-class SVMs (one-vs-all) which is trained by LIBSVM [20].

4 Experiments

Five classes are collected for the trajectory recognition. After the initial segmentation and spatial correlations using the calibrated camera models, the tracking is accomplished using Kalman filter. The five classes consist by "Jump", "Left", "Right", "Circle" and "Forward". Figure 1 shows the example trajectories for each class. Eight participants (two male and six female) are asked to perform the trajectories, 5 for each class. 200 trajectories are collected. 3/4 of the data set is used for training the SVM while the rest of the data set are used for testing.

For posture recognition, six classes of postures are collected. For our two camera set-up, both front view and side view of each class are included. Figure 2 shows six classes of postures. The front views are shown in Fig. 2a while the side views are shown in Fig. 2b. The training set contains 402 well-segmented postures clipped from the gesture videos made by four participants. The testing set contains 432 postures made by different four participant than the original participants used for the training data. For each camera there are total of 216 segmented postures which represents 36 repetitions for each trained class.

For gesture recognition, seven class of gesture are collected. Figure 3 shows examples of each gesture class. 70 gestures are collected among four people for the training stage (10 for each class) and another 70 gestures (10 for each class) collected form the other four are for testing.

Before conducting trajectory recognition results, a trajectory instance "Circle" (Fig. 1d) is selected from the collected trajectories in order to visualize the weight vector and learned DMP models. The acquired hand trajectory is in 3D space where DMP is utilized along each projected directions of the world coordinate system. The SVM with a linear kernel is trained for trajectory recognition. Table 1 gives the 5-fold cross-validation recognition accuracy based different weight vector dimensions. As the dimension increases, for the same number of training data, the accuracy decreases. This is because the bigger the weight vector dimension is, the more parameters in the SVM need to be decided and more training data is needed. The highest recognition rate is obtained at $D_\mathbf{w} = 5$. This is also because the classes of trajectories we are collected are relatively simple and

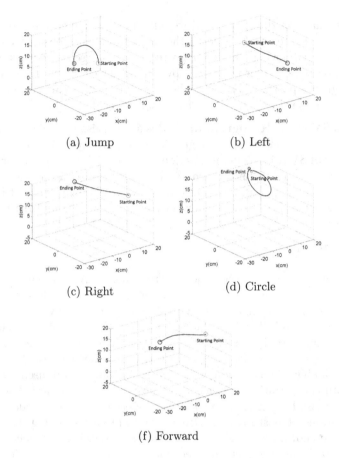

Fig. 1. Trajectory samples of five classes. The starting point is marked in green, and the ending point is in red. (Color figure online)

distinctive from each other. For complex trajectories, a weight vector with higher dimensions is needed. Applying the trained SVM on the testing data set, we obtain the accuracy of 88.0 %. The recognition results are shown in Table 2.

For each of the postures in the training and testing data sets, SIFT is adopted for feature detection. The bag of words [18] method is employed to generate a 1×150 vector for each posture. The multi-class SVM is trained base on such vectors. A SVM with linear kernel trained on 402 template postures achieved an 5-fold cross-validation accuracy of 98.8 %. To test the performance of the trained posture classier, we collected the image of each posture taken from both camera views and recognized them individually. The recognition result shows in Table 3. Each column refers to a posture instance that is classified into the corresponding class. An accuracy of 78.7 % is obtained.

In posture recognition, the hand posture is viewed in both cameras which are further processed for posture recognition. If the recognition results of the

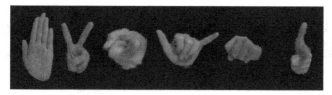

(a) Front view of 6 posture classes

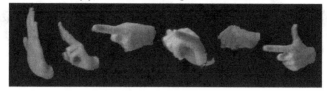

(b) Side view of 6 posture classes

Fig. 2. The front view and side view of six posture classes.

Table 1. Recognition accuracy with different number of $D_\mathbf{w}$

	Weight vector dimension $D_\mathbf{w}$			
Training trajectory number	5	10	20	40
150	86.67 %	74.12 %	59.17 %	51.67 %

postures taken from two cameras are different, then this posture is taken to be ambiguous and discarded. The postures that are recognized to belong to the same class from both camera views are kept for further gesture recognition in follow-up step. For each posture class, the recognition and abandon rate are shown in Table 4. Here, abandon rate corresponds to the cases where there is ambiguity in recognition of postures from two cameras. Although nearly thirty percent of the testing data is abandoned, this scheme increases the recognition accuracy from 78.7 % to 96.6 %.

In our experimental study, since our test trajectories involve one or no changes in postures, we have divided each gesture into two separate segments with two equal durations of time associated with the whole gesture. The dimension of our gesture vector in then defined as $2 \times 6 + 1 = 13$ dimension for each hand gesture. A linear kernel SVM is trained which obtained a 5-fold cross-validation accuracy at 94 %. The trained SVM model recognizes each hand gesture based on the gesture vector. Table 5 shows the recognition results and accuracies for each class. The blanks stand for zero.

The lower recognition accuracy for gesture "Grab" is due to the lower recognition rate for posture "Fist". It has a high chance for misclassifying the posture "Fist" into the posture "Point" (Table 3) due to the similarity of these two postures. With a high recognition rate of posture "Push", "Grab" also has a higher chance to be recognized as "Push".

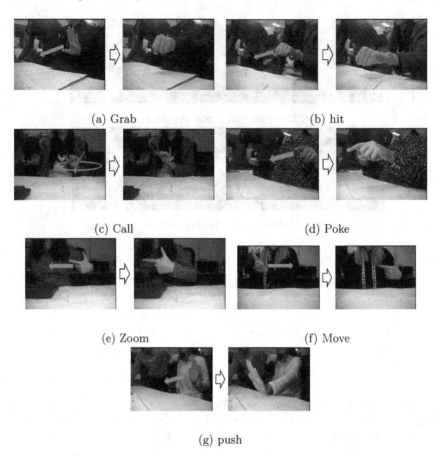

(a) Grab   (b) hit

(c) Call   (d) Poke

(e) Zoom   (f) Move

(g) push

Fig. 3. Example of seven gesture classes. The beginning and ending frames are shown, and the moving direction is shown in blue arrow.

Table 2. Recognition result of testing trajectories.

Class	Jump	Left	Right	Circle	Push
Jump	10	0	0	0	0
Left	0	8	0	0	2
Right	0	0	10	0	0
Circle	0	0	0	10	0
Push	1	3	0	0	6
Accuracy	88.0 % (44/50)				

Table 3. Confusion matrix for posture recognition on the testing set

Posture	Eight	Fist	Palm	Point	Six	V
Eight	71	0	1	4	2	0
Fist	0	46	0	15	9	0
Palm	0	0	61	0	0	2
Point	0	17	3	52	7	2
Six	1	9	0	1	43	1
V	0	0	7	0	11	67
Accuracy	98.6%	63.9%	84.7%	72.2%	59.7%	93.06%
78.7% (340/432)						

Table 4. Recognition result and abandon rate for the testing set

Posture	Recognition accuracy	Abandon rate
Eight	35/35	1/36
Fist	16/20	13/36
Palm	27/27	9/36
Point	17/18	18/36
Six	17/17	19/36
V	31/31	5/36
Accuracy	96.6%	30.1%

Table 5. Confusion matrix for gesture recognition on the testing set

Gesture	Grab	Hit	Call	Poke	Zoom	Move	Push
Grab	6			1			3
Hit		8		2			
Call			10				
Poke				10			
Zoom					10		
Move						10	
Push							10
Accuracy	60%	80%	100%	100%	100%	100%	100%
91.4% (64/70)							

5 Discussions And Conclusions

This paper proposed a vision-based approach for gesture recognition (which combines both the hand posture and its trajectory) using two calibrated cameras. DMP models the reconstructed 3D gesture path into a weight vector which preserves the trajectory topology which is robust to speed and scale variation. For the proposed camera set-up and illumination conditions, SIFT is adopted for hand posture recognition. Associating the posture recognition results from both camera views shows superior than a single view recognition. A gesture vector which includes both posture and path information is used for describing a gesture. With such vector, gestures can be recognized with an accuracy of 91.4 %. In this paper, only 5 classes of trajectory and 6 classes of postures are collected, and 7 classes of gesture with different posture and trajectory combination are trained and recognized. Our approach allows the users to add trajectory and posture classes with a few training data based on their demands.

At the current implementation, application of SIFT feature detector poses a considerable computational challenge for offering real-time performance. A faster and more efficient feature detector need to be explored in order to make the system applicable to real-time cases involving fast movements with less restricted illumination. A classifier such as neural network could be applied, but a considerable amount of training data would be needed. Also, the gesture recognition in this paper is focusing on one-hand gestures and does not involve hand position and orientation. The research of two-hand gestures needs to be explored and more robust hand segmentation and tracking method needs to be studied.

References

1. Stokoe, W.C.: Sign language structure (1978)
2. Elmezain, M., Al-Hamadi, A., Appenrodt, J., Michaelis, B.: A hidden markov model-based continuous gesture recognition system for hand motion trajectory. In: 19th International Conference on Pattern Recognition, 2008. ICPR 2008, pp. 1–4. IEEE (2008)
3. Bashir, F., Qu, W., Khokhar, A., Schonfeld, D.: Hmm-based motion recognition system using segmented pca. In: IEEE International Conference on Image Processing, 2005. ICIP 2005. vol. 3, pp. III-1288. IEEE (2005)
4. Bashir, F.I., Khokhar, A.A., Schonfeld, D.: Object trajectory-based activity classification and recognition using hidden markov models. IEEE Trans. Image Process. **16**(7), 1912–1919 (2007)
5. Mitra, S., Acharya, T.: Gesture recognition: A survey. IEEE Trans. Syst., Man, Cybern., Part C: Appl. Rev. **37**(3), 311–324 (2007)
6. Hong, P., Turk, M., Huang, T.S.: Gesture modeling and recognition using finite state machines. In: 2000 Proceedings of the Fourth IEEE International Conference on Automatic Face and Gesture Recognition, pp. 410–415. IEEE (2000)
7. Schaal, S.: Dynamic movement primitives -a framework for motor control in humans and humanoid robotics. In: Kimura, H., Tsuchiya, K., Ishiguro, A., Witte, H. (eds.) Adaptive Motion of Animals and Machines, pp. 261–280. Springer, Heidelberg (2006)
8. Li, Y.: Hand gesture recognition using kinect. In: 2012 IEEE 3rd International Conference on Software Engineering and Service Science (ICSESS), pp. 196–199. IEEE (2012)
9. Letessier, J., Bérard, F.: Visual tracking of bare fingers for interactive surfaces. In: Proceedings of the 17th Annual ACM Symposium on User Interface Software and Technology, pp. 119–122. ACM (2004)

10. Fang, Y., Wang, K., Cheng, J., Lu, H.: A real-time hand gesture recognition method. In: 2007 IEEE International Conference on Multimedia and Expo, pp. 995–998. IEEE (2007)
11. Phung, S.L., Bouzerdoum, A., Chai Sr, D.: Skin segmentation using color pixel classification: analysis and comparison. IEEE Trans.Pattern Anal. Mach. Intell. 27(1), 148–154 (2005)
12. Chai, D., Ngan, K.N.: Face segmentation using skin-color map in videophone applications. IEEE Trans. Circ. Syst. Video Technol. 9(4), 551–564 (1999)
13. Welch, G., Bishop, G.: An introduction to the kalman filter (1995)
14. Bradski, G., Kaehler, A.: Learning OpenCV: Computer vision with the OpenCV library. O'Reilly Media, Inc. (2008)
15. Ijspeert, A.J., Nakanishi, J., Hoffmann, H., Pastor, P., Schaal, S.: Dynamical movement primitives: learning attractor models for motor behaviors. Neural Comput. 25(2), 328–373 (2013)
16. Liu, Z., Hu, F., Luo, D., Wu, X.: Visual gesture recognition for human robot interaction using dynamic movement primitives. In: 2014 IEEE International Conference on Systems, Man and Cybernetics (SMC), pp. 2094–2100. IEEE (2014)
17. Lowe, D.G.: Distinctive image features from scale-invariant keypoints. Int. J. Comput. Vis. 60(2), 91–110 (2004)
18. Csurka, G., Dance, C., Fan, L., Willamowski, J., Bray, C.: Visual categorization with bags of keypoints. In: Workshop on Statistical Learning in Computer Vision, ECCV. vol. 1, pp. 1–2. Prague (2004)
19. Bishop, C.M., et al.: Pattern recognition and machine learning. Springer, New York (2006)
20. Chang, C.C., Lin, C.J.: LIBSVM: A library for support vector machines. ACM Trans. Intell. Syst. Technol. 2, 27:1–27:27 (2011)

Pattern Recognition

Automatic Verification of Properly Signed Multi-page Document Images

Marçal Rusiñol[⊠], Dimosthenis Karatzas, and Josep Lladós

Computer Vision Center, Department Ciències de la Computació, Edifici O,
University Autònoma de Barcelona, 08193 Bellaterra, Barcelona, Spain
{marcal,dimos,josep}@cvc.uab.es

Abstract. In this paper we present an industrial application for the automatic screening of incoming multi-page documents in a banking workflow aimed at determining whether these documents are properly signed or not. The proposed method is divided in three main steps. First individual pages are classified in order to identify the pages that should contain a signature. In a second step, we segment within those key pages the location where the signatures should appear. The last step checks whether the signatures are present or not. Our method is tested in a real large-scale environment and we report the results when checking two different types of real multi-page contracts, having in total more than 14,500 pages.

1 Introduction

Nowadays, big bank corporations tend to release their local branches from all the paperwork load by providing efficient ways to digitize and forward the paper documents to their central services for processing. However, bank customers contract daily tens of thousands of financial products such as loans, mortgages, insurances, investments, etc. yielding to huge volumes of document images to be processed. This document processing usually requires manual intervention in tasks such as document classification, information extraction, verification, etc. Document Image Analysis research provides solutions for automating some of these processes with minimal human intervention.

One of the tasks requiring a tedious manual intervention is the verification of properly signed contracts. Before providing a certain service to a customer, the institution has to confirm that the contract has been properly signed. Since the dawn of Document Image Analysis research, many works dealing with signature verification [1–3] have been proposed. However, before verifying that the signature from a given customer is genuine, there should be a step verifying that the document is properly signed, i.e. it contains the correct amount of signatures in the right places.

The method proposed by Zhu et al. in [4,5], is one of the few contributions in signature localization in document images. The authors propose a detection framework based on analyzing the curvature of contour fragments over multiple scales. The proposed method performs well at both the localization and matching

© Springer International Publishing Switzerland 2015
G. Bebis et al. (Eds.): ISVC 2015, Part II, LNCS 9475, pp. 327–336, 2015.
DOI: 10.1007/978-3-319-27863-6_30

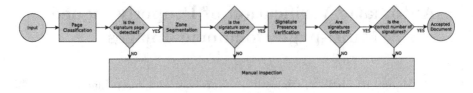

Fig. 1. Overview of the proposed architecture.

tasks. Other methods working on signature segmentation are [6,7]. However, in most scenarios, the incoming documents are usually structured and there is no need to look at signatures over the whole of the document image, instead, focusing at a particular location would be enough, which recasts the problem into a sequential process of segmentation and detection each much simpler than the holistic approach.

In addition, although many works dealing with single-page document representations can be found in the literature (e.g. [8,9]), works dealing with variable-length multi-page documents, representing a most realistic scenario, are scarce. To our best knowledge, the only works handling a multi-page scenario are devoted to document classification. Examples are the works by Frasconi et al. [10] where a hidden Markov model categorizes documents by looking at sequences of pages. Gordo and Perronnin proposed in [11] a bag-of-pages approach that treats multi-page documents as unordered sets of pages. Finally, Rusiñol et al. proposed in [12,13] a multi-page document classification system that takes into account both textual and visual cues to categorize incoming documents.

We present in this paper a method that allows the automatic verification of properly signed multi-page structured documents. Our method is tested in a real large-scale environment (more than 14,500 pages have been tested) and we report the results when checking two different types of real multi-page contracts. The remainder of the paper is organized as follows. In Sect. 2 we overview the proposed method and detail the use cases we focus in. Section 3 gives the details on the page classification step and in Sect. 4 we present the signature detection strategy. We present the experimental results in Sect. 5 and we conclude in Sect. 6.

2   Problem Definition and Method Overview

In our particular scenario, the incoming documents to check are multi-page documents. Although we do have knowledge of the document flow structure, i.e. we know when a new document starts and when it ends, the pages can come in a different order and orientation depending on how the operator fed the physical document to the scanner. For a particular document type, we know which pages have to be signed and where on these pages the customer or the bank representative has to sign the document. Given a multi-page document image as input, we want to accept the document if the correct amount of signatures is found in the proper

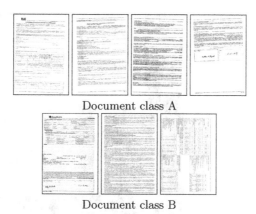

Document class A

Document class B

Fig. 2. Example of the multi-page document images considered as use cases.

pages at the correct places. Whenever the obtained evidences are too weak, the whole document should be rejected and forwarded to the manual process.

We can see in Fig. 1 an overview of the proposed architecture. We divide our proposed industrial application in three separate stages. First, a page classification step identifies from a page sequence, corresponding to a multi-page document, which are the pages that should be signed. In a second step, we focus on the detected pages and we perform a segmentation step aimed at localizing the particular region on those pages where a signature has to be found. The last step decides whether there is a signature in the document or the signature zone was left blank.

In our scenario it is quite important that the system does not provide false positives. Accepting a non-signed document has a much higher cost than rejecting and sending to manual inspection a signed one. Thus, each of the above steps have associated rejection criteria. In case of doubt, we prefer to forward the document to manual inspection rather than accepting non-properly signed contracts. Hence rejection is configured to work in a conservative fashion.

In this paper we report the obtained results when dealing with two different types of contracts that arrive daily at the central services. We can see an example of those documents in Fig. 2. Since those documents contain private information, we have retouched sensitive parts for presentation here, preserving the overall look and structure of the original documents.

3   Page Classification

Within the document image analysis literature, many descriptors encoding the visual appearance of document images have been proposed. In this work we have used a simple description of document images presented by Héroux et al. in [14], that encodes pixel densities at different scales. In order to remove small details and noise from the incoming images, a Gaussian smoothing operator

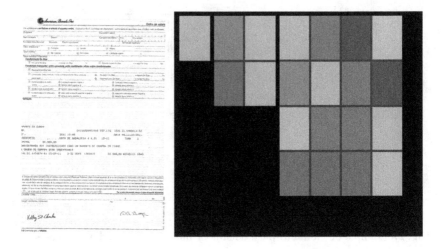

Fig. 3. Example of the multi-scale density descriptor.

is used to blur the images before computing the visual descriptor. Then, the multi-scale descriptor recursively splits each document image into rectangular regions forming a pyramid. In each of the regions the pixel density is computed and stored in the corresponding position of the feature vector. We can see an example of the first levels of the pyramid in Fig. 3. In our experimental setup, we use four scale levels, yielding to an 85-dimensional page descriptor.

Page feature vectors are then $L2$-normalized. The similarity between two pages is assessed by the cosine distance computed using the dot product between both feature vectors. We then use the k-NN classification method over a set of labeled pages in order to decide whether the given page is a page that is expected to bear a signature or not and to which document type the incoming image belongs to. A threshold for the most frequent class in the neighborhood is set as a rejection criteria in order to filter cases where the evidence obtained is weak.

Given a multi-page document, the system either returns a single page where the signature should be found, a negative answer when none of the pages in the document are similar to a signature page or has low confidence on the decision and rejects the document. In the first case, these candidate pages continue the process whereas the rejected or negative documents are forwarded to manual inspection.

4   Signature Detection

Given that the pages that should be signed from the document flow have been identified in the previous step, we aim the following step at detecting the signatures. First we segment the zone of the image in which the signature should be located and then we check whether the zones or interest actually contain a signature or not.

4.1 Zone Segmentation

In all the contracts, the zone within the multi-page document where the customer and the bank clerk have to sign is delimited by some layout structure, e.g. by a box, bold lines, white spaces, etc. In addition, it is usually the case that the zone is also defined by some text indicating where the customers have to sign. Our segmentation framework takes advantage of such graphical (box and line detection) and textual (patterns) layout characteristics, that can be configured for as many different types the system might receive. In the two different types of contracts we deal in our use case we found that these zones of interest are either delimited by a framing box or a bold straight line (Fig. 4).

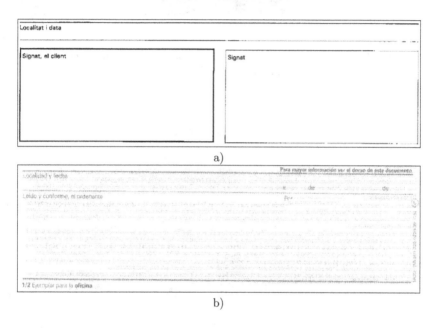

Fig. 4. Example of the zones of interest where the user should sign for the two document types A and B respectively.

First, a set of preprocessing steps aim at enhancing the image quality and get rid of the text. A Gaussian smoothing filter is first applied to the document to get rid of punctual noise introduced during the scanning process. The document images are then binarized by applying the Otsu method after a contrast enhancement aimed at stretch the intensity histogram within the 0-255 range. Afterwards, a connected component analysis aims at pruning small objects corresponding to textual characters. A final run-length smearing algorithm is applied to obtain an image where just long straight horizontal and vertical lines are maintained. In the processed image, having only line elements, we perform an horizontal and vertical projection profile analysis [15] in order to locate the zone

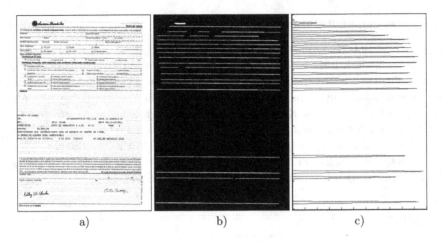

Fig. 5. Original image and graphical part extraction. (a) Original image, (b) filtered image, (c) horizontal projection profiles.

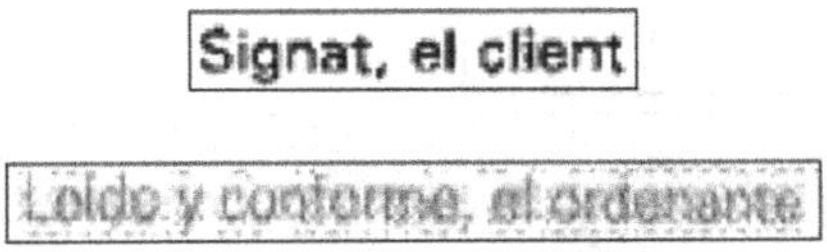

Signat, el client

Leido y conforme, el ordenante

Fig. 6. Example of textual patterns used as rejection criteria for the zone segmentation.

of the document corresponding to the signature zone layout. We can see an example of the result of such steps in Fig. 5.

Rotated pages are handled as well by looking at all the possible configurations of the sought graphical configurations. Since the localization of the zone of interest is based on projection profiles, the method tends to fail if the pages present severe skew deformations, but tolerates well slightly skewed images.

Finally, in order to ensure that the detected zone is really the location where the signatures should appear, we check whether within the segmented zones we can find certain textual patterns. In all document types, some standard text such as "Customer's signature" or "Read and agree" appears in the zone of interest (see Fig. 6), its existence in the candidate zone can therefore be used as a supporting evidence. A pattern matching implemented through a normalized cross correlation [16] is used as rejection criterion. If in the candidate zone we do not have enough confidence to find those standard texts, we reject the whole image to be processed manually. The locations of those textual patterns are used as well as anchors to refine the zone segmentation.

4.2 Signature Presence Verification

Finally, the last step of the system is to verify whether in the zones of interest there is actually a signature or not. Again some preprocessing steps devoted to reduce the noise are applied to these zones. From the original image we enhance the contrast by stretching the intensity histogram and then threshold the region by applying the Otsu algorithm. Small connected components are pruned in order to get rid of noise provoked by the bleed-through effect. Then, the zones' classification is done by looking at the following features of the remaining connected components.

- **Area:** number of pixels of the connected component.
- **Aspect ratio:** ratio between the height and the width of the connected component.
- **Eccentricity:** ratio of the distance between the foci of the ellipse having the same second-moments as the connected component and its major axis length.
- **Stroke's width:** computed by means of the distance transform.

Here, some experimentally set thresholds established through validation over a training set of 300 pages determine whether a zone contains a signature or not, or rejects the document if we are not confident on the decision.

Given a candidate page, the system either returns a verdict on the signature verification or rejects the page and thus the whole document is forwarded to manual inspection. In our use case, the contracts should contain two signatures each, so the system's answer can be either that both signatures are present, that one of the two is missing or that both are missing.

5 Experiments

Our test dataset consists of 3300 multi-page document images sampled from a real banking workflow consisting of around 14,500 pages. The dataset contains two different types of contracts denoted here as class A and B. The two documents arrive randomly in the document flow. We can see an example of those documents in Fig. 2. We can see in Fig. 7 a qualitative result of the proposed signature verification process. Zones where a signature has been detected are framed in green whereas zones that should contain a signature but it is missing are framed in red.

Concerning the classification step, we can see some results in Tables 1 and 2 for documents of class A and B respectively. The documents labeled as "Not signature page" are forwarded to manual inspection in order to determine whether the document did not contain a signature page or if the system was unable to recognize it. On the other hand, all the document pages labeled as probable signature containers continue the process of zone segmentation and signature presence verification. In that case the proposed system presents an 1.91 % and 1.89 % of false positives (pages that should not be signed but are labeled as if they should) for class A and B respectively.

Table 1. Page classification for documents type A

	System outcome		
	Signature page	Not signature page	Total
Signature page	820	35	855
Not signature page	16	97	113
Total	836	132	968

Table 2. Page classification for documents type B

	System outcome		
	Signature page	Not signature page	Total
Signature page	2118	51	2169
Not signature page	41	130	171
Total	2159	181	2340

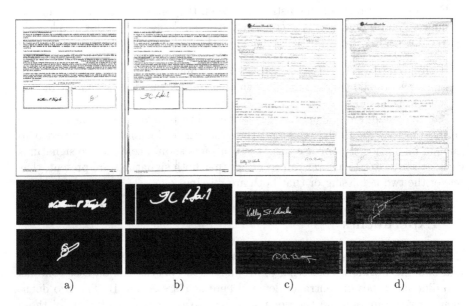

a)   b)   c)   d)

Fig. 7. Qualitative signature verification results. (a) and (b) properly signed and missing signature documents of type A respectively, (c) and (d) properly signed and missing signature documents of type B respectively (Color figure online).

After the classification step, the documents with non signature pages have been rejected and only 836 and 2159 pages from classes A and B respectively continue the pipeline. Tables 3 and 4 present the signature verification confusion matrices for documents of class A and B respectively. All the contracts used in the test should contain both the customer and the bank clerk signatures. A document is valid when it contains both signatures and invalid otherwise. In order to reach an accuracy of 92.94 % and 94.34 % the system rejects 20.26 % of documents of class A and 21.37 % of documents of class B respectively. How-

Table 3. Signature verification for documents type A

	System outcome				
	Both signatures ok	Only first signature	Only second signature	Neither signature ok	Total
Both signatures ok	698	21	17	2	738
Only first signature	2	66	0	1	69
Only second signature	0	0	5	0	5
Neither signature ok	0	0	0	8	8
Not signature page	3	9	2	2	16
Total	703	96	24	13	836

Table 4. Signature verification for documents type B

	System outcome				
	Both signatures ok	Only first signature	Only second signature	Neither signature ok	Total
Both signatures ok	1541	13	22	11	1587
Only first signature	2	85	0	6	93
Only second signature	10	1	187	2	200
Neither signature ok	3	8	3	224	238
Not signature page	19	6	9	7	41
Total	1575	113	221	250	2159

ever, not all the errors produced by the system have the same effect. Negative answers, that is pages missing one or more signatures, are forwarded to manual inspection, so false negatives (i.e. incorrectly reporting a document as invalid) are not really critical since the human observer will report them as valid ones. Our system delivers a 4.78 % and a 2.13 % of false negatives per class respectively. Misclassifying an invalid document (e.g. saying that it misses one signature when it misses both) is a recoverable error as well since it is still an invalid image that goes through manual processing. This is the case of a 1.67 % and 1.94 % of the documents respectively. On the other hand, a false positive (i.e. accepting a document as properly signed when it really lacks one of the signatures) is a critical mistake as it will go through undetected. In that case, our system just yields 0.59 % and 1.57 % of the images that are incorrectly accepted by the system.

6   Conclusions

In this paper we have presented an industrial application for the automatic screening of incoming multi-page documents in a banking workflow aimed at determining whether the documents were properly signed or not. We have tested our method in a real large-scale environment, handling more than 14,500 pages. The reported results were obtained using two different types of real multi-page contracts. The proposed system is able to automatically verify the signature presence while merely accepting an 1 % of critical false positives.

References

1. Plamondon, R., Lorette, G.: Automatic signature verification and writer identification - the state of the art. Pattern Recogn. **22**, 107–131 (1989)
2. Leclerc, F., Plamondon, R.: Automatic signature verification: the state of the art 1989–1993. Int. J. Pattern Recogn. Artif. Intell. **8**, 643–659 (1994)
3. Hou, W., Ye, X., Wang, K.: A survey of off-line signature verification. In: Proceedings of the International Conference on Intelligent Mechatronics and Automation, pp. 536–541 (2004)
4. Zhu, G., Zheng, Y., Doermann, D.: Signature-based document image retrieval. In: Proceedings of the Tenth European Conference on Computer Vision, pp. 752–765 (2008)
5. Zhu, G., Zheng, Y., Doermann, D., Jaeger, S.: Signature detection and matching for document image retrieval. IEEE Trans. Pattern Anal. Mach. Intell. **31**, 2015–2031 (2009)
6. Ahmed, S., Malik, M., Liwicki, M., Dengel, A.: Signature segmentation from document images. In: Proceedings of the, pp. 425–429 (2012)
7. Mandal, R., Roy, P., Pal, U.: Signature segmentation from machine printed documents using contextual information. Int. J. Pattern Recogn. Artif. Intell. 26 (2012)
8. Doermann, D.: The indexing and retrieval of document images: A survey. Comput. Vis. Image Underst. **70**, 287–298 (1998)
9. Chen, N., Blostein, D.: A survey of document image classification: problem statement, classifier architecture and performance evaluation. Int. J. Doc. Anal. Recogn. **10**, 1–16 (2006)
10. Frasconi, P., Soda, G., Vullo, A.: Hidden Markov models for text categorization in multi-page documents. J. Intell. Inform. Syst. **18**, 195–217 (2002)
11. Gordo, A., Perronnin, F.: A bag-of-pages approach to unordered multi-page document classification. In: Proceedings of the International Conference on Pattern Recognition, pp. 1920–1923 (2010)
12. Rusiñol, M., Karatzas, D., Bagdanov, A., Lladós, J.: Multipage document retrieval by textual and visual representations. In: Proceedings of the International Conference on Pattern Recognition (2012)
13. Rusiñol, M., Frinken, V., Karatzas, D., Bagdanov, A., Lladós, J.: Multimodal page classification in administrative document image streams. Int. J. Doc. Anal. Recogn. **17**, 331–341 (2014)
14. Héroux, P., Diana, S., Ribert, A., Trupin, E.: Classification method study for automatic form class identification. In: Proceedings of the Fourteenth International Conference on Pattern Recognition, pp. 926–928 (1998)
15. Likforman-Sulem, L., Zahour, A., Taconet, B.: Text line segmentation of historical documents: A survey. Int. J. Doc. Anal. Recogn. **9**, 123–138 (2007)
16. Lewis, J.: Fast normalized cross-correlation. Vis. Interface **10**, 120–123 (1995)

CRFs and HCRFs Based Recognition
for Off-Line Arabic Handwriting

Moftah Elzobi[1]([✉]), Ayoub Al-Hamadi[1], Laslo Dings[1], and Sherif El-etriby[2]

[1] Institute for Information Technology and Communications (IIKT),
Otto-von-Guericke-University Magdeburg,
P.O. Box 4210, 39016 Magdeburg, Germany
{Moftah.Elzobi,Ayoub.Al-Hamadi}@ovgu.de
[2] Umm Al-Qura University, Makkah 21421, Saudi Arabia

Abstract. This paper investigates the application of the probabilistic discriminative based Conditional Random Fields (CRFs) and its extension the hidden-states CRFs (HCRFs) to the problem of off-line Arabic handwriting recognition. A CRFs- and A HCRFs- based classifiers are built on top of an explicit word segmentation module using two different set of shape description features. A simple yet effective taxonomization technique is used to reduce the number of the class labels, and 3000 letter samples from IESK-arDB database are used for the training and 300 words are used for the evaluation. Experiments compare the performance of the CRFs to the HCRFs as well as to that of a generative based HMMs. Results indicate superiority of discriminative based approaches, where HCRFs achieved the best performance followed by CRFs.

1   Introduction

In general, probabilistic models used for recognition of off-line handwriting can be categorized into two broad categories, generative models and discriminative models. According to generative based approaches, recognition starts firstly by calculating the joint distribution $p(\mathbf{x}, \mathbf{y})$ and then any new value of the observed data $\mathbf{x}$ is assigned to the most likely label $\mathbf{y}$ by estimating the conditional probability $p(\mathbf{y}|\mathbf{x})$ using the popular Bayes rule. Alternatively, discriminative based prediction directly models the posterior conditional distribution $\mathbf{p}(\mathbf{y}|\mathbf{x})$ of an input sequence $\mathbf{x}$ and a label sequence $\mathbf{y}$, i.e. solely molding the classification condition and avoid computing any unnecessarily distributions e.g. $p(\mathbf{x})$. CRFs and HCRFs are discriminative based classification models that have been recently successfully applied in a number of sequences labeling problems, e.g. natural language processing, computer vision, and bioinformatic [1]. CRFs- and HCRFs-based approaches are advantageous because, firstly, they make no dependence assumptions among the input data, and secondly allow representation of complex relations between $\mathbf{y}$ and $\mathbf{x}$. Several works suggest the use of CRFs for the recognition of off-line Chinese- and Latin- based handwritings. Zhou et al. [2], propose a method for the recognition of Chinese/Japanese text based on semi-Markov CRFs. They start by defining a semi-CRF on a lattice of all possible

© Springer International Publishing Switzerland 2015
G. Bebis et al. (Eds.): ISVC 2015, Part II, LNCS 9475, pp. 337–346, 2015.
DOI: 10.1007/978-3-319-27863-6_31

segmentation-recognition hypotheses of a string to directly estimate their posteriori probabilities. CRFs' features functions are defined on top of geometric and linguistic information of character recognition, and negative log-likelihood is used for model parameters optimization. At the level of characters, recognition rates of 95.20 % and 95.44 % are reported for Chinese and Japanese respectively. In an early related work, Feng et al. [3], investigate the performance of CRFs and compared it to HMMs for the task of word recognition in historical handwritten documents. A set of discrete features is extracted from 20 pages of George Washington's manuscripts and used to train and evaluate the CRFs- as well as HMMs based classifiers. They experiment different beam search methods in order to speed up the training of CRF classifier and prove that CRFs superior to HMMs. However, to boost performance, they found it is necessary to reduce the state space by applying CRFs at the characters' level (which we adopted in this paper). By reviewing the literature on the subject, we could not find any previous work proposing CRFs for the recognition of off-line Arabic handwriting; therefore, we believe this is the first work addressing the issue. In this work, we introduce the standard conditional random fields (CRF) and the hidden conditional random field classifiers (HCRF) for the recognition of Arabic handwritten words. The proposed system is built on top of an explicit segmentation approach introduced in Elzobi et al. [4]. Moreover, and in order to reduce the state domain of CRFs classifiers, we taxonomie the segmented letters into four basic groups using some primitive shape features. Additionally, robust yet simple approach is suggested to extract two different sets of shape representative features that used for training, evaluation, and testing.

2   Shape Descriptor Based Features for Arabic Handwriting

The most popular sequential features used in handwriting recognition systems, are those extracted using the principle of the so-called sliding-window [5,6]. When using sliding-window, a large number of features are firstly extracted, then to get rid of redundant and irrelevant features, feature selection and/or reduction algorithms are usually needed to turn the recognition process computationally tractable. Inspired by the work of [7], we propose a robust yet simple approach for extracting two sets of shape descriptor features. Which besides avoiding the drawbacks of the sliding-window based features, have a number of desirable characteristics, e.g., less expensive to extract and to process, capturing letter global shape characteristics, invariant to stroke width and less sensitive to handwriting distortions (e.g., skew and slant). Since our objective is to build a robust and an unconstrained recognition system for the handwritten words, our feature extraction approach is built on top of an explicit segmentation approach for off-line Arabic handwriting proposed in [4]. Providing that writing styles differ greatly with respect to height, width, skew, and slant, feature extraction starts by normalizing the handwritten word against handwriting deformations, i.e., skew and slant. Then, in order to further minimize the within-class variations, segmented

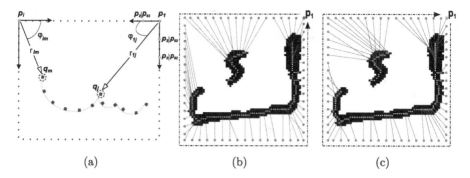

(a)   (b)   (c)

Fig. 1. Features extraction, (a) proposed method, (b) anti-clockwise features, (c) clockwise features.

images are size normalized while preserving the segment aspect ratio. For size normalization, a backward linear normalization method is employed to map the pixels coordinates of all segmented images (usually of different size) into a standard plane of fixed $N \times N$ dimension where $N = 64$ is found to be optimal [8]. The core idea of our approach is illustrated in Fig. 1(a), where feature extraction starts by uniformly distributing a set $\mathbb{P} = \{p_1, p_2, \ldots, p_m\}$, $p_i \in \mathbb{R}^2$, of m reference points along a rectangle that tightly contains the segment skeleton image. Typically, m can be any natural number less than or equal to n, where n is the total number of pixels constituting the segment skeleton image. In practice, m should be proportional to the size of the normalized images, therefore $m = 64$ is chosen. And since the focus of this work is mainly on Arabic handwriting, the first reference point p_1 is positioned on the rectangle's upper-right corner. Additionally, Let $\mathbb{Q} = \{q_1, q_2, \ldots, q_n\}$, $q_i \in \mathbb{R}^2$ the set of pixels coordinates constituting skeleton segment. Starting at the reference point p_1, and for every $p_i \in \mathbb{P}$, we search for the nearest corresponding $q_j \in \mathbb{Q}$. Upon identifying q_j, and by using p_i as the pole, we estimate the radial distance r_{ij} and the angle φ_{ij} according to Eq. 1. Then, we exclude the pixel q_j from the segment pixel set $\mathbb{Q}$.

$$r_{ij} = \sqrt{(\acute{x}_j - x_i)^2 + (\acute{y}_j - y_i)^2}, \quad \varphi_{ij} = \tan^{-1}\left(\frac{\acute{y}_j - y_i}{\acute{x}_j - x_i}\right), \tag{1}$$

where $p_i = (x_i, y_i)$, $q_j = (\acute{x}_j, \acute{y}_j)$.

Algorithm 1, Illustrates the features extraction process, where it is applied twice in anti-clockwise direction (Fig. 1(b)) as well as in clockwise direction (Fig. 1(c)), resulting in two different feature descriptors $\boldsymbol{\chi_a} = (\chi_1, \chi_2, \ldots, \chi_{64})$ and $\boldsymbol{\chi_c} = (\chi_1, \chi_2, \ldots, \chi_{64})$, respectively, where $\chi_i = (r_i, \varphi_i)$. In this context, it is also important to mention that computation is performed with the assumption that $n \geq m$, i.e., the number of pixels m in the segment skeleton image are greater than or equal to the number of reference points m.

As a final step in the feature extraction process and to map the huge number of different input values into a far smaller set of discrete values, we performed

---

Algorithm 1. Extraction of features descriptors

Data: $\mathbb{P}$ the set of reference points, $\mathbb{Q}$ the set of segment skeleton pixels.

Result: χ

begin

    $p_1 \longleftarrow$ *Start Point*

    for $\forall p_i \in \mathbb{P}$ **do**

        for $\forall q_j \in \mathbb{Q}$ **do**

$$q_j^* = \arg\min_{q_j} \|q_j - p_i\|$$

            $r_{ij} = \|q_j^* - p_i\|$

            $\varphi_{ij} = \arctan(q_j^*, p_i)$

            $\chi_{ij} = (r_{ij}, \varphi_{ij})$

        $\chi \mathrel{+}= \chi_{ij}$

        $q_j^* \notin \mathbb{Q}$

---

an efficient vector quantization approach suggested in [9]. The approach proposes the affinity propagation algorithm to initialize a k-means based quantizer. Consequently, χ_a and χ_c are quantized into $\mathbf{f_a} = (f_{a_1}, f_{a_2}, \ldots, f_{a_{64}})$, $f_{ai} \in \{1, 2, \ldots, 16\}$, and $\mathbf{f_c} = (f_{c_1}, f_{c_2}, \ldots, f_{c_{64}})$, $f_{ci} \in \{1, 2, \ldots, 16\}$, respectively, where $k = 16$ is found to be the optimal number of quantization levels.

2.1 Letters Taxonomization

The Arabic alphabet consists of 28 letters, and a letter may appear in two to four distinct shapes according to its position in a word. As a consequence, there are 96 different shapes that have to be learned. CRFs and HCRFs, typically use a single exponential model to represent the joint probability of a sequence of labels given an observation sequence. Learning parameters for models with such large number of labels is intractable problem. A straightforward method to alleviate this problem, is through taxonomizing letters according to very primitive shape characteristics. In this work, we categorize letters according to the number of segment and whether they contain a loop(s) or not. Consequently, letters of one segment and no loop e.g. د, ر, ا, ح will be grouped under one group (Tax.1), letters of one segment and a loop(s) e.g. و, ح, ه will fall under another (Tax.2), letters consist of multi segments e.g. أ, ث, ك will be grouped in a third group (Tax.3), and the fourth group is made of letters such as ض, ة, غ that contain multi strokes and a loop(s) (Tax.4). This simple step, reduced the number of labels to a maximum of 36 across different categories. Moreover, it is important to notice that the total number of labels is 115 labels. This is because several letters appear in two different categories, since they may be written with or without loops depending on the personal writing style and the used font.

3   CRFs Based Recognition of Arabic Handwriting

3.1   Linear-Chain CRFs Based Recognition

In our approach, starting with the assumption that labels of the letter classes $\mathbf{y}$ are fully observed where each $y_i \in \mathbf{y}$ represents a label of a basic shape a given letter may appear in, and by using vector of observations $\mathbf{x}$ (i.e. $\mathbf{f_a}$ or $\mathbf{f_c}$), two different exponential linear-chain CRFs models are created for each taxonomy (one for each direction). Where every letter shape represented by a state in the corresponding taxonomy model. Furthermore, the proposed models are built using two different types of feature functions, i.e. transition feature functions $t(y_{i-1}, y_i, \mathbf{x}, i)$ and state or emission feature functions $s(y_i, \mathbf{x}, i)$, where those functions take as input an entire sequence of observations $\mathbf{x}$, the current position within the sequence i, the current class label y_i, the previous class label y_{i-1}, and output a real-valued number. Transition functions are typically dedicated to estimate the dependency of neighboring class labels given the value of the current position in $\mathbf{x}$. Whereas the state/emission functions are employed to estimate a real values represent the possibility that the current label emits the current value in $\mathbf{x}$.In our approach, to calculate the sequence overall likelihood, firstly, the transition and emission functions are assigned weights $\boldsymbol{\lambda}$ and $\boldsymbol{\mu}$ respectively, that learned from the training data. Then they combined together to form a potential function as follows:

$$F_\theta(y_{i-1}, y_i, \mathbf{x}, i) = \sum_f \lambda_f t_f(y_{i-1}, y_i, \mathbf{x}, i) + \sum_g \mu_g s_g(y_i, \mathbf{x}, i), \qquad (2)$$

where $\theta = (\lambda_1, \lambda_2, ..., \lambda_{N_f}; \mu_1, \mu_2, ..., \mu_{N_g})$, $\lambda_i \in \boldsymbol{\lambda}$, $\mu_i \in \boldsymbol{\mu}$, and N_f and N_g are the total number of feature functions and state/emission functions, respectively. Finally, to convert the outputs of F_θ into proper probabilities, the F_θ is summed over all $x_i \in \mathbf{x}$ and the result is exponentiated and normalized as follows:

$$p_\theta(\mathbf{y}|\mathbf{x}) = \frac{\exp\left(\sum_{i=1}^{n} F_\theta(y_{i-1}, y_i, \mathbf{x}, i)\right)}{Z_\theta(\mathbf{x})}, \qquad (3)$$

where the normalization factor $Z_\theta(\mathbf{x})$ is given by:

$$Z_\theta(\mathbf{x}) = \sum_{\mathbf{y}} \exp\left(\sum_{i=1}^{n} F_\theta(y_{i-1}, y_i, \mathbf{x}, i)\right). \qquad (4)$$

To investigate the effect of dependency range on the performance of the proposed system, and In addition to θ parameter, the potential function F_θ can be also parameterized by ω a window size parameter that define the number of previous and subsequent observations used when predicting a class label at the current position i,(e.g., for a window size ω, the observation from $i - \omega$ to $i + \omega$ will be used to calculate the outputs of F_θ). For each taxonomy, we trained seven linear-chain CRFs models corresponding to seven different window-sizes ($\omega = 0, \omega = 1, ..., \omega = 6$).

Learning CRFs Parameters. To learn the CRFs feature functions weights θ, we applied the popular gradient ascent algorithm on fully labeled training sequences $D = \{(\mathbf{x}^t, \mathbf{y}^t)\}_{t=1}^T$, where $\mathbf{x}^t$ is a sequence of observed features, $\mathbf{y}^t$ is the corresponding sequence of labels, and T is the total number of training samples. The learning process starts by randomly initializes θ, then for each training sample and for each potential function, the gradient of the log probability with respect to θ is calculated as follows:

$$\frac{\partial L(\theta)}{\partial \theta} = \sum_{t=1}^T \left(\sum_{i=1}^n \frac{\partial F_\theta(y_{i-1}^t, y_i^t, \mathbf{x}^t, i)}{\partial \theta} - \sum_{\mathbf{x}} p_\theta(\mathbf{y}|\mathbf{x}^t) \sum_{i=1}^n \frac{\partial F_\theta(y_{i-1}, y_i, \mathbf{x}^t, i)}{\partial \theta} \right),$$

(5)

where $L(\theta) = \sum_{t=1}^T \log p_\theta(\mathbf{y}^t|\mathbf{x}^t)$, and the first term in the gradient is the contribution of F_θ under the true label, whereas the second term is the expected contribution of F_θ under the current model. The well known L-BFGS algorithm [10], is used for the gradient calculation while convergence is assumed at 300 iterations.

Class Label Prediction. To predict the class label of segment, we start first by individually labeling every element of the sequences $\mathbf{f_a}$ and $\mathbf{f_c}$ through calculating the corresponding optimal Viterbi path using the segment trained taxonomy model. Then, separately, each sequence will be assigned to the class label that most frequently occurring along the sequence of predicted labels. A letter is considered recognized if both sequences are assigned to the same class label, otherwise it is rejected.

3.2   HCRFs Based Recognition

To explore the performance of the probabilistic discriminative based models with hidden states in the field, we introduce the recently proposed HCRFs model [11]. HCRFs is an extension of the discriminative fully observed linear-chain CRFs model, where HCRFs model is equipped with an intermediate set of hidden variables (states) $\mathbf{h} = \{h_1, h_2, .., h_n\}$ (between the observations and labels), globally conditioned on the observation vector $\mathbf{x}$. To optimize the number of hidden states for each HCRFs model, a similar approach to that of the HMMs-based approach in [12] is adopted. Where the number of hidden states are chosen to be proportional to the letter shape complexity. Accordingly, the optimized hidden states numbers were found to be 5, 8, 10, and 10, for taxonomy 1 to 4, respectively. Analogous to the formulation of CRFs, HCRFs models the conditional probability of a class label given a sequence of observations as follows:

$$p_\theta(\mathbf{y}|\mathbf{x}) = p_\theta(\mathbf{y}, \mathbf{h}|\mathbf{x}) = \frac{\sum_{\mathbf{h}} \exp\left(\sum_{i=1}^n F_\theta(y_{i-1}, y_i, \mathbf{h}, \mathbf{x}, i)\right)}{\sum_{\mathbf{y}, \mathbf{h}} \exp\left(\sum_{i=1}^n F_\theta(y_{i-1}, y_i, \mathbf{h}, \mathbf{x}, i)\right)},$$

(6)

where the denominator is a normalization factor similar to $Z_\theta(\mathbf{x})$ in Eq. 3, and the potential function $F_\theta(y_{i-1}, y_i, \mathbf{h}, \mathbf{x}, i)$ computes the similarity between a class

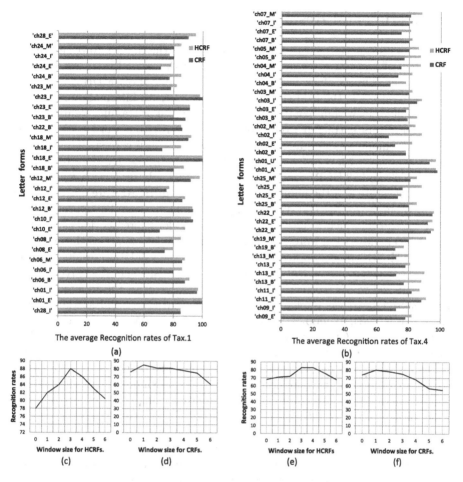

Fig. 2. Validation results, (a) CRFs vs. HCRFs on letter Tax.1, (b) CRFs vs. HCRFs on letter Tax.4, (c) ω effect on HCRFs for Tax.1, (d) ω effect on CRFs for Tax.1, (e) ω effect on HCRFs for Tax.4, (f) ω effect on CRFs for Tax.4.

label, a sequence of observations, and a configuration of the hidden states. As in CRFs case, to estimate the HCRFs optimal weights θ for a given taxonomy, we calculate the gradient ascent of its training samples set $D = \{(\mathbf{x}^t, \mathbf{y}^t)\}_{t=1}^T$ using the L-BFGS algorithm, while assuming the corresponding number of hidden states. Furthermore, convergence is also assumed to be reached at 300 iterations and seven different HCRFs models built for each taxonomy each with different window-size value (i.e., $\omega = 0, \omega = 1, ..., \omega = 6$). The prediction of the elements of the two different observation sequences and the estimation of the final recognition results are computed just like in the CRFs case.

4 Experimental Results

Experiments are performed to fulfill two purposes, the first, is to optimize the parameters of respective recognition system, and the second is to test the system efficiency. The database used is the freely available IESK-arDB [4], which is the only database in the field that contains ground-truth information regarding letters boundaries within handwritten Arabic words. For training we used more than 3000 letters extracted from around 800 handwritten words. We trained seven CRFs models as well as seven HCRFs models each corresponding to different taxonomy with a different window-size ($\omega = 0, \omega = 1, .., \omega = 6$) (i.e. 28 CRFs models and 28 HCRFs models). In each model, a letter form is represented with a single state and the number of hidden states in HCRFs are optimized for every taxonomy. As for HMMs training, we adopted a similar approach suggested in [12], in which a left-to-right banded model is created for each letter form. A 300 handwritten words contains around 1000 letters are used to validate CRFs and HCRFs models, where Fig. 2 illustrates the validation results. Testing is carried out using unseen 300 randomly drawn words, where the percentage of fully recognized words (correctly recognizing every letter) is the adopted metric. The proposed system is implemented in Matlab and C++, and Matlab HCRF library [13] is used. Wondows 7 professional 64-bit and Matlab R2013a, on an Intel(R) Xenon(R) CPU server machine with 2.67 GHz and 64.0 GB installed memory ran the task. Because of space limitations, we will limit our discussion here to the results of experiments conducted on taxonomies Tax.1 and Tax.4. Where experiments on taxonomies Tax.2 and Tax.3 exhibit similar characteristics. Also notice that letters on charts are indexed using their respective order in the alphabet and a letter stands for the form (i.e. isolated (I), begin (B), middle (M), end (E)). Figure 2(a), summarizes the validation results of HCRFs and CRFs on samples of taxonomy Tax.1, where the achieved recognition rates "in average" are around 88 % and 85 %, respectively. Letters with distinctive and simple shapes are expected to be recognized relatively easy; experiments confirmed such a claim, where the best recognition rates (i.e. 100 %) are reached for letters such as xxx and xxx ("ch01_E" and "ch18_E" in Fig. 2(a), respectively). Figure 2(c) and (d), show the affects of modeling the dependency range through the window-size (ω) on HCRFs and CRFs. Experiments indicate that HCRFs performance improves as ω increases, reaching its peak at $\omega = 3$, whereas CRFs' best performance is reached at $\omega = 1$. Such tendency implies that incorporating dependencies positively influence performance, especially when hidden pattern is also considered like in HCRFs case. For taxonomy Tax.4, on average, HCRFs and CRFs achieved less recognition rates 84 % and 80 %, respectively (see Fig. 2(b)). This can be referred to the fact that letters fall under Tax.4 have usually complex shape charact'eristics (e.g. multi strokes and loops), that may not be fully accommodated in the respective model. In Fig. 2(e) and (f), optimization results of ω parameter for both models, confirm their previous counterparts obtained using Tax.1. In summary, HCRFs models perform better across the four taxonomies, with an overall average rate of 84.7 %, whereas CRFs obtained 80.5 %. Since letter's borders within a word are known, the problem of word recognition

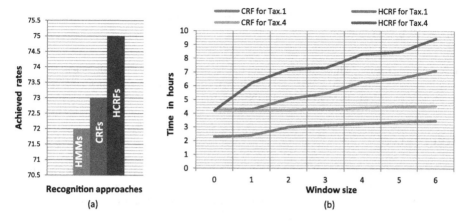

Fig. 3. Word based recognition rates and training costs, (a) HMMS vs. CRFs vs. HCRFs, (b) training costs in terms of time.

can be formulated as the problem of recognizing a sequence of letters. Starting from the right, features of segments within a word are extracted and individually labeled, then a lookup table is used to convert labels into a sequence of equivalent Unicode. When testing HMMs, CRFs, and HCRFs on the samples of test set, the overall recognition rates obtained were 72 %, 73 %, and 75 %, respectively (see Fig. 3(a)). Where a word is considered recognized only, if all of its segments are correctly recognized. Even though, such results are still imperfect, however, they confirm the superiority of discriminative based approaches against generative based ones. Although, CRF and HCRF show a strong performance compared to HMM, however, they are very expensive in terms of training costs. Figure 3(b), summarize the time costs on Tax.1 and Tax.4, and further indicating that the cost is proportional to the window size and the number of hidden states in case of HCRFs.

5   Conclusion and Future Works

To the best of our knowledge, this work is the first to propose conditional random fields based recognition for off-line Arabic handwriting. The linear-chain CRFs and the HCRFs are trained on letters taken from segmented handwritten Arabic words using shape descriptive features. In order to reduce the labels space we performed a pre-classification step which significantly reduced the labels number. Results were satisfactory and there is significant space for future improvements. We expect that more complex pre-classification steps are achievable by including diacritics, as several Arabic letters sharing the same basic shape and are only distinguishable through diacritics. Hence, a further reduction in class labels is a future improvement idea. On the prediction level, A more complex CRFs based algorithm, namely, Latent-Dynamic Conditional Random Fields (LDCRF), is demonstrating a strong performance in fields such as NLP and bioinformatics [2].

Therefore, we plan to investigate its performance in the field of off-line Arabic handwriting recognition.

Acknowledgment. This work is supported by the National Plan for Science, Technology and Innovation (MAARIFAH) - King Abdulaziz City for Science and Technology(KACST) - KSA. Project code: 13-INF604-10.

References

1. Sutton, C., McCallum, A.: An introduction to conditional random fields (2010)
2. Zhou, X., Wang, D., Tian, F., Liu, C., Nakagawa, M.: Handwritten chinese/-japanese text recognition using Semi-Markov conditional random fields. IEEE Trans. Pattern Anal. Mach. Intell. **35**, 2413–2426 (2013)
3. Feng, S., Manmatha, R., Mccallum, A.: Exploring the use of conditional random field models and HMMs for historical handwritten document recognition. In: the Proceedings of the 2nd IEEE International Conference on Document Image Analysis for Libraries (DIAL), pp. 30–37
4. Elzobi, M., Al-Hamadi, A., Al Aghbari, Z., Dings, L.: IESK-arDB: A database for handwritten Arabic and an optimized topological segmentation approach. Int. J. Doc. Anal. Recogn. (IJDAR) **16**, 1–14 (2012)
5. Likforman-Sulem, L., AlHajj Mohammad, R., Mokbel, C., Menasri, F., Bianne-Bernard, A.L., Kermorvant, C.: Features for HMM-Based Arabic handwritten word recognition systems. In: Maergner, V., El Abed, H. (eds.) Guide to OCR for Arabic Scripts, pp. 123–143. Springer, London (2002)
6. Ploetz, T., Fink, G.: Markov models for offline handwriting recognition: A survey. Int. J. Doc. Anal. Recogn. (IJDAR) **12**, 269–298 (2009)
7. Saabni, R., El-Sana, J.: Keywords image retrieval in historical handwritten Arabic documents. J. Electron. Imaging **22**, 013016–013016 (2013)
8. Liu, C.L., Nakashima, K., Sako, H., Fujisawa, H.: Handwritten digit recognition: investigation of normalization and feature extraction techniques. Pattern Recogn. **37**, 265–279 (2004)
9. Zhu, Y., Yu, J., Jia, C.: Initializing k-means clustering using affinity propagation. In: 2009 Ninth International Conference on Hybrid Intelligent Systems, HIS 2009, vol. 1, pp. 338–343 (2009)
10. Liu, D.C., Nocedal, J.: On the limited memory BFGS method for large scale optimization. Math. Program. **45**, 503–528 (1989)
11. Quattoni, A., Wang, S., Morency, L., Collins, M., Darrell, T., Csail, M.: Hidden-state conditional random fields. In: IEEE Transactions on Pattern Analysis and Machine Intelligence (2007)
12. Elzobi, M., Al-Hamadi, A., Dings, L., Elmezain, M., Saeed, A.: A hidden Markov model-based approach with an adaptive threshold model for off-line Arabic handwriting recognition. In: ICDAR, pp. 945–949 (2013)
13. Morency, L.P., Quattoni, A., C.M.C., Wang, S.: Hidden-state conditional random field library (2012). http://pt.sourceforge.jp/projects/sfnethcrf/

Classifying Frog Calls Using Gaussian Mixture Models

Dalwinderjeet Kular[1]($\boxtimes$), Kathryn Hollowood[3], Olatide Ommojaro[4],
Katrina Smart[1], Mark Bush[2], and Eraldo Ribeiro[1]

[1] Computer Vision Laboratory, Department of Computer Sciences,
Florida Institute of Technology, Melbourne, FL, USA
dkular2009@my.fit.edu
[2] Department of Biological Sciences, Florida Institute of Technology,
Melbourne, FL, USA
[3] Department of Computer Science, Mathematics, and Physics,
Roberts Wesleyan College,Rochester, NY, USA
[4] Engineering, Mathematics, and Computer Science, Georgia Perimeter College,
Clarkston, GA, USA

Abstract. We focus on the automatic classification of frog calls using
shape features of spectrogram images. Monitoring frog populations is a
means for tracking the health of natural habitats. This monitoring task
is usually done by well-trained experts who listen and classify frog calls,
which are tasks that are both time consuming and error prone. To automate this classification process, our method treats the sound signal of
a frog call as a texture image, which is modeled as Gaussian mixture
model. The method is simple but it has shown promising results. Tests
performed on a dataset of frog calls of 15 different species produced an
average classification rate of 80 %, which approximates human performance.

1 Introduction

The conservation of natural ecosystems is a concern of modern society [1]. Among
ecosystems, wetlands - areas that are inundated or saturated with water periodically - have high degree of biodiversity. In the U.S., preservation of wetlands is
particularly relevant to the economy of southern and coastal states [2]. Monitoring the health in these regions is key to conservation. One way ecologists monitor
wetlands is by listening to frog calls to estimate frog populations [3,4,9]. These
type of data-collection procedure is time consuming and require expert training to improve classification ability. Automating frog-call classification can help
ecologists to streamline data collection and analysis of habitat health.

Early work on automating frog-call classification aimed at identifying an invasive species of Cane toad [10] in northern Australia. Their classification method
use decision trees. The training data consisted of twenty-two species of frog and
a few cricket species that sounded similar to frogs. The species were classified
using the peaks of energies on the spectrograms, which are spatio-temporal representations of acoustic signals. Recent advances in machine learning have motivated the development of a number of new methods for frog-call recognition.

© Springer International Publishing Switzerland 2015
G. Bebis et al. (Eds.): ISVC 2015, Part II, LNCS 9475, pp. 347–354, 2015.
DOI: 10.1007/978-3-319-27863-6_32

Classification techniques such as Support Vector Machines and Hidden Markov Models are able to recognize frog calls with good accuracy [11,12]. However, most automatic methods for frog-call classification rely on the extraction of syllabus from the acoustic signal. While automatic syllable-segmentation methods do exist [13], they can be unreliable when applied to signals obtained in the field. Other representative works on frog-call classification include [15,16].

In this paper, we propose a method for classifying frog calls by treating the frog call as a texture image. Our classification method models each frog species as a mixture of Gaussian densities over texture visual words and Mel Scale spectral features (MFCCs). Here, we create feature vectors from filter-response maps describing the local shape of a spectral acoustic feature obtained by convolving the spectrogram of the audio signal with a bank of multi-scale multi-orientation filters. We tested our methods on a dataset containing calls of 15 different species of frog. Classification rates for our method produced promising results, reaching classification rates of 80 %, which is close to human performance.

2   Method

The basic approach for our method begins by taking the audio data and creating a spectrogram. The spectrogram of an audio signal maps the frequency as a function of time. There are different colors on a spectrogram that represent the different levels of energy for the different frequencies. The highest levels of energy are in red and these are what you would hear when listening to a frog call. These high peaks of energy are our landmarks and from these landmarks we extract features. These features are then used to classify the frog call using Gaussian Mixture Modeling. This process is shown in Fig. 1.

2.1   Feature Extraction

The audio data and sampling rate of each frog call were extracted into the program and key features were selected to be used for training. The sampling rate and audio data were used to find the fingerprint of the data, the mapping of the frequencies and their energies. This fingerprint was a two-dimensional array. This array was used to create a spectrogram. A spectrogram is a visual representation of the audio sample with the y-axis containing the frequencies, the x-axis containing the time, and different colors represented the amounts of energies each frequency had.

We convolved the spectrograms with a subset of the bank of filters used for texture classification by Varma and Zisserman [17]. The bank contains 48 filters with multiple scales and orientations. We used 12 of these filters for feature extraction. These were the twelfth through the seventeenth filters and the thirtieth through the thirty-fifth of these filters. These filters were selected because they included the most orientations. Figure 2 shows the full filter bank. The 12 filters we used for the feature extraction are enclosed by the red rectangles.

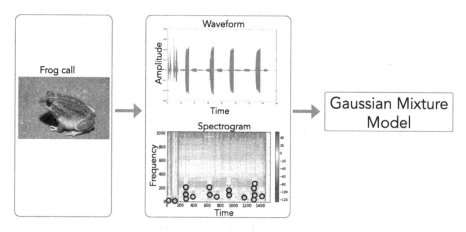

Fig. 1. Basic methodology of our classification methods. Feature extraction uses peaks detected from the spectrogram of the audio signal. For our approach using Gaussian mixture model, MFCC coefficients located at detected high-energy peaks are converted into feature vectors. Gaussian mixture models are learned for each frog species. Classification is based on a maximum-likelihood approach. (Color figure online)

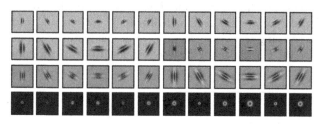

(a) The complete filter bank as in Varma and Zisserman [17].

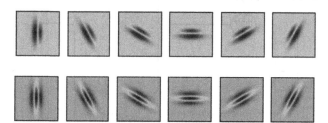

(b) The subset of filters that we used for feature extraction.

Fig. 2. The Leung-Malik filter bank. (Color figure online)

Each of these filters were convolved with the spectrogram. After the convolution, a threshold was applied to the filtered spectrograms, this threshold removed the lower-energy peaks, leaving only the highest peaks of energy that have been filtered through. At each of these peaks, the frequency and the shape

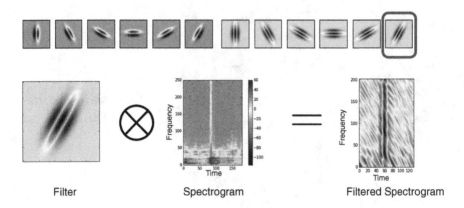

Filter   Spectrogram   Filtered Spectrogram

Fig. 3. Filtered spectrogram. From left to right: filter bank is convolved with the spectrogram of the audio signal. The result of the convolution is a map of filter responses (i.e., convolution coefficients). We threshold the filter response to keep only the locations with the highest energy. (Color figure online)

of the peak were used as features. Figure 3 shows a spectrogram being filtered by one of these filters and then a threshold being applied to the same spectrogram. The remaining red peaks were used as features.

In addition to using the filter response values as features, we also used Mel-Frequency Cepstral Coefficients extracted at locations of prominent peaks in

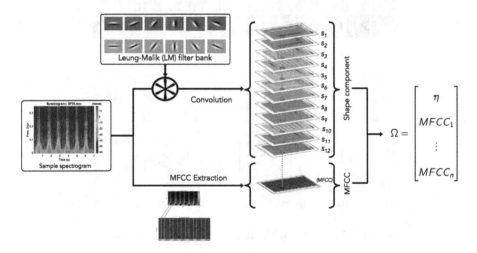

Fig. 4. Feature Extraction. The label of filters with the maximum response at a spectral peak and the MFCC column at the peak location are concatenated to form a feature vector. (Color figure online)

the spectrogram. Mel-Frequency Cepstral Coefficients, or MFCCs, are commonly used in speech recognition. MFCCs are designed to extract data that reflects what the human ear actually hear. This helps extract the most relevant features of the signal that define the sound. Using MFCCs extracts another important features of the sound. The sound is split up into time frames and at each time frame a certain amount of coefficients are extracted. For the purpose of this experiment, the only MFCCs that were used were ones extracted from time frames that contained a high-energy peak to assure that the MFCCs corresponded to key features of the frog call. Once these MFCCs were found and the features from the filtered spectrograms were found, they were combined into one vector of features. This feature vector contained all of the data points used for training and classification. The diagram in Fig. 4 shows the process of feature extraction for Gaussian Mixture Modeling.

2.2 Training and Classification

Once these features were extracted, the training process began. Each species would be fit to its own GMM. Then these models would be tested with a frog call that was not used in training. In order to do this the leave one out method was used. The leave one out method involves using every data point except one data point to train the data with. In this specific experiment, one frog call out of every frog call recorded was taken out to be used as a test call. Then every other call was put in categories with the corresponding species and features were extracted. These features were fit to GMM. Once this process was completed, features were extracted from the test call a test was run to determine to which model the test call was most likely to belong. Each frog species was represented by a GMM. Data points are determined to belong to certain GMM based on probability. The probability density function for a given species is:

$$p(\mathbf{x}|c) = \sum_{k=1}^{K} r_k \mathcal{N}(\mathbf{x}, \mu_k, \Sigma_k). \tag{1}$$

The $\mathcal{N}(\mathbf{x}, \mu_k, \Sigma_k)$ represents the Gaussian density over the data $\mathbf{x}$. The mean is represented by μ_k and the co-variance matrix is represented by Σ_k. r_k represents the mixture weights, these are the proportions of the density for the different components in the mixture model. The posterior probability of a frog species is then calculated using:

$$
\begin{aligned}
p(c|\mathbf{x}) &= \frac{p(\mathbf{x}|c)\,p(c)}{p(\mathbf{x})} \\
&= \frac{p(\mathbf{x}|c)\,p(c)}{\sum_{c=1}^{C} p(\mathbf{x}|c)\,p(c)}.
\end{aligned}
\tag{2}
$$

Here, C represents the number of frog species, or classes. The species with the maximum posterior probability is assumed to be the correct species for that frog call, i.e.:

$$\hat{c} = \arg\max_{c \in C} \left\{ p(c|\mathbf{x})p(c) \right\}. \tag{3}$$

For simplicity, we assume that each species is equally likely and set $p(c) = 1$.

3   Experiments

We tested our method on a dataset containing 226 audio samples of frog calls supplied by the Florida Tech's Paleo Ecology Laboratory. The dataset has calls from 15 frog species. A list of the frog species are shown in Table 1. We adopted the leave-one-out training approach. The Leave-one-out approach defined the training data set with all observation data with known labels, but one that was left out. The left-out observation was the test sample.

Table 1. Table of frog species data

Species Name	Common Name	Acronym	No. of Calls
Lithobates catesbeianus	Bull Frog	BF	17
Hyla gratiosa	Barking Tree Frog	BTF	10
Pseudacris nigrita	Southern Chorus Frog	CF	11
Gastrophyne carolinesis	Eastern Narrow Mouth Toad	ENMT	9
Lithobates clamitans	Green Frog	GF	19
Hyla cinear	Green Tree Frog	GTF	23
Pseudacris ocularis	Little Grass Frog	LGF	12
Anaxyrus quercics	Oak Toad	OT	7
Rana grylio	Pig Frog	PF	18
Hyla femoralis	Pine Woods Tree Frog	PWTF	9
Acris Gryllus	Southern Cricket Frog	SCF	21
Hyla squirella	Squirrel Tree Frog	SF	9
Lithobates phenocephalus	Southern Leopard Frog	SLF	16
Pseudacris cruicifer	Spring Peeper	SP	23
Anaxyrus terrestris	Southern Toad	ST	22

3.1   GMM Experiments

The confusion matrix in Fig. 5 shows the classification rates of our method. Average classification is about 80 %, which is close to rates achieved by humans.

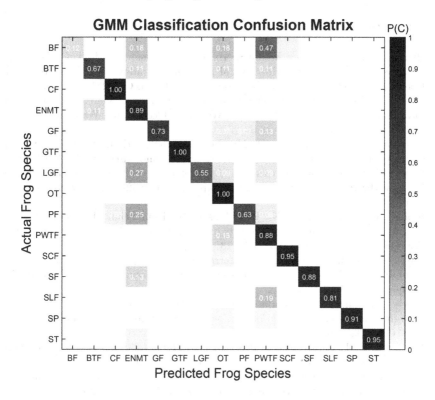

Fig. 5. Confusion matrix of classification results with GMM

4   Conclusion

We described a method for classifying frog calls using texture features and Gaussian mixture models. The texture features here are simply the label of the maximum-response filter representing the local shape of the spectral region around a salient peak on the spectrogram.

The classification procedure based on GMM produced promising results. In the future, we plan to modify the filter component of the feature extraction. Instead of using a bank of pre-designed filters, we would like to create our own filters tailored to frog vocalizations.

Acknowledgments. The authors acknowledge support from National Science Foundation (NSF) grants No. 1263011 and No. 1152306. Any opinions, findings, and conclusions or recommendations expressed in this material are those of the authors and do not necessarily reflect the views of the NSF.

References

1. Hsu, A., Emerson, J., Levy, M., de Sherbinin, A., Johnson, L., Malik, O., Schwartz, J., Jaiteh, M.: The 2014 environmental performance index. Technical report,Yale Center for Environmental Law and Policy, New Haven, CT (2014)
2. Agency, U.E.P.: America's wetlands : our vital link between land and water. Technical report, Office of Water, Office of Wetlands, Oceans, and Watersheds, Washington, DC (1995)
3. Knutson, M., Sauer, J., Olsen, D., Mossman, M., Hemesath, L., Lannoo, M.: Landscape associations of frog and toad species in iowa and wisconsin, USA. (2000)
4. Papadimitriou, E., Loumbourdis, N.: Copper kinetics and hepatic metallothionein levels in the frog rana ridibunda, after exposure to cucl2. Biometals **16**, 271–277 (2003)
5. Stolyar, O., Loumbourdis, N., Falfushinska, H., Romanchuk, L.: Comparison of metal bioavailability in frogs from urban and rural sites of western ukraine. Arch. Environ. Contam. Toxicol. **54**, 107–113 (2008)
6. Carey, C., Bryant, C.J.: Possible interrelations among environmental toxicants, amphibian development, and decline of amphibian populations. Environ. Health Perspect. **103**, 13–17 (1995)
7. Searle, C.L., Biga, L.M., Spatafora, J.W., Blaustein, A.R.: A dilution effect in the emerging amphibian pathogen batrachochytrium dendrobatidis. Proc. Nat. Acad. Sci. **108**, 16322–16326 (2011)
8. Relyea, R.A.: Trait-mediated indirect effects in larval anurans: reversing competition with the threat of predation. Ecology **81**, 2278–2289 (2000)
9. Mossman, M.J., Hartman, L.M., Hay, R., Sauer, J.R., Dhuey, B.J.: Monitoring long-term trends in wisconsin frog and toad populations. In: Status and Conservation of Midwestern Amphibians, pp. 169–198 (1998)
10. Grigg, G., Taylor, A., Mc Callum, H., Watson, G.: Monitoring frog communities: an application of machine learning. In: Proceedings of Eighth Innovative Applications of Artificial Intelligence Conference, Portland Oregon, pp. 1564–1569 (1996)
11. Acevedo, M.A., Corrada-Bravo, C.J., Corrada-Bravo, H., Villanueva-Rivera, L.J., Aide, T.M.: Automated classification of bird and amphibian calls using machine learning: A comparison of methods. Ecol. Inform. **4**, 206–214 (2009)
12. Han, N.C., Muniandy, S.V., Dayou, J.: Acoustic classification of australian anurans based on hybrid spectral-entropy approach. Appl. Acoust. **72**, 639–645 (2011)
13. Graciarena, M., Delplanche, M., Shriberg, E., Stolcke, A., Ferrer, L.: Acoustic front-end optimization for bird species recognition. In: 2010 IEEE International Conference on Acoustics Speech and Signal Processing (ICASSP), pp. 293–296 (2010)
14. Belyaeva, N.A., Yash-yee, K.L., Smartc, K.M., Ribeirod, E.: Whatfrog: A comparison of classification algorithms for automated anuran recognition
15. Chen, W.P., Chen, S.S., Lin, C.C., Chen, Y.Z., Lin, W.C.: Automatic recognition of frog calls using a multi-stage average spectrum. Comput. Math. Appl. **64**, 1270–1281 (2012). Advanced Technologies in Computer, Consumer and Control
16. Xie, J., Towsey, M., Yasumiba, K., Zhang, J., Roe, P.: Detection of anuran calling activity in long field recordings for bio-acoustic monitoring. In: 2015 IEEE Tenth International Conference on Intelligent Sensors, Sensor Networks and Information Processing (ISSNIP), pp. 1–6. IEEE (2015)
17. Visual Geometry Group: Department of Engineering Science, U.o.O. Texture classification (2007)

Ice Detection on Electrical Power Cables

Binglin Li[1], Gabriel Thomas[1(✉)], and Dexter Williams[2]

[1] Electrical and Computer Engineering Department, University of Manitoba,
Winnipeg, MB, Canada
lib34568@myumanitoba.ca, Gabriel.Thomas@umanitoba.ca
[2] Manitoba HVDC Centre, Winnipeg, MB, Canada
willimas@hvdc.ca

Abstract. In northern countries, ice storms can cause major power disruptions such as the one that occurred on December 2013 that left more than 300,000 customers in Toronto with no electricity immediately after such an ice storm. Detection of ice formation on power cables can help on taking actions for removing the ice before a major problem occurs. A computer vision solution was developed to detect ice on difficult imaging scenarios such as images taken under fog conditions that reduces the image contrast, passing cars that are within the field of view of the camera as well as different illumination problems that can occur when taking images during different times of the day. Based on a neural network for classification and six image features that can deal with these difficult images, we reduced the errors on a set of images that was previously yielding 20 errors out of 50 images to only one error.

Keywords: Power lines · Ice detection · Neural networks · Co-occurrence matrix · Hough transform

1 Introduction and Motivation

Extreme cold conditions combined with humidity as well as ice storms contribute to ice formation on power cables. Manitoba Hydro removes ice from power lines as quickly as possible to prevent equipment breakage and loss of power. In windy conditions, icy lines with the extra ice weight can swing violently, causing wires to break, wood poles to snap, and even steel towers to crumple. Ice storms can cause major power disruptions such as the one that occurred on December 2013 that left more than 300,000 customers in Toronto with no electricity [1]. Power blackouts can be very costly; one can cite the major 2003 event that affected the US and Canada that reduced the gross domestic product in this country by 0.7 % [2]. Thus detection of ice formation on power cables can help on taking actions for removing the ice before a major problem occurs.

Innovative ways have been proposed to address this problem using meteorological data with a support vector machine to predict the ice accumulations [3] but as indicated in [3] different learning algorithms can be used to improve on the ice accumulation cases. Based on a computer vision system, Manitoba Hydro has a frost detection system that uses binary images of the power line as an input as reported in [4] that uses the

© Springer International Publishing Switzerland 2015
G. Bebis et al. (Eds.): ISVC 2015, Part II, LNCS 9475, pp. 355–364, 2015.
DOI: 10.1007/978-3-319-27863-6_33

topological features of the area where the cable is located within the image to assess the amount of ice. The Manitoba Hydro system takes images from the overhead line conductors. The system includes 23 ice detection stations installed in a number of ice prone locations throughout the province. It consists of a data collection server that acquires high resolution images and a web based interface. Information is available on a 24-7 basis to the appropriate ice storm staff using the corporate WAN infrastructure. The system, as it is currently implemented, yields a number of false positives, an alarm indicating ice formation in a cable that does not present a problem as well as false negatives. As it can be seen in Fig. 1(a), challenging cases such as moving objects (car) and different background illumination such as a much darker background in (b) as compared to (c) make the detection of ice difficult.

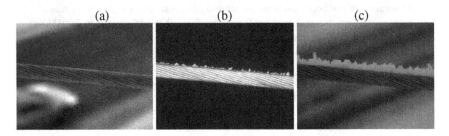

Fig. 1. Examples of images obtained for three different cases: (a) Alarm issued when there is no ice. (b) No alarm even though ice started to accumulate, and (c) no alarm for considerable ice buildup scenario.

Thus the following sections describe the methodology used to develop a computer vision system that can take on images such as the ones shown in Fig. 1. We targeted fifty images for which twenty ice detection classification errors occurred. Based on six image features and the use of a neural network as our classifier, our new system reduced this number of errors to only one.

2   Image Analysis and Feature Extraction

We start this section describing how we solved the different background scenarios followed by the rationale used for selecting the six image features that were successfully used as input to a neural network that makes the final classification of ice problems. All images were high resolution ones of size 1536 by 2048 pixels.

One of the first issues to address was the very different gray intensity levels that can appear on the background as can be seen in Fig. 1(b) and (c). Not only does the background appear much darker in (b) but within the power cable that intensity can change as well as the cable appears much brighter in (b) than (c). One way to avoid these differences is to work with only the edges of the images as shown in Fig. 2.

(a)   (b)

Fig. 2. Edge detection of images corresponding to Fig. 1(b) and (c).

The algorithm used to detect the edges was Canny [5]. The algorithm assigns pixels as edges at local maxima of the gradient of a Gaussian-smoothed image. Comparisons have been made with various edge detection techniques [6, 7], and it has been proven that Canny edge detection achieves a good performance in many image processing applications [8, 9].

The Canny edge detection scheme follows these steps:

(1) Smooth the image by convolving a Gaussian filter to reduce the noise.
(2) Compute the gradient magnitude to estimate the edge strength.
(3) Apply non-maxima suppression to the gradient magnitude to avoid spurious maximum.
(4) Utilize two thresholds to the gradient in order to reduce the false edge points and link the gaps in the detected edges. An upper threshold ignores all edges above that value and a lower threshold does the same for small intensity edge pixels.

As mentioned before, working with the image edges alleviates the problem of having different background and illumination. Another advantage of using edge detection based on Canny is that the thresholds mentioned in point (4) above also allow to alleviate the problem of having low contrast images. By estimating the contrast of the image using the global standard deviation of the original image σ_G, we assigned the Canny threshold T_c according to the following table:

This threshold T_c refers to the upper level one, the lower threshold is assigned as $0.4T_c$. The values on Table 1 were found empirically after careful examination of fifty images that included all the scenarios mentioned in the abstract.

Table 1. Assignment of threshold used in the Canny edge detection technique

Global standard deviation	$\sigma_G < 10$	$10 \leq \sigma_G < 15$	$\sigma_G \geq 15$
Canny threshold	$T_c = 0.18$	$T_c = 0.46$	$T_c = 0.13$

Figure 3 shows an example using the calculated thresholds as shown in Table 1. The value of the standard deviation associated to the Gaussian filter was set to square root of two for all the cases. Note how different contrast between the cable images yield edges around the cable only.

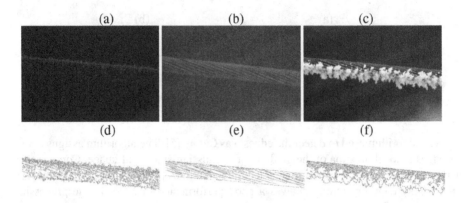

Fig. 3. Original images (a) to (c). Images obtained using Canny edge detection for the cases of $\sigma_G = 8.889$, $\sigma_G = 13.59$ and $\sigma_G = 46.35$ are shown from (d) to (f) respectively.

3   Image Features for Classification

The most important aspect to consider was the selection of the features to be used for classification. A reduced number of features was required as the system was meant to analyze images at many different locations and albeit a real time signal processing system was not mandatory, we wanted it to be fast enough so that not to depend on an expensive computing set up. This section describes the six features selected for this job and the criteria that were followed for this selection. Because the position of the cables is mostly constant, we set the very top and bottom areas on the edge images as zeros and concentrated on the central areas.

3.1   Features Based on the Gray Level Co-occurrence Matrix (GLCM)

The Gray Level Co-occurrence Matrix (GLCM) is a popular feature extraction technique that has been widely used in image processing [10–14]. GLCM calculates how often the gray-level value a of a pixel occurs conformed to certain spatial relationships with a pixel that has a value b. It can reveal that the images have fine or coarse textures, or how they are texturally uniform. The final features based on GLCM are calculated as second order statistical textures [11]. A total of fourteen textural features were proposed in [15] based on GLCM properties. In general, to reduce the computational complexity, only a number of relevant features are preferred. For example in [10] only four features (energy, entropy, contrast and inverse difference moment) were applied and achieved better performance than Gabor wavelet features. The work presented in [13] applied only seven features. In addition, some properties of the GLCM are proportional or inversely proportional to other features [10] and not all the fourteen features contribute equally to the classification performance. So in this project, we only added the correlation and contrast features of GLCM that gave good results based on the available 50 images. For the GLCM energy feature, it was found that it has a similar performance trend as that of the 2D standard deviation feature. So GLCM based energy features were excluded in the project.

Correlation of Co-occurrence Matrix. The correlation feature is a measure of linear dependency between two pixels over the whole image [15]. It indicates how a pixel is correlated to another pixel in the particular direction defined in the predicate that specifies the direction between pixels a and b mentioned above. Correlation is commonly applied to measure changes in images [12]. Let $p(i, j)$ be the $(i, j)^{\text{th}}$ entry of the GLCM, $p_x(i)$ is the ith entry in the marginal-probability matrix defined as $p_x(i) = \sum_{j=0}^{L} p(i,j)$, where $L + 1$ is the total number of gray level values. Similarly, $p_y(j) = \sum_{i=0}^{L} p(i,j)$. Then the feature is calculated as: $\text{Correlation} = \frac{\sum_{i=0}^{L} \sum_{j=0}^{L} (i,j)p(i,j) - \mu_x \mu_y}{\sigma_x \sigma_y}$ where μ_x, μ_y, σ_x and σ_y are the means and standard deviations of p_x and p_y. For the edge images, the correlation values for images with ice are lower than those without ice in the particular direction which the inner structure of the cables generate line edges within the top and the bottom of the cable (offset = [1 7], which means 7 pixels to the right and 1 pixel below). Some ice may cover the bottom and the top of the cable, so the correlation features of ice images are lower, as shown in Fig. 4(a) and (b) with values of 0.147 and 0.11. The images without ice have more line edges at the particular angle specified by the offset and hence have relatively higher values, as shown in Fig. 4(c) with 0.24. However, for the image in Fig. 4(d), the correlation value can also be high, because there is little ice on the visible surface of the cable. Note that some background objects (cars and lights) can also be detected when using the Canny method, but we observed that the objects have different edge angles. So the background points do not bring much trouble to the correlation feature in this application.

| (a) | (b) | (c) | (d) |

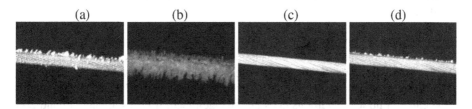

Fig. 4. (a) Correlation = 0.147 (b) Correlation = 0.11 (c) Correlation = 0.24 (d) Correlation = 0.2162.

Contrast of Co-occurrence Matrix. The contrast feature measures the amount of local variations in an image [15]. For a constant image, the GLCM contrast is 0. The contrast is quantified as: $\text{Contrast} = \sum_{i=0}^{L} \sum_{j=0}^{L} (i - j)^2 p(i,j).$

This feature was also calculated using the edge images. To obtain variations, we observed that when setting the offset at [1 3] (3 pixels to the right and 1 pixel below), the images with ice have more local variations than those without ice and therefore have higher contrast features. Consider Fig. 4(b) which has a contrast value of 0.03284 and compare it with the contrast value obtained in Fig. 5(a) and (b).

(a)   (b)

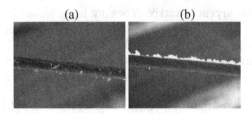

Fig. 5. (a) Contrast = 0. 01267 (b) Contrast = 0.01147.

3.2 Statistical Features

Central Standard Deviation Feature. The standard deviation can be used as a measure of texture. It was found in [16] that in general, first order statistics (mean, variance, etc.) perform about as well as second-order statistics. Quite different areas of application such as forest classification and melanoma detection have successfully used these statistical features [17, 18].

As accumulated ice on the cable yields high variance values compared to cables with no ice, we calculated the standard deviations. Let $I(x,y)$ denote the gray-level value in image I. The standard deviation σ_C can be obtained by:

$$\sigma_C = \sqrt[2]{\frac{1}{N*M-1}(\sum_{x=1}^{N}\sum_{y=1}^{M}|I(x,y)|^2 - \mu_C^2)} \text{ where } \mu_C = \frac{1}{N*M}\sum_{x=1}^{N}\sum_{y=1}^{M}I(x,y) \text{ and } N, M \text{ are the}$$

size of the matrix used to compute the standard deviation.

In Fig. 4(a), the central standard deviation is 0.1493, while in (c) the value is lower with 0.1144.

Measuring the Width of the Cable. The width changes of the cable were calculated for each image column where the measurement of width is that as shown in Fig. 6. The maximum and minimum vertical coordinates for each column pixel was found and the difference of these two values was considered to be the width of the cable. The width of the cable in no ice conditions does not change much. It keeps a value around 200 pixels, while in images with ice is larger than 200 pixels. When the computed width for a certain column is less than 200, we set the width equal to that from the previous column. This step can reduce the width error which results from the detected discontinuous edges present on some patterns on the surface of the cables. Isolated points in the background can result in a sharp increase of width change between two successive columns. As the ice accumulating between two adjacent columns is fairly continuous, the difference of width change between two neighboring columns which is larger than 50 pixels can be regarded as isolated points, thereby we also set the width as that found in the previous column.

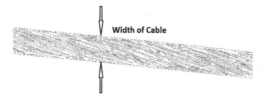

Fig. 6. Figure that highlights the width of the cable. We refer to width not to the actual width of the cable but the distance from top to bottom as indicated by the arrows.

The standard deviation of the width changes was the feature used in this application. Assume $d(i)$ represents the ith value of cable width along the horizontal axis. The standard deviation σ_W is calculated as: $\sigma_W = \sqrt[2]{\frac{1}{N-1}\sum_{i=1}^{N}(d(i)-\mu_W)^2}$ where $\mu_W = \frac{1}{N}\sum_{i=1}^{N}d(i)$ and N is the horizontal dimension of image I.

Figure 7 shows the width changes for the images shown in Fig. 8. The standard deviations of the width changes calculated are 542.9248 for the ice case and 3.3192 for the no ice scenario.

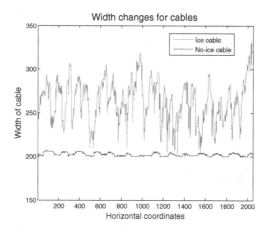

Fig. 7. Width values for every column shown in Fig. 8 below.

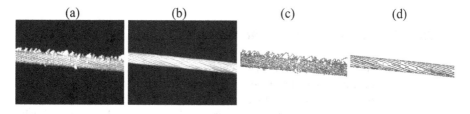

Fig. 8. Original images (a) and (b) and their corresponding edges (c) and (d) respectively.

3.3   Features Based on the Hough Transform

The Hough transform is a popular technique that detects lines in pictures [19]. For every point x and y on a binary image, every value of m and b are calculated from $y = mx + b$ to form and accumulator matrix where the maximum value will indicate the total number of pixels associated with a line for those m and b values. These maximum values will be higher on images such as the one shown in Fig. 8(d) than the ones shown in part (c). Another benefit is that if the only lines to be computed have the angles of the lines corresponding to the cable with no ice, then these high Hough values would indicate no ice and all the edges corresponding to the car shown in Fig. 1(a) would be mostly omitted. These maximum values in the accumulator matrix form the fifth feature extracted from the images.

As some of the ice accumulates on for example, the bottom of the cable, please refer to Fig. 3(c), edge detection would capture the edges on the top of the cable and the Hough transform would tend to yield higher values. A very difficult case is shown in Fig. 9(a) where the surface of the cable would yield a high value in the accumulator matrix indicating no ice where in fact there is considerable ice on the upper left part. In order to alleviate these problems, the pixels in the original image with intensity values above 80 % of the maximum value were segmented with a simple threshold and then edge detection was performed to eliminate the edges from the cable. The Hough transform would yield a feature value that corresponds to the ice scenario as the edges after segmentation present almost no lines in the right angle as shown in Fig. 9(b). This modification then forms the sixth and last feature extracted from the images.

(a)   (b)

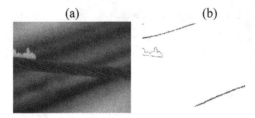

Fig. 9.   (a) Original image with a high Hough feature value indicating no ice. (b) Edge detection after segmentation. Hough features in this image would reduce the value of pixels with lines corresponding to the cable indicating no ice.

4   Classification Using a Neural Network

A two layer feed-forward back propagation neural network was used updating the weight and bias values according to the Levenberg-Marquardt optimization training algorithm. Hyperbolic tangent sigmoid transfer functions were used in the architecture. The size of the training, validation and testing sets were 70 %, 15 % and 15 % respectively. Figure 10 shows the final architecture used as well as the mean squared error values obtained during the training. Of the fifty images used, twenty were given Manitoba Hydro International misclassification cases, ten of those twenty misclassification cases were false positives and the other ten were false negatives. In this work those errors were reduced to only *one*, an image with very little ice on the cable that was misclassified as having no ice.

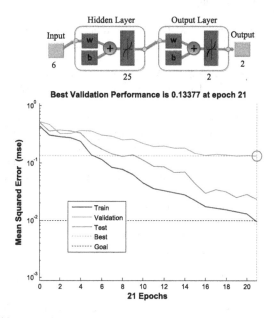

Fig. 10. Top: Architecture of the neural network used for the final classification. An output value of [1 0] indicated ice and [0 1] indicated no ice on the cable. Bottom: Mean squared error values during the training of the network.

5 Conclusions

We were given a series of images to detect the cases of ice on electrical power transmission cases in which scenarios such as passing cars, different illumination and poor contrast caused problems for twenty out of fifty images. Using six features based on Canny edge detection and a combination of features extracted from the GLCM matrix, the width of the cable and the Hough transform, a neural network could successfully reduce the number of errors to one.

References

1. Tens of thousands in U.S., Canada without power days after ice storm. http://www.cnn.com/2013/12/25/us/winter-weather/
2. U.S.-Canada Power System Outage Task Force, Final Report on the August 14, 2003 Blackout in the United States and Canada: Causes and Recommendations, April 2004
3. Zarnani, A., Musilek, P., Shi, X., Ke, X., He, H., Greiner, R.: Learning to predict ice accretion on electric power lines. Eng. Appl. Artif. Intell. **25**(3), 609–617 (2012)

4. Wachal, R., Stoezel, J.S., Peckover, M., Godkin, D.: A computer vision early-warning ice detection system for the smart grid. In: Transmission and Distribution Conference and Exposition (T&D), IEEE PES, pp. 1–6, May 2012
5. Canny, J.: A computational approach to edge detection. IEEE Trans. Pattern Anal. Mach. Intell. **PAMI-8**(6), 679–698 (1986)
6. Sharifi, M., Fathy, M., Mahmoudi, M.T.: A classified and comparative study of edge detection algorithms. In: International Conference on Information Technology: Coding and Computing, pp. 117–120, April 2002
7. Shrivakshan, G.T., Chandrasekar, C.: A comparison of various edge detection techniques used in image processing. Int. J. Comput. Sci. Issues **9**(5), 272–276 (2012)
8. Ramamurthy, B., Chandran, K.R.: Content based image retrieval for medical images using canny edge detection algorithm. Int. J. Comput. Appl. **17**(6), 0975–8887 (2011)
9. Cheng, H.Y., Weng, C.C., Chen, Y.Y.: Vehicle detection in aerial surveillance using dynamic bayesian networks. IEEE Trans. Image Process. **21**(4), 2152–2159 (2012)
10. Partio, M., Cramariuc, B., Gabbouj, M., Visa, A.: Rock texture retrieval using gray level co-occurrence matrix. In: Proceedings of the 5th Nordic Signal Processing Symposium, vol. 75, October 2002
11. Mohanaiah, P., Sathyanarayana, P., GuruKumar, L.: Image texture feature extraction using GLCM approach. Int. J. Sci. Res. Publ. 3(5) (2013)
12. de Siqueira, F.R., Schwartz, W.R., Pedrini, H.: Multi-scale gray level co-occurrence matrices for texture description. Neurocomputing **120**, 336–345 (2013)
13. Tripathi, N., Panda, S.P.: A review on textural features based computer aided diagnostic system for mammogram classification using GLCM & RBFNN. Int. J. Eng. Trends Technol. **17**(9), 462–464 (2014)
14. Beura, S., Majhi, B., Dash, R.: Mammogram classification using two dimensional discrete wavelet transform and gray-level co-occurrence matrix for detection of breast cancer. Neurocomputing **154**, 1–14 (2015)
15. Haralick, R.M., Shanmugam, K., Dinstein, I.: Textural features for image classification. IEEE Trans. Syst. Man. Cybern. **SMC-3**(6), 610–621 (1973)
16. Weszka, J.S., Dyer, C.R., Rosenfeld, A.: A comparative study of texture measures for terrain classification. IEEE Trans. Syst. Man. Cybern. **SMC-6**(4), 269–285 (1976)
17. Coburn, C.A., Roberts, A.C.B.: A multiscale texture analysis procedure for improved forest stand classification. Int. J. Remote Sens. **25**(20), 4287–4308 (2004)
18. Barata, C., Ruela, M., Francisco, M., Mendonça, T., Marques, J.S.: Two systems for the detection of melanomas in dermoscopy images using texture and color features. IEEE Syst. J. **8**(3), 965–979 (2014)
19. Duda, R.O., Hart, P.E.: Use of the hough transformation to detect lines and curves in pictures. Commun. ACM **15**(1), 11–15 (1972)

Facial Landmark Localization
Using Robust Relationship Priors
and Approximative Gibbs Sampling

Karsten Vogt$^{(\boxtimes)}$, Oliver Müller, and Jörn Ostermann

Institut Für Informationsverarbeitung (TNT),
Leibniz Universität Hannover, Hanover, Germany
{vogt,omueller,ostermann}@tnt.uni-hannover.de

Abstract. We tackle the facial landmark localization problem as an inference problem over a Markov Random Field. Efficient inference is implemented using Gibbs sampling with approximated full conditional distributions in a latent variable model. This approximation allows us to improve the runtime performance 1000-fold over classical formulations with no perceptible loss in accuracy. The exceptional robustness of our method is realized by utilizing a L_1-loss function and via our new robust shape model based on pairwise topological constraints. Compared with competing methods, our algorithm does not require any prior knowledge or initial guess about the location, scale or pose of the face.

1 Introduction

Accurate facial landmark localization algorithms play a vital role for many applications, such as biometric authentication [1] or human-machine-interfaces [2]. The goal of these algorithms is to estimate the pixel-coordinates of a configuration of predefined facial landmarks in an input image. Research has been mostly focused on three different approaches. First are methods based on a global appearance model [3]. Next are the part-based methods [4,5]. There, the individual landmarks are detected separately, while a shape model, typically based on a *point-distribution-model* (PDM), acts as a prior that constrains the space of valid configurations. The third class of methods are based on shape regression [6–9]. Regressors are trained to predict improved location estimates of all landmarks based on a previous guess. The relationship between landmarks is typically not explicitly modelled, but either learned from training data or enforced via regularization.

Past works usually focused on fast and accurate localization, while robustness was more of an afterthought. For example, methods that utilize an *Active Shape Model* (ASM) [10] require the approximate position and scale of the face to be known a-priori, as the landmark coordinates need to be aligned with a mean shape. Methods based on shape regression also require an initial guess. Arashloo et al. [11] proposed a method to perform global optimization in a Markov Random Field (MRF) formulation of the landmark localization problem.

© Springer International Publishing Switzerland 2015
G. Bebis et al. (Eds.): ISVC 2015, Part II, LNCS 9475, pp. 365–376, 2015.
DOI: 10.1007/978-3-319-27863-6_34

This approach is promising, but its implementation has two issues. First, the use of an ASM still retains the aforementioned shape alignment issues. Secondly, for each landmark only a small number of heuristically selected locations are ever considered. We wish to present a new landmark localization framework which is based on Bayesian principles that will solve the aforementioned problems and prioritizes robustness, while also achieving competitive performance and near real-time speed.

In this work, we tackle the facial landmark localization problem as an inference problem over an MRF. Efficient inference is implemented in the *Markov Chain Monte Carlo* MCMC framework, namely using Gibbs sampling. Our approach does not require a favorable initial estimate of the landmark locations. In fact, a completely random initialization will suffice. Our Gibbs sampler has the capability to rapidly cover the entire configuration space. We achieve large speedups over classical Gibbs sampling formulations by decomposing the full conditional distributions into sets of discrete latent variables. To solve the resulting sampling problem, we propose to approximate their probability distributions by exploiting the factorization of the posterior. We also propose a new PDM based shape model that is, in contrast to ASM and its variants, translation and scale invariant. This shape model consists of two components. The first component models the topology of the landmark configuration by imposing a set of simple pairwise relationship rules. The second one is inspired by *Shape-Indexed-Features* [12] and models the exact landmark locations in relation to nearby landmarks. Its main task is to fine-tune the results after the optimization has already mostly converged.

Our main contributions can be briefly stated as follows:

1. We present an approximation of the full conditional distributions that allow for speed-ups by a factor > 1000 over classical Gibbs sampling.
2. This approximation works in conjunction with a new translation and scale invariant shape model.
3. We always optimize over the full configuration space without requiring sub-sampling while also being extremely robust to bad initializations.

Section 2 introduces factor graphs. After formulating the landmark localization problem as an MRF inference problem, we present our new Gibbs sampling algorithm in Sect. 3 and our face model in Sect. 4. Section 5 ties everything together into a complete landmark localization framework. We evaluate our work in Sect. 6 and finish with conclusions in Sect. 7.

2   Graphical Models

The posterior probability of a facial landmark configuration $c = (x_1, y_1, \ldots, x_L, y_L)$ with L landmarks can be modeled as a factor graph $G = (V, F, E)$, where the set of vertices $V = \{v_1, \cdots, v_L\}$ represent the individual landmarks. The factors $F = \{f_1, \ldots, f_{|F|}\}$ define the relationships between vertices and are connected to them via undirected edges $E = \{e_1, \ldots, e_{|E|}\}$. A set of vertices $\mathcal{N}_f$

connected to the same factor $f \in F$ is called a *clique* and we will denote $\mathcal{F}_v$ as
the set of factors that are connected to vertex v. We also define the probability
of a configuration c conditioned on the observed image data I as

$$p(c \,|\, I(x,y)) \propto \prod_{f \in F} p_f^{\alpha_f}(c(\mathcal{N}_f) \,|\, I), \qquad (1)$$

where $c(\nu)$ are the coordinates for a subset of landmarks $\nu \subseteq V$, $p_f(c(\mathcal{N}_f) \,|\, I)$ is
the clique potential for a factor $f \in F$ and α_f is a tuning parameter that adjusts
the influence of said factor.

Our objective is to find a landmark configuration which maximizes the posterior in Eq. (1) via inference on the graphical model G. Efficient inference
on general graphical models is a notably complex problem. In the past, several different approaches have been proposed. The most successful ones have
been *Gibbs sampling* [13], *belief propagation* [14] and *dual decomposition* [15].
While Gibbs sampling has been mostly succeeded by competing approaches in
convergence speed, it still allows for a more natural handling of large clique
sizes than either belief propagation or dual decomposition. Furthermore, all
these methods struggle to perform well as the configuration space becomes
larger. Assuming a full-HD image, each landmark can be situated in one of
1920×1080 different locations. The full configuration space is therefore comprised
of $2.073.600^L$ different configurations. Efficient solutions can still be achieved
either by sub-sampling the configuration space [11] or via particle-sampling [16].
Conceptually, both approaches achieve their runtime gains by considering only
part of the full configuration space. This can be problematic, since the global
optimum may not even be among the candidate configurations.

In this paper, we propose to solve the inference problem via *Gibbs sampling*.
In contrast to competing methods, we will always consider the full configuration space. Our landmark detector will therefore be significantly more robust
with regards to its initialization. Large speed-ups will be gained by introducing
appropriate latent variables into the Gibbs sampling formulation.

3   Approximative Gibbs Sampling

Sampling based detectors first draw a representative random sample of configurations $\{c_0, \ldots, c_N\}$ from the posterior $p(c \,|\, I)$. Different types of estimates can
then be derived from this sample to find the solution that is best supported
by the observed data. Gibbs sampling generates such a sample by sequentially
generating new configurations c_{t+1} from c_t by sampling from each variable $c(v_i)$
in turn, while keeping all other variables fixed. By exploiting the factorization of
the posterior, we can simplify the full conditionals by discarding clique factors
that are conditionally independent from the target variable:

$$c(v_1) \sim p(c(v_1)\,|\,c(V \setminus v_1), I) \propto \prod_{f \in \mathcal{F}_{v_1}} p_f^{\alpha_f}(c(\mathcal{N}_f)\,|\,I)$$

$$\vdots \tag{2}$$

$$c(v_L) \sim p(c(v_L)\,|\,c(V \setminus v_L), I) \propto \prod_{f \in \mathcal{F}_{v_L}} p_f^{\alpha_f}(c(\mathcal{N}_f)\,|\,I).$$

At each Gibbs sampling step, we have to draw a variate from one of these discrete distributions. Even though there are well known algorithms to sample from arbitrary discrete distributions in constant time [17], these still require a linear time preprocessing step. As the configuration space becomes very large, the computational requirements of exact sampling can become prohibitive.

We solve this problem with a sampling strategy that can be best described as divide-and-conquer sampling. As shown in Fig. 1, we recursively subdivide the configuration space until we end up with elementary events. At each split, we have to sample from a latent variable $\phi_{v,i}$ with M possible outcomes. Instead of sampling from one variable with K outcomes, we sample from $\lceil \log_M(K) \rceil$ variables, each with only M outcomes.

Next, we have to define the probability distributions of the latent variables. Since the landmark locations are inherently two-dimensional, sampling is actually performed with a quad-tree structure ($M = 4$). The probabilities for landmark v to be in one of the four quadrants $Q_{i,j}$ ($j \in \{TL, TR, BL, BR\}$) can be directly stated as:

$$p_{\phi_{v,i}}(j) \propto \sum_{(x,y) \in Q_{i,j}} \prod_{f \in \mathcal{F}_v} p_f^{\alpha_f}(c(\mathcal{N}_f)\,|\,I). \tag{3}$$

This formulation requires us to evaluate the sum by computing the landmark location probability for each valid coordinate, which has a runtime complexity linear to the number of pixels in quadrant $Q_{i,j}$. If possible, we want to transform the problem such that the summation can be computed in constant time. Here is where our approximation comes into play. By upper-bounding the quadrant probabilities in Eq. (3) using the generalized *Hölder* inequality and renormalizing, we get the following approximated quadrant probabilities:

$$\tilde{p}_{\phi_{v,i}}(j) = \frac{1}{Z} \prod_{f \in \mathcal{F}_v} \left(\sum_{(x,y) \in Q_{i,j}} p_f^{\alpha_f \cdot |\mathcal{F}_v|}(c(\mathcal{N}_f)\,|\,I) \right)^{1/|\mathcal{F}_v|}, \tag{4}$$

where Z is a normalizing constant. This greatly simplifies the complexity of each sum and allows us to directly compute their result in constant time, for some well chosen families of factor distributions, independent of the size of the quadrant. Next, we will present clique potentials which are suitable for facial landmark detection and which also fulfill the required constant time complexity.

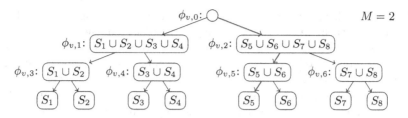

Fig. 1. Sampling in the latent variables formulation on a discrete variable v with states $S_1, \ldots, S_8$. The red path shows an exemplary sampling path resulting in state S_5.

4 Face Model

4.1 Appearance Model – Unary Potentials

Unary potentials are only conditioned on the image data and describe the location probabilities for individual landmarks. Recent publications propose very diverse strategies to define such distributions. This broad range of approaches include *template matching* [18], MOSSE *filters* [4], *geometric blur features* [19] and HOG *features* [20]. In this paper, we chose to implement dense HOG *features* as described in [21]. They are invariant to changes in brightness, robust to changes in contrast, and robust to affine transformations by their very design. Location probabilities can be derived from these HOG features using any supervised classifier with soft outputs. We chose to use a multi-class linear SVM with calibrated outputs [22], as it is fairly robust to outliers and noise while also having a low amount of hyperparameters that need to be hand-tuned.

The summation over location probabilities in Eq. (4) can be efficiently computed in constant time by transforming it into a simple look-up operation. Since the unary potentials provide a single location probability for each landmark/pixel, independent of the location of neighboring landmarks, we can transfer the work to a preprocessing step, e.g., via summed-area-tables.

4.2 Shape Model – Higher Order Local Gaussian Potentials

The higher order potentials (≥ 2) in our factor-graph model should describe the spatial relationship between different landmarks. We only draw new coordinates for one landmark at a time during each Gibbs sampling step, as all other landmarks remain fixed. Thus, we can directly model these relationships with bivariate probability distributions that are conditioned on the fixed coordinates. Due to the way our approximate sampling scheme is set up, and based on our own observations, these distributions should have the following properties:

1. To increase the expressiveness of the shape model, local shape components should be independently deformable.
2. If possible, we want to achieve invariance to translation and scale.
3. The sums in Eq. (4) must be evaluable in constant time.

Property 1 leads us to consider a local model that obeys the Markov property, i.e., each landmark v_i is only dependent on the positions of its direct spatial neighbors. Property 2 strongly favors models that operate in an appropriately chosen local coordinate system, which is constructed from these neighbors. Of course, the local coordinates would not be very informative if the neighborhood system is allowed to change during the course of the simulation. Therefore, the neighborhood system will first be extracted from a mean face shape $\bar{S}$, which is estimated from a training set via *Generalized Procrustes Analysis*. After constructing the *Delaunay* triangulation of all landmarks $V \setminus v_i$ on $\bar{S}$, the three neighbors $n_1(v_i)$, $n_2(v_i)$ and $n_3(v_i)$ are simply the vertices of the encompassing triangle of v_i. We can now construct a non-orthogonal local basis for v_i from any landmark configuration as shown in Fig. 2(a). We first select $\boldsymbol{o} = \boldsymbol{c}(n_1(v_i))$ as the origin of this coordinate system and $\boldsymbol{x} = \boldsymbol{c}(n_2(v_i)) - \boldsymbol{o}, \boldsymbol{y} = \boldsymbol{c}(n_3(v_i)) - \boldsymbol{o}$ as its basis vectors. Transforming a coordinate vector $\boldsymbol{c}(v_i)$ from the global basis to the local basis can be achieved as follows: $\boldsymbol{c}_{local}(v_i) = A \cdot (\boldsymbol{c}(v_i) - \boldsymbol{o})$, where $A = \left[\frac{\boldsymbol{x}}{\boldsymbol{x}^T \boldsymbol{x}} \ \frac{\boldsymbol{y}}{\boldsymbol{y}^T \boldsymbol{y}} \right]^T$. The bivariate location distribution for each landmark can now be defined as a Gaussian distribution parametrized in the local coordinate system.

$$p_f(\boldsymbol{c}(\mathcal{N}_f) \,|\, I) \propto p(\boldsymbol{c}(v_i) \,|\, \boldsymbol{c}(n_1(v_i)), \boldsymbol{c}(n_2(v_i)), \boldsymbol{c}(n_3(v_i))) \tag{5}$$

$$\boldsymbol{c}_{local}(v_i) \sim \mathcal{N}(\mu_{local}, \Sigma_{local}) \tag{6}$$

$$\Leftrightarrow \boldsymbol{c}(v_i) \sim \mathcal{N}(A^{-1}\mu_{local} + \boldsymbol{o}, A^{-1}\Sigma_{local}A^{-T}). \tag{7}$$

The local distribution parameters μ_{local} and Σ_{local} can be estimated from the training set. Since the transformation between the local and global bases is simply a linear relationship, we can also reproject the local distribution back into the global space (Eq. (7)). For landmarks situated on the convex hull of $\bar{S}$, we will have to employ a different procedure to select an appropriate local basis. It turned out that the robustness of the location estimate is more important than its accuracy. For these landmarks we will therefore select K uniformly distributed bases at random, each inducing an independent estimate of their location. The final distribution for $\boldsymbol{c}(v_i)$ represents the consent between all K estimates and can than be derived by multiplying their normal distributions in the global space, which again produces a single bivariate normal distribution.

The product of Gaussians is itself proportional to a Gaussian distribution [23]. As required for Property 3, the sum in Eq. (4) can therefore be evaluated using the bivariate cumulative distribution function (CDF) of the Gaussian distribution in Eq. (7), e.g. using the algorithm described in [24].

4.3   Shape Model – Rule-Based Binary Potentials

The local Gaussian relationship model is prone to slow convergence and can be unstable if the initial solution is not chosen well. We tackle this problem by augmenting our graphical model with additional robust binary factors. While the local Gaussian factors model the geometry of the shape, these binary factors

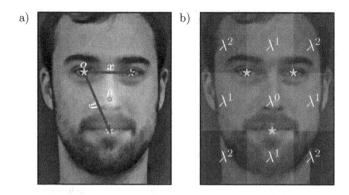

Fig. 2. Higher-order potentials for the nose landmark. (a) Gaussian distribution with mean μ and its local coordinate system and (b) penalty regions for the rule-based prior with penalty term λ.

should only model its topology. Each binary clique represents a simple relationship rule. These rules may be of the form $r(f, v_i, v_j) \in \{\text{is-left-of, is-right-of, is-above, is-below}\}$. The clique distribution of a single rule is then defined as

$$p_f(\boldsymbol{c}(v_i), \boldsymbol{c}(v_j)) \propto p_f(\boldsymbol{c}(v_i)\,|\,\boldsymbol{c}(v_j)) = \lambda^{1_{(\text{rule } r(f, v_i, v_j) \text{ is violated})}}, \qquad (8)$$

where λ is a user-adjustable penalty term and 1_A is the indicator function. Multiple rules operating on the same landmark v_i are statistically independent and their conditional distribution may be jointly calculated by counting how many rules are being violated by the proposed configuration. The higher-order clique distribution over all landmarks is then defined as

$$p(\boldsymbol{c}(v_i)\,|\,\boldsymbol{c}(V \setminus v_i)) = \lambda^{\sum_{f \in \mathcal{F}_{v_i}} 1_{(\text{rule } r(f, v_i, v_j) \text{ is violated})}}. \qquad (9)$$

Allowing only the specified types of rules, the probability distribution in Eq. (9) is always piecewise-constant and axis-aligned. Evaluation of the sum for the quadrant probabilities in Eq. (4) can therefore be implemented such that the runtime complexity is invariant to the image size and only depends on the number of rules. This relationship model has the advantage of being fairly simple and robust, yet also invariant to translation and scale. Robustness to rotations can be improved by augmenting the training set with slightly rotated versions of the input faces. We automatically select rules by including all relationship rules that hold true for at least 95 % of all images in the training dataset.

5   Landmark Localization Algorithm

Our facial landmark localization algorithm uses the Gibbs sampling scheme (Sect. 3) as its core component. Yet, there are a few details that could not be covered in the previous sections. Algorithm 1 presents our algorithm AGS (*Approximative Gibbs Sampling*) in pseudo code. We explain the individual components of our algorithm step-by-step:

Algorithm 1. AGS landmark localization algorithm.

---

function LOCALIZELANDMARKS(Image I_{RGB})
  Parameters:
    Sample-Chains C, Burn-In Samples B, Sample-Size N, Loss-Function $\mathcal{L}$
  Preprocessing:
    $I_{YUV} \leftarrow$ RGB2YUV(I_{RGB})
    for *pose* $\in$ {*ProfileLeft, Frontal, ProfileRight*} **do**
      $U_{pose} \leftarrow$ UNARYPOTENTIALS(I_{YUV}, *pose*)
      $T_{pose} \leftarrow$ CONSTRUCTSUMMEDAREATABLE(U_{pose})
  Sampling:
    $S \leftarrow$ Empty List   ▷ Initialize list of samples
    for each Sample-Chain $\in [1 \ldots C]$ **do**
      for *pose* $\in$ {*ProfileLeft, Frontal, ProfileRight*} **do**
        $c_{pose,0} \leftarrow$ Random Configuration
        for $i = 1 \ldots B/2$ **do** $c_{pose,i} \leftarrow$ GIBBSSAMPLING($c_{pose,i-1}, T_{pose}$, *pose*)
      $\theta \leftarrow \arg\max_\theta p(c_{\theta,B/2} \mid I)$ ▷ Select *pose* θ which maximizes the posterior
      for $i = B/2 \ldots B + N$ **do**
        $c_{\theta,i} \leftarrow$ GIBBSSAMPLING($c_{\theta,i-1}, T_\theta, \theta$)
        if $i > B$ **then** Append $c_{\theta,i}$ to S
  Estimation:
    switch $\mathcal{L}$ **do**
      case MAP: **return** $\arg\max_{c \in S} p(c \mid I)$
      case L_1-loss: **return** MEDIAN(S)
      case L_2-loss: **return** MEAN(S)

Pose is discretized into three possible states: frontal, left looking and right looking, and for each of these poses a learned shape model is available. During the preprocessing step, we will first precompute the unary potentials. The HOG-features are computed separately for each color channel in the YUV color space. We accelerate the computation of the quadrant probabilities in Eq. (4) by storing the unary potentials in a summed-area-table. The preprocessing step of the algorithm is actually the most performance critical one, because we have to evaluate the location probabilities for each landmark and pixel coordinate at the original image resolution. Therefore it is paramount to use optimized implementations for the dense HOG-feature extraction and classification. We then create multiple independent sampling chains, each initialized with a random landmark configuration. Pose estimation is handled in a very straight-forward manner. For each chain, we simply try all three pose states and advance the sampling chain for a few iterations using our approximate Gibbs sampler. The pose that results in the maximum a-posteriori configuration will then be selected for this chain. Following this, we draw the remaining samples from the sampling chain.

6 Evaluation

Here, we evaluate our landmark detector with respect to its landmark localization error and convergence properties. We compare our results with the DBASM algorithm [4] and the npBCLM algorithm [25], since both are recent part-based landmark detector grounded in Bayesian methodology. Unless otherwise noted, all experiments are performed with 8 independent sampling chains and a sample-size of 500 per chain, of which the first 100 burn-in samples are discarded. Other

Table 1. Parameters for the AGS landmark detector.

HOG block_size	36×36	HOG normalization	L_2	α_{unary}	1.0
HOG cell_size	9×9	SVM C	1.0	α_{Gaussian}	0.025
HOG num_bins	8	rule-based penalty λ	0.1	$\alpha_{\text{rule-based}}$	0.25

parameters were tuned empirically and set as shown in Table 1. Each chain is always initialized with a random landmark configuration.

IMM Dataset: The IMM dataset [26] contains 240 annotated images of 40 different subjects with a resolution of 640×480 pixels. The annotations for each face image include 58 different landmarks and the sex of the subject. All face images were captured indoors under studio condition but vary in pose, size, facial expression and lighting.

Localization Error: The automatically detected landmark locations should be as close as possible to a manually created ground-truth. Additionally, the detector should also generalize well to unseen faces. To this end, we perform a 40-fold cross-validation by partitioning the dataset by subject. Errors are measured as the interocular distance normalized error averaged over all images in the dataset. Figure 3 shows the results for our AGS algorithm with three different loss functions. The MAP estimator simply selects the single best landmark configuration that achieves the highest posterior probability from all sampling chains. The Bayesian estimator with L_2-loss assumes a normally distributed localization error, while the Bayesian estimator with L_1-loss assumes longer tails for the posterior distribution and is therefore more robust to outliers. As can be clearly seen, a simple MAP estimate is too noisy to get satisfactory results. In comparison, both L_2 and L_1 loss functions will always generate significantly improved

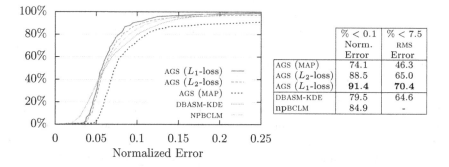

Fig. 3. Cross-validation results for the IMM dataset. The graph shows percentile plots over normalized landmark localization errors. The table compares the performance of different algorithms and presents the proportion of images in the dataset, for which the localization error falls below a predefined threshold (higher is better).

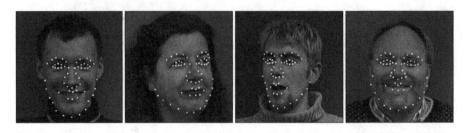

Fig. 4. Example results for the IMM dataset.

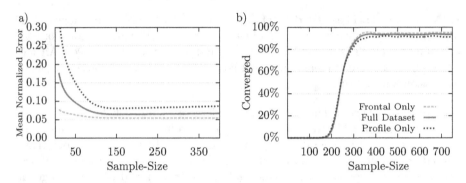

Fig. 5. Convergence results for the IMM dataset. (a) Mean normalized error over sample-size and (b) percent of converged images over sample-size.

location estimates. Our method also improves on the results of the DBASM algorithm using a kernel density estimate and the npBCLM algorithm. Figure 4 shows some typical results for this dataset.

Convergence: Convergence can be measured in multiple ways. Our method has fast convergence in both a theoretical and a practical sense. Figure 5(a) shows the average normalized landmark location error as a function of the sample size. We show that the sampling chains typically show fast mixing behavior by calculating the multivariate scale reduction factor $\hat{R}^p$ as in [27]. For mixing chains, $\hat{R}^p$ should start to approach 1.0 with increasing sample sizes. Figure 5(b) shows the percentage of input images for which the sampling converged ($\hat{R}^p < 1.2$) within a specified sampling size. As can be seen in Fig. 5(a), less than 100 iterations will usually suffice to get close to optimal results. Frontal faces show slightly lower estimation errors than profile views, but on average the results are. Mixing behavior is almost impeccable. For more than 95 % of the dataset, less than 300 iterations are sufficient to demonstrate convergence to the target distribution. Only a few particularly uncommon facial expressions or poses exhibit slow convergence.

Runtime Performance: We give the runtime for all three phases of our algorithm separately in Table 2. The preprocessing step includes the computation

Table 2. Runtime performance measurements. The experiments were done on a *Intel(R) Xeon(R) E5-2690* CPU at 3 GHz using 8 cores.

Preprocessing		Sampling		Estimation
HOG Features	Unary Potentials	Approx. Sampling	Exact Sampling	
270 ms	3000 ms	0.4 ms/draw	633 ms/draw	$\ll 1$ ms

of HOG features, pixelwise evaluation of the appearance model and preparation of the integral images. Sampling speed is shown for our approximative sampling scheme and for exact discrete Gibbs sampling. The measurements were averaged over a large number of draws. Currently, the major bottleneck of our algorithm is the precomputation of the appearance model. In case a high image throughput is required, we suggest to pipeline the three phases of our localization algorithm. The preprocessing phase could than be distributed over multiple networked computers. The runtime improvements due to our approximation scheme are usually on the order of a 1600-fold reduction in sampling time for a 640×480 pixel image. Larger images will of course amplify the gain. Given the long compute time for exact Gibbs sampling, we do not present cross-validation results for this method.

7   Conclusion

This work presents a facial landmark localization algorithm based on Gibbs sampling with approximated full conditional distributions. Compared with competing methods, the presented algorithm does not require an initial guess and improves on their localization errors. A new robust shape model allows for translation and scale invariant landmark localization while generally achieving fast convergence for a variety of poses and facial expressions.

References

1. Jain, A.K., Ross, A., Prabhakar, S.: An introduction to biometric recognition. TCSVT **14**, 4–20 (2004)
2. Mulligan, J.B.: A software-based eye tracking system for the study of air-traffic displays. In: ETRA, ACM, pp. 69–76 (2002)
3. Cootes, T.F., Edwards, G.J., Taylor, C.J.: Active appearance models. TPAMI **23**, 681–685 (2001)
4. Martins, P., Caseiro, R., Henriques, J.F., Batista, J.: Discriminative bayesian active shape models. In: Fitzgibbon, A., Lazebnik, S., Perona, P., Sato, Y., Schmid, C. (eds.) ECCV 2012, Part III. LNCS, vol. 7574, pp. 57–70. Springer, Heidelberg (2012)
5. Zhou, F., Brandt, J., Lin, Z.: Exemplar-based graph matching for robust facial landmark localization. In: ICCV, IEEE, pp. 1025–1032 (2013)
6. Dollár, P., Welinder, P., Perona, P.: Cascaded pose regression. In: CVPR, IEEE, pp. 1078–1085 (2010)

7. Dantone, M., Gall, J., Fanelli, G., Van Gool, L.: Real-time facial feature detection using conditional regression forests. In: CVPR, pp. 2578–2585. IEEE (2012)
8. Cao, X., Wei, Y., Wen, F., Sun, J.: Face alignment by explicit shape regression. Int. J. Comput. Vis. **107**, 177–190 (2014)
9. Ren, S., Cao, X., Wei, Y., Sun, J.: Face alignment at 3000 fps via regressing local binary features. In: CVPR, pp. 1685–1692. IEEE (2014)
10. Cootes, T.F., Taylor, C.J., Cooper, D.H., Graham, J.: Active shape models-their training and application. Comput. Vis. Image Underst. **61**, 38–59 (1995)
11. Arashloo, S.R., Kittler, J., Christmas, W.J.: Facial feature localization using graph matching with higher order statistical shape priors and global optimization. In: BTAS, pp. 1–8. IEEE (2010)
12. Burgos-Artizzu, X.P., Perona, P., Dollár, P.: Robust face landmark estimation under occlusion. In: ICCV, pp. 1513–1520. IEEE (2013)
13. Walsh, B.: Markov chain monte carlo and gibbs sampling (2004)
14. Bishop, C.M., et al.: Pattern Recognition and Machine Learning, vol. 4. Springer, New York (2006)
15. Komodakis, N., Paragios, N., Tziritas, G.: MRF energy minimization and beyond via dual decomposition. TPAMI **33**, 531–552 (2011)
16. Müller, O., Yang, M.Y., Rosenhahn, B.: Slice sampling particle belief propagation. In: ICCV, pp. 1129–1136. IEEE (2013)
17. Walker, A.J.: An efficient method for generating discrete random variables with general distributions. TOMS **3**, 253–256 (1977)
18. Yuille, A.L., Hallinan, P.W., Cohen, D.S.: Feature extraction from faces using deformable templates. Int. J. Comput. Vis. **8**, 99–111 (1992)
19. Berg, A.C., Malik, J.: Geometric blur for template matching. In: CVPR, pp. 607–614. IEEE (2001)
20. Albiol, A., Monzo, D., Martin, A., Sastre, J., Albiol, A.: Face recognition using hog-ebgm. Pattern Recogn. Lett. **29**, 1537–1543 (2008)
21. Dalal, N., Triggs, B.: Histograms of oriented gradients for human detection. In: CVPR, pp. 886–893. IEEE (2005)
22. Fan, R.E., Chang, K.W., Hsieh, C.J., Wang, X.R., Lin, C.J.: Liblinear: a library for large linear classification. J. Mach. Learn. Res. **9**, 1871–1874 (2008)
23. Bromiley, P.: Products and convolutions of gaussian distributions. Medical School, Univ. Manchester, Manchester, UK, Technical report 3 (2003)
24. Genz, A.: Numerical computation of rectangular bivariate and trivariate normal and t probabilities. Stat. Comput. **14**, 251–260 (2004)
25. Martins, P., Caseiro, R., Batista, J.: Non-parametric bayesian constrained local models. In: CVPR, pp. 1797–1804. IEEE (2014)
26. Nordström, M.M., Larsen, M., Sierakowski, J., Stegmann, M.B.: The IMM face database - an annotated dataset of 240 face images. Technical report, Informatics and Mathematical Modelling, Technical University of Denmark, DTU, Richard Petersens Plads, Building 321, DK-2800 Kgs. Lyngby (2004)
27. Brooks, S.P., Gelman, A.: General methods for monitoring convergence of iterative simulations. J. Comput. Graph. Stat. **7**, 434–455 (1998)

Recognition

Off-the-Shelf CNN Features
for Fine-Grained Classification of Vessels
in a Maritime Environment

Fouad Bousetouane[(⊠)] and Brendan Morris

Electrical and Computer Engineering Department,
University of Nevada, Las Vegas, NV 89154, USA
{fouad.bousetouane,brendan.morris}@unlv.edu
http://rtis.oit.unlv.edu/

Abstract. Convolutional Neural Networks (CNNs) have recently achieved spectacular performance on standard image classification benchmarks. Moreover, CNNs trained using large datasets such as ImageNet
have performed effectively even on other recognition tasks and have
been used as generic feature extraction tool for off-the-shelf classifiers.
This paper, presents an experimental study to investigate the ability of
off-the-shelf CNN features catch discriminative details of maritime vessels for fine-grained classification. An off-the-shelf classification scheme
utilizing a linear support vector machine is applied to the high-level
convolution features that come before fully connected layers in popular deep learning architectures. Extensive experimental evaluation compared OverFeat, GoogLeNet, VGG, and AlexNet architectures for feature
extraction. Results showed that OverFeat features outperform the other
architectures with a $mAP = 0.7021$ on the nine class fine-grained problem
which was almost 0.02 better than its closest competitor, GoogLeNet,
which performed best on smaller vessel types.

1 Introduction

Fine grained classification (FGC) is a fundamental operation in computer vision.
This operation is defined as the task of distinguishing sub-ordinate categories of
the same class such as boats, aircraft, bird or car models [1,2]. The major challenge of FGC is object description with relevant and discriminative appearance
features to differentiate fine details in appearance between visually very similar object class. Generating powerful and discriminative features to catch subtle
appearance differences between subcategories is a challenging problem and is
currently of particular importance within the computer vision community.

Recently, a sequence of results has demonstrated that off-the-shelf features
generated from CNNs have powerful interclass discrimination [3,4] in contrast to
hand-engineered features (e.g., SIFT [5] or HOG [6]). The major goal of this work
is to present an experimental study to investigate the ability of generic CNN-
features extracted from state-of-art topologies to catch fine appearance details
in a maritime environment for vessel FCG. An off-the-shelf classification scheme

© Springer International Publishing Switzerland 2015
G. Bebis et al. (Eds.): ISVC 2015, Part II, LNCS 9475, pp. 379–388, 2015.
DOI: 10.1007/978-3-319-27863-6_35

is adopted (Fig. 1) where features are extracted from the last convolutional layer of popular CNN architectures and classified using a one-vs-all support vector machine (SVM) with linear kernel. This study can help better understand CNN topology effects for fine-grained classification, especially when objects are small and there is high intra-class appearance and shape variability as found in maritime vessel data from the Annapolis Harbor.

The overall structure of the paper is organized as follows: Sect. 2 is dedicated to description of background information and related work. Section 3 presents the off-the-shelf classification scheme. Experimental comparison with various state-of-the-art CNN architectures is provided in Sect. 4 for the Annapolis Harbor with a discussion of results and concluding remarks are provided in Sect. 5.

2   Background

Over last 10 years, researchers have developed a wide spectrum of different hand-engineered descriptors for object classification which encode shape and structure. Dalal and Triggs proposed an object detector based on histogram of oriented gradients (HOG) [6]. This detector set the standard for pedestrian detection and was the winner of the PASCAL object detection challenge in 2006. HOG features then become popular and widely used in different visual recognitions tasks. The HOG-based deformable parts model (DPM) provided more fine-level description of an object and better handled appearance variation [7]. Several approaches based on DPM have been proposed for multi-part localization and object-part description in fine-grained domains [8].

In 2012, Chai et al. [9] showed that the Fisher vector was more well suited for describing the appearance of an object's parts than HOG for FGC. Intensity based Local features, such as SIFT [5,11] and kernel based descriptors [10,12,13] have been proven as effective descriptors to catch low-level appearance details of objects, such as color, edge, or context. Engineered features have been well exploited for FGC through sophisticated machine learning models, such as Exemplar-SVM (E-SVM) [8] and latent SVM (L-SVM) with deformable part models [14]. However, comparison between HOG, E-SVM, and L-SVM showed significant room for improvement for maritime vessel classification [15].

More recently, many records of computer vision challenges have been shattered using deep CNNs such as large-scale image/video classification [3]. Some vision researchers believe that the spectacular results obtained by CNNs on many visual recognition benchmarks will likely lead to a complete replacement engineered features for challenging visual recognition tasks, such as fine-grained recognition, classification, and detection [3,4]. The most popular deep CNN networks are the OverFeat network proposed by LeCun et al. [16], AlexNet network proposed by Krizhevsky et al. in [17], GoogLeNet [18] and VGG [19]. These CNN networks all had impressive classification results on the ImageNet Large Scale Visual Recognition Challenge. Recently, Razavian et al. [3] have report that simple off-the-shelf CNN-features from OverFeat have obtained state-of-the-art results in many visual classification tasks on different datasets compared to more

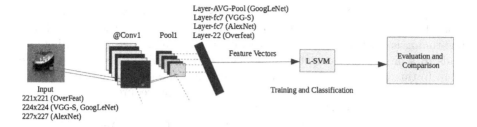

Fig. 1. Off-the-shelf Fine Grained classification scheme

complicated tuned algorithms for a particular application. A detailed comparison between different CNN architectures for FCG of maritime vessels in the Annapolis Harbor can then be used as a high baseline.

3 Off-the-Shelf FGC Scheme

In order to classify maritime vessels, an off-the-shelf FGC scheme with three stages was developed as illustrated in Fig. 1. The first stage takes an image as input to a {OverFeat, AlexNet, VGG, GoogLeNet} CNN to extract features from the highest convolutional layer. The second stage uses the feature to train multi-class linear kernel SVM classifiers with a one-vs-all approach. In the final stage, the SVM classifiers are evaluated for maritime vessel classification using receiver operating characteristic (ROC) curves and recall/precision techniques.

3.1 CNN-Generic Features Extraction

Convolution neural networks consists of many sequential convolutional/pooling layers. The sequential layers allows CNNs to learn a deep hierarchy of filters to catch the most relevant and discriminative appearance-based characteristics of objects. Hence, the power behind CNNs is their ability to extract discriminative features from pixels directly and reducing the number of connection through the sparse connectivity between neurons. Different layers correspond to a hierarchy of features; earlier layers have more low-level features. Generic CNN features are extracted at the highest level (or last convolution/pooling layer) to encode appearance relationships. In this work, OverFeat, AlexNet, VGG, and GoogLeNet are used as black-box feature extractors.

AlexNet CNN: AlexNet [17] is one of the well known CNN-topology which has been very successful on the ImageNet classification challenge. The AlexNet CNN topology consists of 8 layers, 5 convolutional layers and 3 fully connected layers. It also minimizes overfitting by using a dropout technique during training. Since training a large CNN from scratch is difficult and leads to overfitting without sufficient data, the common strategy of fine-tuning is utilized. Fine-tuning consists of the initialization of the CNN in training stage by models (i.e. weights) of the pre-trained CNN on a very large dataset (e.g. ImageNet, which contains more than 1 million images with 1000 categories). Many researchers showed that fine-tuning is very effective for CNN training after supervised pre-training and leads to a large improvement in different visual recognition tasks performance [3,4]. In this work, we follow previously used fine-tuning procedure for AlexNet after initializing with pre-trained weights [4]. This procedure consists of three main steps: 1. The CNN's classification layer (fully connect layer) should be removed, because it is specific to the pre-trained dataset. 2. Add another classification layer with the desired number of classes. 3. Run the back-propagation algorithm, gradient descent or stochastic gradient descent to learn the hierarchy of the filters. The training stage is initialized by the pre-trained model on a large scale dataset (e.g. ImageNet). To deal reduce overfitting, it is recommended to start the training with very small learning rate 0.001. In this work, AlexNet CNN is used as a blackbox for features extraction from fc7 layer after fine-tuning of the network. These generic features are used for fine-grained classification.

OverFeat CNN: OvereFeat [16] is a CNN which has two different topologies; fast-Overfeat and accurate Overfeat. The fast-Overfeat CNN topology is composed of 21 layers (Convolution,max-pooling, ReLU) and one output layer (softmax function). However, the accurate Overfeat is more deep (large) CNN network which is composed of 24 layers (Convolution,max-pooling, ReLU) and one output layer (softmax function). Unlike the AlexNet topology, pooling and rectification operations are considered as a separate layers [16]. Razavian et al. [3] reported that features off-the-shelf extracted from layer 22 of the large Over-Feat network have achieved results compared to the highly tuned state-of-the-art systems. The layer 22 of the Overfeat Network is used in this work for generic features off-the-shelf extraction. The output of this layer consists of a vector of 4096 dimensions.

GoogLeNet CNN: GoogLeNet [18] is a deep CNN. The topology of this CNN is composed of 27 layers including pooling layers. In this work, the average pooling layer (AVG-Pool) is used for generic features extraction from a network pretrained on ImageNet dataset. The output features vector of this layer is only of 1024 dimensions, which is the smallest of all the CNN architectures studied though it has the highest-level features.

VGG CNN: VGG [19] is a CNN which consists of 5 convolutional layers (conv-1 – conv-5) and three fully-connected layers (full-1 – full-3). There are 3 different topologies of VGG CNN based on the parameters of the convolution kernel, pooling and padding. In this work, layer fc7 of the pretrained S-VGG topology on ImageNet is used for generic features extraction. The output of this layer is features vector of 4096 dimensions.

3.2   Classification

It is common to use the softmax function for classification problems using convolutional neural network techniques. This function is usually used as the last layer of the CNN to produce a probability distribution over the N classes labels through the minimization of the cross-entropy loss. For example, given nine classes, the softmax layer has 9 nodes denoted by n_i, where $i = 1, \ldots, 9$. n_i is a discrete probability distribution, in which $\sum_{i=1}^{9} n_i = 1$.

However, Many researcher have reported that replacing the softmax function by a linear support vector machine improved the classification results [3,20]. Hence, minimizing a margin-based loss in the classification layer of the CNN is more effective than the cross-entropy loss or maximizing the log-likelihood. In this work, SVMs with linear kernels are used as a classifier of the CNN generic features.

SVM is one of the most popular supervised machine learning methods due to margin maximization and can be applied to classification or regression [20]. This method is originally developed for two-classes classification problems, i.e. binary classification. Let, (X_j, Y_j), $j = 1, ..., m$ denotes the training data and its corresponding class. $X_j \in \mathbb{R}^D$, $D = 4096$ dimension of the CNN-generic features vector, $Y_j \in \{\pm 1\}$. The unconstrained optimization problem of the binary SVM is defined by the following equation:

$$\min_{\mathbf{w}} \frac{1}{2} \|\mathbf{w}\|^2 + C \sum_j \max(1 - y_j \mathbf{w}^T \mathbf{x}_j, 0) \tag{1}$$

Further information about the SVM with different norms for CNN classification can be found in this publication [20]. To perform multi-class classification, in this work we have used the extended SVMs version for multi-class problems. This version is based on the $one - vs - rest$ or $one - vs - all$ approach. For an N class problem, N SVMs with linear kernel are trained independently, where the negative data of one class is composed of positive data of other classes. In our case $N = 9$ and we have trained 9 SVMs with linear kernel classifiers from generic CNN features using cross validation to obtain the best SVM parameters.

4   Experimental Setup

For all the experiments, generic off-the-shelf CNN features are obtained from various publicly available CNN architectures. Training features are used to learn linear SVM classifiers for recognition of maritime vessels.

4.1   Annapolis Dataset Overview

The Annapolis-Maritime vessel dataset consists of a collection of wide field-of-view videos captured by high-resolution (2032×1072) video camera. Images were grabbed from a camera overlooking the Annapolis Harbor at 1 Hz over the course of a week from 19:40 Friday August 13, 2010 through 03:00 Saturday August 21,

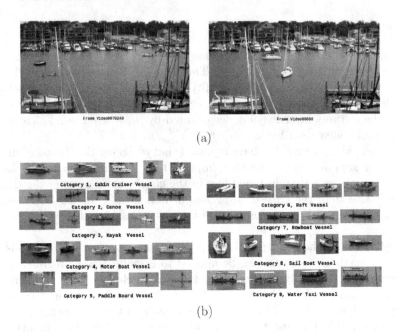

Fig. 2. Annapolis Maritime Vessel Dataset. (a) Example of wide view of Annapolis Harbor. (b) Example of images of annotated Annapolis Maritime vessel categories.

2010. A total of 58,365 images were obtained during the 180 h period for evaluation. Figure 2b illustrates an example of Annapolis dataset images. This dataset contains nine maritime vessel classes, which are: cabin cruiser, canoe, kayak, motorboat, paddle-board, raft, row-boat, sail-boat and water taxi. Figure 2b gives images of each of the nine maritime vessel classes. The main challenge of this dataset is wide variability in appearance of vessels from the same class (e.g. cabin cruiser). More details about the dataset can be found in [15].

4.2   Implementation Details

The CNN features are extracted from the first fully connected layer (layer 22) of the Overfeat. The second fully connected layer (layer fc7) AlexNet. Layers AVG-Pool and fc7 for GoogLeNet and VGG respectively. The feature vectors are of 4096 dimensions for Overfeat, AlexNet and VGG and 1024 dimensions for GoogLeNet. Feature vectors are normalized using the L2-norm [20]. All the networks are pre-trained versions that used the ImageNet dataset. To evaluate the effectiveness of fine-tuning on performance, the AlexNet was fine-tuned (AlexNet-FT) using images from the Annapolis dataset using the procedure described in [20].

Before the generation of generic CNN features, the annotated Annapolis dataset images were re-sized to fit Overfeat kernel (221 × 221), AlexNet CNN kernel (227 × 227), GoogLeNet kernel (224 × 224) and VGG kernel (224 × 224).

To enrich the training and test samples for each Annapolis-maritime vessel's category we performed different transformations on the images: rotation with three degrees ($15°, 30°$ and $45°$). These transformations allow also the evaluation of the intra-class invariance ability of the generic CNN features. The training dataset for each Annapolis vessel category was split into two tests: the SVM classifier is trained using 70 % of the training observations and then evaluated on the remaining 30 %. The implementation is derived from the publicly available GPU-version of the Caffe toolbox [21] for AlexNet, VGG, and GoogLeNet. Off-the-shelf features are extracted from the OverFeat network using the publicly available precomputed GPU-version [16]. The LibSVM toolbox is used to ensure training and classification using linear SVM.

4.3   Results and Discussion

The performance of the off-the shelf classification scheme is evaluated using ROC and recall/precision evaluation techniques. Technically, for each trained CNN-SVMs classifier results we have computed the Area Under Curve (AUC) from ROC curves and average precision (AP) from recall/precision curves. The mean average precision (mAP) metric is also computed. The evaluation considered ImageNet pre-trained models for OverFeat, GoogLeNet, VGG, and AlexNet as well as a fine-tuned version of AlexNet-FT.

Table 1 provides a comparison between the CNN architectures for classification with the AUC metric. OverFeat and GoogLeNet have the best performance. Overfeat was the best performing for six of the nine categories with GoogLeNet first in the remaining three. Example ROC curves for four of the Annapolis vessels types with various shape and size are plotted in Fig. 3(a)–(d). The general trend shows that OverFeat, GoogLeNet, and VGG tend to perform better than the AlexNet variants. This is reinforced by the mAUC scores in the 0.7 for those and much less for AlexNet. Yet, all of the classifier systems perform effectively for the nine class problem.

Results in Table 2 gives a comparison using the Pascal VOC evaluation criteria of AP. The associated precision/recall curves are shown in Fig. 3(e)–(f). OverFeat has the top AP score for five of the nine classes and two each belong with GoogLeNet and VGG. Again, the performance of OverFeat, GoogLeNet, and VGG tend to out pace AlexNet. OverFeat had a large gap ($>0.3AP$) over others for the Rowboat, Kayak, and Paddle board. GoogLeNet's gap was apparent for Motorboat and VGG for Canoe.

Astonishingly, off-the-shelf OverFeat features had the highest performance of 0.7021 mAP for the Annapolis dataset by a somewhat large margin. In general, the features extracted from deeper CNNs (e.g. Overfeat, GoogLeNet, etc.) performed better than less-deep (e.g. AlexNet). These results confirm the intuition that deeper CNN architecture results in more discriminative features and are better suited for the FGC task. It is noted that while CNNs are effective for the Annapolis dataset, fine-tuning of the AlexNet resulted in a significant improvement in mAUC and mAP suggesting that off-the-shelf is a good baseline but adaption is required for a complete implementation.

Table 1. Area Under Curve Results

CNN	Cabin Cruiser	Motorboat	Sailboat	Canoe	Paddle Board	Kayak	Raft	Rowboat	Water Taxi	mAUC
OverFeat	**0.9620**	0.8454	0.8439	**05573**	**0.4616**	0.8149	**0.7324**	**0.6562**	**0.9272**	**0.7556**
GoogLeNet	0.9195	**0.8499**	**0.8644**	0.5516	0.4130	**0.8786**	0.6925	0.6315	0.8993	0.7444
VGG	0.9350	0.7879	0.8387	0.5490	0.4171	0.7743	0.6132	0.5895	0.9002	0.7116
AlexNet	0.6278	0.5264	0.5045	0.3014	0.2379	0.4658	0.4255	0.3820	0.6065	0.4530
AlexNet-FT	0.7844	0.6920	0.6436	0.4447	0.3775	0.7292	0.5609	0.5609	0.7311	0.6138

Table 2. Average Precision Results

CNN	Cabin Cruiser	Motorboat	Sailboat	Canoe	Paddle Board	Kayak	Raft	Rowboat	Water Taxi	mAP
OverFeat	0.9396	0.6647	0.8488	0.4734	**0.4387**	**0.7841**	**0.6895**	**0.5726**	**0.8832**	**0.7021**
GoogLeNet	**0.9512**	**0.7862**	0.8287	0.4816	0.3714	0.7397	0.6580	0.5085	0.8710	0.6884
VGG	0.9178	0.6893	**0.8552**	**0.5952**	0.2844	0.7318	0.6052	0.4511	0.8573	0.6652
AlexNet	0.5888	0.4567	0.6041	0.3152	0.2649	0.4807	0.3980	0.3833	0.6168	0.4565
AlexNet-FT	0.7047	0.5370	0.6910	0.3714	0.3044	0.5838	0.4348	0.3833	0.7328	0.5270

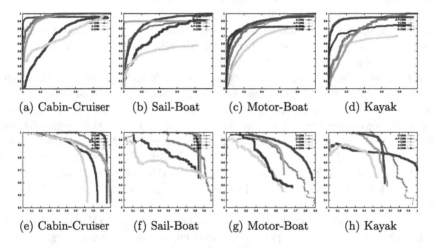

(a) Cabin-Cruiser   (b) Sail-Boat   (c) Motor-Boat   (d) Kayak

(e) Cabin-Cruiser   (f) Sail-Boat   (g) Motor-Boat   (h) Kayak

Fig. 3. (a)–(d) are Roc curves of the classification results of the trained CNNs-SVMs of four Annapolis categories. (e)–(f) are Recall/Precision curves of the classification results of the trained CNNs-SVMs of four Annapolis categories. The dashed coloured lines are labeled as follow: Red for the pretrained Overfeat-SVMs, Blue for the pretrained GoogLeNet-SVMs, Green for the pretrained VGG-SVMs, Black for the fine-tuned AlexNet-SVMs and finally yellow for the pre-trained AlexNet-SVMs (Color figure online).

5   Conclusion

In this paper, we have presented an experimental study to investigate the ability of off-the-shelf CNN features to catch details of maritime vessels for fine-grained classification. CNN features were extracted from the popular CNN-topologies and classified using simple linear SVM classifiers. We found that OverFeat features off-the-shelf outperform other state-of-art CNN architectures for the Annapolis-maritime dataset. Thus, it can be concluded that CNN off-the-shelf features are very powerful appearance based object description in the fine grained domains.

Acknowledgments. Thanks to ONR 311 and NRL for supporting this research.

References

1. Krause, J., Gebru, T., Deng, J., Li, L.J., Fei-Fei, L.: Learning features and parts for fine-grained recognition. In: 2014 22nd International Conference on Pattern Recognition (ICPR), pp. 26–33. IEEE (2014)
2. Xiao, T., Xu, Y., Yang, K., Zhang, J., Peng, Y., Zhang, Z.: The application of two-level attention models in deep convolutional neural network for fine-grained image classification (2014). arXiv preprint arXiv:1411.6447
3. Razavian, A.S., Azizpour, H., Sullivan, J., Carlsson, S.: Cnn features off-the-shelf: an astounding baseline for recognition. In: 2014 IEEE Conference on Computer Vision and Pattern Recognition Workshops (CVPRW), pp. 512–519. IEEE (2014)
4. Agrawal, P., Girshick, R., Malik, J.: Analyzing the performance of multilayer neural networks for object recognition. In: Fleet, D., Pajdla, T., Schiele, B., Tuytelaars, T. (eds.) ECCV 2014, Part VII. LNCS, vol. 8695, pp. 329–344. Springer, Heidelberg (2014)
5. Yang, S., Bo, L., Wang, J., Shapiro, L.G.: Unsupervised template learning for fine-grained object recognition. In: Advances in Neural Information Processing Systems, pp. 3122–3130 (2012)
6. Dalal, N., Triggs, B.: Histograms of oriented gradients for human detection. In: CVPR 2005, IEEE Computer Society Conference on Computer Vision and Pattern Recognition, vol. 1, pp. 886–893. IEEE (2005)
7. Felzenszwalb, P.F., Girshick, R.B., McAllester, D., Ramanan, D.: Object detection with discriminatively trained part-based models. IEEE Trans. Pattern Anal. Mach. Intell. **32**, 1627–1645 (2010)
8. Zhang, N., Farrell, R., Iandola, F., Darrell, T.: Deformable part descriptors for fine-grained recognition and attribute prediction. In: 2013 IEEE International Conference on Computer Vision (ICCV), pp. 729–736. IEEE (2013)
9. Chai, Y., Rahtu, E., Lempitsky, V., Van Gool, L., Zisserman, A.: TriCoS: a tri-level class-discriminative co-segmentation method for image classification. In: Fitzgibbon, A., Lazebnik, S., Perona, P., Sato, Y., Schmid, C. (eds.) ECCV 2012, Part I. LNCS, vol. 7572, pp. 794–807. Springer, Heidelberg (2012)
10. Bousetouane, F., Dib, L., Snoussi, H.: Improved mean shift integrating texture and color features for robust real time object tracking. Visual Comput. **29**, 155–170 (2013)
11. Shirazi, M.S., Morris, B.: Contextual combination of appearance and motion for intersection videos with vehicles and pedestrians. In: Bebis, G., et al. (eds.) ISVC 2014, Part I. LNCS, vol. 8887, pp. 708–717. Springer, Heidelberg (2014)
12. Bo, L., Ren, X., Fox, D.: Kernel descriptors for visual recognition. In: Advances in Neural Information Processing Systems, pp. 244–252 (2010)

13. Bousetouane, F., Vandewiele, F., Motamed, C.: Occlusion management in distributed multi-object tracking for visual-surveillance. Pattern Recognition and Image Analysis **25**, 295–300 (2015)
14. Duan, K., Parikh, D., Crandall, D., Grauman, K.: Discovering localized attributes for fine-grained recognition. In: 2012 IEEE Conference on Computer Vision and Pattern Recognition (CVPR), pp. 3474–3481. IEEE (2012)
15. Chua, M., Aha, D.W., Auslander, B., Gupta, K.M., Morris, B.: Comparison of object detection algorithms on maritime vessels (2014)
16. Sermanet, P., Eigen, D., Zhang, X., Mathieu, M., Fergus, R., LeCun, Y.: Overfeat: Integrated recognition, localization and detection using convolutional networks (2013). arXiv preprint arXiv:1312.6229
17. Krizhevsky, A., Sutskever, I., Hinton, G.E.: Imagenet classification with deep convolutional neural networks. In: Advances in neural information processing systems, pp. 1097–1105 (2012)
18. Szegedy, C., Liu, W., Jia, Y., Sermanet, P., Reed, S., Anguelov, D., Erhan, D., Vanhoucke, V., Rabinovich, A.: Going deeper with convolutions (2014). arXiv preprint arXiv:1409.4842
19. Chatfield, K., Simonyan, K., Vedaldi, A., Zisserman, A.: Return of the devil in the details: Delving deep into convolutional nets (2014). arXiv preprint arXiv:1405.3531
20. Tang, Y.: Deep learning using linear support vector machines (2013). arXiv preprint arXiv:1306.0239
21. Jia, Y., Shelhamer, E., Donahue, J., Karayev, S., Long, J., Girshick, R., Guadarrama, S., Darrell, T.: Caffe: convolutional architecture for fast feature embedding (2014). arXiv preprint arXiv:1408.5093

Joint Visual Phrase Detection
to Boost Scene Parsing

Keke Tang[(⊠)], Zhe Zhao, and Xiaoping Chen

University of Science and Technology of China, Hefei, China
{kktang,zhaozhe}@mail.ustc.edu.cn, xpchen@ustc.edu.cn

Abstract. Scene parsing is a very challenging problem which attracts increasing interests in many fields such as computer vision and robotics. However, occluded or small objects which are difficult to parse are always ignored. To deal with these two problems, we integrate visual phrase into our joint system, which has been proved to have good performance on describing relationships between objects. In this paper, we propose a joint model which integrates scene classification, object and visual phrase detection, as well as scene parsing together. By encoding them into a Conditional Random Field model, all tasks mentioned above could be solved jointly. We evaluate our method on the MSRC-21 dataset. The experimental results demonstrate that our method achieves comparable and on some occasions even superior performance with respect to state-of-the-art joint methods especially when there exist partially occluded or small objects.

1 Introduction

Scene parsing, or image labeling, whose goal is to label each pixel in the image with its semantic category, is a very important problem in computer vision. Many applications such as automatic driver assistant, task planning of a robot in an unknown environment, all benefit from it.

There exist diverse scene parsing methods in the literature. [1,2] propose using nonparametric approaches. To parse an image, these methods first retrieve a small set of similar images from the pre-labeled database and then transfer their associated labels in pixels or superpixels. These data-driven approaches do not do any training at all, thus are suitable for open-universe datasets. However, these methods always fail when there are no similar images in the pre-labeled sets. Other approaches for scene parsing always begin by specifying many pre-trained category models using kinds of discriminative features [3–5]. After getting the evidence of semantic label for a pixel or superpixel, they formulate the problem as inference in a conditional random field or markov random field to enforce neighboring pixels (superpixels) smooth. To leverage the benefits of both nonparametric and parametric methods, [6] tries to integrate the nonparametric method with per-exemplar sliding window detectors.

With significant progress in tasks such as object detection [8], scene classification [9], and scene parsing [3], many researchers try to joint them together [7,10].

© Springer International Publishing Switzerland 2015
G. Bebis et al. (Eds.): ISVC 2015, Part II, LNCS 9475, pp. 389–399, 2015.
DOI: 10.1007/978-3-319-27863-6_36

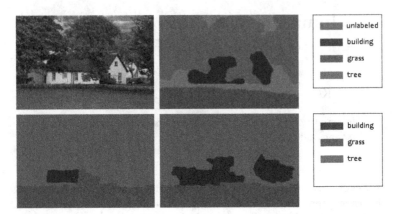

Fig. 1. Top left: a query image, top right: the human annotation, button left: the result of [7] which tends to miss the building partially occluded by trees, button right: with the cue of building-tree visual phrase, our approach correctly segments the occluded building.

Knowledge of other tasks is typically incorporated as context evidence, encouraging the result of scene parsing to agree with other tasks. These joint systems have been proved to work much better than single ones.

However, most conventional approaches only focus on clear scene with a few salient objects, such as a cow on the grass, a chair on the floor. Actually, the scene in the real world is more complex. We could see occlusion everywhere. For example, tables and chairs are always placed nearby and several cows may group together. It's hard for segmentation and object detection algorithms to deal with them. In addition, many classes in relatively small sizes are also ignored [11].

To solve these two problems, we use visual phrases. [12] demonstrates that "visual phrases often display significant reduced visual complexity". When a building is partially occluded by trees (see Fig. 1), it is much easier to detect the visual phrase "a building with trees around it" other than to detect these two classes individually. On the other hand, the size of a small object that is grouped with others should be larger than the individual one. To the best of our knowledge, this is the first work that tries to solve these two problems together.

Visual phrase or object group, attracts many researchers in recent years. [13] builds a 4-dimensional transform space to discover visual phrases. [14] tries to use the Minimum Description Length principle to discover high-level object groups. They add this information either to boost scene categorization or object detection. However, few of them use the information of visual phrase to boost scene parsing. As [7] demonstrates that scene categorization and object detection could boost the performance of scene parsing. We are interested in knowing whether visual phrase detection could indirectly boost scene parsing.

In this paper, we propose a novel joint system which integrates scene classification, object detection, visual phrase detection, scene parsing together. Probably the most similar approach to ours is [15], which uses 3D geometric phrases to reason about the semantic and geometric relationships between objects in

Fig. 2. The illustration of some visual phrases generated by our method. Three phrases in the above are pruned due to low occurrence frequency.

the indoor scene. We frame the joint system as a structure prediction problem in a graphical model defined over different constituent tasks. Beyond the joint system, we propose a simple method of retrieving and training a set of visual phrase models. We demonstrate our joint system on the MSRC-21 dataset [16]. The results show that the proposed algorithm achieves comparable and on some occasions even superior scene parsing performance than state-of-the-art algorithms, which suggests that visual phrases can truly boost the performance of scene parsing.

2   Generating Visual Phrases

Choosing Visual Phrases. Visual phrases choosing is a critical step for our joint system. A good visual phrase should reveal the high-level relationship between constituent objects. In this paper, we simply group nearby objects together to compose visual phrase candidates. For each object in the image, we slightly enlarge its bounding box. Then we exhaustively search all other bounding boxes in this image. If there is an overlap, we group them. For example, a man sitting on a chair can compose a man-chair two-words visual phrase. In this way, we generate a large amount of candidates.

Pruning Visual Phrases. In the previous step, we get a lot of visual phrase candidates. It would cost too much time for training all of them. On the other hand, some phrases may only occur on very exceptional occasions. We calculate the frequency of each visual phrase $Freq(P_i)$ and the frequency of constituent words $Freq(W_j^{P_i})$. Any visual phrase will be pruned if $Freq(P_i)/\prod Freq(W_j^{P_i}) < th$, where P_i is the i-th visual phrase, $W_j^{P_i}$ is the

j-th constitute word of P_i, and th is the threshold set by us. Figure 2 shows some examples of both origin and pruned visual phrases.

Training Visual Phrases. To train the appearance model of visual phrases, we follow the per-exemplar framework of [17]. As the limited training data in the MSRC-21 dataset and high intra-class variation, per-exemplar detectors are better suited for our task than traditional SVM methods. We learn these models for each of the generated phrases using estimated bounding box which covers most parts of its constituent words. For each visual phrase, all training images without any constituent words of it are regarded as negative examples. We then do hard negative mining over these hard negatives. Due to the limited size of the MSRC-21 dataset, we do not calibrate our detectors.

3   Our Joint Model

In this section, we describe our joint model of scene parsing. We address the multi-class scene parsing problem from a holistic perspective, in which we jointly assign a category label to every pixel, infer the type of scene and class occurrence, as well as locating the objects and visual phrases. Our main focus is to explore the consistency within each task told above.

To this end, we formulate the problem as one of Conditional Random Field (CRF). The random fields contain variables representing the preliminary results of constituent tasks. Figure 3 gives an overview of our model.

Following [7], we use two-level hierarchy segments. Formally, let $sp_i \in \{1, \cdots, C\}$ be a random variable representing the category class of the i-th segment in the low hierarchy while let $ss_j \in \{1, \cdots, C\}$ be a random variable representing the category class of the j-th segment in the high hierarchy. For objects, we use Deformable Part Model [8] to generate candidates of class c_k.

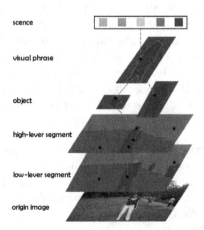

Fig. 3. The illustration of our model.

For visual phrases, we use Examplar-SVM to generate candidates p_k. We also add an auxiliary variable $z_k \in \{1, 0\}$ to indicate whether class k exists in the image. For the scene, we use the variable $s \in \{1, \cdots, S\}$ to represent it.

A connected CRF model is defined as the sum of unary and pairwise potentials over cliques of the same or different tasks. The energy function is as follows

$$E(sp, ss, c, p, s, z) = \underbrace{\sum_{v \in V} \omega_v \psi_v}_{\text{Unary Potentials}} + \underbrace{\sum_{(p,q) \in E} \omega_q^p \phi_q^p}_{\text{Pairwise Potentials}} \qquad (1)$$

The first term in Eq. 1 represents all the unary potentials: potentials for two hierarchy segments, object detection, visual phrase detection, scene classification and class occurrence. The other term represents all the pairwise potentials: potentials between two level segments, potentials between object detection and high-level segments, and so on. Each potential is associated with a weight shared across cliques, which can be learned from training data by the method proposed in [18]. Then we describe the potentials in details.

3.1 Unary Potentials

In this section, we describe the potentials over five constituent tasks of our joint model.

Unary potentials for two hierarchy segments. We use the method of [19] with two different thresholds to generate a fine segmentation sp and a coarse one ss. Then we average the pixel potentials calculated by TextonBoost [5] in each region as its potential.

Unary potential for object detection. For each detection, we define the unary potential using the score of the object detectors

$$\psi^{class}(c) = -\log(\sigma(r_c - thres_c)) \qquad (2)$$

where $thres_c$ is the threshold of class c, r_c is the detector's score, and σ is a logistic function.

Unary potential for class occurrence. We use the class occurrence statistics calculated from the training data as the unary potential.

Unary potential for visual phrase detection. Like object detection, the unary potential over visual phrase detection is as follows

$$\psi^{phrase}(p) = -\log(\sigma(r_p - thres_p)) \qquad (3)$$

where $thres_p$ is the threshold of visual phrase p, r_p is the detector's score, and σ is also a logistic function.

Unary potential for scene information. In order to get global information about the scene, we compute the global feature and train a SVM classifier follow the method described in [9]. The unary potential over the scene is defined as

$$\psi^{scene}(s) = -\log(\sigma(r_s)) \tag{4}$$

where r_s is the score of the scene classifier for the scene label s and σ is again the logistic function.

3.2   Pairwise Potentials

In this section, we will describe all pairwise potentials used in our joint model. All $\beta_i (i \in 1, 2, 3...)$ in this section are positive numbers which could be implicitly learned during training.

Pairwise potential between two-level hierarchy segments. Following [20], we use the p^n potentials, which can be written as

$$\phi_{ss}^{sp}(i,j) = \begin{cases} \beta_1 \ if \ sp_i \neq ss_j \\ 0 \ otherwise \end{cases} \tag{5}$$

Pairwise potential between object detection and high-level segment. This term penalizes the situation when the segment and the object detection in the same region share different labels.

$$\phi_c^{ss}(i,j) = \begin{cases} \beta_2 \ if \ ss_i \neq c_j \\ 0 \ otherwise \end{cases} \tag{6}$$

Pairwise potential between class occurrence and high-level segment. This term penalizes the situation when segment is labelled as class k while k does not exist in this scene.

$$\phi_z^{ss}(i,k) = \begin{cases} \beta_3 \ if \ z_k = 0 \wedge ss_i = k \\ 0 \ otherwise \end{cases} \tag{7}$$

Pairwise potential between the detection of class and visual phrase. This potential is designed to enforce the consistency between the detection of classes and visual phrases. For example, the detection of class head could boost the detection of visual phrase head-body around it and the detection of visual phrase head-body could also boost the detection of class head inside it. We define the potential as

$$\phi_p^c(i,j) = \begin{cases} -\beta_4 \ if \ c_i \in p_j \wedge Overlap(c_i, p_j) > \epsilon \\ 0 \ otherwise \end{cases} \tag{8}$$

where $Overlap(c, p) = A(c \cap p)/A(c)$, ϵ is the threshold and A is the area function.

Pairwise potential between class occurrence and scene. The pairwise potential between class occurrence and scene is defined as

$$\phi_s^z(i,j) = \begin{cases} -\log(Cooc(z_i, s_j)) \ if \ z_i = 1 \wedge Cooc(z_i, s_j) > 0 \\ \beta_5 \ if \ z_i = 1 \wedge Cooc(z_i, s_j) = 0 \\ 0 \ otherwise \end{cases} \tag{9}$$

where $Cooc(z_i, s_j) = (p(z_i|s_j) + p(s_j|z_i))/2$, which represents the probability of occurrence of class i in scene j.

Pairwise potential of class co-occurrence. Like Eq. 9, the pairwise potential between classes is defined as

$$\phi_z^z(i,j) = \begin{cases} -\log(Cooc(z_i, z_j)) \ if \ Cooc(z_i, z_j) > 0 \\ \qquad\qquad 0 \ otherwise \end{cases} \tag{10}$$

where $Cooc(z_i, z_j)$ represents the probability of occurrence of class i and class j.

Pairwise potential between class occurrence and object detection. This term makes a penalty if we find some candidates of a class while the class does not exist in the scene.

$$\phi_z^c(i,k) = \begin{cases} \beta_6 \ if \ c_i = k \wedge z_k = 0 \\ \ 0 \ otherwise \end{cases} \tag{11}$$

3.3 Learning and Inference

Our energy is defined in terms of a linear model. Following [7], we use the loss function considering all constitute tasks. As this is a structure learning problem, all parameters could be learned using the primal dual learning framework of [18]. For inference, we compute the minimal energy by computing

$$\min_{sp,ss,c,p,s,z} E(sp, ss, c, p, s, z) \tag{12}$$

which can be approximated efficiently by the algorithm of [21].

4 Experiment Results

4.1 Experiment Details

We test our joint system on the task of scene parsing on the MSRC-21 dataset with bounding boxes of 17 classes and scene annotations provided by [7]. After pruning, we get 10 visual phrases, such as building-tree, body-face. The dataset contains 591 images from 21 different categories. Following the standard train/test spilt [16], the dataset is divided into a training set of 335 images and a testing set of 256 images. Our results are tested on the standard error measure [3].

To get the two-layer segments, we employ the method of [19] on each image, which produces some regions according to the threshold of the boundary. Balancing efficiency and accuracy, we follow the thresholds used in [7], and get average 65 and 19 regions for the two-layers in the hierarchy. We then use the latest version of Deformable Part Model [8] to train object models. In the period of detection, we slightly lower the threshold to produce over-detections. Higher threshold may lose some objects, while lower threshold may produce too many false detections. Both situations may affect our performance. We test different thresholds and find that lowering the threshold $thres_c$ about 0.1 always produces a fairly good result. As our visual phrase detectors have no calibration, we simply set $thres_p = -1.0$ as their thresholds.

Table 1. Scene parsing results on the MSRC-21 dataset

	building	grass	tree	cow	sheep	sky	aeroplane	water	face	car	bicycle	flower	sign	bird	book	chair	road	cat	dog	body	boat	average	global
Dense CRF [23]	**75**	**99**	91	**84**	82	**95**	82	71	89	**90**	94	95	**77**	48	96	61	**90**	78	48	**80**	22	78.3	86.0
Holistic [7]	71	98	91	78	**86**	93	88	**84**	90	84	94	**98**	75	**53**	97	68	**90**	83	55	67	17	78.9	86.0
Ours	73	98	**93**	79	85	93	**89**	83	**94**	87	94	**98**	76	**53**	97	68	89	83	55	72	16	**79.8**	**86.7**

4.2 Baselines

To demonstrate the effectiveness of our joint model, we choose two baselines:

Dense CRF [22], which is supposed to be the best non-joint model. We simply use the results reported in their paper.

Holistic [7], which is supposed as the best joint model. We run their codes ourselves and use all 17 object detectors trained using the annotations provided by them.

4.3 Comparison on the Whole Dataset

Table 1 reports the accuracy on the MSRC-21 dataset. Comparing these results, the performance of our method is the best on both the average accuracy and the global accuracy.

Note that our approach outperforms holistic [7] on the accuracy of many classes. Some classes like boat, road, without training visual phrase detectors, fall behind just a little.

We present some qualitative results in Fig. 4. Face, due to its small size, is easily ignored. The two results on the left show the detection of body-face could boost the locating of it. The result on the right shows the detection of body-body boosts the locating of two persons standing nearby with a little occlusion. The result in Fig. 1 shows the detection of building-tree could boost the locating of partly occluded building.

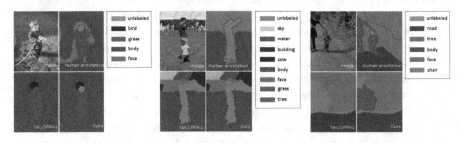

Fig. 4. Some representative scene parsing results on the MSRC-21 dataset generated by [7] and our approach.

4.4   Comparison on the Chosen Sub-Dataset

As most images in the MSRC-21 dataset are quite easy, for example, only a tree on the grass or just a cow, which makes our model not discriminative with others. In order to show the powerful improvement due to adding the information of visual phrase, we choose two visual phrases (body-face, building-tree) to demonstrate it. We manually select the images which contain the visual phrase from the test set of the MSRC-21 dataset and apply our method on them. These images are much more difficult and contain at least two or three objects in each one.

In the following, we will demonstrate the experiments on two visual phrase sub-datasets. The experiment on the body-face sub-dataset is to demonstrate the effectiveness of our method in segmenting small objects, while the experiment on the building-tree sub-dataset is to show the effectiveness in segmenting objects in occlusion.

Experiment of Body-Face Visual Phrase. To prevent over-fitting, the model is still trained on the full training set. We then manually select the images containing the body-face visual phrase from the test set, and stitch other images out. In this sub-dataset, most heads in the images are fairly small.

Table 2 shows the scene parsing results on this body-face sub-dataset. As most of the faces in this sub-dataset are in fairly small sizes, the accuracy rate of segmenting face using the method of [7] is only 13.04 %. This probably suggests that small objects are easily ignored by [7]. However, with the information of body-face visual phrase, our method achieves the accuracy of 41.51 %, obtaining nearly 30 % improvement. Besides, the accuracy of segmenting body also obtains 3 % improvement. This proves the effectiveness of our method in segmenting small objects.

Table 2. Scene parsing results on the body-face sub-dataset

	building	grass	tree	sky	water	face	flower	road	body	average	global
Holistic [7]	42.00	97.77	99.00	88.71	94.51	13.04	0.00	70.46	64.06	63.28	80.21
Ours	42.00	98.49	98.91	88.71	94.59	**41.72**	0.00	70.46	**67.55**	**66.94**	**81.67**

Experiment of Building-Tree Visual Phrase. Like the experiment of body-face visual phrase, we manually select the images containing the building-tree visual phrase. In this sub-dataset, some buildings are occluded partly by the surrounding trees.

Table 3 shows the scene parsing results on the building-tree sub-dataset. Our method outperforms the method of [7] in segmenting the occluded buildings and obtains 2 % improvement. This proves the effectiveness of our method in dealing with occlusion.

Table 3. Scene parsing results on the building-tree sub-dataset

	building	grass	tree	sky	road	body	average	global
Holistic [7]	78.60	97.28	96.19	94.97	95.02	47.30	84.90	87.94
Ours	**80.86**	97.28	**96.32**	94.97	95.02	47.30	**85.29**	**88.81**

5 Conclusion

In this paper, we firstly present a simple approach to generate visual phrases and then integrate it with scene parsing, object detection, scene classification together. With the information from visual phrases, our approach achieves comparable and on some occasions superior performance on the MSRC-21 dataset especially when there is occlusion or small objects. To the best of our knowledge, this is the first work that solves these two problems together.

In the future, we plan to integrate more information to further boost the performance of scene parsing and test on more challenging datasets.

References

1. Liu, C., Yuen, J., Torralba, A.: Nonparametric scene parsing: Label transfer via dense scene alignment. In: IEEE Conference on Computer Vision and Pattern Recognition, CVPR 2009, pp. 1972–1979. IEEE (2009)
2. Tighe, J., Lazebnik, S.: SuperParsing: scalable nonparametric image parsing with superpixels. In: Daniilidis, K., Maragos, P., Paragios, N. (eds.) ECCV 2010, Part V. LNCS, vol. 6315, pp. 352–365. Springer, Heidelberg (2010)
3. Ladický, L., Sturgess, P., Alahari, K., Russell, C., Torr, P.H.S.: What, where and how many? combining object detectors and CRFs. In: Daniilidis, K., Maragos, P., Paragios, N. (eds.) ECCV 2010, Part IV. LNCS, vol. 6314, pp. 424–437. Springer, Heidelberg (2010)
4. Ren, X., Bo, L., Fox, D.: Rgb-(d) scene labeling: Features and algorithms. In: 2012 IEEE Conference on Computer Vision and Pattern Recognition (CVPR), pp. 2759–2766. IEEE (2012)
5. Shotton, J., Winn, J., Rother, C., Criminisi, A.: Textonboost for image understanding: Multi-class object recognition and segmentation by jointly modeling texture, layout, and context. Int. J. Comput. Vis. **81**, 2–23 (2009)
6. Tighe, J., Lazebnik, S.: Finding things: Image parsing with regions and per-exemplar detectors. In: 2013 IEEE Conference on Computer Vision and Pattern Recognition (CVPR), pp. 3001–3008. IEEE (2013)
7. Yao, J., Fidler, S., Urtasun, R.: Describing the scene as a whole: Joint object detection, scene classification and semantic segmentation. In: 2012 IEEE Conference on Computer Vision and Pattern Recognition (CVPR), pp. 702–709. IEEE (2012)
8. Felzenszwalb, P., McAllester, D., Ramanan, D.: A discriminatively trained, multiscale, deformable part model. In: IEEE Conference on Computer Vision and Pattern Recognition, CVPR 2008, pp. 1–8. IEEE (2008)

9. Xiao, J., Hays, J., Ehinger, K.A., Oliva, A., Torralba, A.: Sun database: Large-scale scene recognition from abbey to zoo. In: 2010 IEEE Conference on Computer Vision and Pattern Recognition (CVPR), pp. 3485–3492. IEEE (2010)
10. Wojek, C., Schiele, B.: A dynamic conditional random field model for joint labeling of object and scene classes. In: Forsyth, D., Torr, P., Zisserman, A. (eds.) ECCV 2008, Part IV. LNCS, vol. 5305, pp. 733–747. Springer, Heidelberg (2008)
11. Yang, J., Price, B., Cohen, S., Yang, M.H.: Context driven scene parsing with attention to rare classes. In: Proceedings of the CVPR (2014)
12. Sadeghi, M.A., Farhadi, A.: Recognition using visual phrases. In: 2011 IEEE Conference on Computer Vision and Pattern Recognition (CVPR), pp. 1745–1752. IEEE (2011)
13. Li, C., Parikh, D., Chen, T.: Automatic discovery of groups of objects for scene understanding. In: 2012 IEEE Conference on Computer Vision and Pattern Recognition (CVPR), pp. 2735–2742. IEEE (2012)
14. Sadovnik, A., Chen, T.: Hierarchical object groups for scene classification. In: 2012 19th IEEE International Conference on Image Processing (ICIP), pp. 1881–1884. IEEE (2012)
15. Choi, W., Chao, Y.W., Pantofaru, C., Savarese, S.: Understanding indoor scenes using 3d geometric phrases. In: 2013 IEEE Conference on Computer Vision and Pattern Recognition (CVPR), pp. 33–40. IEEE (2013)
16. Shotton, J., Johnson, M., Cipolla, R.: Semantic texton forests for image categorization and segmentation. In: IEEE Conference on Computer Vision and Pattern Recognition, CVPR 2008, pp. 1–8. IEEE (2008)
17. Malisiewicz, T., Gupta, A., Efros, A.A.: Ensemble of exemplar-svms for object detection and beyond. In: ICCV (2011)
18. Hazan, T., Urtasun, R.: A primal-dual message-passing algorithm for approximated large scale structured prediction. In: Advances in Neural Information Processing Systems, pp. 838–846 (2010)
19. Arbelaez, P., Maire, M., Fowlkes, C., Malik, J.: Contour detection and hierarchical image segmentation. IEEE Trans. Pattern Anal. Mach. Intell. **33**, 898–916 (2011)
20. Kohli, P., Kumar, M.P., Torr, P.H.S.: P3 and beyond: Solving energies with higher order cliques. In: Proceedings of IEEE Conference on Computer Vision and Pattern Recognition (2007)
21. Schwing, A.G., Hazan, T., Pollefeys, M., Urtasun, R.: Distributed Message passing for large scale graphical models. In: Proceedings of the CVPR (2011)
22. Krähenbühl, P., Koltun, V.: Efficient inference in fully connected crfs with gaussian edge potentials. In: Advances in Neural Information Processing Systems, pp. 109–117 (2011)

If We Did Not Have ImageNet: Comparison of Fisher Encodings and Convolutional Neural Networks on Limited Training Data

Christian Hentschel[✉], Timur Pratama Wiradarma, and Harald Sack

Hasso Plattner Institute for Software Systems Engineering, Potsdam, Germany
{christian.hentschel,harald.sack}@hpi.de,
pratama.wiradarma@student.hpi.uni-potsdam.de

Abstract. This work aims to compare two competing approaches for image classification, namely Bag-of-Visual-Words (BoVW) and Convolutional Neural Networks (CNNs). Recent works have shown that CNNs (Convolutional Neural Networks) have surpassed hand-crafted feature extraction techniques in image classification problems. Their success is partly attributed to the fact that benchmarking initiatives such as ImageNet in a massive crowd sourcing effort gathered sufficient data necessary to train deep neural networks with a very large number of model parameters. Obviously, manually annotated training datasets on a similar scale cannot be provided in every classification scenario due to the massive amount of required resources and time. In this paper, we therefore analyze and compare the performance of BoVW- and CNN-based approaches for image classification as a function of the available training data. We show that CNNs benefit from growing datasets while BoVW-based classifiers outperform CNNs when only limited data is available. Evidence is given by experiments with gradually increasing training data and visualizations of the classification models.

1 Introduction

Recently, approaches for image classification based on the Bag-of-Visual-Words (BoVW) model as well as its more powerful successors (e.g. Vector of Locally Aggregated Descriptors, VLAD [1] and Fisher Vector encodings, FV [2]) have been significantly outperformed by approaches based on convolutional neural networks. The fact that BoVW encodings are largely based on handcrafted image descriptors was identified as a major drawback: Typically, a vector space representation of an image is computed by extracting local features (usually gradient based, e.g. SIFT [3]) at densely sampled image regions and summarizing these features into a global image descriptor (e.g. a histogram of vector quantized local features). Quantization of the local region descriptors (e.g. by using KMeans or Gaussian Mixture Models) is actually the only step where the features are adjusted to the training data. All other parameters (e.g. the number of bins

© Springer International Publishing Switzerland 2015
G. Bebis et al. (Eds.): ISVC 2015, Part II, LNCS 9475, pp. 400–409, 2015.
DOI: 10.1007/978-3-319-27863-6_37

and orientations in the SIFT gradient histogram) are kept fixed. Visual concept models are learned on top of these feature representations (typically linear and non-linear support vector machines are applied).

On the other hand, Convolutional Neural Networks combine several layers of non-linear feature extractors whose weights are trained directly on the image data at hand. Feature extraction and visual concept model training is performed in a single step of training one neural network. The large number of model parameters allows for a more fine-grained adjustment of the image features but also comes at the cost of increased training complexity: deep neural networks can only be reasonably trained on highly parallelized hardware (GPUs are exploited in most cases) and a large number of model parameter demands for large training data in order to avoid overfitting of the model. Especially the latter aspect represents a significant limitation: Assembly of (manually annotated) training data is considered a costly and time consuming process. On the other hand, BoVW-based approaches have shown reasonable classification accuracy even in scenarios with very little training data available.

In this paper, we therefore analyze the impact of varying training dataset size on the achieved classification performance using either BoVW and CNNs. By gradually increasing the number of available training data we are able to estimate a decision threshold based on which users can decide, which method to favor. Furthermore, we analyze the learned models by visualizing their classification accuracy in selected scenarios. The results give insights into the differences of the respective approaches in terms of adaptation to the training data.

This paper is structured as follows: In Sect. 2 we briefly review the related work. We describe the setup of our experiments, the employed BoVW descriptors as well as the architecture of the CNNs used in Sect. 3. Furthermore, we present the various training and test datasets used throughout our experiments. Section 4 provides a detailed analysis of the obtained results. Heat map visualizations computed for some of the trained models give further insights into how increased number of training images is used by the respective approach to learn the depicted concept. Finally, Sect. 5 concludes our paper and gives a short outlook to future work.

2 Related Work

In the first two years of the ImageNet Large Scale Visual Recognition Challenge (ILSVRC) [4] the leading participating teams all used Bag-of-Visual-Words derived approaches such as Fisher Vector (or the closely related Super Vector) encodings. In 2012, the authors in [5] proposed an approach based on Deep Convolutional Neural Networks that outperformed the BoVW competitors by a large margin. The neural network architecture presented has more than 60 million parameters which made training on a GPU a necessity (training on two GTX 580 GPUs took between 5 to 6 days) and which makes the approach prone to overfitting – only attenuated by the large number of training images available in ILSVRC (1.2 million images were manually assigned to 1,000 categories).

By 2014 almost all participating teams had adopted CNN-based approaches. One straightforward way of improving the performance of deep neural networks is by increasing their size – the winning team in the 2014 ILSVRC used a CNN with up to 144 million parameters [6] – which, however, likewise increases the risk of overfitting. Hence, several efforts have focused on exploring approaches that work in low training data scenarios as well.

One promising idea are CNN models *pre*-trained on a larger dataset (e.g., ILSVRC) and *fine-tuned* on the new target outputs [7–9]. Another approach uses the pen-ultimate layer of the pre-trained CNN as a powerful feature descriptor and then applies machine learning (e.g. linear Support Vector Machines) to train the target models (e.g. [10,11]). However, both approaches rely on the assumption that the data used to train the initial CNN exhibits features similar to the data that is actually supposed to be classified. In [9], Wei et al. compare both methods to a Fisher Vector based implementation and report the superior performance of the CNN approaches in a multi label classification experiment conducted on the PASCAL VOC-2007 dataset [12]. Similarly, the authors in [7] conclude that using CNNs as feature descriptors (pre-trained on ILSVRC2012 data) and SVMs as linear predictors outperform Improved Fisher Encodings when tested on the VOC and Caltech [13] datasets. Both datasets – PASCAL VOC and Caltech – show real world objects and scenes and should be considered as visually similar to the ILSVRC dataset (with some images from the Caltech datasets being also present in the ILSVRC 2012 dataset).

Using a CNN without pre-training immediately on a comparatively small dataset such as the Caltech datasets, leads to significantly worse results than BoVW-based classifiers as reported in [11], which underlines the above mentioned necessity of large datasets for CNN training. Furthermore, this makes BoVW like implementations a competitive candidate for scenarios with low amounts of training data.

In this paper, we therefore analyze the impact of incrementing training set sizes on the classification performance of FV and CNN based approaches for image classification. We directly train both approaches on the datasets, i.e. without relying on pre-trained CNN models, in order to avoid a bias induced by dataset similarities. Thus, the reported results are valid even in scenarios where the data to be classified differs strongly from the ImageNet datasets typically used to pre-train CNNs. Our assumption is that FV are better candidates when limited training data is at hand.

3   Experimental Setup

In this section we will detail the two image representation and training approaches that we compared in our experiments: linear Support Vector Machines trained on (improved) Fisher Vector encodings as well as Convolutional Neural Networks. Furthermore, we describe the dataset employed and the different experiments conducted.

3.1   Improved Fisher Encodings and Linear Predictors

In [14] the authors compared different local feature encodings in a large scale experiment and conclude the superior performance of FV encodings which we therefore adopted in our experiments as well. Consistent with the FV implementations proposed in [2], our approach starts by extracting SIFT descriptors [3] at a dense grid with a stride of 4 pixels at 7 different scales. We use the implementation provided by [15] which uses triangular feature reweighting (as opposed to Gaussian feature weighting proposed by Lowe). Following [2] we decorrelate and reduce the original feature dimensions from $d = 128$ to $d = 80$ by means of Principal Component Analysis (PCA). We further enhance the local descriptors by spatially extending the features with the (normalized) sampling point's coordinates, yielding a $d = 82$ dimensional local descriptor

A FV encoding is then obtained by first computing the Gaussian Mixture Model (GMM) with $k = 256$ components on a random subset of $n = 256,000$ local descriptors equally selected from all training images. Subsequently, each local descriptor of an image is soft-quantized using the obtained mixtures and first and second order statistics between the descriptor and its Gaussian cluster are accumulated. Finally, the *improved* version of FV (IFV, as suggested by the authors in [2]) applies signed square-rooting to the individual components of the encoding followed by a $\| \cdot \|_2$ normalization.

Usually, visual concept models are trained based on these global feature representations using Support Vector Machines ([7,14]). Our implementation learns a linear SVM per image class (using a *one-vs-rest* pattern) by minimizing the hinge loss function. While in theory the regularization C hyperparameter should be optimized using cross validation, we fix it to $C = 10$ in order to reduce training time. Empirical results in small test scenarios have shown no significant disadvantage incurred from this simplification, however, clearly this leaves room for future improvements.

3.2   Convolutional Neural Networks

The CNN-based classifiers follow the architecture as proposed by Krizhevsky, et al. in [5] with some minor modifications. These modifications address the sequence of pooling and normalization layers (compared to the original model we flip the order, i.e. pooling is applied before normalization) and mainly help to speed up the forward run without sacrificing the accuracy. The remainder of the architecture is left unchanged: The network consists of five convolutional layers activated by a Rectified Linear Unit (ReLU) and followed by a max pooling layer (applied to $1^{st}, 2^{nd}$ and 5th convolutional layer). Local Response Normalization is applied after the 1st and 2nd pooling layer. Layers 6 to 8 are fully connected layers and a softmax layer computes a probability for each target class.

Our implementation uses the Caffe framework [16]. Different from [5], we train the model on a Tesla K20X GPU with 6GB of memory (instead of using two independent GPUs with less memory). Following Krizhevsky, et al., every image is resized to 256×256 pixels and the center crop (224×224) is used as input image for the model. Additionally, mean subtraction – obtained by averaging the pixel values from all training images – is carried out for each input image. No further data augmentation is applied in this experiment since it was reported to contribute only slightly to the results.

3.3   Dataset

Since our primary goal is to analyze the impact of trainingset size on the two competing approaches – IFV and CNN – we had to make sure, to provide enough training data for CNN to be able to show its true power. We therefore opted for the ILSVRC 2012 training and validation datasets, which likewise provides comparability to other experiments.

In order to reduce the overall training time, we decided to limit our experiments to train and test models for only 10 out of the entire 1,000 classes provided in the ILSVRC. Considering the mean error from the top 5 predictions from all submissions to the 2012 ILSVRC[1], we took the 5 best and worst performing classes respectively yielding a total number of 12,424 training and 500 test images. Figure 1 shows example images for each class.

(a) Best performing classes   (b) Worst performing classes

Fig. 1. Example images for best and worst performing classes according to the 2012 ILSVRC submissions. From left to right: (a) *geyser, odometer, canoe, website, yellow lady's slipper*; (b) *hook/claw, muzzle, spatula, hatchet, ladle*.

While keeping the test data fixed, we conducted 7 individual training runs by selecting between 5 % and 100 % of the original data, uniformly distributed over all classes. In order to test the impact of additional negative data, in two further tests, we added training images from 90 and 190 classes randomly selected from the remaining categories provided in the ILSVRC 2012 dataset. Since our linear predictors in the IFV approach were trained in a one-vs.-rest fashion, we simply added the additional images to the negative sets. In the CNN-based training scenario, however, in order to avoid problems arising from imbalanced datasets, we actually trained a total of 100 and 200 classes. Test results were always evaluated based on the achieved scores for the initial set of 10 classes. Just like in the original 10-class scenario, we generated uniformly sampled subsets of the

---

[1] See http://image-net.org/challenges/LSVRC/2012/ilsvrc2012.pdf for more information.

100- and 200-class training sets as well. Table 1 provides and overview over the respective number of classes and available training images. In total, we trained and tested models based on $7 \times 3 = 21$ ILSVRC 2012 subsets.

Table 1. Training sets generated by taking the top and least performing classes from the ILSVRC 2012 dataset (10 classes) and subsampling the obtained train images. Further sets are obtained by adding additional 90 and 190 randomly selected classes.

No. Classes	No. of images per subset						
	5 %	10 %	20 %	40 %	60 %	80 %	100 %
10 classes	622	1,242	2,485	4,969	7,455	9,939	12,424
100 classes	6404	12,808	25,615	51,230	76,846	102,461	128,076
200 classes	12,866	25,728	51,456	102,911	154,368	205,823	257,279

4   Analysis

Based on the trained models and the separated test set we have computed Average Precision (AP) scores for all 10 test classes and each individual run. Figure 2 shows the mean AP scores for both approaches, IFV and CNNs.

The plots show that adding more *positive* samples (i.e. going from 5 % to 100 % of the individual subsets) increases the performance for both approaches – most significant improvements occurring between 5 % to 40 %. However, incrementing the number of training samples from 80 % to 100 % contributes little to the achieved MAP score (i.e., less than 1 % MAP increase) for both models. On the other hand, while IFV based models seem to saturate at around 80 % of the entire dataset sizes (MAP even drops slightly), the CNN models seem to continue growing if going beyond 100 %.

Interestingly, when increasing the number of *negative* samples (i.e. going from 10 to 200 classes) we observe a clear drop in the achieved accuracy for the IFV based model (MAP score dropping from 76 % to 71 %) whereas CNNs benefit from the increased number of (negative) examples. In fact, the best performance by the IFV model (MAP = 76.5 %) is achieved when using the initial set of 10 classes for training whereas the best CNN-based model reaches an MAP score of 78.6 % when using 100 % of the data of the 200 class scenario. One reason may be that IFV models do not learn enough features to be able to separate classes on a more fine-grained level. When analyzing the individual per class AP scores (see Fig. 3) we observe that the best performing classes can be mostly predicted correctly by both approaches whereas the worst performing classes are equally hard to capture for CNNs as well as IFV.

Considering our initial hypothesis, we observe that our assumption of IFV outperforming CNNs in low training set scenarios holds. Especially when considering the individual per class AP scores of the best performing classes we see that IFV-based models achieve high accuracies even when provided with as little

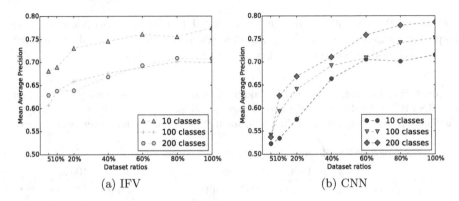

Fig. 2. A comparison of IFV and CNN MAP scores on the different ILSVRC subsets.

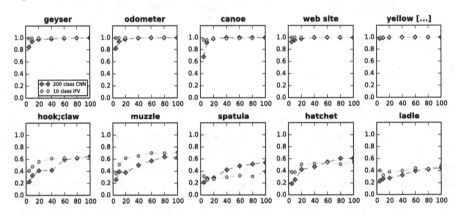

Fig. 3. Comparison of per class Average Precision scores obtained by the best performing IFV and CNN models.

as 5 %–20 % of the original data. Similarly, IFV achieves better results for most of the least performing classes, whereas CNNs need up to 60 % of the training data to catch up. This makes IFV a valid candidate when the effort to manually label large amounts of training images cannot be taken.

4.1   Model Visualization

In order to better understand what the individual models learn and how they evolve over different number of training images, we computed heat map visualizations. The method we use follows the one presented in [11], and works by stepwise occluding parts of a test image prior to classification. By that, we are able to visualize, which regions of the image have the highest impact on the overall classification score. A sliding window of 64×64 pixels is moved over the image pane partially setting the occluded pixel values to 0. The models trained

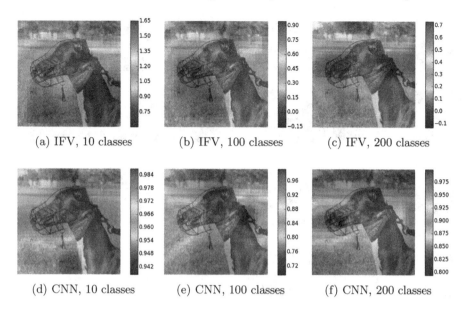

(a) IFV, 10 classes   (b) IFV, 100 classes   (c) IFV, 200 classes

(d) CNN, 10 classes   (e) CNN, 100 classes   (f) CNN, 200 classes

Fig. 4. An test image taken from the *muzzle* class superposed by heat maps depicting the impact of individual image regions. Blue color denotes higher impact. All heat maps are based on models trained with the maximum number of available images when using 10, 100 and 200 classes of the 2012 ILSVRC dataset (Color figure online).

in Sect. 3 are used to compute a prediction score for the true class label of that test image – each partially occluded image will give a different score. Finally, all scores are aggregated to compute the heat map.

Figure 4 presents heat maps computed using IFV and CNN models trained on 10, 100 and 200 classes (using the entire sets, i.e. 100 % of the data as we have shown that both approaches benefit from increased number of positive samples). The figure shows an image from the class *muzzle* – one of the hardest classes to be correctly predicted – superposed with the computed heat maps. Colors represent the achieved classification score when obstructing the respective region. A blue color here denotes a lower score meaning that the region is more important for the overall classification than a region superposed with a red color (denoting a higher classification score).

When comparing the heat maps of IFV and CNN based models, one can clearly see, that with increasing number of training examples, the CNN focuses more on the region of the object to be classified (*muzzle*): while in the smallest example (10 classes) the grassy region in the lower part of the image is considered important, in the 200 class setting, the CNN almost perfectly concentrates on the region depicting the *muzzle*. In contrast, the IFV-based approach right from the beginning covers the object (which might explain the higher accuracy in the 10 class scenario), however, likewise large parts of the background as well.

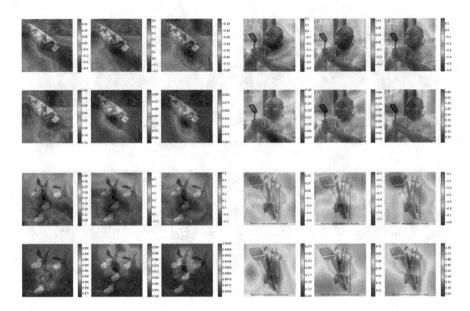

Fig. 5. Heat maps computed for some of the easier classes (*canoe* [top left]) and *yellow lady's slipper* [bottom left]) show that CNNs focus better on the depicted object when adding more (negative) data. Example images taken from one of the harder classes (*spatula* [right]) convey that both approaches have difficulties in locating meaningful regions. IFV models (top rows) and CNN model (bottom rows) have been trained on 10, 100 and 200 classes datasets (Color figure online).

While this changes slightly with increased number of classes, the IFV-based approach never reaches the precision of the CNN. Similar observations can be made when analyzing heat maps from other classes (see Fig. 5).

5   Conclusions

In this paper we evaluated the impact of growing trainingset sizes on the classification performance of Convolutional Neural Networks and Improved Fisher Vector-based image predictors. In line with our initial hypothesis, we have shown, that while CNNs largely benefit from bigger datasets, IFV is a competitive candidate when limited amounts of training data are available. Furthermore, we have presented that CNNs may use negative images to learn better feature representations. On the other hand, the precision of IFV-based models suffer from the increased diversity. Computed heat map visualizations underline our findings.

Future work will target the comparison of CNNs and IFV with feature representations obtained from CNNs pre-trained on large datasets. We aim to explore whether these representations can generalize well to significantly smaller and visually completely different datasets and how these representations compare to BoVW-like representations.

References

1. Jegou, H., Douze, M., Schmid, C., Perez, P.: Aggregating local descriptors into a compact image representation. In: 2010 IEEE Computer Society Conference on Computer Vision and Pattern Recognition (2010)
2. Perronnin, F., Sánchez, J., Mensink, T.: Improving the fisher kernel for large-scale image classification. In: Daniilidis, K., Maragos, P., Paragios, N. (eds.) ECCV 2010, Part IV. LNCS, vol. 6314, pp. 143–156. Springer, Heidelberg (2010)
3. Lowe, D.G.: Object recognition from local scale-invariant features. In: Proceedings of the Seventh IEEE International Conference on Computer Vision, vol. 2. IEEE (1999)
4. Russakovsky, O., Deng, J., Su, H., Krause, J., Satheesh, S., Ma, S., Huang, Z., Karpathy, A., Khosla, A., Bernstein, M., Berg, A.C., Fei-Fei, L.: ImageNet Large Scale Visual Recognition Challenge. Int. J. Comput. Vision **115**(3), 211–252 (2015)
5. Krizhevsky, A., Sutskever, I., Hinton, G.E.: Imagenet classification with deep convolutional neural networks. In: Advances in Neural Information Processing Systems 25. Curran Associates, Inc. (2012)
6. Simonyan, K., Zisserman, A.: Very deep convolutional networks for large-scale image recoginition. In: International Conference on Learning Representations (ICLR) (2015)
7. Chatfield, K., Simonyan, K., Vedaldi, A., Zisserman, A.: Return of the devil in the details: Delving deep into convolutional nets (2014). CoRR abs/1405.3531
8. Oquab, M., Bottou, L., Laptev, I., Sivic, J.: Learning and transferring mid-level image representations using convolutional neural networks. In: Proceedings of the 2014 IEEE Conference on Computer Vision and Pattern Recognition, Washington, DC, USA (2014)
9. Wei, Y., Xia, W., Huang, J., Ni, B., Dong, J., Zhao, Y., Yan, S.: CNN: single-label to multi-label (2014). CoRR abs/1406.5726
10. Razavian, A.S., Azizpour, H., Sullivan, J., Carlsson, S.: CNN features off-the-shelf: an astounding baseline for recognition (2014). CoRR abs/1403.6382
11. Zeiler, M.D., Fergus, R.: Visualizing and understanding convolutional networks (2013). CoRR abs/1311.2901
12. Everingham, M., Van Gool, L., Williams, C., Winn, J., Zisserman, A.: The pascal visual object classes (VOC) challenge. Int. J. Comput. Vis. **88**, 303–338 (2010)
13. Fei-Fei, L., Fergus, R.P.: One-shot learning of object categories. IEEE Trans. Pattern Anal. Mach. Intell. **28**, 594–611 (2006)
14. Chatfield, K., Lempitsky, V., Vedaldi, A., Zisserman, A.: The devil is in the details: an evaluation of recent feature encoding methods (2011)
15. Vedaldi, A., Fulkerson, B.: VLFeat: An open and portable library of computer vision algorithms (2008). http://www.vlfeat.org/
16. Jia, Y., Shelhamer, E., Donahue, J., Karayev, S., Long, J., Girshick, R., Guadarrama, S., Darrell, T.: Caffe: Convolutional architecture for fast feature embedding (2014). arXiv preprint arXiv:1408.5093

Investigating Pill Recognition Methods for a New National Library of Medicine Image Dataset

Daniela Ushizima[1,2]([envelope]), Allan Carneiro[3], Marcelo Souza[3],
and Fatima Medeiros[3]

[1] BIDS, University of California, Berkeley, CA 94720, USA
dushizima@lbl.gov
[2] CRD, Lawrence Berkeley National Laboratory, Berkeley, CA 94720, USA
[3] LABVIS, Federal University of Ceara, Fortaleza, CE 60020-181, Brazil
allanccarneiro@gmail.com, marcelo.mssouza@gmail.com, fsombra@ufc.br

Abstract. With the increasing access to pharmaceuticals, chances are that medication administration errors will occur more frequently. On average, individuals above age 65 take at least 14 prescriptions per year. Unfortunately, adverse drug reactions and noncompliance are responsible for 28 % of hospitalizations of the elderly. Correctly identifying pills has become a critical task in patient care and safety. Using the recently released National Library of Medicine (NLM) pill image database, this paper investigates descriptors for pill detection and characterization. We describe efforts in investigating algorithms to segment NLM pills images automatically, and extract several features to assembly pill groups with priors based on FDA recommendations for pill physical attributes. Our contributions toward pill recognition automation are three-fold: we evaluate the 1,000 most common medications in the United States, provide masks and feature matrices for the NLM reference pill images to guarantee reproducibility of results, and discuss strategies to organize data for efficient content-based image retrieval.

Keywords: Segmentation · Pill detection · NLM dataset

1 Introduction

Identifying the right pill has become a critical task in daily patients care and their safety. Nine out of ten US citizens over age 65 take more than one prescription pill, in average 14 pills per year, misidentifying medications is a significant concern. Better medication identification could prevent 6,000 to 8,000 deaths each year due to adverse events [1]. The importance of the identification of unknown medications in disaster and emergency situations has motivated the development of automated algorithms and software tools for recognizing drugs, particularly prescription tablets and capsules.

© Springer International Publishing Switzerland 2015
G. Bebis et al. (Eds.): ISVC 2015, Part II, LNCS 9475, pp. 410–419, 2015.
DOI: 10.1007/978-3-319-27863-6_38

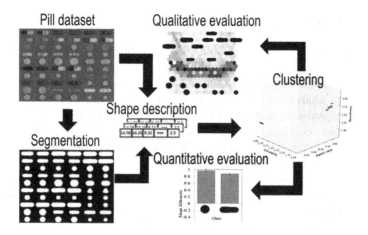

Fig. 1. Toward an automated pill image recognition system: from raw DR files to qualitative and quantitative results.

In order to advance pill identification, the National Library of Medicine (NLM) issued a call for participation in a contest entitled the "Pill Image Recognition" (PIR) Challenge, in January 2015. This NLM project targets the creation of a software tool (content-based image retrieval) to help users identify medications accurately from supplied smartphone pictures by matching the unknown medication to the prototypes in the NLM high-resolution image database. The NLM made two image datasets available: high-resolution as well as smartphone-acquired pictures.

Our paper investigates segmentation, descriptors and clustering methods (Fig. 1) for pill detection and recognition from the images provided by the NIH NLM PIR challenge. The main contributions of this paper are the release of the first, to the best of our knowledge, recognition results using the NLM PIR reference samples, including publication of the segmentation masks, feature matrices and code to improve reproducibility of experiments.

Section 2 describes previous work and details about NLM PIR image database. Section 3 specifies both methods and corresponding results using the high-resolution images from the NLM PIR reference dataset: first, it delineates how our segmentation code, based on multiband edge detection, automatically selects the pill from the picture; second, it shows how the applied morphometric image descriptors and clustering of the reference images can organize the NLM PIR images. All these algorithms are deployed as software for Fiji [2], an open-source Java image processing program, inspired by NIH Image. Section 4 discusses how our results match prior knowledge about FDA approved solid medication, it summarizes our contributions and sketches upcoming developments using NLM PIR data.

2   Background

There has been much progress in pill identification, with a few helpful public tools, such as the National Library of Medicine (NLM) Pillbox [3] and WebMD pill identification tool [4]. Both sites have been developed to aid in the recognition of solid dosage pharmaceuticals. These systems show high-resolution images of tablets and capsules to enable users to visually identify FDA approved pills. Another common characteristic among these sites is to offer user-interface tools where the user can manually enter medication specifications that matches the following attributes: imprint, color and shape. Although useful, these sites are cumbersome to use in dynamic conditions often observed in clinical settings [5].

In the United States, drugs approved for consumption receive a National Drug Code and a Structured Product Labeling (SPL), which characterizes the medication. Table 1 summarizes the analysis of public SPL records reported in [5], and illustrates the complexity in identifying pills.

Table 1. Shape and color distribution of drugs approved for consumption in United States [5]

Shape	Round	Oval	Capsule	Others
%	46.76	26.30	21.60	5.34
Color	White	Yellow	Orange	Others
%	44.29	12.90	9.10	66.29

Our challenge is to enable automatic identification of pills using smartphone-acquired pictures through a customized application. This software should reliably match the unknown product picture to an image in the NIH NLM reference image library within 15 s. One of the hurdles associated to the development of image-based pill identification has been the unavailability of open source pill image datasets with representative amount of samples, appropriate resolution pictures and associated metadata. Fortunately, the NIH NLM PIR database is a new public image set, containing JPEG images, with standards that are promising to enable pill recognition. This image database consists of two main sets: (a) a directory with reference images (a.k.a. DR), containing 2,000 high quality image files (front and back of each of 1,000 pills); (b) a directory (a.k.a. DC), containing 3,000 files of consumer images taken with cell phone and consumer grade digital cameras. A spreadsheet with correspondence between DR and DC items is also available, but lacks product label and pill physical property descriptions.

The NLM PIR challenge requires that the proposed pill recognition software identifies which consumer image corresponds to which reference image. According to NIH researchers [5] and other pill recognition investigators [6–8], there are three main characteristics used to identify pills: shape, color and imprint. Lee [6] developed an automatic pill matching system based on edge information to characterize 2,116 pill images based on imprint patterns only. This work considered image acquisition with uniform background for capturing pictures of illicit and legal pills. Their experimental results showed 76.74 % (93.02 %) rank-1 (rank-20) matching accuracy. Recent work on automatic identification of prescription drugs using shape distribution models [5] showed successful evaluation of the 568 most prescribed drugs in the United States. The proposed method relies on the identification of pill centers and the calculation of statistics based on the absolute difference between the center and the equally spaced boundary points. The advantages of the method were the invariance to rotation, translation, and classification accuracy of 91.13 %.

Most previous work, including [5–8], share common reproducibility hindrances: (a) the assumption that pill masks are available, i.e., the processed images are pre-segmented, (b) the image datasets are not available to reproduce or continue investigations, and/or (c) segmentation masks and feature matrices are not public. In order to evaluate the accuracy of upcoming algorithms and software tools in pill recognition, particularly using the NLM PIR challenge image database, we propose algorithms and provide results for NLM PIR DR images as described in the next sections.

3   Pill Recognition: Methods and Results

Our goal is to develop algorithms and share the first standardized results based on the NLM PIR challenge datasets. Henceforth, other researchers working with the same data may reuse our work to advance recognition systems toward matching unknown pills to one or more samples in a reference picture dataset. Given the importance of the information in the reference dataset, this paper focuses on the DR dataset segmentation, characterization, clustering and classification as it follows.

3.1   Segmentation: Edge Map and Texture

The NLM PIR reference or DR dataset consists of cross-polarized pictures of 2,400 × 1,600 pixels, with uniform background, as shown in the montage of Figs. 2 and 3. The apparent homogeneity of each image proved to be difficult to segment using a single cut-off or channel-based thresholding algorithms, both considering single channels and the RGB or HSV color spaces. We observed that the texture variation is much higher within the pill than in the background. This information led us to adopt multiband Sobel edge detection, followed by sharpening iterations for the segmentation of the pills. The input to our algorithm is the raw image (0.5 GB), only modified by an affine transformation to rescale the data to 20 % using linear interpolation, which both speeds up computing and minimizes noise. We propose a very efficient algorithm for the segmentation step, fully coded for Fiji [2], that detects 100 % of the pills, and within 1.8 s running in a 1.3 GHz Macbook Air.

3.2   Feature Extraction

FDA recommendations [9] on physical attributes for pill design offer guidance for industry to manufacture drugs in a way that minimizes patient medication intake errors. This report focuses on two main pill characteristics: shape and size of the pills. The importance of shape descriptors motivates our paper to focus on

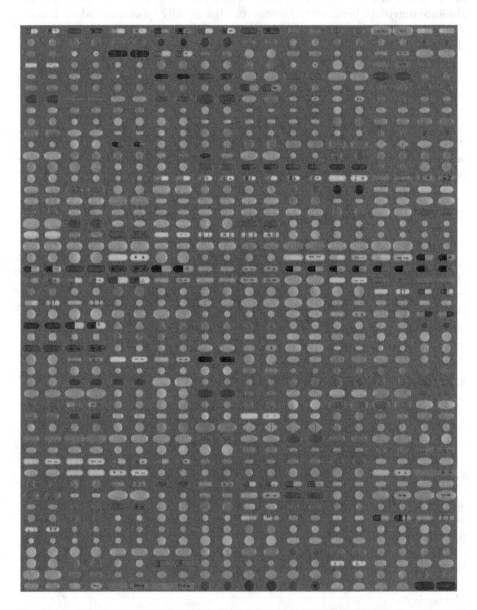

Fig. 2. Automatic segmentation of 1–1000 images of NLM pills using the proposed algorithm: dark border around pills indicates detection result (Color figure online).

capturing such pill properties, e.g., circularity (C), aspect ratio (AR), roundness (R) and solidity (S). Their definitions are as it follows: $C = \frac{4\pi Area}{Perimeter^2}$, $AR = \frac{majorAxis}{minorAxis}$, $R = \frac{4Area}{\pi majorAxis^2}$ and $S = \frac{Area}{ConvexHullArea}$ (Fig. 4).

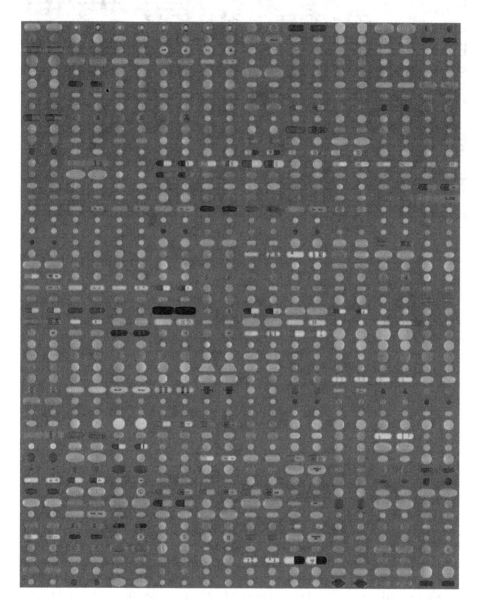

Fig. 3. Automatic segmentation of 1001–2000 images of NLM pills using the proposed algorithm: dark border around pills indicates detection result (Color figure online).

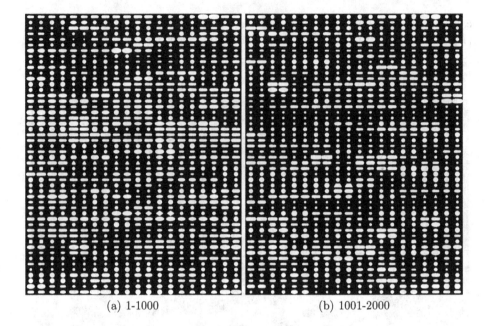

(a) 1-1000   (b) 1001-2000

Fig. 4. Automatically calculated masks from NLM PIR reference image dataset.

In addition, we also consider other shape, size and intensity descriptors such as the pills area, perimeter, width and height, major and minor axes, including intensity statistical measurements such as mean, standard deviation, mode, minimum, maximum, kurtosis, skewness, raw integrated density of the gray-levels within the pill. A detailed explanation of the software tools necessary to extract all these attributes is available in [10].

3.3 Clustering and Performance Evaluation

Because the NLM PIR dataset omits labels, e.g. tablet or capsule, medication name, we manually labeled all the 2,000 samples into four shape classes, following the shape standards presented in Table 1. After visual inspection, we listed that there are: 454 round tablets, 166 capsules, 336 oval tablets and 44 oddly shaped tablets in the NLM PIR dataset. We also noticed that these numbers are proportionally in agreement with the public SPL records, as reported in [5], although slightly skewed toward capsules.

Figure 5 shows the results of running k-means [11] for $k = 4$, where circles mostly correspond to round tablets, crosses to capsules, triangles to oval tablets and x's to oddly shaped forms. Using our labels from the visual inspection, we also verified that segregating the pills into four classes resulted in 100 % of the round pills to be grouped together, the same occurring to the capsules. About 51 % of the oval-shaped pills were clustered together, mostly mixed with oddly-shaped (others) pills. The pills with "other" shapes were scattered among the

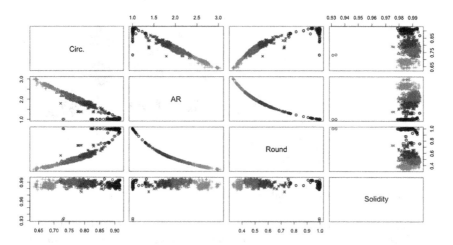

Fig. 5. K-means cluster for pill shape descriptors: circularity, aspect ratio, roundness and solidity for the whole reference database - circles mostly correspond to round tablets, crosses to capsules, triangles to oval tablets and x's to oddly shaped pills.

4 classes, being 37 % grouped together in a separate cluster. We performed similar analysis using all the features described in the previous section, however the features depicted in Fig. 5 presented better cluster separation.

SOM Network and U-matrix: A self-organizing map, or SOM network [12], is a neural network that performs nonlinear dimensional reduction of high-dimensional data by projecting it onto a more compact space. It is a useful tool for exploratory data analysis that has been used for image visualization [13], cluster learning [14], among others. A SOM network represents structural features of the input data in a low-dimensional lattice structure. These features are spatialy expressed in the lattice using a neighborhood similarity criterion, e.g. Euclidean distance, of the input data. Figure 6(a) shows the result for the whole image dataset, although some masks may overlap in this matrix. Notice that the frontiers indicate three well-defined clusters at the top, middle-right and bottom-left of the diagram.

Quantitative Evaluation Using Silhouette: The silhouette [15, 16] is a cluster quality metric that measures how well an entity y_i fits into the cluster to which it is assigned. It compares the within-cluster cohesion, based on the distance to all entities in the same cluster, to the cluster separation. In other words, the mean silhouette considers $s(y_i) = \frac{b(y_i) - a(y_i)}{max(a(y_i), b(y_i))}$, where $a(y_i)$ is the average dissimilarity of $y_i \in S_k$ to all other $y_j \in S_k$, and $b(y_i)$ is the minimum dissimilarity over all clusters S_l, to which y_i is not assigned, such as $y_j \in S_l, l \neq k$ and $-1 \leq s(y_i) \leq 1$. If $s(y_i)$ is approximately zero, the entity y_i could be assigned to another cluster without making the cluster cohesion or separation any worse.

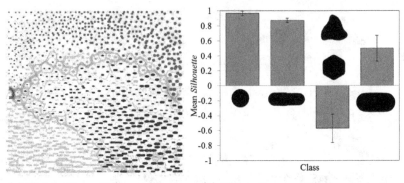

(a) Matrix-U: qualitative analysis   (b) Silhouette: quantitative analysis

Fig. 6. Analysis of cluster quality of pill dataset using shape descriptors: (a) matrix-U darker frontiers indicate higher dissimilarity among shapes; (b) higher mean silhouette values indicate lower intra-class and high inter-class relationship.

A negative $s(y_i)$ suggests that y_i's cluster assignment is damaging the cluster cohesion and separation, whereas an $s(y_i)$ closer to 1 means the opposite. Figure 6(b) shows the mean silhouette for the whole dataset when using only shape descriptors.

4   Conclusions

This paper releases the first results of using NIH NLM Pill Image Recognition datasets that were part of the Request for Information regarding software for the identification of prescription pills. We investigated descriptors that can be calculated from NIH-based tools, and anticipate that future investigators will benefit of building upon these results to progress toward pill recognition. We verified that the NLM PIR image database follows a pill shape distribution quite similar to the distribution found in the public SPL records (Table 1). We organized the dataset using visual inspection of pill shape, and classified 45.4 % as round tablets, 16.6 % as capsules, 36.6 % as oval tablets and 4.4 % as oddly shaped pills that serves as our ground-truth.

Our quantitative and qualitative metrics indicate that the shape descriptors support pill image organization in 4 main clusters, however other descriptors, such as those considering convexity points, will play a major role in improving results, particularly separating oddly shape pills from other classes. While categorization of pills into 4 main groups allowed us to understand how the feature selection was impacting clusters, our algorithms still require much advancement before enabling patients to detect the right pill using photography. New developments will include descriptors that better represent contour variations such as those based on contour bending energy [17] and information theoretic measures [18]. Future work will also include lower resolution image samples acquired with standard mobile cameras, as those in the NLM PIR DC image sets.

We acknowledge that this research was partially supported by the Sloan-Moore Foundation through BIDS - UC Berkeley, the Office of Science, U.S. Department of Energy, under Contract No. DE-AC03-76SF00098/DE-AC02-05CH11231, and PVE-CAPES/CNPq 401442/2014-4.

References

1. Hale, D.: Pill meets poka to enhance patient safety. Medicine 2.0 Congress (2009)
2. Fiji: Fiji is just imagej. (http://fiji.sc/Fiji). Accessed 27 August 2015
3. NIH NLM: Pillbox rapid identification, reliable information. (http://pillbox.nlm.nih.gov/pillimage/search.php). Accessed 27 August 2015
4. WebMD: Pill identification tool. (http://www.webmd.com/pill-identification/). Accessed 27 August 2015
5. Caban, J., Rosebrock, A., Yoo, T.: Automatic identification of prescription drugs using shape distribution models. In: ICIP, pp. 1005–1008 (2012)
6. Lee, Y.B., Park, U., Jain, A.K., Lee, S.W.: Pill-id: matching and retrieval of drug pill images. Pattern Recogn. Lett. **33**, 904–910 (2012)
7. Lee, H.C., Chang, T.: Pill recognition using image processing and neural networks. Gerontechnology **13**, 335 (2014)
8. Yu, J., Chen, Z., Kamata, S.I.: Pill recognition using imprint information by two-step sampling distance sets. In: International Conference on Pattern Recognition, pp. 3156–3161 (2014)
9. FDA: Size, shape, and other physical attributes of generic tablets and capsules guidance for industry. (http://www.fda.gov/downloads/drugs/guidance complianceregulatoryinformation/guidances/ucm377938.pdf). Accessed 27 August 2015
10. Fiji-plugins: Analysis tools. (http://rsb.info.nih.gov/ij/docs/menus/analyze.html). Accessed 27 August 2015
11. Hartigan, J.A., Wong, M.A.: A k-means clustering algorithm. Appl. Stat. **28**, 100–108 (1979)
12. Kohonen, T., Schroeder, M.R., Huang, T.S. (eds.): Self-Organizing Maps, 3rd edn. Springer-Verlag New York Inc., Secaucus (2001)
13. Strong, G., Gong, M.: Similarity-based image organization and browsing using multi-resolution self-organizing map. Image Vis. Comput. **29**, 774–786 (2011)
14. Kuroiwa, J., Inawashiro, S., Miyake, S., Aso, H.: Self-organization of orientation maps in a formal neuron model using a cluster learning rule. Neural Netw. **13**, 31–40 (2000)
15. Rousseeuw, P.: Silhouettes: a graphical aid to the interpretation and validation of cluster analysis. J. Comput. Appl. Math. **20**, 53–65 (1987)
16. de Amorim, R.C., Hennig, C.: Recovering the number of clusters in data sets with noise features using feature rescaling factors. Inf. Sci. **324**, 126–145 (2015)
17. Backhaus, A., Kuwabara, A., Bauch, M., Monk, N., Sanguinetti, G., Fleming, A.: LEAFPROCESSOR: a new leaf phenotyping tool using contour bending energy and shape cluster analysis. New Phytol. **187**, 251–261 (2010)
18. Vinh, N.X., Epps, J., Bailey, J.: Information theoretic measures for clusterings comparison: variants, properties, normalization and correction for chance. J. Mach. Learn. Res. **11**, 2837–2854 (2010)

Realtime Face Verification with Lightweight Convolutional Neural Networks

Nhan Dam[1], Vinh-Tiep Nguyen[2], Minh N. Do[3], Anh-Duc Duong[4],
and Minh-Triet Tran[2(✉)]

[1] AI Lab, University of Science, VNUHCM, Ho Chi Minh, Vietnam
dntnhan@apcs.vn
[2] Faculty of IT, University of Science, VNUHCM, Ho Chi Minh, Vietnam
{nvtiep,tmtriet}@fit.hcmus.edu.vn
[3] Department of ECE, University of Illinois at Urbana-Champaign,
Champaign, USA
minhdo@illinois.edu
[4] University of Information Technology, VNUHCM, Ho Chi Minh, Vietnam
ducda@uit.edu.vn

Abstract. Face verification is a promising method for user authentication. Besides existing methods with deep convolutional neural networks to handle millions of people using powerful computing systems, the authors aim to propose an alternative approach of a lightweight scheme of convolutional neural networks (CNN) for face verification in realtime. Our goal is to propose a simple yet efficient method for face verification that can be deployed on regular commodity computers for individuals or small-to-medium organizations without super-computing strength. The proposed scheme targets unconstrained face verification, a typical scenario in reality. Experimental results on original data of Labeled Faces in the Wild dataset show that our best CNN found through experiments with 10 hidden layers achieves the accuracy of $(82.58 \pm 1.30)\%$ while many other instances in the same scheme can also approximate this result. The current implementation of our method can run at 60 fps and 235 fps on a regular computer with CPU-only and GPU configurations respectively. This is suitable for deployment in various applications without special requirements of hardware devices.

Keywords: Unconstrained face verification · Convolutional neural network · Lightweight

1 Introduction

User authentication is one of the most important steps for various applications to ensure that only authorized users can access systems, and to provide appropriate information and services corresponding to the users. Face recognition or verification can be used as a natural yet efficient method for user authentication.

For face verification, we simply answer whether a pair of photos are of the same person or not. This is appropriate for applications in which we cannot

© Springer International Publishing Switzerland 2015
G. Bebis et al. (Eds.): ISVC 2015, Part II, LNCS 9475, pp. 420–430, 2015.
DOI: 10.1007/978-3-319-27863-6_39

know or do not need to know all persons in advance but only to match people in various scenarios from their faces, such as in parking system, video surveillance, etc. Face verification can also be used to recognize a user from a face photo by comparing it with registered photos of known persons.

Face verification under unconstrained conditions is still a challenging problem because of various obstacles, such as partial occlusion, illumination and especially pose variation, etc [12]. Different methods have been proposed for face verification in unconstrained environment, most of which use hand-crafted local or global features based on the known structure of a human face.

Instead of manually constructing a new type of hand-crafted features, features can be learned from training samples in deep learning with convolutional neural networks. However, most of these methods usually use complex networks with very deep structures and require powerful computing systems that may not be available for individuals or small-to-medium organizations. This motivates our proposal to use lightweight CNNs with less complex structures for face verification that can be deployed on a regular commodity computer. Specifically, our goal is to propose a novel method that can balance the following criteria: (i) achieve reasonable high accuracy, (ii) process in realtime, and (iii) can be deployed in regular commodity computers.

As there are different ways to devise a CNN, we first define a scheme with a common structure of CNN instances for consideration. Through experiments, we find the best instance in this family of CNNs according to the above criteria can achieve the accuracy of $(82.58 \pm 1.30)\%$ on the standard dataset Labeled Faces in the Wild (LFW [11]) following the "Image Restricted, No Outside Data" protocol, higher than the results of existing methods on original images with the same protocol.

The main contributions of our paper are as follows:

- We propose a scheme of lightweight Convolutional Neural Networks for face verification that can be deployed on a popular commodity computer. Not only the best CNN but many other instances of our proposed lightweight CNN scheme can achieve the accuracy of up to 80 % on LFW.
- To speed up the processing time, we eliminate the funnel step [8,10] to transform an input face image to an upright one. Experiments show that multiple CNNs in our method can achieve the accuracy up to 80 % even with original images without transformation.
- The best found CNN instance conforming our proposed lightweight CNN scheme can perform face verification task at 60 fps and 235 fps with CPU-only and GPU implementations on a regular computer respectively. Therefore, it will be suitable to integrate face verification component into various applications for individuals and small-to-medium organizations without special requirements on hardware devices.

The rest of the paper is organized as follows. In Sect. 2, we briefly review the background and related works. Our proposed scheme of lightweight CNNs for face verification is in Sect. 3. Experimental results on LFW dataset with various instances of our proposed lightweight CNN scheme and the best found CNN are discussed in Sect. 4. The final section is for conclusion and future work.

2   Background and Related Works

Although face verification with images captured in constrained conditions has been well studied and achieves high accuracy, it is still a challenging problem to handle face images captured in unconstrained environment because of various factors, such as illumination, partial occlusion, and especially pose variation [13].

One of the common trends for face verification is to extract features or facial components from a face image, then perform different techniques and metrics [2] to evaluate the matching score between faces in a pair of photos. Several methods use local features, such as SIFT [4,5], LBP [4,5], Walsh LBP [12], Gabor [21], LE [4]. These features can also be combined to boost the overall performance. Structures of a face are also extracted, such as hierarchical-probabilistic elastic parts [13]. Some methods also exploit 3D structure of a face from an image for face frontalization [7] or face alignment with explicit 3D modeling of faces [20].

While most methods for face verification use hand-crafted features, using deep neural networks has become a new trend for face verification [6,9], especially DeepFace [20]. In these methods, instead of manually constructing a specific feature type, features can be learned from trained data.

Most of existing methods with deep neural networks for face verification use complex networks with very deep structures and process large data to train such networks. These tasks require super-computing systems with high computational resources to handle huge computation workload. These methods and systems provide high accuracy to process millions of people. However, it would be necessary to devise lightweight neural networks that can utilize the structures and techniques in common deep neural networks for face verification. Such lightweight versions can be deployed on regular commodity computers and may be appropriate for individuals or small-to-medium organizations to realize their own solutions for face verification.

3   Proposed Method

3.1   Overview

Face verification is solved as a binary classification, in which input is a correlation representation of a pair of images and output is a binary label determining whether the pair is matched or mismatched.

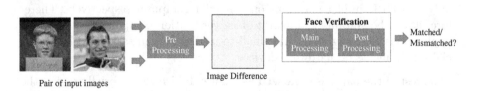

Pair of input images   Image Difference

Fig. 1. Overview of face verification process

Figure 1 illustrates the overview of our proposed process of face verification. An image difference is generated from a pair of face images and fed into Face Verification module, which contains two main parts of *Main processing* and *Post-processing*. The first part is for feature extraction and learning while the second part improves the overall performance and guarantees the desired format of output. The final output is "Matched" or "Mismatched" which means that the network predicts the 2 images to be of the same person or to capture different faces respectively.

In the pre-processing step, each input image is transformed into grayscale, resize to a specific size, then normalized so that its histogram follows the zero-mean Gaussian distribution $N(0, \sigma)$ to reduce the impact of illumination. In our system, we choose the image size of 100×100 pixels and $\sigma = 1$.

To speed up the whole process of face verification, we intentionally skip all pre-processing techniques that may require extra/high computational cost, such as Face Frontalization [7] or even the funnel transformation [8,10].

There are several ways to calculate element-wise differences between two normalized images [15], such as squared, absolute-value, and square-root absolute-value. In our proposed method, we take the pixel intensity difference between 2 normalized images as the correlation representation to feed into the face verification process. In the training phase, for a given pair of input images (I_1 and I_2), we generate two different training inputs corresponding to $I_1 - I_2$ and $I_2 - I_1$ with the same label (matched/mismatched) so that the CNN can learn this pair in two scenarios. In the testing phase, we only use either $I_1 - I_2$ or $I_2 - I_1$ as the input.

There are different ways to construct the initial structure and modify the structure of a convolutional neural network for a specific problem. Although there are best practices or common approaches to design the structure of a CNN, there are not concrete guidelines for this task. Therefore, to define the search space for a promising structure of CNN for face verification that satisfies our objectives (c.f. Sect. 1), we propose a scheme to construct a family of CNNs with reference to common design patterns of existing CNNs.

3.2 Main Processing

Figure 2 shows the first part in our proposed scheme for CNNs. A face has some features that can be used to distinguish different people: forehead, eye, nose, mouth, chin, cheek, overall face shape, etc. The network is designed to first extract and describe those features separately, then combine them to form a complete feature vector of the difference of faces. In other words, local features of individual parts are used to distinguish the difference of faces. Then, those features are combined in a relatively spatial manner, which means that spatial information of the facial parts is considered but the image transformation (such as translation, reflection, rotation) has little effect on the network performance.

Main processing consists of K rounds, each consists of a convolutional layer and a pooling layer. Since the face has some small parts (eyes, nose, mouth, chin, cheek, ears, etc.) that can be used to distinguish different people, a *convolutional*

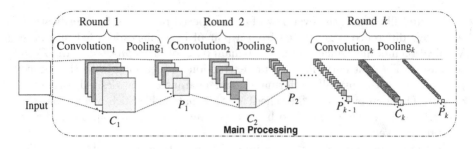

Fig. 2. Structure of main processing module.

layer applies a filter bank with multiple filters of the same size to the image to focus on amplifying local features while considering only a neighboring area for each pixel. A ReLU (rectified linear unit) layer is always embedded right after each convolutional layer to add non-linearity property to the network, otherwise the learning algorithm only works for linear models. Besides, ReLU layers help to ensure the sparsity for features learned in the network. For simplicity of presentation, we do not mention ReLU layers in the following sections as they implicitly exist in convolutional layers.

After a convolutional layer, a *pooling layer* with a kernel of 2×2 makes the image representation more compact and more abstract by downsampling and eliminating redundant information. In the proposed scheme, we use the max-pooling, which keeps only the maximum value in a region. The underlying assumption to use pooling layer is that the probability that any neighboring pixels of a feature point is also a feature point is low. The deeper the network is, the more detailed and abstract the learned features are.

The ith convolutional layer has the following parameters: $kernel_i$ and n_i^C are the size of the kernel and the number of filters in its filter banks, respectively. The number of filters in filter banks are in ascending order, i.e. $n_i^C < n_{i+1}^C$ for $1 \leq i < K$. We set the value of the stride in all convolutional layers to 1.

In practice, we define two *strategies for a filter bank* in each convolutional layer in a round: (A) small filters and (B) large filters. We consider small filters with $3 \leq kernel_i \leq 7$ and large filters with $17 \leq kernel_i \leq 23$. In the first convolutional layer, the number of filters n_1^C starts at 10 and increases with the step of 5. In subsequent convolutional layer, we always ensure that $10 \leq n_{i+1}^C - n_i^C \leq 50$.

3.3   Post Processing

Figure 3 shows the detailed structure of the post-processing part. Similar to those in main processing part, two more convolutional layers are used to extract local features from the feature map. To generate a flat feature vector the last layers in the network, we gradually decrease the number of filters in filter banks in post processing part, i.e. $n_K^C > n_{K+1}^C > n_{K+2}^C$. We also apply Dropout [18] as a simple technique to prevent the network from overfitting.

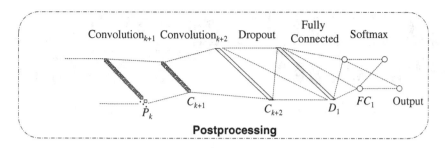

Fig. 3. Structure of post processing module.

Several fully connected layers are usually used as the last layers of a neural network, e.g. GooLeNet [19], VGG network [3,17], fully convolutional semantic segmentation model [14]. However, we use convolutional layers instead of fully connected ones to follow the pure idea of CNN: each feature can be extracted from a local area. Only one fully connected layer is used right before the softmax layer the feature dimension is small then. Moreover, the use of convolutional layers reduces complexity and the number of weights for the network due to shared weight and spatial subsampling properties of CNN. In test (or practical use) phase, the very last layer computes the probabilities that the correlation difference belongs to each label ("matched" or "mismatched") and stores them as a vector. In training, softmax layer is replaced by log-loss softmax, which first computes the probabilities that the correlation representation belongs to each label and then the error penalty for wrong prediction, which is the complement of the probability the correlation representation is categorized to the correct label.

With the mentioned above structure of main processing and post processing parts of CNN in our proposed scheme, we can define a family of CNN instances with similar structures. For each value of K, we can instantiate multiple convolutional neural networks with $2K + 4$ hidden layers, depending on different choices for parameters in $K + 2$ convolutional layers. As we only consider lightweight CNNs, we choose $K \leq 5$ to keep CNN not to be too complicated with a very deep structure.

4   Experiments and Results

4.1   Labeled Faces in the Wild

Experiments are conducted on Labeled Faces in the Wild (LFW) [11], a standard dataset for unconstrained face verification. LFW contains $13,233$ images of $5,749$ people collected from the Internet. There are numerous variations in images: 2-D rotation, inconsistent illumination, non-frontal poses, major change in expression, accessories obscuring, inconsistent localization, etc. The only constraint is that faces must be detected by Viola-Jones face detector. Figure 4 shows samples of matched pairs (left) and mismatched pairs (right) of LFW.

Fig. 4. Samples of matched pairs (left) and mismatched pairs (right) in LFW dataset.

There are 2 main groups of data in LFW: the original dataset and 3 aligned datasets, including funneled images [8], LFW-a , and deep funneled images [10]. All these aligned methods try to standardize the images under the criteria: the face is rotated to be straight vertically, scaling is applied in order to ensure an equivalent proportion that face occupy over images.

In this paper, we target the original dataset without any transformations to evaluate the efficiency of CNN in face verification even with raw original inputs. Furthermore, we also want to skip extra computation for alignment in real applications to speed up the whole face verification process.

Among 6 common protocols to conduct experiments with LFW, we follow the protocol of "Image Restricted, No Outside Data". In the restricted configuration, the pairs of matched and mismatched images have been given beforehand by configuration files on LFW dataset homepage, whereas the unrestricted configuration allows the researchers to freely create matched and mismatched pairs. In both cases, the identity of each image is not allowed to be used to support classification. Besides, we do not use any extra outside data for training.

4.2 Experiment Set 1: Model Selection for Main Processing Part

We use View 1 of LFW to quickly evaluate different CNN instances conforming our proposed CNN scheme with different number of rounds K and different kernel sizes in convolutional layers. View 1 of LFW only contains $1,100$ matched and mismatched pairs of photos for training and 500 pairs of photo for testing.

As mentioned in Sect. 3.2, there are two strategies to select the kernel size for each convolutional layer. Small filters with kernel size $3 \leq kernel_i \leq 7$, denoted by Strategy A, can be used to extract small parts/features on the face, such as nose, mouth, eye, chin, or cheek. Large filters with kernel size $17 \leq kernel_i \leq 23$, denoted by Strategy B, can be appropriate to capture the overall features in a face. For each convolutional layer in the main processing part, we try either of the two strategies independently from other convolutional layers. In general, there are 2^K different configurations for each value of K.

For $K = 1$, as there is only one convolutional layer, there are 2 configurations, A or B. For $K = 2$, A-B is a possible configuration with small filters in the first convolutional layer and large filters in the second convolutional layer. There are 3 other possible configurations for $K = 2$: A-A, B-A, and B-B.

For each configuration, we empirically evaluate on View 1 of LFW to select the best instance of CNN conforming to our proposed CNN scheme.

Table 1. Results of model selection for main processing part on dataset LFW.

K	1		2			3
Configuration	A	B	A-B	A-A	B-A	A-B-A
Accuracy	79.5 %	79.2 %	78.8 %	81.5 %	80.1 %	**83.7 %**

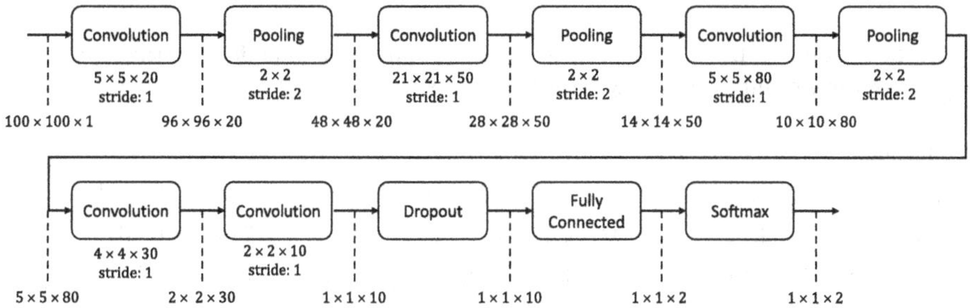

Fig. 5. Structure of the chosen CNN instance with $K = 3$ (feature dimensions are connected with dotted lines)

Table 1 demonstrates the results of model selection for main processing part with several configurations. For $K - 1$, configuration B leads to the lowest accuracy. In case $K = 2$, configuration A - B has less significant performance than other settings. This may be because of using a large window size kernel at the end may reduce the chance of learning detail information. For $K = 4$ and 5, the results are lower than that of $K = 3$, as the networks become too complicated and there are not enough samples to train such networks.

The best CNN instance found empirically (Fig. 5) in our proposed scheme has $K = 3$ and follows the configuration A-B-A, which means that the second convolutional layer uses a large kernel size while the other two convolutional layers use small kernel sizes. This CNN instance achieves the accuracy of 83.7 % on View 1 of LFW. We select this CNN instance for experiment set 2.

4.3 Experiment Set 2: Benchmark Result on LFW Dataset

View 2 of LFW is used for final performance report. There are 10 disjoint sets divided by the dataset creatorsto be used in 10-fold cross validation learning algorithm. At each time, a set is considered as the test set and the other 9 sets are fed to training phase. Table 2 shows the results of face verification on LFW with "Image Restricted, No Outside Data" protocol. We learn 10 models and have 10 corresponding accuracy results.

Experiments show that the best CNN instance found in Sect. 4.2 achieves the accuracy of (82.58 ± 1.30) % on 10 sets of LFW. Table 3 shows the results of other works on dataset LFW with the same setting as our experiments (image-restricted and no outside data). The first 3 rows are results of methods on the original dataset, while other methods use funneled dataset in which faces have

Table 2. Results of face verification on 10-fold dataset LFW.

Test set	1	2	3	4	5
Accuracy	85.500 %	81.833 %	82.500 %	83.667 %	80.833 %
Test set	6	7	8	9	10
Accuracy	82.667 %	82.167 %	82.333 %	81.333 %	83.000 %

Table 3. Results on LFW with image-restricted and no outside data setting. *(Source: http://vis-www.cs.umass.edu/lfw/results.html. Accessed on October 1, 2015)*

Methods	Accuracy
Matthew A. Turk et al., CVPR 1991, *original*	(60.02 ± 0.79) %
Eric Nowak et al., CVPR 2007, *original*	(72.45 ± 0.40) %
Conrad Sanderson et al., ICB 2009, *original*	(72.95 ± 0.55) %
Gary B. Huang et al., ICCV 2007, *funneled*	(73.93 ± 0.49) %
Lior Wolf et al., ECCV 2008, *funneled*	(78.47 ± 0.51) %
Nicolas Pinto et al., CVPR 2009, *funneled*	(79.35 ± 0.55) %
Shervin Rahimzadeh Arashloo et al., BTAS 2013, *funneled*	(79.08 ± 0.14) %
Our proposed method, *original*	**(82.58 ± 1.30) %**
Haoxiang Li et al., CVPR 2013, *funneled*	(84.08 ± 1.20) %
Karen Simonyan et al., BMVC 2013, *funneled*	(87.47 ± 1.49) %
Haoxiang Li et al., ACCV 2014, *funneled*	(88.97 ± 1.32) %
Haoxiang Li et al., CVPR 2015, *funneled*	(91.10 ± 1.47) %
Shervin Rahimzadeh Arashloo et al., TIFS 2014, *funneled*	(95.89 ± 1.94) %

been transformed and aligned for better results. Our method provides the best result on the original data, outperform nearly 10 % higher than the current best result of (72.95 ± 0.55) % on original data [16]. Besides, the result of our method is also higher than those of several methods that take advantages of funneled data [1, 15].

4.4   Performance

As one of the criteria for our proposed method is that it can be deployed on a regular computer, we evaluate the processing speed of our method with two implementations. Using CPU-only implementation, our method runs at 60 fps on a computer with Intel Core i7 2.4 GHz and 8 GB RAM. With GPU-enabled implementation, our method runs up to 235 fps on a computer with Intel Core i7 2.4 GHz, 16 GB RAM, and Nvidia GT750M.

5 Conclusion

We propose a lightweight Convolutional Neural Network for face verification with images captured in unconstrained environment. We first define the structure of a CNN scheme with the parameter K, the number of convolution-pooling pairs in the main processing part of the network, and other necessary parameters for each layer in a network. By this way, we can define the search space for a lightweight CNN instance conforming to our proposed scheme that can provide the best accuracy for face verification.

Experiments on LFW dataset show that the best found CNN instance (with $K = 3$ and 10 hidden layers) achieves the best accuracy of $(82.58 \pm 1.30)\%$, higher than the current best accuracy of $(72.95 \pm 0.55)\%$ on the original data following the protocol "Image Restricted, No Outside Data" [16]. Besides, not only the best instance but many other CNN instances in our proposed CNN scheme can achieve up to 80 % of accuracy on the original data. This supports the idea that it is possible to devise a not-very-complex CNN that can provide sufficient accuracy for face verification, even with photos without special transformation or alignment.

Because of the simplicity of our chosen CNN structure, face verification can be performed in realtime with up to 60 fps (CPU only) and 235 fps (GPU-enabled). Thus, our method can be applied into various practical applications, such as smart house, video surveillance, robotics, or platforms for Internet-of-Things to provide appropriate services and information to users in realtime.

As current successful methods with deep convolutional neural networks usually have complex structures with very deep layers to process huge amount of data, it might lead to the consideration that CNN may not be applicable for regular commodity computers. It is obvious that a lightweight CNN cannot achieve the high accuracy as deep networks for face verification or other tasks. However, it may be useful for individuals and small-to-medium organizations in various applications without special requirements of computing resources and devices.

Acknowledgement. This article was funded in part by a grant from the Vietnam Education Foundation (VEF). The opinions, findings, and conclusions stated herein are those of the authors and do not necessarily reflect those of VEF.

References

1. Arashloo, S.R., Kittler, J.: Efficient processing of mrfs for unconstrained-pose face recognition. In: IEEE Sixth International Conference on Biometrics: Theory, Applications and Systems, BTAS 2013, pp. 1–8 (2013)
2. Cao, Q., Ying, Y., Li, P.: Similarity metric learning for face recognition. In: IEEE International Conference on Computer Vision, ICCV 2013, pp. 2408–2415 (2013)
3. Chatfield, K., Simonyan, K., Vedaldi, A., Zisserman, A.: Return of the devil in the details: delving deep into convolutional nets. In: British Machine Vision Conference, BMVC 2014 (2014)

4. Chen, D., Cao, X., Wang, L., Wen, F., Sun, J.: Bayesian face revisited: a joint formulation. In: Fitzgibbon, A., Lazebnik, S., Perona, P., Sato, Y., Schmid, C. (eds.) ECCV 2012, Part III. LNCS, vol. 7574, pp. 566–579. Springer, Heidelberg (2012)
5. Chen, D., Cao, X., Wen, F., Sun, J.: Blessing of dimensionality: high-dimensional feature and its efficient compression for face verification. In: IEEE Conference on Computer Vision and Pattern Recognition, CVPR 2013, pp. 3025–3032 (2013)
6. Chopra, S., Hadsell, R., LeCun, Y.: Learning a similarity metric discriminatively, with application to face verification. In: IEEE Computer Society Conference on Computer Vision and Pattern Recognition, CVPR 2005, pp. 539–546 (2005)
7. Hassner, T., Harel, S., Paz, E., Enbar, R.: Effective face frontalization in unconstrained images. In: IEEE Conference on Computer Vision and Pattern Recognition, CVPR 2015 (2015)
8. Huang, G.B., Jain, V., Learned-Miller, E.G.: Unsupervised joint alignment of complex images. In: IEEE International Conference on Computer Vision, ICCV 2007, pp. 1–8 (2007)
9. Huang, G.B., Lee, H., Learned-Miller, E.G.: Learning hierarchical representations for face verification with convolutional deep belief networks. In: IEEE Conference on Computer Vision and Pattern Recognition, CVPR 2012, pp. 2518–2525 (2012)
10. Huang, G.B., Mattar, M.A., Lee, H., Learned-Miller, E.G.: Learning to align from scratch. In: 26th Annual Conference on Neural Information Processing Systems, NIPS 2012, pp. 773–781 (2012)
11. Huang, G.B., Ramesh, M., Berg, T., Learned-Miller, E.: Labeled faces in the wild: a database for studying face recognition in unconstrained environments. Technical report 07-49, University of Massachusetts, Amherst, October 2007
12. Juefei-Xu, F., Luu, K., Savvides, M.: Spartans: single-sample periocular-based alignment-robust recognition technique applied to non-frontal scenarios. IEEE Trans. Image Process. 24(12), 4780–4795 (2015)
13. Li, H., Hua, G.: Hierarchical-PEP model for real-world face recognition. In: IEEE Conference on Computer Vision and Pattern Recognition, CVPR 2015, pp. 4055–4064 (2015)
14. Long, J., Shelhamer, E., Darrell, T.: Fully convolutional networks for semantic segmentation. CoRR abs/1411.4038 (2014)
15. Pinto, N., DiCarlo, J.J., Cox, D.D.: How far can you get with a modern face recognition test set using only simple features? In: IEEE Computer Society Conference on Computer Vision and Pattern Recognition, CVPR 2009, pp. 2591–2598 (2009)
16. Sanderson, C., Lovell, B.C.: Multi-region probabilistic histograms for robust and scalable identity inference. In: Tistarelli, M., Nixon, M.S. (eds.) ICB 2009. LNCS, vol. 5558, pp. 199–208. Springer, Heidelberg (2009)
17. Simonyan, K., Zisserman, A.: Very deep convolutional networks for large-scale image recognition. Computing Research Repository - CoRR abs/1409.1556 (2014)
18. Srivastava, N., Hinton, G.E., Krizhevsky, A., Sutskever, I., Salakhutdinov, R.: Dropout: a simple way to prevent neural networks from overfitting. J. Mach. Learn. Res. 15(1), 1929–1958 (2014)
19. Szegedy, C., Liu, W., Jia, Y., Sermanet, P., Reed, S., Anguelov, D., Erhan, D., Vanhoucke, V., Rabinovich, A.: Going deeper with convolutions. Computing Research Repository-CoRR abs/1409.4842 (2014)
20. Taigman, Y., Yang, M., Ranzato, M., Wolf, L.: Deepface: closing the gap to human-level performance in face verification. In: IEEE Conference on Computer Vision and Pattern Recognition, CVPR 2014, pp. 1701–1708 (2014)
21. Yi, D., Lei, Z., Li, S.Z.: Towards pose robust face recognition. In: IEEE Conference on Computer Vision and Pattern Recognition, CVPR 2013, pp. 3539–3545 (2013)

Virtual Reality

Relighting for an Arbitrary Shape Object Under Unknown Illumination Environment

Yohei Ogura$^{(\boxtimes)}$ and Hideo Saito

Keio University, 3-14-1 Hiyoshi, Kohoku, Yokohama, Kanagawa 223-8522, Japan
{y.ogura,saito}@hvrl.ics.keio.ac.jp

Abstract. Relighting techniques can achieve the photometric consistency in synthesizing a composite images. Relighting generally need the object's shape and illumination environment. Recent research demonstrates possibilities of relighting for an unknown shape object or the relighting the object under unknown illumination environment. However, achieving both tasks are still challenging issue. In this paper, we propose a relighting method for an unknown shape object captured under unknown illumination environment by using an RGB-D camera. The relighted object can be rendered from pixel intensity, surface albedo and shape of the object and illumination environment. The pixel intensity and the shape of the object can simultaneously be obtained from an RGB-D camera. Then surface albedo and illumination environment are iteratively estimated from the pixel intensity and shape of the object. We demonstrate that our method can perform relighting for a dynamic shape object captured under unknown illumination using an RGB-D camera.

1 Introduction

Relighting is a technique to change appearance of objects in images captured in arbitrary illumination. We can achieve the photometric consistency in composite images synthesized from different images taken under different illuminations by using the relighting technique. Users can change the lighting condition for the object as if they are in other illumination environment. We consider that this technique can be applied for the entertainment experience or lighting simulation.

Relighting technique generally requires the shape and the surface albedo of the target object, and the illumination environment of the scene. There are some relighting approach with off-line processing or on-line processing with known geometry owing to the difficulty of exacting the object's shape and the surface albedo. Therefore, these methods generally can handle a single image or a static object in a movie. Not only shape and surface albedo of the target object, but also illumination environment are indispensable to relighting. Illumination estimation has been one of the important research issues in computer vision field. Illumination environment can be estimated based on inverse rendering, without any light probes. When we implement the illumination estimation method based on inverse rendering, the pixel intensity, the shape, and the surface albedo of the object are needed. However, the surface albedo of arbitrary object is unknown

© Springer International Publishing Switzerland 2015
G. Bebis et al. (Eds.): ISVC 2015, Part II, LNCS 9475, pp. 433–442, 2015.
DOI: 10.1007/978-3-319-27863-6_40

in many cases. It is not easy to estimate the illumination environment without any known surface albedo or light probes such as a mirror ball or cast shadows. In other words, achieving relighting methods which can be used for arbitrary shape object captured under unknown illumination is still a challenging task.

In this paper, we propose a relighting method for an arbitrary shape object captured under unknown illumination environment by using an RGB-D camera. A relighted object is rendered from pixel intensity, surface albedo and shape of the object and illumination environment. The pixel intensity and shape of the object are obtained from RGB-D camera. The surface albedo and the illumination environment are iteratively estimated from the pixel intensity and shape of the object. All of those properties can be estimated via on-line process for each frame in a movie sequence so that our implementation can be used for a dynamic shape object.

2   Related Work

Debevec et al. proposed a method to acquire the reflectance field of a target object using Light Stage, which has one spotlight which spirals around a target [1]. They take 2048 images under the different light conditions to estimate the reflectance function of the object. Not only they can get relighting result, but also they can change the viewpoint from the reflectance function. Wenger et al. implement newer Light Stage with high speed cameras to take larger number of images [2]. Their method can be applied to the dynamic shape object by using their new Light Stage with high speed cameras. These methods can obtain high quality relighting results thanks to their special equipment.

Zhen et al. proposed a relighting technique for human faces using 3D morphable face model [3]. This method builds on a ratio-image based technique. The advantage of this method is that single face image is required as input. However, the normal map should be estimated from generic human face model in off-line process. Aldrian et al. proposed a face relighting method considering not only diffuse component but also specular component so that more natural relighting results can be obtained [4]. These techniques don't consider acquisition of the illumination environment but simply assume that Illumination environment should be obtained in an arbitrary manner.

Wang et al. proposed a method that relights human faces from a single image under unknown lighting conditions [5]. They integrate spherical harmonics into the 3D morphable model to represent a human face captured under arbitrary unknown lighting and pose by representing them with three dimensional parameters.

However, those methods [3–5] assume that the input is a static human face image. Obtaining shape of the target from 3D morphable models, it is difficult to apply to the whole body of human or other objects.

3   Proposed Method

Our goal is to relight an arbitrary shape object recorded under unknown illumination environment. Our system consists from two parts: illumination estimation
part and relighting part. Input data are color images and depth images from an
RGB-D camera. Illumination environment is estimated from pixel intensity, surface albedo, and normal map of an arbitrary object existing in the image based
on an inverse rendering technique without any light probes that has been proposed by Ramamoorthi and Hanrahan [6]. The normal map of a target object
is obtained from depth images. After the normal map estimation, we segment
the input color image into some regions. On each region, we set the average
pixel intensity as initial albedo. After estimating illumination environment from
the normal map and the initial albedo, we re-estimate the surface albedo from
the estimated illumination and the normal map. We repeat this process until
the improvement of the estimation is lower than a threshold (Fig. 1).

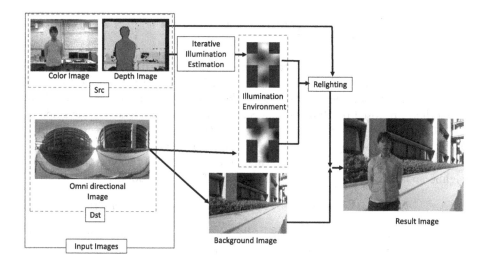

Fig. 1. System overview

After estimating illumination environment, we relight the object to match
the illumination of the object with the background scene. Relighting process is
done with pixel intensity, normal map and estimated illumination environment.
We implement Image based rendering and Spherical Harmonics approximation.
We calculate the ratio of the inverse rendering equation of the both scene to get
the relighting result. Finally, the target object is superposed to the background
image to get the final result. From Sects. 3.1 to 3.4, we introduce the theory of
normal map estimation and spherical harmonics relighting.

3.1   Normal Map Estimation

Our relighting method uses normal map in the illumination estimation section and the relighting part. Simply calculating the normal map from two vectors calculated from neighbor pixels, the result may not be good because of noises on depth images. Before the normal map estimation, we apply bilateral filter and temporal filter to depth images. The bilateral filtered depth map D_b is obtained from raw depth map D by using following equation:

$$D_b(\boldsymbol{u}) = \frac{1}{k'(\boldsymbol{u})} \sum_{v \in \Omega_{g'}} g_{s'}(\boldsymbol{u}, \boldsymbol{v}) g_d(D(\boldsymbol{u}), D(\boldsymbol{v})) D(\boldsymbol{v}) \tag{1}$$

Note that $g_{s'}(\boldsymbol{u}, \boldsymbol{v})$ is the spatial Gaussian weight and $g_d(I(\boldsymbol{u}), I(\boldsymbol{v}))D$ is the color similarity Gaussian weight. $k'(\boldsymbol{u})$ is a normalize factor and $\Omega_{g'}$ is a square window whose center is $\boldsymbol{u}$. After applying bilateral filter, we also apply the temporal filter [7]. A current depth image is denoised by using a current frame and a previous frame.

$$D_{tf}(\boldsymbol{u}) = \frac{(D_b(\boldsymbol{u})(w + 1) + D_{b-1}(\boldsymbol{u})w)}{2w + 1} \tag{2}$$

w is a constant weight term. After denoising the depth image, we can obtain a vertex map corresponding to a camera coordinate since we assume that the camera's intrinsic parameters are known. We estimate the normal map based on the method proposed by Holzer et al. [8]. This method can generate the smooth normal map. However, this method cannot estimate the normal vector around the boundary of the object where the difference of the depth value is too large. We obtain normal vectors by calculating a cross product of two vectors from the neighbor points on these areas. Normal vector $\boldsymbol{N}(\boldsymbol{u})$ at a point $\boldsymbol{u} = (u, v)$

$$\boldsymbol{N}(\boldsymbol{u}) = (V(u + 1, v) - V(u, v)) \times (V(u, v + 1) - V(u, v)) \tag{3}$$

$V(u, v)$ is a vertex map that corresponds to a camera coordinate. Normal map for illumination estimation and relighting is obtained by merging Holzer's method [8] and Eq. (3).

3.2   Spherical Harmonics Lighting

The purpose of this section is to explain the illumination estimation theory using Spherical Harmonics. Relationship between pixel intensity, albedo and normal vector is presented by Ramamoorthi et al. [6]. This is called "Inverse Rendering". We assume that the light source is distant and objects in the scene have lambertian surfaces.

The irradiance $E(\boldsymbol{x})$ observed at a point $\boldsymbol{x}$ is given by an integral on the distant sphere Ω

$$E(\boldsymbol{x}) = \int_\Omega L(\boldsymbol{\omega}) max((\boldsymbol{\omega} \cdot \boldsymbol{n}(\boldsymbol{x})), 0) d\boldsymbol{\omega} \tag{4}$$

$L(\boldsymbol{\omega})$ is incoming light intensity along the direction vector $\boldsymbol{\omega} = (\theta, \varphi)$ and $\boldsymbol{n}(\boldsymbol{x})$ is normal vector at a point $\boldsymbol{x}$. Then, $\boldsymbol{\omega} \cdot \boldsymbol{n}(\boldsymbol{x})$ is a dot product of a normal vector and incoming light direction. $max((\boldsymbol{\omega} \cdot \boldsymbol{n}(\boldsymbol{x})), 0)$ means that if a normal vector and incoming light vector has the same direction, the light from that direction is fully considered, but if the angle of these two vector is more than 90 degree, we exclude the light from that direction.

It takes too much cost to calculate Eq. (4) regarding the illumination as a aggregate of point sources. Spherical Harmonics(SH) approximation is good way to reduce the calculating cost. The illumination is shown with SH basis function y and the coefficients c The Eq. (4) will be represented in the following equation.

$$E(\boldsymbol{x}) = \sum_{l=0}^{\infty} \sum_{m=-l}^{l} A_l(\theta) L_{l,m} Y_{l,m}(\boldsymbol{\omega}) \tag{5}$$

l denotes the SH band. There are $2l + 1$ functions in band l, and m denotes the index in a band. $A_l(\theta)$ is the SH projection of $max((\boldsymbol{\omega} \cdot \boldsymbol{n}(\boldsymbol{x})), 0)$. It is obtained by rotating the standard cosine term A_l^{std} which is equal to $A_l(0)$ [9]. $Y_{l,m}$ is the basis function of SH and $L_{l,m}$ is the coefficient of each SH basis function. The pixel intensity $I(\boldsymbol{x})$ from a point $\boldsymbol{x}$ on a image is written as

$$I(\boldsymbol{x}) = R_d \sum_{l=0}^{\infty} \sum_{m=-l}^{l} A_l(\theta) L_{l,m} Y_{l,m}(\boldsymbol{\omega}) \tag{6}$$

R_d represents albedo. Since we consider lambertian surface uniformly reflects the incoming light to all direction, R_d is set to constant values.

3.3   Albedo and Illumination Estimation

The purpose of this part is to estimate illumination environment by employing inverse rendering method represented as Eq. (6). Illumination environment can be estimated from the pixel intensity, the normal map and the surface albedo. However, albedo is also unknown property in most cases. Estimating these two properties at the completely same time is ill-posed problem. So, we employ an iterative method to obtain both albedo and illumination environment.

Objects in real scene generally have varying albedo on each point. However, precisely adopting this assumption takes too much cost and makes our method hard to be done. We assume that the region with the same color and the same material, such like a monochromatic T-shirt, have uniform albedo value. We subtract the target region by thresholding depth values. After that, we segment a color image by k-means segmentation to obtain divided regions depending on objects as a first step as shown in Fig. 2. For example in case of that the target is a human, a shirt, a jacket, pants, etc., that can be regions with uniform albedo. For each region, we set an average pixel value as an initial albedo of each region.

After that, we iteratively solve Eq. (6) by using the least-square method to obtain illumination environment. Such iterative approach to estimate albedo

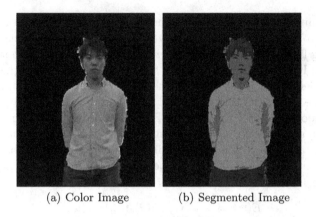

(a) Color Image   (b) Segmented Image

Fig. 2. Image segmentation

under unknown illumination is based on Zou et al. [10]. We implement this method to estimate illumination with unknown albedo values. The iterative illumination estimation is performed as the following procedure. First, initial illumination is estimated with initial albedo value. Next, albedo is updated by fixing the illumination that was estimated in previous step. Then, illumination is also updated with updated albedo. This procedure is repeated until SH coefficients values are converged. We assume that the illumination color is white, so we handle Eq. (6) in gray scale.

3.4   Relighting Using Spherical Harmonics

In this part, we explain the relighting theory. At first, we define the name of each scene. "Src" is the scene where has the object to be relighted and "Dst" is the other illumination environment that is different from Src. We can calculate the pixel intensity of the relighted object fitting to Dst illumination environment from the ratio of Eq. (6) [3]. Relighting result is obtained from following equation.

$$I_{Dst}(x) = I_{Src}(x)\frac{\sum_{l=0}^{2}\sum_{m=-l}^{l}A_l(\theta)L_{l,m}^{Dst}Y_{l,m}(\omega)}{\sum_{l=0}^{2}\sum_{m=-l}^{l}A_l(\theta)L_{l,m}^{Src}Y_{l,m}(\omega)} \tag{7}$$

$I_{Dst}(x)$ is a pixel intensity of the relighted object and $I_{Src}(x)$ is an original pixel intensity of the object to be relighted. We can get the final result image by superposing the relighted target object on Dst background image. Ramamoorthi et al. showed that nine SH coefficients are needed to approximate diffuse reflectance [11]. Based on this method, we use nine SH coefficients and set the supremum of l to 2.

Illumination of Dst scene can be obtained by applying the same illumination estimation method as Src, by using SH projected Omnidirectional images, or synthesized cubemap images. In the case of using the same way as Src to get Dst illumination, we need at least one lambertian object in Dst scene.

4   Experiments

In this section, we present experiments performed to demonstrate the effectiveness of the proposed method. We capture an image of a person under unknown illumination environment by using an RGB-D camera. Albedo of objects such as a blue shirt in the scene are also unknown. We iteratively estimate albedo and illumination environment, and relight the person subsequently. Finally, we get the final result by superposing the person to the background images.

4.1   Experiment Condition

We relight a person who is in a room but the illumination environment is unknown. He is wearing a blue shirt. We estimate the illumination from the region of a target person. Note that 3D reconstruction is hard to be applied because the person moves his body. We also need the illumination environment of Dst scene. In this experience, we obtained the illumination from omnidirectional images captured by omnidirectional camera. The background for resulting composite images is also obtained from these omnidirectional images. Dst scene has sunlight casted from left side. Both images are shown in Fig. 3. We use Microsoft Kinect as an RGB-D Camera and assume that camera intrinsic parameters for converting depth image to vertex map are known.

(a) Src scene   (b) Dst scene

Fig. 3. Src and Dst scenes

4.2   Experiment Result

Src scene is indoor scene that has fluorescent lamps but Dst scene is outdoor scene. As shown in Fig. 4, superposing a person without relighting causes unnatural appearance, because his body does not have any shadow area, while the building wall behind the person strongly shaded. Note that our method doesn't have off-line process because we estimate the surface albedo and illumination iteratively, and obtain the normal map from depth images on each frame. Therefore, our method can also be applied for dynamic shape target. Focusing on front

| 15st frame | 182th frame | 222th frame | 319th frame |

Fig. 4. Relighting results. Each caption shows frame number. Upper column shows composite images without relighting. Lower column shows relighting results.

of his body, the brightness are varying according to the angle of the body. Our method can relight dynamic shape targets under unknown illumination to fit other scene which has different illumination and can realize the photometric consistency on composite images.

4.3   Quantitative Evaluation

To evaluate the performance of our method, we compare the results with a ground truth data. We calculate root mean square error values between each result and the ground truth image. The ground truth is captured in the same room with Src scene, but consists of the same illumination with Dst scene. We relight the mannequin and calculate RMS error values on each of pixels of the mannequin region and get average values. Root mean square error values are shown in Table 1 with other two results. The one is relighting result with illumination estimation method proposed by Gruber et al. [12]. Since this method doesn't provide the way of estimating albedo, we estimate the albedo of mannequin in Src scene and a blue shirt in Dst scene in off-line process. The other one is the result of a composite image without relighting. Those two relighted result images are using the same relighting algorithm [3]. The difference is the method of illumination estimation. Input scenes, the ground truth image and result images are shown in Fig. 5.

Table 1. Root mean square error value comparing with Ground Truth

	RMS Error(pixel value)
Proposed method	13.8
Illumination estimation by [12]	13.7
Without relighting	18.6

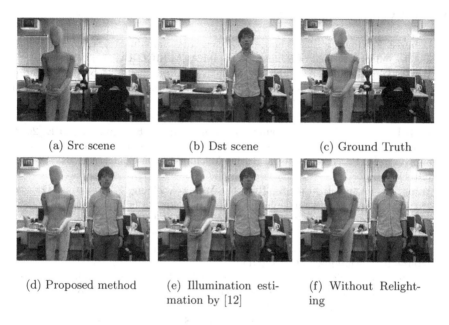

(a) Src scene   (b) Dst scene   (c) Ground Truth

(d) Proposed method   (e) Illumination esti-   (f) Without Relight-
                       mation by [12]   ing

Fig. 5. Comparison with ground truth

An advantage of our method is that we don't need obtain any albedo beforehand. The error value of the proposed method was similar to the result of illumination estimation by Gruber et al. [12]. We can say that we could estimate the illumination environment appropriately from this point. On the other hand, the error value of proposed method was lower than "Without relighting". Our method could also relight an unknown shape object appropriately.

5   Conclusion

In this paper, we proposed a relighting method for arbitrary shape objects captured under unknown illumination by using an RGB-D camera. We can obtain illumination environment without any light probes by the iterative illumination and albedo estimation. This means that we don't need to estimate albedo in off-line process. Thus our method can be used for arbitrary shape objects under unknown illumination. After the iterative illumination estimation, relighting process is done with estimated illumination data. Since our method obtains the normal map from depth images on each frame, a dynamic shape object can be applied to our method. In experiment, we tested our method to show the utility and quantitative evaluation. We will improve our method more robust, and also apply relighting to other purpose such as object tracking.

References

1. Debevec, P., Hawkins, T., Tchou, C., Duiker, H.P., Sarokin, W., Sagar, M.: Acquiring the reflectance field of a human face. In: ACM SIGGRAPH. ACM (2000)
2. Wenger, A., Gardner, A., Tchou, C., Unger, J., Hawkins, T., Debevec, P.: Performance relighting and reflectance transformation with time-multiplexed illumination. ACM Trans. Graph. **24**(3), 756–764 (2005)
3. Zhen, W., Liu, Z., Huang, T.S.: Face relighting with radiance environment maps. In: IEEE Conference on Computer Vision and Pattern Recognition. IEEE (2003)
4. Aldrian, O., Smith, W.: Inverse rendering with a morphable model: A multilinear approach. In: Proceedings of the British Machine Vision Conference. BMVA Press (2011)
5. Wang, Y., Zhang, L., Liu, Z., Hua, G., Wen, Z., Zhang, Z., Samaras, D.: Face relighting from a single image under arbitrary unknown lighting conditions. IEEE Trans. Pattern Anal. Mach. Intell. **31**, 1968–1984 (2009)
6. Ramamoorthi, R., Hanrahan, P.: A signal-processing framework for inverse rendering. In: Proceedings of the 28th Annual Conference on Computer Graphics and Interactive Techniques. ACM (2001)
7. Matsumoto, K., Song, C., de Sorbier, F., Saito, H.: Joint upsampling and noise reduction for real-time depth map enhancement. In: Proceedings of IS&T/SPIE Electronic Imaging, SPIE (2014)
8. Holzer, S., Rusu, R.B., Dixon, M., Gedikli, S., Navab, N.: Adaptive neighborhood selection for real-time surface normal estimation from organized point cloud data using integral images. In: IEEE/RSJ International Conference on Intelligent Robots and Systems. IEEE (2012)
9. Nowrouzezahrai, D., Simari, P., Fiume, E.: Sparse zonal harmonic factorization for efficient SH rotation. ACM Trans. Graph. **31**, 23 (2012)
10. Zou, X., Kittler, J., Hamouz, M., Tena, J.R.: Robust albedo estimation from face image under unknown illumination. In: SPIE Defense and Security Symposium, SPIE (2008)
11. Ramamoorthi, R., Hanrahan, P.: An efficient representation for irradiance environment maps. In: ACM SIGGRAPH. ACM (2001)
12. Gruber, L., Richter-Trummer, T., Schmalstieg, D.: Real-time photometric registration from arbitrary geometry. In: IEEE International Symposium on Mixed and Augmented Reality. IEEE (2011)

Evaluation of Fatigue Measurement Using Human Motor Coordination for Gesture-Based Interaction in 3D Environments

Neera Pradhan[1], Angela Benavides[1]([✉]), Qin Zhu[1], and Amy Ulinski Banic[1,2]

[1] Department of Computer Science, University of Wyoming, Laramie, USA
{neera,abenavi3,qzhu1}@uwyo.edu, abanic@cs.uwyo.edu
[2] Idaho National Laboratory, Idaho Falls, USA

Abstract. Benefits of immersive three-dimensional (3D) applications are enhanced by effective 3D interaction techniques. While gesture-based interaction provides benefits in these types of environments users commonly report higher fatigue than with other interaction solutions. Typically, fatigue is measured subjectively but may lack precision, consistency and depth. Our research proposes a novel, more consistent and predictable measure of fatigue. This paper presents the details of our technique based on human motor coordination, the results of an experimental study on gesture-based interaction, identifies attributing causes for fatigue and outlines design guidelines to reduce fatigue for gesture-based interaction techniques. These results have implications for gesture-based or mid-air interaction techniques for 3D environments, such as virtual environments and immersive visualizations.

1 Introduction and Motivation

Gesture-based interaction is an input mechanism used to interact with computer systems, it is intuitive to the users and effective for large display interactions [1]. There are several advantages expected when using hand-based gestures for interaction [2], for example:

- *Natural interaction:* Natural means for manipulation of virtual objects.
- *Direct interaction:* Controller-free environment where no additional devices are required to interact with the system.
- *Multiple uses of interaction:* For a spatial context, a hand movement can provide both positional values as well as the intended action by just using one hand gesture.

Regardless of its advantages, researchers still face the problem of creating applications that have all the former benefits. The designers of gesture-based interaction work based on application requirements to create new interaction gestures, this limits the exploration stage to understand about limitations and represent a drawback in this field. Therefore, the process of implementing a gesture for an application becomes rigid and may result in unfavorable designs.

© Springer International Publishing Switzerland 2015
G. Bebis et al. (Eds.): ISVC 2015, Part II, LNCS 9475, pp. 443–452, 2015.
DOI: 10.1007/978-3-319-27863-6_41

User studies about hand gesture interaction have revealed that hand/arm fatigue is one of the major drawback in this kind of interaction [3]. They did not addressed how the fatigue measuring tool was used, neither given suggestions to alleviate the fatigue experienced by the users. Our research study aims to use the understanding of human motor coordination to find a new approach of measuring and to explain the contributors to fatigue in gesture interaction.

We propose a novel measurement technique for contributing factors of fatigue. Combining a 3D environment with gestures interaction we present an experimental study to evaluate our measurement technique and validate it against others.

In this study we answer the following questions: (a) Is Continuous Relative Phase (CRP) measure a good way to find precise information on fatigue for hand gestures? (b) Based on the calculated CRP values, what is the pattern that can be used for hand coordination while the users perform the gestures? (c) Can CRP values validate the subjective data reported by users through questionnaires and NASA-TLX?. Our hypotheses were: (a) Our technique will report similar levels of fatigue as standard NASA-TLX. (b) Our technique will provide some information and help to determine the cause of fatigue in the field of gesture design.

2   Related Work

2.1   Measurements of Fatigue

In studies where fatigue was reported users were interviewed, asked to fill out questionnaires and rate their fatigue level according to predefined Borg's scales [4] or NASA-TLX. These metrics lack details about the contributing factors that could cause fatigue and are subjective because they could have variations between users according to their tolerance level. Chung et al. [5] designed a computer based body postural stress evaluation system that take the body joint range of motion and uses the neural network results to predict the workload. In a research conducted by Kölsch et al. [6], a comfort function for gestures design was created. However, this function had a limitation of area of application: horizontal area in front of the body. An objective measurement for muscle movements is electromyography (EMG). It can be used to measure muscle fatigue by recording signals and monitoring changes of different frequencies. The problem with this method is that gesture interactions don't usually have a long duration [1,2], therefore, it reduces the chances of getting enough muscle movement activity to record EMG signals for evaluating muscle fatigue. A similar issue results from the Cybex Text [7], which measures muscle contractions while subjects perform repetitive task until the movement force falls below 50 %.

2.2   Coordination Patterns

For any two joints, a coordination pattern can be represented in continuous relative phase (CRP). This describes the mechanical behavior of two joints using a single variable representing elements related to coordination. It is a function of time showing relative phase angles of a pair of joints throughout the entire movement cycle [8]. The phase angle is calculated using Eq. 1, where $x'(t)$ is the velocity and $x(t)$ is position at time t.

$$\phi(t) = \tan^{-1}\left(\frac{x'(t)}{x(t)}\right). \tag{1}$$

After normalizing the displacement and velocity data obtained from each joint, CRP can be calculated by subtracting the phase angle of one joint from the angle of the other joint, at the same instant throughout the movement cycle [9,10]. The formula to calculate the CRP angle of the lefthand/righthand joint pair is:

$$CPR(t) = \phi_{lefthand}(t) - \phi_{righthand}(t). \tag{2}$$

where $\phi_{lefthand}(t)$ and $\phi_{righthand}(t)$ are the normalized phase angles of the wrist and elbow respectively.

Coordinated movements can be either in-phase (simultaneous contraction of homologous muscles) or anti-phase (simultaneous contraction of non-homologous muscle) when the CRP value is 0° or 180° respectively. The stability of the wrist movement can be represented by the standard deviation for CRP values close to 0, which suggests how well the bimanual coordination pattern can be maintained. As the joint movement approach out-of-phase movements, the instability increases, which may affect the ability of muscle joints to withstand the movement impact [11]. These research findings have only been used in kinematics.

3   Fatigue Measurement Technique

3.1   Relation Between Fatigue and CRP

Calculating CRPs during the entire movement cycle for any two joints, we can evaluate the variability of CRPs from in-phase or anti-phase. Our work focuses on hand coordination by looking at the mean relative phase for the *lefthand - righthand* (distal) joint pair as a function during the movement cycle. The reason to consider this is because distal joints adjust the motion in order to continue with the movement cycle. During fatigue, coordination has been mostly characterized by a strong increase in the wrist movement to compensate proximal joint control impairment [12]. CRP values are represented in a range between 0° to 180° [8]. Our research aims at finding whether the variability measure of CRP for a gesture indicates fatigue during the movement. Using the commonly subjective tools to gather information on fatigue from users we can validate our approach on finding fatigue factors.

4   Experimental Study

To validate our technique, we conducted a research study at the 3-Dimensional Interaction and Agents (3DiA) Laboratory, using a 2 ×2 × 5 mixed study design.

The study encompassed a between-subjects condition, where participants were assigned to one of two display-types: monoscopic large screen display (MD) (stereopsis was turned off) or stereoscopic large screen display (SD).

The within-subjects conditions are the gesture sets and gesture types. These conditions are further detailed in the following sections.

4.1   Gesture Sets and Types (Within-Subjects Conditions)

Participants were assigned to perform the task using two gesture sets:

- *Dynamic:* Refers to dynamic hand gestures based on hand motions and point-ing gestures using hand or arm location. In this case, the user had to perform the gestures in the most open-ended way that they felt comfortable with and were asked to maintain consistency along all the trials.
- *Static or train gestures:* Refers to static hand gestures based on hand postures. In this case, the user learned how to perform the gestures and were asked to maintain consistency along all the trials.

To determine the types of gestures performed in each set and define how the users should execute them five investigators participated in a preliminary data collection session. They performed a selection task on lightweight physical objects to identify how users would interact with these objects using natural gestures they would use in the real world. These actions were tracked, recorded and analyzed to define the types of gestures and instructions for the static set. Based on the analysis we found that most of the gestures performed by users were:

- *Push:* Fully extend the arms in front of the body.
- *Pull:* Bring the hands inwards, towards the body.
- *Divide:* Separate the hands sideways.
- *Hold:* Extend the arms to hold and object in the air.
- *Volume Selection:* Define an arbitrary shape of an object to select it.

Each user performed the gestures in 5 sets of trials in both a dynamic and a static manner. Each set of trials was performed in a random order, which is given in Table 1.

4.2   Apparatus

We used OptiTrack Tracking Tools (OTT) [13] to capture 6-degrees of freedom (6DoF) position and orientation values of wrist, elbow and shoulder joints. Hand joint movement values collected from OTT were recorded using the modified Nat-Net SDK. We also used CyberGlove II [14] to record the angles of the bending of finger joints. In order to record the position and orientation of the forearm in space we attached infrared trackers to the wristband of the CyberGlove. Additionally, VRUI [15] was used to create a 3D visualization that was displayed to the participants. However, there was no visual feedback to their actions. This was done in order to motivate the users to perform actions freely and without any application specific restrictions or cause/effect of their interactions.

Table 1. Random order of tasks for a trial

Type of gestures				
Order 1	Push	Pull	Divide	Hold	Volume selection
Order 2	Pull	Hold	Volume selection	Divide	Push
Order 3	Volume Selection	Divide	Push	Pull	Hold
Order 4	Divide	Push	Volume selection	Hold	Pull
Order 5	Hold	Pull	Divide	Push	Volume selection

4.3   Programming Approach

For this research, the NatNetClient [16] was used collaboratively with the VirtualHand SDK [17]. A C++ program was created to collect position and orientation values from NatNetClient. It also creates virtual hands to display the user's movement using the values received through the CyberGlove. The positional values from OptiTrack were also saved in a separate .dat file for data analysis.

4.4   Data Collection

For calculating CRP values we classified the direction of movement occurred during a gesture according to body planes. Human movements are described in 3D based on a series of planes and axes. There are three planes of motion called body planes that pass through the human body: sagittal plane, frontal plane and transverse (horizontal) plane (Fig. 1a). A body joint movement occurs in all three planes, however, when describing the dominant plane movement, it is classified in just one of the three planes, the one where most movement occurs [18]. For any gesture, we analyzed the dominant plane by plotting the average distance over all trials for each participant. Each gesture was represented by its dominant plane, which also was used for computing the CRP values.

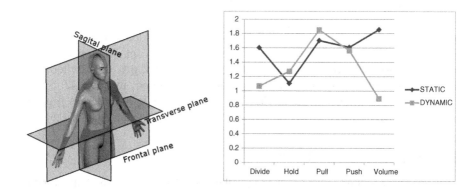

Fig. 1. (a) Body planes [19] and (b) Phase by Gesture Set

Phase angles were calculated using Eq. 2, where position and velocity were computed based on the dominant axis for left hand and right hand joints during the entire movement cycle. Movement values were calculated in an experimental setup where participants' data were different in terms of movement range. The data values were normalized across all trials to minimize all the influence of different movement amplitudes. CRP values were calculated at each frame and then normalized so that 0 % specifies the beginning and 100 % specifies the end of the gesture movement. Using this new range CRP values were aligned and then averaged.

4.5   Participants

A total of 20 participants (10 females, 10 males, mean age = 25.05 and SD = 3.28) participated in the study, either using a stereoscopic display (N = 10) or a monoscopic display (N = 10). All of them were right handed. Before participation, all of them completed a range of motion assessment without any difficulty.

4.6   CRP Results

We determined the average variability of CRP values for different experimental conditions. The higher the variability the higher the coordination between hand movements. This gave insight on how well hand movements were coordinated as in-phase or out-of-phase and whether the movements were randomly shifting causing more muscle usage. The coordinated movements for in-phase and out-of-phase range by gesture set are plotted in Fig. 1b.

4.7   NASA-TLX Overall Workload

TLX overall workload, TLX temporal and TLX effort measures were independently averaged for each participant across testing sessions. A one-way ANOVA on NASA-TLX showed there were no significant differences between participants that were grouped by gesture type. An ANOVA testing for TLX overall workload showed a main effect of display type $F(1, 18) = 11.81$, p = 0.003, whereas p using stereoscopic display reported higher overall workload (mean = 39.73, SD = 3.32). A one-way ANOVA found there is a significant difference among display types for TLX temporal demand $F (1, 18) = 6.08$, p = 0.24 (whereas p using stereoscopic display showed higher temporal demand, mean = 42, SD = 12.07) and for TLX effort demand $F(1, 18) = 6.545$, p = 0.20 (whereas p using stereoscopic display reported higher effort demand, mean = 44, SD = 5.16).

4.8   Completion Time Results

Mean completion times showed a main effect of display type on pull gesture $F(1, 2) = 20.782$, p = .045 (whereas p using stereoscopic display showed higher completion time, mean = 1.18, SD = .13) and on push gesture $F(1, 2) = 29.886$, p = .032 (whereas p using stereoscopic display showed higher completion time, mean = 1.18, SD = .24) (Fig. 2). These values were computed within each of the experimental conditions and averaged over all trials for each participant. One-way ANOVA showed there was no significant differences in mean completion times for the participants that were grouped by order of gesture type, with each F<1.

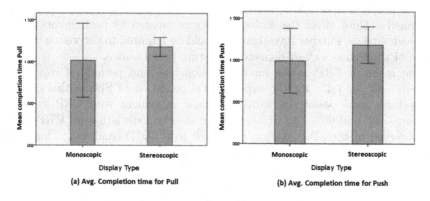

Fig. 2. Average completion time by display types

4.9   Users Perceptions

After the exercise a qualitative open-ended questionnaire was given to each participant to debrief and interview them about fatigue, task completion satisfaction, preference of one-hand vs two-hand gestures and system predicting the gestures and other issues. The mean results of responses are summarized in Table 2.

Table 2. Post-questionnaire responses

	Mean	SD	Liker scale
Task was easy	6.35	0.933	1 = Very difficult
			7 = Very easy
Feelings of tiredness in arms at the end	5.6	1.40	1 = Very tired
			7 = Normal and relaxed
Feelings about own performance	5.9	1.12	1 = Very unsatisfied
			7 = Very satisfied
Perception towards gesture performance	6.1	0.97	1 = Very unsatisfied
			7 = Very satisfied
Suitability of gestures	5.75	0.91	1 = Not suited at all
			7 = Very suitable

5 Discussion and Design Guidelines

In this section we discuss our interpretation of the results and produce guidelines for design of interaction techniques to reduce fatigue in bimanual gesture-based interaction.

For NASA-TLX overload, the Stereoscopic Display (SD) condition showed significantly higher overall workload, temporal demand and effort demand than for Monoscopic Display (MD) condition. The order of the nature of the gestures performed did not affect the choice of gestures created by participants for the open-ended trials. Further investigation would be required to determine the true source of what causes effort demand in gesture performance.

The range of CRP values for all relationships and periods of stance was between $-45°$ and $45°$ for the experimental condition of SD. In this case, all the gestures were performed with anti-phase movement with CRP values in between $-90°$ and $90°$. The higher median standard deviation for CRP values was observed in the SD condition compared to the MD condition. This shows that participants who performed in the SD condition showed higher variation in maintaining the coordination pattern across all trials.

In the SD condition subjective measures showed significant values for overall workload, temporal demand and effort demand. This suggests that higher workload requirement can likely cause fatigue. With CRP analysis higher variability in coordination pattern was visible in the SD condition, which suggests that frequent change in coordination patterns can likely induce fatigue. However, higher variability in CRP values can also suggest inter-trial and inter-person inconsistent performance throughout the trials. More specific gestures to perform the task can be implemented in the future to test the validity of CRP variation.

For user comfort feelings of tiredness were found to be not statistically significant, however, from follow-up interviews it was revealed that discomfort may have been caused because of the wearing of gloves. In the SD condition, both pre-defined and open-ended gestures showed higher variability during volume selection gesture throughout the movement cycle.

6 Conclusion

Statistical analysis showed that when there was significant effect of stereoscopic displays on total workload, temporal and effort demand, CRP values also showed higher variability during the display condition. There was no significant effect of nature of gesture and order of display type chosen for a trial.

Our results show that CRP analysis can be a cheaper alternative to objective clinical measures like EMG to analyze level of fatigue. CRP variability can report fatigue-causing factors with precision and as a measurable quantity. By using the understanding of fatigue as a measurable factor, hand gestures can be designed. We expect that this study will aid in the understanding of human motor coordination and provide more detailed information on attributing causes, leading to design guidelines for future gesture-based interaction.

6.1   Future Work

Further studies on our research work can be done to improve the quality of results produced in this research. We think that some of the areas that need more research to understand hand coordination and fatigue are: (a) Compare with other standard subjective measures to verify the results from our study, and (b) Create gesture designs using our fatigue reporting method and validate the user experience feedback with CRP variability.

References

1. Lee, S.S., Chae, J., Kim, H., Lim, Y.k., Lee, K.p.: Towards more natural digital content manipulation via user freehand gestural interaction in a living room. In: Proceedings of the 2013 ACM International Joint Conference on Pervasive and Ubiquitous Computing, UbiComp 2013, pp. 617–626. ACM, New York (2013)
2. Baudel, T., Beaudouin-Lafon, M.: Charade: remote control of objects using free-hand gestures. Commun. ACM **36**, 28–35 (1993)
3. Hinckley, K., Pausch, R., Goble, J.C., Kassell, N.F.: A survey of design issues in spatial input. In: Proceedings of the 7th Annual ACM Symposium on User Interface Software and Technology, UIST 1994, pp. 213–222. ACM, New York (1994)
4. Wikipedia: Borg scale – wikipedia, the free encyclopedia (2015). [Accessed 26 August 2015]
5. Chung, M.K., Lee, I., Kee, D., Kim, S.H.: A postural workload evaluation system based on a macro-postural classification. Hum. Fact. Ergonomics Manuf. Serv. Ind. **12**, 267–277 (2002)
6. Kölsch, M., Beall, A.C., Turk, M.: The postural comfort zone for reaching gestures. Proc. Hum. Fact. Ergonomics Soc. Ann. Meet. **47**, 787–791 (2003)
7. Binder-Macleod, S.A., Snyder-Mackler, L.: Muscle fatigue: clinical implications for fatigue assessment and neuromuscular electrical stimulation. Phys. Ther. **73**, 902–910 (1993)
8. Bartlett, R.: Introduction to Sports Biomechanics. E & FN Spon, London (1997)
9. Hamill, J., Haddad, J.M., McDermott, W.J.: Issues in quantifying variability from a dynamical systems perspective. J. Appl. Biomech. **16**, 407–418 (2000)
10. Glazier, P., Davids, K., Bartlett, R.: Dynamical systems theory: a relevant frame-work for performance-oriented sports biomechanics research. Sportscience **7**, 85–92 (2003)
11. Dierks, T.A., Davis, I.: Discrete and continuous joint coupling relationships in uninjured recreational runners. Clin. Biomech. **22**, 581–591 (2007)
12. Huffenus, A.F., Amarantini, D., Forestier, N.: Effects of distal and proximal arm muscles fatigue on multi-joint movement organization. Exp. Brain Res. **170**, 438–447 (2006)
13. NaturalPoint, I.: (Optitrack). http://www.optitrack.com/. [Accessed 25 August 2015]
14. Systems, C.G.: (Cyber glove ii). [Accessed 25 August 2015]
15. Kreylos, O.: (Vrui vr toolkit (vrui)). http://idav.ucdavis.edu/okreylos/ResDev/Vrui/. [Accessed 25 August 2015]
16. OptiTrack: (Natnet sdk). http://www.optitrack.com/products/natnet-sdk/. [Accessed 25 August 2015]

17. Systems, C.: (Virtualhand sdk). http://www.cyberglovesystems.com/products/virtual-hand-sdk/overview. [Accessed 25 August 2015]
18. Solutions, P.: (Understanding planes and axes). http://www.physical-solutions.co.uk/wp-content/uploads/2015/05/Understanding-Planes-and-Axes-of-Movement.pdf. [Accessed 25 August 2015]
19. Panjabi, M.M., Goel, V., Oxland, T., Takata, K., Duranceau, J., Krag, M., Price, M.: Human lumbar vertebrae: quantitative three-dimensional anatomy. Spine **17**, 299–306 (1992)

JackVR: A Virtual Reality Training System for Landing Oil Rigs

Ahmed E. Mostafa[1]([✉]), Kazuki Takashima[2], Mario Costa Sousa[1],
and Ehud Sharlin[1]

[1] University of Calgary, Calgary, Canada
aezzelde@ucalgary.ca
[2] Tohoku University, Sendai, Japan

Abstract. We propose JackVR, an interactive immersive simulation
prototype aiming to train domain experts to land jackup oil rigs. Jackup
rigs are among the most common offshore drilling units for extracting oil,
but the process of landing the rigs is mostly challenging because of the
unpredictable sea and weather conditions, lack of clear vision, and the
possible risk of damaging the ocean floor. We designed JackVR to support oil engineers and technicians by allowing them to practice landing
the oil rig within a safe and semi-realistic training environment. Furthermore, the design explores various superimposed spatial indicators that
provide visual warnings on unexpected task conditions. The implemented
prototype supports two modes for training, and utilizes the ray-casting
interaction technique to enable seamless and direct control of the rig. . . .

1 Introduction

Virtual reality (VR) training applications are widely used to prepare users for
tasks that may be too costly or dangerous to practice in real world settings [1].
Within the oil-and-gas domain, some VR training systems were proposed and
implemented in order to support domain experts who train for challenging oil-
and-gas processes and tasks (e.g. [2,3]).

One such task, which is the focus of this paper, is the process of landing
a jackup oil rig. This process involves many environmental challenges including unpredictable weather conditions, varying (deep) ocean forces, and uneven
seabed topography. During the landing process operators are provided with a
very simple user interface to control the oil rig (e.g., move its legs up/down) with
minimal visual feedback, which is arguably insufficient to satisfy the complete
set of task requirements. During the landing the operator must stably and correctly position the rig to previously defined positions at the seabed, while being
aware of dynamically changing environmental conditions, and carefully avoiding risks of damaging the topography of the seabed. These challenges motivated
the design of JackVR, a training system that provides domain users with an
immersive environment that helps them understand the multifaceted challenges
of landing the oil rig, and practice landing it in a semi-realistic environment.

© Springer International Publishing Switzerland 2015
G. Bebis et al. (Eds.): ISVC 2015, Part II, LNCS 9475, pp. 453–462, 2015.
DOI: 10.1007/978-3-319-27863-6_42

Fig. 1. JackVR's Concept: immersive landing of oil rigs with Ray selection technique

Bin He [4] proposed a virtual prototyping system to validate the design of an offshore drilling platform including its jacking systems. However, to our knowledge, there is no immersive system that supports practicing offshore oil rig landing.

We propose JackVR, an interactive training system that supports domain experts train in landing jackup oil rigs in a variety of simulated scenarios. We implemented JackVR as an immersive virtual reality simulation of jackup oil rig landing that allows control of various environmental aspects, such as ocean waves and wind, enabling spatial interaction using ray-casting interaction technique (Figs. 1 and 4). This paper details the main components of JackVR, its usage in landing simulation tasks, and our plans for future improvements.

2   Jackup Rigs

Drilling is one of the main processes in hydrocarbon extractions with onshore or offshore drilling units, the latter are mounted mostly in the middle of the ocean. Jackup rigs [5] are among the most common offshore drilling units due to their low operation and maintenance cost. A jackup rig usually consists of the rig itself, the legs, spud cans (heavy objects attached to the legs to facilitate seabed penetration), and the hull. The legs and the hull of most jackup rigs usually can be lifted down/up through the rigs' jacking system. The weight of the jackup rig impacts the landing at certain points and proper utilization of the rig's hull with the water level inside the hull must be maintained. The stability of the rig mostly depends on the rig's weight and the environmental forces applied (e.g. wind and ocean pressure).

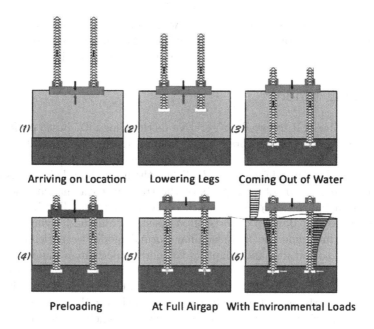

Fig. 2. Steps involved in the process of landing jackup oil rigs.

Many steps are associated with the operation of landing a jackup rig (Fig. 2). The jackup rig is pulled (moved and oriented) to the target location. Then, the jacking system is used to lower the legs with careful attention to the environmental forces and the seabed depth until the legs touch the seabed. Next, the hull is raised out of the water prior to filling it with water to increase the total weight of the rig, in order to enable penetration of the seabed. Following the penetration, the hull is further raised leaving an air gap for more stability, with the rig fully landed and almost ready as a fixed platform for drilling. JackVR allows the user to train and practice all the landing phases with the exception of the pulling to the target location (the 1st phase), which reflects on towage task components that are arguably external to the core landing process.

3   Why JackVR?

Operators of oil rigs may need to control the rig through a simplified desktop interface that only supports basic interaction capabilities to land the rig along with standard 2D images of the under-ocean terrain topography. Due to the apparent lack of simulating the surrounding environment, a more realistic, safe, and engaging training environment is needed to better educate the operators about landing the oil rig, especially with the various weather and environmental challenging conditions. In this project, we propose JackVR, an immersive simulation and semi-realistic training environment for the process of landing oil rigs.

In our early meetings and consultations with our domain collaborators, they highlighted the importance and the potential of having an immersive environment, such as JackVR, to support the training of landing oil rig.

4   Training Modes

The design of JackVR followed an iterative methodology [6], based on feedback from domain expert collaborators. One of the main design requirements was that JackVR will allow users to experience two training modes, "normal" mode, and what we termed a "superhuman" mode. When the normal-mode is set, users are able to completely control all aspects of the training with the same low level of insight that can be expected from the current oil rig landing interfaces. It is worth noting that domain experts often resist learning and trying new tools (e.g., [7]). By including the more familiar normal mode, we anticipate a simpler transition to learning and adapting the new, superhuman, mode. While in the "normal" mode, the visualization is lacking and only few low-level insights can be gained. The "superhuman" mode would allow us to explore intuitive interactions directly as additional features provided over the simpler "normal" mode. The "superhuman" mode enables the user to benefit from empowerment of "supervision" abilities and intuitive feedback including superimposed visualizations of the simulation and awareness of the surrounding objects in the environment. Furthermore, the superhuman mode has potential to extend user engagement beyond the use of natural interaction techniques, with a vision for future use of drones (or small submarines) that would aid in such complex simulation processes. In essence, this unrealistic mode provides a more engaging experience and better understanding of the simulation process and its various attributes, with the goal of enabling rich task awareness and as a result better and more efficient training and learning experience.

5   Simulation Attributes

The design of JackVR simulation included the following simulation elements: (1) a module that simulates the environmental forces, (2) a notification and warning messages generation module, (3) a state-machine module that continuously evaluates the possible outcome of the training, and (4) interactive visualization module which renders the 3D rig model and supports the various spatial VR interactive techniques.

JackVR continuously simulates the environmental conditions impact on the rig's tilt and movement. We use simplified physics simulation with two parameters that affect the rig tilt, ocean condition and depth, as a simplified representation of any horizontal or vertical environmental forces such as wave current and hull weight. We assume that the sea condition is the main (angular) variation that affects the rig tilt, and modeled it using a simplified 1D noise function. The output from the noise function is a random angle that is generated periodically as the new rig orientation, and is interpolated over time to smoothly reflect the

Table 1. Training difficulty levels.

		Smooth	Rough
		Seabed terrain topography	
	Calm	Easy	Normal
Sea Condition	**Normal**	Normal	Hard
	Stormy	Hard	Very Hard

new tilt status. Ocean depth is the secondary parameter that affects the tilt according to the current depth of the rig's legs. When the legs arrive at the seabed, we consider the internal rig weight to be at the highest and stop the tilt motion.

Table 2. Training outcome according to JackVR difficulty levels

If	**Easy**	**Then**	**Failure never occurs**
If	**Normal**	**Then**	**Failure if mistakes ¿ 2**
If	**Hard**	**Then**	**Failure if mistakes ¿ 1**
If	**Very Hard**	**Then**	**Failure for any mistake**

The warning messages module is integrated as a visual notification reporting the status of the landing. A warning value is issued by continuously taking into account the following simulation parameters: sea wave conditions (e.g. calm, normal, or stormy), the depth of the rig's legs relative to the total depth of the ocean floor, the status of the hull (either raised or lowered), and the type of the seabed terrain topography (either smooth or rough). We designed the warning notifications to allow trainees that miss early notifications continuous indication of the simulation status and better chances of recovering from problems that might have been ignored initially.

A state-machine component was designed to continuously evaluate the overall simulation status, and determine landing success or failure. Prior to the simulation start, the user customize a set of parameters that determine the difficulty level, which is later translated into a set of simulation variables (Table 1). This customization phase allow the simulation to better fit the varying expertise of potential users. Afterwards, the training outcome is evaluated based on the simulation difficulty level in combination with the total number of users' errors (Table 2). For instance, in the 'very hard' difficulty level the user can fail the entire task if the rig's hull is raised in the VR immersive environment earlier than it should.

The visual user interface of JackVR was designed to be rendered in stereo head-mounted display (HMD). Text messages were rendered only within the center of the user's view, while other indicators such as some of the system statistics were rendered via alternative cues such as sound and graphics elements.

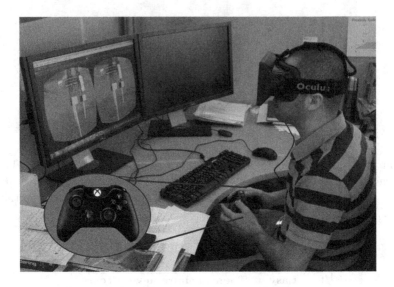

Fig. 3. Landing a jackup oil rig with JackVR

Along this line, we explored superimposed visualization as on-demand visuals which are attached to certain objects within the immersive environment. Such superimposed visuals are scaled relative to the user's eye (or camera location), allowing to simplify the user interface while still providing awareness of the simulation variables and status.

6   Interaction

Interacting with JackVR requires the user to wear an Oculus Rift HMD enabling immersive depiction of the 3D environment and its surroundings (Fig. 3). Within the JackVR immersive environment the user is interacting with a rich set of simulation graphical representations including animated 3D ocean waves (with a simple buoyancy), sky and clouds (through a skybox), a set of floating ships over the sea surface, a jackup rig 3D model, and underwater effects (implemented using Unity3D, Fig. 4). Furthermore, basic audio support has been integrated to further enhance the immersion of the experience.

In the actual landing process, the operator needs to observe the rig and its legs from different perspectives during the landing, and to be able to interact with the rig by moving its legs up and down, or by raising the rig's hull as needed. We mapped these task interactions to fit the design of VR simulation. We realized that the simplicity of the ray-casting interaction technique [8] (Fig. 4) makes it a seamless and a suitable choice for such mapping.

We decided to enable interaction with the JackVR immersive environment using the ray-casting technique, as it fits the navigation and control around the rig. First, with regards to navigation, we wanted to support users with a first-person flying experience to be able to directly see the 3D rig from any point

Fig. 4. JackVR and the ray-casting interaction technique in two ocean conditions: high stormy waves (left), and calm waves (right)

during the simulation. Second, to control the rig, it must be "selected" first, and the design of the ray supports that as follows: the ray is rendered as a virtual screen-inward laser pointer reflecting the user direction. When the ray intersects objects within the 3D world, it can cause specific visualizations to appear and can enable direct interaction with objects. For example, when the user is oriented towards the rig model, the ray will enable a specialized visual element attached to the rig which indicates its current status (Fig. 5). The ray casting approach and the superimposed visualization provide details-on-demand, which prevents over-population of the user's view, and allows direct access to faraway objects.

JackVR uses a Xbox controller to support a gaming-like interaction within the simulated environment. The user can navigate the environment using both the left-analog stick and HMD movement. The controller's buttons are used to control the rig itself. For instance, the "X" and "B" buttons are used to control the rig's legs while the "A" and "Y" are dedicated to raising and lowering the rig's hull.

Sound is utilized as a second sensory element in addition to visual elements. Our initial implementation only supports sound-based feedback when the user is traveling within the virtual environment. For instance, the sound of air while moving above surface differs depending on the user's traveling speed, and moving under the sea level will result in ambient underwater sound effects. Sound can be used beyond the aforementioned effects, e.g., as auditory warnings that aim to notify the user of sever issues regarding the landing process, which we are exploring as future work.

JackVR design is focusing the user's attention on the immersive experience. For example, when a landing problem is indicated it will not be reported numerically or as text, but rather via visual notification cue, such as a red overlay that is blended over the final rendering frame with a transparency that is relative to the current warning value (Fig. 6). While JackVR can integrate the more

Fig. 5. JackVR's "superhuman" mode is active

typical notifications (e.g. textual-based messages, or 2D graphs), it enables with
rich superimposed visualization that provides more direct seamless information
superimposed on the main rendered simulation objects.

The JackVR visualization status varies based on the selected mode. When the
normal-mode is active, the user interface relies on basic textual status indicators,
while still allowing the user to fully control all simulation aspects. When the
superhuman-mode is enabled, the user is empowered with a set of superimposed
visualizations including visual bars, indicators and meters which are integrated
within the immersive environment components. For instance, a visual meter
(similar to cars speedometers) reflecting the current warning level would appear
when facing the rig (Fig. 6).

In the superhuman mode, various graphical interface elements have been
implemented aiming to assist users and inform them of the simulation's status.
For instance, the parameter representing the water level inside the rig is simulated through an icon reflecting two states of the water level inside the rig's tank
(either empty or full). Similarly, the leg depth parameter is represented by a
floating 3D indicator showing the current depth relative to total depth of seabed
(Fig. 6).

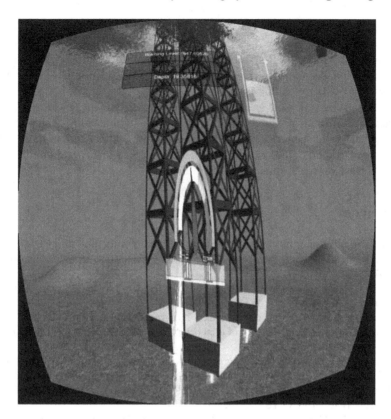

Fig. 6. Warning level representation in superhuman mode

Upon completion of the training, a graphical notification of either success or failure is shown. The user can still navigate around the environment after finishing the training, but at this point cannot change or interact with any object within the virtual world. We enabled this post-task view in order to provide the user with an opportunity to reflect on the task performed and on the reasons for landing success or failure.

7 Discussion and Future Work

JackVR is still a proof-of-concept prototype, and while it was designed iteratively with domain experts, we still did not evaluate it as a practical training tool. Current preliminary feedback from a senior domain expert points to JackVR potential help to engineers who are learning the oil-rig landing process, its ability to provide a sense of difficulty and unexpected hazards, which might happen and cost companies millions of dollars.

We plan to run a user-study of JackVR to gather in-depth feedback from its potential users to refine the representation and better support the domain users expectations.

8   Conclusions

We presented JackVR, a VR simulation aimed at training oil-and-gas practitioners in landing offshore oil rigs, focusing on the process of landing a jackup rig as a widely used offshore drilling unit. JackVR incorporates features that simulate the landing process as well as the surrounding environment, including seabed topography and ocean waves. Users can change the simulation parameters, and practice a variety of scenarios. JackVR also supports a superhuman mode which superimposes 3D indicators with the immersive environment, and enables a ray casting interaction technique.

References

1. Bricken, W.: Training in virtual reality. In: Virtual Reality, pp. 46–48 (1991)
2. Evans, F., et al.: Future trends in oil and gas visualization. In: Visualization, pp. 567–570 (2002)
3. Slatt, R.M., et al.: Visualization technology for the oil and gas industry: today and tomorrow. AAPG Bull. **80**, 453–458 (1996)
4. He, B., et al.: Virtual prototyping-based integrated information modeling and its application in the jacking system of offshore platform. Int. J. of Hybrid Inf. Technol. **6**, 135–140 (2013)
5. Ridehalgh, J., Edwards, A.: Design Operation and Towage of Jack Up Rigs. RINA (1982)
6. Sedlmair, M., et al.: Design study methodology: reflections from the trenches and the stacks. IEEE TVCG **18**, 2431–2440 (2012)
7. Mostafa, A.E., Carpendale, S., Brazil, E.V., Eaton, D., Sharlin, E., Sousa, M.C.: FractVis: Visualizing Microseismic Events. In: Bebis, G., Boyle, R., Parvin, B., Koracin, D., Li, B., Porikli, F., Zordan, V., Klosowski, J., Coquillart, S., Luo, X., Chen, M., Gotz, D. (eds.) ISVC 2013, Part II. LNCS, vol. 8034, pp. 384–395. Springer, Heidelberg (2013)
8. Bowman, D., et al.: 3D User Interfaces: Theory and Practice. Addison-Wesley, Redwood city (2004)
9. Baldonado, M., Chang, C.C., Gravano, L., Paepcke, A.: The stanford digital library metadata architecture. Int. J. Digit. Libr. **1**, 108–121 (1997)

DAcImPro: A Novel Database of Acquired Image Projections and Its Application to Object Recognition

Aleksandr Setkov[1]([✉]), Fabio Martinez Carillo[2,3], Michèle Gouiffès[1], Christian Jacquemin[1], Maria Vanrell[4], and Ramon Baldrich[4]

[1] LIMSI, CNRS, University Paris-Sud, Université Paris-Saclay, Orsay, France
aleksandr.setkov@limsi.fr
[2] LIMSI, CNRS, Université Paris-Saclay, Orsay, France
[3] U2IS, ENSTA ParisTech, Université Paris-Saclay, Palaiseau, France
[4] Computer Vision Center, Universitat Autonoma de Barcelona, Cerdanyola, Spain

Abstract. Projector-camera systems are designed to improve the projection quality by comparing original images with their captured projections, which is usually complicated due to high photometric and geometric variations. Many research works address this problem using their own test data which makes it extremely difficult to compare different proposals. This paper has two main contributions. Firstly, we introduce a new database of acquired image projections (DAcImPro) that, covering photometric and geometric conditions and providing data for ground-truth computation, can serve to evaluate different algorithms in projector-camera systems. Secondly, a new object recognition scenario from acquired projections is presented, which could be of a great interest in such domains, as home video projections and public presentations. We show that the task is more challenging than the classical recognition problem and thus requires additional pre-processing, such as color compensation or projection area selection.

Keywords: Projector-camera systems · Feature descriptors · Object recognition

1 Introduction

Spatial Augmented Reality (SAR) [2] applications have gained a lot of success over the last decades and currently they are applicable in various domains, such as entertainment, prototyping, training simulators and so forth. Projector-camera systems, typically used in SAR applications, consist of one or several projection and acquisition devices. These systems are responsible for providing good visual projection quality which, however, requires compensating various distortions inherently produced during the acquisition process. To estimate these distortions, the methods essentially seek to match initial (reference) images with their acquired projections. Due to high color and geometric divergences in the

© Springer International Publishing Switzerland 2015
G. Bebis et al. (Eds.): ISVC 2015, Part II, LNCS 9475, pp. 463–473, 2015.
DOI: 10.1007/978-3-319-27863-6_43

acquisitions the matching problem becomes even more complicated as compared to classical computer vision problems [17].

Compensation tasks can be divided into geometric correction and color compensation. Often, both tasks are mutually dependent, i.e., the color correction methods assume that per-pixel coordinate mapping between the reference and the acquired images is available [1,7], while geometric compensation methods work well for low color variations between the images in order to facilitate geometric transform extraction [16].

The problem of geometric compensation is typically solved through Structured-Light methods [7,15], in which a set of projected markers are matched by (semi)-dense strategies which further allow extracting the surface shape and performing a precise projection geometric compensation. However, Structured-light methods in most of the cases modify the projected content if any spatial coding is used. Additionally, in the case of temporal coding, the interleaved projected patterns infer with the human perception [15]. In a different model, feature matching-based methods perform non-invasive compensation even if it is dependent of the content or acquisition conditions [6,18]. Temporal tracking of salient feature points [8] provides a good trade-off between the computation time and compensation accuracy. Another approach combines Structured-light and content-based approaches in one compensation system [19] which shows promising results in case of untextured projected images which are difficult for feature matching.

Color compensation methods estimate photometric differences between the reference projected and the acquired images to introduce necessary corrections to the projected content to enhance the visualization once projected back. This is a complex task because it can include several components: background color, ambient illumination, and camera and projector responses. When the conditions remain constant, off-line color calibration can be performed to recover the color transformations [14]. On the other hand, if lighting conditions and background change, the color compensation requires real-time adaptation [7]. Some methods project additional patterns of a known color that facilitates the mapping between projected and acquired colors [7,15]. For instance, in [15] a real-time compensation system is proposed to correct both color and geometric variations. Other approaches have used different strategies such as color-mixing matrices [14], Light-Transport [3], stratified inverses [11], or both of them [12] to project images.

These methods however are evaluated using particular and private projection and acquisition setups. This fact introduces bias because there is no evaluation reference to compare the different strategies. In this paper we present two main contributions. Firstly, this work introduces a novel database of acquired image projections (DAcImPro) that can serve as test data to evaluate different SAR state-of-the-art proposals which work on projected acquired image pairs[1], for instance geometric compensation methods through feature matching [6,18] or tracking [8]. By addressing different geometric and illumination conditions, and equipped with ground-truth data, this database can be considered as the baseline for assessing SAR algorithms. The second major contribution illustrates

---

[1] The database is available on: http://dacimpro.limsi.fr.

a new object recognition application on acquired projections. Contrary to the classical recognition problem [9,13], in this application the model is trained on reference (projected) images, while the recognition is performed for acquired projections. Such an application might be of interest in various domains that make use of projector-camera systems, for example smart home video projections [1], or public presentations. Concretely, the user could take a picture of the movie or the presentation with a smartphone or a tablet, and then this picture could be matched with a database in order to get more information on the object present in the captured projection.

The rest of the paper is organized as follows: Sect. 2 introduces the database of acquired image projections, Sect. 3 is devoted to the description of the object recognition framework used to evaluate the descriptor performance on a part of the database, Sect. 4 sums up the obtained results, and in Sect. 5 we draw conclusions.

2   Database of Acquired Projections

The typical pipeline for a projector-camera system is illustrated in Fig. 1. First, an image I is projected by the projector that introduces color distortions $P(I)$. Then, the camera acquires the geometrically warped projection $G{\circ}C{\circ}P(I)$ where C stands for the color distortions due the camera response and the illumination and G defines the geometric transformation. Steps 3 and 4 consist in geometric and color compensations that can be performed in any order. Finally, the system projects the computed compensation $P^{-1}{\circ}C^{-1}{\circ}G^{-1}{\circ}G{\circ}C{\circ}P(I)$. Additionally to the two first pipeline steps addressed in this work, the proposed database provides some data that help to address color and geometric compensations (Steps 3 and 4).

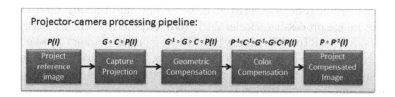

Fig. 1. A sequence of steps performed inside a projector-camera system

2.1   Acquisition Setup

The presented database aims at covering various conditions that typically occur in a projector-camera system. Figure 2 illustrates the acquisition setup. An LCD projector and a CMOS firewire camera were rigidly fixed one with respect to another (Fig. 2a and b). The projections and acquisitions were made in series for each acquisition configuration. Herein, several sources of distortions are covered:

- **Illumination Conditions:** 4 light sources were alternated in the acquisitions;
- **1- and 2-Homography Transforms:** projections were made onto one or two planar surfaces that form the projection scene (Fig. 2c);
- **3 Degrees of Homography Transforms:** for single- or double-homography, projections and acquisitions were performed from three setup positions. Different transforms were obtained by changing the angle and the distance of the system to the projection scene. (Fig. 2e);
- **Color Background:** 4 backgrounds were addressed. The trivial case is a white homogeneous projection surface. (Fig. 2d).

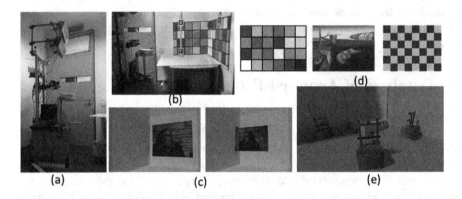

Fig. 2. Acquisition setup description: (a) - global view on the projection-acquisition system with the mounted fluorescent and incandescent bulb lamp; (b) - projection scene and "colorchart" poster; (c) - projections on one and two planar surfaces; (d) three color background posters; (e) - three setup positions (Color figure online).

The acquisition code was written in C++ and additionally used OpenCV library[2] for image processing, Basler API[3] to manipulate on-the-fly with camera parameters, and ARToolKit library[4] to detect special markers present in the scene (more details are given later in this section). The resolution of the reference images were equal to the projector resolution, which is 1024×768. The size of the acquired images were 1624×1234.

Illumination Sources. The four illumination sources are the following: (a) *Dark room:* the only light source in the scene is the projector; (b) *Fluorescent:* an example of white light; (c) *Incandescent lamp:* a source of reddish illumination; (d) *Daylight:* difficult to characterize. The light entered through an open window, while most acquisitions were made in a sunny weather with the clear sky at around midday which allows its approximation with D60 illumination.

---

[2] http://opencv.org/.

[3] http://www.baslerweb.com/en/products/software.

[4] http://www.hitl.washington.edu/artoolkit/.

For each illumination, except for the first one, additional data are included in the database, namely, acquired colorchart and white board images, which allow to estimate camera response matrix and illumination in the scene.

For each geometric configuration, a black and white checkerboard image was projected and used to obtain ground-truth homography transforms. Note, that whatever the illumination, setup position and background poster, we tried to keep the homography transformation constant. To that end, a simple automatic setup alignment was implemented on the basis of ARToolKit marker detection. The markers are, therefore, present in the acquired images outside the projection zone.

2.2 Reference Projected Images

Three categories of projected images were chosen in the database so that to model real applications where projector-camera system can be used.

(1) *Natural Images.* To make the database as generic as possible, 48 images of 10 different categories were selected from Flickr[5] database and used in acquisitions;
(2) *Video Conferences.* From McGill Real-World Face Video Database [5] 8 video sequences were chosen according to the presence of complex background, various human faces as well as different poses. From each video stream 10 frames were randomly selected and included in the set of projected images.
(3) *Powerpoint Presentations.* 5 publicly available educational presentations were selected from Coursera[6] and MIT webpages[7]. In total, 34 slides were selected that represent the following cases: title slides, slides with plots, text and images.

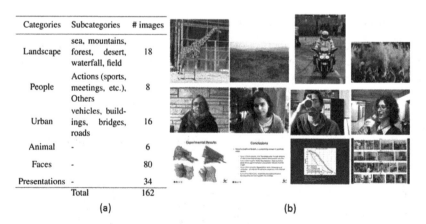

(a)   (b)

Fig. 3. Projected reference images. (a) Summary of 162 projected images; (b) Examples of projected images.

---

[5] https://www.flickr.com/.
[6] https://www.coursera.org/.
[7] http://ocw.mit.edu/.

In total the database includes 162 reference images. Figure 3 shows some samples of the projected images. Figure 4 illustrates acquired projection examples.

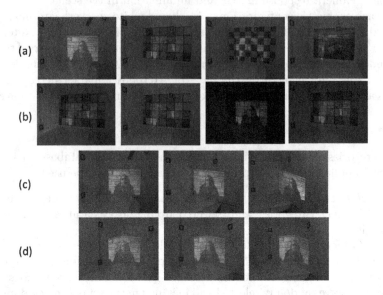

Fig. 4. Examples of acquired images: (a) - 4 color background posters used in the acquisitions; (b) - 4 illumination conditions addressed in the acquisitions; (c) and (d) - 3 setup positions for single- and double-homography transforms, respectively (Color figure online).

2.3   Acquisition Pipeline and Ground-Truth Preparation

Figure 5 illustrates the sequence of processing steps to build the database. Besides the reference images and their acquired projections, we include additional data that can be used to obtain ground-truth data. For color calibration, we gather such data as acquired colorchecker and white board images for camera calibration and scene illumination estimation. An RGB-cube of 256^3 different color was also projected to obtain a precise projector response mapping. For geometric compensation, the database includes acquired projections of a checkerboard image.

Fig. 5. Acquisition pipeline. Blocks in red denote the acquisition process of data used for geometric and color compensation. The green block is the projection/acquisition loop. Blocks in blue describe the preparation steps for acquisitions (Color figure online).

3   Object Recognition Framework

In this section we use the previous dataset to introduce a new application for object recognition from captured projections. While the recognition model is trained on original images, it recognizes objects from acquired projections. Human faces were chosen as recognized categories in this work.

3.1   Training and Test Images

For object recognition we chose the part of the database that corresponds to the human face images (video conferences scenario in Sect. 2.2). Each category is, therefore, represented by a sequence of video frames (300 for each video) in which one person appears. In total, it leads to a training set of 2400 images.

A subset of this database (10 images in each of 8 categories) was projected and acquired in different configurations (geometric and photometric transformations). As a test set for recognition, only a part of the projection database was exploited. For each category, 10 images were projected with three different illumination conditions, which makes the total number of test images equal to $8 \times 10 \times 3 = 240$ images.

Fig. 6. Three scenarios tested in the object recognition framework: (a) - an rgb image provided in the database; (b) - the cropped image; (c) - the cropped and photometrically corrected image (Color figure online).

The comparison between projection acquisitions and reference images is, in general, a complicated task due to color discrepancies and surrounding background. These problems were herein addressed by preprocessing the acquired projections and comparing the obtained results with those obtained on unprocessed images. Therefore, for each tested descriptor, three scenarios were examined (examples of images can be found in Fig. 6): (1) without any compensation of the test images; (2) the projection area was manually identified and cropped so that to reduce the amount of the background in the test images.[8] As a result, projections occupied approximately 1/2 of the image. (3) color compensation was performed on the cropped test images, which includes gamma correction, white balance and compensation for the camera gain. Projector response

---

[8] Since in the acquired images the setup was fixed, it was possible to performed this preselection using a batch script.

and ambient illumination were not compensated in this scenario. The gamma parameter 2.2 was roughly estimated from the acquired colorchecker images (provided in the database) and averaged for the R, G and B channels. White balance correction consisted in multiplying the green channel by 0.8, which corresponds to the Balance Ratio parameters in the camera settings. Finally, the resulted images were multiplied by intensity factor of 0.8 in order to avoid saturations. The formula below presents the transformation as a whole:

$$\begin{pmatrix} O_R \\ O_G \\ O_B \end{pmatrix} = 0.8 \left(\begin{pmatrix} 1 & 0 & 0 \\ 0 & 0.8 & 0 \\ 0 & 0 & 1 \end{pmatrix} \begin{pmatrix} I_R \\ I_G \\ I_B \end{pmatrix} \right)^{1/2.2} \tag{1}$$

where $I_{R,G,B}$ and $O_{R,G,B}$ correspond to the initial and the corrected images.

Local Feature Characterization. The following descriptors were used in the evaluation: I-SIFT [10], its color extension RGB-SIFT, RGB-LHE-SIFT [16] and PHOW-SIFT. All of them were implemented in Matlab by means if VLFeat library[9]. PHOW-SIFT, a multi-scale version of Dense-SIFT, was set to compute descriptors at every 32-th pixel. It was done in order to keep the total size of feature descriptors compared with the other methods and to reduce the execution time.

3.2   Learning Codebook for Acquired Projections

In the training step, the recovered salient features from the acquired projections are used to learn a codebook. Such codebook is made up by a set of salient features in projected images coded as words, formed by the output of a classical k-means algorithm computed using a random selection of 10 % of the salient features extracted over the whole training video set.

Herein different feature words were tested independently, such as I-SIFT, RGB-SIFT, RGB-LHE-SIFT and PHOW-SIFT. The size of each codebook k_l is herein set according to the number of input samples at each time scale, as: $k_l = \sqrt{(S/2)}$, with S the number of input samples computed.

Then, a projected image is represented as a histogram of word occurrences, by counting the number of visual words that belong to each centroid k_l, based on the Euclidean distance on $\mathbb{R}^{3n}$, where $3n$ is the description vector length. Finally, the recognition is carried out by a Support Vector Machine (SVM) using the standard LIBSVM [4] implementation, using the *one-against-one* multi-class SVM classification with a Radial Basis Function (RBF) kernel.

4   Results

A classification test framework was used to evaluate the performance of classical descriptors over the proposed strategy. An average accuracy of the multi-class

---

[9] http://www.vlfeat.org/.

recognition tasks is shown in Fig. 7. Moreover, a confusion matrix was recovered to detail the performance of different descriptors on the dataset. First of all, PHOW-SIFT performs very poor because dense SIFT point sampling is very sensitive to the projection scene background. Even in the cropped images (scenario 2) there remains some background around the projection area which introduces a noise in the recognition. For the sake of compactness and due to its poor performance, this descriptor performance is not depicted in Fig. 8.

The results clearly show the improvement that can be achieved by preselecting the object area and performing a simple color correction. In the best case we achieve almost 90 % recognition accuracy with 8 categories.

If we compare the considered local descriptors, LHE-RGB-SIFT outperforms the descriptors in scenario 1, when images are not corrected photometrically.

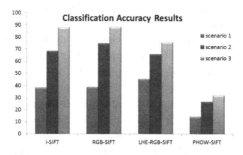

Fig. 7. Classification accuracy results of the BoW method with different descriptors.

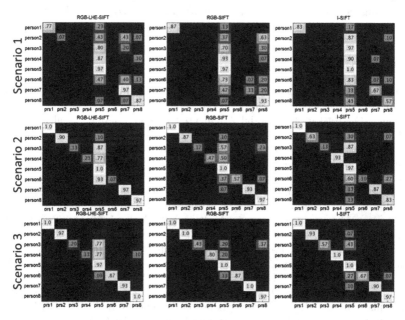

Fig. 8. Confusion Matrices. I-SIFT, RGB-SIFT and LHE-RGB-SIFT

It means that this descriptor is more robust to the present background. When image correction is introduced (scenarios 2 and 3), I-SIFT and RGB-SIFT yield better results. When color correction is not performed (scenario 2), RGB-SIFT outperforms all the methods, which can be explained by its ability to better cope with color distortions than I-SIFT.

As seen in the matrices in Fig. 8, images of most classes tend to be misclassified as "person5" category. Sometimes classes "person7" and "person8" also represent difficulties for the model. It can be due to similar backgrounds in the acquired images and in the reference "person5" images. Color correction and projection area selection can solve this problem.

5   Conclusions

This paper presents a database of acquired image projections that covers various photometric conditions and homography transformations. This database can be useful for different research communities that deal with image projections. Further, we show an example of new object recognition application in which objects from acquired projections are recognized through a BoW model trained on a dataset of reference images. Moreover, by examining different experimental scenarios, we show that a significant classification improvement can be achieved by pre-processing test images in terms of color and projection area selection. Depending on whether test images will be corrected or not, we recommend to use of RGB-SIFT or LHE-RGB-SIFT for BoW-based object recognition.

Acknowledgments. This project has been partially funded by MINECO (TIN2014-61068-R).

References

1. Bimber, O., Emmerling, A., Klemmer, T.: Embedded entertainment with smart projectors. IEEE Comput. **38**(1), 48–55 (2005)
2. Bimber, O., Raskar, R.: Spatial Augmented Reality: Merging Real and Virtual Worlds. CRC Press, New York (2005)
3. Chandraker, M., Bai, J., Ng, T.-T., Ramamoorthi, R.: On the duality of forward and inverse light transport. IEEE TPAMI **33**, 2122–2128 (2011)
4. Chang, C.-C., Lin, C.-J.: LIBSVM: a library for support vector machines. ACM TIST **2**(3), 27:1–27:27 (2011). http://www.csie.ntu.edu.tw/cjlin/libsvm
5. Demirkus, M., Clark, J.J., Arbel, T.: Robust semi-automatic head pose labeling for real-world face video sequences. Multimed. Tools Appl. **70**(1), 495–523 (2014). doi:10.1007/s11042-012-1352-1
6. Drouin, M.-A., Jodoin, P.-M., Premont, J.: Camera-projector matching using an unstructured video stream. In: 2010 IEEE Computer Society Conference on CVPR Workshops (CVPRW), vol. 33, p. 40 (2010)
7. Fujii, K., Grossberg, M.D., Nayar, S.K.: A projector-camera system with real-time photometric adaptation for dynamic environments. In: 2005 IEEE Computer Society Conference on (CVPR 2005), San Diego, CA, USA, 20–26 June 2005, pp. 814–821 (2005)

8. Kooi, T., de Sorbier, F., Saito, H.: Colour descriptors for tracking in spatial augmented reality. In: Park, J.-I., Kim, J. (eds.) ACCV Workshops 2012, Part II. LNCS, vol. 7729, pp. 387–399. Springer, Heidelberg (2013)

9. Kumar, V., Namboodiri, A.M., Jawahar, C.V.: Face recognition in videos by label propagation. In: 22nd ICPR 2014, Stockholm, Sweden, 24–28 August 2014, pp. 303–308 (2014)

10. Lowe, D.G.: Distinctive image features from scale-invariant keypoints. Int. J. Comput. Vis. **60**(2), 91–110 (2004)

11. Ng, T.-T., Pahwa, R.S., Bai, J., Quek, T.Q.S., Tan, K.-H.: Radiometric compensation using stratified inverses. In: IEEE 12th ICCV 2009, Kyoto, Japan, 27 September – 4 October 2009, pp. 1889–1894 (2009)

12. Ng, T.-T., Pahwa, R.S., Bai, J., Tan, K.-H., Ramamoorthi, R.: From the rendering equation to stratified light transport inversion. Int. J. Comput. Vis. **96**(2), 235–251 (2012)

13. Ortiz, E.G., Wright, A., Shah, M.: Face recognition in movie trailers via mean sequence sparse representation-based classification. In: 2013 IEEE Conference on CVPR, Portland, OR, USA, 23–28 June 2013, pp. 3531–3538 (2013)

14. Park, H., Lee, M.-H., Kim, S.-J., Park, J.-I.: Contrast enhancement in direct-projected augmented reality. In: Proceedings of the 2006 IEEE International Conference on Multimedia and Expo, ICME 2006, Toronto, Ontario, Canada, 9–12 July 2006, pp. 1313–1316 (2006)

15. Park, H., Lee, M.-H., Seo, B.-K., Park, J.-I., Jeong, M.-S., Park, T.-S., Lee, Y., Ryong Kim, S.: Simultaneous geometric and radiometric adaptation to dynamic surfaces with a mobile projector-camera system. IEEE Trans. Circ. Syst. Video Technol. **18**(1), 110–115 (2008)

16. Setkov, A., Gouiffès, M., Jacquemin, C.: Color invariant feature matching for image geometric correction. In: 6th International Conference on Computer Vision/Computer Graphics Collaboration Techniques and Applications, MIRAGE 2013, Berlin, Germany, 06–07 June 2013, pp. 7:1–7:8 (2013)

17. van de Sande, K.E.A., Gevers, T., Snoek, C.G.M.: Evaluating color descriptors for object and scene recognition. IEEE TPAMI **32**(9), 1582–1596 (2010)

18. Yamanaka, T., Sakaue, F., Sato, J.: Adaptive image projection onto non-planar screen using projector-camera systems. In: 20th ICPR 2010, Istanbul, Turkey, 23–26 August 2010, pp. 307–310 (2010)

19. Zollmann, S., Langlotz, T., Bimber, O.: Passive-active geometric calibration for view-dependent projections onto arbitrary surfaces. JVRB J. Virtual Reality Broadcast. **4**(6), 10 (2007)

Deformable Object Behavior Reconstruction Derived Through Simultaneous Geometric and Material Property Estimation

Shane Transue[✉] and Min-Hyung Choi

University of Colorado Denver, Denver, USA
shane.transue@ucdenver.edu

Abstract. We present a methodology of accurately reconstructing the deformation and surface characteristics of a scanned 3D model recorded in real-time within a Finite Element Model (FEM) simulation. Based on a sequence of generated surface deformations defining a reference animation, we illustrate the ability to accurately replicate the deformation behavior of an object composed of an unknown homogeneous elastic material. We then formulate the procedural generation of the internal geometric structure and material parameterization required to achieve the recorded deformation behavior as a non-linear optimization problem. In this formulation the geometric distribution (quality) and density of tetrahedral components are simultaneously optimized with the elastic material parameters (Young's Modulus and Possion's ratio) of a procedurally generated FEM model to provide the optimal deformation behavior with respect to the recorded surface.

1 Introduction

Replicating realistic deformation behaviors of elastic objects is an extensively studied domain with a broad array of applications in countless fields including soft-tissue medicine, structural engineering, and animation. The ability to reliably reproduce physically plausible deformations exhibited by elastic objects has been widely developed and several techniques have been introduced in an effort to automate the process of reproducing realistic deformations of recorded real-world objects. Physically-based Finite Element Methods (FEM) define the standard for emulating realistic elastic behaviors demonstrated by deformable objects and provide accurate representations of physically plausible behaviors. While methods of estimating material properties, such as Young's modulus and Poisson's ratio for FEM-based physical models from scanned objects have been introduced [1,2], these techniques rely on the assignment of an arbitrary internal geometric structure contained by a scanned surface or use artistically created template meshes. The assumption that these techniques present is that the ideal material properties will accurately reproduce the observed deformation. The premise of our methodology deviates from this standard process through the introduction of a technique for simultaneously extracting both the ideal internal

© Springer International Publishing Switzerland 2015
G. Bebis et al. (Eds.): ISVC 2015, Part II, LNCS 9475, pp. 474–485, 2015.
DOI: 10.1007/978-3-319-27863-6_44

geometric density and homogeneous isotropic material properties to provide an accurate reconstruction of a recorded deformation.

In this work we present an automated extension of modeling physically simulated deformable objects and present a means to optimizing FEM-based physical models to behave consistently with their real-world counterparts. However, unlike most prior work within this field, we base our approach on the notion that strictly identifying the material parameters of a deformable object may not provide optimal behavioral correspondence to these real-world deformations. Typically the geometric composition of the simulated object has a large impact on the deformation behaviors that are exhibited by the object and can degrade the accuracy of the exhibited deformations. The images in Fig. 1 illustrate the impact of geometric density on the resulting deformation behavior of an FEM mesh with increasing geometric densities simulated using VegaFEM [3].

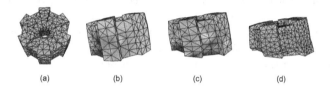

(a)   (b)   (c)   (d)

Fig. 1. Impact of the geometric density on an FEM for the deformation of a model (a) dropped onto a flat surface for resolutions of: (b) $n = 688$, (c) $n = 868$, and (d) $n = 4070$ nodes. Young's modulous (1.0e5) and Poisson's ratio (0.4) are held constant.

Based on this geometric density influence, we build upon existing methods for replicating an elastic-based deformation within a physical simulation, emphasizing the process of extracting the characteristics of the geometric density that facilitate the generation of the exhibited deformation. The objective of our approach is to procedurally generate both the geometric and material properties that provide the closest approximation of the deformation defined by the recorded surface. Additionally, our technique aims to provide a practical implementation that can automatically obtain the physical simulation template model and replicate the recorded deformation behavior without manual intervention.

2   Related Work

The extraction and estimation of the material properties of FEM-based objects has been extensively studied and several forms of material optimization techniques have been developed [4,5]. Multi-view tracking systems for iteratively reconstructing the deformation of an initial surface mesh have also been developed [6]. In relation to the methodology we present, independent aspects of our approach have been validated with respect to the optimization of deformable object animations. This includes the optimization of the underlying geometry of deformable mass-spring systems and associated material properties [7] and the extraction of a physical surface model from a multi-view configuration [8]. Recently, template-based techniques have been developed [1] to eliminate the use of several scanning devices to improve the practicality of the scanning process; however our technique differs from this in that we continue to extract the simulation mesh (template) from the animated surface sequence (set of temporally coherent, aligned 3D scans) for our proposed optimization process.

Recent techniques propose an approach that utilizes a similar process for the extraction and optimization of the material parameters of the elastic material properties to match a provided input deformation [2]. In this work the observed deformation behavior is defined through signed distance fields that define the required deformation of the provided watertight surfaces. While the input to our method is consistent with this approach, our technique varies from this work in two aspects: (1) we continuously regenerate the rest-state internal geometry of the simulated FEM mesh to improve the consistency between the simulated surface and the provided animation and (2) the objective of our method is to provide a process for reconstructing the generalized deformation behavior without introducing undesirable secondary behaviors. These secondary behaviors can be introduced as surface artifacts within noisy surface scans and should not be incorporated within the reproduced deformation behavior.

3   Method

In the objective of reconstructing scanned deformation behaviors, we propose a methodology of capturing and emulating the deformations exhibited by a surface scanned by a set of high-speed depth-imaging devices. In this process we present an automated method of extracting the surface of the scanned object over time, the procedural generation of the tetrahedral simulation mesh, and the material coefficients used to recreate the observed deformation. We then optimize both the geometric and elastic material properties used to define the simulation mesh that provides the optimal approximation of the recoded deformation behavior.

The implemented method of reproducing the observed deformation behavior of the scanned surface model is based on a five-stage process: (1) The recorded deformation is provided as a set of surface scans (point-clouds) that defines the objects surface state at incremental instances in time. (2) The reconstruction of the objects surface at these discrete time steps are stored into a reference animation A as surface states $S_0, S_1, ..., S_n$. (3) These discrete states are then used to generate temporally aligned displacement fields $D = \{d_1, d_2, ..., d_{n-1}\}$. These displacement fields are obtained using the non-rigid Coherent Point Drift (CPD) algorithm [9] between each set of consecutive surface states S_i, S_j. This allows us to identify topological correspondence regions as the topology of the reconstructed surface varies over time. (4) The optimization of both the geometric and material properties of the simulated object are obtained using the following formulation: $P(q, v, y, p)$ where (q) is the tetrahedral quality, (v) is the maximal tetrahedral volume (density coefficient), (y) is Young's Modulous and (p) is Possion's Ratio. (5) Based on the instance of the simulation mesh defined by $P(q, v, y, p)$, we then optimize the forces inferred by the displacement fields to reconstruct the recorded animation of the form: $P(f, q, v, y, p)$ (Fig. 2).

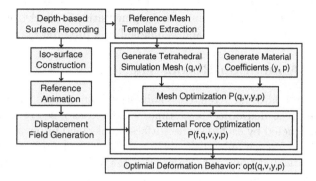

Fig. 2. Implemented methodology for the reconstruction of a surface-based recorded animation using simultaneous geometric and material parameter optimization. The enclosed regions illustrate the optimization hierarchy.

This process is defined as a compound non-linear optimization that identifies the optimal forces to apply to the deformable mesh in each simulation time-step for each unique parameterization of the geometric construction and elastic material parameters of the model. The objective function of this optimization process is to minimize the displacements generated between the simulated surface and the scanned reference animation during each simulation time-step. The output of this approach is the set of geometric and material parameters $opt(q, v, y, p)$ that can be used with the set of optimal external forces f for each simulation time-step to accurately reconstruct the recorded deformation.

4 Deformable Surface Tracking

The process of recording the behaviors of a deformable object's surface state has been well explored [10, 11] and approaches using high-speed video as well as and depth-based surface scanning techniques [12] have begun to emerge as a large segment within the field of elastic material property extraction. In our approach, we can utilize the improved sampling rates of modern depth-imaging devices to capture the deformation behaviors of real-world objects in real-time.

As a prerequisite for our optimization procedure, we assume that the surface of the deforming object can be acquired through a sequence of depth-images or point-clouds that can be obtained using synchronized depth-imaging devices. These scans are then automatically aligned using existing techniques [13–15] to form an accurate approximation of the deforming object's surface over time. We define this sequence of aligned depth images as the input of one animation frame within A, which will be used to represent the topology of the recorded object.

Our approach assumes the point-cloud representing the objects entire surface at each discrete time increment is defined as a set of unordered point set with estimated surface normals [16]. To reduce the density of the provided point clouds we employ a simple standard-deviation outlier filter and voxel-based reduction filters provided within PCL [17]. Point-cloud segmentation for background removal is beyond the scope of this work and is handled in a pre-processing stage.

5   Surface Animation Reconstruction

The process of establishing a reference animation that defines the deformation behavior of the recorded object over time is defined as a set of surface states that are reconstructed by each discrete surface point-cloud into a set of n frames to generate the reference animation A. For each scanned surface state $S_i \in A$ containing an oriented point set, we employ the volumetric iso-surface reconstruction algorithm presented in [18] to reconstruct the surface of the object in each recorded frame. This algorithm takes the oriented point set in each state and utilizes a Marching Cubes (MC) [19] variant to extract a surface mesh that accurately represents the surface of the scanned object. Due to the high resolution of the resulting mesh that contains a large number of sliver triangles due to the MC surface extraction process, we utilize a filtering process to eliminate these defects and any degenerate triangles [20].

The result of this process is a high-quality discrete surface animation that describes the deformation behavior of the recoded object. We explicitly note the requirement of generating a high-quality mesh for the objects rest-state. Due to our template-free approach, we extract the simulation surface from the surface state S_0 to generate the tetrahedral mesh used in our simulation optimization. An example of the input surface point clouds and their corresponding surface reconstructions are illustrated in Fig. 3. This allows us to automatically generate the physical representation of our simulation mesh which is required for the process of optimizing the internal geometry of the recorded object.

6   Rest-State Mesh Generation

The basis of our automated optimization process relies on the ability to extract a high quality tetrahedral mesh from the first frame of the reference animation

Fig. 3. Surface reconstruction of two scanned surface states: the input surface point-clouds (top) and their corresponding surface reconstructions (bottom) define the reference surface animation A.

A. This tetrahedral mesh represents the geometric construction of the mesh used within the simulation to reconstruct the observed deformation behavior. Ensuring that the quality of the surface state $S_0 \in A$ allows us to provide a reliable procedural regeneration process for creating a high quality simulation mesh. Based on the triangulated surface provided by the iso-surface extraction algorithm, we employ the Delaunay-based tetrahedral construction algorithm implemented within Tetgen [21]. The images in Fig. 4 illustrate the change in distribution and density of the tetrahedral components that compose the internal geometry of the simulated object. The quality and density of the generated tetrahedra are varied based on the parametrization $P(q, v)$. The images in Fig. 4(a) illustrate the slight difference in geometric density that contribute to a different deformation characteristic and the comparison presented in Fig. 4(b) illustrates a slight variance in the tetrahedral quality leading to elongated tetrahedral components.

(a)   (b)

Fig. 4. Tetgen generated tetrahedral mesh based on the surface state S_0. The images in (a) and (b) illustrate two different potential geometric densities generation during the optimization process.

The quality of a tetrahedral element within the simulation mesh is defined as the radius-edge ratio of the generated tetrahedra. The objective of this parameter is to ensure that a minimal number of sliver tetrahedra (those with excessively small or large dihedral angles) are introduced into the internal structure of the object. These forms can introduce unwanted artifacts within the surface of the deforming object and alter the distribution of the homogeneous material we aim to model. Similarly the restriction of the maximal tetrahedral volume alters the observed deformation behavior of the object under a constant set of external forces. The density of the geometric structure defined by this parameter impacts the magnitude of the external forces applied to reach the desired deformation.

7   Deformation Force Optimization

The collection of surface states used to determine the displacements of the deformation behavior of the object are also used to generate the required external forces that imposed the deformation. Since our approach assumes that no information about the external forces imposed on the reference object, we present an automated method of estimating these forces using the displacement fields generated from the discrete differences in the input surface sequence defined within the generated reference animation.

Based on the surface displacement exhibited by the scanned object between discrete states within the reference animation, we form the generated external forces by calculating an approximate linear velocity and acceleration of the surface. This pseudo acceleration is defined through the linearization of the displacement given the discrete time-step dt. From this we derive the linear velocity $v = d_i/dt$ and pseudo acceleration $\alpha = dv/dt$ where the mass coefficient for each discrete surface node is one and the external force is defined as $\mathscr{F} = m\alpha$.

The error metric presented in the objective function of this optimization is based on the magnitude of the displacement field generated between the current simulation mesh surface consisting of p_n points and the alignment of these points to the scanned surface using the CPD algorithm. Equation 1 defines the objective function as the norm of the displacement field for this set of points between their current position and their resulting alignment with the next deformation surface state S_{i+1}.

$$|D| = \sum_{i=1}^{n} \| \, align(\boldsymbol{p}_i, S_{i+1}) - \boldsymbol{p}_i \| \tag{1}$$

Scanned surface reconstructions typically contain artifacts due to the depth measurement error associated with the employed depth-imaging device. Therefore rather than optimizing the simulated surface to match exact the topology changes and potential artifacts in the scanned surface within some tolerance, our approach generates the forces required to emulate the generalized deformation behavior to avoid artifacts. Using this approach we impose an approximation of the overall deformation behavior while maintaining physically plausibility.

8   Deformation Optimization

The optimization process that we utilize to derive the optimal deformation behavior of the simulated tetrahedral mesh with respect to the reference animation is composed of a two-phase process: (1) the geometric construction of the simulated object is generated using the current geometric parameters for tetrahedral quality and maximal tetrahedral volume along with the elastic material coefficients. (2) The forces required to achieve the deformation behavior observed in the reference animation are optimized with respect to this current parameterization as introduced in Sect. 7. This optimization is performed with a derivative-free optimization: Constrained Optimization by Linear Approximations [22] implemented within the NLopt numerical library [23].

As the elastic material properties and geometric construction of the tetrahedral mesh are updated, the new external forces required to impose the same deformation behaviors are naturally addressed through this optimization hierarchy. The objective of this non-linear optimization is to obtain the parameterization that provides the closest matching deformation behavior observed in the reference animation. This optimization is driven by the minimization between the simulated mesh surface with respect to the scanned surface state. From this

Table 1. Optimization parameterization $P(q, v, y, p)$ to generate the optimal deformation defined within the provided set of surface states $S_0, S_1, ..., S_n$.

Optimization parameter	q	v	y	p	$|D_{anim}|$
Geometric Opt $|D_{anim}|_{max}$	2.13242	1.01930e-4	1.000e6	0.300000	72.2366
Geometric Opt $|D_{anim}|_{opt}$	2.12500	1.01321e-4	1.000e6	0.300000	53.0002
Material Opt $|D_{anim}|_{max}$	2.00000	1.00000e-4	5.000e5	0.287500	75.4573
Material Opt $|D_{anim}|_{opt}$	2.00000	1.00000e-4	4.634e5	0.221214	53.3687
Geo + Mat $|D_{anim}|_{max}$	2.13601	8.01578e-5	6.55411e5	0.294876	75.7081
Geo + Mat $|D_{anim}|_{opt}$	2.12799	8.03421e-5	6.52945e5	0.297177	51.6883

metric we define the total deformation error of a given animation A as the sum of norms for each displacement field generated between the simulated surface and the scanned surface state as defined in Eq. 2.

$$|D_{anim}| = \sum_{S_i \in A} \sum_{j=1}^{n} \| \, align(\boldsymbol{p}_j, S_i) - \boldsymbol{p}_j \| \qquad (2)$$

In this approach, we identify three configurations of this optimization process that illustrate the complex relationship between the geometric composition of the object, the elastic material properties, and the forces required to impose the recorded deformation. We formulate these configurations as follows: (1) Geometric optimization of the parameters $P(q, v)$ with the associated force optimization required to impose the recorded deformation. (2) The optimization of the FEM-based material parameters $P(y, p)$ with associated force optimization. The internal geometric composition of the object in this case is based on the optimal quality parameters defined by Tetgen ($q = 2.0$, $v = 1.0$). (3) The last configuration performs a complete optimization of the parameterization $P(q, v, y, p)$ to obtain the optimal uniform internal geometric structure and material properties required to match the recorded behavior.

9   Results

The results of our proposed simulation framework are illustrated through a synthetic deformation example of a scanned surface that represents a bend deformation imposed on a deformable object composed of a isotropic, homogeneous, elastic material. This object is scanned during the deformation process to capture the imposed deformation and our implementation extracts the external forces, geometric construction and elastic material properties through the optimization $P(f, q, v, y, p)$ to reproduce a simulation that matches the reference animation (Fig. 5).

The parameters provided in Table 1 define the parameterization of this optimization based on three configurations (1) optimizing only the internal geometric

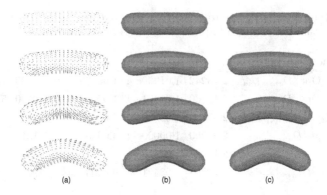

Fig. 5. Resulting deformation imposed on the simulated tetrahedral mesh based on the scanned deformation behavior. The image sequence displays the displacement fields of the observed deformation (a), the reconstructed scan surface over time (b), and the resulting simulated deformation (c).

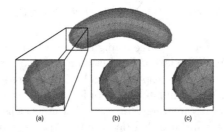

Fig. 6. Resulting alignment of the optimal parameterization (a) of the FEM mesh superimposed onto the scanned surface. The Geometric Opt (b) and Material Opt (c) results illustrate the misalignment obtained during the independent optimizations.

structure of the object with optimized external forces, (2) Fixing the ideal internal tetrahedral quality and density and optimizing only the material parameters, and (3) the process of optimizing all parameters of the form: $P(q, v, y, p)$.

The results provided through our optimization process indicates that simultaneously optimizing the external forces with respect to the internal geometric structure $P(q, v)$ and the elastic material properties of the FEM-based model $P(y, p)$ provides a closer approximation of the input animation A. Based on the set of external forces estimated from the displacement fields generated from the input surface states, we illustrate the complex relationship between the applied forces, geometric structure, and material properties exacerbate the difficulty in extracting an exact deformation replication. This is indicated by the slight numerical adjustments that made to the optimization parameters provided in Table 1 drastically alter the resulting animation displacement score $|D_{anim}|$. The result in Fig. 6 provides the optimized superimposed alignment between the simulated mesh and the scanned surface, illustrating the improvement obtained through optimizing both geometric composition and elastic material properties.

10   Evaluation

The core of our methodology provides an accurate behavioral alignment however, it relies on the ability to extract the physical model from the reference frame S_0 within the animation A. For complex geometric structures there are instances where surface occlusions will degrade the quality of the generated surface. These occlusions can not only reduce the quality of the simulation mesh but also interfere with the recorded deformation behavior.

The process of optimizing the forces required to drive the deformation behavior defined by the subsequent displacement fields generated from the original surface animation can introduce drift between the simulated FEM model and the recorded surface. This can introduce an unintended translation and net torque to be imposed during the force optimization process. Due to the generalized nature of the external forces generated in our approach, exact surface deformations may not be obtained. While our technique avoids the propagation of any imperfections in the scanned surface through the generalized deformation behavior, we still may not obtain an exact replica of the observed deformation. However this process can be directed towards reproducing physically plausible deformation behaviors based on a guided reference model.

11   Conclusion

In this work we have present an automated methodology for extracting the deformation behavior for a scanned surface of an object composed of a homogeneous elastic material that exhibits isotropic characteristics. We present an effective method for extracting a quality tetrahedral simulation mesh from an input scanned surface and illustrated the behavioral impact of optimizing both the geometric structure and material properties with respect to a set of unknown external forces. Based on the implementation of our approach, our method was able to generate a high-quality animated surface sequence, procedurally generate the physical representation of the object, derive the eternal forces and optimal material properties required to impose the deformation observed in the scanned surface. Furthermore, our method validates the process of simultaneously optimizing the geometric construction of the simulated object with the FEM-based elastic material properties to improve the deformation generated in the force optimization process, thus providing an objective analysis of our displacement-driven simulation and illustrating the impact of geometric density with FEM-based simulations.

References

1. Wuhrer, S., Lang, J., Tekieh, M., Shu, C.: Finite element based tracking of deforming surfaces. Graph. Mod. **77**, 1–17 (2015)

2. Choi, J., Szymczak, A.: Fitting solid meshes to animated surfaces using linear elasticity. ACM Trans. Graph. **6**(1–6), 10 (2009)
3. Barbic, J.: Vega fem library. A physics library for three-dimensional deformable object simulation (2015). http://run.usc.edu/vega/
4. Becker, M., Teschner, M.: Robust and efficient estimation of elasticity parameters using the linear finite element method. In: Schulze, T., Preim, B., Schumann, H. (eds.) SimVis. SCS Publishing House e.V., Erlangen (2007)
5. Frank, B., Schmedding, R., Stachniss, C., Teschner, M., Burgard, W.: Learning the elasticity parameters of deformable objects with a manipulation robot. In: IEEE/RSJ International Conference on Intelligent Robots and Systems, pp. 1877–1883 (2010)
6. Cagniart, C., Boyer, E., Ilic, S.: Iterative deformable surface tracking in multi-view setups. In: 5th International Symposium on 3D Data Processing, Visualization and Transmission, 3DPVT (2010)
7. Bianchi, G., Solenthaler, B., Szekely, G., Harders, M.: Simultaneous topology and stiffness identification for mass-spring models based on fem reference deformations. In: Medical Image Computing and Comptuer-Assisted Intervention (2004)
8. Furukawa, Y., Ponce, J.: Dense 3d motion capture from synchronized video streams. In: IEEE Conference on Computer Vision and Pattern Recognition (2008)
9. Myronenko, A.: Point-set registration: coherent point drift. IEEE Trans. Pattern Analy. Mach. Intell. **32**, 2262–2275 (2010)
10. Schulman, J., Lee, A., Ho, J., Abbeel, P.: Tracking deformable objects with point clouds. In: International Conference on Robotics and Automation (2013)
11. Hur, J., Lim, H., Ahn, S.C.: 3D deformable spatial pyramid for dense 3D motion flow of deformable object. In: Bebis, G., Boyle, R., Parvin, B., Koracin, D., McMahan, R., Jerald, J., Zhang, H., Drucker, S.M., Kambhamettu, C., El Choubassi, M., Deng, Z., Carlson, M. (eds.) ISVC 2014, Part I. LNCS, vol. 8887, pp. 118–127. Springer, Heidelberg (2014)
12. Li, Y., Chen, C.F., Allen, P.K.: Recognition of deformable object category and pose. In: International Conference on Robotics and Automation (2014)
13. Chen, Y., Medioni, G.: Object modeling by registration of multiple range images. In: Proceedings of IEEE International Conference on Robotics and Automation, vol. 3, pp. 2724–2729 (1991)
14. Besl, P., McKay, N.D.: A method for registration of 3-d shapes. IEEE Trans. Pattern Anal. Mach. Intell. **14**(2), 239–256 (1992)
15. Rusu, R., Blodow, N., Beetz, M.: Fast point feature histograms (fpfh) for 3d registration. In: IEEE International Conference on Robotics and Automation, ICRA 2009, pp. 3212–3217 (2009)
16. Rusu, R.B.: Semantic 3D Object Maps for Everyday Manipulation in Human Living Environments Ph.D. thesis, Computer Science, Universitaet Muenchen, Germany (2009)
17. Source, O.: Pcl: the point cloud library is a standalone, large scale, open project for 2d/3d image and point cloud processing (2015). http://pointclouds.org/
18. Kazhdan, M.: Reconstruction of solid models from oriented point sets. In: Proceedings of the Third Eurographics Symposium on Geometry Processing, SGP 2005, Eurographics Association (2005)
19. Lorensen, W.E., Cline, H.E.: Marching cubes: a high resolution 3d surface construction algorithm. In: Proceedings of the 14th Annual Conference on Computer Graphics and Interactive Techniques, SIGGRAPH 1987, pp. 163–169. ACM (1987)
20. Curless, B., Levoy, M.: A volumetric method for building complex models from range images. In: Proceedings of SIGGRAPH 1996, pp. 303–312 (1996)

21. Si, H.: TetGen: a quality tetrahedral mesh generator and three-dimensional delaunay triangulator (2007). http://wias-berlin.de/software/tetgen/
22. Powell, M.: A direct search optimization method that models the objective and constraint functions by linear interpolation. In: Gomez, S., Hennart, J.-P. (eds.) Advances in Optimization and Numerical Analysis. Mathematics and Its Applications, vol. 275, pp. 51–67. Springer, Heidelberg (1994)
23. Johnson, S.G.: Nlopt: the nlopt nonlinear-optimization package (2015). http://ab-initio.mit.edu/nlopt

Poster

Accidental Fall Detection Based on Skeleton Joint Correlation and Activity Boundary

Martha Magali Flores-Barranco[1],
Mario-Alberto Ibarra-Mazano[1], and Irene Cheng[2]([✉])

[1] Universidad de Guanajuato, Guanajuato, Mexico
[2] University of Alberta, Edmonton, Canada
locheng@ualberta.ca

Abstract. We propose a system to detect accidental fall from walking or sitting activity in a nursing home. Differing from the trajectory tracing techniques, which detects periodic movements, our algorithm explores secondary features (angle and distance), focusing on the correlation between joints and the boundary of this correlation. We generated skeleton joint data using the Kinect sensor because it is affordable and supports sufficiently large capture space. However, other similar smart sensors can also be used. The angle feature denotes the correlation between the normal vector of the floor and the vector formed by linking the knee and ankle (on the left and right leg separately). The distance feature denotes the correlation between the floor and each of several important joints. A fall is reported when the angle is greater than and the distance is less than the respective threshold value. We created an activity database to evaluate our technique. The activities simulate elderly people walking, sitting and falling. Experimental results show that our algorithm is simple to implement, has low computational cost and is able to detect 36/37 falling events, and 57/57 walking and sitting activities accurately.

1 Introduction

Due to the increasing number of elderly people and limited hospital beds, along with economic reasons and a labour market that demands younger family members to spend longer hours on the job than taking care of the older generation, nursing homes have become necessary for vulnerable individuals. However, in a nursing home, individual occupants do not always have care-givers to monitor their safety. In particular, when an elderly person falls down, (s)he often requires immediate attention in order to avoid life threatening casualty. Common factors associated with falling include heart attack, gait and balance problems. It is estimated that approximately 25 to 35 % of older people fall each year, and the incidence of falling is the highest among elderly people in nursing facilities [1]. Falls may result in soft tissue injury, head trauma, internal hemorrhage, severe laceration, hip fracture, and loss of other bodily functions. Therefore, an automatic monitoring system capable of detecting falls and alerting care-givers immediately will benefit not only to save life but also to reduce healthcare cost in the long term.

© Springer International Publishing Switzerland 2015
G. Bebis et al. (Eds.): ISVC 2015, Part II, LNCS 9475, pp. 489–498, 2015.
DOI: 10.1007/978-3-319-27863-6_45

Videos are commonly used for surveillance, but individuals especially elderly people often do not like their faces and bodies being monitored on surveillance screens. With the recent technological advances in the development of smart sensors, researchers have proposed less intrusive methods to track human activities. There are systems that analyze data collected from wearable sensors. The problem with this approach is that accidental falls can only be detected if the person wears the sensor [2]; elderly people may forget to wear or reject the idea of wearing a sensing device. Another approach makes use of a smart-phone to detect the change of acceleration and orientation when a fall occurs [3]. Again, the assumption is that the person must always carry the smart-phone. There are systems that propose detecting falls through audio signal analysis but the detection performance is not satisfactory [4].

Among the vision-based methods, Kinect sensors are popular because it is non-intrusive and affordable. Kinect sensors are commonly used in games for detecting hand and leg motions so that the player(s) can interact with the virtual scene in the game. Simple limb movements, e.g., kicking and throwing, are sufficient in game applications. However, more complex joint detection and analysis is necessary for recognizing accidental falls in real-life. Kinect is a RGB-D sensor, providing synchronized color and depth images, as well as skeleton joint data in 3D space. The RGB camera delivers three basic color components of the video. The 3D depth sensor system consists of an infrared (IR) laser projector and an IR camera, for generating a depth map, which provides the distance information between an object and the camera [5]. Kinect also detects human skeleton joints. The IR camera can locate the joints up to two users accurately and track their movements over time. Since only skeleton joints are tracked for motion analysis, much privacy is preserved compared to video display. The depth-sensing technology of Kinect has been used for detecting falls [6–9]. In these approaches, the target skeleton is segmented from the background and tracked. For example, in the approach of Rougier et al. [7], the target is segmented and localized from the depth image. Their system uses the body centroid to calculate the velocity. A fall is reported when the velocity is above a certain threshold while the distance of the body centroid from the floor is below another threshold. The floor is recognized by using histogram analysis on a V-disparity image. From the V-disparity image, Hough transform is applied to determine straight lines and to choose the depth of the floor plane in the scene. This method can detect floor-plane but does not give stable result and is time consuming due to the complexity of Hough transform [10].

We propose an automatic fall detection system based on joint correlation and boundary threshold. In this paper, we focus on monitoring the activities of elderly people in nursing facilities. We select six important joints as controlling parameters. By analyzing the correlation of these joints, our algorithm is able to detect whether the elderly is walking, sitting or has fallen. The angle and distance features generated from these joints are computed. The distance ise measured between the floor and each of the Shoulder Center, Pelvis, Left Knee and Right Knee. The angle is calculated between the normal vector of the floor and the vector linking the Left (Right) Knee and Left (Right) Ankle joints. A fall is reported when the feature angle is greater than a threshold while the joint distance from the floor is less than another threshold. Our system can perform in real-time, is affordable and computationally efficient. User privacy is respected because only skeleton joints are used.

The rest of this paper is organized as follows. Some existing vision-based fall detection systems are described in Sect. 2. The proposed algorithm is explained in Sect. 3. Experimental results and discussion are given in Sect. 4 and the conclusion is given in Sect. 5.

2   Related Works

Many algorithms have been proposed to detect a person falling down. For example, Miaou et al. [11] proposed a system which applies MapCam-based image processing on personal information. A clean background model is first established without a person or moving object in the scene. When a person enters into the scene, the person's information is analyzed and the person's silhouettes is extracted. The ratio of the person's height and weight, and an associated threshold are used to determine the state of falling. The body mass index is used to automatically adjust the threshold. However, their results show many false positives. Another interesting method proposed by Töreyin et al. [12] is based on video and audio analysis. Bounding boxes of the moving regions are obtained and aspect ratios are estimated. A time-series signal describing the motion of a person in the video is extracted. The wavelet transform of this signal is calculated and used as a feature signal in the Hidden Markov Models (HMMs) for classification. They observed that the wavelet coefficients can characterize moving regions. The authors set two thresholds (T1 and T2) for defining Markov states in the wavelet domain. T1 is used to determine the wavelet signal close to zero, which indicates that a fall has occurred. If the wavelet coefficients exceed the higher threshold T2, the object is likely exhibiting periodic behavior. To reach a final decision of a fall state, an audio analysis is performed to discriminate between falling and sitting down. This method requires mapping between the wavelet signal and the audio signal. Features are extracted by determining the variance and number of zeros. Three 3-state Markov models are then used to classify the walking, talking and falling sounds. An alarm is issued only when both the video and audio data yield the required probability in their model. A major limitation of this work is the necessity of audio information in order to distinguish a falling from a sitting event. Furthermore, audio signal is often interfered by external sources like television, music and talking.

Auvinet et al. [13] proposed a system to detect falls based on a multi-camera network used to reconstruct the 3D shapes of people. Fall events are detected by analyzing the vertical volume distribution ratio along the vertical axis. This ratio is obtained by dividing the volume that is below a given height by the total volume. For people lying on the ground, this ratio is high compared to when they are standing up. An alarm is triggered when the major part of this distribution is substantially near the floor during a predefined time duration, which implies that someone has fallen on the floor. Nevertheless, this method is expensive and difficult to set up accurately, involving multi-camera calibration and synchronization.

Video cameras are traditionally used but found invasive because they graphically display personal appearance. Recent work indicates that older adults are willing to compromise certain level of privacy to keep safe and independent [14]. Their more accommodating attitudes lead researchers to explore smart sensors, e.g., Kinect, and skeleton joint analysis. For example, Alazrai et al. [15] proposed an anatomical-plane-based representation of human-body, which is used in a hierarchical classification framework for detecting a person's fall. This framework consists of a representation layer and two classification layers. RGBD images are acquired from a Kinect sensor, and the 3D joint positions are estimated. A view-invariant Motion Pose Geometric Descriptor (MPGD) representation for each input frame is constructed, which consists of two profiles describing the motion and pose of the human body parts. A support-vector-machine (SVM) is used to classify each frame into one of the various states in the first classification layer. The state describes the spatio-temporal configuration of the person in each frame. In the second classification layer, constrained dynamic time warping is used to classify the whole frame sequence of states generated from the SVM classifier into a falling or non-falling activity. The evaluation of the framework was performed through computer simulation scenario. They used 66 datasets, which contained 14,400 frames and 180 activity sequences. To test the fall detection accuracy, five-fold cross validation was used. The precision, recall, and F1-measure results were 98.01 %, 97.13 %, and 97.57 %, respectively. However, this method requires processing 20 joints and a huge feature space, which has high computational cost and noise level.

Many methods choose joint velocity as a feature to detect fall. This approach assumes that the velocity will be lower when a person is doing regular activities than falling. For example, Kawatsu et al. [16] proposed two steps. A single frame is first used to detect a possible fall by computing the distance from the floor to each joint. The second step calculates the velocity of each joint along the normal vector of the floor plane. The velocities are averaged over all joints and many frames. If the average velocity exceeds some threshold value, a fall is reported. One disadvantage of this approach is that the velocity is not always a suitable indicator because a person can fall down slowly by grabbing onto some nearby support. Le et al. [10] also proposed using velocity and distance with the addition of an angle. Their fall detection system computes the velocity, distance of a chosen joint J to the floor plane, angle between the normal vector of the floor plane and the vector linking the head joint and joint J. In order to distinguish fall from the other regular activities, such as lying down to sleep, a SVM classifier is used. Their results show 3.3 % false positive (FP) and 0 % false negative (FN). The limitation of this approach is that their feature set only takes into account of the upper body ignoring the leg movements. In contrast, our features focus on joints on the leges, which are more correlated to a fall and thus generated no FP in our experiments.

3   Proposed Automatical Fall Detection Algorithm

Our goal is to detect accidental falls in nursing homes. We propose selecting feature joints that can be used to accurately detect falling from walking or sitting. An overview of our framework is depicted in Fig. 1. Kinect sensor can acquire twenty skeleton joints. Each joint position is represented in 3D (X,Y,Z). We selected six joints, which are then smoothed across frames. The angle and distance features are computed and compared with pre-defined thresholds. The detail of the algorithm is explained below.

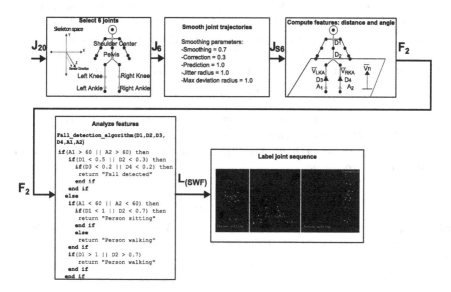

Fig. 1. An overview of our monitoring system to detect accidental falls in senior homes. (J_{20}) denotes the skeleton joints acquired by Kinect. Six joints (J_6) are selected. Their trajectories are smoothed (J_{S6}) and used to compute the angle and distance features (F_2), which are then compared with the predefined thresholds (L_{SWF}).

3.1   Skeleton Joints Acquisition and Selection

After a person's skeleton joints are captured, our algorithm applies temporal smoothing on the 3D trajectories of the target joints across the frame sequence in order to remove local jittering and noises. A smoothing filter based on the Holt Double Exponential Smoothing method is used. The filter is controlled via five parameters: smoothing, correction, prediction, jitter radius, and max deviation radius as shown in the Fig. 1.

Although 20 joints are captured by Kinect, not all of them are useful to detect falls. For example, hand movements are often random while a person is walking or sitting, and are not robust to detect falling. Based on our experimental observations, we have selected 6 joints: Shoulder Center, Pelvis, Right Knee,

Left Knee, Right Ankle, and Left Ankle. These joints maintain a relative stable pattern and, to a large extent, correlate to each other when a person is walking or sitting. When a fall occurs, these joints and their correlation changes rapidly compared to normal activities. We propose the distance and angle features computed from the 6 selected joints, to detect falling from walking or sitting.

3.2   Secondary Feature Extraction

Angle Feature: the vectors formed between the Right Knee and Right Ankle, and Left Knee and Left Ankle are denoted by V_{RKA} and V_{LKA} respectively (see Fig. 2). Note that when a fall has occurred, these vectors often cast an angle close to 90 degrees to the normal vector of the floor, V_n. To compute these angles we need to estimate the floor plane coordinates. Each skeleton frame generated by the Kinect Software Development Kit (SDK) provides a floor-clipping-plane vector, which contains the coefficients of an estimated floor-plane equation:

$$ax + by + cz + d = 0. \tag{1}$$

The Kinect sensor is located at the origin of this Cartesian coordinate system, and d is the height of the sensor from the floor. The normal vector of the floor plane is given by $V_n = [a, b, c]$. The angle between the Knee-Ankle vector and the floor normal vector is:

$$A_{1,2} = cos^{-1}(\frac{V_{KA} \cdot V_n}{|V_{KA}| \cdot |V_n|}). \tag{2}$$

where V_{KA} is the vector formed by the Knee and Ankle joints (either on the left or right leg). A_1 and A_2 are the angles formed between V_{LKA}-V_n and V_{RKA}-V_n respectively.

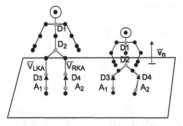

Fig. 2. Joints used to compute distance features (D_1, D_2, D_3, D_4) and vectors for computing angle feature (V_{LKA} and V_{RKA}).

Distance Feature: When a person is walking or sitting, the 6 selected joints maintain a relative stable distance from the floor plane, but such pattern changes when a fall occurs. To calculate the distance between the floor and each selected joint, we need the coefficients of the floor plane given by Eq. 1 and the 3D position of the joint. We then apply the point to plane computation:

$$D = \frac{ax + by + cz + d}{\sqrt{a^2 + b^2 + c^2}}. \tag{3}$$

3.3 Activity Detection

The algorithm reports a fall based on comparing the angle and distance features with threshold values. The angle threshold is determined using Linear Discriminant Analysis(LDA) and the distance thresholds are determined based on data analysis presented in Figs. 3 and 4 of the Experimental Section. To determine whether a person has fallen, or is walking or sitting, the algorithm first verifies the angle feature. If it exceeds the established threshold, the algorithm continues to check whether a fall has occurred by evaluating the distance features.

4 Experimental Setting, Result and Analysis

A large number of angle and distance features can be extracted from the 20 captured joints. However, not all are useful in term of fall detection. Our selected features are well defined by *activityboundaries*, which are used as thresholds to distinguish walking, sitting and falling.

4.1 Dataset

Since there is no publicly available database for accidental fall detection, we collected the test data using Kinect Studio, We recorded accidental fall events at 30 FPS. The database contains (a) 17 falls from walking, (b) 20 falls from sitting, (c) 31 sitting, and (d) 26 walking events. We simulated motions of older people, including the use of a walking stick. The Kinect sensor was placed one meter above the floor with an inclination angle of $-7°$.

4.2 Thresholds Selection

The trajectories of five joints were recorded in two falling scenarios: (a) from walking, and (b) from sitting. The recorded joint sequences were used to calculate the angle and distance features. In our method, Linear Discriminant Analysis (LDA) is applied on the training data to locate the angle threshold. LDA provides the acceptable boundary for each activity class as depicted in Fig. 3. The x-axis represents the angle between the normal vector V_n and the vector linking the Left Knee and Left Ankle joints V_{LKA} of walking, sitting and after falling respectively. The y-axis represents the angle generated by the counterpart V_{RKA} linking the Right Knee and right Ankle. Based on the training data, larger than a threshold angle of 60° alerts a possible fall and less than 60° indicates walking or sitting. We found that our angle threshold is sufficiently discriminative to classify a fall but is inadequate to discriminate between walking and sitting. Hence, the distance feature is necessary to distinguish these two activities. The 4 thresholds for the distance feature is set based on experimental observation as shown in the Fig. 4. We note that the shoulder center distance (red) is between 0.9 m and 1 m when sitting; thus the threshold for sitting is set to 1 m. The pelvis (green), left knee (blue) and right knee (black) thresholds are set similarly. The left- and right- knee distances are useful to describe the relative pose of the two legs.

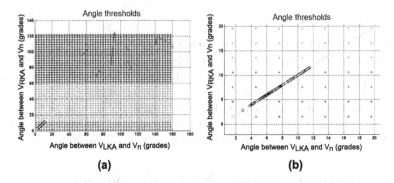

Fig. 3. (a) Motion boundaries obtained with LDA: Blue (60–120) depicts a fall event, green (< 60) represents walking , and red (> 0) shows sitting. Pink denotes the angle pattern for the fall event. Yellow denotes the angle pattern for walking. Black denotes the angle pattern for sitting. (b) Enlargement of the walking and sitting regions in (a): Note that the angle boundary is very similar between walking and sitting, and thus the distance feature is needed to distinguish the two motions (Color figure online).

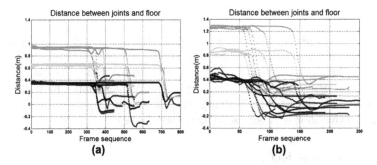

Fig. 4. Joint distance from the floor when a fall occurs (a) during sitting, and (b) during walking. (Top) Red, (middle) green, (bottom) blue and black lines represent distance of the shoulder center, left knee and right knee respectively (Color figure online).

4.3   Algorithm Performance Analysis

37 falls from sitting or walking, 31 sitting, and 26 walking activities were tested. Figure 5 shows some selected frames. Our method correctly detected 36/37 of the falls. One fall was not recognized because the person was too close to the Kinect sensor and thus the knee-joints and ankle-joints could not be tracked. For the walking and sitting activities, 57/57 were detected correctly even when a walking stick was used. Despite this encouraging outcome, the current algorithm has limitations. First, if the lower joints (knees and ankles) are hidden, Kinect cannot detect the fall. However, this can be overcome by incorporating the velocity feature in our algorithm. Second, the distance thresholds may need to adjust if a person is far from the regular height (currently defined as 1.58 m). In future work, we plan to record the height of a person when (s)he first appears and normalize it so as to adjust the distance feature adaptively. Third, the algorithm may not

Fig. 5. Depth maps: (a) Fall from sitting, and (c) Fall from walking and sitting. Snapshots of joints detected by Kinect are shown in (b) and (d).

work on staircase because the floor plane reference varies on each step. However, in a nursing facility, the staircase locations are known. Such information can be incorporated for accurate fall detection.

5 Conclusion

Compared to video-based monitoring, using smart sensors to detect accidental falls in nursing facilities provides more privacy to individuals. Smart sensors generate valuable 3D joint information for activity analysis. Although joint velocity can be used as a clue to detect falls, misses can occur due to an elderly's slow motion or an individual's resistance to fall, e.g., grab onto a railing. We propose the angle and distance features from selected skeleton joints, together with activity boundaries, to address this deficiency. We tested our algorithm by creating a test database, which contains walking, sitting and falling motions. Experimental results show that our method achieves 100 % accuracy when detecting walking and sitting, and only 1/37 fall detection was missed due to occlusion. Our system is easy to implement and the algorithm has low computational cost. In future work, we will study stereo sensing, which is anticipated to overcome occlusion.

References

1. Kiely, D.K., Kiel, D.P., Burrows, A.B., Lipsitz, L.A.: Identifying nursing home residents at risk for falling. J. Am. Geriatr. Soc. **46**, 551–555 (1998)

2. Winters, J.: Emerging rehabilitative telehealthcare anywhere was the homecare technologies workshop visionary. Emerging and Accessible Telecommunications, Information and Healthcare Technologies, pp. 95–111 (2002)
3. Viet, V.Q., Lee, G., Choi, D.: Fall detection based on movement and smart phone technology. In: 2012 IEEE RIVF International Conference on Computing and Communication Technologies, Research, Innovation, and Vision for the Future (RIVF), pp. 1–4. IEEE (2012)
4. Zhuang, X., Huang, J., Potamianos, G., Hasegawa-Johnson, M.: Acoustic fall detection using gaussian mixture models and gmm supervectors. In: IEEE International Conferenceon Acoustics, Speech and Signal Processing (ICASSP), pp. 69–72. IEEE (2009)
5. Han, J., Shao, L., Xu, D., Shotton, J.: Enhanced computer vision with microsoft kinect sensor: a review. IEEE Trans. Cybern. **43**, 1318–1334 (2013)
6. Bevilacqua, V., Nuzzolese, N., Barone, D., Pantaleo, M., Suma, M., D'Ambruoso, D., Volpe, A., Loconsole, C., Stroppa, F.: Fall detection in indoor environment with kinect sensor. In: 2014 IEEE International Symposium on Innovations in Intelligent Systems and Applications (INISTA) Proceedings, pp. 319–324. IEEE (2014)
7. Rougier, C., Auvinet, E., Rousseau, J., Mignotte, M., Meunier, J.: Fall detection from depth map video sequences. In: Abdulrazak, B., Giroux, S., Bouchard, B., Pigot, H., Mokhtari, M. (eds.) ICOST 2011. LNCS, vol. 6719, pp. 121–128. Springer, Heidelberg (2011)
8. Nghiem, A.T., Auvinet, E., Meunier, J.: Head detection using kinect camera and its application to fall detection. In: 2012 11th International Conference on Information Science, Signal Processing and their Applications (ISSPA), pp. 164–169. IEEE (2012)
9. Stone, E.E., Skubic, M.: Fall detection in homes of older adults using the microsoft kinect. IEEE J. Biomed. Health Inform. **19**, 290–301 (2015)
10. Le, T.L., Morel, J.M., et al.: An analysis on human fall detection using skeleton from microsoft kinect. In: 2014 IEEE 5th International Conference on Communications and Electronics (ICCE), pp. 484–489. IEEE (2014)
11. Miaou, S.G., Sung, P.H., Huang, C.Y.: A customized human fall detection system using omni-camera images and personal information. In: 1st Transdisciplinary Conference on Distributed Diagnosis and Home Healthcare, D2H2 2006, pp. 39–42. IEEE (2006)
12. Töreyin, B.U., Dedeoğlu, Y., Çetin, A.E.: HMM based falling person detection using both audio and video. In: Sebe, N., Lew, M., Huang, T.S. (eds.) HCI/ICCV 2005. LNCS, vol. 3766, pp. 211–220. Springer, Heidelberg (2005)
13. Auvinet, E., Multon, F., Saint-Arnaud, A., Rousseau, J., Meunier, J.: Fall detection with multiple cameras: an occlusion-resistant method based on 3-d silhouette vertical distribution. IEEE Trans. Inf. Technol. Biomed. **15**, 290–300 (2011)
14. Demiris, G., Oliver, D.P., Giger, J., Skubic, M., Rantz, M.: Older adults' privacy considerations for vision based recognition methods of eldercare applications. Technol. Health Care **17**, 41 (2009)
15. Alazrai, R., Zmily, A., Mowafi, Y.: Fall detection for elderly using anatomical-plane-based representation. In: 2014 36th Annual International Conference of the IEEE Engineering in Medicine and Biology Society (EMBC), pp. 5916–5919. IEEE (2014)
16. Kawatsu, C., Li, J., Chung, C.J.: Development of a fall detection system with microsoft kinect. In: Kim, J.-H., Matson, E., Myung, H., Xu, P. (eds.) Robot Intelligence Technology and Applications. AISC, vol. 208, pp. 623–630. Springer, Heidelberg (2013)

Generalized Wishart Processes for Interpolation Over Diffusion Tensor Fields

Hernán Darío Vargas Cardona[✉], Mauricio A. Álvarez, and Álvaro A. Orozco

Faculty of Engineering, Universidad Tecnológica de Pereira, Pereira, Colombia
{hernan.vargas,malvarez,aaog}@utp.edu.co

Abstract. Diffusion Magnetic Resonance Imaging (dMRI) is a non-invasive tool for watching the microstructure of fibrous nerve and muscle tissue. From dMRI, it is possible to estimate 2-rank diffusion tensors imaging (DTI) fields, that are widely used in clinical applications: tissue segmentation, fiber tractography, brain atlas construction, brain conductivity models, among others. Due to hardware limitations of MRI scanners, DTI has the difficult compromise between spatial resolution and signal noise ratio (SNR) during acquisition. For this reason, the data are often acquired with very low resolution. To enhance DTI data resolution, interpolation provides an interesting software solution. The aim of this work is to develop a methodology for DTI interpolation that enhance the spatial resolution of DTI fields. We assume that a DTI field follows a recently introduced stochastic process known as a generalized Wishart process (GWP), which we use as a prior over the diffusion tensor field. For posterior inference, we use Markov Chain Monte Carlo methods. We perform experiments in toy and real data. Results of GWP outperform other methods in the literature, when compared in different validation protocols.

1 Introduction

Diffusion Magnetic Resonance Imaging (dMRI) is a non-invasive procedure to find connections into biological mediums such as fiber nerves and muscle tissue. From dMRI it is possible to estimate the apparent diffusivity coefficient (ADC) of water particles within tissue by solving the Stejskal-Tanner formulation [1]. A 2-rank diffusion tensor (D) is employed to modeling ADC in each specific voxel, where D is a symmetric and positive definite 3×3 matrix. Following this notion, a diffusion tensor imaging (DTI) field is understood as a grid of individual but related diffusion tensors. Although a DTI field shows how some nerve fiber bundles are interconnected, there are some limitations in MRI scanners. For example, dMRI is sensitive to the difficult compromise between spatial resolution and signal to noise ratio (SNR). This leads to data acquisitions with low resolution [2].

To enhance DTI data resolution, interpolation provides an interesting and feasible methodological solution. Diffusion tensor fields belong to a Riemannian space, where the Riemannian metric is defined by the inner product assigned

© Springer International Publishing Switzerland 2015
G. Bebis et al. (Eds.): ISVC 2015, Part II, LNCS 9475, pp. 499–508, 2015.
DOI: 10.1007/978-3-319-27863-6_46

to each point of this space. With this metric, one can to compute geodesic distances between diffusion tensors and to calculate different statistics in this space [3]. An important condition is keeping the smooth transition of anisotropic features inherent in the given tensor fields (i.e. Fractional anisotropy-FA maps), especially around degenerate points, where at least two of three eigenvalues are equivalent [4]. Interpolation of the diffusion tensor fields have many applications. For example, registration of DTI datasets will require resolution enhancement when a registration transformation is applied to a tensor field. Other examples that require DTI interpolation include segmentation, atlas construction, diagnosis of neurological diseases, etc. [5]. Currently, the clinical acquisition protocols of dMRI data allow one or two millimeters resolution for each voxel. The problem here, is that brain tissue fiber bundles are in micrometers scale. Therefore, the tractography models developed from DTI data can be imprecise due to the current low resolution in acquired images. Normally, visualization of DTI is discrete, where it is used ellipsoids or glyphs for graphic representation. Tractography is the search of fiber connection among neighboring voxels. The basic idea is to generate a continuous data representation [6]. According to this, an accurate interpolation approach may improve the spatial resolution of diffusion tensor fields in a considerable factor. Therefore, the tractography process will describe with more detail the fiber tissue connection.

Some recent works have proposed interpolation methods for tensor fields in DTI. They developed a variety of mathematical approaches, such as: direct smooth approximation [7] and euclidean approaches, but they do not retain the principal properties of a DTI, i.e. positive definite tensors. For this reason, the scientific community has been looking alternative methods for estimating tensor fields that keep the symmetric positive definite (SPD) constraint inside the grid of tensors. [8] presented a Log-euclidean approximation and [3] developed a Riemannian framework achieving important advances in tensor fields geometry, but they lack in smoothness property in presence or high level of noise. [5] presented a b-spline scheme that interpolates SPD tensors with high accuracy using the Riemannian metric. The authors introduced a tensor product of B-splines that minimizes the Riemannian distance between tensors. Following the Riemannian framework, [9] presented Geodesic-loxodromes that can identify isotropic and anisotropic components of the tensor and interpolates each component separately. Finally, alternative methodologies have been posited: a tensor field reconstruction based on eigenvector and eigenvalue interpolation [6], location of degenerated lines in 2-D planar [4], and a feature-based interpolation [10]. However, those methods do not achieve an adequate representation of a DTI field obtained from noisy real data.

As previously was pointed out, most of the methods for DTI interpolation are based on Riemannian geometry. While they preserve the main properties of DTI data, and solve limitations of the Euclidean approaches, they lead to rigid interpolations that fail to fully adapt to the variety of diffusion patterns in biological tissues [11]. In this work, we present a novel methodology for interpolation of DTI fields. Instead of a Riemann geometry framework, we propose a stochastic modeling of DTI. We assume that a DTI field follows a generalized Wishart process (GWP). A GWP is a collection of symmetric positive definite random matrices indexed by an arbitrary dependent variable [12], i.e. the x, y, z position. In this context, we use it to model the entire DTI field $D(x, y, z)$. Then, through approximate Bayesian inference (i.e. Elliptical slice sampling and Markov Chain Monte Carlo methods), we estimate the optimal parameters of the model. Stochastic modeling of DTI fields has some advantages: positive definite matrices, robustness to noise, smooth transition among nearby tensors and good accuracy for estimating new data. We compare our approach with linear interpolation [7] and a Riemannian method known as log-euclidean interpolation [8]. We perform experiments in toy and real DTI data. Results of GWP improve to the comparison methods in different validation protocols.

2   Materials and Methods

2.1   DTI Estimation from dMRI and DTI Fields

Diffusion Magnetic Resonance Imaging (dMRI) studies the diffusion of water particles in the human brain. Diffusion can be described by a symmetric positive definite 3×3 matrix proportional to the covariance of a Gaussian distribution [1,13].

$$\mathbf{D} = \begin{bmatrix} D_{xx} & D_{xy} & D_{xz} \\ D_{yx} & D_{yy} & D_{yz} \\ D_{zx} & D_{zy} & D_{zz} \end{bmatrix}$$

For water, the diffusion tensor (DT) is symmetric, so that $D_{ij} = D_{ji}$, where $i, j = x, y, z$. The diffusion tensor for each voxel of the dMRI is calculated using the Stejskal-Tanner formulation [1]:

$$S_k = S_0 e^{-b\hat{\mathbf{g}}_k^\top D \hat{\mathbf{g}}_k}, \tag{1}$$

where S_k is the kth dMRI, S_0 is the reference image, $\hat{\mathbf{g}}_k$ is the gradient vector and b is the diffusion coefficient. At least 7 dMRI measurements are necessary for each slice ($k = 0, 1, ..., 7$). Usually, DTI fields are estimated from (1) using least squares [14]. However, there are robust methods for DT estimation. In this work, we use the RESTORE algorithm [15] for solving the DTs.

Traditionally, rank-2 DTs have been visualized by constructing the ellipsoid given by:

$$\mathbf{r}^\top D^{-1} \mathbf{r} = C \tag{2}$$

where $\mathbf{r}^\top = [x, y, z]$ is the position vector, and C is a constant with the units of time. Therefore, the resulting shape is a level surface of the expression on the left side of (2), and it is possible to show by diagonalization that these surfaces are ellipsoids.

2.2   Generalized Wishart Process (GWP)

We begin with the Wishart distribution, which defines a probability density function over a symmetric positive definite matrix. Let S be a $p \times p$ symmetric positive definite matrix of random variables. Let V be a (fixed) positive definite matrix of size $p \times p$. Then, if $\nu \geq p$, S has a Wishart distribution with ν degrees of freedom if it has a probability density function given by:

$$S = \frac{|S|^{\nu-p-1}}{2^{\nu p/2}|V|^{\nu/2}\Gamma_p(\frac{\nu}{2})} e^{-\frac{1}{2}\operatorname{trace}(V^{-1}S)},$$

where $|\cdot|$ is the determinant and Γ_p is the multivariate gamma function:

$$\Gamma_p\left(\frac{\nu}{2}\right) = \pi^{\frac{p(p-1)}{4}} \prod_{j=1}^{p} \Gamma\left(\frac{\nu}{2} + \frac{1-j}{2}\right)$$

Following this notion and according to the definition given in [12], a generalized Wishart process (GWP) is as a collection of symmetric positive definite random matrices indexed by an arbitrary and high dimensional dependent variable $\mathbf{z}$. In DTI fields, the dimension is $p = 3$ because diffusion tensors are represented by 3×3 matrices, and the indexed variable refers to position coordinates $\mathbf{z} = [x, y, z]^{\top}$. Assume 3ν independent Gaussian process functions $u_{id}(\mathbf{z}) \sim \mathcal{GP}(0, k)$, for $i = 1, ..., \nu$ and $d = 1, 2, 3$, where $k(\mathbf{z}, \mathbf{z}')$ is the covariance or kernel function for the GP. Given a set of input vectors $\{\mathbf{z}\}_{n=1}^{N}$, the vector $(u_{id}(\mathbf{z}_1), u_{id}(\mathbf{z}_2), ..., u_{id}(\mathbf{z}_N))^{\top} \sim \mathcal{N}(\mathbf{0}, K)$, being K an $N \times N$ Gram matrix with entries $K_{ij} = k(\mathbf{z}_i, \mathbf{z}_j)$. If we define $\hat{\mathbf{u}}_i(\mathbf{z}) = (u_{i1}(\mathbf{z}), u_{i2}(\mathbf{z}), u_{i3}(\mathbf{z}))^{\top}$ and L as the lower Cholesky decomposition of a $p \times p$ scale matrix V, such that $LL^{\top} = V$, for each input position $\mathbf{z} = [x, y, z]^{\top}$, the diffusion tensor $D(\mathbf{z})$ follows a Wishart distribution,

$$D(\mathbf{z}) = \sum_{i=1}^{\nu} L\hat{\mathbf{u}}_i(\mathbf{z})\hat{\mathbf{u}}_i^{\top}(\mathbf{z})L^{\top} \sim \mathcal{GWP}_p(\nu, V, k(\cdot, \cdot)), \qquad (3)$$

In this work, we use the squared exponential kernel $k(\mathbf{z}, \mathbf{z}')$,

$$k(\mathbf{z}, \mathbf{z}') = \exp\left(-0.5\frac{\|\mathbf{z} - \mathbf{z}'\|^2}{\theta^2}\right),$$

where θ is the length-scale hyperparameter.

2.3   Bayesian Inference for DTI Field Learning

In order to perform DTI interpolation, we first need to compute the posterior distribution for the variables in the model. For a DTI field, we assume a prior given by a Generalized Wishart process

$$p\left(D(\mathbf{z})\right) \sim \mathcal{GWP}_3(\nu, V, k(\cdot, \cdot)) = \sum_{i=1}^{\nu} L\hat{\mathbf{u}}_i(\mathbf{z})\hat{\mathbf{u}}_i^{\top}(\mathbf{z})L^{\top}. \qquad (4)$$

For the likelihood function, we assume each element from the diffusion tensor data follows an independent Gaussian distribution with the same variance σ^2. This leads to a likelihood with the following form:

$$p(S|u, L, \nu) \propto \prod_{i=1}^{N} \exp \left(-\frac{1}{2\sigma^2} \|S(\mathbf{z}_i) - D(\mathbf{z}_i)\|_{frob}^2 \right),$$

where $S(\mathbf{z})$ is the known initial DTI field with low resolution, $D(\mathbf{z})$ is constructed from Eq. (4), and Frobenius norm is given by

$$\|\mathbf{X}\|_{frob}^2 = \text{trace}\left(\mathbf{X}^T\mathbf{X}\right).$$

The purpose is to infer the posterior probability of $D(\mathbf{z})$ given a known tensorial data set $S(\mathbf{z}) = \{S(\mathbf{z}_1), S(\mathbf{z}_2), ..., S(\mathbf{z}_N)\}$, being N the number of data in the initial DTI field. We first compute the posterior of the relevant variables in Eq. (4) including the vector of all GP function values $\mathbf{u}$, length-scale hyperparameter of the GP kernel function θ, the lower Cholesky decomposition of the scale matrix L, such that $LL^T = V$, and the degrees of freedom ν. Given a GWP prior for the model and the likelihood function, the posterior distributions can be computed by

$$p(\mathbf{u}|\theta, L, S) \propto p(S|\mathbf{u}, L, \nu)p(\mathbf{u}|\theta), \tag{5}$$

$$p(\theta|\mathbf{u}, L, S) \propto p(\mathbf{u}|\theta, L, D)p(\theta), \tag{6}$$

$$p(L|\mathbf{u}, \theta, S) \propto p(S|\mathbf{u}, L, \nu)p(L). \tag{7}$$

We use Markov chain Monte Carlo algorithms to sample in cycles. We employ Metropolis-Hastings to sample θ from (6), and the elements of scale matrix L from (7). To sample $\mathbf{u}$ from (5), we employ elliptical slice sampling [16]. We choose $\nu = 5$ through cross-validation. We set a log-normal prior on θ, a spherical Gaussian prior on elements of L and the prior $p(\mathbf{u}|\theta) \sim \mathcal{N}(\mathbf{0}, K_B)$ is a Gaussian distribution with $3\nu N \times 3\nu N$ block diagonal covariance matrix K_B, formed using 3ν of the K matrices.

2.4   DTI Field Interpolation Through GWP Modeling

Once we find the posterior distributions over all relevant variables for the model, we can compute the posterior distribution for $D(\mathbf{z}_*)$ in a new spatial position $\mathbf{z}_* = [x_*, y_*, z_*]^\top$. First, we have to infer the distribution over all unknown GP function values $\mathbf{u}_*$ in $\mathbf{z}_*$, where $\mathbf{u}_*$ is a vector with elements given by $u_{id}(\mathbf{z}_*)$. The joint distribution over $\mathbf{u}$ and $\mathbf{u}_*$ is given by,

$$\begin{bmatrix} \mathbf{u} \\ \mathbf{u}_* \end{bmatrix} \sim \mathcal{N} \left(\mathbf{0}, \begin{bmatrix} K_B & A^\top \\ A & I_p \end{bmatrix} \right)$$

If $\mathbf{u}_*$ and $\mathbf{u}$ have p and q elements respectively, A is a $p \times q$ matrix that represents the covariances between $\mathbf{u}_*$ and $\mathbf{u}$ for all pairs of training and validation data, this is $A_{ij} = k_i(\mathbf{z}_*, \mathbf{z}_j)$ for $i + (i-1)N \leq j \leq iN$, and 0 otherwise. I_p is a $p \times p$ identity matrix. Using the properties of a Gaussian distribution, and conditioning on $\mathbf{u}$, we obtain:

$$p(\mathbf{u}_*|\mathbf{u}) \sim \mathcal{N}\left(AK_B^{-1}\mathbf{u}, I_p - AK_B^{-1}A^\top\right) \tag{8}$$

From values of $\mathbf{u}_*$ obtained from (8), and using Eq. (3), we can construct $D(\mathbf{z}_*)$.

2.5   Validation Procedure and Datasets

As ground truth (gold standard) we employ three different types of data. The first one corresponds to a synthetic DTI field. The second one corresponds to a simulation of crossing fibers using the algorithm of the fanDTasia toolBox [17], available at http://www.cise.ufl.edu/abarmpou/lab/fanDTasia/. The third one, corresponds to a DTI dataset estimated from real dMRI through the RESTORE method [15]. dMRI data of the head were acquired from a healthy subject on a General Electric Signa HDxt 3.0T MR scanner using the body coil for excitation, and an 8-channel quadrature brain coil for reception. We employ 25 gradient directions with a value for b equal to $1000\,S/mm^2$. The study contains $128 \times 128 \times 33$ images in axial plane. For the three datasets, we downsample the DTI field by a factor of two. The downsampled field is the input data for the GWP. After we perform inference over the GWP, we interpolate the DTI field, and calculate two error metrics, having the gold standards as our references. Also, we repeat the same procedure for linear [7] and log-euclidean interpolation [8] for a comparison with two commonly used methods in the state of the art. We use two metrics to measure the differences between the interpolated fields and the ground truth, the Frobenius norm, and the Riemman distance, defined by

$$\mathrm{Frob}(T_1, T_2) = \sqrt{\mathrm{trace}\left[(T_1 - T_2)^\top (T_1 - T_2)\right]},$$

$$\mathrm{Riem}(T_1, T_2) = \sqrt{\mathrm{trace}\left[\log(T_1^{-1/2}T_2T_1^{-1/2})^\top \log(T_1^{-1/2}T_2T_1^{-1/2})\right]},$$

where T_1 and T_2 are the estimated and the ground truth tensors, respectively. The error metrics are computed for each *voxel*. We report the mean and standard deviation for the errors over the predicted data.

3   Experimental Results and Discussion

In this section, we present the interpolation results for the different DTI datasets. We compare with linear [7] and log-euclidean interpolation [8].

3.1   Synthetic Data

We generate noisy random DTI data to construct a $2D$ field of 37×37 tensors. We assume 25 gradient directions for generating DTs, and b value of $1000\,s/mm^2$. In Fig. 1 we can see the initial downsampled DTI field, linear and log-euclidean interpolation, the interpolated field with GWP, and the ground truth respectively. Table 1 shows the error metrics.

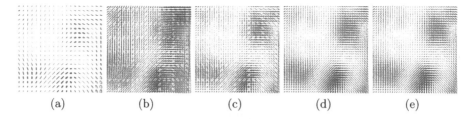

(a)   (b)   (c)   (d)   (e)

Fig. 1. Graphic results for DTI interpolation (2×) applied in synthetic data. (a) Downsampled DTI field (data used for estimation). (b) Linear interpolation. (c) Log-euclidean interpolation. (d) Interpolation with GWP. (e) Ground truth.

Table 1. Metric results for synthetic DTI field

	Frobenius distance ($\times 10^{-5}$)	Riemman distance
GWP	7.06 ± 1.51	0.160 ± 0.125
Linear interpolation	50.11 ± 4.26	8.54 ± 1.36
Log-euclidean	35.25 ± 3.92	6.34 ± 1.22

3.2 DTI from Crossing Fibers

One of the most critical DTI datasets correspond to crossing fibers. We generate this type of DTI field through FanDTasia toolbox [17]. This datset describes a $2D$ crossing fiber field with 31×31 tensors. Figure 2 and Table 2 show the comparative results.

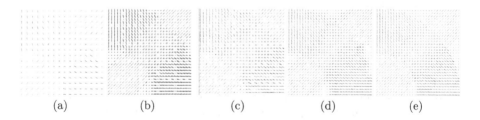

(a)   (b)   (c)   (d)   (e)

Fig. 2. Graphic results for DTI interpolation (2×) applied in crossing fibers field. (a) Downsampled DTI field (data used for estimation). (b) Linear interpolation. (c) Log-euclidean interpolation. (d) Interpolation with GWP. (e) Ground truth.

Table 2. Error measures for crossing fibers in a DTI field.

	Frobenius distance ($\times 10^{-5}$)	Riemman distance
GWP	18.11 ± 11.82	0.184 ± 0.114
Linear interpolation	73.12 ± 8.26	11.14 ± 2.65
Log-euclidean	61.09 ± 6.15	9.74 ± 1.67

3.3   Real DTI Field Estimated from dMRI

Finally, we test our method in real DTI data estimated from dMRI acquired in a human subject. The field corresponds to an axial slice with 49×55 tensors. Figure 3 and Table 3 show the comparative results.

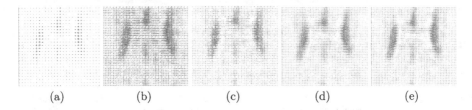

(a)   (b)   (c)   (d)   (e)

Fig. 3. Graphic results for DTI interpolation ($2\times$) applied in real DTI data. (a) Downsampled DTI field (data used for estimation). (b) Linear interpolation. (c) Log-euclidean interpolation. (d) Interpolation with GWP. (e) Ground truth.

Table 3. Error measures for the real DTI field example

	Frobenius distance ($\times 10^{-5}$)	Riemman distance
GWP	6.26 ± 3.20	0.146 ± 0.080
Linear interpolation	45.76 ± 7.21	7.25 ± 2.10
Log-euclidean	$31.67 \pm\ \ 6.10$	6.89 ± 1.86

3.4   Discussion

Linear and log-euclidean interpolation seek to minimize geodesic distances. The geometric (Riemann and Euclidean) approaches work well in smooth DTI fields. However, they reduce their performance in presence of high level of noise. For example, when we interpolate the synthetic noisy data (Fig. 1 and Table 1), we can observe a swelling effect for the estimated tensors in the new input locations, when using linear interpolation. This is a critical issue, because this effect modifies the fractional anisotropy maps. Another big problem with linear interpolation is the possibility of obtaining non-positive definite tensors. Although, non-positive definite tensors are avoided with log-euclidean interpolation, the accuracy of tensor estimation in new input locations is not satisfactory. On the other hand, GWP guarantees positive definite tensors because of its mathematical construction. Also, this probabilistic model is more robust to noise, and it can keep the smooth transition among spatially nearby data. This property avoids the swelling effect. If we look at the results for more complex data like crossing fibers DTI, and the real DTI field (Figs. 2 and 3 and Tables 2 and 3), we can see better accuracy results for GWP. Both average distances (Frobenius and Riemann) are smaller in GWP than Linear and Log-Euclidean methods. Similarly, graphic results of DTI fields are smoother for GWP interpolation. The GWP takes into account the global spatial behavior of the DTI data, while geometric approaches estimate tensors only with the nearest tensors.

A drawback in GWP happens when there are strong changes in tensor orientation (i.e. crossing fibers). The GWP cannot capture extreme modifications in data of crossing fibers. Recall that a GWP is a superposition of Gaussian processes (GP) and GPs are modeled with smooth kernels functions. The best alternative to model crossing fibers is through Higher Order Tensors (HOT). Nevertheless a GWP does not describe HOT. However, for the geometric approaches, strong changes in tensor orientation generate worse results than GWP.

In summary, the probabilistic modeling of DTI fields that we employ here allows a better description of global spatial transition in tensorial imaging. Geometric approaches are fine for simple DTI fields. However in real applications, DTI data are very complex (High level of noise, Heterogeneous data, Non-positive definite tensors, etc.). The GWP has many advantages, for example: it guarantees positive definite tensors, robustness to noisy data, smooth transition among nearby data, no swelling effect, it keeps important properties of DTI (FA maps) and excellent accuracy.

4   Conclusions and Future Work

In this paper we developed a probabilistic methodology to interpolate Diffusion Tensor Imaging (DTI) data. We model a DTI field as a Generalized Wishart process (GWP). We employ approximate Bayesian inference for optimizing the relevant variables in GWP. Results obtained with GWP in synthetic and real DTI data outperform to commonly used geometric methods. Also, our proposed method guarantees positive definite tensors, excellent accuracy and it avoids an issue in tensorial interpolation known as swelling effect. As future work, we would like to extend this concept to modeling Higher Order Tensors. The above approach would be relevant when describing tensorial data from crossing fibers.

Acknowledgments. H.D. Vargas Cardona is funded by Colciencias under the program: *formación de alto nivel para la ciencia, la tecnología y la innovación - Convocatoria 617 de 2013*. This research has been developed under the project financed by Colciencias with code 1110-657-40687.

References

1. Basser, P., Mattiello, J., Le Bihan, D.: Estimation of the effective self-diffusion tensor from the nmr spin echo. J. Magn. Reson. **103**, 247–254 (1994)
2. Yang, F., Zhu, Y.M., Luo, J.H., Robini, M., Liu, J., Croisille, P.: A comparative study of different level interpolations for improving spatial resolution in diffusion tensor imaging. IEEE J. Biomed. Health Inform. **18**, 1317–1327 (2014)

3. Fletcher, P., Joshi, S.: Riemannian geometry for the statistical analysis of diffusion tensor data. Signal Process. **87**, 250–262 (2007)
4. Bi, C., Takahashi, S., Fujishiro, I.: Interpolating 3D diffusion tensors in 2D planar domain by locating degenerate lines. In: Bebis, G., Boyle, R., Parvin, B., Koracin, D., Chung, R., Hammoud, R., Hussain, M., Kar-Han, T., Crawfis, R., Thalmann, D., Kao, D., Avila, L. (eds.) ISVC 2010, Part I. LNCS, vol. 6453, pp. 328–337. Springer, Heidelberg (2010)
5. Barmpoutis, A., Vemuri, B., Shepherd, T., Forder, J.: Tensor splines for interpolation and approximation of dt-mri with applications to segmentation of isolated rat hippocampi. IEEE Trans. Med. Imaging **26**, 1537–1546 (2007)
6. Hotz, I., Sreevalsan-Nair, J., Hamann, B.: Tensor field reconstruction based on eigenvector and eigenvalue interpolation. In: Scientific Visualization: Advanced Concepts, pp. 110–123 (2010)
7. Pajevic, S.: A continuous tensor field approximation of discrete dt-mri data for extracting microstructural and architectural features of tissue. J. Magn. Reson. **154**, 85–100 (2002)
8. Arsigny, V., Fillard, P., Pennec, X., Ayache, N.: Log-euclidean metrics for fast and simple calculus on diffusion tensors. Magn. Res. Med. **56**, 411–421 (2006)
9. Kindlmann, G., Estepar, R., Niethammer, M., Haker, S., Westin, C.: Geodesic-loxodromes for diffusion tensor interpolation and difference measurement. Med. Image Comput. Comput. Assist. Interv. **10**, 1–9 (2007)
10. Yang, F., Zhu, Y.M., Magnin, I., Luo, J.H., Croisille, P., Kingsley, P.: Feature-based interpolation of diffusion tensor fields and application to human cardiac dt-mri. Med. Image Anal. **16**, 459–481 (2012)
11. Chang, I.S., Shun-Ren, X.: Diffusion tensor interpolation profile control using non-uniform motion on a riemannian geodesic. Comput. Electron. **13**, 90–98 (2012)
12. Wilson, A., Ghahramani, Z.: Generalised wishart processes. In: UAI 2011, pp. 736–744 (2011)
13. Tanner, J., Stejskal, E.: Spin diffusion measurements: spin echoes in the presence of a time-dependent field gradient. J. Chem. Physiol. **42**, 288–292 (1965)
14. Basser, P.: Inferring microstructural features and the physiological state of tissues from diffusion weighted images. NMR Biomed. **8**, 333–344 (1995)
15. Lin-Chin, C., Jones, D., Pierpaoli, C.: RESTORE: robust estimation of tensors by outlier rejection. Magn. Reson. Med. **53**, 1088–1095 (2005)
16. Murray, I., Adams, R., Mackay, D.: Elliptical slice sampling. JMLR **9**, 541–548 (2010)
17. Barmpoutis, A., Vemuri, B.: A unified framework for estimating diffusion tensors of any order with symmetric positive-definite constraints. In: Proceedings of ISBI 2010: IEEE International Symposium on Biomedical Imaging, pp. 1385–1388 (2010)

Spatio-Temporal Fusion for Learning of Regions of Interests Over Multiple Video Streams

Samaneh Khoshrou[1,2]([✉]), Jaime S. Cardoso[1,2],
Eric Granger[3], and Luís F. Teixeira[1,2]

[1] INESC TEC, Porto, Portugal
{skhoshrou,jaime.cardoso}@inescporto.pt, lft@fe.up.pt
[2] Faculdade de Engenharia da Universidade do Porto (FEUP), Porto, Portugal
[3] Laboratoire d'Imagerie, de vision et d'intelligence artificielle,
École de technologie supérieure, Université du Québec, Montreal, Canada
eric.granger@etsmtl.ca

Abstract. Video surveillance systems must process and manage a growing amount of data captured over a network of cameras for various recognition tasks. In order to limit human labour and error, this paper presents a spatial-temporal fusion approach to accurately combine information from Region of Interest (RoI) batches captured in a multi-camera surveillance scenario. In this paper, feature-level and score-level approaches are proposed for spatial-temporal fusion of information to combine information over frames, in a framework based on ensembles of GMM-UBM (Universal Background Models). At the feature-level, features in a batch of multiple frames are combined and fed to the ensemble, whereas at the score-level the outcome of ensemble for individual frames are combined. Results indicate that feature-level fusion provides higher level of accuracy in a very efficient way.

1   Introduction

Video surveillance applications, such as activity recognition, are increasingly making use of multiple sensors and modalities. The fusion of multiple diverse sources of information is expected to benefit the system for the recognition of objects, persons, activities and events captured in an array of cameras.

Networks of video cameras are commonly employed to monitor large areas for a variety of applications. A central issue in such networks is the tracking and recognition of individuals of interests across multiple cameras. These individuals must be recognized when leaving the Field of View (FoV) of one camera and re-identified when entering the FoV of another camera. Systems for video-to-video recognition are typically employed for person re-identification (PR). In a FoV, the appearance of an individual may be captured in reference RoIs and representative models may be learned from RoI trajectories. Then, the probe RoI may be matched against the reference model in either live (real-time monitoring) or archived (post-event analysis) [1]. In this paper, we address a PR system over wide network of cameras where no target individual enrolled to the system in advance.

© Springer International Publishing Switzerland 2015
G. Bebis et al. (Eds.): ISVC 2015, Part II, LNCS 9475, pp. 509–520, 2015.
DOI: 10.1007/978-3-319-27863-6_47

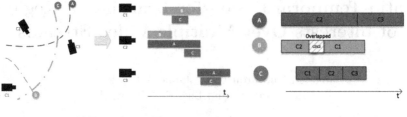

(a) Typical surveillance scenario   (b) An example of time-line graph

Fig. 1. A surveillance scenario including three persons A, B, and C, moving in the scene, crossing the FoV of 3 cameras: c_1, c_2, and c_3.

In such environments, where objects move and cross in the FoV of multiple cameras, it is likely to have multiple streams, recorded at different starting points with various lengths, for the same RoI of individuals (see Fig. 1a). The surveillance system must track that person across all cameras whose FoV overlap the person's path. Thus, a suitable outcome for this system could be a time-line graph assigning streams from each camera to an identity for the indicated presence period, as illustrated in Fig. 1b. Environmental challenges such as the variation in appearance of individuals due to changes in illumination, contrast, positioning of acquisition devices, motion blur as well as occlusion lead to noisy and/or partial RoI captures. These challenges have previously been addressed by a batch divisive strategy in [2], that views a batch of RoI as a unique element to classify, since learning from these batches may reduce noise and fill the gaps caused by dropped-tracks. A batch includes a fixed number of consecutive RoIs (sources of information) of a given stream and a single label is assigned to the batch of same person in time. Each batch of RoI can be learnt using a one-class classifier, and the pool of classifiers generated in one or more FoVs can be combined into an ensemble of classifiers. Fusion of multiple sources into an ensemble have been addressed by three main approaches in the literature: early, mid-level, and late [3]. Early fusion combines the information in the first possible level (so called signal level fusion in image processing), whereas late fusion combines the information as late as possible (decision level fusion) [4,5]. Mid-level fusion is an interesting compromise that combines the information in an intermediate abstraction level [6].

Score-level fusion is the most popular way of fusion. A quantitative similarity measure disseminates valuable information about the input, and yet it is still easy to process compared to sensor-level or feature-level data. However the score space is subject to considerable flexibilities, e.g. different normalization methods may lead to different decision boundaries. Furthermore, small number of scores in a batch might easily overfit the data [7]. On the other hand, feature-level fusion schemes derive the most abstract form of original multiple feature set by eliminating redundant information. The advantages of this scheme are the use of only one learning stage to combine the information (instead of running individual learning stage for every single feature set) for rapid decisions.

In this paper, two feature-level abstraction schemes that represent the entire batch with a single descriptor are proposed. These descriptors are obtained by combining features of individual frames in different ways. To the best of our knowledge, this work is the first attempt to explore spatial-temporal fusion schemes for RoI batches captured from video streams generated in a multi-camera surveillance scenario. We compare their performances with two score-level fusion schemes.

Next Sect. 2 provides an overview of the framework. Section 3 briefly reviews the employment of fusion schemes and introduces algorithms. Section 4 discusses the experimental methodology. In Sect. 5, we experimentally compare the effectiveness of different levels of fusion on several real-world videos.

2   Background on the NEVIL.ubm Approach

A surveillance system should track and recognize the object from the first moment it is captured by a camera and across all cameras whose fields of view overlap the path. In this section, the Never Ending Visual Information Learning with UBM (NEVIL.ubm) framework is briefly presented. NEVIL.ubm [8] is designed for learning in non-stationary environments in which no labelled data is available but the learning algorithm is able to interactively query the user to label the desired outputs at carefully chosen data points (Fig. 2).

The system receives multiple visual streams, generated by a typical tracking algorithm, which analyses sequential video frames and tracks RoIs over time. For each RoI the features corresponding to some pre-selected object representation (e.g. bag of words) are extracted ($v[l]$ $l = 1, ..., B$). A batch $v_t^{m_t}$ is a temporal sequence of frames $v_{t,f}^{m_t}$, where f runs over 1 to the batch size B. Initially, the composite model is initialized to yield the same probability to every class

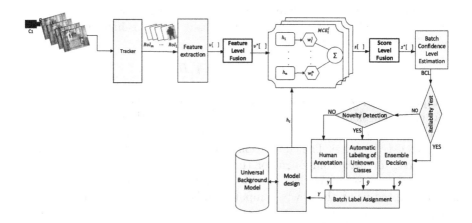

Fig. 2. Block diagram of NEVIL.ubm. (The diagram shows both possible level of fusion applied in the framework)

(uniform prior). When the features of batches of RoIs $v_{t,f}^{m_i}$ in time slot t become available, the framework starts computing the scores $\mathscr{S}(v_t^{m_i}|C_k, H_{t-1})$ for each batch $v_t^{m_i}$ in the time slot. The scores are obtained from the likelihood ratio test of the batch data obtained by the individual class model C_k and the UBM.

The composite model H_t is an ensemble of Micro-classifiers ensembles $(MCE_t^j, j = 1, ..., k)$. Each MCE_t^j includes classifiers that are incrementally trained (with no access to previous data) on incoming batches of j_{th} class at t, h_t^j. The individual models h_t^j are combined using a weighted majority voting, where the weights are dynamically updated with respect to the classifiers' time of design.

The prediction output by the composite model MCE_t^j for a given ROI $(v_{t,f}^{m_i})$ is

$$p(C_k|v_{t,f}^{m_i}, MCE_t^j) = \sum_{\ell=1}^{t} W_\ell^t h_\ell(C_K|v_{t,f}^{m_i}) \tag{1}$$

where $h_\ell^j(.)$ is the classifier trained from batches of j_{th} at TS ℓ, W_ℓ^t is the weight assigned to classifier ℓ, adjusted for time t. The weights are updated and normalised at each time slot and chosen to give more credit to more recent knowledge. After combining the decisions of classifiers inside every MC-ensemble, the ensemble will assign a batch to the label of MC-ensemble with highest score $(\mathscr{S}(v_t^{m_i}|C_k, H_{t-1}))$.

Such on-line learning may suffer if labelling errors accumulate, which is inevitable. To help mitigate this issue, the system is designed to interact wisely with a human. Once $\mathscr{S}(v_t^{m_i}|C_k, H_{t-1})$ is obtained, a batch confidence level (BCL) is estimated. In NEVIL.ubm framework, if the scores associated to all observed classes are significantly low (below a predetermined threshold), it is very likely that this class has not been observed before and it is considered novel and a new label $(\ddot{y})$ is automatically assigned to this batch(es). Having decided that the batch data belongs to an existing class, one needs to decide if the automatic prediction is reliable (the reliability test is positive) and accepted or rather a manual labelling needs to be requested. If BCL is high enough (above a predefined threshold), the predicted label

$$\hat{y} = \arg\max_{C_k} \mathscr{S}(v_t^{m_i}|C_k, H_{t-1}) \tag{2}$$

is accepted as correct; otherwise the user is requested to label (y) the data batch.

At each time slot, the batches predicted to belong to the same class are used to generate the class model by *tuning the UBM parameters* in a maximum a *posteriori* (MAP) sense. The adaptation process consists in two main estimation steps. First, for each component of the UBM, a set of sufficient statistics is computed from a set of M class specific feature vectors. Each UBM component is then adapted using the newly computed sufficient statistics, and considering diagonal covariance matrices.

Note that the UBM is trained offline, before the deployment of the system. It is designed from a large pool of streams aimed to be representative of the complete set of potentially observable 'objects'.

3   Spatial-Temporal Fusion Schemes over Frames

Although in many real visual applications different sources of information are available, learning from multiple sources is a less explored area. Early fusion has been applied to define whether the audio signal is consistent with the speaker video file [9]. Pixel-level fusion has shown promising performance in video-based biometric recognition [10] as well as multiple object tracking [11]. Some authors demonstrate [12,13] demonstrated the effectiveness of the decision level fusion strategies on object tracking, video segmentation, and video event detection. Feature-level fusion has gained much importance over the past few years, and various approaches have been introduced in the literature [5,14,15]. Most approaches combined the information of multiple modalities (sensors), while some methods used the complementary descriptors. The former requires multiple sensors (visible light cameras combined with depth or infra-red camera), and the latter adds more complexity to the system specially in an online application. To the best of our knowledge, the employment of feature-level techniques over frames in a PR scenario has not been addressed before.

Fusion schemes have been successfully used in large-scale recognition systems to address multiple issues confronting these systems such as accuracy, practicality, and efficiency. Inspired by the rationale behind such systems, two fusion schemes to combine the information in a PR system are proposed. Each frame can be considered as an independent source of information and combining such information in different levels could be beneficial for a PR system. The batch score $(\mathscr{S}(v_t^{m_i}|C_k, H_{t-1}))$ can be obtained in two ways: either by combining the scores of individual RoIs in a batch (score-level fusion), or by combining the patterns of M RoIs in a batch (feature-level fusion).

3.1   Feature-Level Fusion

Finding a joint representation for a group of frames is a challenging problem in visual applications. There is a considerable body of research works that addressed this problem by choosing a key frame, which represents the entire batch. As the quality of the batch representation relies heavily on the representative sample and an inappropriate choice may lead to unreliable results, such methods seem impractical for challenging environments. This is the main rationale behind approaches exploiting fusion schemes. In this paper, two feature-level fusion that aggregate descriptors of all the frames in a given batch are proposed. Let $v_{t,f}^{m_t}$ be the descriptor of $f - th$ frame in a batch, the average histogram that combines the information of entire batch in a single histogram defined by

$$v^{*m_i}_t = \frac{1}{B} \sum_{f=1}^{B} v_{t,f}^{m_i}(b) \quad where \quad b = 1, ..., M \tag{3}$$

Where M is the number of histogram bins (Fig. 3).

In our scenario, it is very likely to obtain outlier values for some frames in a batch due to occlusion or miss tracking. The median might be seen as a better

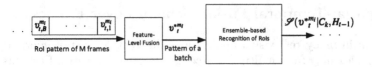

Fig. 3. Block diagram of feature-level fusion (performed before micro-ensembles recognition)

indication of central tendency than the arithmetic mean in such cases, since it is less susceptible to the exceptionally large or small values in data. Hence, as an alternative option we consider estimating the descriptor of a given batch by:

$$v^{*m_i}_t = \mathcal{M}edian\ v^{m_i}_{t,f}(b)\ \ where\ \ b = 1, ..., M \tag{4}$$

Given the single representation, a score $\mathcal{S}(v^{m_i}_t | C_k, H_{t-1})$ is calculated for the batch.

3.2   Score-Level Fusion

The composite model, H_{t-1}, can be used to predict directly $p(v^{m_i}_{t,f} | C_k, H_{t-1})$ but not $p(v^{m_i}_t | C_k, H_{t-1})$. The individual scores per frame $\mathcal{S}(v^{m_i}_{t,j} | C_k, H_{t-1})$ can then be immediately obtained as $\mathcal{S}(v^{m_i}_{t,j} | C_k, H_{t-1}) = \frac{p(v^{m_i}_{t,j} | C_k, H_{t-1})}{p(v^{m_i}_{t,j} | UBM)}$. The batch label prediction can be analysed as a problem of combining information from multiple (B) classification decisions. Considering that, per frame, the composite model produces approximations to the likelihoods/scores for each class, different combination rules can be considered to build the batch prediction from the individual frame predictions. Applying arithmetic mean, the score per batch is obtained as (Fig. 4):

$$\mathcal{S}^*(v^{m_i}_t | C_k, H_{t-1}) = \frac{\sum_{j=1}^{B} \mathcal{S}(v^{m_i}_{t,j} | C_k, H_{t-1})}{B} \tag{5}$$

As an alternative choice, the median of the scores were also evaluated, since it may be more robust to the outliers. The batch score is defined by:

$$\mathcal{S}^*(v^{m_i}_t | C_k, H_{t-1}) = \mathcal{M}edian\mathcal{S}(v^{m_i}_{t,j} | C_k, H_{t-1}) \tag{6}$$

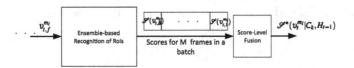

Fig. 4. Block diagram of score-level fusion (performed after micro-ensembles recognition)

Although other robust statistics could be considered from the individual frame scores, experimentally we will only compare the two options. In the end, NEVIL.ubm assigns each batch to the class maximizing $\mathscr{S}(v_t^{m_i}|C_k, H_{t-1})$.

4   Experimental Methodology

4.1   Datasets

In order to explore the properties of the proposed framework, we evaluated it on multiple datasets covering various possible scenarios in a multi-camera sur-veillance system. We conducted our experiments on a number of CAVIAR video clips including: OneLeave ShopReenter1, Enter ExitCrossingPaths1, OneSho-pOneWait1, OneStop Enter2 and WalkBy Shop1front as well as PETS2009. These sequences present challenging situations with cluttered scenes, high rates of occlusion, different illumination conditions as well as different scales of the person being captured. We employ an automatic tracking approach to track objects in the scene and generate streams of bounding boxes, which define the tracked objects' positions. As the method may fail to perfectly track the targets, a stream often includes frames of distinct objects. A hierarchical bag-of-visterms method is applied to represent the tracked objects, resulting in a descriptor vec-tor of size 11110 for each frame (refer to [16] for more information). In order to avoid the curse of dimensionality that system may suffer from, PCA is applied to the full set of descriptor features as a pre-processing step. Hence, the number of features in each stream is reduced to 85.

4.2   Confidence Measure

Various criteria have been introduced as uncertainty measures in literature for a probabilistic framework.

Most Confident Measure (MC): Perhaps the simplest and most commonly used criterion relies on the probability of the most confident class, defining the confi-dence level as

$$\max_{C_k} \mathscr{S}(C_k|v_t^{m_i}, H_{t-1}) \tag{7}$$

Modified Margin Measure (MM): MC only considers information about the most probable label. Thus, it effectively "throws away" information about the remain-ing label distribution [17]. To correct this, an option is to adopt a margin con-fidence measure based on the first and second most probable class labels under the model. We evaluate experimentally the BCL base on the *ratio* of the first and second most probable class labels:

$$\mathscr{S}(C^*|v_t^{m_i}, H_{t-1})/\mathscr{S}(C_*|v_t^{m_i}, H_{t-1}), \tag{8}$$

where C^* and C_* are the first and second most probable class labels, respectively.

4.3   Evaluation Criteria

Active learning aims to achieve high accuracy using as little annotation effort as possible. Thus, a trade-off between accuracy and proportion of labelled data can be considered as one of the most informative measures.

Accuracy. In a classical classification problem the disparity between real and predicted labels explains how accurately the system works. However, in our scenario the labels do not carry any semantic meaning (it is not a person recognition problem). The same person should have the same label in different batches, whichever the label. As such, when evaluating the performance of our framework we are just comparing the partition of the set of batches as defined by the reference labelling with the partition obtained by the NEVIL labelling. We adopted a generic partition-distance method for assessing set partitions, initially proposed for assessing spatial segmentations of images and videos [18]. Thus, the accuracy of the system is formulated as:

$$Accuracy = \frac{N - Cost}{N} \tag{9}$$

where N denotes the total number of batches, and $Cost$ refers to the cost, yielded by the assignment problem.

Annotation. Assume MLB and TB denote the manually labelled batches and all the batches available during a period (includes one or more time slots), respectively. The *Annotation Effort* is formulated as:

$$Annotation\ effort = \frac{\#MLB}{\#TB} \tag{10}$$

It is expected that the accuracy increases with the increase of the annotation effort.

Area Under the Learning Curve (ALC). [19] is a standard metric in active learning research that combines *accuracy* and *annotation effort* into a single measurement. ALC, which provides an average of accuracy over various budget levels, seems to be a more informative metric. Herein, the learning curve is the set of accuracy plotted as a function of their respective annotation effort, a, $Accuracy = f(a)$. The ALC is obtained by:

$$ALC = \int_0^1 f(a)da \tag{11}$$

Table 1. ALC of fusion at feature-level on videos. The rank of each setting in a given dataset is presented next to the ALC between parentheses. Highlighted row indicates the optimal design. Values in bold indicate better performance than score-level fusion for optimal setting.

Confidence Measure	Combination Rule	Reenter1	Reenter2	front	Paths1	Enter2	Wait1	Enter1	PETS09
MC	Median	0.96(1)	0.96(2)	0.91(4)	0.87(2)	0.96(1)	0.90(4)	0.90(1)	0.85(2)
MC	Mean	0.94(3)	**0.97(1)**	**0.96(1)**	**0.88(1)**	**0.96(1)**	**0.91(3)**	**0.90(1)**	**0.86(1)**
MM	Median	0.95(2)	0.94(3)	0.93(3)	0.86(3)	0.96(1)	0.93(1)	0.88(2)	0.79(3)
MM	Mean	0.96(1)	0.96(2)	0.95(2)	0.87(2)	0.96(1)	0.92(2)	0.90(1)	0.73(4)

5   Results

Table 1 shows the ALC performance of the proposed fusion techniques using all datasets along with the mean of ALC rank averaged over all the experiments (the std of the results is always below ± 0.01). The table shows that settings in where sum rule have been applied for combining the information occupy the two top spots for both feature-level and score-level fusion. The results indicate that the most confident class as batch confidence measure selects more informative batches than modified margin, as settings employing the former have better mean rank. Based on the average rank, we conclude that the arithmetic mean as fusion rule and the most confident as selection criterion presents the optimal design. Comparing the ALC of identical designs of two fusion schemes (highlighted rows in Tables 1 and 2) for every dataset, we observe that for 6 out of 8 datasets feature-level fusion attains better performance (higher ALC) than score-level fusion.

Figure 5 presents the results of optimal design (arithmetic mean as fusion rule and the most confident as selection criteria) for two fusion levels on all video clips. Since ALC measures the average performance over various budget levels, it does not give detailed information for every single budget level. We chose the point obtained by labelling 20 % of batches for a more detailed analysis. Given that budget while employing mid-level fusion, we obtain 100 % accuracy for four scenarios (OneLeaveShopReenter2, OneLeaveShopReenter1, OneStopEnter2, and WalkByShop1front). For more complex scenarios, such as OneStopMoveEnter1 (in where 42 streams from 14 classes are available) 88 % of

Table 2. ALC of fusion at score-level on videos. The rank of each setting in a given dataset is presented next to the ALC between parentheses. Highlighted row indicates the optimal design. Values in bold indicate better performance than score-level fusion for optimal setting.

Confidence Measure	Combination Rule	Datasets							
		Reenter1	Reenter2	front	Paths1	Enter2	Wait1	Enter1	PETS09
MC	Median	0.96(1)	0.93(3)	0.93(2)	0.86(2)	0.95(2)	0.88(4)	0.87(3)	0.79(2)
MC	Mean	**0.96(1)**	**0.97(1)**	0.90(3)	0.87(1)	0.95(2)	0.90(3)	0.90(1)	0.85(1)
MM	Median	0.96(1)	0.91(4)	0.95(1)	0.87(1)	0.93(3)	0.91(2)	0.87(3)	0.71(4)
MM	Mean	0.96(1)	0.95(2)	0.93(2)	0.85(3)	0.96(1)	0.92(1)	0.89(2)	0.75(3)

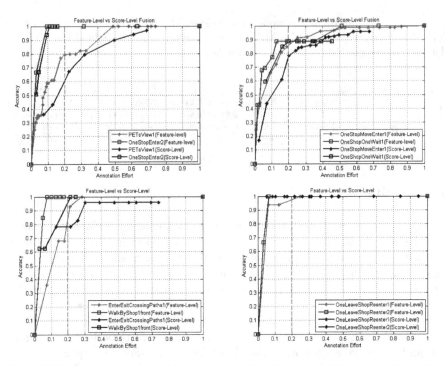

Fig. 5. ALC vs annotation effort for feature-level with score-level fusion on the various videos. _ _ _ highlights 20 % budget.

batches are correctly classified, showing an improvement over score-level fusion results (80 % accuracy). The results indicate the better performance of feature-level over score-level fusion.

Complexity. Improving the accuracy is not the only advantage of feature-level fusion. In real-time learning, when massive amount of information is available, efficiency is equally important. In contrary to score-level fusion, where an independent recognition process is applied to every single RoI (of M RoI in a batch) and then the results are mathematically combined, feature-level fusion employs a single learning stage on the joint representation of a batch of M frames. Thus, the time and complexity of the framework decrease dramatically. Since the framework was developed in MATLAB without any efficiency concerns, a straightforward assessment of the time efficiency is not adequate. Nevertheless our experiments shows that combining the information at feature-level is able to process the streams almost twice as fast as score-level fusion, for a framerate of 25 fps (running in an Intel Core i7 at 3.2 GHz).

6   Conclusions

In this paper, two spatio-temporal fusion strategies to combine the patterns of RoIs in various streams captured in a multi-camera surveillance system are presented. We experimentally investigated the impact of feature-level and score-level fusion on the performance of the PR system. Experiments indicate the potential of feature-level fusion for on-line applications, as they attained the best performance with much lower time complexity. For future work, we plan to exploit descriptors that are specifically designed to represent video shots.

References

1. Dewan, M.A.A., Granger, E., Marcialis, G.L., Sabourin, R., Roli, F.: Adaptive appearance model tracking for still-to-video face recognition. Pattern Recogn. **49**, 129–151 (2016)
2. Khoshrou, S., Cardoso, J.S., Teixeira, L.F.: Active learning of video streams in a multi-camera scenario. In: 22nd International Conference on Pattern Recognition (2014)
3. Dietrich, C., Palm, G., Schwenker, F.: Decision templates for the classification of bioacoustic time series. Inf. Fusion **4**, 101–109 (2003)
4. Jiang, B., Martínez, B., Valstar, M.F., Pantic, M.: Decision level fusion of domain specific regions for facial action recognition. In: 22nd International Conference on Pattern Recognition, ICPR 2014, Stockholm, Sweden, 24–28 August 2014, pp. 1776–1781 (2014)
5. Abouelenien, M., Wan, Y., Saudagar, A.: Feature and decision level fusion for action recognition, pp. 1–7 (2012)
6. Schels, M., Glodek, M., Meudt, S., Scherer, S., Schmidt, M., Layher, G., Tschcchne, S., Brosch, T., Hrabal, D., Walter, S., Traue, H.C., Palm, G., Schwenker, F., Rojc, M., Campbell, N.: Multi-Modal Classifier-Fusion for the Recognition of Emotions. In: Converbal Synchrony in Human-Machine Interaction, pp. 73–97. CRC Press (2013)
7. Tao, Q., Veldhuis, R.: Hybrid fusion for biometrics: combining score-level and decision-level fusion. In: 2008 IEEE Computer Society Conference on Computer Vision and Pattern Recognition, Workshop on Biometrics, Los Alamitos, pp. 1–6. IEEE Computer Society Press (2008)
8. Khoshrou, S., Cardoso, J.S., Teixeira, L.F.: Learning from evolving video streams in a multi-camera scenario. Mach. Learn. **100**, 609–633 (2015)
9. Fisher, J.W., Darrell, T.: Signal level fusion for multimodal perceptual user interface. In: Workshop on Perceptive User Interfaces, pp. 1–7 (2001)
10. Colores-Vargas, J.M., García-Vázquez, M., Ramírez-Acosta, A., Pérez-Meana, H., Nakano-Miyatake, M.: Video images fusion to improve iris recognition accuracy in unconstrained environments. In: Carrasco-Ochoa, J.A., Martínez-Trinidad, J.F., Rodríguez, J.S., di Baja, G.S. (eds.) MCPR 2012. LNCS, vol. 7914, pp. 114–125. Springer, Heidelberg (2013)
11. Cvejic, N., Nikolov, S., Knowles, H., Loza, A., Achim, A., Bull, D., Canagarajah, C.: The effect of pixel-level fusion on object tracking in multi-sensor surveillance video. In: ICVPR, pp. 1–7 (2007)
12. Krishna Mohan, C., Dhananjaya, N., Yegnanarayana, B.: Video shot segmentation using late fusion technique. In: ICMLA, pp. 267–270 (2008)
13. Kamishima, Y., Inoue, N., Shinoda, K.: Event detection in consumer videos using GMM supervectors and SVMs. EURASIP J. Image Video Process. **51** (2013)
14. Sharma, V., Davis, J.W.: Feature-level fusion for object segmentation using mutual information. In: Hammoud, R.I. (ed.) Augmented Vision Perception in Infrared, pp. 295–320. Springer, London (2009)

15. Chen, C., Jafari, R., Kehtarnavaz, N.: Improving human action recognition using fusion of depth camera and inertial sensors. IEEE Trans. Hum.-Mach. Syst. **45**, 51–61 (2015)

16. Teixeira, L.F., Corte-Real, L.: Video object matching across multiple independent views using local descriptors and adaptive learning. Pattern Recogn. Lett. **30**, 157–167 (2009)

17. Settles, B.: Active learning literature survey. Technical report 1648, University of Wisconsin-Madison (2009)

18. Cardoso, J.S., Corte-Real, L.: Toward a generic evaluation of image segmentation. IEEE Trans. Image Process. **14**, 1773–1782 (2005)

19. Cawley, G.C.: Baseline methods for active learning. In: Active Learning and Experimental Design@ AISTATS, pp. 47–57 (2011)

Patch Selection for Single Image Deblurring Based on a Coalitional Game

Jung-Hsuan Lin$^{(\boxtimes)}$, Rong-Sheng Wang, and Jing-wei Wang

The Innovative DigiTech-Enabled Applications and Services Institute (IDEAS),
Institute for Information Industry (III), Taipei, Taiwan
imjustaup@gmail.com

Abstract. Most single-image deblurring methods estimate the blur kernel using whole image, however, that may lead to incorrect estimation and more computations. In this paper, we focus on accelerating the blind deconvolution algorithm and increasing the accuracy of kernel estimation by using only a small region in image to perform the process of kernel estimation. Then, the problem now is to find the most proper region. At first, we found informative pixels to locate useful patches. Inspiring by game theory, we propose a coalitional game based patch selection method to choose a group of patches for kernel estimation. In this game, each patch represents a player, and our purpose is to find a coalition that has the maximal payoff. Shapley Value is applied to fairly distribute the utility to each player. We show the speed-up and the quality improvement of our method both on real-world and synthetic images.

1 Introduction

A motion blur is a common artifact that produces an undesirable blurry image. It is usually caused by camera shake during long exposure time, particularly under low-light conditions. The observed blurry image B can be modeled as the convolution of a latent image l with a motion blur kernel k plus the noise n

$$B = l \otimes k + n \tag{1}$$

The motion blur can be due to camera motion or objects motion within the scene. In objects motion, the blur kernel is spatially-variant. Estimating a spatially-variant blur kernel imposes a significant computational cost, because we have to repeatedly estimate the kernel all over the image. On the contrary, the blur kernel in camera motion is spatially-invariant and it is more simple to estimate.

The task of deblurring an image is image deconvolution. If the blur kernel is unknown, the problem is said to be "blind". The typical blind deconvolution approach is to estimating the latent image with the gradient of the image to find a combination $(l'; k')$ that minimizes $\left\| B - l' \otimes k' \right\|$. A straightforward approach of deconvolution is to solve the maximum-a-posteriori (MAP) problem. One can either estimate the kernel k while marginalizing over the latent image l or jointly estimate a kernel and latent image. The MAP approach may fail without well constraint for either the kernel or the

© Springer International Publishing Switzerland 2015
G. Bebis et al. (Eds.): ISVC 2015, Part II, LNCS 9475, pp. 521–531, 2015.
DOI: 10.1007/978-3-319-27863-6_48

latent image, and it is usually computationally complex. However, the MAP approach is still useful in blind image deconvolution [2, 3, 9, 10].

In this paper, we discovered that using whole image to estimate the blur kernel is not necessary, and may lead to more computations and errors at estimation step. Our idea is to find a patch which has the enough information for estimating kernel. Main contribution of our method is to find a proper patch for kernel estimation, which can reduce the computational time and increase accuracy. We first find the informative pixels using the gradient magnitude of the blurry image and a usability map proposed by [3] then we form patches centered at these pixels. To find the most proper patches, we apply a coalitional game which treats every small patch as a player. We compute the Shapley Value [12] of each patch in the coalitional game to fairly distribute the payoff for each player in the coalition. In experiment, we modified two blind deconvolution method from [1, 5] using the patch selected by our method. The result shows the feasibility and effectiveness of our proposed method.

2 Related Work

Hyeoungho Bae et al. [6] constructed the patch mosaic by tiling informative image patches to synthesize a new, compact blurry image which significantly reduce the computational time. Jiaya Jia suggested using small fractions of an image for blind deconvolution [5]. However, their work needs user intervention for selecting appropriate areas and foreground and background colors for alpha-blending calculation. Gupta et al. [7] estimated the camera trajectory by a patch-based analysis of the scene. However, they applied the blind deconvolution algorithm of [2] for the entire image area first. Then, they used RANSAC-based approach to select appropriate results among initially deblurred image patches to estimate the camera motion.

There are several works tried to reduce the computational time of blind motion-deblurring algorithms. Cho and Lee [8] took the MAP approach and used discrete Fourier transform to reduce the computational time. However, they still needed to use the whole image area to estimate the blur kernel.

Although the accuracy of the method of Xu and Jia [3] is remarkable, due to the sophisticated masking and the larger number of iterations, it's still computationally complex.

3 Patch Selection Method with Game Theory

The main concern of our approach is how to reduce the amount of data to be processed without sacrificing the accuracy of kernel estimation. Since our method uses a patch instead of the whole image to predict the sharp image and estimate the blur kernel, the amount of data to be processed has been reduced, as a result, the computational time will be saved.

In game theory, the Shapley Value is a solution concept in cooperative game (coalitional game). To each cooperative game, it assigns a unique distribution (among the players) of a total surplus generated by the coalition of all players. In this paper, at

first, we locate the informative pixels, considering a small patch centered at each of them. Next, we try to find a set of these small patches which has the most significant information. To find the best combination of the set, Coalition game is applied, we treat each small patch as a player. Shapley Value is then used to fairly distribute the contribution of patches. Once the value of patches is defined, we simply find the set which has the best payoff.

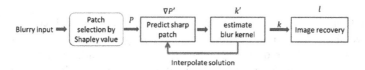

Fig. 1. Image deblurring using Patch Selection Method with Game Theory

Our patch selection method automatically finds a proper patch for kernel estimation from blurry input. Figure 1 shows the overall framework. Given a blurry input, our method finds a patch P by calculating Shapley Value and predicts its sharp patch $\nabla P'$ to estimate the kernel k' in coarse-to-fine iteration. Then we use the estimated blur kernel k to recover the latent image l.

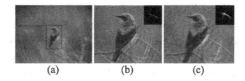

(a)   (b)   (c)

Fig. 2. The effect of using different patches to estimate blur kernel by Fergus, R. et al. [1]. (a) The blurry input, yellow rectangle is the patch selected by our manually and red rectangle is the patch selected by our method. (b)(c) The magnified deblurring results using yellow patch and red patch, respectively, and Top-right are the estimated blur kernels (Color figure online)

Although we focus on spatially-invariant cases, insignificant patch would lead to error in kernel estimation. We show the example of this issue in Fig. 2. We take two different patches as input of the deblurring algorithm by Fergus, R. et al. [1]. The red rectangle in Fig. 2(a) is the patch selected by our method, and the yellow rectangle is selected manually. We can clearly observe that the ringing artifact in (b) is more serious than (c), and the estimated kernel is different because we used different patch to predict the latent patch. Therefore, we proposed a patch selection method based on Shapley Value in order to find the patch which is significant to kernel estimation process.

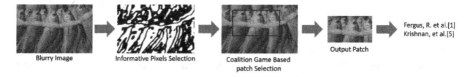

Fig. 3. The block diagram of the proposed patch selection method

The overall process of our patch selection method is shown in Fig. 3. It is composed of two parts: (1) Informative pixels selection, and (2) Coalition game based patch selection.

In the informative pixel selection, we first locate the usable part of the blurry image by finding the sharp edges by gradient magnitude, and apply the usability map proposed by Xu et al. [3] to rule out the useless edges. To find the pixels with more information, we proposed an algorithm to find the locally importance without loss of spatiality.

In the Coalitional game based patch selection, we form a patch centered at each informative pixel. We treat every small patch as a player who seeks to join a coalition to improve its payoff. We design the utility of the coalition and compute the Shapley Value for each player to fairly distribute the utility generated by the coalition. Players in this game are allowed to join a coalition only if the Shapley Value of each player is improved. In the end of the game, we will have a coalition such that no other player can obtain an outcome better than the current assignment. Finally, the region formed by the coalition is the most proper patch.

3.1 Informative Pixels Selection

Given a blurry image, our goal in this section is to find the useful pixels of image. Edges of an image usually provide useful information for image deblurring. Joshi et al. [9] assumed that the blur kernel is unimodal to predict the location of sharp edges in a blurry photo and estimate the blur kernel accordingly. Cho et al. [10] estimated camera shake by analyzing edges in the image, effectively constructing the Radon transform of the kernel. We use the gradient magnitude of a down sampled image to locate the sharpest edges.

However, Xu and Jia [3] has discovered that salient edges do not always improve kernel estimation; on the contrary, if the scale of an object is smaller than that of the blur kernel, the edge information could damage kernel. They proposed a new metric to measure the usefulness of gradients which is defined as follow

$$R(x) = \frac{\left\| \sum_{y \in N_h(x)} \nabla B(y) \right\|}{\sum_{y \in N_h(x)} \nabla B(y) + 0.5} \tag{2}$$

Where B is the blurry image, $N_h(x)$ is a $h \times h$ window centered at pixel x in the image B, and $\nabla B(y)$ is the gradient of blurry image at pixel y. The function $R(x)$ (see Eq. (2)) is the usability map which calculate the usefulness of a pixel x in image B, 0.5 is to prevent producing a large R in flat regions. The value of each pixel in usability map R is normalized between zero and one. A small R implies that either spikes or a flat region is involved within the window.

To secure sufficient information about kernel estimation, we start with a pixel-wise measure of informativeness that incorporates our desired criteria. We use the gradient magnitude and usability map to locate the informative pixels: We first rule out the pixels belonging to small R-value and small gradient magnitude by using threshold,

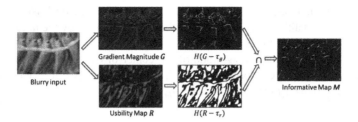

Fig. 4. Block diagram of our Informative pixels selection process

and then we take the intersection of these two measurements. We specify the informativeness of a pixel by

$$M = H\big(G - \tau_g\big) \cap H(R - \tau_r) \tag{3}$$

Where $H(.)$ is the Heaviside step function, outputting zeros for negative values and ones otherwise, G is the gradient magnitude of the blurry image, τ_g and τ_r are threshold values for filtering low-intensity edges and less-usable edges, respectively, and R is the usability map defined in Eq. (2). The function M (see Eq. (3)) is a binary map we denote as informative map and the value of informative pixels in M will be ones or zeros for non-informative pixels. We show the block diagram of our informative pixels selection process in Fig. 4.

3.2 Coalitional Game Based Patch Selection

With the informative map, we try to find a patch that contains all the informative pixels. To find the patch more accurate, we take the neighbors of each informative pixel into account. For each informative pixel in M, we form a small patch centered at it.

However, single patch has not enough information for kernel estimation; we attempt to find a group of the patches which has the most utility. We apply a coalitional game to help us to find these patches. We treat every small patch as a player who seeks to join a coalition to improve its payoff. We compute the Shapley Value for each coalition to fairly distribute the payoff. In game theory, the Shapley Value is a solution concept in cooperative game (coalitional game) theory. To each cooperative game it assigns a unique distribution (among the players) of a total surplus generated by the coalition of all players. In this game, our goal is to find a coalition such that no other player can obtain an outcome better than the current assignment. After the game, we use the patches in the coalition we found at last to form a large patch surrounded by the minimal and the maximal coordinates in these patches. We specify our game model as follow:

Players: Given the n patches, each patch $p_i(1, \cdots, n)$ is regard as a player in the game.

Coalitions: Given n players, we denote $N = \{p_1, p_2, \cdots, p_n\}$ is the set of all players. S is a coalition that belongs to all the non-empty subset of N which means there are $2^n - 1$ possible coalitions.

Utility of a coalition: Utility of a coalition is calculated by following guidelines.

A. The size of patch may have significant effect on kernel estimation. We consider the size of given coalition S denoted as $size(S)$ which is defined by the union region of each patch in a coalition S.

B. We consider the gradient magnitude of the center of each patch. Strong gradient magnitude gives more information for kernel estimation. Given a coalition S, we denote the gradient magnitude of S as $G_m(S)$

$$G_m(S) = \sum_{i=1}^{n} |\nabla B(C_{p_i})| \tag{4}$$

Where C_{p_i} denotes the central pixel of patch p_i, and $|\nabla B(C_{p_i})|$ denotes the gradient magnitude at pixel C_{p_i}. The function G_m(see Eq. (4)) is the sum of gradient magnitude of each center of patch in coalition S.

C. Given a coalition S, we compute the usability map R for the region composed by union each patches in S. Considering the number of usable pixels in R, We compute the percentage of the usable pixels of the coalition S as $P(S)$

$$P(S) = \frac{\sum_{x \in S} H(R(x) - \tau_r)}{size(S)} \tag{5}$$

Where x presents each pixel inside the region consisted by given coalition S, and $H(R(x) - \tau_r)$ determine whether the pixel x is usable or not by using threshold τ_r. If x is usable, then the function returns 1 and vice versa. The function P (see Eq. (5)) is the value of percentage of how many usable pixels in a coalition.

D. The distance between centers of patches should be as close as possible. If the patches are far away from each other, the region of S may contain too many unusable pixels. We denote $D(S)$ as the deviation of S. We calculate the standard deviation of the coordinates of patch centers in the coalition, higher standard deviation should yield less utility and vice versa

$$D(S) = \frac{1}{std(\{C_{p_i} | i \in S\})} \tag{6}$$

Where C_{p_i} is the center of patch p_i, and $std(.)$ returns the standard deviation of given set of central pixels according to the coordinates. The function D is invert of standard deviation of patch centers in S.

Utility function: In summary, we define the utility function of the coalition S as

$$U(S) = w_a size(S) + w_g G_m(S) + w_p P(S) + D(S) \tag{7}$$

Where w_a, w_g and w_p are the weight for $size(S)$, $G_m(S)$ and $P(S)$, respectively. In experiment, we set $w_a = 10^{-4}$, $w_g = 1$, and $w_p = 5$.

3.3 Ordered Coalitional Game

Calculating the Shapley Value of all possible coalitions may take a while. Here, we proposed an ordered coalitional game. In this game, we compute the Shapley Value follow by a specified order. The order is defined by the sequence that we find the local maximum count pixel. The front sequence of patch means it is more important. Following the order, we check each player by computing the Shapley Value; a player is allowed to join the coalition only if the Shapley Value after he joined of each player is improved. After checking all players, we will find a coalition such that no other player can obtain an outcome better than the current assignment. The Shapley Value of player i in the coalition S is denoted as v_i

$$v_i = \sum_{c \in comb_i(S)} \frac{(n-k)!(k-1)!}{n!} [U(c) - U(c-i)] \tag{8}$$

Where S is the coalition which has n players and $comb_i(S)$ means the all possible non-empty combinations of subsets of $\{P_1, P_2, \cdots, P_n\}$ which include player i. For example, considering a three players coalition (P_1, P_2, P_3), the $comb_{P_i}(S)$ for P_1 is $\{\{P_1\}, \{P_1, P_2\}, \{P_1, P_3\}, \{P_1, P_2, P_3\}\}$. $k = \{1, 2, \cdots, n\}$ is the possible numbers of players. $\frac{(n-k)!(k-1)!}{n!}$ represents the weight for different k, $[U(c) - U(c-i)]$ means the marginal contribution for player i. v_i presents the Shapley Value of play i in S.

4 Experimental Results

To demonstrate our method, we modified two single-image blind deconvolution algorithms [1, 4]. Their Matlab codes are available online. Without loss of generosity, we just modified the patch axis that used to estimate blur kernel to fit the patch selected by our method. The kernel size is set to 25 and other parameters are set to be the same as original code to all images. However, the only difference is the patch that used to estimate the blur kernel.

Sample images for our experiment have been tested in many researches [1–3, 4, 8, 10]. For comparing the quality of our results, we calculate the PSNR between the synthetic images which are synthesized by the ground truth kernels in [11] and our experimental results. Moreover, processing time has also been considered. Experiment shows that the algorithms applying with our selected patch can effectively speed up. Our testing environment was a PC running MS Windows 7 64bit version with Intel Core i5-2400 CPU 3.10 GHz, 4 GB RAM, and no additional graphics card. It takes only about a second to run our game theoretic patch selection algorithm in Matlab implementation. This section is divided into three parts. In part A, we test the synthetic images in order to quantify the quality of the results. In part B, we test the real-world blurry image. The result shows that the deblurring algorithms applying our patch selection method can effectively reduce the ringing artifacts. In part C, the accuracy and processing speed have been quantitatively analyzed, showing significant speed-up.

For the quantitative analysis of the accuracy of deblurring result, we use the six synthetic images shown in Fig. 5. We compute the PSNR between the original algorithm and our modified algorithm with ground truth images shown in Fig. 7(a). Generally, our modified algorithm yields higher PSNR.

Our method effectively suppressed ringing artifacts as shown in Fig. 6(c) and (e) although the difference between (e) and (f) is not obvious, the computational time is reduced. Our method can yield a same or even better deblurring result with less computations. As shown in Fig. 7(b) our method effectively reduces the amount data to be processed. The average reduced percentage of computational time of Fergus, *et al.* is 47 % and 61 % to Krishnan, *et al.*

A. Synthetic blurry benchmark image

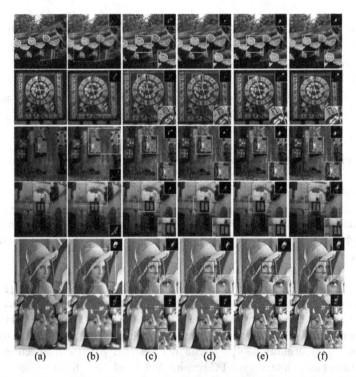

| (a) | (b) | (c) | (d) | (e) | (f) |

Fig. 5. Comparison of deblurring result with synthetic blurry image. The top-right block in (b) ~ (f) are the predicted blur kernels. The bottom-right block in (c) ~ (f) are the zoomed region for red rectangle. (a) Ground truth image. (b) Synthetic blurry image form (a), yellow rectangle is the patch used to kernel estimation selected by our method. (c) Deblurring results by Fergus, et al. [1]. (d) Deblurring result by Fergus, et al. [1] with our method. (e) Deblurring results by Krishnan, et al. [4]. (f) Deblurring results by Krishnan, et al. [4] with our method (Color figure online)

B. **Real-world blurry image**

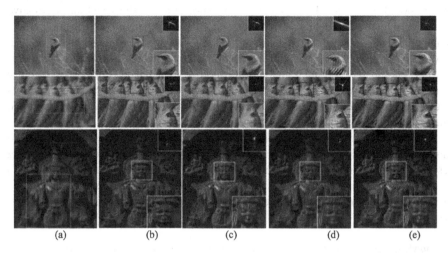

(a)   (b)   (c)   (d)   (e)

Fig. 6. Comparison of deblurring result with real-world blurry image. The top-right block in (b) ∼ (f) are the predicted blur kernels. The bottom-right block in (c) ∼ (f) are the zoomed region for red rectangle. (a) Real blurry image, yellow rectangle is the patch used to kernel estimation selected by our method. (b) Deblurring results by Fergus, et al. [1]. (c) Deblurring result by Fergus, et al. [1] with our method. (d) Deblurring results by Krishnan, et al. [4]. (e) Deblurring results by Krishnan, et al. [4] with our method (Color figure online)

C. **Analysis of experiments**

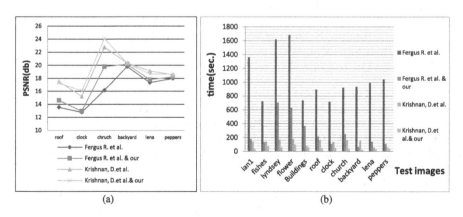

(a)   (b)

Fig. 7. Quantitative analysis of our method. (a) PSNR comparison, (b) Computational time comparison

5   Conclusion

We proposed a coalitional game based framework to automatic select an informative patch for single-image deblurring. We first find the informative pixels using gradient magnitude and usability map. For more precisely, we form a patch centered at each informative pixels. Then, we apply the coalitional game which treats these patches as players. Players seek to join a coalition to improve their payoff. For fairness, we use Shapley Value to compute the marginal contribution for each player making all players get the maximum payoff. At the end of this game, we will have a stable coalition and our final patch can be formed by the members in the coalition.

Our method can accelerate the blind deconvolution algorithm by reduce the amount data to be processed in kernel estimation. We also increase the quality of deblurring result because we increase the accuracy of the kernel estimation. Our method select a proper patch avoiding user supervision. Experimental results demonstrate that the two deblurring algorithms with the patch selected by our method in most of cases outperform those with whole image.

Acknowledgment. This study is conducted under the "Online and Offline integrated Smart Commerce Platform (2/4)" of the Institute for Information Industry which is subsidized by the Ministry of Economy Affairs of the Republic of China

References

1. Fergus, R., Singh, B., Hertzmann, A., Roweis, S.T., Freeman, W.T.: Removing camera shake from a single photograph. ACM Trans. Graph. (TOG) **25**(3), 787–794 (2006). ACM
2. Shan, Q., Jia, J., Agarwala, A.: High-quality motion deblurring from a single image. ACM Trans. Graph. (TOG) **27**(3), 73 (2008). ACM
3. Xu, L., Jia, J.: Two-phase kernel estimation for robust motion deblurring. In: Daniilidis, K., Maragos, P., Paragios, N. (eds.) ECCV 2010, Part I. LNCS, vol. 6311, pp. 157–170. Springer, Heidelberg (2010)
4. Krishnan, D., Tay, T., Fergus, R.: Blind deconvolution using a normalized sparsity measure. In: 2011 IEEE Conference on Computer Vision and Pattern Recognition (CVPR), pp. 233–240. IEEE, June 2011
5. Jia, J.: Single image motion deblurring using transparency. In: CVPR 2007 IEEE Conference on Computer Vision and Pattern Recognition, pp. 1–8. IEEE, June 2007
6. Bae, H., Fowlkes, C.C., Chou, P.H.: Patch mosaic for fast motion deblurring. In: Lee, K.M., Matsushita, Y., Rehg, J.M., Hu, Z. (eds.) ACCV 2012, Part III. LNCS, vol. 7726, pp. 322–335. Springer, Heidelberg (2013)
7. Gupta, A., Joshi, N., Lawrence Zitnick, C., Cohen, M., Curless, B.: Single image deblurring using motion density functions. In: Daniilidis, K., Maragos, P., Paragios, N. (eds.) ECCV 2010, Part I. LNCS, vol. 6311, pp. 171–184. Springer, Heidelberg (2010)
8. Cho, S., Lee, S.: Fast motion deblurring. ACM Transactions on Graphics (TOG) **28**(5), 145 (2009). ACM
9. Joshi, N., Szeliski, R., Kriegman, D.J.: PSF estimation using sharp edge prediction. In: Computer Vision and Pattern Recognition, CVPR 2008, pp. 1–8. IEEE, June 2008

10. Köhler, R., Hirsch, M., Mohler, B., Schölkopf, B., Harmeling, S.: Recording and playback of camera shake: benchmarking blind deconvolution with a real-world database. In: Fitzgibbon, A., Lazebnik, S., Perona, P., Sato, Y., Schmid, C. (eds.) ECCV 2012, Part VII. LNCS, vol. 7578, pp. 27–40. Springer, Heidelberg (2012)

11. Levin, A., Weiss, Y., Durand, F., Freeman, W.T.: Understanding and evaluating blind deconvolution algorithms. In: IEEE Conference on Computer Vision and Pattern Recognition, CVPR 2009, pp. 1964–1971. IEEE, June 2009

12. Shapley, L.S.: A Value for n-person Games. In: Contributions to the Theory of Games, vol. II, by H.W. Kuhn and A.W. Tucker, editors. Annals of Mathematical Studies v. 28, pp. 307–317. Princeton University Press (1953)

A Robust Real-Time Road Detection Algorithm Using Color and Edge Information

Jae-Hyun Nam, Seung-Hoon Yang, Woong Hu, and Byung-Gyu Kim[✉]

Department of Computer Engineering, Sun Moon University, Asan, Korea
{jh.Nam,sh.Yang,hoewoong}@mpcl.sunmoon.ac.kr, bg.kim@ieee.org

Abstract. A vision-based road detection technique is important for implementation of a safe driving assistance system. A major problem of vision-based road detection is sensitivity to environmental change, especially illumination change. A novel framework is proposed for robust road detection using a color model with a separable brightness component. Road candidate areas are selected using an adaptive thresholding method, then fast region merging is performed based on a threshold value. Extracted road contours are filtered using edge information. Experimental results show the proposed algorithm is robust in an illumination change environment.

1 Introduction

Road recognition and lane departure warning systems have been developed over the last decade but have not been commercialized. A system mounted on a vehicle generally produces a constant error due to mechanical movement that needs to be cancelled to enable robust road detection. Hence, development of an algorithm is necessary to provide reliable information for robust feature extraction and development of tracking technologies [1, 2]. The automotive industry is changing from traditional mechanical to electronics and intelligence-based products to increase driving comfort and safety. Also, by setting the speed intelligently using radar is equipped with adaptive cruise control, to adjust the distance between the vehicles [3]. Another system is provided a method that gives a driving system by confirming the fatigue state of the driver, it will help to be able to make safe driving [4].

Road region detection techniques have been reported in the last few years [5]. A road environment detection technique must change and be adaptable to different road conditions. In order to process a large amount of data in real time, detection of the road from a video sequence requires use of fast algorithms [6]. On the other hand, vision-based road detection and object tracking techniques are beneficial for driver safety [7].

A major problem of vision-based road detection is sensitivity to environmental change, especially illumination change. In outdoor environments shadows are common. Peter Corke described how to convert a color image of the scene to a greyscale invariant image where pixel values are a function of underlying material property not lighting [8]. To cope with this problem, a novel color-based road detection technique is proposed herein. Also, an edge-based outline filtering method is also suggested to increase the robustness of the proposed algorithm.

© Springer International Publishing Switzerland 2015
G. Bebis et al. (Eds.): ISVC 2015, Part II, LNCS 9475, pp. 532–541, 2015.
DOI: 10.1007/978-3-319-27863-6_49

This paper is organized as follows: In Sect. 2, the proposed road detection algorithm including fast region merging scheme and edge-based smoothing, are introduced. The experimental results and discussion will be given in Sect. 3. Section 4 presents concluding remarks

2   Robust Road Detection Algorithm

2.1   Road Candidate Extraction Based on Robust Color Model

A major challenge for road detection techniques is brightness change based on environmental changes. The RGB color model, which includes a brightness (luminance) component, is not robust enough to overcome illumination changes. For this reason, a color model is needed that can separate the luminance component. Lab and HSV color models can all separate the luminance component.

Weaknesses and robustness for environment changes are shown in Table 1 for different color models. Here, 'W' and 'R' indicate 'weakness' and 'robustness', respectively. The HSV model and Lab model are promising due to robustness in all three indicated environments.

Table 1. Weakness and robustness against environmental changes of the RGB, Lab, and HSV color models

	R	G	B	ab	H	S
Change brightness	W	W	W	R	R	R
Shadow region	W	W	W	R	R	R
Road type	W	W	W	R	R	R

The HSV color model was used in this study for extraction of the road area. The HSV color model and the color region of the road are shown in Fig. 1. The H value is not significant when the S value is less than a predetermined value. If the S value is less than the predetermined value, the model devolves to similar saturation of the urban road.

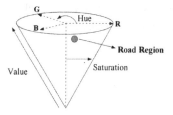

Fig. 1. HSV color model (Color figure online).

The histogram value of S in a sample image of an urban road is shown in Fig. 2(b). The histogram indicates a peak value of approximately 25 % of the S value. This property can be useful for development of an adaptive thresholding scheme for separation of the road region from the background.

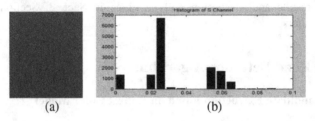

(a)   (b)

Fig. 2. (a) A road sample image and (b) histogram of S component in the HSV color model.

The extracted S component of the road image is shown in Fig. 3. The S component of the general road area is shown in Fig. 3(a) and the road region with shadow is shown in Fig. 3(b). Road and non-road areas are clearly delineated. However, determination of the road region is not acceptable. Based on color model analysis, an adaptive thresholding scheme that can cope with this kind of brightness change is needed.

(a)   (b)

Fig. 3. Extracted S component of road sample images.

In this paper, the Otsu thresholding method [9] was used based on a histogram of the S component. In the Otsu method, the total variance (σ^2) of input for an image can be represented by the sum of the inter-class variance and the intra-class variance, as:

$$\sigma^2 = \sigma_w^2 + \sigma_b^2, \tag{1}$$

where σ_w^2 is the intra-class variance and σ_b^2 is the inter-class variance. In order to obtain a good threshold value, the smaller inter-class variance value of the two classes can be obtained as:

$$\text{Min}_{t^*}\sigma_b^2(t) = \sigma^2 - \sigma_w^2(t), \tag{2}$$

where t^* is the optimal threshold value.

Results in binary form using the adaptive threshold t^* with input images are shown in Fig. 4. Even though the input image has shadow on both road and non-road regions, it is possible to distinguish the road region (Fig. 4b). A road candidate region was, thus, identified using the above procedure.

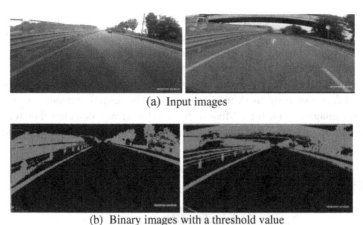

(a) Input images

(b) Binary images with a threshold value

Fig. 4. Segmented results by Otsu method [6].

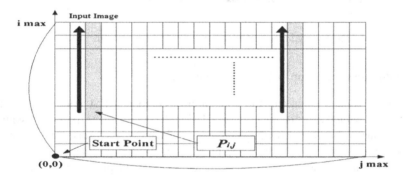

Fig. 5. Boundary detection for road region segmentation.

A region-growing mechanism based on boundary detection is shown in Fig. 5. Usually, the road area is located from the wide-bottom to the narrow-top in front of the driver. Using this factor, a region-growing mechanism was designed based on the initial segmented result shown in Fig. 5. From the first bottom pixel $p_{i,j}$, the intensity S value is compared with the next pixel $p_{i+1,j}$ vertically. If the S value of $p_{i,j}$ is less than one of $p_{i+1,j}$ by an amount of the pre-defined threshold (t^*), then this is confirmed as road region. Otherwise, all remained pixels are set as non-road area from $p_{i+1,j}$ to $p_{i_{max},j}$ without any further examination. For all vertical column lines, the above procedure is performed iteratively. For all j, the merged candidate road region ($R_c(i,j)$) can be expressed as the following:

$$\Delta S_{i,j} = \left| S_{i,j} - S_{i+1,j} \right|, \quad \text{for all } i,$$

$$R_c(i,j) = \begin{cases} \text{non - road}, & \Delta S_{i,j} \geq t^*, \\ \text{road}, & \text{otherwise.} \end{cases} \tag{3}$$

A value larger than the threshold value t^* is interpreted as a different color area (non-road area). Pixel values are compared with vertically neighboring pixels and merged into the road area or background (non-road) area. As mentioned in Eq. (2), the optimal threshold value is computed in each frame and used to separate regions.

Road detection results are shown in Fig. 6. The input image is shown in Fig. 6(a) and the road detection result is shown in Fig. 6(b). The detection result was overlapped for the input image in Fig. 6(c). Most of the road area was detected, but some errors were observed. In order to solve this problem, an edge-based filtering method was introduced.

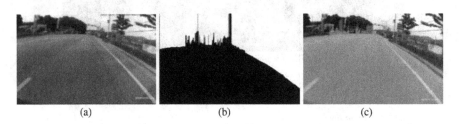

(a)   (b)   (c)

Fig. 6. The detected road area based on the proposed region merging scheme.

2.2 Edge-Based Outline Filtering Mechanism

The road candidate region can still contain error components. A filtering method is, therefore, presented to refine the road area. Using the binary image, points are sampled that are equally distributed on the contour line of the extracted road. The number of sampling points (β) can be determined by the user. With each sampling point as the center, a rectangular area is set that can include α other sampling points to check the smoothness.

The concept of setting rectangular area on the contour line of the extracted road is shown in Fig. 7(a) and the actual rectangles for outline filtering when $\alpha = 5$ and $\beta = 30$ are shown in Fig. 7(b). With high α and β values the accuracy is increased but the complexity is also increased. Rectangular areas usually contain noise and road regions together. The region of the rectangle is used to distinguish road and non-road areas based on edge smoothing processing for removal of noise.

To smooth the outline of the candidate road area, band-pass type edge magnitude filtering is used so that if the examined edge pixel has a value larger than TH_h, it is verified as a noise component. Otherwise, if the magnitude of the examined edge pixel is less than TH_l, it is also classified as noise. Edge pixels are considered with a magnitude between TH_l and TH_h as credible edges. In this study, $TH_l = 200$ and $TH_h = 100$ were set for processing of edge smoothing. It is because it indicates the most accurate results when the road contour detecting experimentally. After edge smoothing, all filtered edge pixel are connected to make an updated contour line of the extracted road area.

Result for connected edge pixels are shown in Fig. 8. The smoothed edge component can be represented as straight lines to define the exact road region. The straight lines can be formulated using Eq. (5):

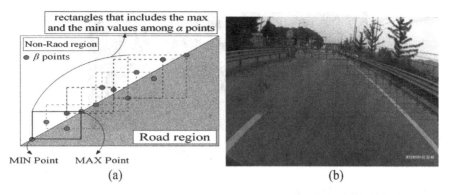

Fig. 7. Outline filtering scheme using local rectangular blocks.

$$P(t) = (1 - t)P_1 + tP_2,$$
$$P(s) = (1 - s)P_3 + sP_4,$$

(5)

where t and s are values from 0 to 1, and $P_k = (x_k, y_k)$. From Eq. (5), an intersection point of the two lines $P(t)$ and $P(s)$ can be computed, as expressed in Eq. (6):

$$(1 - t)P_1 + tP_2 = (1 - s)P_1 + sP_2,$$
$$x_1 + t(x_2 - x_1) = x_3 + s(x_4 - x_3),$$
$$y_1 + t(y_2 - y_1) = y_3 + s(y_4 - y_3.)$$

(6)

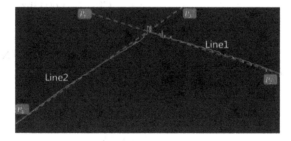

Fig. 8. Outlier filtering result.

After obtaining the intersection point p_{xy}, which is called the banishing point, the upper part vertically of this point is defined as the non-detected road area (area of no interest). The intersection (crossing) point from Eq. (6) is shown in Fig. 9. Based on this crossing point, the upper horizontal area is identified as the non-road region. Thus, the inner part of two lines and down region of intersection point p_{xy} is recognized as the final road region.

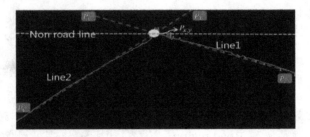

Fig. 9. Crossing point detection and the final road region.

3 Experimental Results

The hardware (HW) platform is shown in Table 2. Different HD video sequences were used. Road detection results are shown Fig. 10. The detected road area in typical environments is shown in Fig. 10(a, b). Results with shadow are shown in Fig. 10(c). An example with vehicles on an urban road is shown in Fig. 10(d). For all environments, the proposed method achieved good detection results and the extraction result with shadow was especially remarkable Fig. 10(c). Also, the proposed algorithm produced good results for a complex urban road environment (Fig. 10c).

Table 2. System Spec. for experiment.

CPU		Intel(R) Core(TM) i7-4790 CPU @ 3.60 Ghz	
OS		Window 7 64 bit	
GRAPHIC DEVICE		NVIDIA GeForce GTX 750	
MEMORY	TOTAL CORE NO.	2 Gb	128

For evaluation of quantitative performance, measurement parameters were defined as:

$$\text{Accuracy: } ACC = \frac{TP_i + TN_i}{N_i}, \tag{8}$$

$$\text{Error Rate: } ERR = \frac{FN_i + FP_i}{N_i}, \tag{9}$$

$$\text{True Positive Rate: } TRR = \frac{TP_i}{TP_i + FN}, \tag{10}$$

$$\text{False Positive Rate: } TRR = \frac{FP_i}{TN_i + FP_i}, \tag{11}$$

(a) Road data1

(b) Road data2

(c) Road data3

(d) Road data4

Fig. 10. Road detection results using the proposed algorithm. First column: Original images. Second column: Detected road area. Third column: Overlay images on the original.

where TP_i indicates the road for pixel I, TN_i is non-road area of pixel I, FP_i is a value for detection as non-road pixel for the given pixel that is actually true road, and FN_i is a measure for detection of a non-road pixel that is actually not road pixel.

Performance of the proposed road detection algorithm is shown in Table 3. High accuracy with a small error rate (1.42 %) using different video sequences was achieved. The proposed algorithm produced credible TPR and FPR performance (Table 3). The FPR value was large due to addition of noise components; however, even with a high FPR value, a high recognition accuracy rate was achieved (Table 3).

Table 3. Accuracy performance of the proposed algorithm.

Items	ACC	ERR	TPR	FPR
Avg. value (%)	96.03	1.42	98.89	17.91

Simple CPU-based implementation achieved approximately 4.51 frames per second (FPS) for processing time of the extracted road area (Table 4). To increase the speed of the process, Compute Unified Device Architecture (CUDA) parallel programming using a graphic processing unit (GPU) on an NVIDIA Jetson TK1 development kit was used. The enhanced speed of the GPU parallel programming while maintaining the accuracy, is shown in Table 4. An approximate 10 times improvement in processing speed was achieved compared with simple CPU processing. Thus, the proposed algorithm can be operated in real-time with GPU implementation.

Table 4. Improved speed of the proposed algorithm using CUDA.

Contents	Processing speed (FPS)
CPU-based Avg. value	4.51
GPU-based Avg. value	41.12

4   Conclusions

An efficient vision-based road detection algorithm is proposed using robust color-based region merging and edge-based filtering mechanisms for autonomous vehicles and safe driving. The proposed algorithm was tested using various road environments. Based on the proposed method, a 98.89 % recognition ratio was achieved. Also, GUP-based system implementation was developed for real-time processing in a driver assistance system (DAS).

Acknowledgment. This work was supported by the Sun Moon University Research Grant of 2014.

References

1. Alvares, J.M.A., Gevers, T., Lopez, A.M.: Vision-based road detection using road models. In: IEEE International Conference on Image Processing (ICIP), pp. 2073–2076 (2009)
2. Alvarez, J.M.A., Gevers, T., Lopez, A.M.: Evaluating color representations for on-line road detection. In: International Conference on Computer Vision Workshops (ICCVW), pp. 594–559 (2013)
3. Hasan, N., Didar-Al-Alam, S.M., Rezwanul Huq, S.: Intelligent car control for a smart car. Int. J. Comput. Appl. **14**, 15–19 (2011)
4. Bhumkar, S.P., Deotare, V.V., Babar, R.V.: Intelligent car system for accident prevention using ARM-7. Int. J. Emerg. Technol. Adv. Eng. **2**(4) (2012)
5. Vipul, H., Ramji, M.: Survey: vision based road detection techniques. Int. J. Comput. Sci. Inf. Technol. **5**, 4741–4747 (2014)
6. Moghadam, P., Starzyk, J.A., Wijesoma, W.S.: Fast vanishing-point detection in unstructured environments. In: IEEE International Conference on Image Processing (ICIP), pp. 425–430 (2011)

7. Kuhnl, T., Fritsch, J.: Vision-spatial road boundary detection for unmarked urban and rural roads. In: IEEE Intelligent Vehicles Symposium Proceedings, pp. 1251–1256 (2014)
8. Corke, P., Paul, R., Churchill, W., Newman, P.: Dealing with shadows: capturing intrinsic scene appearance for image-based outdoor localization. Intelligent Robots and Systems (IROS), pp. 2085–2092 (2013)
9. Sezgin, M., Sankur, B.: Survey over image thresholding techniques and quantitative performance evaluation. J. Electron. Imaging **13**, 146–165 (2004)

SeLibCV: A Service Library for Computer Vision Researchers

Ahmad P. Tafti[1], Hamid Hassannia[2], Dee Piziak[1], and Zeyun Yu[1]($\boxtimes$)

[1] Department of Computer Science, University of Wisconsin-Milwaukee,
Milwaukee, WI 53211, USA
yuz@uwm.edu
[2] IEEE Member, Uppsala, Sweden

Abstract. Image feature detectors and descriptors have made a big advance in several computer vision applications including object recognition, image registration, remote sensing, panorama stitching, and 3D surface reconstruction. Most of these fundamental algorithms are complicated in code, and their implementations are available for only a few platforms. This operational restriction causes various difficulties to utilize them, and even more, it makes different challenges to establish novel experiments and develop new research ideas. SeLibCV is a Software as a Service (SaaS) library for computer vision researchers worldwide that facilitates Rapid Application Development (RAD), and provides application-to-application interaction by tiny services accessible through the Internet. Its functionality covers a wide range of computer vision algorithms including image processing, features extraction, motion detection, visualization, and 3D surface reconstruction. The present paper focuses on the SeLibCV's routines specializing in local features detection, extraction, and matching algorithms which offer reusable and platform independent components, leading to reproducible research for computer vision scientists. SeLibCV is freely available at http://selibcv.org for any academic, educational, and research purposes.

1 Introduction

Efficient detection and description of image features such as edges, corners, and blobs has been a longstanding topic in computer vision and multimedia systems. Scale Invariant Feature Transform (SIFT) [28,29], Speed Up Robust Features (SURF) [16], and Binary Robust Independent Elementary Features (BRIEF) [17] are among the most popular feature extractors which have made a big advance in a wide scope of computer vision applications including object recognition and tracking [20,21,25], image forgery detection [23], remote sensing [27,30], image registration [13,14,24,36], and 3D surface reconstruction [10,32,34,35].

These fundamental computer vision strategies are essential for almost every computer vision problem, but their implementations are often available in binary (executable) format and only for a few special platforms. Unavailability of multi platform implementations for such computer vision algorithms causes operational restrictions to examine new experiments and develop modern applications especially in the Internet of Things (IoT) era.

© Springer International Publishing Switzerland 2015
G. Bebis et al. (Eds.): ISVC 2015, Part II, LNCS 9475, pp. 542–553, 2015.
DOI: 10.1007/978-3-319-27863-6_50

Concerning the software engineering, "Software as a Service" (SaaS) as the basic idea behind the centralized computing, is a design pattern as well as a delivery model in which a software could be accessed by human users through a web browser or by an application using an application-oriented interface [1,11,22,37]. To our best knowledge, a SaaS based architecture has not yet been explored for highly demanded computer vision algorithms and our work is one of the first to design and develop a set of World Wide Web services for the computer vision community.

The present work is mainly focused on a novel design and development of a SaaS library specializing in local features detection, description, and matching algorithms including SIFT [28,29], SURF [16], and BRIEF [17] to provide reusable, platform independent, and highly available components for such basic computer vision strategies. To find a best set of the matching points, we also utilize RANSAC [19], and KNN [12] techniques. We named this library as **SeLibCV** which stands for Service Library for Computer Vision Researchers. SeLibCV developed for convenience of use and high availability, and it aims Rapid Application Development (RAD) and fast prototyping for computer vision researchers, students, and scientists all around the world. SeLibCV is an easy-to-use service which is freely available at [9] for any academic, educational, and research purposes.

This contribution will target the following objectives: (1) To assist fast prototyping and Rapid Application Development (RAD) for computer vision community by tiny services available on the Internet, (2) To provide application-to-application interaction for highly demanded computer vision algorithms, (3) To make the fundamental computer vision algorithms available through both human-oriented and application-oriented interfaces, (4) To provide better scalability in which the SeLibCV can spread all requests between different parallel data sources without having details of the resource to be implemented by service requesters, and (5) To create better portability and interoperability by imposing no programming language or operating system limitations in using the services. We summarized our **main contributions** as follows:

- We initiate the study of SaaS based architectures for essential computer vision techniques. To the best of our knowledge, our work is one of the first attempt to design and develop a set of World Wide Web services for highly demanded computer vision algorithms.
- With the current work, we novel design and develop a SaaS based architecture to implement reusable, platform independent, and extensible software components for the SIFT, SURF, and BRIEF algorithms. This contribution makes these algorithms available from any computers or smart phones, anytime, anywhere. An Internet connection is all we need!
- As an important contribution, the present work is expected to bridge the gap between computer vision, World Wide Web services, and SaaS architectural design, and open the doors for many interesting directions from the computer vision community to the IoT technology, the latest industry buzzword used as an advanced technology.

The rest of the paper is arranged as follows. The proposed system design along with its underlying technologies come in Sect. 2. Experimental validations are reported in Sect. 3. Conclusion and possible directions for future works are presented in Sect. 4.

2 Methods

In this section we shall begin with a brief introduction to a set of computer vision algorithms implemented in the SeLibCV. We then demonstrate our proposed SaaS based model as well as the system implementation.

2.1 Feature Detection Algorithms Implemented in the SeLibCV

This section gives a short introduction to a selection of algorithms implemented in the SeLibCV.

SIFT: The Scale Invariant Feature Transform (SIFT) [28,29,33] is one of the most commonly feature descriptors utilized in computer vision applications which is able to detect image features invarinat to scaling, rotation, and translation. SIFT is basically divided into four main components, namely: (1) Scale-space construction using Difference-of-Gaussian (DoG), (2) Stable features localization, (3) Gradient orientation computation and magnitude assignment, and (4) Feature descriptors extraction [29]. The original executable SIFT implementation is available at [3]. An open source C implementation among with MATLAB [2] interfaces are also available at [5].

SURF: The Speed Up Robust Features (SURF) [16] is another well-known local feature detector and descriptor algorithm which employs an integer approximation of the determinant of Hessian blob detector [16] to detect feature points. The standard version of SURF algorithm is faster than the SIFT algorithm. SURF is available in the MATLAB Computer Vision System Toolbox [8]. An open source OpenCL [4] implementation is also available at [6].

BRIEF: The Binary Robust Independent Elementary Features (BRIEF) [17] has been a popular feature extractor which is proper for real-time applications. BRIEF is based on comparisons. Let's say we have a patch (40 pixels by 40 pixels), we then choose two feature points and compare the intensities of those points, if the intensity of the first feature point is larger than the second feature point, we assign the value '1', otherwise '0', and we do that for a number of pairs and we finally end up with a string of boolean values. An open source C++ implementation is available at [7].

In order to find the best corresponding points (inliers), we employ KNN [12] as a way to discover nearest points, and we then utilize RANSAC [19] to compute a perspective transform to get rid of the outliers.

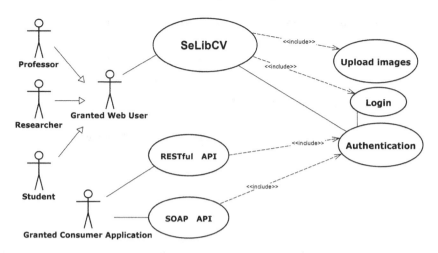

Fig. 1. SeLibCV: Use Case View. This figure shows the main functionalities of the system and how different users can interact with those functionalities.

2.2   System Architecture

The "4+1" view model [15, 18, 22, 26] as the most commonly used software architectural model template describes the anatomy of a software system using multiple concurrent views including Use Case view, Logical view, Process view, Development view, and Physical view [26]. Here, we use the "4+1" view model to present and explain the Use Case and Process diagrams of the proposed system.

Use Case View: This view is one of the fundamental perspectives of the "4+1" view model which represents the main functionalities of the system along with their users who can interact with the system [26]. The Use Case view of the SeLibCV is shown in Fig. 1. Different users can login into the system as a previously granted user, and then they can utilize the SeLibCV through the web. In case they intend to use the web based functionalities, they have to upload their own images. The other option is that the SeLibCV services may be called as a RESTful or SOAP based API [31] for a granted consumer application developed with any programming language. In this case, the services will work on the local images which are located in the consumer application's machine.

Process View: The process view deals with the dynamic aspects of a system to show how objects will integrate into the complete system based on a time sequence [26]. The sequence diagram of the process view is presented in Fig. 2.

2.3   Service Architecture

The high level service architecture of the SeLibCV is shown in Fig. 3. The service compromises two disparate machines: a service consumer (Client) and the

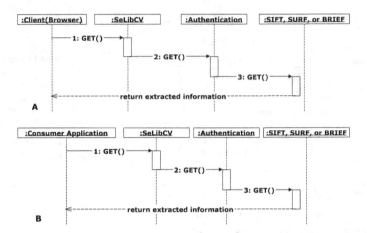

Fig. 2. SeLibCV: Process View. The sequence diagram models the collaboration of objects based on a time sequence. GET() defines a particular communication between lifelines of an interaction. (A) A client by using an Internet browser can get into the system once after authentication. (B) A consumer application can also use the system once after authentication.

service provider (Application Server). Main components of each section are presented within the subsystem blocks. The abstract view of the system is also shown in Fig. 4.

2.4   How to Use the SeLibCV Services

A complete user manual of the system is fully documented and available at [9]. In order to use the SeLibCV services, users should send a request for a valid username/password through [9]. We will then create and send them back the username/password to use the SeLibCV services either through an Internet browser or a consumer application. Using a consumer application, the following sample codes is all that is necessary to call the SIFT algorithm inside the SeLibCV. This sample code is written in Java to get the SIFT description vectors.

Sample Code 1. How to call the SIFT algorithm implemented in SeLibCV:

```
try {
  SiftResult sift = proxy.sift(USER, PASS, fileContent);
  BufferedImage img =
  ImageIO.read(new ByteArrayInputStream(sift.getImageDate()));
  ImageUtil.saveAsPNG(img, args[1]);
  System.out.println("Results SIFT Image saved as:"+ args[1]);
  System.out.println("SIFT result:");
  for(int i=0; i< sift.getPoints().length(); i++) {
    System.out.println("X:" + sift.getPoints[i].getX()+
    "Y:" + sift.getPoints[i].getY());}
    } catch (Exception e) {
    e.printStackTrace();
  }
```

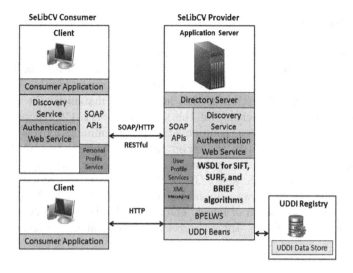

Fig. 3. SeLibCV: service architecture. **WSDL** (Web Service Description Language) is such an XML-based document for describing the SeLibCV services and how to access it over the standard Internet protocols. **UDDI** is a specification for a distributed type registry of the SeLibCV services which can communicate through SOAP, or Java RMI Protocols. **Discovery Service** permits the discovery of the SeLibCV services. **SOAP** (Simple Object Access Protocol) is a XML-based Internet protocol for exchanging structural information in the implementation of web services between computers and applications. **BPELWS** (Business Process Execution Language for Web Service) aims to model the main behaviors and operation of both executable and abstract implementation of the web service processes.

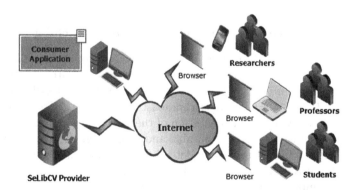

Fig. 4. SeLibCV: abstract view of the system. Different group of users can get into the system using a browser, and developers may build a consumer application to call the SeLibCV services without employing any browsers.

3   Experimental Validations

To examine the general performance, accuracy, reliability, and speed of the proposed system discussed in Sect. 2, extensive experiments on real digital images are carried out in this section. In Sect. 3.1, we compare the accuracy of the system in feature points detection as well as features matching. We then study and compare the time efficiency of the system in Sect. 3.2. All parts of the SeLibCV including its classes, components, and communication packages were implemented by Java SE 8. In the server side (Application Server), we used a 64-bit Linux CentOS operating system on a virtual server with 2 v cores processor, 100 Mbits/s bandwidth, and 2 GB of RAM. In the client side (Client), we employed 64-bit MS Windows 8 operating system with 3.00 GHz Intel Dual core CPU, 2 MB cache, and 4 GB of RAM. The bandwidth between server and client was established as 5 Mbit/s. Table 1 shows a dataset which used to examine the behavior of the SeLibCV.

Table 1. The proposed dataset.

Image Set	Image	Image Size	File Size	File Type
Books	books01	1152*864	329 KB	JPEG
Books	books02	1152*864	320 KB	JPEG
Building	b01	922*692	943 KB	PNG
Building	b02	922*692	1.02 MB	PNG
Hall	hall01	807*605	720 KB	TIFF
Hall	hall02	807*605	681 KB	TIFF

3.1   Accuracy in Feature Points Detection and Matching

Here, we bring up a validation summary on the accuracy of the proposed system for feature points detection in a single image, and features matching between an image pair. We compare the accuracy of the SeLibCV services with the original executable SIFT [3], the SURF algorithm exists in MATLAB Computer Vision Toolbox [8], and the C++ implementation of the BRIEF algorithm [7] (Table 2). As you can see the results are comparable and promising. Figures 5 and 6 show the visual accuracy in feature points detection and matching, respectively. We can see that the error threshold for both detection and matchning processes are less than 2 % comparing with the three other implementations.

3.2   Runtime Efficiency

This part gives a validation summary on the runtime efficiency of the SeLibCV services. Table 3 compares the runtime efficiency of the SeLibCV services with

Table 2. Features detection and matching accuracy. We labeled our proposed system as "SeLibCV".

System	Image	Number of detected features
Original executable SIFT	books01	2166
SeLibCV - SIFT Service	books01	2148
Original executable SIFT	books02	2221
SeLibCV - SIFT Service	books02	2213
SURF in MATLAB Computer Vision Toolbox	b01	1612
SeLibCV - SURF Service	b01	583
SURF in MATLAB Computer Vision Toolbox	b02	656
SeLibCV - SURF Service	b01	1660
C++ implementation of the BRIEF algorithm	hall01	309
SeLibCV - BRIEF Service	hall01	311
C++ implementation of the BRIEF algorithm	hall02	244
SeLibCV - BRIEF Service	hall02	207
System	Image #1	Image #2	Number of matches
Original executable SIFT	hall01	hall02	166
SeLibCV	hall01	hall02	158
SURF in MATLAB Computer Vision Toolbox	books01	books02	1781
SeLibCV	books01	books02	1764
C++ implementation of the BRIEF algorithm	b01	b02	301
SeLibCV	b01	b02	305

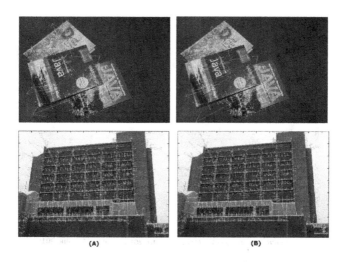

Fig. 5. Visual accuracy in feature points detection. (A) Results obtained by the original executable SIFT software [3]. (B) Results obtained by the SeLibCV. Image Sets: Books and Building.

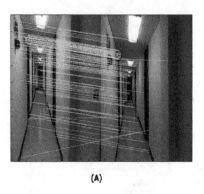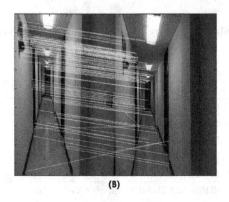

(A)   (B)

Fig. 6. Visual accuracy in feature points matching. (A) Results obtained by the original executable SIFT software [3]. (B) Results obtained by the SeLibCV. Image Set: Hall.

three other implementations. The experiment shows that the runtime efficiency of the SeLibCV services are better, and the reason could be associated to the nature of the Java programming language which is a little faster than C and a quite faster than MATLAB Toolbox.

As you have seen in the experiments, the time efficiency, accuracy, and reliability of the SeLibCV services are promising. In addition, the proposed system includes platform independent and reusable components which can provide both human-to-application and application-to-application interaction. The SeLibCV is also going to be popular in the computer vision community (Fig. 7). In the near future, we would very much like to increase the number of users to the project.

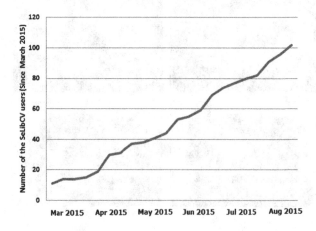

Fig. 7. Impact of the SeLibCV. Number of users of the SeLibCV services since March 2015.

Table 3. The comparison of runtime efficiency. We labeled our contribution system as "SeLibCV".

System	Image	# of features	Elapsed time
Original executable SIFT	books01	2166	15.7 s
SeLibCV - SIFT Service	books01	2148	14.8 s
Original executable SIFT	books02	2221	17 s
SeLibCV - SIFT Service	books02	2213	15.3 s
SURF in MATLAB Computer Vision Toolbox	b01	1612	12.11 s
SeLibCV - SURF Service	b01	1583	11.09
SURF in MATLAB Computer Vision Toolbox	b02	1656	12.74 s
SeLibCV - SURF Service	b01	1660	12.18 s
C++ implementation of the BRIEF algorithm	hall01	309	8.64 s
SeLibCV - BRIEF Service	hall01	311	7.05 s
C++ implementation of the BRIEF algorithm	hall02	244	8.13 s
SeLibCV - BRIEF Service	hall02	207	6.59 s

4 Conclusions and Future Works

In this contribution, we design and develop a Software as a Service (SaaS) library specializing in local features detection, description, and matching algorithms. The proposed service based library offers promising results. The services are available at http://selibcv.org from any computers or smart phones, anytime, anywhere. An Internet connection is all we need!

Based on the unique architecture of SaaS and its functional attributes, it has a remarkable impact on computer vision community by either bringing a modern architectural model which is distinguished from traditional software architectures, or allowing application-to-application communication through World Wide Web services. Employing SaaS architecture for computer vision algorithms will provide following quality attributes: (1) Accessibility and availability, anywhere, anytime, (2) Interoperability, (3) Portability, (4) Less deployment time, and (5) Scalability. Future works will be focused on adding more practical computer vision algorithms to the SeLibCV, plugging the library into a Cloud infrastructure, such as Google Cloud or Amazon EC2. The main idea of the work is that the SeLibCV could be considered as an initial step towards the next generation of computer vision applications.

References

1. World wide web consortium. http://www.w3.org/
2. Matlab (2000). http://www.mathworks.com/products/matlab/
3. Lowe's executable sift software (2005). http://www.cs.ubc.ca/lowe/keypoints/
4. Opencl (2008). https://www.khronos.org/opencl/

5. Vlfeat (2010). http://www.vlfeat.org/index.html
6. Clsurf (2011). https://code.google.com/p/clsurf/
7. Cvlab (2013). http://cvlab.epfl.ch/research/detect/brief
8. Computer vision system toolbox^{TM} matlab (2014). http://www.mathworks.com/products/computer-vision/
9. Selibcv (2015). http://selibcv.org
10. Agarwal, S., Furukawa, Y., Snavely, N., Simon, I., Curless, B., Seitz, S.M., Szeliski, R.: Building rome in a day. Commun. ACM **54**(10), 105–112 (2011)
11. Almorsy, M., Grundy, J., Ibrahim, A.: Adaptable, model-driven security engineering for saas cloud-based applications. Autom. Softw. Eng. **21**, 187–224 (2014)
12. Altman, N.S.: An introduction to kernel and nearest-neighbor nonparametric regression. Am. Stat. **46**, 175–185 (1992)
13. Baghaie, A., D'souza, R.M., Yu, Z.: Sparse and low rank decomposition based batch image alignment for speckle reduction of retinal oct images (2014). arXiv preprint arXiv:1411.4033
14. Baghaie, A., Yu, Z.: Curvature-based registration for slice interpolation of medical images. In: Zhang, Y.J., Tavares, J.M.R.S. (eds.) CompIMAGE 2014. LNCS, vol. 8641, pp. 69–80. Springer, Heidelberg (2014)
15. Bass, L., Clements, P., Kazman, R.: Software Architecture in Practice. Addison-Wesley, Boston (2013)
16. Bay, H., Ess, A., Tuytelaars, T., Gool, L.V.: Surf: speed up robust features. Comput. Vis. Image Underst. (CVIU) **110**, 346–359 (2008)
17. Calonder, M., Lepetit, V., Strecha, C., Fua, P.: BRIEF: binary robust independent elementary features. In: Daniilidis, K., Maragos, P., Paragios, N. (eds.) ECCV 2010, Part IV. LNCS, vol. 6314, pp. 778–792. Springer, Heidelberg (2010)
18. Eriksson, H.-E., Penker, M.: Business modeling with UML. Wiley, Hoboken (2000)
19. Fischler, M.A., Bolles, R.C.: Random sample consensus: a paradigm for model fitting with applications to image analysis and automated cartography. Commun. ACM **24**(6), 381–395 (1981)
20. Girod, B., Chandrasekhar, V., Chen, D.M., Cheung, N.-M., Grzeszczuk, R., Reznik, Y., Takacs, G., Tsai, S.S., Vedantham, R.: Mobile visual search. IEEE Signal Process. Mag. **28**(4), 61–76 (2011)
21. Hamidi, S.A., Bravo, H.R., Klump, J.V., Waples, J.T.: The role of circulation and heat fluxes in the formation of stratification leading to hypoxia in Green Bay, Lake Michigan. J. Great Lakes Res. (2015)
22. Hatch, R.: SaaS Architecture, Adoption and Monetization of SaaS Projects using Best Practice Service Strategy, Service Design, Service Transition. Service Operation and Continual Service Improvement Processes. Brisbane, Emereo Pty Ltd (2008)
23. Huang, H., Guo, W., and Zhang, Y.: Detection of copy-move forgery in digital images using sift algorithm. In: Computational Intelligence and Industrial Application, PACIIA 2008, vol. 2 (2008)
24. Huijuan, Z., Qiong, H.: Fast image matching based-on improved surf algorithm. In: 2011 International Conference on Electronics, Communications and Control (ICECC), pp. 1460–1463. IEEE (2011)
25. Kalal, Z., Mikolajczyk, K., Matas, J.: Tracking-learning-detection. IEEE Trans. Pattern Anal. Mach. Intell. **34**(7), 1409–1422 (2012)
26. Kruchten, P.B.: The 4+ 1 view model of architecture. IEEE Softw. **12**, 42–50 (1995)

27. Li Song, Z., Li, S., George, T.F.: Remote sensing image registration approach based on a retrofitted sift algorithm and lissajous-curve trajectories. Opt. Express **18**(2), 513–522 (2010)
28. Lowe, D.G.: Object recognition from local scale-invariant features. In: International Conference on Computer Vision, Corfu, Greece, pp. 1150–1157 (1999)
29. Lowe, D.G.: Distinctive image features from scale-invariant keypoints. Int. J. Comput. Vision **60**(2), 91–110 (2004)
30. Song, Z.L., Zhang, J.: Remote sensing image registration based on retrofitted surf algorithm and trajectories generated from lissajous figures. IEEE Geosci. Remote Sens. Lett. **7**(3), 491–495 (2010)
31. Srinivasan, L., Treadwell, J.: An overview of service-oriented architecture, web services, and grid computing. HP Software Global Business Unit (2005)
32. Tafti, A.P., Baghaie, A., Kirkpatrick, A.B., Owen, H.A., D'Souza, R.M., Yu, Z.: A comparative study on the application of sift, surf, brief and orb for 3d surface reconstruction of electron microscopy images. In: Computer Methods in Biomechanics and Biomedical Engineering: Imaging & Visualization (2015)
33. Tafti, A.P., Hassannia, H., Yu, Z.: siftservice. com-turning a computer vision algorithm into a world wide web service (2015a). arXiv preprint arXiv:1504.02840
34. Tafti, A.P., Kirkpatrick, A.B., Owen, H.A., Yu, Z.: 3D microscopy vision using multiple view geometry and differential evolutionary approaches. In: Bebis, G., Boyle, R., Parvin, B., Koracin, D., McMahan, R., Jerald, J., Zhang, H., Drucker, S.M., Kambhamettu, C., El Choubassi, M., Deng, Z., Carlson, M. (eds.) ISVC 2014, Part II. LNCS, vol. 8888, pp. 141–152. Springer, Heidelberg (2014)
35. Tafti, A.P., Kirkpatrick, A.B., Alavi, Z., Owen, H.A., Yu, Z.: Recent advances in 3d sem surface reconstruction. Micron **78**, 54–66 (2015b)
36. Teke, M., Temizel, A.: Multi-spectral satellite image registration using scale-restricted surf. In: 2010 20th International Conference on Pattern Recognition (ICPR), pp. 2310–2313. IEEE (2010)
37. Yu, J., Lin, J.: An architecture for cloud-based consumer support software-as-a-service. In: IEEE International Symposium on Computer, Consumer and Control (IS3C) (2014)

Bicycle Detection Using HOG, HSC and MLBP

Farideh Foroozandeh Shahraki, Ali Pour Yazdanpanah[✉],
Emma E. Regentova, and Venkatesan Muthukumar

Electrical and Computer Engineering Department,
University of Nevada, Las Vegas, NV 89154, USA
pouryazd@unlv.nevada.edu

Abstract. Due to the growing number of bicycles on roads, safety of bicyclists is drawing the increasing attention of transportation departments. Intelligent Transportation Systems (ITS) use automated tools for processing and analysis of traffic video data to plan and implement safety measures. One of important factors that influence the planning and safety countermeasures for bicyclists is the bicycle count. In this paper, we develop a bicycle detection method that can be used in a bicycle counting system. We strive to improve the efficiency of detection by looking for classification features that deliver more versatile information to automatic classifiers. We explore a combination of Histograms of Oriented Gradients (HOG), Histogram of Shearlet Coefficients (HSC) and Multi-scale Local Binary Pattern (MLBP) to improve detection and count of bicycles in video data. It is shown that the combination of the above features secures a higher detection accuracy.

1 Introduction

The number of general traffic accidents is currently decreasing because of the influence of numerous Intelligent Transportation System applied to transportation safety. But the rate of bicycle's accidents is gradually increasing due to the increased number of bicyclists on roads. To prevent traffic accidents, it is necessary to evaluate risk metrics at roadways and intersections [1]. The main factor in evaluating risk factor is determining accurate vehicle, pedestrian and bicycle counts. Therefore, bicycle detection in video based system is of the great significance to the research and application of Intelligent Transportation Systems (ITS). Also analysis of existing video recordings can deliver valuable information about the nature of the problem, i.e., behavior, road and traffic condition so conclusions can be drawn and safety measures developed. As Department of Transportation develops infrastructure to stream live video data from roadways and intersections, there is a need for automated detection and analysis. In spite of a remarkable progress in computer vision, bicycle detection is a challenging problem due to variation in appearances and poses, cluttered environment, and occlusion. Machine learning methods based on feature extraction and classification have achieved good results in bicycle detections over the last decade. This work focuses on determining classification features and efficient classifiers for high detection accuracy. Different key appearance features have been analyzed. Recent methods use

© Springer International Publishing Switzerland 2015
G. Bebis et al. (Eds.): ISVC 2015, Part II, LNCS 9475, pp. 554–562, 2015.
DOI: 10.1007/978-3-319-27863-6_51

Histogram of Oriented Gradients (HOG) feature descriptor introduced by Dalal and Triggs [2] for human detection. HOG feature captures edges by analyzing the gradient magnitudes and directions at different resolutions. H. Cho et al. [3] suggest part-based model bicycle detection using Histogram of Gradient (HOG)-PCA features and Linear Support Vector Machine (SVM). In [4], authors developed a bicycle detector using HOG, linear SVM and focus on pedaling movement. J. Heewook et al. [5] proposed a method based on the HOG features and Real-Adaboost meta-algorithm to detect a person on the bicycle. H. Jung et al. [6], propose a method based on MSC-HOG and Real-Adaboost.

Figure 1 shows few examples of a bicycle appearance at different orientations and rotations around z-axis. As the distance from the camera to the image pane varies and orientation changes, features have to be able to capture details at different directions. For safety studies, street cameras are set at the fixed locations with respect to the expected route of pedestrians, cars and bicycles, they can appear at the variable scales. HOG is robust under local intensity variations and is rotation invariant, if rotation is smaller than the orientation of the bin interval [7]. However, HOG is not efficient to analyze images at different scales. In [8], Multi-scale Local Binary Pattern (MLBP) has been introduced as an extension of well-known LBP features [9] for addressing the problem of texture characterization at different scales. Wavelet transform is known for its power to analyze signals at different scales, and it has been used for last decades for extracting edges and texture features for numerous applications. However, wavelet approach ability to analyze edges and textures in multiple directions beyond horizontal and vertical directions is limited. Recently, ridgelets, curvelets and shearlets have been introduced as an analytical tool for extracting information about structures in images more efficiently. Shearlet transform is the most advanced in delivering information about edges in various orientations and at multiple scales [7]. Therefore, in this paper, we explore detection of bicycles by augmenting a combination of Histogram of Shearlet Coefficients (HSC) [7] and HOG feature. Two basic orientations of bicycles are considered, frontal and the side views, w.r.t. the camera for training a linear SVM. The results show improvements when HOG features are complemented by HSC and MLBP. The paper is organized as follows. Section 2 explains the implemented method and gives a brief definition of feature descriptors and a classifier. In Sect. 3, we introduce the dataset and discuss experimental results. Conclusion is discussed in Sect. 4.

2   The Method

In this paper, we present a novel bicycle detection method based on combination of various features. We study HOG, HSC and MLBP features in combination to achieve the best classification accuracy. Subsequently, we choose the best of HOG, HSC and MLBP, normalize them and then combine. Finally, we compare the results of HOG, HSC, MLBP, HOG-HSC, HOG-MLBP and HOG-HSC-MLBP.

Fig. 1. Bicycle rotation

2.1 HOG Feature

HOG is a shape-based feature descriptor for finding edge information of objects. For this feature extractor, the image is divided into blocks and each block is divided into cells. In each cell, the histogram of gradients is computed. Histograms of cells are concatenated and then normalized with L2-norm normalization to form a feature vector. The authors of HOG use a 64×128 detection window for scanning the images. Each window is divided into 16×16 pixels size blocks with 50 % overlap and each block consists of 4 cells each of 8×8 pixels. Four histograms of four cells make a 1D feature vector of length 3780. Overall, each detection window has $7 \times 15 = 105$ overlapped blocks.

2.2 HSC Feature

The continuous shearlet transformation [10] of an image is defined as below.

$$SH_\varphi(a, s, t) = \int f(x)\psi_{a,s,t}(t - x)dx$$

Where a, s, t are the scale, orientation, and location in spatial domain respectively and $f(x)$ is a two dimensional image. Shearlets $\psi_{a,s,t}$ are given by

$$\psi_{a,s,t}(x) = \left|\det K_{a,s}\right|^{\frac{-1}{2}} \psi\left(K_{a,s}^{-1}(x - t)\right)$$

$$K_{a,s} = \begin{pmatrix} a & \sqrt{a}s \\ 0 & \sqrt{a} \end{pmatrix} = BA = \begin{pmatrix} a & 0 \\ 0 & \sqrt{a} \end{pmatrix}\begin{pmatrix} 1 & s \\ 0 & 1 \end{pmatrix}$$

Where A is an anisotropic scaling matrix and B is a shear matrix. $f(x)$ can be reconstructed back using the following formula:

$$f = \sum_{a,s,t} \langle f, \psi_{a,s,t}\rangle \psi_{a,s,t}$$

Due to the good localization properties in both time and frequency of Meyer wavelet, this wavelet is used as mother wavelet for the shearlet transformation implementation.

The HSC introduced in [7] uses statistics, a histogram of shearlet coefficients for a compact representation instead of coefficients themselves. The HSC features are calculated at different scales and orientations, since the shearlet coefficients of large magnitude come from edges [11]. In HSC method, image is divided into blocks. Each block is divided into 4 cells. In each cell, we have a number of decomposition levels and a number of orientations. For each decomposition levels, we estimate a histogram where the number of its bins is equal to the number of orientations on that decomposition level. Entry of each bin is computed as the absolute value of the shearlet coefficients where $H_{dl}(s)$ shows the s-th bin of the histogram is calculated for the dl-th decomposition level.

$$H_{dl}(s) = \sum |SH_\varphi(a, s, t)|$$

Finally, the histograms computed for all levels, cells and blocks are concatenated respectively and L2-norm normalization is employed. In this method, we have used 8×8, 16×16, 32×32 and 64×64 block sizes with 50 % overlap to calculate HSC. Each block has 4 cells. Histogram of shearlet coefficient is computed for each cell. The number of decomposition levels at each cell has been changed from 1 to 5, and each level has 8 orientations. After testing all possible states of block-size and level-number, we found 8×8 block with 50 % overlap and 4 cells in 2 scales with 8 orientations is best HSC feature to describe the image. The histograms of scale 1 and scale 2 are concatenated; L2-norm normalization is applied and a higher dimension feature is generated. The final feature vector length is (number of blocks in the image × number of cells in each block × number of levels in each cell × number of orientations) $15 \times 15 \times 4 \times 2 \times 8 = 14400$.

2.3   MLBP Feature

Local Binary Pattern (LBP) is a powerful feature for texture classification [9] and face recognition [12]. In this paper, we use Multi-scale LBP (MLBP) proposed in [8]. For MLBP, the LBP is calculated at three scales with radii of 1, 3 and 5. Number of directions in each scale is 8. Thus, each scale has a $2^8 = 256$-bin histogram. Histograms of scales are concatenated and a $3 \times 256 = 768$-bin histogram is generated. Figure 2

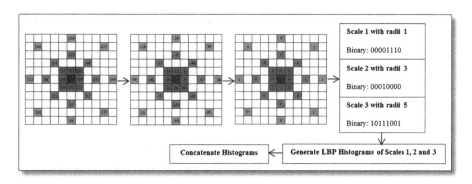

Fig. 2. Calculation of MLBP feature

shows the chosen pattern for MLBP. For all scales, the marked pixels participated in calculations lie along same directions to yield a more robust feature.

2.4   Combined Features

In order to get a more accurate and robust detector, we implement HOG-HSC, HOG-MLBP, HSC-MLBP and HOG-HSC-MLBP for our dataset and use linear SVM to train them. First, we extract HOG, MLBP and HSC features. For the HOG feature, we have set the window size to be 64×64 pixels and use blocks and cells of variable sizes. Our research shows that blocks with size 16×16 pixels and 50 % overlap and cells with size 2×2 pixels give the best result. Therefore, each detection window consists of 49 blocks for total and feature vector of length 28224. The MLBP feature is calculated per block. We divide each training image into 32×32-size blocks with 50 % overlap. For our dataset, each training image is divided into 9 blocks. So, the size of feature vector for each image is $9 \times 768 = 6912$. For the combined features, we use smaller feature vector of HSC than that used for only HSC implementation for bicycle training. We use blocks of 32×32 pixels with 50 % overlap in three scales with eight orientations.

The flowchart of the design and testing is shown in Fig. 3. HOG, HSC and MLBP features are extracted, normalized, combined and are fed into linear SVM. To test the classifier, the test images are scanned by a 64×64 sliding windows, and features are calculated.

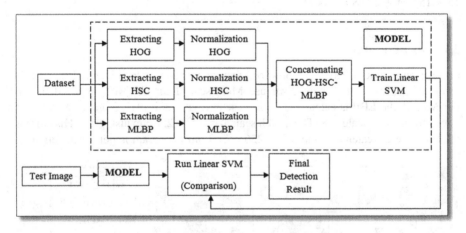

Fig. 3.   The framework of the bicycle detector

2.5   Classification

For classification, we use a linear Support Vector Machine (SVM) [13].

3 Experiment

3.1 Dataset

The data collection has three subsets: «Horizontal» includes images that are mostly side views of bicycles and bicycles with human legs (pedaling). The dataset «Vertical» has mostly includes image with frontal or rear view of bicycles or bicycles with human legs and finally the third subset «Wheels» includes only single bicycle wheels. The reason for the latter choice is to make the detector capable to handle partial views that are more common in the real world that is on the road with cars and pedestrians present. Totally, the collection contains 1250, 668 and 800 positive training samples and 1250, 668 and 800 negative training fragments respectively for horizontal, vertical and wheel sets. The size of all training patches is 64 × 64. For testing, we have 312, 168 and 200 images respectively for testing the "horizontal", "vertical" and "wheel" sets. We have conducted separate tests on each part of the collection to find how to proceed with building the system and fine-tuning it for implementation. As shown in Figs. 4, 5 and 6, we included all types of bicycle models, positions, under various illuminations, and possible backgrounds into positive training set. Figure 7 shows that the negative training set contains street, tree, building, sky, human, cars, motorcycles etc.

Fig. 4. «Vertical» Dataset

Fig. 5. «Horizontal» Dataset

Fig. 6. «Wheel» Dataset

Fig. 7. «Negative» Dataset

3.2   Results

The percentages of true and false positive detections for features and their combinations on the test datasets are listed in Table 1. We have evaluated the performance using the area under the ROC (AUC) as presented in Table 2. The highlighted values in Table 1 indicate highest true positive and lowest false positive counts and highest AUC in Table 2, respectively. The combination of HSC and MLBP features secures highest detection rates in true cases, whereas combining all three features allows for reducing false positives. Overall, combinations of HOG with any of texture features, i.e., HSC or MLBP, or inclusion of both of them increases the performance of the detector. Depending on how the complete system is projected either HOG-HSC – MLBP or HOG-MLBP can be used. Apparently, detecting the complete bicycle is advantageous for accurate counting, otherwise, provisions are to be made to count the bicycles based

Table 1. Percentage of true and false positive detections per feature

Dataset	HOG		HSC		MLBP		
	TP	FP	TP	FP	TP	FP	
Horizontal	96.1	0.9	96.4	3.8	97.1	10.5	
Vertical	97.0	10.7	96.4	18.4	97.6	26.7	
Wheel	99.5	**0**	**100**	11.0	**100**	10.5	
Dataset	HOG-HSC		HSC-MLBP		HOG-MLBP		HOG-HSC-MLBP
TP	FP	TP	FP	TP	FP	TP	FP
Horizontal	96.7	2.5	**97.7**	3.8	**97.1**	1.9	97.1	**0.9**
Vertical	97.6	10.1	**98.2**	8.3	**98.2**	10.1	**98.2**	**8.3**
Wheel	99.5	**0.5**	**100**	6.5	100	10.0	82.0	13.5

Table 2. Area under the ROC curve (AUC)

Dataset	HOG	HSC	MLBP	HOG-HSC	HSC-MLBP	HOG-MLBP	HOG-HSC-MLBP
Horizontal	0.9965	0.9953	0.9836	0.9974	0.9962	0.9984	**0.9989**
Vertical	0.9781	0.9669	0.9341	0.9874	0.9798	0.9809	**0.9875**
Wheel	0.9998	0.9994	0.997	0.9999	0.9997	1	0.9031

on detecting a single wheel, and that requires analyzing additional clues for differentiating between two wheels belonging to the same bicycle or two wheels of two different bicycles.

4 Conclusions

We have proposed combined bicycle descriptors, i.e., we have augmented the shape-based feature descriptor (HOG) with two texture-based feature descriptors (HSC and MLBP). Using the AUC, it was shown than the HOG-HSC and HOG-MLBP are more robust detectors than the sole HOG for all orientations of bicycles. The HOG-HSC-MLBP achieves a highest accuracy for horizontal and vertical views of bicycles, but has a diminished performance for a single wheel detection.

References

1. Shirazi, M.S., Morris, B.: Observing behaviors at intersections: a review of recent studies & developments. In: IEEE Intelligent Vehicle Symposium(IV 2015), pp. 1258–1263 (2015)
2. Dalal, N., Triggs, B.: Histograms of orientation gradients for human detection. In: Proceedings of Computer Vision and Pattern Recognition (CVPR 2005)
3. Cho, H., Rybski, P.E., Zhang, W.: Vision-based bicycle detection and tracking using a deformable part model and an EKF algorithm. In: 13th International IEEE Conference on Intelligent Transportation Systems (ITSC 2010), pp. 1875–1880 (2010)
4. Takahashi, K., Kuriya, Y., Morie, T.: Bicycle detection using pedaling movement by spatiotemporal gabor filtering. In: IEEE Region 10 Conference TENCON, pp. 918–922 (2010)

5. Jung, H., Tan, J.K., Ishikawa, S., Morie, T.: Applying HOG feature to the detection and tracking of a human on a bicycle. In: 11th International Conference on Control, Automations and Systems (ICCAS 2011), pp. 1740–1743 (2011)

6. Jung, H., Ehara, Y., Tan, J.K., Kim, H., Ishikawa, S.: Applying MSC-HOG feature to the detection of a human on a bicycle. In: 12th International Conference on Control, Automation and Systems (ICCAS 2012), pp. 514–517 (2012)

7. Schwartz, W.R., da Silva, R.D., Davis, L.S., Pedrini, H.: A novel feature descriptor based on the shearlet transform. In: 18th IEEE International Conference on Image Processing (ICIP 2011), pp. 1033–1036 (2011)

8. Cao, Y., Pranata, S., Yasugi, M., Zhiheng, N., Nishimura, H.: Staggered Multi-scale LBP for Pedestrian Detection. In: 19th IEEE International Conference on Image Processing (ICIP 2012), pp. 449–452 (2013)

9. Ojala, T., Pietikainen, M., Harwood, D.: A comparative study of texture measures with classification based on feature distributions. Pattern Recogn. **29**(1), 51–59 (1996)

10. Yi, S., Labate, D., Easley, G.R., Krim, H.: A shearlet approach to edge analysis and detection. IEEE Trans. Image Process. **18**(5), 929–941 (2009)

11. Easley, G.R., Labate, D., Colonna, F.: Shearlet-based total variation diffusion for denoising. IEEE Trans. Image Process. **18**(2), 260–268 (2008)

12. Ahonen, T., Hadid, A., Pietikainen, M.: Face description with local binary patterns: application to face recognition. IEEE Trans. Pattern Anal. Mach. Intell. **28**(12), 2037–2041 (2006)

13. Cortes, C., Vapnik, V.: Support-vector networks. J. Mach. Learn. **20**(3), 273–297 (1995)

On Calibration and Alignment of Point Clouds in a Network of RGB-D Sensors for Tracking

George Xu$^{(\boxtimes)}$ and Shahram Payandeh

Experimental Robotics and Imaging Laboratory, Simon Fraser University,
Burnaby, BC V7E 4N4, Canada
{georgex,payandeh}@sfu.ca

Abstract. This paper investigates the integration of multiple time-of-flight (ToF) depth sensors for the purposes of general 3D tracking and specifically of the hands. The advantage of using a network with multiple sensors is in the increased viewing coverage as well as being able to capture a more complete 3D point cloud representation of the object. Given an ideal point cloud representation, tracking can be accomplished without having to first reconstruct a mesh representation of the object. In utilizing a network of depth sensors, calibration between the sensors and the subsequent data alignment of the point clouds poses key challenges. While there has been research on the merging and alignment of scenes with larger objects such as the human body, there is little research available focusing on a smaller and more complicated object such as the human hand. This paper presents a study on ways to merge and align the point clouds from a network of sensors for object and feature tracking from the combined point clouds.

1   Introduction

With a single stationary depth camera, the coverage of the scene is limited by the field of view and positioning of the camera. Objects captured from a single depth camera are also susceptible to occlusion. For a complex and self-occluding object such as the human hand, this is particularly problematic as the occluded data is crucial in creating a 3D reconstruction of the hand. In tracking applications, self-occlusion is also a major problem as it causes the hand or body to appear disjoint. This can be addressed by using methods such as the joint-evidence approach [1], regressions forests [2], and Dijkstra's [3] to join the disjoint areas of the hand or body by estimating and connecting the skeletal model. Such approaches attempt to fit a skeletal model over the point clouds by relying on training sets. In some cases, with an estimate of the skeletal model, the occluded data may no longer be required to track the object. However, these setups are computationally demanding and due to the reliance on training sets, it makes these algorithms unsuitable for dynamic scenes or where the reconstruction of the grasping and manipulation of objects by hand is required.

A single depth camera system has been used to capture a wider view of a static scene, as long as it can be moved to scan the scene so that each of the separate depth images can be stitched together to create a panoramic depth map of the scene. For RGB cameras, the equivalent is panorama stitching, which relies on stitching and aligning adjacent

G. Bebis et al. (Eds.): ISVC 2015, Part II, LNCS 9475, pp. 563–573, 2015.
DOI: 10.1007/978-3-319-27863-6_52

images using feature detectors and matchers such as SURF [4] or FAST [5]. Given a static scene, some examples for 3D point clouds can be the iterative closest point algorithm proposed by the KinectFusion [6] or multi-view stereo techniques of [7] to merge all the depth maps into a single one. However, such solutions impose two limitations, first, there must exist the ability to move the camera. Second, such systems are not ideal for real-time reconstruction. As the reconstructed scene is a merged composite of past and live data obtained through the movements of the camera as done by [7] or [8], where the camera is moved by hand or [9], where the camera is moved about the scene on a mobile platform. An alternative application for dynamic scenes is to have multiple depth cameras networked and integrated as one system. Having multiple cameras increases the total viewing volume while addressing the problem of occlusion without the dependency on a moving camera which can introduce additional errors to the measurements.

A point cloud of an object sampled from different viewpoints can make existing research in feature detection, tracking and gesture recognition more effective. However, there exist many challenges in combining the point clouds of each sensor together to form a composite of the scene as well as in sensor integration and placement. This paper will present an investigation into how the aforementioned problems can be addressed, namely the calibration and merging of point clouds.

The camera network used in this paper is introduced in Sect. 2, as is the approach of its integration. For Sect. 3, the technique for calibrating the depth cameras is discussed, followed by the merging of each camera's point cloud in Sect. 4. In Sect. 5, some preliminary results from hand tracking are presented followed by the discussions highlighting the key findings of this study and some future research goals.

2   Sensor Network

In our proposed setup, two SoftKinetic DepthSense 325 sensors [10] are arranged in order to capture a common object between them from separate angles (Fig. 1).

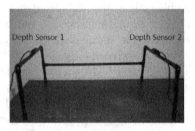

Fig. 1. A network of two SoftKinetic DepthSense sensors are setup to view the area in front of them from separate angles.

For a time-of-flight sensor, modulated beams of light are sent from a laser emitter on the sensor and then based on the round trip travel time of that beam of light from the sensor to the scene and back, distances can be resolved into a depth map. Time-of-flight sensors however suffer from small depth map resolutions and are also susceptible to noise from light reflectivity or interference from other sensors. Figure 2 shows an example of such interference which occurs when multiple camera signals overlap on a single object.

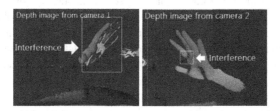

Fig. 2. Interference from multiple sensors viewing the hand. The area that appears in the view of both cameras shows interference.

Due to the fundamental differences between infrared sensors and time-of-flight sensors, techniques for addressing interference for one camera type are often incompatible with the other. Time-multiplexing is a more general technique which schedules each sensor in the network to be activated separately from one another so that only one camera is on at one time [11]. However, as only one camera can be on at a given time, for a two camera network, only 50 % of the information is usable. One technique particular to depth sensors is to send the signal from each camera in the network at a different frequency so that the signals don't overlap [12, 13]. However, as frequency correlates to sensor range and only one sensor can run on one frequency without causing interference in the system, it constrains the total number of sensors that a given network can support.

The DepthSense 325 supports laser frequencies of 30 MHz and 50 MHz. The default frequency of 50 MHz has an effective range of 0.15 m to 1.5 m while the 30 MHz has an effective range of 0.45 m to 2.0 m. The proposed two camera system running 50 MHz and 30 MHz in the two cameras results in an effective range of 0.45 to 1.5 m. Thus, while modulating the signal frequencies removes interference from the sensor network, closer attention needs be paid in defining the effective coverage of overlapping range of the sensors. Figure 3 shows the refined depth images which are free of interference captured by our system in comparison to the depth images in Fig. 2.

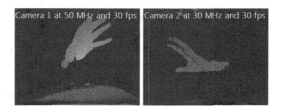

Fig. 3. Minimal interference is exhibited on both frame rates by using different laser frequencies for each sensor in the network. Sensor 1 sends a laser signal at 50 MHz at a frame rate of 30. Sensor 2's laser signal is set to 30 MHz with the same frame rate as sensor 1.

3   Multi-sensor Calibration

To merge and align the point clouds of each sensor in the network, they have to be transformed to the same reference frame in space. Thus, the primary objective is to find the transformation matrices of each point cloud to the origin of the chosen frame. RGB camera calibration is commonly performed to adjust for distortion in the images captured through the pinhole camera model [14]. As depth images are essentially grayscale images and have no special properties, RGB calibration can be lent to depth images as well. Camera calibration can also be used to determine the transformation matrix from the camera to the frame of the checkerboard when it is moved around the field of view of the camera.

The transformations found during calibration only transforms the RGB image to that of the checkerboard. However, the RGB and depth images are not aligned with each other as shown in Fig. 4. This is due to an offset that exists between the RGB and depth sensors that creates a visible difference in the view of both images. Given such differences it is required to find the transformation of the depth frame to that of the RGB frame, $^{Depth}_{RGB}T$. In order to transform the depth data to the checkerboard frame, we use the transform, $^{Depth}_{World}T$, which can be found using the equation $^{Depth}_{World}T = {}^{Depth}_{RGB}T{}^{RGB}_{World}T$. The transform, $^{RGB}_{World}T$ can be found through calibration.

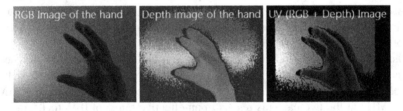

Fig. 4. For the same frame, the RGB image shows the hand in the lower right corner of the frame and the depth image shows the hand in the center of the frame. The RGB image in the left is mapped to the depth image in the middle using a UV map to create a combined RGB and depth model shown on the right (Color figure online).

To calibrate the sensors, the world coordinate system is set to the corner of the checkerboard so that all the points of the checkerboard lies on one plane using a homography transform. The homography matrix can be split into two separate matrices of intrinsic and extrinsic parameters. The intrinsic parameter matrix can be solved by supplying the layout of the checkerboard and the size of each square as input parameters to the calibration algorithm. The unknowns of the extrinsic matrix are solved by positioning a black and white checkerboard in various views in front of the camera and solving a system of linear equations for each unknown.

The intrinsic parameters include c_x, c_y which are the coordinates of the optical center (typically the center pixel of the RGB image), which is the point where all light rays intersect inside the lens of the RGB sensor as well as f_x and f_y which are the focal lengths, defined as the distance from the camera lens to the camera sensor. The extrinsic parameters include the rotation matrix and translation vector which represents the transformation matrix from the RGB image to the frame of the checkerboard. Thus, by positioning the checkerboard in the view of multiple cameras, (Fig. 11), we obtain the transformations of every camera to the frame of the checkerboard which can later be used to align the point clouds together.

The DepthSense 325 provides a UV map (where U and V are the coordinates of a pixel in the image plane), which is a pixel-to-point map that maps RGB pixels to its corresponding point in the depth image as seen in Fig. 4. By converting the UV map to a transformation matrix, we are able to find the transformation of the depth frame to the RGB frame, $^{Depth}_{RGB}T$. Combining the Depth-to-RGB transformation and the RGB-to-World transformation, $^{RGB}_{World}T$, found earlier through calibration, we have the transformation of the point cloud to the world frame, $^{RGB}_{World}T$.

The transformation that maps the depth images to the RGB image frame can be represented by a 3×3 affine transformation matrix. As the RGB image does not have a 3rd dimension, the Z value for each depth point is set to 1 and later modified with the proper depth values in actual depth alignment transformation. This transformation matrix can be solved for the three unknown values as defined in the transformation matrix (namely, θ, x, y). Given a system with three unknowns, three equations from the system can be used in order to solve for them. Given one pair of points where one is obtained from the RGB image and one obtained from the depth image (found via the UV map), we can obtain two equations from the separate X and Y values. Hence, two pairs of points can be used to create three equations to solve for the three unknowns in the transformation matrix.

$$\begin{bmatrix} X_{RGB} \\ Y_{RGB} \\ 1 \end{bmatrix} = \begin{bmatrix} \cos\theta & -\sin\theta & x_{trans} \\ \sin\theta & \cos\theta & y_{trans} \\ 0 & 0 & 1 \end{bmatrix} \begin{bmatrix} X_{Depth} \\ Y_{Depth} \\ 1 \end{bmatrix} \tag{1}$$

$$\begin{aligned} X_{RGB1} &= \cos\theta \cdot X_{Depth1} - \sin\theta \cdot Y_{Depth1} + x_{trans} \\ Y_{RGB1} &= \sin\theta \cdot X_{Depth1} + \cos\theta \cdot Y_{Depth1} + y_{trans} \\ X_{RGB2} &= \cos\theta \cdot X_{Depth2} - \sin\theta \cdot Y_{Depth2} + x_{trans} \\ Y_{RGB2} &= \sin\theta \cdot X_{Depth2} + \cos\theta \cdot Y_{Depth2} + y_{trans} \end{aligned} \tag{2}$$

A different transformation matrix (Eq. 1) can be found for each pair of points as the UV map is not uniform. This matrix can be broken into a system of equations in (Eq. 2). By solving for the system of equations in (Eq. 2), we can solve for the unknowns in the matrix. However, to transform the point cloud to the frame of the RGB image, only one transformation matrix can be used. The RANSAC algorithm [15], can be used to find that transformation matrix. The RANSAC algorithm is an iterative method that finds the most appropriate transformation matrix $^{Depth}_{RGB}T$ from the observed data set (found using the UV map) with the most inliers, which are the pairs of RGB and depth points that fits the transformation matrix.

4 Point Cloud Alignment and Merging

Before aligning the point clouds, some pre-processing can be done prior to alignment in order to improve the accuracy and performance by segmenting and filtering the depth images. The point cloud at each frame of each camera can be segmented along its depth values (Z-axis of the depth image) in order to separate the foreground from the background by removing depth points that are above or below an upper or lower threshold respectively. Additional segmentation along the X-axis and Y-axis can be done to separate specific objects from the rest of the foreground.

4.1 Noise Filtering

Having resolved multi-sensor interference, noise can still appear in the depth images which can be the result of sensor noise or any other environmental noise. The Depth-Sense 325 has an inherent device noise of less than 1.4 cm at distances of up to 1 meter with 50 % reflectivity [16]. This noise can be removed by using a statistical outlier filter. For every point in the point cloud, a statistical outlier filter finds the mean distance from it to all its neighbors; the amount of neighbors can be defined beforehand. Neighbors with a distance greater than a defined standard deviation of the mean distance will be removed from the point cloud as shown in Fig. 5 below.

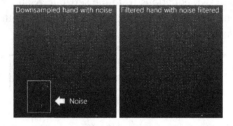

Fig. 5. Applying the statistical outlier filter removes the noise in the point cloud by iterating through points and removing neighbors with a standard deviation greater than the mean distance of neighbouring points.

Other examples of noise filtering methods can include smoothing and Gaussian filters, where each point is replaced by the mean of its neighbors. However, these filters are applied globally to all depth points so it may remove valid points from the point cloud erroneously. Furthermore, they are not ideal for removing the noise which is often minor and far away from neighboring points.

Other implementations such as the KinectFusion [6, 13] can remove noise by averaging the depth points in the current frame with the corresponding depth points in previous frames so that the noise is smoothed out by the corresponding values from other frames. However, such implementations are only used for static reconstruction and are not suited for real-time applications, due to the heavy influence of previous frames from the past on the current frame in time.

4.2   Point Cloud Alignment

The point clouds are transformed and concatenated together to create a single point cloud. Without any additional adjustments, the merging of the point clouds remains imperfect due to offsets in the final point cloud as seen in Fig. 6. For an object such as the hand where there are many contours in a small surface area, the offsets in the point clouds prevent any feasible means of performing tracking or feature extraction from the point cloud.

Fig. 6. The left image shows the point cloud from camera 1, the middle image shows the point cloud from camera 2, and the right image shows the combined point cloud with errors due to offsets between the point clouds as a result of poor alignment.

Errors in the point cloud can be a result of device noise, if the object is not the same size in the view of both cameras or if homography does not exist between the two camera views meaning that the depth images for each point cloud don't share the same 2D plane [17]. These factors will result in a mismatched point cloud when each point cloud is merged and aligned. Furthermore, ensuring that the objects are the same size in the cameras and minding view angles limits the flexibility of the system. By only applying a rigid transformation, it is difficult to align the point clouds due to the objects being viewed from varying angles and distances. In [18], these factors are referred to biases and are proposed to be compensated for by applying a rigid body transformation, which aligns the point clouds, a directional compensation, which corrects misalignments between the viewing rays and angles for each camera in the system, and also correction of ray distance, to account for varying object distances from the sensors. By compensating for system bias, better aligned point clouds can be created as seen in Fig. 7 compared to Fig. 6.

Fig. 7. The left image shows the point cloud from camera 1, the middle image shows the point cloud from camera 2, and the right image shows the merged point cloud which appears more aligned in comparison to Fig. 6 as the fingers in the merged point cloud are clearly separated and defined.

5   Hand Tracking

To explore the viability of tracking in a multi-sensor system, we applied 2D feature detection and tracking techniques to the depth images as 3D feature detection and tracking is still an area of research and thus, outside the scope of this paper. Only a single camera will be used in this section as despite the alignment of the point clouds from each camera, the imperfections in the point cloud make the point cloud still not reliable enough for accurate feature detection.

2D frames can be extracted from the 3D depth image by ignoring the Z-axis on each 3D depth image. Once a feature has been detected the 2D coordinate of the feature can be retranslated to its approximate 3D coordinate by adding back the Z-axis and finding the feature's Z-coordinate by looking up the depth image using the X and Y coordinate of the feature found using 2D detection. From the 3D coordinate, a 3D shape can then be reconstructed in the centroid of the point cloud representing the feature to represents its location in the model of the point cloud. There are obvious limitations to this methodology, but this was done as an evaluation of the system for future research.

In Fig. 8, the hand is positioned in such a way so that the palm is facing the sensor of only one of the cameras. This depth image is treated as a 2D image and the contours of the hand are located. A convex hull of the hand can then be enclosed by fitting a convex polygon to the extremities of the contour [19]. The convexity defects, which are the start and end points of the edges in the convex hull are saved and tracked. The convexity defects, symbolized by the red points in Fig. 8, can then be filtered to display only the five fingertips. Following this, the locations of the fingertips can now be visualized on the image by drawing green circles at the locations of the defects.

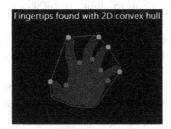

Fig. 8. The palm of the hand captured from one camera is placed in front of one of the depth sensor so that a convex hull can be enclosed around the hand. The fingertips can then be found by locating (Color figure online)

The locations of the fingertips denoted by the green points are found on the convex hull of the hand shown by the red lines in Fig. 8. By connecting the points of the fingertips to the convexity defects as shown by the red points, we are able to segment out the hand from the depth image [19]. These areas can then be segmented out and supplied as inputs to the tracking algorithm to be tracked. Once tracked, a green sphere can be drawn at the centroid of the point cloud of the tracked fingertips to show their locations as the hand moves as shown in Fig. 9.

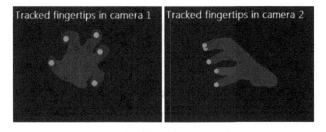

Fig. 9. The tracked fingertips are drawn on the hand in the view of the two cameras as the hand moves in real-time. The convexity defects no longer exists at the fingertips, but the tracking algorithm from the Point Cloud Library using the initial frame found using the convexity defects ensures that the fingertips are tracked as the hand moves (Color figure online).

We utilized the Point Cloud Library [20] to track sections of the point cloud by estimating potential object poses through calculating their likelihoods based on the combined weighting of point coordinates, color data and surface normals. The most likely position of the tracked object in the current frame is determined from its position, color and normal from the previous frame. The library is optimized for real-time applications using Monte Carlo sampling techniques, multi-core optimization and adaptive particle filtering.

6 Discussions and Future Work

Using a network of depth sensors, we can expand the viewing coverage of the object inside the view of the system. With this, objects susceptible to self-occlusion such as the hand can be better represented by the merged point cloud. A comprehensive point cloud provides more data which will allow for higher accuracy feature detection and tracking. One of the overall goals of this research is to create a system which can be used to model the hand grasping and manipulating objects in real-time.

To merge the point clouds, the transformation matrices are first obtained using checkerboard calibration. Calibrating a checkerboard while it's in the view of the two cameras allows the determination of the transformation between the cameras and the checkerboard. To merge the point clouds accurately, the images in the sensor should have a compensation for directional alignment as well as correction for ray distance, otherwise, the point clouds may not align unless in specific camera and object placements.

Feature detection is done on the point cloud without first constructing a surface mesh due to the clarity of the cloud point itself making the mesh reconstruction unneeded. Signification computational resources can be saved by not constructing the mesh which will improve the overall speed of the system. As discussed, there are numerous algorithms that support feature detection and tracking for point clouds directly. Features detected can then be drawn and visualized on the point cloud to obtain a geometrical entity that is more representative of the object. This can then be used to develop the interaction model between the fingers and objects by highlighting the fingertips and the edges of the objects being grasped.

References

1. Oikonomidis, I., Kyriazis, N., Argyros, A.A.: Efficient model-based 3d tracking of hand articulations using kinect. In: Proceedings of the 2011 British Machine Vision Conference, pp. 101.1–101.11 (2011)
2. Kıraç, F., Kara, Y.E., Akarun, L.: Hierarchically constrained 3D hand pose estimation using regression forests from single frame depth data. Pattern Recogn. Lett. Spec. Issue Depth Image Anal. (2013)
3. Baak, A., Müller, M., Bharaj, G., Seidel, H.-P., Theobalt, C.: A data-driven approach for real-time full body pose reconstruction from a depth camera. In: Proceedings of the 2011 IEEE International Conference on Computer Vision, pp. 1092–1099 (2011)
4. Bay, H., Ess, A., Tuytelaars, T., Van Gool, L.: SURF: speeded up robust features. Comput. Vis. Image Underst. (CVIU) **110**(3), 346–359 (2008)
5. Rosten, E., Drummond, T.: Machine learning for high-speed corner detection. In: Proceedings of the 2005 IEEE International Conference on Computer Vision, pp. 1508–1515 (2010)
6. Newcombe, R., Izadi, S., Hilliges, O., Molyneaux, D., Kim, D., Davison, A., Kohli, P., Shotton, J., Hodges, S., Fitzgibbon, A.: KinectFusion: real-time dense surface mapping and tracking. In: Proceedings of the 2011 International Symposium on Mixed and Augmented Reality, pp. 127–136 (2011)
7. Newcombe, R., Davison, A.J.: Live dense reconstruction with a single moving camera. In: Proceedings of the 2010 IEEE Conference on Computer Vision and Pattern Recognition, pp. 1498–1505 (2010)
8. Tanskanen, P., Kolev, K., Meier, L., Camposeco, F., Saurer, O., Pollefeys, M.: Live metric 3D reconstruction on mobile phones. In: Proceedings of 2013 IEE International Conference on Computer Vision, pp. 65–72 (2013)
9. Marton, Z., Rusu, R., Beetz, M.: On fast surface reconstruction methods for large and noisy point clouds. In: Proceedings of 2009 IEEE International Conference on Robotics and Automation, pp. 3218–3233 (2009)
10. SoftKinectic DepthSense Cameras. http://www.softkinetic.com/en-us/products/depthsense cameras.aspx
11. Berger, K., Ruhl, K., Brümmer, C., Schröder, Y., Scholz, A., Magnor, M.: Markerless motion capture using multiple color-depth sensors. In: Proceedings of the 2011 Vision, Modeling and Visualization, pp. 317–324 (2011)
12. Xu, G., Payandeh, S.: Sensitivity study for object reconstruction using a network of time-of-flight depth sensors. In: Proceedings of the 2015 IEEE International Conference on Robotics and Automation (2015)
13. Kim, Y., Theobalt, C., Diebel, J., Kosecka, J., Miscusik, B., Thrun, S.: Multi-view Image and ToF sensor fusion for dense 3D reconstruction. In: Proceedings of the 2009 IEEE 12th International Conference on Computer Vision Workshops, pp. 1542–1549 (2009)
14. Zhang, Z.: A flexible new technique for camera calibration. IEEE Trans. Pattern Anal. Mach. Intell. **22**(11), 1330–1334 (2000)
15. Fischler, M., Bolles, R.: Random sample consensus: a paradigm for model fitting with applications to image analysis and automated cartography. Commun. ACM **24**(6), 381–395 (1981)
16. SoftKinetic DS325 Datasheet. http://www.softkinetic.com/Portals/0/Documents/PDF/WEB_20130527_SK_DS325_Datasheet_V4.0.pdf
17. Agarwal, A., Jawahar, C.V., Narayanan, P.J.: A survey of planar homography estimation techniques. Technical Reports: International Institute of Information Technology Hyderabad (2005)

18. Kim, Y., Chan, D., Theobalt, C., Thrun, S.: Design and calibration of a multi-view TOF sensor fusion system. In: Proceedings of the 2009 IEEE Computer Vision and Pattern Recognition Workshops, pp. 1–7, June 2009
19. Dhawan, A., Honrao, V.: Implementation of hand detection based techniques for human computer interaction. Int. J. Comput. Appl. **72**(17), June 2013
20. Rusu, R., Cousins, S.: 3D is here: point cloud library (PCL). In: Proceedings of the 2011 IEEE International Conference on Robotics and Automation, May 2011

Semantic Web Technologies for Object Tracking and Video Analytics

Benoit Gaüzère[1], Claudia Greco[2], Pierluigi Ritrovato[2(✉)], Alessia Saggese[2], and Mario Vento[2]

[1] Laboratoire d'Informatique,
du Traitement de l'Information et des Systmes (LITIS), Universit de Rouen,
Avenue de l'universit, 76800 Saint-Étienne-du-Rouvray, France
benoit.gauzere@insa-rouen.fr
[2] Department of Information Engineering,
Electrical Engineering and Applied Mathematics, University of Salerno,
Via Giovanni Paolo II, 132, 84084 Fisciano, SA, Italy
{pritrovato,asaggese,mvento}@unisa.it

Abstract. As demonstrated in several research contexts, some of the best performing state of the art algorithms for object tracking integrate a traditional bottom-up approach with some knowledge of the scene and aims of the algorithm. In this paper, we propose the use of the Semantic Web technology for representing high-level knowledge describing the elements of the scene to be analysed. In particular, we demonstrate how to use the OWL ontology language to describe scene elements and their relationships together with a SPARQL based rule language to infer on the knowledge. The proof of the implemented concept prototype is able to track people even when occlusions between persons and/or objects occur, only using the bounding box dimensions, positions and directions. We also demonstrate how the Semantic Web Technology enables powerful video analytics functions for video surveillance applications.

1 Introduction

The increasing interest in security, contributes to expanding deployments of video surveillance systems, which constituted 50 % of the universal big data production in 2012 and are expect to rise to 65 % in 2015 [1]. Availability of this large amount of data contributes to the continuous development of methods devoted to automatically process the video sequences for tracking objects (humans, cars) and recognize actions or violations (fight, robberies, access to forbidden areas, speed limits) to name but a few. In particular, the object tracking is among the most investigated research topics in video analysis. Given a video sequence, the tracking problem consists in labeling the objects in the scene and in identifying, frame by frame, the trajectories associated to these objects while they move in the scene.

As explained in [2], object tracking is a very complex task due the following problems: *(i)* loss of information caused by projection of the 3D world on a 2D

© Springer International Publishing Switzerland 2015
G. Bebis et al. (Eds.): ISVC 2015, Part II, LNCS 9475, pp. 574–585, 2015.
DOI: 10.1007/978-3-319-27863-6_53

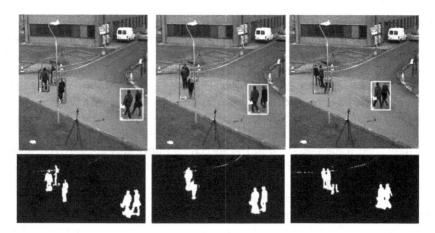

Fig. 1. Frames 22,29,36 (1 frame per second resolution) extracted from view 1 of PETS 2009 dataset. The first row shows the tracking output, while the second one shown the *foreground mask*, where white pixels correspond to pixels moving in the scene.

image; *(ii)* noise in images; *(iii)* complex object motion and shapes; *(iv)* nonrigid or articulated nature of objects; *(v)* partial and full object occlusions; *(vi)* scene illumination changes and shadows; *(vii)* real-time processing requirements.

A system for tracking objects, namely a tracker, essentially executes two main steps: a *detection* step, aiming at recognizing everything is moving in the scene at a given time instant; a *tracking* step, which uses the output of the detection step for building the trajectories of the objects.

Due to the relevance of the problem in the computer vision domain, several contexts and datasets have been developed for evaluating the performance of different tracking methods [3]. The difficulty of objects tracking is clearly visible after looking at the performance results of the algorithms analysed in [3] even in videos lasting few minutes (Fig. 7 in [3]). What generally happens in many object tracking algorithms is that they try to describe the scene content following a bottom-up approach, starting therefore from the simplest available information: namely the pixels. This guarantees a very good computational performance but limits the accuracy level of identifying and describing elements of the scene. As evidenced by the analysis of the performance of different tracking algorithms, good results are often achieved by trackers that are able to combine (fuse) information extracted at different levels (pixel, regions, scene areas, etc.) together with some object attributes like appearance or shape [4].

These kinds of approaches are more demanding from a computational point of view, meaning that we are able to elaborate only a few frames per second. But do we really need a high-frame rate for understanding what happens in a scene? Our mind is able to track objects in situation like those depicted in Fig. 1 with a frame rate of 1 frame per second, realising that there is an occluding object in the middle of the scene (even from the black and white sequence that is the output of classical background substraction algorithm) hiding a part of the person in the scene and that the dots in the pictures are due to the movement of the ribbon.

Nonetheless, pattern matching and machine learning scholars involved in visual and image analysis know very well that as soon as we are able to bring together a high-level contextual knowledge of a domain with low level (pixel-based) feature information of objects in the domain, the understanding process of the meaning of the information available becomes more precise [5]. This is what happened some years ago in the speech and optical character recognition. Indeed, in both cases, by recognizing only basic elements (to reduce the amount of information) like phonemes and syllables and by reducing the space of possible interpretation hypotheses according to a dictionary, a breakthrough of the recognition capabilities in both systems has been achieved.

This is the path we intend to follow, with the belief that semantic web technologies can be used to describe the high-level knowledge of scene and aims of the algorithm (tracking objects) and contribute to the definition of contextual knowledge which can be used by different algorithms composing the tracker to quickly and accurately give the right meaning to the information extracted from the video sequence.

The paper is structured as follows. First, we present in Sect. 2 the rationale of passing from a classical bottom-up approach to a new generic top-down one. Second, in Sect. 2.1, we introduce the ontology framework. The framework is powerful in order to manage the knowledge associated to our top-down approach. Then, in Sect. 3, we present our knowledge-based approach dedicated to track humans traversing a scene. Finally, after presenting in Sect. 3.2 some preliminary results, we draw the conclusions and future work.

2   Knowledge-Based Top-Down Approach

As described in the previous section, many trackers follow a bottom-up approach by first identifying low level features and combining them to obtain a higher level description of objects to track. This bottom-up scheme is newly reproduced to identify tracks from the set of objects identified in the lower level. Using such an approach, parts of each level are associated according to some heuristics defined on the data available in frames such as position, shape, number of object pixels, pixel's colors and so on. Unfortunately, since the detection phase is computed on an imperfect video acquisition, the detection phase of bottom-up approach may suffer from some problems [6] and produce an incorrect set of blobs.

Moreover, existing bottom-up approaches do not include an explicit and global idea of what must be identified in the sequence, such as humans walking in the scene. Also, adding such knowledge into a bottom-up approach may be quite complicated since each single part processed at a low level provides local information. This lack of information may lead to tracking errors. For instance, errors during the detection phase (due to noise or spurious blobs) may be avoided or checked by considering that a human can only appear in the scene in limited areas (or when a group is close to the new identified object) and can not appear suddenly in the middle of the scene. In the same way, splitting problems can be anticipated by knowing that some specific blobs are generated in an area subject to this phenomenon.

As evidenced in several studies [7,8], the visual object recognition is a hierarchical task. In the visual perception, the human brain uses a high-level knowledge of forms, structures, colors and directions to give a meaning to the low level information (signals that arrive to different brain zones involved in the identification and recognition process). Zhaoping explains in [7] that an early vision is subject to two bottlenecks of the visual pathway, namely the transmission capacity of the optic nerve and the limited human attention. Both aspects impose a data compression (at the low level) and a data deletion (at the upper level) in the visual process. This is what exactly several tracking algorithms do, for instance when applying a background substraction algorithm or when they use high-level features like shape or appearance for classifying an object.

We attempt to reduce the number of low level data to process and improve the recognition capability by providing a high-level knowledge of the context. We have started demonstrating the feasibility of our approach by considering only sizes, positions and directions of blobs' bounding boxes identified during the detection phase of a classical background-substraction based algorithm described in [4].

Adding high level knowledge may allow to check whether the information extracted from the scene is consistent with a set of constraints associated to the task of tracking trajectories. For instance, using a global point of view of the scene, we can include as knowledge that the number of humans on the scene is consistent with exits and entries detected in the video sequence. Using simple rules, we can perform reasoning on different hypotheses behind group formation (people joining/leaving a group) and many other events.

2.1   Ontology and Rule Based Systems

In the Knowledge Representation domain, the research is focused on providing high level descriptions of facts, hierarchies of terms and conceptual networks in order to make them available to intelligent applications that could infer implicit consequences from the available information. While creating a representation language, it is important to strike a balance between expressive power and computational efficiency. Description Logic (DL) [9] is a family of formal logics which corresponds to a compromise between requirements of expressiveness and efficiency.

An Ontology is an explicit and formal specification of a conceptualization of a specific domain [10]. The current standard for Ontology definition and instantiation is Web Ontology Language (OWL) [11] which is defined as a family of markup languages based on the Resource Description Framework (RDF). OWL2 [12] is the successor of OWL and has been developed for overcoming some OWL limitations and for improving computability in specific applications through language profiles. An ontology is composed of a series of axioms that assign restrictions to sets of individuals and relationships between individuals.

Axioms can be stored in RDF Triple stores (semantic database systems) and searched using the SPARQL query language (SPARQL Protocol and RDF Query Language). Ontology axioms can be analysed by *DL inference engines* which infer new information from explicitly asserted data using a deductive process called *Reasoning*: one or more logical premises lead to a specific conclusion. Conversely, rule-based reasoning acts on the semantic knowledge by applying one or more predefined rules to add new information.

The problem with ontology based reasoners is that they support only deductive reasoning, *i.e.* simple IF THEN ELSE statements that express certainty in a sequence of events. However, Scene Interpretation Algorithms need abductive reasoning [13], *i.e.* taking a set of facts as input and finding a suitable hypothesis that explains them. Considering this, our methodology follows one of the most recent approaches in rule languages, the W3C member submission SPARQL Inferencing Notation (SPIN)[1]. SPIN is a SPARQL based rule language. SPIN rules correspond to SPARQL queries which can be used to assert new facts, create new individuals or compute the probability of a certain event (abductive reasoning). Moreover, the SPIN Modeling Vocabulary [14] defines a collection of properties and classes that can be used to link OWL classes with SPARQL queries. For example, a set of rules can be linked to a specific OWL class using the property *spin:rule* and during reasoning they are automatically applied to each individual of the class. Having a rule language based on SPARQL, the standard query language of the Semantic Web, allow to create very powerful query for retrieving information from the knowledge base. Availability of these languages allow to solve one of the most relevant problems in the video surveillance applications. Human operators have to analyse several hours of video for the identification of those events that the video analysis algorithms are not programmed to recognise.

2.2   Related Work

Several works have demonstrated use of ontology for managing the so called semantic gap in video and image analysis and video surveillance applications [5, 15–17] Practical implementations of ontology-based systems for people tracking are still scarce. SanMiguel [18] uses domain knowledge (knowledge about a specific location) and semantic reasoning only as a high-level classifier to determine the best procedures, algorithms and system reactions to use during video analysis. The actual detection is performed by the Visual Analysis Framework built with the selected items. Gomez *et al.* [13] follows a rule-based approach using standard (deductive) and non-standard (abductive) ontology-based reasoning with nRQL (new RACER Query Language) and the RACER reasoner. Simple blob-objects associations are performed by the General Tracking Layer, while the Context Layer based on semantic technologies performs further analysis. This is similar to our hybrid approach (described in the following paragraph), but it is worth observing that Gomez *et al.* uses a non-standard query language

---

[1] http://www.w3.org/Submission/spin-overview/.

(nRQL) and related reasoner (RACER), while we seek to reuse semantic web standards to improve interoperability. Moreover, vocabulary and structure of their ontology reflect how their detection algorithm works. We aim to develop an ontology using a vocabulary which can be accepted by the object tracking community to facilitate application development and reuse. None of these works have considered the potentiality offered by the use of standard ontology languages for developing advanced analytics functions that are relevant in video surveillance applications in order to quickly detects specific events or query the system for retrieving events not codified *a priori* by the algorithm, like identifying all persons with a yellow shirt that stationed in a specific area of the scene. Several interesting works demonstrating the application of Semantic Web technology stack like OWL/OWL2 and SPARQL for knowledge representation and reasoning have been produced in the area of human and activity recognition [19,20] our approach is strongly related to these works as explained in the following sections.

3   Approach Implementation and Preliminary Results

3.1   The Proposed Technological Stack

Our knowledge layer uses the OWL2/SPARQL/SPIN Semantic Technology stack: OWL2 is used to create classes and properties, SPARQL as a query language and SPIN to define rules. Using OWL2, we created the **Tracking Ontology**. The Tracking Ontology (Fig. 2) serves four main purposes: *(i)* to convert information coming from the lower levels (detection algorithm) in a semantic format (with OWL Classes like Frame, Blob, BoundingBox); *(ii)* to codify knowledge about the scene; *(iii)* to describe Objects which have been identified in each frame (Noise, TrackedObject, GhostObject); *(iv)* to maintain information about the status of tracked objects (JoinAGroup, LeavingAGroup, EnteringScene, LeavingScene, Stationary). TrackedObjects can be people or groups. We have reused the FOAF Ontology [21] to express people, groups and group membership. The Tracking Ontology also includes a set of properties relating ontology classes (i.e. TrackedObject *hasBoundingBox* BoundingBox, Blob *isContainedIn* Area, TrackedObject *isOccludedBy* OccludingObject, etc.). The scene Knowledge consists of entry and exit areas (EntryArea and ExitArea classes in the model), areas where occlusion may occur (OArea), perspective areas (SpatialPerspectiveArea), background objects (OccludingObjects). OccludingObjects are classified according to their shape and, consequently, to the type of occlusion they cause: ThinVertical objects cause horizontal splitting, Scattered objects (e.g. a tree) cause scattering, and so on. This kind of classification helps the system identify the proper actions to take to confront these problems.

The Tracking Ontology is used as a basis for a SPIN rule system that we named **Tracking Rules**. In Tracking Rules, there are series of functions (right

Fig. 2. Tracking ontology: class tree (first 2 left), object properties tree (middle), SPIN functions (right)

side tree in Fig. 2), which take in input ontology individuals and perform calculus on their properties. As an example, the function *hasHumanSize* in Listing 1.1 checks if a blob has the size of a human according to the PerspectiveArea it is included in.

These functions are applied to ontology instances using rules. It is possible to govern the order in which spin rules are executed by grouping rules in sets and ordering these sets. The system fires all the rules according to the predefined order and each time new information is generated, iterates the rules on the new information [14].

Listing 1.1. Function spin:hasHumanSize

```
ASK WHERE {
    ?spThing a tracking:Blob .
    ?spThing tracking:width ?w .
    ?spThing tracking:height ?h .
    OPTIONAL {
        ?spThing spin:hasPerspectiveArea ?area .
        ?area tracking:maxHumanWidth ?wMax .
        ?area tracking:maxHumanHeight ?hMax .
        ?area tracking:minHumanWidth ?wMin .
        ?area tracking:minHumanHeight ?hMin .
    } .
    OPTIONAL {
        tracking:DefaultPerspectiveArea tracking:maxHumanWidth ?wMax .
        tracking:DefaultPerspectiveArea tracking:maxHumanHeight ?hMax .
        tracking:DefaultPerspectiveArea tracking:minHumanWidth ?wMin .
        tracking:DefaultPerspectiveArea tracking:minHumanHeight ?hMin .
    } .
    FILTER ((((?w <= ?wMax) && ( ?h <= ?hMax)) && (?w >= ?wMin)) && (?h >= ?hMin)) .
}
```

3.2   Preliminary Results

People Tracking. The ontology, rules and technology stack described in the previous section have been used to implement a semantic prototype. This prototype has been tested using view 1 of the S2 (people tracking) PETS 2009 dataset[2], that proposes problems like the presence of background moving objects (the ribbon) generating spurious blobs and occluding objects (the central pole with the sign) that cause blob splitting. Our approach aims at utilizing as little low level information as possible. For this reason, we used as input only 1 frames per second (instead of the seven frame per second of the dataset), bounding box coordinates of the blobs identified by the detection algorithm described in [22] and we calculated size, direction and velocity using our rule system. Frames, blobs, bounding box coordinates and the position of occluding background objects were translated into instances of the Tracking Ontology, using an *ad-hoc* application developed using OWL API. Instances of SpatialPerspectiveArea class represent regions where, according to the distance from the Camera, a specific multiplying factor is applied to compute the real size of blobs. Persons recognition and tracking are achieved considering a series of hypotheses that the semantic rule system assumes during the inference. Some examples of hypotheses are: *(H1)* Person cannot disappear in the middle of a scene. If a person disappears very far from the entry/exit areas, he has entered a group or is hidden by an occluding object of a suitable size; *(H2)* people moves at a common speed so that position, direction and current speed can help predict if two tracked objects will collide in the next frame; *(H3)* occluding objects can cause different types of blob splitting depending on their shape (scattering, vertical splitting, horizontal splitting and so on); *(H4)* background moving object (like ribbon) generates blobs with a dimension not compatible with human or group sizes (according to the perspective area where they are generated). Since we had not fully implemented all the rules needed for managing all possible situations, we restricted our experiment to 3 sets of frames of view 1 (001 – 015; 176 – 274; 400 – 470, for a total of 57 frames) that highlight typical tracking problems. The detection algorithm running on these frames set identified 16, 84 and 65 blobs respectively. As can be seen in Fig. 3(a), our system is able to easily track people in simple cases. Figure 3(b) describes the behaviour of our algorithm when blob splitting occurs: the system make uses of Scene Knowledge (the position of pole and the sign on it) to automatically merge blobs around OccludingObjects. Errors still occur in presence of people leaving groups, because the Tracking Rules does not cover all the cases yet (Fig. 3(c)).

The Tracking Rules currently consists of 21 SPIN rules, which refer to 30 custom SPIN functions to execute calculus on semantic instance properties, for a total of 3476 lines of code. As an example of SPIN rule, in Listing 1.2, let us see a rule that assigns a blob (*?this*) to a person identified in the previous frame: a blob is assigned if it has a human size and is the nearest non-assigned blob to the bounding box of persons identified in the previous frame.

---

[2] Available at http://www.cvg.reading.ac.uk/PETS2009/a.html.

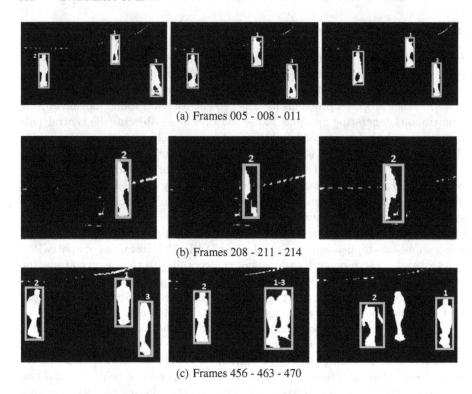

(a) Frames 005 - 008 - 011

(b) Frames 208 - 211 - 214

(c) Frames 456 - 463 - 470

Fig. 3. Behaviour of the algorithms on the PETS 2009 dataset

Listing 1.2. Rule for people tracking

```
INSERT {
    ?previousPerson tracking:blobMatch ?this .
    ?previousPerson tracking:hasBoundingBox ?bb .
}
WHERE {
    ?previousPerson a foaf:Person .
    ?previousPerson tracking:blobMatch ?b2 .
    ?this tracking:hasBoundingBox ?bb .
    ?b2 tracking:belongTo ?frame1 .
    ?this tracking:belongTo ?frame2 .
    ?frame1 tracking:hasNext ?frame2 .
    BIND (tracking:blobDistance(?b2, ?this) AS ?dist) .
    FILTER ((( spin:hasHumanSize(?this)) && (?dist = spin:getMinHumanDistance(?b2))) .
}
```

Semantic Video Analytics. As mentioned in the previous sections, a relevant advantage of our approach with respect to the traditional one lies in the analytics capabilities. In general, commercial video surveillance applications are developed to identify and store specific events, like intrusion detections or access violation, car transit in restricted areas, to mention a few of them. In these "codified" cases human operators are able to quickly analyse the events without looking at several hours of recorded video. Unfortunately, even in these situations if we are looking for slightly different information not fully codified by the algorithms, like people that stationed close to a restricted areas, the operators are forced to watch hours of video unless a new version of the algorithm for detecting the new specific event is released. With our approach we are not only able to create new knowledge reasoning on the existing one, but also to infer new information thanks to the use of SPARQL query language and SPIN language. The Listing 1.3 provides an example of query for identifying stationary persons. Similarly, we can identify frames showing persons that cross a specific area of the scene even at a specific velocity. As in our case velocity and directions of blobs identified by the detection algorithm are calculated by using a simple rule. In case of missing information, it is possible to calculate it or add a new property to the ontology and run a rule for calculating the property on the fly.

Listing 1.3. Example of Query for identifying frames with people stopping in a specific area

```
SELECT DISTINCT ?frame ?person
WHERE {
    ?person a foaf:Person .
    ?person tracking:blobMatch ?blob .
    ?blob tracking:belongTo ?frame .
    ?person tracking:hasAverageSpeed ?speed .
    FILTER (?speed = 0) .
# this first part of the query identify all stationary tracked person.
    ?blob tracking:hasBoundingBox ?bb .
    ?bb tracking:bottomLeftVertex ?bl .
    ?bl tracking:x ?xcoord .
    ?bl tracking:y ?ycoord .
    FILTER ((?xcoord>500) && (?xcoord<800) && (?ycoord>400) && (?ycoord<600)) .
#The second part of the query select just the blobs with the right position.
}
```

4   Conclusions and Future Work

In this paper a possible approach for carrying out person tracking and analytics using high-level knowledge described using standard ontology and rule-based system has been presented. This first proof of concept has been designed as a self consistent system. this means that the inferencing for person tracking has been implemented using rules based on the SPIN language and the TopBraid Composer tool[3]. An experimentation has been carried out on view 1 of the S2 PETS 2009 dataset. Even if the system is in its infancy, and not all the rules for managing the different situations have been fully implemented, the achieved results seem to be promising, with successful tracking of people in situation where they join/leave a group, cross behind an occluding object, traverse the scene close to background moving objects. It is worth noticing that the system works using very few information: bounding box coordinates, directions, velocity and only 1 frame per second. In spite of the promising results in terms of people tracking, the performance from the computational point of view are not, at the moment, comparable with actual solutions. The analysis of each blob detected in a frame requires several seconds (average time is 2,5 s per blob) and the computational time growth is not linear with respect to the number of blobs to

---

[3] http://www.topquadrant.com/tools/modeling-topbraid-composer-standard-edition/.

analyse. This is mainly due to two aspects: (i) control over SPIN rules firing (in some situations we are forced to delete entities from the OWL structure) and (ii) execution of mathematical calculus for evaluating several hypotheses (like a person hiding other persons, who joined another person or group). Our idea is to move toward the development of a hybrid approach where the traditional bottom-up solutions are extended with our top-down knowledge-based system to be only consulted in specific situations where the detection and tracking algorithms identify conditions (presence of high level of uncertainty) that could lead to a mistake or wrong decisions.

References

1. Tiejun, H.: Surveillance video: the biggest big data. Comput. Now 7 (2014)
2. Yilmaz, A., Javed, O., Shah, M.: Object tracking: a survey. Comput. Surv. **38**, 1–45 (2006)
3. Ferryman, J., Ellis, A.L.: Performance evaluation of crowd image analysis using the PETS2009 dataset. Pattern Recogn. Lett. **44**, 3–15 (2014). Pattern recognition and crowd analysis
4. Lascio, R.D., Foggia, P., Percannella, G., Saggese, A., Vento, M.: A real time algorithm for people tracking using contextual reasoning. Comput. Vis. Image Underst. **117**, 892–908 (2013)
5. Li, L.J., Socher, R., Fei-Fei, L.: Towards total scene understanding: classification, annotation and segmentation in an automatic framework, pp. 2036–2043 (2009)
6. Toyama, K., Krumm, J., Brumitt, B., Meyers, B.: Wallflower: principles and practice of background maintenance. In: The Proceedings of the Seventh IEEE International Conference on Computer Vision, vol. 1, pp. 255–261. IEEE (1999)
7. Zhaoping, L.: Theoretical understanding of the early visual processes by data compression and data selection. Netw. Comput. Neural Syst. (Bristol, England) **17**, 301–334 (2006)
8. DiCarlo, J., Zoccolan, D., Rust, N.: How does the brain solve visual object recognition? Neuron **73**, 415–434 (2012)
9. Baader, F.: The Description Logic Handbook: Theory, Implementation, and Applications. Cambridge University Press, Cambridge (2003)
10. Gruber, T.R.: Toward principles for the design of ontologies used for knowledge sharing? Int. J. Hum. Comput. Stud. **43**, 907–928 (1995)
11. McGuinness, D.L., Van Harmele, F., et al.: Owl web ontology language overview. W3C Recommendation **10**, 2004 (2004)
12. Hitzler, P., Krötzsch, M., Parsia, B., Patel-Schneider, P.F., Rudolph, S.: Owl 2 web ontology language primer. W3C Recommendation **27**, 1–123 (2009)
13. Gomez-Romero, J., Patricio, M.A., Garca, J., Molina, J.M.: Ontology-based context representation and reasoning for object tracking and scene interpretation in video. Expert Syst. Appl. **38**, 7494–7510 (2011)
14. Knublauch, H.: Spin-modeling vocabulary. W3C Member Submission 22 (2011)
15. Bloehdorn, S., Petridis, K., Saathoff, C., Simou, N., Tzouvaras, V., Avrithis, Y., Handschuh, S., Kompatsiaris, Y., Staab, S., Strintzis, M.G.: Semantic annotation of images and videos for multimedia analysis. In: Gómez-Pérez, A., Euzenat, J. (eds.) ESWC 2005. LNCS, vol. 3532, pp. 592–607. Springer, Heidelberg (2005)

16. Wang, H., Liu, S., Chia, L.T.: Does ontology help in image retrieval?: a comparison between keyword, text ontology and multi-modality ontology approaches, pp. 109–112 (2006)
17. Snidaro, L., Belluz, M., Foresti, G.: Representing and recognizing complex events in surveillance applications, pp. 493–498 (2007)
18. SanMiguel, J., Martinez, J., Garcia, A.: An ontology for event detection and its application in surveillance video. In: Sixth IEEE International Conference on Advanced Video and Signal Based Surveillance, AVSS 2009, pp. 220–225 (2009)
19. Riboni, D., Bettini, C.: Owl 2 modeling and reasoning with complex human activities. Pervasive Mobile Comput. **7**, 379–395 (2011)
20. Meditskos, G., Dasiopoulou, S., Efstathiou, V., Kompatsiaris, I.: SP-ACT: a hybrid framework for complex activity recognition combining owl and sparql rules, pp. 25–30 (2013)
21. Brickley, D., Miller, L.: Foaf vocabulary specification 0.98. Namespace document 9 (2012)
22. Conte, D., Foggia, P., Percannella, G., Tufano, F., Vento, M.: An experimental evaluation of foreground detection algorithms in real scenes. EURASIP J. Adv. Sig. Process. **2010**, 7 (2010)

Home Oriented Virtual e-Rehabilitation

Yogendra Patil[1]([⊠]), Iara Brandão[2], Guilherme Siqueira[3], and Fei Hu[1]

[1] University of Alabama, Tuscaloosa, USA
yjpatil@crimson.ua.edu
[2] Universidade Federal da Bahia, Salvador, Brazil
[3] Universidade de Saõ Paulo, Saõ Paulo, Brazil

Abstract. We present a collaborative framework to provide a simple solution for home-oriented rehabilitation of post-stroke patients. Our final aim is to build a system that will act like a therapist, giving sound advice to the patient and also improve patient's confidence to perform their daily routine activities independently. In this study, we discuss our rehab system, with strong emphasis on techniques implemented for the integration of rehab robot with the virtual reality games. Experimental observations proves the feasibility of our system.

1 Introduction

A combination of Virtual Reality (VR) and rehab robot has the potential to provide efficient home-oriented rehabilitation service for post-stroke patients [1]. According to U.S. Centers for Disease Control and Prevention, approximately 795,000 people suffer from stroke problems, resulting in a need for rehabilitation in the remainder part of their lives. Almost one-fourth of the total stroke patients have reported recurrent attacks [9–11]. Studies have revealed that, chances of recurrent attack can be reduced by performing regular physical activities [10]. Hence, the stroke rehabilitation research work is aimed at implementing system, that could motivate patients to perform exercises on daily basis and could also assist patients to acquire mobility skills.

2 Related Work

In recent years, the field of stroke rehabilitation using VR is gathering immense attention [1–3,5–8]. The main aim is directed towards helping patients to acquire independent mobility skills by performing tasks in VR environments. Although many practitioners may disagree with the VR rehabilitation approach, various VR-based rehabilitation studies have indicated improvements in patient's motor capabilities [1]. A home-oriented virtual rehab system was developed by Kizony et al. [5,6] for patients with post stroke and spinal stenosis. The system uses a PC and web cam to capture the hand motions of the patients and allows them to interact with the objects in virtual world. Similary, Lozano et al. [7] introduced a similar low-cost, home-oriented rehabilitation system, which captures the patient's body movement using KINECT. Although these systems are

© Springer International Publishing Switzerland 2015
G. Bebis et al. (Eds.): ISVC 2015, Part II, LNCS 9475, pp. 586–596, 2015.
DOI: 10.1007/978-3-319-27863-6_54

low cost, it can only be used either to perform sedentary activities or for static balance training. Kuttuva et al. [2] and Cameirao et al. [4] introduced a VR-based rehabilitation system using robot, for patients suffering from upper extremity weaknesses. More recently, S. Mottura [8] introduced a robot-aided rehabilitation system, for post-stroke patients to perform exercises. These systems are aimed at improving patient's hand gesture skills or to perform static exercises. Therefore, most of the VR world based stroke rehabilitation systems require stroke patients to perform exercises either by sitting or standing. Hence, our goal is directed towards developing a VR-based home-oriented rehabilitation system for stroke patients requiring dynamic balance training with no external support.

3   Design Features and Challenges

We implement our home oriented rehabilitation system for dynamic balance of stroke patient's with basic features such as -

i. **Easy Accessibility**: The system is developed for home oriented rehabilitation purposes. This allows patients to use our system whenever they need it.
ii. **Fun to Use**: Conventional rehabilitation techniques are considered to be repetitive and may tend to reduce patients motivation. One of the key features of our system is, it is fun to use. We develop our virtual environment with interactive games to increase patient's commitment.
iii. **Cognitive Capability**: The proposed system provides basic functionalities such as assisting in harmful situations (patient falling), recording patient activities, etc.

Conventional rehab techniques require scheduling appointments, frequent visits to clinics, plus the therapy style is repetitive in nature - which may not seem motivational to the patient. Our system is intended to address these key issues. Although the features make the proposed system ideal for home-oriented stroke rehabilitation, certain key challenges need to be addressed before implementing the actual system. The key challenges include-

a. **VR-Robot Integration**: The main aim is to create a VR world experience using a robot, i.e., simulate real behaviors in virtual world and *vice-versa*. The main challenge is to set up communication or consistent synchronization, between the virtual environment and the rehab robotic system.
b. **Automated Assistance**: At early stages, patients tend to fall very often, while performing various tasks in virtual world. In absence of a caretaker or therapist, patients might need assistance during their therapy sessions. The challenge in this case is to provide automated assistance, if the patient falls.
c. **User-Friendly Interface**: As the system is home-oriented and no therapist is available to observe the performance of the patient, the challenge is to present patient's performance with interpretable data.

To answer the first key challenge, we design a single `.dll` file to continuously read and interprets the sensor values. These values are then passed on to the VR development engine, to imitate the patient's real world behavior in VR world. To ensure safety of the user, we program our robot in such a way that it first detects harmful situations (patient falling) and then immediately supplies external force (support) to lift up the patient. Finally, a user friendly interface is created to present the robot's sensor signals into meaningful parameters.

4   System Design

The three basic components of our Home Oriented Virtual electronic Rehabilitation (HOVeR) system are: a personal computer (PC)- with NVIDIA GeForce GTX 970 graphics card to handle the 3D virtual environment computation overload, virtual world - displayed by head mounted unit (Oculus®), and a robotic system called as KineAssist® or KAMX® provided by HDT technologies (Fig. 1). Unreal Engine 4 (version 4.72) or UE4 platform is utilized to develop the virtual world. The PC (client) uses 64-bit Windows OS to run the UE4 and also to communicate with the KAMX (server) over LAN simultaneously.

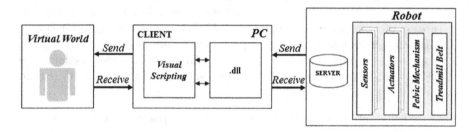

Fig. 1. System Design Framework showing three main modules: (1) *Robot or Server*, (2) *PC or Client*, and the (3) *Virtual World*

4.1   Robot Module

Figure 2 shows the robot module, primarily designed to improve the dynamic balance skills of post stroke patients. The KAMX provides 3 degrees of freedom (DoF), i.e. in X, Y and Z axis direction (Fig. 2), for patient's movements. The movement in X axis (forward and backward) direction is controlled using a treadmill. The movement along Y axis (left and right) is controlled by a pelvic mechanism (PM). The PM also has the ability to move in the up and down direction, along the Z-axis. The PM is equipped with a harness belt (not shown in Fig. 2) and ensures that the patient does not fall off the treadmill.

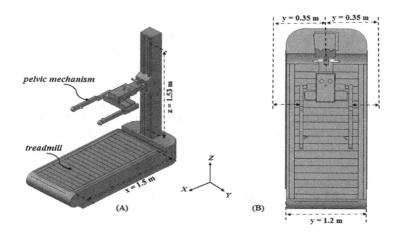

Fig. 2. Rehabilitation robot: (A) *Side View*, (B) *Top View*

Table 1. List of variables used during networking to receive and send data to KAMX.

Name	Role	Location	Usage
YPot	Receive	Pelvic mechanism	Detect left, middle and right
AbsPosWrtFloor	Receive	Pelvic mechanism	Detect fall of subject
XAxisPosition	Receive	Treadmill	Record the distance traveled
XAxisVelocity	Receive	Treadmill	Record the velocity
Xf	Send	Treadmill	Apply force on treadmill
Xv	Send	Treadmill	Stop the treadmill movement
Zf	Send	Pelvic mechanism	Assist when patient falls

4.2   Client Module

The basic networking interface between the client and server (KAMX) was provided by HDT technologies, but was modified by our team for the development of our proposed system. Table 1 details the list of variables used for sending and receiving sensor data to and from KAMX, and also states the location of the sensors from where the values are received or send. Based on the sensor values received from each sensor, we devise a method to determine the user's motion and position. For example, YPot values between -0.1 and $+0.1$, the user's position is consider in the middle portion (MIDDLE) of the treadmill. The user is considered to be on right hand side (RIGHT), if the value is less than -0.1, and on left hand side (LEFT) if greater than $+0.1$. These range of sensor values for defining a particular state of motion was found out empirically (by simulating each state of motion). The 95 % confidence interval was used to set up the interval threshold values. The information from sensor 'YPot' is necessary to control the left and right movements of the virtual world character. The value of AbsPosWrtFloor varies between 0.45 to 1.53, which represents the distance traveled by the pelvic mechanism along Z-axis in meters. This information is used to determine whether the patient is walking normally or fallen down. The variable 'XAxisVelocity' is an important one, as it determines the state of user's motion. For values between 0.3 and −0.005, the state is considered as stationary (STOP). In case of values between 0.3 and 0.5 the user is considered to be limping (LIMP), and for values less than −0.005 it is considered as walking backwards (BACK). Finally, when the values are between 0.5 and 0.1 then the user is considered to be walking (WALK), and if greater than 1.0, then the user is walking fast or running (RUN). This information is necessary to control the forward and backward movements of the virtual world character. The 'Xf' and 'Zf' are force-related variables used to exert force on treadmill (mild brakes on treadmill), and to exert upward force (when the patient looses it's balance) respectively.

Table 2. Look Up Table (LUT) for assigning a value to user's state of motion

	STOP	LIMP	WALK	RUN	BACK
LEFT	0	1	2	3	4
RIGHT	5	6	7	8	9
MIDDLE	10	11	12	13	14

Dynamic Linking Library. In order to integrate the KAMX machine with the VR software (UE4), a single .dll file was created. The .dll file allows to setup connection with the KAMX, read sensor values as well as send commands to the KAMX machine consistently. Due to complexity of the code, only important parts of the code are discussed here. The code consists of two main modules as shown in Algorithms 1 and 2. First module (Algorithm 1) consists of a function implemented to interpret sensor values of KAMX, so as to decide the user's state of motion. Second module (Algorithm 2) is the main CONTROLKAMX.dll file that calls the function from Algorithm 1 to learn the current user's state of motion. It also sends commands to the KAMX, so as to simulate the VR world behavior in real world and also implements safety measures. In case of Algorithm 1, the connection with KAMX is first initiated. After establishing the connection, the data from sensor 'YPot' is interpreted to determine whether the user is on LEFT, RIGHT or MIDDLE of the treadmill. Then the user's state of motion and position(defined in each IF-ELSE) is determined by using values of 'XAxisVelocity' and 'YPot', and eventually an equivalent value is assigned to a variable *status* defined by the Look Up Table (Table 2). For example, if the value received from the variable 'XAxisVelocity' is 0.567 and that for 'YPot' is −0.134, then the algorithm detects that the user is walking on right side of the KAMX. Hence, according to the LUT (Table 2) the value assigned to *status* is 7. This information is very useful to control the VR world character's behavior. It should be noted that the values in Table 2 are arbitrarily assigned values. For Algorithm 2, the main aim is to get information (using the variable

Algorithm 1. To read and interpret data from KAMX sensors

1: Initialize: Socket   ▷ Only Once
2: **function** READSENSORDATA( ) ▷ First Check if subject is on left, middle or right
3:     **if** READ(YPot) == LEFT **then**
4:         $status \leftarrow$ [READ(XAxisVelocity) ∩ READ(YPot)] ∈ VALUE(LUT)
5:     **else if** READ(YPot) == RIGHT **then**
6:         $status \leftarrow$ [READ(XAxisVelocity) ∩ READ(YPot)] ∈ VALUE(LUT)
7:     **else**
8:         $status \leftarrow$ [READ(XAxisVelocity) ∩ READ(YPot)] ∈ VALUE(LUT)
9:     **end if**
10:     **return** $status$
11: **end function**

$GetInfo$) from the VR world (software) such as - whether the user is walking on plain surface or slope. Based on the received information from the VR software, Algorithm 2 then sends control commands to the KAMX to initiate the appropriate actions. For example, if the user encounters an ascent in the VR world, then the algorithm will receive an external command from the VR software to exert a force (mild brakes) on treadmill. The algorithm also returns the user's state of motion at the same time. The Algorithm 2 checks if the user has not fallen down while walking. The 'FALL' condition is determined by continuously monitoring the value of 'AbsPosWrtFloor'. For example, the initial value of 'AbsPosWrt-Floor' is recorded as the patient's normal posture and if this values falls below a user-defined value, then the 'FALL' condition has occurred. During this period, the algorithm immediately sends a command to stop the treadmill motion and picks up the user by exerting an upward force on the PM. The force exerted is set to $value2$ based on the maximum weight (350 lbs. or 150 kg) the PM can carry. The CONTROLKAMX.dll file is imported by the VR software (UE4) so as to control the character or avatar's behavior in VR world.

Algorithm 2. To control and read sensor values from KAMX

1: **function** CONTROLKAMX(GetInfo)
2:     **if** GetInfo == NORMAL **then**
3:         SEND(Xf ← 0)
4:         SEND(Zf ← 0)
5:     **else if** GetInfo == CLIMB **then**
6:         SEND(Xf ← $value1$)   ▷ Exert a force on treadmill
7:     **else if** GetInfo == FALL **then**
8:         SEND(Zf ← $value2$)   ▷ Exert an upward force on pelvic mechanism
9:         SEND(Xv ← 0)   ▷ Stop treadmill movement
10:     **end if**
11:     $motion \leftarrow$ READSENSORDATA( )   ▷ value returned by the function
12:     **return** $motion$
13: **end function**

4.3 Virtual Reality Module

The VR environment interaction was created using visual scripting in Unreal
Engine 4 (UE4). Visual scripting is a programming language that allows user
to create programs using graphical modules (known as blueprint nodes in case
of UE4), so as to create and control different gaming scenarios. The UE4 consists of two main modules blueprint editor and VR level editor. In blueprint
editor, visual scripting is performed by connecting various blueprint nodes. In
VR level editor, character, various scenes and elements are constructed. We
develop the patient's character in the VR world as first person view. The communication between the UE4 platform (or VR world) and KAMX (or Robot)
is set up by importing the `ControlKAMX.dll` file to the UE4 blueprint editor
as 'CONTROLKAMX' blueprint node (Fig. 3(1)). At every instance, the value
returned by 'CONTROLKAMX' blueprint node is one of the value from the LUT
(Table 2), which corresponds to the user's motion and position status. Therefore,
another blueprint node is created (not shown in Fig. 3) to interpret these values and initiate appropriate actions to control the VR character movements. In
addition to this, 'CONTROLKAMX' blueprint node is also useful to change the
force applied to KAMX treadmill belt. Figure 3 shows an example on how to
change the force applied on the treadmill belt of the KAMX when the patient
tries to climb the hill in VR world. In order to do that, an element called as
'Box Trigger' is imported to the VR world editor and placed at the location
where the ascent begins. The 'Box Trigger' module is shown in Fig. 3(2) as a
yellow border cube. After placing the cube, its by default functions or Blueprint
modules are called in the blueprint editor (Fig. 3(1)), namely - 'OnActorBegi-
nOverlap(TriggerBox)' and 'OnActorEndOverlap(TriggerBox)'. These modules
are then connected to the blueprint node 'GetInfo', which is responsible for
changing the variable's value to either 'CLIMB' or 'NORMAL'. Hence, if the
character in VR world enters or overlaps with the Box Trigger zone, the value of

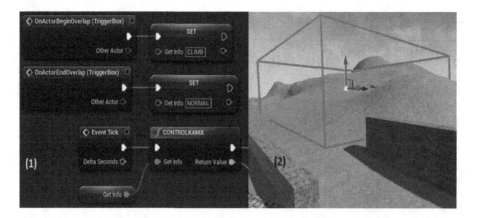

Fig. 3. VR design module showing: (1) *Blueprint layout to send control command to
KAMX*, (2) *Box Trigger module shown as a yellow bordered cube (Color figure online)*

'GetInfo' is set to 'CLIMB' by the 'OnActorBeginOverlap(TriggerBox)' blueprint node. This value is immediately passed on to the KAMX via blueprint module 'CONTROLKAMX', which is the imported `CONTROLKAMX.dll` file. Hence, according to Algorithm 2, *value1* is assigned to treadmill force variable 'Xf', and the user experiences a behavior as if he or she is ascending on a hill. Reverse behavior is observed when the character leaves the 'TriggerBox' zone. More complex behaviors are left for future studies.

5   Demo Experiments and Observations

In order to test our system, we conducted an experiment on 5 subjects (4 males, 1 female). Our main aim was to observe, whether the subjects were able to consistently control their motions in VR world (Fig. 4(1)). The main task of the subject's character in the VR game was to collect maximum possible golden rings. Subjects were asked to control the character with two states of motion walk with and without limp. It was observed that, no subjects reported difficulty in controlling their character in the VR world. Although, they did report some difficulty in taking sharp turns on corners. This is due to the degrees of freedom available on the KAMX left or right, and front or back. So we introduced Box Trigger module (as discussed before) on sharp turns. When overlapped by the character, changes the character's line of sight in the required direction. Various performance parameter were also recorded from raw sensor data and presented via a user friendly interface (Fig. 4(2)).

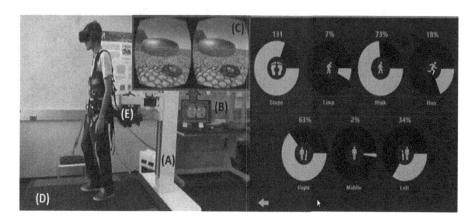

Fig. 4. (1) Integration of VR and Robot: (A) *Robot or Server*, (B) *PC or Client*, (C) *Virtual World*, (D) *Treadmill* and the (E) *Pelvic Mechanism*, (2) User friendly interface displaying user's performance data

In order to test the automated assistance feature, a simulated test was carried out during which the subject was asked first to walk in forward and backward directions, then fall while walking backwards, wait for the PM assistance and then continue to walk in forward direction. For experimental observations, various sensor values from the KAMX were plotted. First plot (titled as 'z axis Position') in Fig. 5 shows the value for variable 'AbsPosWrtFloor', which is the relative distance of PM from the floor in meters. The second plot (titled as 'Distance traveled') is the distance traveled by the subject (in meters), followed by velocity in m/s. During the interval of τ_1 secs, subject starts walking at 4sec, with 'AbsPosWrtFloor' equal to 0.92 m. Figure 5(1) shows one instance during the interval τ_1. Time interval τ_2 shows the period during which the subject falls while walking backwards. The minimum value reached by 'AbsPosWrtFloor' is around 0.7 m at 16.7 sec. Figure 5(2) shows an instance during τ_2, when the subject falls. According to Algorithm 2, the system detects the subject has fallen down and therefore tries to exert an upward force via the PM (Fig. 5(3)). The upward force is exerted until the value of 'AbsPosWrtFloor' reaches around the original position. Interval τ_3 demonstrates the automated assistance period. Finally, during interval τ_4, the subject retains the normal posture and continues to walk for few more seconds. This experiment demonstrates that our system has cognitive capability (i.e. fall detection plus safety lift-up feature) and hence safe to use.

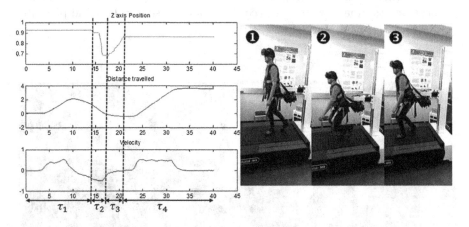

Fig. 5. Plots demonstrating the system's automated assistance feature: (1) *Subject walking in normal posture*, (2) *Subject falls and is assisted by the PM*, (3) *Subject is assisted until the original position is reached*

6   Conclusions and Future Work

In this study we presented our home-oriented, VR-based rehabilitation system, designed for dynamic balance training of post-stroke patients. Although our system is equipped with safety lift mechanism, there is no way to determine the patients health status during the session. Hence, our current work is focused on including low-cost biomedical sensors, to continuously monitor patient's health status. This will also help us to record new parameters to determine and compare patients performance. The initial configuration to our VR game is done manually based on the known health status of the patient. Our future work is also focused on introducing adaptive VR games, similar to Ma et al. [3], that presents game scenarios to the patient based on the health status determined by the biomedical sensors. Although this study introduces the feasibility of home-oriented rehabilitation system, one of the main drawback is the current cost of the rehabilitation robot. We would like to work with HDT Technologies KAMX team to reduce the cost of the robot. In conclusion, this study validated the feasibility of home-oriented rehabilitation system for dynamic balance training of stroke patients. Our research is centered around implementing various safety features for the user. With the introduction of the biomedical sensors to monitor the patients health status, we are confident that our rehab system can be made available to the general public and clinics.

References

1. Holden, M.K.: Virtual environments for motor rehabilitation: review. Cyberpsychology Behav.: Impact Internet Multimedia Virtual Reality Behav. Soc. 8(3), 187–211 (2005). (Discussion 212–219)
2. Kuttuva, M., Boian, R., Merians, A., Burdea, G., Bouzit, M., Lewis, J., Fensterheim, D.: The Rutgers Arm, a rehabilitation system in virtual reality: a pilot study. Cyberpsychology Behav.: Impact Internet Multimedia Virtual Real. Behav. Soc. 9(2), 148–151 (2006)
3. Ma, M., McNeill, M., Charles, D., McDonough, S., Crosbie, J., Oliver, L., McGoldrick, C.: Adaptive virtual reality games for rehabilitation of motor disorders. In: Stephanidis, C. (ed.) UAHCI 2007 (Part II). LNCS, vol. 4555, pp. 681–690. Springer, Heidelberg (2007)
4. Cameirao, M.S., i Badia, S.B., Zimmerli, L., Oller, E.D., Verschure, P.F.M.J.: The rehabilitation gaming system: a virtual reality based system for the evaluation and rehabilitation of motor deficits. In: Virtual Rehabilitation, pp. 29–33 (2007)
5. Kizony, R., Raz, L., Katz, N., Weingarden, H., Weiss, P.L.T.: Video-capture virtual reality system for patients with paraplegic spinal cord injury. J. Rehabil. Res. Dev. 42(5), 595–608 (2005)
6. Kizony, R., Weiss, P.L., Shahar, M., Rand, D.: TheraGame: a home based virtual reality rehabilitation system. In: Disability, Virtual Reality and Associated Technologies (2006)
7. Lozano-Quilis, J.A., Gil-Gómez, H., Gil-Gómez, J.A., Albiol-Pérez, S., Palacios, G., Fardoum, H.M., Mashat, A.S.: Virtual reality system for multiple sclerosis rehabilitation using KINECT. In: 2013 7th International Conference on Pervasive Computing Technologies for Healthcare (PervasiveHealth), pp. 366–369 (2013)
8. Mottura, S., Fontana, L., Arlati, S., Zangiacomi, A., Redaelli, C., Sacco, M.: A virtual reality system for strengthening awareness and participation in rehabilitation for post-stroke patients. J. Multimodal User Interfaces, 1–11 (2015)

9. Stroke Facts — cdc.gov. http://www.cdc.gov/stroke/facts.htm
10. Stroke.org. http://www.stroke.org/we-can-help/survivors/stroke-recovery/first-steps-recovery/preventing-another-stroke
11. AHA/ASA Guidelines on Prevention of Recurrent Stroke - American Family Physician. http://www.aafp.org/afp/2011/0415/p993.html

WHAT2PRINT: Learning Image Evaluation

Bohao She and Clark F. Olson[✉]

University of Washington Bothell, Bothell, USA
cfolson@uw.edu

Abstract. The popularity of digital photography has changed the way images that are taken, processed, and stored. This has created a demand for systems that can evaluate the aesthetic quality of images. Applications that auto-assess image aesthetic quality and modify images to raise their aesthetic quality are widely available, but applications that automatically select aesthetic images from a given image collection are limited. The goal of this project is to create a portable application that can recommend user-given images from a given image collection, using criteria learned from user preferences. We train a Support Vector Machine on seven extracted image features. This system achieves a correct prediction rate of 70 % on a public image dataset. The use of additional or improved features should yield increased prediction rates.

1 Introduction

The popularity of digital photography has changed the way images that are taken, processed, and stored. Images are no longer placed in albums, but on hard drives and cloud services for easy access and sharing with others. Unfortunately, this doesn't mean that the quality of the images has improved. Images uploaded onto photo.net and dpchallenge.com are good examples. The website dpchallenge.com alone has many thousands of images uploaded by professional and amateur photographers for participating in theme-based photography contests. After these images are uploaded, they are then rated by other photographers on aesthetic-quality scoring from 1 to 10, with 1 being the lowest and 10 being the highest. Interestingly, when Datta et al. [1] did their study on algorithmic inference of aesthetics and emotion in natural images, out of the 16,509 images gathered from dpchallenge.com, the mean score was only 5.5. Because of this result, machine-vision researchers like Murray et al. [2], Su et al. [3], Datta and Wang [4], and Lo and Chen [5] have independently tried to create automated image aesthetic-quality assessment systems to give photographers a sense of what score their image would most likely receive, if it were uploaded to a photography challenge website.

At the same time, professional and amateur photographers themselves have tried to improve the aesthetic quality of their images by various post-image processing tools before submitting their works into the contest. Programs that are well-known and popular among photographers, such as Adobe Photoshop, Adobe Darkroom, and Google Picasa, are designed to meet this demand by

© Springer International Publishing Switzerland 2015
G. Bebis et al. (Eds.): ISVC 2015, Part II, LNCS 9475, pp. 597–608, 2015.
DOI: 10.1007/978-3-319-27863-6_55

aiding the user through auto editing and fine-tuning images to a higher aesthetic-quality. Other machine-vision researchers, such as Li et al. [6] and Liu et al. [7], have attempt to tackle this problem through image-cropping to increase the aesthetic quality of a given image. On one hand, the solution provided by Li et al. [6] is only targeted at images with human faces by looking at the color variation, image composition, and facial characteristics of the original image. On the other hand, the solution given by Liu et al. [7] is not only targeted towards human portraits but also landscape images, as long as an obvious target subject is in the image; by analyzing the original image's aesthetical quality based on its composition governed by the principles such as Rule of Thirds, shapes and lines, amputation avoidance, visual balance, and diagonal dominance [8]. Bhattacharya et al. [9] go one step farther, where they reposition the target subject to a more aesthetically pleasing location on a given image without any changes to the original image's background. As result, the authenticity of the original image is not preserved.

Although image processing tools have been examined by many researchers, existing image management programs that can automatically distinguish high-quality images from low-quality ones in a given collection are limited. Yeh et al. [10] and Ryu et al. [11] have independently tried to create such systems with some success. However, both of them are far from perfect. Yeh et al. [10] tried to pick out images from a given image-collection based on the images composition, whereas Ryu et al. [11] organized the image collection based on their relative capture time and location. In our project, we not only leverage the work done by Yeh et al. and Ryu et al., but also improve on them to create an image recommendation system that not only selects high-aesthetic-quality images from a given image-collection, but also recommends images tailored towards the user's own image aesthetic preferences. We will also use this opportunity to evaluate a number of existing image composition guidelines described in photography books by Krages [8] or Barnbaum [12] for what constitutes an aesthetically pleasing image.

This paper is organized as follows: Sect. 2 reviews relevant previous works on image-aesthetic-quality evaluation system using machine-vision techniques as well as existing image selection systems using both machine-vision and machine-learning techniques. Section 3 describes our approach. Section 4 presents the results and evaluation our system. Finally, Sect. 5 gives our conclusions and discusses future directions.

2   Related Work

This section discusses previous work on the assessment of image aesthetic quality and systems for performing this assessment.

2.1   Assessing Aesthetic Quality

Rule of Thirds. The most famous rule is the Rule of Thirds. This rule states that an image should be evenly divided into nine equal parts by two equally spaced horizontal lines and two equally spaced vertical lines, and that important compositional elements placed along these lines or their intersections, called power points, with a short depth of field will create a more aesthetically pleasing image than if those elements were placed in the center or on the side of the image. This technique has been used in systems by Luo et al. [13], Su et al. [3], Liu et al. [7], and Yeh et al. [10].

Color Balance. Another technique that has been used regularly to increase aesthetic quality of an image is the use of color balance. Professional photographers have used various methods to control the color patterns in their image and thus raise specific emotion of the viewer. Warm colors such as red, orange, and yellow may create excitement or even joyful responses in an image. In contrast, cool colors such as violet, blue, and green may create calming and peaceful effects. Furthermore, the use of complementary colors can increase the intensity of an image and the use of combination of colors can trigger particular kinds of emotions [14]. This response to a particular color arrangement may make an image more aesthetically pleasing depending on the viewer's cultural background [15]. This technique has been used by Tang et al. [16], Nishiyama et al. [17], Ke et al. [18], Yeh et al. [10], and Lo et al. [19].

Use of Lines. The use of lines to increase the aesthetic quality of an image is also popular among photographers. Vertical lines tend to impart a feeling of strength and stability to an image [20]. Horizontal lines, on the other hand, tend to impart a restfulness and calm feeling to the viewer [20]. Diagonal lines leading toward a vanishing point give an inherent kinetic energy to an image, as if a vertical line is in the process of rising or falling. The use of these lines give an image a strong sense of dynamism and activity [20]. This technique has been used by Yeh et al. [10], and Su et al. [21].

Visual Balance. A visually balanced image can bring a sense of harmony to the viewer [8]. In some cases, a visually balanced image will have its salient objects distributed evenly around the center of the image [7]. In other cases, an image is partitioned into the golden ratio proportions to bring a sense of visual balance to the image [21]. Bhattacharya et al. [9] and Liu et al. [7] use this idea in their systems.

2.2   Quality Assessment Systems

Image assessment techniques systems are relatively recent developments [7,9,10,16,18,19,21]. For example, Datta et al. [4] have developed a system called ACQUINE that allows the user to upload one image to their web-based application and evaluate it. However, image-management system design with image-quality-assessment support is still a relatively new field.

Our image evaluation system leverages content-based image quality assessment research done by Tang et al. [16], Datta et al. [22], and Yeh et al. [10],

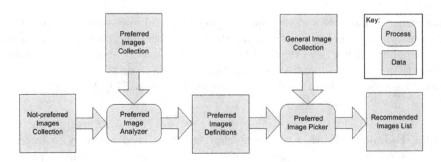

Fig. 1. System dataflow

which provide useful references and serve as a starting point for the feature-extraction portion of our system. In addition, research done by Datta et al. [23] using weighed-least-squares-regression and a naive Bayes' classifier, to distinguish high-quality images from low-quality ones, provides a useful reference for the machine-learning portion of our system.

3 System Design

In this section, we discuss the design of our image evaluation system, including the overall system architecture and the image features used in the system.

3.1 System Architecture

Overall System Architecture. The system is implemented in C++ using OpenCV. It consists of two parts, the Preferred Image Analyzer and the Preferred Image Picker. This system requires three input datasets: the Preferred Image Collection, the Not-Preferred Image Collection, both of which are used for learning, and the General Image Collection, to which the learned preferences are applied.

The system dataflow is shown in Fig. 1. The first process, the Preferred Image Analyzer, takes in the Preferred/Not-Preferred Image Collections and outputs the Preferred Image Definitions file which is saved as a YAML file. After that, the Preferred Image Picker takes this file as its input, along with the General Image Collection and produces the Recommended Images List. The images in this list should share the identifiable aesthetic features extracted from the Preferred Image Definitions file.

Preferred Image Analyzer. The Preferred Image Analyzer comprises two major parts. The first, the Image Features Extractor, is a process that uses various machine-vision techniques to extract potential aesthetic features from the Preferred Image Collection and Not-Preferred Image Collection. After all images in both collections have been analyzed by the Image Features Extractor, a table of aesthetic scores for all the images is generated. We use this table, called the Image Feature Scores List, as input to the second process of the system, named Image Pattern Learner, which uses machine-learning techniques to segregate preferred images from not-preferred ones. After the Image Pattern Learner is done analyzing the image features, it generates and saves a Preferred Image Definitions file for later use in the Preferred Image Picker portion of the overall system.

Preferred Image Picker. The Preferred Image Picker is similar to the Preferred Image Analyzer in the sense that it also consists of two major processes. One is called the Image Features Extractor, like the one in the Preferred Image Analyzer. The only difference in this, however, is that it takes in one image collection, namely the General Image Collection. However, it performs the same function, generating an Image Feature Scores List for the second portion of the Preferred Image Picker subsystem, the Image Picker. Combining the Preferred Images Definitions file with the Image Feature Scores List, the Image Picker then can create the Recommended Images List.

3.2 Image Feature Extraction

Finding the focused subject of an image is an important step to distinguishing high-quality images from low-quality ones. We use the technique of Achanta et al. [24] to generate a binary saliency map. This map is used in calculating scores for the rule of thirds, region of interest, image contrast. and color simplicity.

Rule of Thirds. Using the binary saliency map, we compute a rule-of-thirds image quality score for each image. First, we divide the binary saliency map into 4 sub-images. Each sub-image contains exactly one rule-of-thirds power point. We then compute the subject's center to power point distance as a ratio of distance between the power point in this sub-image and the farthest image corner. A weight factor of the ratio between the subject and the overall sub-image size is then added, in order to avoid false positive cases, in which a small subject is closer to its power point in a sub image than a larger subject is from its power point on a different subimage. The subimage score, f_{rot_sub}, is formulated as:

$$f_{rot\_sub} = w_{o/b} \frac{||o(x,y) - p(x,y)||}{\max(||c(x,y) - p(x,y)||)}, \tag{1}$$

where $w_{o/b}$ is the weight factor ratio of subject over its background, $o(x,y)$ is the subject center, $p(x,y)$ is the power point on this sub image, and $\max(||c(x,y) - p(x,y)||)$ is the maximum distance between an image corner and the power point. With all four f_{rot_sub} scores calculated, the total f_{rot} score is computed as the sum of the four subscores.

Region of Interest. Regions of interest are also extracted from the binary saliency map. The equation to calculate the region-of-interest score for each image, f_{roi}, is:

$$f_{roi} = \sum_{k=1}^{n} \frac{\text{ROI}_k}{WH}, \tag{2}$$

where ROI_k is the area of the k-th region-of-interest in the image, W is the width of the entire image, H is the height of the entire image, and n is the total number of ROI.

Image Contrast. Using the previously extracted binary saliency map, we calculate the image contrast score, $f_{weber\_contrast}$, using the Weber contrast formula [10], which is defined as:

$$f_{weber\_contrast} = \frac{1}{WH} \sum_{x=1}^{W} \sum_{y=1}^{H} \frac{|I(x,y) - I_{avg}|}{I_{avg}}, \tag{3}$$

where $I(x,y)$ is the intensity at location (x,y) on the grayscale image and I_{avg} is the average intensity of the grayscale image.

Color Simplicity. The image background color simplicity can also be an important factor in terms of distinguishing high-quality images from low-quality ones. High-quality images tend to have plain backgrounds and bring the viewer's attention to the focused subject in the image. After locating the background information (again using the saliency technique of Achanta et al. [24]), we created a color histogram of the image background, based on work by Luo et al. [13]. This quantizes each RGB color channel into 16 values and then creates a histogram with 4096 bins to cover all possible color combinations in this set. Once we have created the color histogram, we then find the maximum histogram value h_{max} of all of the color bins. The simplicity feature score, $f_{simplicity}$, is calculated as:

$$f_{simplicity} = \frac{1}{4096} \sum_{i=1}^{4096} H_{filter}(i), \tag{4}$$

where H_{filter} is defined as:

$$H_{filter}(i) = \begin{cases} H(i) & \text{if } H(i) \geq 0.01 \cdot h_{max} \\ 0 & \text{otherwise} \end{cases} \tag{5}$$

and $H(i)$ is total count of i-th bin in the color histogram.

Line Distribution. To calculate a feature based on the line distribution, we first detect edges using the Canny edge detector [25]. We apply the Hough-transform [26] to the resulting edge image to generate a list of lines for evaluation. To avoid false positives due to color variation on an image, we count an edge as a line only if its length was at least a quarter of the image width or height (whichever is smaller). Next, we compute a histogram of the line orientations in an 180 bin histogram. The line distribution score, $f_{line\_distribution}$, is calculated as:

$$f_{line\_distribution} = \sum_{i=1}^{180} H_{line\_angle}(i), \tag{6}$$

where $H_{line\_angle}(i)$ is defined as:

$$H_{line\_angle}(i) = \begin{cases} H(i) & \text{if } i \leq 90 \\ -H(i) & \text{otherwise} \end{cases} \tag{7}$$

where $H(i)$ is the total number of line counts in i-th bin of the line angle histogram. By making the total line count for bins greater than 90, we can have the lines that have angles less than 90 degree canceled out the lines that have angles greater than 90 degree. Hence, the remaining lines are more inclined toward one direction of tilts. The $f_{line\_distribution}$ score will be close to zero when lines are evenly distributed at all angles of tilts in an image.

Golden Ratio. The method we use to compute the image quality score in terms of the golden ratio partition is by first computing where the two hypothetical golden ratio partition dividing lines would be on a given image (dividing the image into two parts with areas related by the golden ratio - 1.618). Then, using the line extraction method discussed above, we count how many lines that are less than 10 degrees of tilt and located within 10% of those two hypothetical golden ratio partition dividing lines are located. The golden ratio partition score, $f_{golden\_ratio\_partition}$, can be formulated as:

$$f_{golden\_ratio\_partition} = \sum_{\theta=1}^{10} H_{line\_location(\theta)}, \tag{8}$$

where $H_{line\_location(\theta)}$ is defined as:

$$H_{line\_location(\theta)} = \begin{cases} H(\theta) & \text{if } |H(\theta)_y - \text{GRL}(y)| \leq (0.1 \cdot \text{GRL}(y)) \\ 0 & \text{otherwise,} \end{cases} \tag{9}$$

where $H(\theta)$ is the count in θ-th bin of the line angle histogram, $H(\theta)_y$ is the y location of that line, and $\text{GRL}(y)$ is the y location of the two hypothetical golden ratio partition dividing lines.

Aspect Ratio. The image aspect ratio can be an important factor that affects the perception of image quality. Some common preferred rations are 4:3, 16:9, and 1:1. The aspect ratio, f_{aspect_ratio}, can be simply calculated as:

$$f_{aspect\_ratio} = W/H. \tag{10}$$

3.3 SVM Classifier

In order to learn the user preferences, we use a Support Vector Machine (SVM) [27]. We separately tested an artificial neural network (ANN) classifier, but found the performance with this classifier significantly lower on our test set.

4 Results

This section discusses the data set and results. All experiments were performed on an Intel i5, 3 GHz quad-core CPU with 12 gigabytes of RAM.

4.1 Data Set

The data set we used our tests is from City University of Hong Kong (CUHK), collected and provided by Ke et al. [18]. It has also been used by Yeh et al. [10], Su et al. [3,21], Luo et al. [13], and Lo et al. [19]. The original 60,000 images, taken in various settings by 40,000 different photographers, came from the website dpchallenge.com and were rated from 1 (the lowest rating) to 10 (the highest rating) by at least a hundred users. The CUHK dataset was created by first sorting these images by their rating then selecting only the top and bottom 10 %, to create the total dataset of 12,000 images. Although the original user rating was not recorded in the dataset and only the general classifications (good or bad) are available, this was adequate for our system, which performs binary classification.

4.2 Classifier Performance

In order test classifier performance, we performed five-fold-cross-validation. We found this more efficient than ten-fold-cross-validation, while yielding similar results. We first collected the image feature data using the Image Features Extractor, along with their ground truth score, with 1 as preferred image and −1 as not-preferred image. We then randomly shuffled the dataset and divided it into five subsets. Of these subsets, a single subset (20 % of the complete set) is used as the validation testing set, and the remaining four subsets (80 % of the complete set) are used as the training set. The cross-validation process is then repeated five times, with each of the five subsets used exactly once as the validation test set. The accuracy results from each run are then averaged to calculate the overall accuracy for the classifier under evaluation. The accuracy rate results for the SVM classifier was 70.4 % on this data.

4.3 Feature Effectiveness

We also performed five-fold-cross-validation on each of the aesthetic features separately, to see if they positively contributed, and to what degree they contributed, to the classification based on their individual Average Accuracy Rate

(AAR). As shown in Table 1, with their AAR and the Standard Deviation (SD) of each test run, certain aesthetic features specifically, the Rule of Thirds (F1), Region of Interest Area on Image (F2), Image Contrast (F4), and Background Color Simplicity (F3) seem to contribute more to the classification than the other features; aesthetic features such as the Image Aspect Ratio (F0), the Golden Ratio (F5), and the Line Distribution (F6) seem to have minimal contributions to the classification.

Table 1. Accuracy rate of individual features

Image feature	AAR	StdDev
Aspect ratio (F0)	0.53	0.020
Rule of thirds (F1)	0.58	0.011
Region of interest (F2)	0.59	0.027
Background simplicity (F3)	0.59	0.040
Image contrast (F4)	0.58	0.056
Golden ratio (F5)	0.46	0.023
Line distribution (F6)	0.47	0.031

When we only used the top four features (F1–F4) for classification, the accuracy rate decreased only to 69.6 %. This indicates that the Image Aspect Ratio, the Golden Ratio, and the Line Distribution make a negligible contribution to the image classification for this data set.

Fig. 2. Example of recommended images canvas

4.4 Example Results

Figure 2 shows an example canvas of recommended images, after the Preferred Image Picker has finished picking the images from the user-given General Image Collection based on the user-given Preferred Image and Not-Preferred Images collections.

The example recommended images in Fig. 2 were picked from 50 images from the General Image Collection shown in Fig. 3(a) based on the Preferred Image Collection shown in Fig. 3(b) and Not-Preferred Images Collection shown in Fig. 3(c).

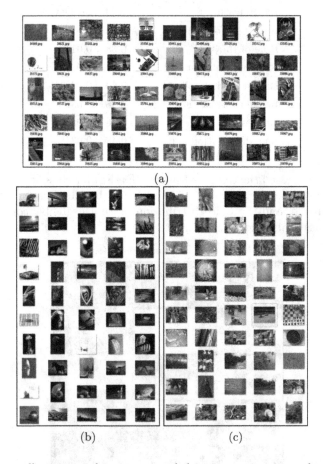

(a)

(b)   (c)

Fig. 3. Image collections used in recommended images canvas example. (a) General image collection. (b) Preferred image collection. (c) Not-Preferred images collection.

5 Conclusions

We have implemented an image-evaluation system that can automatically evaluate and recommend images from a user provided collection of images, based on user preferences. We found that good performance (70 % accuracy) in picking high-quality images can be achieved using four image features (rule of thirds, region of interest, background simplicity, and image contrast) when trained using a support vector machine.

We have also looked into the contribution of the specific aesthetic features of our classifiers and found that image aesthetic features, such as the Rule of Thirds, Image Contrast, Region of Interest Area, and Background Color Simplicity, contributed more to classifying an image as preferred or not-preferred than other image aesthetic features, such as the Image Aspect Ratio, the Golden Ratio, and Line Distribution.

A future direction for study is to improve on our current image features, specifically, the Golden Ratio and Line Distribution, but also experiment with adding more aesthetic image features, such as vanishing point position detection and color variation of the entire image (instead of just background), to see if we can further improve the accuracy rate on predictions. Experiments with additional classifiers might also improve prediction performance.

References

1. Datta, R., Li, J., Wang, J.Z.: Algorithmic inferencing of aesthetics and emotion in natural images: an exposition. In: Proceedings of the IEEE International Conference on Image Processing, pp. 105–108 (2008)
2. Murray, N., Marchesotti, L., Perronnin, F.: AVA: a large-scale database for aesthetic visual analysis. In: Proceedings of the IEEE Conference on Computer Vision and Pattern Recognition, pp. 2408–2415 (2012)
3. Su, H.H., Chen, T.W., Kao, C.C., Hsu, W.H., Chien, S.Y.: Scenic photo quality assessment with bag of aesthetics-preserving features. In: Proceedings of the 19th ACM International Conference on Multimedia, pp. 1213–1216 (2011)
4. Datta, R., Wang, J.Z.: Acquine: aesthetic quality inference engine real-time automatic rating of photo aesthetics. In: Proceedings of the ACM International Conference on Multimedia Information Retrieval, pp. 421–424 (2010)
5. Lo, L.Y., Chen, J.C.: A statistic approach for photo quality assessment. In: Proceedings of the International Conference on Information Security and Intelligence Control (ISIC), pp. 107–110 (2012)
6. Li, C., Loui, A.C., Chen, T.: Towards aesthetics: a photo quality assessment and photo selection system. In: Proceedings of the International Conference on Multimedia, pp. 827–830 (2010)
7. Liu, L., Chen, R., Wolf, L., Cohen-Or, D.: Optimizing photo composition. Comput. Graph. Forum (Proc. Eurograph.) 29, 469–478 (2010)
8. Krages, B.: Photography: The Art of Composition. Allworth Press, New York (2005)
9. Bhattacharya, S., Sukthankar, R., Shah, M.: A framework for photo-quality assessment and enhancement based on visual aesthetics. In: Proceedings of the International Conference on Multimedia, pp. 271–280 (2010)
10. Yeh, C.H., Barsky, B.A., Ouhyoung, M.: Personalized photograph ranking and selection system considering positive and negative user feedback. ACM Trans. Multimed. Comput. Commun. Appl. 10, 1–20 (2014). Article No. 36

11. Ryu, D.S., Kim, K.H., Park, S.Y., Cho, H.G.: A web-based photo management system for large photo collections with user-customizable quality assessment. In: Proceedings of the ACM Symposium on Applied Computing, pp. 1229–1236 (2011)
12. Barnbaum, B.: The Art of Photography: An Approach to Personal Expression. Rocky Nook, Santa Barbara (2010)
13. Luo, Y., Tang, X.: Photo and video quality evaluation: focusing on the subject. In: Forsyth, D., Torr, P., Zisserman, A. (eds.) ECCV 2008, Part III. LNCS, vol. 5304, pp. 386–399. Springer, Heidelberg (2008)
14. Manav, B.: Color-emotion associations and color preferences: a case study for residences. Color Res. Appl. **32**, 144–150 (2007)
15. Gao, X.P., Xin, J.H., Sato, T., Hansuebsai, A., Scalzo, M., Kajiwara, K., Guan, S.S., Valldeperas, J., Lis, M.J., Billger, M.: Analysis of cross-cultural color emotion. Color Res. Appl. **32**, 223–229 (2007)
16. Tang, X., Luo, W., Wang, X.: Content-based photo quality assessment. IEEE Trans. Multimed. **15**, 1930–1943 (2013)
17. Nishiyama, M., Okabe, T., Sato, I., Sato, Y.: Aesthetic quality classification of photographs based on color harmony. In: Proceedings of the IEEE Conference on Computer Vision and Pattern Recognition, pp. 33–40 (2011)
18. Ke, Y., Tang, X.: The design of high-level features for photo quality assessment. In: Proceedings of the IEEE Conference on Computer Vision and Pattern Recognition, pp. 419–426 (2006)
19. Lo, K.Y., Liu, K.H., Chen, C.S.: Assessment of photo aesthetics with efficiency. In: Proceedings of the IAPR International Conference on Pattern Recognition, pp. 2186–2189 (2012)
20. Barnbaum, B.: The Essence of Photography: Seeing and Creativity. Rocky Nook, Santa Barbara (2014)
21. Su, H.H., Chen, T.W., Kao, C.C., Hsu, W.H.: Preference-aware view recommendation system for scenic photos based on bag-of-aesthetics-preserving features. IEEE Trans. Multimed. **14**, 833–843 (2012)
22. Datta, R., Joshi, D., Li, J., Wang, J.Z.: Studying aesthetics in photographic images using a computational approach. In: Leonardis, A., Bischof, H., Pinz, A. (eds.) ECCV 2006. LNCS, vol. 3953, pp. 288–301. Springer, Heidelberg (2006)
23. Datta, R., Li, J., Wang, J.Z.: Learning the consensus on visual quality for next-generatoin image management. In: Proceedings of the 15th International Conference on Multimedia, pp. 533–536 (2007)
24. Achanta, R., Hemami, S., Estrada, F., Susstrunk, S.: Frequency-tuned salient region detection. In: Proceedings of the IEEE Conference on Computer Vision and Pattern Recognition, pp. 1597–1604 (2009)
25. Canny, J.: A computational approach to edge detection. IEEE Trans. Pattern Anal. Mach. Intell. **8**, 679–697 (1986)
26. Duda, R.O., Hart, P.E.: Use of the Hough transformation to detect lines and curves in pictures. Commun. ACM **15**, 11–15 (1972)
27. Vapnik, V.N.: Statistical Learning Theory. Wiley, New York (1998)

Use of a Large Image Repository to Enhance Domain Dataset for Flyer Classification

Payam Pourashraf[✉] and Noriko Tomuro

DePaul University, 243 S. Wabash Ave, Chicago, IL 60604, USA
ppourash@cdm.depaul.edu, tomuro@cs.depaul.edu

Abstract. This paper describes our exploratory work on supplementing our dataset of images extracted from real estate flyers with images from a large general image repository to enhance the breadth of the samples and create a classification model which would perform well for totally unseen, new instances. We selected some images from the Scene UNderstanding (SUN) database which are annotated with the scene categories that seem to match with our flyer images, and added them to our flyer dataset. We ran a series of experiments with various configurations of flyer vs. SUN data mix. The results showed that the classification models trained with a mixture of SUN and flyer images produced comparable accuracies as the models trained solely with flyer images. This suggests that we were able to create a model which is scalable to unseen, new data without sacrificing the accuracy of the data at hand.

1 Introduction

Flyers are a popular form of advertising material, intended to be distributed to a wide audience. Flyers are often multimodal – description of the subject matter in text is accompanied by related images. For example, a flyer of a commercial real estate typically indicates the listing information such as address, square footage, price, property type (industrial, office, etc.) and broker's name in text, and includes some pictures of the property. Figure 1 shows an example flyer of an industrial property. In recent years, with the wide spread of desktop publishing tools and the prevalence of the internet, flyers have become more popular, especially in the electronic file format. For example, brokers of commercial real estate create a flyer for each property they sell in the (commonly) pdf format and post it on their website; they also collect flyers of other brokers or in the public domain (again commonly in the pdf format) to create a large, online searchable database of available properties to attract customers. However, manually searching and indexing flyers in building a database is a tedious, time-consuming task – a better solution would be to automatically extract the relevant information.

In this paper, we describe our work on classifying images embedded in real estate flyers by the property types (retail, office, industrial and multi-family). It is a part of a larger project which aims to develop a high-performance multimodal system which

© Springer International Publishing Switzerland 2015
G. Bebis et al. (Eds.): ISVC 2015, Part II, LNCS 9475, pp. 609–617, 2015.
DOI: 10.1007/978-3-319-27863-6_56

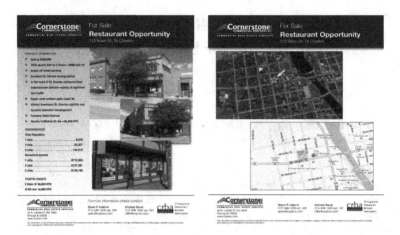

Fig. 1. A commercial real estate flyer of a retail property (© Cornerstone Commercial Partners II LLC)

extracts information from real estate flyers by using both texts and images. In this paper, we focus on the image classification component. In the full system, the results of the image classifier will be combined with the textual information to extract various information from a flyer regarding the property and listing.

Recently, there has been a lot of effort, in both research and industry, on how to mine from multimodal or multimedia data and utilize the extracted content for various purposes [1–3]. However, there are several major difficulties in the pursuit. First, automatic media/image extraction/cropping tools (from various file formats) are generally not very accurate. Not only are they unable to differentiate irrelevant graphic elements such as banners and border strips from real content images, most tools also have difficulty with images which are overlapping or laid out in an unusual or creative way. In fact, the latter situation occurs commonly with flyers because they are essentially free-form. Consequently the 'ratio of successful automatic harvesting' of good, clean content images from embedded files is usually only moderate. For example, the dataset we used in our previous work showed about 30 % of the automatically cropped images were fragments, and around 15–20 % were failures (where typically an original file with multiple embedded images is returned without cropping). This will become a problem when one wants to collect sample images of many categories. The second major difficulty is the problem of semantic gap. It refers to the difficulty in image analysis in relating the low-level pixel data with a high-level concept which the image represents (e.g. 'sunset') [4, 5]. In our case, the gap is particularly deep because of the subtle similarity or overlaps between the property types. For example, retail and industrial spaces often have an office (room) inside, while an office may have a storage room which looks like a small warehouse.

The work we present in this paper is an attempt to address the first problem (of obtaining clean content images), by supplementing the images extracted from the flyers with images from a large publicly available image repository, in particular the Scene

UNderstanding (SUN) database[1] [6]. The SUN database contains over 130,000 images, which are annotated with over 900 scene categories. Also the scene categories are organized in a hierarchy. We selected several scene categories from SUN which seem to match with our flyer images, and added the images of those categories in our dataset. Actually, enhancing the dataset would also help narrow the semantic gap as well in our case, because generally a real estate property consists of spaces of several types (for different usage) and our flyer dataset did not have enough samples of various space types (partly due to the low cropping accuracy).

However, there is a critical assumption in adding SUN images in our dataset, that SUN images are similar to our flyer images. Since SUN is essentially an external source, mixing in with our primary dataset, which is from a particular domain of real estate flyers, might not bring improvement in classification. To test the assumption and see the usefulness of the SUN dataset in our research, we added a selected subset of the SUN images to our flyer image dataset and ran a series of experiments. The results showed that the classification models trained with a mixture of SUN and flyer images produced comparable accuracies as the models trained solely on flyer images. This is a promising result, which indicates that we would be able to utilize the SUN database to boost our data and create a larger training dataset without sacrificing classification accuracy.

2 Related Works

Preliminary works have been done on text and image independently. In one of the previous works for the text side various techniques in Information Extraction and Text Categorization has been used to do the task. The combination of textual (e.g. token and token kind) and visual features (e.g. font color, size, position in the flyer) were used to extract various information about the property, such as property type (e.g. retail, industrial, office, land), address, space size (square footage, acres), and the name and contact information of the broker [7]. The visual features which have been used in that work included: font size and Y coordinate [7].

In a recent work on the image side, the images embedded in real estate flyers were classified into five genres (map, schematic drawing, aerial photo, indoor-building and outdoor-building) [8]. At the start of this experiment, the features were extracted from over 3000 images from publicly available online real estate flyers, including Autocorrelogram, Tamura, Local Binary Patterns, Histogram of Oriented Gradients, number of lines (by using Hough Transform) and the number of points with high cornerness (by using Harris corner detection). A two-level ensemble classifier model was built in which the first (Tier-1) consisted of several binary classifiers, each of which was trained to classify data for a given genre, and the second (Tier-2) classifier combined the output of the Tier-1 classifiers to produce the final output. The result showed that the model has a significant out performance in comparison to the baseline the classifiers (Naïve Bayes, Decision Tree and KNN) [8].

---

[1] http://groups.csail.mit.edu/vision/SUN/.

The SUN database from which we have selected the images has been widely used as a benchmark dataset for different tasks such as scene understanding [9] and object recognition [10]. SUN-397 database is a subset of this dataset which has also been used as a benchmark dataset [6, 11]. In [12] the authors selected over 500 images from the SUN categories of "bedroom" and "living room" for the task of 3D model scene understanding.

3   Methodology and Experimental Design

3.1   Image Dataset

In this work, we used the real estate flyer dataset used in [7]. From the original dataset we selected 144 flyers that included indoor images. We focused on indoor images in this work, because we thought indoor images were more suggestive of the property types than other kinds of images (such as maps and outside images). Then from each flyer, we extracted images by using software tools[2] and wrote our own code to filter 'noisy' non-content images (such as image fragments, color borders and company logos). Finally, we got 686 indoor images out of the original flyers. The distribution of the number of flyers and their proportion in the dataset are shown in Table 1.

Table 1.   Distribution of the number of flyers and their proportion in the dataset

Property type	Number of flyers	Proportion of flyers	Number of images	Proportion of images
Industrial	37	21 %	143	26 %
Multi-family	7	8 %	60	5 %
Office	58	41 %	280	40 %
Retail	42	30 %	203	29 %
Total	144	100 %	686	100 %

We supplemented the flyer images with images from the SUN database. SUN has 3 categories in its' third level of hierarchy, which are indoor, outdoor man-made, and outdoor natural. We selected some categories from the SUN indoor categories which seem to match with our flyer images. We tagged each of the selected categories by one property type (industrial, multi-family, office and retail). In order to have a balanced dataset, we randomly selected the images inside each property type and made the same number of images for each of them. The total number of images was 5140 (1285 for each of the four property type). Selected SUN categories with their assigned property types are shown in Table 2.

---

[2] We used pdf to html (http://sourceforge.net/projects/pdftohtml/) to convert pdf to html, and Gimp (http://www.gimp.org/) to crop individual images.

Table 2. Selected SUN categories

Property type	Number of images	SUN categories
Industrial	1285	Basement, Assembly line, Furnace room, Parking garage, Storage room, Warehouse
Multi-family	1285	Kitchen, Bedroom
Retail	1285	Cafeteria, Coffee shop, Lobby, Restaurant, Shopping mall
Office	1285	Computer room, Conference room, Cubicle, Office, Office cubicles

Figures 2 and 3 show examples of the indoor-building genre from our flyers and example images of the SUN indoor categories, respectively.

Fig. 2. Example images of the indoor-building genre from our flyers

Fig. 3. Example images of the SUN indoor categories

3.2 Feature Extraction

The indoor-building genre images of our work and the selected list of categories from SUN database have then been incorporated to extract the GIST and color features. The details of the extracted features are described as follows:

GIST: Some prior studies [13, 14] have proposed that the scene recognition is initiated from the getting of the global configuration of the scene. The task Scene recognition can be done by looking at their GIST. Thus, the general 512 dimensions' GIST [14, 15] feature has been extracted from the images.

Color features: For color features, we have extracted 6 color moments (the 2 first color moments from each R,G,B) and 32 color Autocorrelogram [16] of 1 and 3 distances. Finally we got 38 color features.

3.3 Experimental Setup

To see the effect of SUN images mixed in with the flyer images to build classification models, we ran several experiments with varying degrees/proportions of SUN vs. flyer mixes in the training set. In particular, we added SUN images in the training set gradually with a 20 % increment, that is, all of the flyer images + 0 % of SUN images, + 20 % of SUN, + 40 % of SUN, and so on until (all of flyer images) + 100 % of SUN – thereby a total of 6 mixture configurations. We also wanted to see the effect of color features (in addition to the GIST features) in the classification for our problem – thus two feature configurations: GIST only or GIST + Color. Putting all these together, we had a total of 12 (= 6 mixture * 2 feature) configurations. Table 3 below summarizes the configurations and provides the breakdown of the training set for each configuration. Note the selection of the SUN images in each configuration was random but stratified (evenly across the property types, because the same number of images were selected from the SUN database for each property type).

Table 3. Various configurations of the training dataset

SUN Mix / Image Features	0%	20%	40%	60%	80%	100%
GIST only (512 features)	686 flyer images	686 flyer images	686 flyer images	686 flyer images	686 flyer images	686 flyer images
GIST + Color (550 features)	+ 0 SUN images	+ 1028 SUN images	+ 2056 SUN images	+ 3084 SUN images	+ 4112 SUN images	+ 5140 SUN images

To build classification models, we used Support Vector Machine (SVM) with a linear kernel. We chose SVM because it has been shown to produce good results in many works (including in several of our previous work).

4   Results and Discussion

To evaluate the classification models, we used (5-fold) cross validation (CV), with two data partition schemes: one by images (but stratified with respect to property types, which is the target category) and another by flyers (from which the images were extracted). We thought of partitioning at the flyer level because it would better simulate the realistic situations - a trained model receives images from a new (totally unseen) flyer.

The accuracy results for various experimental configurations are demonstrated in Table 4 and Fig. 4. Note that cross validation was done for flyer images only, not for

SUN images. For each scheme (1 through 4), for each fold the training was done using all of the SUN images (always), plus a 4/5 of the flyer images partitioned based on the images/property types or the original flyers from which the images came from, and the testing was done using the remaining 1/5 of the flyer images. Also the p-values were computed between the 0 % SUN vs. each of the SUN mix for each scheme.

Table 4. Accuracy results for various experimental configurations

Scheme	CV Partition	SUN Mix / Image Features	0%	20%	40%	60%	80%	100%
1	Image level	GIST only	51.45	37.75	32.78	31.77	30.01	30.89
		(p-value)	(--)	(.009)	(.002)	(.001)	(.002)	(.001)
2		GIST + Color	52.62	51.73	48.97	47.65	47.08	47.07
		(p-value)	(--)	(.663)	(.177)	(.111)	(.032)	(.010)
3	Flyer level	GIST only	43.86	41.94	43.91	42.31	43.15	42.45
		(p-value)	(--)	(.568)	(.989)	(.663)	(.863)	(.758)
4		GIST + Color	45.68	45.66	42.50	41.15	42.24	42.94
		(p-value)	(--)	(.995)	(.290)	(.315)	(.366)	(.514)

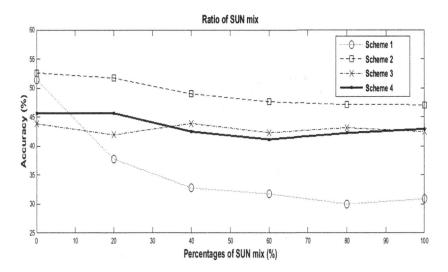

Fig. 4. Accuracies of various experimental schemes

For all schemes, accuracy went down as more SUN images were added in the training set (except for Scheme 3: accuracy went up slightly for 40 % SUN mix). It was an expected result because SUN images were essentially 'out-of-domain' data from the perspective of flyer images. Also the decrease was more drastic when the test data was partitioned on the image level (Scheme 1 and 2) – the decrease was statistically signif-icant (p-value < .05 for all non-zero SUN mixes for Scheme 1, and 80 and 100 % SUN mixes for Scheme 2). A notable result is that, for Scheme 3 and 4, accuracy didn't go down significantly, or yet stayed about the same, as more SUN images were added in the training set (as evidenced by the p-values $> .05$ for all non-zero SUN mixes). This is an encouraging result – Since Scheme 3 and 4 used the partition based on flyers (and simulate realistic and predictive situations), maintaining approximately the same level of accuracy as more SUN data was injected means that these models are scalable to unseen, new data.

As for the effect of color features, there was a large effect in the classification accuracy when the test data was partitioned at the image level (Scheme 1 vs. 2), but not when it was partitioned at the flyer level (Scheme 3 vs. 4).

5   Conclusions and Future Work

In conclusion, in this work we presented our work on classifying the indoor images embedded in the real estate flyers by the property type. The proposed model could be scaled to new data without compromising the accuracy of data classification.

For future work, we plan to bridge the problem of semantic gap by building a multi-modal system using both texts and images. Other future studies with focus on the other image genres besides indoor-building (map, schematic drawing, aerial photo and outdoor-building) would also be helpful for improvement of flyers classification.

References

1. Manjunath, T.N., Hegadi, R.S., Ravikumar, G.K.: A survey on multimedia data mining and its relevance today. IJCSNS **10**(11), 165–170 (2010)
2. Bhatt, C.A., Kankanhalli, M.S.: Multimedia data mining: state of the art and challenges. Multimedia Tools Appl. **51**(1), 35–76 (2011)
3. Guillaumin, M., Verbeek, J., Schmid, C.: Multimodal semi-supervised learning for image classification. In: 2010 IEEE Conference on Computer Vision and Pattern Recognition (CVPR), pp. 902–909 (2010)
4. Dorai, C., Venkatesh, S.: Bridging the semantic gap with computational media aesthetics. IEEE Multimedia **10**(2), 15–17 (2003)
5. Zhao, R., Grosky, W.I.: Bridging the semantic gap in image retrieval. In: Distributed Multimedia Databases: Techniques and Applications, pp. 14–36 (2002)
6. Xiao, J., Ehinger, K.A., Hays, J., Torralba, A., Oliva, A.: SUN database: Exploring a large collection of scene categories. Int. J. Comput. Vision 1–20 (2014)
7. Apostolova, E., Tomuro, N.: Combining visual and textual features for information extraction from online flyers. In: Empirical Methods in Natural Language Processing (EMNLP) (2014)
8. Pourashraf, P., Tomuro, N., Apostolova, E.: Genre-based image classification using ensemble learning for online flyers. In: Seventh International Conference on Digital Image Processing (ICDIP) (2015)
9. Li, C., Parikh, D., Chen, T.: Automatic discovery of groups of objects for scene understanding. In: 2012 IEEE Conference on Computer Vision and Pattern Recognition (CVPR), pp. 2735–2742 (2012)
10. Manen, S., Guillaumin, M., Van Gool, L.: Prime object proposals with randomized prim's algorithm. In: 2013 IEEE International Conference on Computer Vision (ICCV), pp. 2536–2543 (2013)

11. Su, Y., Jurie, F.: Improving image classification using semantic attributes. Int. J. Comput. Vis. **100**(1), 59–77 (2012)
12. Satkin, S., Lin, J., Hebert, M.: Data-driven scene understanding from 3D models. In: BMVC (2012)
13. Biederman, I.: Aspects and extensions of a theory of human image understanding. In: Computational Processes in Human Vision: An Interdisciplinary Perspective, pp. 370–428 (1998)
14. Khosla, A., Das Sarma, A., Hamid, R.: What makes an image popular?. In Proceedings of the 23rd International Conference on World Wide Web, pp. 867–876 (2014)
15. Oliva, A., Torralba, A.: Modeling the shape of the scene: A holistic representation of the spatial envelope. Int. J. Comput. Vis. **42**(3), 145–175 (2001)
16. Huang, J., Kumar, S.R., Mitra, M., Zhu, W., Zabih, R.: Image indexing using color correlograms. In: Proceedings of the 1997 IEEE Computer Society Conference on Computer Vision and Pattern Recognition, pp. 762–768 (1997)

Illumination Invariant Robust Likelihood Estimator for Particle Filtering Based Target Tracking

Buti Al Delail$^{(\boxtimes)}$, Harish Bhaskar, M. Jamal Zemerly,
and Mohammed Al-Mualla

Khalifa University of Science, Technology and Research (KUSTAR),
Abu Dhabi, United Arab Emirates
{buti.aldelail,harish.bhaskar,jamal.zemerly,almualla}@kustar.ac.ae

Abstract. Tracking visual targets under illumination changes is a challenging problem, especially when the illumination varies across different regions of the target. In this paper, we solve the problem of illumination invariant tracking during likelihood estimation within the particle filter. Existing particle filter based tracking frameworks mainly deal with changes in illumination by the choice of color-space or features. This paper presents an alternate likelihood estimation algorithm that helps dealing with illumination changes using a homomogrphic filtering based weighted illumination model. That is, a homomorphic filter is first used to separate the illumination and reflectance components from the image, and further by associating an appropriate weight to the illumination, the target image is reconstructed for the accurately measuring the likelihood. The proposed algorithm is implemented using a simple particle filter tracking framework and compared against other tracking algorithms on scenarios with large illumination variations.

1 Introduction

Visual target tracking in dynamic environments serves many applications in surveillance and human-computer interfaces. Tracking algorithms attempt to estimate a moving target position in a sequence of video frames. The problem becomes more complex to solve when the target appearance varies over time due to the nature of the target or changes in the environment. The main challenges that impact the tracking performance are changes in illumination, camera motion, background, full-partial target occlusion, non-rigid shape, blurring or noise [1]. These changes hugely degrade the tracker performance or cause it to fail. Much work has already been done in visual tracking to address these issues though there is still no visual tracking algorithm robust enough to all kinds of variations. Illumination variation is an important aspect for targets trackers to cope-with in different environments. The target illumination is mainly affected by the ambient lighting, which greatly varies between different times and locations. Hence, the target appearance is highly influenced by these changes. For visual

© Springer International Publishing Switzerland 2015
G. Bebis et al. (Eds.): ISVC 2015, Part II, LNCS 9475, pp. 618–627, 2015.
DOI: 10.1007/978-3-319-27863-6_57

trackers to work in real environment, there is need for it to be robust to different lighting conditions.

In the last decade, a lot of research have contributed to visual target tracking, being one of the main topics in computer vision. In this context, particle filter based approaches have captured a fair amount of attention. This is mainly due to the fact that, it effectively deals with solving non-linear and non-gaussian probability distribution, and its ability to recover from occlusions, thus making it suitable for many tracking problems. However, current particle filtering methods do not perform well under complex conditions, specifically when the target appearance changes significantly. Allowing the particle filter to deal with large variations in pose, shape and illumination increases the risk of losing the tracked target to another but yet similar target present in the visual scene. Furthermore, homomorphic filtering is an image enhancement technique that is a popular in illumination invariant face recognition used to normalise the amount of illumination present in the image [2]. Since illumination is dominantly represented by low frequency components, many techniques are based on either Fast Fourier Transform (FFT), Discrete Cosine Transform (DCT) or Discrete Wavelet Transform (DWT).

2   Related Work

2.1   Particle Filters

Particle filtering based visual tracking methods have received a considerable amount of attention in the literature with many applications in surveillance [3], hand tracking [4], vehicle tracking [5], sports [6], aquaculture [7]. The idea behind particle filtering is to use a set of random samples with associated weights in order to represent the posterior density function given a noisy or partial observation. Various particle filters exist in the literature, these include Auxiliary Particle Filter (Aux-PF) [8], Unscented Particle Filter [9], Rao-Blackwellized Particle Filter (RBPF) [10], PF-Doucet [11], and Gaussian PF [12]. Most of the highlighted work in particle filtering for visual tracking under varying illumination conditions can be classified into appearance based and similarity based approaches.

Appearance modelling methods are feature-based approaches that define a feature extraction procedure based on color, edges, texture, etc. Robust features enable the particle filter to track targets in complex conditions. The work in [13] proposed a joint spatial-color appearance model in order to utilize more information than what the conventional color histogram provides. It deals with illumination invariance through the choice of color space. Although such an approach has proven to deal with small variations in illumination, a large amount of illumination change continues to present problems. Multi-cue based particle filter tracking achieves better stability under sudden illumination change than any single-cue method [14]. The authors in [15] proposed a dynamic multi-cue appearance model that describes the target appearance using color, texture and edge information. Similarly, an appropriate choice of color space is said to make the tracker robust against illumination changes. The HSV color-space is often used, where the value (V component) is given less weight. Such approaches can handle small-to-medium variations in illumination. However, illumination-invariant tracking requires targets to be tracked under large illumination changes. Thus, the choice of color space alone is not enough. More recently, an enhanced RBPF tracking algorithm based on background modelling for multi-target tracking in a dynamic scene has been proposed in [16], where illumination variation between consecutive frames is modelled using a forward-reverse procedure to compute the associated cost.

Similarity measure and likelihood estimation are very important aspects that contribute to the performance of particle filters. The authors in [17] proposed a particle filter with incremental likelihood calculation that includes histogram and Bhattacharyya similarity calculations to improve the performance of the tracker. Illumination tolerance is achieved by the choice of feature descriptors, for example, Edge Orientation Histogram (EOH) and Local Binary Pattern (LBP). The combination of these features with color and edge information results can produce more accurate target tracking. However, the work mainly focused on the aspect of tracking performance. The work in [18] analysed various distance measures including Bhattacharyya, Matusita and X^2 distances, and concluded that there is no fundamental reason to prefer one over the other.

The choice of the proposal distribution for importance sampling in particle filters remains a tough problem to solve. The authors in [19] proposed a swarm intelligence based particle filter algorithm to overcome the sample impoverishment problem of particle filters. The appearance model and the similarity measure are very important to the performance of particle filters. The color histogram, so far, is widely used in particle filters for its simplicity. However, its main disadvantage is that it lacks the information about the spatial layout of the target, and is very sensitive to illumination changes.

2.2   Likelihood Estimators

Likelihood estimation is an important step in particle filtering that measures the difference between the initial target appearance and its current observation. The basic function of the likelihood estimator is to give higher importance scores (weight) to predicted states that are close to the target while suppress predictions that are far from the target. Most particle filters in visual target tracking employ a Gaussian likelihood function due to the common assumption of a normally distributed measurement error. However, a heavier tailed Cauchy distribution might be better in some other situations [20].

The authors in [21] proposed an algorithm to detect occlusions by learning the observation likelihoods. The algorithm is integrated within a particle filter tracking and implemented using $l1$ minimization, where a SVM classifier is trained on the observation likelihoods to be able to detect occlusions. The work in [22] proposed a multimodal particle filter based tracking algorithm to track moving objects across illumination changes. Their particle filter algorithm models illumination variance, where the illumination change is detected using generalised Expected (negative) Log Likelihood (gELL) statistic [23]. The authors proposed an auxiliary particle filter with a likelihood model that combines the similarity measure obtained from the intensity gradient information with a color model of the tracked elliptical region.

3   Proposed Methodology

3.1   Particle Filtering - Problem Formulation

In a simple 2D particle filter based target tracking, the target is represented by the following elliptical model (note this model assumes constant target size).

$$\{x, y, w, h, \delta x, \delta y, i, \delta i\} \tag{1}$$

where (x,y,w,h) the center and scale of the target's elliptical region, $(\delta x, \delta y)$ represents the acceleration parameters. While i is the weight of illumination and δi is the variation in illumination.

The initialisation of the tracker is done based on given coordinates of the target, where the first frame is used to extract the template of the target. Suppose that $Z_t = \{z_1, ..., z_t\}$ denotes the observation at time t, and the corresponding states are $X_t = \{x_1, ..., x_t\}$ the tracking problem can be formulated as

$$p(x_t|Z_t) = \sum_{i=1}^{n} \hat{w}_t^i \delta(x_t - x_t^i) \tag{2}$$

$$w_t^i = w_{t-1}^i * \frac{p(z_t|x_t^i)\, p(x_t^i|x_{t-1}^i)}{q(x_t^i|x_{t-1}^i, Z_t)} \tag{3}$$

Weights are determined by evaluating the likelihood function $p(z_t|x_t^i)$ for each sample. In order to construct an illumination invariant likelihood measurement algorithm, we should first construct the histogram for the particles with various weights of illumination. To do so, we employ homomorphic filtering to separate the illumination component from the image, and then reconstruct the image using weighted illumination (refer to Sect. 3.3 for details). The obtained weights are then normalised to calculate $\hat{w}_t^i$ as follows

$$\hat{w}_t^i = \frac{w_t^i}{\sum_{i=1}^{n} w_t^i} \tag{4}$$

The weighting of the particles provides an estimate of the target state and determines the evolution of the particles. In particle filtering, if the particles slowly start to propagate away from the target, the likelihood measures low scores, and thus becoming less important to the tracking. Therefore the process of resampling removes these particles, and replaces them with those stationed on the target (i.e. have high likelihood score). The effective number of particles is estimated as

$$N_{eff} = \frac{1}{\sum_{i=1}^{n} (\hat{w}_t^i)^2} \tag{5}$$

3.2   Controlled Illumination Using Homomorphic Filtering

The intensity in a video frame of Lambertian object surfaces is modelled as

$$f(x, y) = i(x, y)r(x, y) \qquad (6)$$

where i is the illumination and r is the reflectance components. Homomorphic filtering attempts to separate these components. First, a natural log transformation is applied on the image. Therefore, the components become additive

$$ln\ f(x, y) = ln\ i(x, y) + ln\ r(x, y) \qquad (7)$$

Then we chose to use the Fourier transform to transform the image into the frequency domain.

$$\mathfrak{F}\{ln\ f(x, y)\} = \mathfrak{F}\{ln\ i(x, y)\} + \mathfrak{F}\{ln\ r(x, y)\} \qquad (8)$$

and therefore, can be expressed as

$$Z(u, v) = F_i(u, v) + F_r(u, v) \qquad (9)$$

where $F_i(u, v)$ is the Fourier transform of the log-illumination $ln\ i(x, y)$ and $F_r(u, v)$ is the Fourier transform of the log-reflectance $ln\ r(x, y)$. Now we apply a highpass filter $H(u, v)$, we chose a high-boost Butterworth filter with a cutoff value of 0.02.

$$R(u, v) = H(u, v).Z(u, v) = H(u, v).F_i(u, v) + H(u, v).F_r(u, v) \qquad (10)$$

Based on the fact that most of the illumination is present in the low frequency, a highpass filter would ideally eliminate the whole illumination component from the image, leaving out reflectance. Therefore, we recover the image using inverse Fourier transform as a close estimate of the reflectance.

$$ln\ r(x, y) = \mathfrak{F}^{-1}\{H(u, v).Z(u, v)\} = \mathfrak{F}^{-1}\{H(u, v).F_r(u, v)\} \qquad (11)$$

Now from the reflectance image and Eq. (6), we compute the illumination as

$$ln\ i(x, y) = ln\ f(x, y) - ln\ r(x, y) \qquad (12)$$

Then the image is recovered again with the illumination being suppressed, where $0 < \alpha < 1$ is the intensity factor. A value of $\alpha = 0$ will remove all illumination and $\alpha = 1$ will result in the same image. While, $0 < \beta < 1$ is the intensity factor for the reflectance image.

$$ln\ f(x, y) = \alpha\ ln\ i(x, y) + \beta\ ln\ r(x, y) \qquad (13)$$

Then finally the illumination weighted image is obtained by the exponential operation

$$f'(x, y) = exp(ln\ f(x, y)) \qquad (14)$$

3.3 Likelihood Estimation Under Variable Illumination

Likelihood estimation gives the tracker a similarity measure of the current observation, thus used to compute the weights w_t^i for each particle x_i. We maximise the likelihood by applying homomorphic filtering with incremental illumination weights across the region of the particle as shown in Fig. 1. Therefore the likelihood estimation can be modelled as follows

$$l_t^i = max\{exp(-\frac{1 - d_t(i,j)}{2\sigma^2})\} \tag{15}$$

where l_t^i is the maximum likelihood at time t for particle i. σ is the standard deviation for the similarity measure, and d_t^i is the distance between the current measure and initial templates. d_t^i is chosen to be the Bhattacharyya distance, calculated as follows

$$d_t^i = \sum_{i=1}^{n} \sqrt{h_t^i * T_a} \tag{16}$$

where h_t^i corresponds to the histogram for particle i at time t, and T_a is the template histogram. Figure 2 shows the improvements that can be achieved in the likelihood confidence by weighting the illumination in 2 scenarios. Improvements are noticeable mainly when there is huge variation in illumination. A normal particle filter likelihood will be operating on the 100 % illumination level (original image). Our results show further improvement can be achieved when the illumination is reduced to levels of 10 % to 30 % of the original illumination. For smaller weights of illumination, the reconstructed image is equivalent to the reflectance (which does not contain color information and, therefore, the color histogram cannot be used).

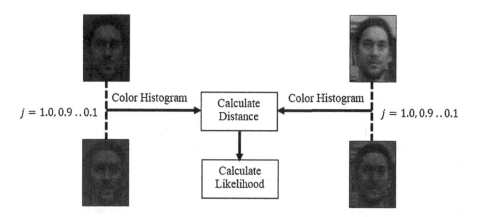

Fig. 1. Various weights of illumination in homomorphic filtering. Left side shows the target template. Right side shows the tracked target under different illumination using homomorphic filtering.

Fig. 2. Illumination invariant likelihood confidence values. (a) shows the likelihood values when small and large illumination variation exists between the template and target with $\sigma = 0.6$. (b) shows the likelihood values of the small variation with $\sigma = 0.2$. (c) shows the likelihood values of the large variation with $\sigma = 0.2$

4   Experimental Method

4.1   Dataset Description

The data set selection was based on the challenges that the proposed algorithm attempts to solve, which is towards illumination invariant tracking. The public dataset includes many videos with various challenging tracking aspects related to pose, occlusion, etc. However, our primary goal is to test the tracker only on those video segments where illumination changes dramatically. Therefore, we experimented mainly on sequences with challenging illumination only. The sequence here is obtained from the public dataset available in [24].

4.2   Results

The tracking results shown in Fig. 3 demonstrates an improved tracking of the visual target when the illumination rapidly changes. Despite that the particle filter eventually loses the tracked target due to the sole reliance on color histogram, it is able to track the target better when the illumination changes rapidly. In comparison, the other baseline tracking algorithms fail to do so. We compare the normal particle filter framework with our improved particle filter, and other algorithms including SPT [25] and IVT [26]. These tracking frameworks perform better on constant illumination than the normal particle filter framework used here; they lose the target when a large illumination variation occurs. The geometric affine parameters $\{x, y, th, scale, aspect, skew\}$ are set as follows. The SPT initialisation parameters in our experiment were set as $\{246, 202, 0.86375, 0, 0, 1.2\}$ while the IVT parameters where set as $\{246, 202, 1.6875, 0, 0, 2.25\}$. Other parameters are kept as default. Figure 3(c) shows the moment these trackers lose tracking. The initialisation parameters only represent the initial target, and has no effect on the tracking results. However, the size of the tracked window may affect the tracking performance.

Future work will evaluate the proposed likelihood method on other trackers than the particle filter used in this experiment. Figure 4 shows the overlapped tracking regions between the trackers. It is worth noting that each tracker performed differently at the moment of large illumination variance.

Fig. 3. Tracking scenario with large change in illumination. (a) is normal particle filter tracking based on color histogram. (b) is the improved particle filter tracking with our proposed algorithm. (c) Frame when illumination variance is large results order [IVT, SPT, LPF (Ours)].

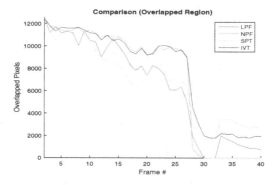

Fig. 4. Comparison of overlapped tracking regions between the trackers

5   Conclusion and Future Work

This paper presented a robust likelihood estimator for illumination invariant particle filter based target tracking. The algorithm calculates the likelihood confidence as the maximum measure of similarity under variable weights of controlled illumination using homomorphic filtering. We implemented the algorithm in a color-based particle filter tracking and shown improvements achieved on a scenario consisting of large illumination variation. The particle filter tracking framework is compared with other state-of-the-art tracking algorithms, where the baselines fail due to the large change in illumination. Our future work will focus on improving the tracking framework by incorporating robustness against other challenges such as occlusion, noisy background, together with the current model for illumination invariance.

References

1. Yang, H., Shao, L., Zheng, F., Wang, L., Song, Z.: Recent advances and trends in visual tracking: a review. Neurocomputing **74**, 3823–3831 (2011)
2. Chen, W., Er, M.J., Wu, S.: Illumination compensation and normalization for robust face recognition using discrete cosine transform in logarithm domain. IEEE Trans. Syst. Man Cybern. Part B Cybern. **36**, 458–466 (2006)
3. Xu, X., Li, B.: Adaptive rao-blackwellized particle filter and its evaluation for tracking in surveillance. IEEE Trans. Image Process. **16**, 838–849 (2007)
4. Shan, C., Tan, T., Wei, Y.: Real-time hand tracking using a mean shift embedded particle filter. Pattern Recogn. **40**, 1958–1970 (2007)
5. Chan, Y.M., Huang, S.S., Fu, L.C., Hsiao, P.Y., Lo, M.F.: Vehicle detection and tracking under various lighting conditions using a particle filter. IET Intell. Transp. Syst. **6**, 1–8 (2012)
6. Lu, W.L., Okuma, K., Little, J.J.: Tracking and recognizing actions of multiple hockey players using the boosted particle filter. Image Vis. Comput. **27**, 189–205 (2009)
7. Pinkiewicz, T., Williams, R., Purser, J.: Application of the particle filter to tracking of fish in aquaculture research. In: Digital Image Computing: Techniques and Applications (DICTA). IEEE, pp. 457–464 (2008)
8. Pitt, M.K., Shephard, N.: Filtering via simulation: auxiliary particle filters. J. Am. Stat. Assoc. **94**, 590–599 (1999)
9. Van Der Merwe, R., Doucet, A., De Freitas, N., Wan, E.: The unscented particle filter. In: NIPS, pp. 584–590 (2000)
10. Khan, Z., Balch, T., Dellaert, F.: A rao-blackwellized particle filter for eigentracking. In: Proceedings of the 2004 IEEE Computer Society Conference on Computer Vision and Pattern Recognition, CVPR 2004, vol. 2, p. II-980. IEEE (2004)
11. Doucet, A., Godsill, S., Andrieu, C.: On sequential monte carlo sampling methods for bayesian filtering. Stat. Comput. **10**, 197–208 (2000)
12. Kotecha, J.H., Djurić, P.M.: Gaussian particle filtering. IEEE Trans. Sig. Process. **51**, 2592–2601 (2003)
13. Wang, H., Suter, D., Schindler, K.: Effective appearance model and similarity measure for particle filtering and visual tracking. In: Leonardis, A., Bischof, H., Pinz, A. (eds.) ECCV 2006. LNCS, vol. 3953, pp. 606–618. Springer, Heidelberg (2006)

14. Yin, M., Zhang, J., Sun, H., Gu, W.: Multi-cue-based camshift guided particle filter tracking. Expert Syst. Appl. **38**, 6313–6318 (2011)
15. Wang, Y., Tang, X., Cui, Q.: Dynamic appearance model for particle filter based visual tracking. Pattern Recogn. **45**, 4510–4523 (2012)
16. Bhaskar, H., Dwivedi, K., Dogra, D.P., Al-Mualla, M., Mihaylova, L.: Autonomous detection and tracking under illumination changes, occlusions and moving camera. Sig. Process. **117**, 343–354 (2015)
17. Liu, H., Sun, F.: Efficient visual tracking using particle filter with incremental likelihood calculation. Inf. Sci. **195**, 141–153 (2012)
18. Dunne, P., Matuszewski, B.: Choice of similarity measure, likelihood function and parameters for histogram based particle filter tracking in cctv grey scale video. Image Vis. Comput. **29**, 178–189 (2011)
19. Zhang, X., Hu, W., Maybank, S.: A smarter particle filter. In: Zha, H., Taniguchi, R., Maybank, S. (eds.) ACCV 2009, Part II. LNCS, vol. 5995, pp. 236–246. Springer, Heidelberg (2010)
20. Ikoma, N., Ichimura, N., Higuchi, T., Maeda, H.: Maneuvering target tracking by using particle filter. In: IFSA World Congress and 20th NAFIPS International Conference, Joint 9th, vol. 4, pp. 2223–2228. IEEE (2001)
21. Kwak, S., Nam, W., Han, B., Han, J.H.: Learning occlusion with likelihoods for visual tracking. In: IEEE International Conference on Computer Vision (ICCV), pp. 1551–1558. IEEE (2011)
22. Kale, A., Vaswani, N., Jaynes, C.: Particle filter with mode tracker (pf-mt) for visual tracking across illumination change. In: IEEE International Conference on Acoustics, Speech and Signal Processing. ICASSP 2007, vol. 1, p. I-929. IEEE (2007)
23. Vaswani, N.: Additive change detection in nonlinear systems with unknown change parameters. IEEE Trans. Sig. Process. **55**, 859–872 (2007)
24. Dataset: (Advanced video and signal based surveillance (2007). www.eecs.qmul.ac.uk)
25. Yang, F., Lu, H., Yang, M.H.: Robust superpixel tracking. IEEE Trans. Image Process. **23**, 1639–1651 (2014)
26. Ross, D.A., Lim, J., Lin, R.S., Yang, M.H.: Incremental learning for robust visual tracking. Int. J. Comput. Vis. **77**, 125–141 (2008)

Adaptive Flocking Control of Multiple Unmanned Ground Vehicles by Using a UAV

Mohammad Jafari$^{(\boxtimes)}$, Shamik Sengupta, and Hung Manh La

Department of Computer Science and Engineering,
University of Nevada, Reno 89557, USA
mohamadj@gmail.com, {ssengupta,hla}@unr.edu

Abstract. In this paper we aim to discuss adaptive flocking control of multiple Unmanned Ground Vehicles (UGVs) by using an Unmanned Aerial Vehicle (UAV). We utilize a Quadrotor to provide the positions of all agents and also to manage the shrinking or expanding of the agents with respect to the environmental changes. The proposed method adaptively causes changing in the sensing range of the ground robots as the quadrotor attitude changes. The simulation results show the effectiveness of proposed method.

Keywords: Hybrid system · Multi-agent · Flocking control · UAVs · UGVs

1 Introduction

Multi-agent control has attracted a lot of interests in recent years. A multi-agent system consists of a group of agents (for instance: robots, vehicles, etc.) cooperating with each other to do a task. The importance of multi-agent systems is shown specifically when a single agent can not easily solve a problem by itself. Utilizing a team of robots has several advantages over a single one such as: (i) there are some intrinsically distributed tasks, (ii) working multiple robots simultaneously, parallelizable problems can be solved in less time, (iii) minimizing the effects of failure by working together to accomplish the same task [1,2].

The applications of multiple robots include but not limited to rescue operations, environmental monitoring, robot-soccer and target tracking [1–6]. Search and rescue operations are definitely among the most important tasks that should be done when natural and/or manmade disasters such as earthquakes, tsunamis or explosions happen. Disaster environments are always hazardous and dangerous for people. In past years, researchers mostly focused on UGVs for employing them in search and rescue operations. Recently, UAVs attract many researchers because of their various capabilities. Employing UAVs and UGVs together as a hybrid system, UAVs could cooperate with UGVs to better accomplish the tasks.

Utilizing hybrid systems raises specific challenges and questions. Why do we need hybrid systems? What are their advantages? What kind of UAVs are more appropriate for hybrid systems? What kind of constraints need to be considered?

G. Bebis et al. (Eds.): ISVC 2015, Part II, LNCS 9475, pp. 628–637, 2015.
DOI: 10.1007/978-3-319-27863-6_58

What are the challenges of using hybrid systems? Because of the challenges of disaster environments (i.e., buildings pose 3-D constraints on visibility, low visibility, communication, GPS, etc.) a network of aerial and ground vehicles working in cooperation is more beneficial. Also in noisy cluttered environments top layer UAVs, by relying on measurements from the GPS/IMU and camera parameters, could help bottom layer UGVs by providing localization data and acting as communication relays. For example UAVs at the higher attitude could localize UGVs using a sequence of images taken which relates the position of the robot in a global coordinate frame with its pixel coordinates in the image or by using a set of known landmarks in the image.

Furthermore, having adaptive formation for multiagent systems seems to be more practical compared with fixed formation of the system. Changing the formation can be caused by several reasons that include but not limited to:

(i) Covering the greater part of the environment: In some application such as mapping an area, it is important that the multiagent system has the ability to spread out or gather. Therefore, by having the ability to change the formation, the agents can successfully accomplish the predefined mission.

(ii) Moving along an obstacle or through obstacles: Changing the formation in case of facing an obstacle is a preferable way to avoid collision.

"Formation control is a multi-agent application where the objective is for the robotic network to move into a desired formation (particular pattern, i.e., triangular) and to accomplish the desired task" [6]. There are diverse approaches for formation control of a multi-agent system such as: the behavior based, virtual structure, leader follower, and graph theoretical approaches [3].

In the leader-follower approach, several mobile robots are chosen as leaders and the rest of robots as followers. The leaders track predefined trajectories while the followers chase after the leaders. The main advantage of this method is its simplicity [3]. The goal of virtual structure method is to force a group of agents to stay in a rigid formation [5]. However, existing approaches have some limitations dealing with the terrain changes, for example when the UGV fleet has to pass a narrow space through a difficult terrain. Formation of the fleet might be changed or the connectivity of them might be lost. Also, the fleet might get stuck and this would affect the tracking performance. Furthermore, it is difficult for each agent to sense the whole area and estimate the size of the obstacles or the passing space among them. Therefore, designing an adaptive flocking control by utilizing a UAV to providing the global information about the environment is an interesting and challenging task.

The main contribution of this paper is to propose a new approach to the adaptive flocking control with the consideration of utilizing a quadrotor to provide the position information for ground agents and to manage their shrinking or expansion through altitude adaptation so the collision and obstacle avoidance could perform more effectively. The proposed hybrid system idea can allow ground agents to shrink or expand to effectively avoid the obstacles due to the difficult terrain. The simulation results for the proposed work is given (Fig. 1).

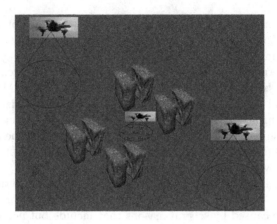

Fig. 1. Agents shrinking and expansion due to the terrain changes.

The rest of the paper is organized as follows. The next section presents modeling of both UGVs and UAV. Section 3 addresses the problem definition and our proposed formulation for hybrid system. The simulation results are presented in Sect. 4. The conclusion of the paper is provided in Sect. 5.

2   Preliminaries

2.1   Modeling Unmanned Ground Vehicles

We consider the flocks topology for modeling the Unmanned Ground Vehicles (UGVs) in this paper. It is based on the model introduced by Olfati-Saber [7] in 2006. Assuming that n agents are moving in an m dimensional space ($m = 2, 3$), the equation of motion of each dynamic agent could be described as follow:

$$\begin{cases} \dot{q}_i = p_i \\ \dot{p}_i = u_i \quad , i = 1, 2, ..., n \end{cases} \tag{1}$$

where $q_i, p_i, u_i \in \mathbb{R}^m$ are position, velocity, and control input of the agent i respectively. Let us consider a dynamic graph $G(v, \varepsilon)$ consisting of a set of edges and vertices as follow: $v = \{1, 2, ..., n\}, \varepsilon \subseteq \{(i, j) : i, j \in v, j \neq i\}$. In topology of flocks each agent is represented by a vertex and each edge shows that there exists a communication link between two agents. The neighborhood set of agent i could be defined by

$$N_i^\alpha = \{j \in v_\alpha :\| q_j - q_i \| < r, j \neq i\} \tag{2}$$

where r is an interaction range between two agents and $\| \cdot \|$ is the Euclidean norm in $\mathbb{R}^m$. By solving the following set of algebraic conditions, a geometric model of flocks (α-lattice) [7] could be described:

$$\| q_j - q_i \| = d \quad \forall j \in N_i^\alpha \tag{3}$$

where d (a positive constant) is the distance between two neighbors i and j. The above equation causes a singularity for the collective potential function at $q_i = q_j$. To resolve the aforementioned problem, (3) could be rewritten as follow [7]:

$$\| q_j - q_i \|_\sigma = d_\alpha \quad \forall j \in N_i^\alpha \tag{4}$$

where $d_\alpha = \| d \|_\sigma$ and $\| \cdot \|_\sigma$ is called σ-norm which is defined as follow:

$\| z \| = \frac{1}{\epsilon}[\sqrt{1 + \epsilon \| z \|^2} - 1], \epsilon > 0$. For a vector z, σ-norm is a map from $\mathbb{R}^m$ to $\mathbb{R} \geq 0$. The new map (i.e., $\| z \|_\sigma$) is differentiable everywhere while the Euclidean norm (i.e., $\| z \|$) is not differentiable at $z = 0$. A smooth collective potential function which is induced by the above constraints is obtain from: $V(q) = \frac{1}{2} \sum_i \sum_{j \neq i} \psi_\alpha(\| q_j - q_i \|_\sigma)$. where $\psi_\alpha(z)$ is a smooth pairwise potential function which is defined by $\psi_\alpha(z) = \int_{d_\alpha}^z \phi_\alpha(s)ds$. Here $\phi_\alpha(z) = \rho_h(z/r_\alpha)\phi(z - d_\alpha)$, $\phi(z) = \frac{1}{2}[(a + b)\sigma_1(z + c) + (a - b)]$ and $\sigma_1(z) = z/\sqrt{1 + z^2}$ while $\phi(z)$ is an uneven sigmoidal function with parameters $0 < a \leq b, c = |a-b|/\sqrt{4ab}$ to guarantee $\phi(0) = 0$. There is a bump function $\rho(z)$ which is scalar function that smoothly varies between $[0,1]$. One possible choice is the following [7]:

$$\begin{cases} 1, & z \in [0, h) \\ \frac{1}{2}\left[1 + cos(\pi \frac{(z-h)}{(1-h)})\right], & z \in [h, 1] \\ 0, & otherwise \end{cases} \tag{5}$$

The main flocking control algorithm ($u_i = u_i^\alpha + u_i^\beta + u_i^\gamma$) in [7] has the capability of controlling all agents to form a lattice configuration (which is called α-lattice) while avoiding the obstacles. The algorithm consists of three terms: u_i^α is the interaction component between two α-agents, u_i^β is the interaction component between α-agent and an obstacle (which is called β-agent), and u_i^γ is a goal component which consist of a distributed navigational feedback.

$$u_i^\alpha = c_1^\alpha \sum_{j \in N_i^\alpha} \phi_\alpha(\| q_j - q_i \|_\sigma)\mathbf{n}_{i,j} + c_2^\alpha \sum_{j \in N_i^\alpha} a_{ij}(q)(p_j - p_i) \tag{6}$$

$$u_i^\beta = c_1^\beta \sum_{k \in N_i^\beta} \phi_\beta(\| \hat{q}_{i,k} - q_i \|_\sigma)\hat{\mathbf{n}}_{i,k} + c_2^\beta \sum_{k \in N_i^\beta} b_{i,k}(q)(\hat{p}_{i,k} - p_i) \tag{7}$$

$$u_i^\gamma = -c_1^\gamma \sigma_1(q_i - q_r) - c_2^\gamma (p_i - p_r) \tag{8}$$

where $c_1^\alpha, c_2^\alpha, c_1^\beta, c_2^\beta, c_1^\gamma$, and c_2^γ are all positive constants. The pair (q_r, p_r) is the virtual leader (i.e., γ-agent) which could be defined as follow:

$$\begin{cases} \dot{q}_r = p_r \\ \dot{p}_r = f_r(q_r, p_r) \end{cases} \tag{9}$$

The vectors $\mathbf{n}_{i,j}$ and $\hat{\mathbf{n}}_{i,k}$ are described as below: $\mathbf{n}_{i,j} = \frac{q_j - q_i}{\sqrt{1 + \epsilon \| q_j - q_i \|^2}}$, $\hat{\mathbf{n}}_{i,k} = \frac{\hat{q}_{i,k} - q_i}{\sqrt{1 + \epsilon \| \hat{q}_{i,k} - q_i \|^2}}$. The $a_{ij}(q)$ and $b_{i,k}(q)$ are the elements of spatial adjacency matrix $A(q)$ and heterogeneous adjacency matrix $B(q)$ respectively which could be defined by: $a_{ij}(q) = \rho_h(\| q_j - q_i \|_\sigma / r_\alpha) \in [0,1], j \neq i$, $b_{i,k}(q) = \rho_h$ $(\| \hat{q}_{i,k} - q_i \|_\sigma / d_\beta)$. where $r_\alpha = \| r \|_\sigma$, $a_{ii}(q) = 0$ for all i and q, $d_\beta = \| d \|_\sigma$, and $r_\beta = \| r' \|_\sigma$. The repulsive action function $\phi_\beta(z)$ is defined as: $\phi_\beta(z) = \rho_h(z/d_\beta)(\sigma_1(z - d_\beta) - 1)$.

Similar to (2) we can define the set of β-neighbors of an α-agent i as follow: $N_i^\beta = \{k \in \upsilon_\beta : \| \hat{q}_{i,k} - q_i \| < r' \}$, where $r' > 0$ is interaction range of an α-agent with obstacles.

2.2   Modeling Unmanned Aerial Vehicles

Nowadays, the knowledge of UAVs is at the cutting edge of researches. Among them, flights which have the capability of Vertical Takeoff and Landing (VTOL), specially Quadrotors, are under our consideration. Various control methods have been employed for stabilizing and controlling of Quadrotors. Coza et al. [8] utilized Adaptive fuzzy control for a quadrotor helicopter. Adaptive neural control of a quadrotor helicopter with extreme learning machine was designed by Zhang et al. [9]. In [10] adaptive sliding mode control design was presented. Attitude control of a quadrotor using brain emotional learning based intelligent controller was proposed in 2013 [11]. Bou-Ammar and his colleagues [12] employed the reinforcement method to control the quadrotor. Nonlinear robust output feedback tracking control of a quadrotor by using quaternion representation was proposed in [13]. Robust attitude controller design for miniature quadrotors has been recently proposed by Liu et al. [14]. To model a quadrotor, a hybrid frame consisting of E-frame and B-frame was utilized [15], where the equations in H-frame (i.e., linear equation WRT E-frame and angular equations WRT B-frame) were defined as follows:

$$\begin{cases} \ddot{X} = (sin\psi sin\phi + cos\psi sin\theta cos\phi)\frac{U_1}{m} \\[2mm] \ddot{Y} = (-cos\psi sin\phi + sin\psi sin\theta cos\phi)\frac{U_1}{m} \\[2mm] \ddot{Z} = -g + (cos\theta cos\phi)\frac{U_1}{m} \\[2mm] \dot{p} = \frac{I_{YY} - I_{ZZ}}{I_{XX}} qr - \frac{J_{TP}}{I_{XX}} q\Omega + \frac{U_2}{I_{XX}} \\[2mm] \dot{q} = \frac{I_{ZZ} - I_{XX}}{I_{YY}} pr - \frac{J_{TP}}{I_{YY}} p\Omega + \frac{U_3}{I_{YY}} \\[2mm] \dot{r} = \frac{I_{XX} - I_{YY}}{I_{ZZ}} pq + \frac{U_4}{I_{ZZ}} \end{cases} \qquad (10)$$

where ψ, ϕ, and θ are Yaw, Roll, and Pitch angles respectively. I_{XX}, I_{YY}, and I_{ZZ} are body moment of inertia around x, y, and z axis respectively. J_{TP} is total rotational moment of inertia around propeller axis, g is the acceleration due to gravity and m is quadrotor mass. Relation between basic movements and the propellers' speed are defined by

$$\begin{cases} U_1 = b_q(\Omega_1^2 + \Omega_2^2 + \Omega_3^2 + \Omega_4^2) \\[2mm] U_2 = b_q l(-\Omega_2^2 + \Omega_4^2) \\[2mm] U_3 = b_q l(-\Omega_1^2 + \Omega_3^2) \\[2mm] U_4 = d_q(-\Omega_1^2 + \Omega_2^2 - \Omega_3^2 + \Omega_4^2) \\[2mm] \Omega = -\Omega_1 + \Omega_2 - \Omega_3 + \Omega_4 \end{cases} \qquad (11)$$

where Ω_1, Ω_2, Ω_3, and Ω_4 are front, right, rear, and left propeller speeds respectively. Also, b_q is trust factor of the quadrotor, d_q is drag factor of the quadrotor and l is distance between center of the quadrotor and center of the propeller.

3   Problem Definition

There are several challenges existing when a fleet of UGVs has to pass through difficult terrain with obstacles. The fleet might get stuck behind the obstacles and this will cause problem in tracking the target [7]. Also because of 3-D constraints on visibility, communication constraints, GPS denied area and so on, the UGV fleet might lost their connectivities and/ or their path. Because of their capabilities, UAVs can be of help in guiding the UGV fleet. In this work, we are proposing methodology for UAV so that it can help the UGV fleet most effectively.

The idea of this work is based on changing of the quadrotor height. As the UAV senses the area is free it flies to the higher attitude so this should force the agents to expand and cover more area. On the contrary, if the UAV sees any obstacle or narrow path it will fly down to a lower height and this will cause the agents to shrink so they could successfully pass the narrow area. We assume that the quadrotor is providing the positional information for the UGVs. As Fig. 2 (a) shows, changing of the quadrotor height will change the Field of View (FOV) of the bottom camera. The higher the quadrotor flies, the bigger FOV will be achieved. The radius of FOV of the quadrotor could be defined as follow:

$$r_{FOV} = H_q tan(\alpha_q) \qquad (12)$$

where H_q is the quadrotor attitude and α_q is the half-angle of view of the quadrotor bottom camera. The quadrotor tracks the same trajectory as the UGV fleet while its x-y position is same as the center of mass of the UGV fleet. The x-y position of the quadrotor can be defined by:

$$\bar{q} = \frac{1}{n} \sum_{i=1}^{n} q_i \qquad (13)$$

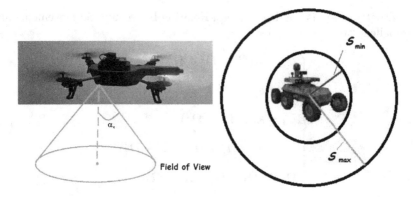

Fig. 2. (a) Field of View of quadrotor bottom camera (Left), (b) Minimum and Maximum sensing range of UGV (Right).

For implementing the above idea we will utilize a function F(h) which will help us insert the above mentioned method to the system. The function F should have the following properties:

1. $F(h) = 0, \qquad \forall 0 < h \le h_{min}$
2. $0 < F(h) < S_{dif}, \forall h_{min} < h \le h_{max}$
3. $F(h)$ is smooth and differentiable

Where h_{min} and h_{max} are minimum and maximum heights of the UAV respectively. S_{dif} could be obtained as follow:

$$S_{dif} = S_{max} - S_{min} \qquad (14)$$

where S_{max} and S_{min} are maximum and minimum sensing ranges of the ground agents respectively (depicted in Fig. 2 (b)).

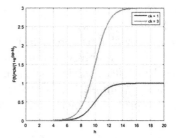

Fig. 3. Function $F(h)$ with two different c_k.

We introduce the following function which satisfies all the above mentioned properties:

$$F(h) = c_k \frac{1}{1 + e^{(c_p - h)}} \qquad (15)$$

where c_k and c_p are two positive constants. c_p affects the smoothness of the $F(h)$ (which defines the speed of shrinking or expansion of the agents) while c_k changes the variation range of the $F(h)$. Figure 3 demonstrates the $F(h)$ with two different $c_k = 3$ and $c_k = 1$, and $c_p = 10$. We then rewrite equation (4) to include the effect of quadrotor height changes to the shrinking or expanding of the flocks. Below is the new equation which is employed in our proposed work:

$$\| q_j - q_i \|_\sigma = d_\alpha^{new} + F(h) \quad \forall j \in N_i^\alpha \qquad (16)$$

where d_α^{new} is the minimum sensing range of the ground agents.

It could be clearly seen as the height of the quadrotor is changing it affects the sensing range of the ground agents and force them to expand or shrink as they are moving towards their goals. The ground agents still have a capability of avoiding the collision or obstacles by themselves.

4   Simulation Results

In this section two different simulation results from 2-D and 3-D view prospective are presented. The following parameters are used through the simulations: $d_\alpha^{new} = 4$, $r = 1.2d_\alpha^{new}$, $d = 0.6d_\alpha^{new}$, $r = 1.2d$, for σ-norm $\epsilon = 0.1$, $a = b = 5$ (for $\phi(z)$), for the bump functions of $\phi_\alpha(z)$ and $\phi_\alpha(z)$, $h = 0.2$ and $h = 0.9$ respectively. Other parameters of the algorithm and the initial positions and velocities of all agents are specified in each case separately.

4.1   3-D View

We employed 10 ground agents which are randomly distributed in the $[-40, 80]^2$. The initial velocities of all agents are equally chosen as zero. We used the obstacles which can be defined from following matrix:

$$OBS_1 = \begin{bmatrix} 220 & 220 & 260 & 260 & 300 & 300 & 340 & 340 \\ -30 & 70 & -30 & 70 & -30 & 70 & -30 & 70 \\ -15 & -15 & -13 & -13 & -15 & -15 & -13 & -13 \\ 20 & 20 & 20 & 20 & 20 & 20 & 20 & 20 \end{bmatrix}.$$

The first three rows of the matrix are the locations (i.e., x-y-z position) of the obstacles, while the last row shows the radius of the obstacles. Figure 4 demonstrates the shrinking of the agents. As the quadrotor senses the obstacles, it flies in a lower attitude and this is forcing the ground agents to reduce the inter-agent distance between them and it results in shrinking. As it could clearly be seen the flock size successfully reduced.

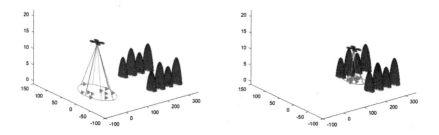

Fig. 4. Shrinking of the agents.

4.2   2-D View

We employed 150 ground agents which are randomly distributed in the $[-40, 80]^2$. The initial velocities of all agents are equally chosen as zero. We used the obstacles which can be defined from following matrix:

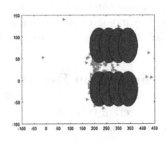

$$OBS_2 = \begin{bmatrix} 220 & 220 & 260 & 260 & 300 & 300 & 340 & 340 \\ -20 & 80 & -20 & 80 & -20 & 80 & -20 & 80 \\ 40 & 40 & 40 & 40 & 40 & 40 & 40 & 40 \end{bmatrix}.$$

The first two rows of the matrix are the locations (i.e., x-y position) of the obstacles, while the last row shows the radius of the obstacles.

Fig. 5. 2D view of Shrinking of the agents- algorithm from [7].

Figure 5 shows the result which is obtained using algorithm in [7]. As it is clearly seen the ground agents get stuck in the narrow path between the obstacles. Figure 6(Left) demonstrates how the ground agents are shrinking and passing the narrow space between obstacles while avoiding any collisions between agents or obstacles. Figure 6(Right) illustrates the closer look of Fig. 6(Left).

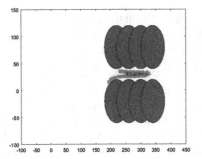
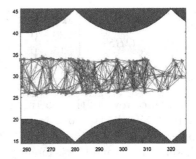

Fig. 6. 2D view of Shrinking of the agents.

All these results are obtained using proposed adaptive flocking method. The results shows the satisfactory performances of the proposed work.

5   Conclusions

Adaptive flocking control of multiple unmanned ground vehicles by using an unmanned aerial vehicle is presented in this paper. We employed a Quadrotor to provide the positions of all agents and also to manage the shrinking or expanding of the agents with respect to the terrain changes. The simulation results show the effectiveness of proposed method.

References

1. Sabattini, L.: Nonlinear Control Strategies for Cooperative Control of Multi-Robot Systems. Ph.D. thesis, Universitá di Bologna (2012)
2. Garg, D.P., Fricke, G.K.: Potential function based formation control of mobile multiple-agent systems. In: 1st International and 16th National Conference on Machines and Mechanisms (iNaCoMM2013) (2013)
3. Dong, W., Guo, Y., Farrell, J.: Formation control of nonholonomic mobile robots. In: American Control Conference, pp. 5602–5607. IEEE (2006)
4. Speranzon, A.: On Control Under Communicaiton Constraints in Autonomous Multi-Robot Systems. KTH, Signals, Sensors and Systems, Stockholm (2004)
5. Arranz, L.B.: Cooperative control design for a fleet of AUVs under communication constraints. Ph.D. thesis, Université de Grenoble (2011)
6. Clark, J.D.: Cooperative hybrid control of robotic sensors for perimeter detection and tracking. Master's thesis, Oklahoma State University (2005)
7. Olfati-Saber, R.: Flocking for multi-agent dynamic systems: algorithms and theory. IEEE Trans. Autom. Control 51, 401–420 (2006)
8. Coza, C., Nicol, C., Macnab, C., Ramirez-Serrano, A.: Adaptive fuzzy control for a quadrotor helicopter robust to wind buffeting. J. Intell. Fuzzy Syst.: Appl. Eng. Technol. 22, 267–283 (2011)
9. Zhang, Y., Xu, B., Li, H.: Adaptive neural control of a quadrotor helicopter with extreme learning machine. In: Cao, J., Mao, K., Cambria, E., Man, Z., Toh, K.-A. (eds.) Proceedings of ELM-2014 Volume 2, PALO, vol. 4, pp. 125–134. Springer, Heidelberg (2014)
10. Islam, S., Faraz, M., Ashour, R., Cai, G., Dias, J., Seneviratne, L.: Adaptive sliding mode control design for quadrotor unmanned aerial vehicle. In: 2015 International Conference on Unmanned Aircraft Systems (ICUAS), pp. 34–39. IEEE (2015)
11. Jafari, M., Shahri, A.M., Shouraki, S.B.: Attitude control of a quadrotor using brain emotional learning based intelligent controller. In: 2013 13th Iranian Conference on Fuzzy Systems (IFSC), pp. 1–5. IEEE (2013)
12. Bou-Ammar, H., Voos, H., Ertel, W.: Controller design for quadrotor uavs using reinforcement learning. In: 2010 IEEE International Conference on Control Applications (CCA), pp. 2130–2135. IEEE (2010)
13. Xian, B., Diao, C., Zhao, B., Zhang, Y.: Nonlinear robust output feedback tracking control of a quadrotor UAV using quaternion representation. Nonlinear Dyn. 79, 2735–2752 (2015)
14. Liu, H., Li, D., Xi, J., Zhong, Y.: Robust attitude controller design for miniature quadrotors. Int. J. Robust Nonlinear Control (2015)
15. Bresciani, T.: Modelling, Identification and Control of a Quadrotor Helicopter. Lund University, Department of Automatic Control, Lund (2008)

Basic Study of Automated Diagnosis of Viral Plant Diseases Using Convolutional Neural Networks

Yusuke Kawasaki[1], Hiroyuki Uga[2],
Satoshi Kagiwada[3], and Hitoshi Iyatomi[1]([✉])

[1] Applied Informatics, Graduate School of Science and Engineering,
Hosei University, Tokyo, Japan
iyatomi@hosei.ac.jp
[2] Saitama Agricultural Technology Research Center, Saitama, Japan
[3] Clinical Plant Science, Faculty of Bioscience and Applied Chemistry,
Hosei University, Tokyo, Japan

Abstract. Detecting plant diseases is usually difficult without an experts' knowledge. Therefore, fast and accurate automated diagnostic methods are highly desired in agricultural fields. Several studies on automated plant disease diagnosis have been conducted using machine learning methods. However, with these methods, it can be difficult to detect regions of interest, (ROIs) and to design and implement efficient parameters. In this study, we present a novel plant disease detection system based on convolutional neural networks (CNN). Using only training images, CNN can automatically acquire the requisite features for classification, and achieve high classification performance. We used a total of 800 cucumber leaf images to train CNN using our innovative techniques. Under the 4-fold cross-validation strategy, the proposed CNN-based system (which also extends the training dataset by generating additional images) achieves an average accuracy of 94.9 % in classifying cucumbers into two typical disease classes and a non-diseased class.

1 Introduction

Plant diseases have a devastating effect on agricultural products. The monetary loss caused by plant diseases is estimated to be \$30–50 billion annually [1]. Viral plant diseases in particular cause significant damage to agriculture. Because there is no treatment for these diseases, infected plants must be removed as quickly as possible to avoid secondary infection; thus, early detection is required. As a result, the number of diagnostic requests to prefectural agricultural agencies in Japan has been increasing. In general, plant diagnosis by experts is expensive, and viral plant diseases are occasionally missed or misdiagnosed because their symptoms are difficult to identify. Thus, plant pathologists have shared their knowledge with farmers through farming communities [2].

To improve diagnostic results, several studies on machine learning-based automated plant diagnosis have been conducted [3–7]. Huang proposed a recognition method based on a multi-layer perceptron to identify bacterial soft rot,

© Springer International Publishing Switzerland 2015
G. Bebis et al. (Eds.): ISVC 2015, Part II, LNCS 9475, pp. 638–645, 2015.
DOI: 10.1007/978-3-319-27863-6_59

bacterial brown spot and Phytophthora black rot appearing on orchids, and reported an average classification accuracy of 89.6 % [4]. Phadikar et al. proposed a system based on self-organized maps; it achieved over 70 % accuracy in distinguishing between rice blast and brown spot appearing on rice leaves [5]. Zhang et al. utilized a support vector machine for distinguishing downy mildew, brown spot, and angular leaf spot on cucumbers, and attained an accuracy of 83.3 % [6]. Xu et al. analyzed nutrient deficiency in tomatoes by means of k-nearest neighbor clustering based on leaf color and texture. They detected nitrogen and potassium deficiencies with probabilities of 90 % and 85 %, respectively [7]. However, these methods are faced with several difficulties, involving the detection of regions of interest (ROIs) for subsequent processing, the design and implementation of efficient diagnosis parameters, and so on. Accordingly, it is quite difficult to apply these techniques to other purposes (i.e. detecting other diseases or detecting diseases on different plants).

On the other hand, convolutional neural networks (CNNs) are widely percieved as one of the most promising classification techniques among machine learning fields. The most attractive advantage of CNN is their ability to acquire requisite features for the classification from the images automatically during their learning processes. Recently, CNN have demonstrated excellent performance in large scale general image classification tasks [8], traffic sign recognition [9], leaf classification [10], and so on.

In this paper, we introduce an innovative technique to enhance the learning ability of a CNN training schema, and propose an automated plant disease diagnosis system based on these enhancements. In this study, we develop a plant disease diagnosis system that uses leaf images to detect two harmful viral infections that afflict cucumber plants (MYSV: melon yellow spot virus, and ZYMV: zucchini yellow mosaic virus). We designed our system to receive, as input, a leaf image from the plant to be investigated, and to yield a diagnostic result as output for practical use. Because CNNs are highly expected to acquire "features of the disease" automatically, our system is designed to perform accurate recognition without troublesome pre-processing or parameter design. We expect that farmers will be able to detect early-stage infections by using our system in conjunction with a common digital camera.

2   Plant Disease Detection System

2.1   Material and Pre-processing

Images of cucumber leaves were supplied by Saitama Prefectural Agriculture and Forestry Research Center, Japan. The dataset consists of 800 cucumber leaf images (300 with MYSV, 200 with ZYMV, and 300 non-diseased); the leaves are situated in the center of the images. OLYMPUS 5P560UZ and SONY DSV-RX100 color digital cameras were used for capturing leaf images with resolutions of 2048×1536 and 2736×1824 pixels, respectively. All pre-processing procedures are shown in Fig. 1.

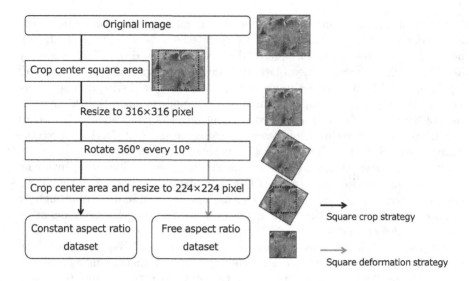

Fig. 1. Schematics of pre-processing.

We designed our system to accept images from various types of digital cameras, to accommodate images captured in different sizes and aspect ratios. In our tests, we used two pre-processing strategies, namely (1) square crop, and (2) square deformation. In the (1) square crop strategy, we select a square region from the center of the photographed image. Thus, the side regions of the images are removed, because we assume that leaf images will be photographed with the area of interest situated in the center. In the (2) square deformation strategy, we forcibly deform the photographed image into a 1:1 aspect ratio with bi-linear interpolation. These pre-processing strategies created two types of datasets: the "constant aspect ratio dataset" and "free aspect ratio dataset", respectively. Subsequently, each leaf image in the training dataset was artificially extended 36 times by rotating the image with 10 degree increments. This rotation process helps CNN to acquire various types of local features, because CNN learn them using convolutional integration. Therefore, we expect this rotation process to help improve classification performance. Finally, we resized these images into 224×224 pixels using bi-linear interpolation.

2.2   Architecture

Our CNN-based plant disease detection system, which uses the Caffe framework [11], includes convolution layers (Conv), pooling layers (Pool), and local contrast normalization layers (Norm). Each neuron in our network is activated by a rectified linear unit (ReLU) function [12]. An illustration and details of our CNN architecture are shown in Fig. 2 and Table 1. In Fig. 2, we omitted "Pool" and "Norm" to enhance clarity.

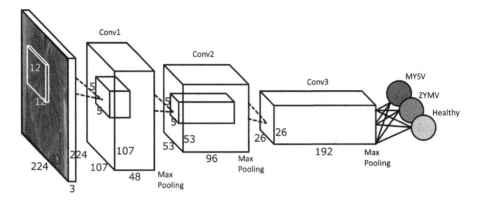

Fig. 2. Illustration of our CNN architecture.

Table 1. Details of our CNN architecture.

Layer name	Function	Weight filter sizes	Output tensor†	Notes
Input	-		$3 \times 224 \times 224$	
Conv1	Convolution	12×12	$3 \times 107 \times 107$	stride = 2, bias = 0
Pool1	Max Pooling	3×3	$48 \times 53 \times 53$	stride = 2
Norm1	Normalization	5×5	$48 \times 53 \times 53$	
Conv2	Convolution	5×5	$96 \times 53 \times 53$	stride=2, bias = 1
Pool2	Max Pooling	3×3	$96 \times 26 \times 26$	stride = 2
Norm2	Normalization	5×5	$96 \times 26 \times 26$	
Conv3	Convolution	5×5	$192 \times 24 \times 24$	stride = 2, bias = 0
Pool3	Max Pooling	3×3	$192 \times 12 \times 12$	stride = 2
Norm3	Normalization	5×5	$192 \times 12 \times 12$	
Dense	Full Connection		$3 \times 1 \times 1$	bias = 0

† : # of color channels (or # of feature maps) × width of the map × height of the map

The convolution layer convolves the input image (or the output of the previous layer) with tunable weight filters (i.e. kernel), in order to extract local features. By applying various types of weight filters, CNN acquire shift invariant and scale invariant local features from the input. The pooling layer summarizes the output of the previous layer and acquires the object's translation invariance. In our system, we applied max-pooling in each pooling layer. The local contrast normalization layer subtracts the means of neighborhood pixels from the target pixels and divides the subtracted pixels by the standard deviations of the neighborhood pixels. This layer enables the system to compensate for variations among images captured under different conditions.

3 Experiments

We performed three experiments for each dataset to estimate the effectiveness of our method. Experimental conditions are shown in Table 2. Experiment-1 uses the normal CNN model for the classification (as a performance baseline). Experiment-2's training epoch is adjusted with our proposed method (Experiment-3). Experiment-3 tests our proposed method. Note again that the proposed method rotates the input image every 10 degrees, to generate a total of 36 training images. All of these experiments were evaluated under the 4-fold cross-validation strategy.

Table 2. Experimental conditions.

	rotation	# of training epochs
Experiment-1	no	40
Experiment-2	no	1440(40 × 36)
Experiment-3(proposed)	yes	40

Tables 3 and 4 summarizes the system's classification performance when differentiating between plant diseases with the constant aspect ratio and free aspect ratio datasets.

Table 3. Differentiation results (Constant aspects ratio dataset).

	Accuracy(%)	Sensitivity for MYSV (%)	Sensitivity for ZYMV (%)	Specificity (%)
Experiment-1	78.5	74.0	76.0	84.7
Experiment-2	83.1	82.7	76.5	88.0
Experiment-3(proposed)	**94.9**	**96.3**	**89.5**	**97.0**

Table 4. Differentiation results (Free aspects ratio dataset).

	Accuracy(%)	Sensitivity for MYSV (%)	Sensitivity for ZYMV (%)	Specificity(%)
Experiment-1	77.0	73.0	74.5	82.7
Experiment-2	80.5	78.6	77.5	84.3
Experiment-3(proposed)	**92.5**	**95.3**	**87.0**	**92.0**

Figure 3 shows examples of correctly classified leaf images (a:MYSV, b:ZYMV and c:Non-diseased) from the constant aspect ratio dataset. Figure 4 shows weight filters obtained by the first convolution layer in Experiments-1, 2 and 3.

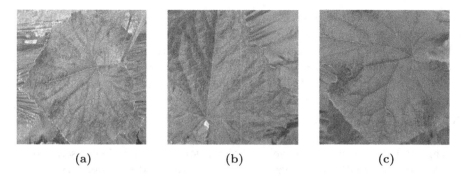

Fig. 3. An example of (a) MYSV, (b) ZYMV, (c) Normal leaves.

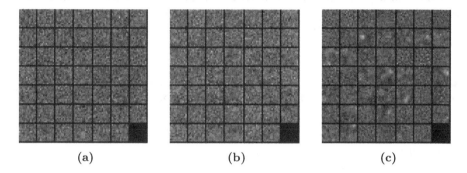

Fig. 4. Weights of kernel obtained in (a) Experiment-1, (b) Experiment-2, (c) Experiment-3.

4   Discussion

The results in Tables 3 and 4 show that CNN could solve difficult differentiation problems without segmenting the ROI (i.e. leaf areas) or extracting the requisite features. For both datasets, the proposed method (Experiment-3) achieved the best performance. From these results, we confirmed that expanding the dataset by rotating the images contributed significantly to enhancing the system's differentiation ability. Furthermore, the rotation created a larger variety of weight maps (Fig. 4(c)) than ordinal cases (Fig. 4(a) and (b)). On the other hand, we can also see the results from the constant aspect ratio dataset showed superior performance compared to the results from the other dataset, while the differences were limited. These facts imply that, as expected, CNN acquired various types of local features as a result of the rotation process. We can conclude that CNN learned effective features of plant diseases that are robust to geometrical distortions; in addition, we conclude that it not necessary to consider the aspect ratio of the camera, because of the effectiveness of the data expansion process.

We note that our dataset was obtained from a single source, and that the disease types were limited in this study. In practical use, there will be undesired effects appearing in the images, such as sunlight, drooping, and others effects caused by various situations. We will continue to improve our system, to enable it to distinguish a greater variety of diseases under various photographic conditions.

5   Conclusion

In this study, we proposed a novel detection system for viral plant disease using CNN and confirmed its effectiveness. We also confirmed that our strategy for training the CNN significantly improved its classification accuracy. We expect that this will free the system's users from paying a extra attention to the details of the photographic conditions. We will develop our next system in near future. We expect that our future system will make a significant contribution to the agricultural field.

Acknowledgement. This research was partially supported by the Japan Science and Technology Agency (JST) (A-STEP Feasibility Study program, AS262Z00664N, 2014-2015).

References

1. Sastry, K.S.: Plant Virus and Viroid Diseases in the Tropics, vol. II. Springer, Heidelberg (2013)
2. Mathews, D.M.: Optimizing detection and management of virus diseases of plants. In: Proceedings of the Landscape Disease Symposium, pp. 10–20 (2010)
3. Barbedo, J.G.A.: Digital image processing techniques for detecting, quantifying and classifying plant diseases. SpringerPlus **2**, 660 (2013)
4. Huang, K.Y.: Application of artificial neural network for detecting Phalaenopsis seedling diseases using color and texture features. Comput. Electron. Agric. **57**, 3–11 (2007)
5. Phadikar, S., Sil, J.: Rice disease identification using pattern recognition techniques. In: Computer and Information Technology, pp. 420–423 (2008)
6. Zhang, J., Zhang, W.: Support vector machine for recognition of cucumber leaf diseases. In: Proceedings of the 2nd IEEE International Conference on Advanced Computer Control, ICACC 2010, vol. 5, pp. 264–266 (2010)
7. Xu, G., Zhang, F., Shah, S.G., Ye, Y., Mao, H.: Use of leaf color images to identify nitrogen and potassium deficient tomatoes. Pattern Recogn. Lett. **32**, 1584–1590 (2011)
8. Krizhevsky, A., Sutskever, I., Hinton, G.E.: ImageNet classification with deep convolutional neural networks. In: Advances In Neural Information Processing Systems, pp. 1–9 (2012)
9. Jin, J., Fu, K., Zhang, C.: Traffic sign recognition with hinge loss trained convolutional neural networks. IEEE Trans. Intell. Transp. Syst. **15**, 1991–2000 (2014)
10. Hall, D., McCool, C., Dayoub, F., Sunderhauf, N., Upcroft, B.: Evaluation of features for leaf classification in challenging conditions. In: 2015 IEEE Winter Conference on Applications of Computer Vision, pp. 797–804 (2015)

11. Jia, Y., Shelhamer, E., Donahue, J., Karayev, S., Long, J., Girshick, R., Guadarrama, S., Darrell, T.: Caffe: Convolutional architecture for fast feature embedding. arXiv preprint arXiv:1408.5093 (2014)
12. Nair, V., Hinton, G.E.: Rectified linear units improve restricted boltzmann machines. In: Proceedings of the 27th International Conference on Machine Learning, pp. 807–814 (2010)

Efficient Training of Evolution-Constructed Features

Meng Zhang[✉] and Dah-Jye Lee

Department of Electrical and Computer Engineering,
Brigham Young University, Provo, UT 84602, USA
mengzhang24@hotmail.com

Abstract. Evolution-Constructed (ECO) features have been shown to be effective for general object recognition. ECO features use evolution strategies to build series of transforms and thus can be generated automatically without human expert involvement. We improved on our successful ECO features algorithm by reducing their dimensions before putting them into the classifier in order to create more effective ECO features. Efficient training of ECO features allows features to be more robust in representing the images.

1 Introduction

Object recognition is a very challenging task in computer vision. Humans recognize a multitude of objects in images easily while object recognition by computers requires a lot of effort. The difficulty lies in the fact that images of objects vary in lots of ways such as sizes, lighting conditions, view-points, and when they are partially occluded.

In our previous work, we proposed Evolution Constructed Features (ECO) [1]. It is a fully automated feature construction method to achieve general object recognition. Rather than relying on human experts to design or construct features, the proposed method uses standard genetic algorithm [2] to get high quality features from the raw pixels of the input images. These high quality features constructed by the ECO features are either good features that would have been selected by the human experts or features that are often overlooked by humans.

In this paper, we improve the effectiveness of ECO features by using feature descriptors. Feature descriptors are well-known to represent the features in a compact way and are widely used in object recognition. We were inspired to use feature descriptors in order to reduce the dimensionality of ECO features. Our original ECO features are represented as an ordered list of pixel values and thus are usually of high dimension. By using feature descriptors in the improved ECO features, feature dimensions can be greatly reduced.

Feature descriptors are also able to extract useful information that is robust to transforms, rotation or lighting change of objects [3]. Representing ECO features using pixel values is sensitive to small variation of object location and orientation although it's very simple to compute. Feature descriptors can help us not only to capture the most robust information of images but also make ECO features training more efficiently.

The rest of this paper is organized as follows. Section 2 describes the ECO features in detail. Feature descriptors details are presented in Sect. 3. Dataset and experimental

G. Bebis et al. (Eds.): ISVC 2015, Part II, LNCS 9475, pp. 646–654, 2015.
DOI: 10.1007/978-3-319-27863-6_60

results and related analysis are discussed in Sect. 4. Section 5 concludes the paper by summarizing our main idea.

2   ECO Features

2.1   What is an ECO Feature?

An ECO feature is generated by employing a series of image transforms onto a sub-region of the input image. The output of one transform is the input of the next transform. The order of image transforms, the corresponding parameters of each transform and the location of the sub-regions are all determined by the genetic algorithm.

This process is shown in Eq. 1, where V is the ECO feature output vector, n is the number of transforms an ECO feature is composed of, $I(x_1, y_1, x_2, y_2)$ defines the sub-region of an image I, T_i represents each transform at step i and ϕ_i is the corresponding parameter vector of each transform at step i.

$$V = T_n \left(V_{n-1}, \phi_n \right)$$
$$V_{n-1} = T_{n-1} \left(V_{n-2}, \phi_{n-1} \right)$$
$$\dots$$
$$V_1 = T_1 \left(I(x_1, y_1, x_2, y_2), \phi_1 \right)$$

(1)

Almost any transform is possible but we are mostly interested in those transforms that can be found in a typical image processing library. The number of transforms used to initially create an ECO feature, n, varies from 2 to 8 transforms. It was found that the average ECO feature has 3.7 transforms with a standard deviation of 1.7 transforms on the datasets we were using. The range 2 to 8 transforms allowed the search to yield good results while being less complicated.

So each ECO feature is composed of two parts: an ordering of transforms and the parameters used in each transform. Figure 1 shows a graphical example of two ECO features.

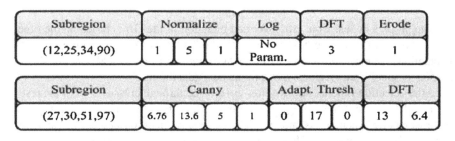

Fig. 1. Two example ECO features. The first example shows an ECO feature where the transforms are applied to the sub-region where $x_1 = 12$, $y_1 = 25$, $x_2 = 34$, and $y_2 = 90$ from Eq. 1. The values below the transforms are the parameter vectors ϕ_i also from Eq. 1

Rather than making any assumptions about where the good features can be found and what the salient regions are, the genetic algorithm is used to search for the sub-region parameters x_1, y_1, x_2, y_2. In this way the saliency of a sub-region is not determined by any experts but in its own ability to help classify objects, after being operated on by those transforms.

A sub-region of the image can range from single pixel area to the whole image. It allows both global and local information to be captured. And thus each ECO feature will specialize at representing different aspects of the object.

2.2 ECO Features Based Feature Construction

ECO features [1] is an effective method for feature construction. ECO features are constructed using a standard genetic algorithm (GA) [2]. Both the transforms and associated parameters are determined by GA. Initially, GA randomly generates a population of ECO features and verifies that each ECO feature consists of a valid ordering of transforms.

A fitness score is assigned to each ECO feature to indicate how good this ECO feature in the current generation is. In order to calculate a fitness score, each ECO feature is associated with a single perceptron which is defined in Eq. 2. The perceptron maps the ECO feature input vector $\mathbf{V}$ to a binary classification, α, through a weight vector $\mathbf{W}$ and a bias term b.

$$\alpha = \begin{cases} 1 & \text{if } \mathbf{W} \cdot \mathbf{V} + b > 0 \\ 0 & \text{else} \end{cases} \tag{2}$$

A fitness score, s, is defined in Eq. 3, which reflects how well the single perceptron classifies a holding set. In Eq. 3, t_p is the number of true positives, f_n is the number of false negatives, t_n is the number of true negatives, and f_p is the number of false positives. The fitness score ranges from 0 to 1000.

$$s = \frac{t_p \cdot 500}{f_n + t_p} + \frac{t_n \cdot 500}{f_p + t_n} \tag{3}$$

The process of selection, crossover and mutation of the GA is based on the fitness score. After a fitness score has been obtained for every ECO feature, a portion of the population is selected to continue to the next generation. A tournament selection method is used to select which features move to the next generation.

After selection is done, new features are created through crossover from the rest of the population. Once the next generation is filled, each of the parameters in the ECO features can be mutated. Examples of crossover and mutation process are shown in Fig. 2. This whole process of finding features is summarized in Algorithm 1.

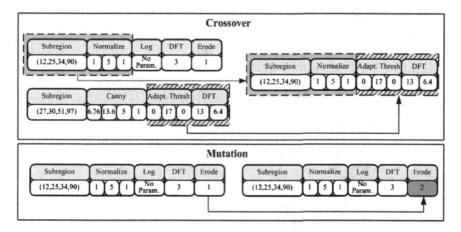

Fig. 2. Examples of crossover and mutation

Algorithm 1. Finding Features.

```
for Size of population do
    Randomly create ECO features and select all
    parameters
end for
for number of generations do
    for every ECO feature do
        for every training image do
            Process image with feature transforms
            Train feature's perceptron
        end for
        for every holding set image do
            Process image with feature transformations
            Use perceptron output to update fitness score
        end for
        Assign fitness score to the feature
        Save feature if fitness score > threshold
    end for
    Select features to go to next generation
    Create new features using crossover
    Apply mutations to the population
end for
```

2.3   Training AdaBoost

Adaboost [6] is an algorithm that combines a number of weak classifiers to get a strong classifier. It has been successfully used in face detection [4], gender classification [5]. How AdaBoost is trained is shown in Algorithm 2. After ECO features have been successfully constructed, Adaboost is used to combine the perceptrons that are associated with ECO features, to make a stronger classifier. The resulting Adaboost model consists of a list of perceptrons and coefficients that indicate how much to trust each perceptron.

Algorithm 2. Training AdaBoost.

```
Set of training images M
for every training image, m do
   Initialize δ_M[m] = 1/|M|
end for
for x = 0 to X do
   for every perceptron, ω, do
      for every training image, m do
         if wrongly classified then
            δ_M+ = δ_M[m]
         end if
      end for
   end for
   Select perceptron with minimum error, Ω
   if δ_M[Ω] ≥ 50% then
      break
   end if
   Calculate coefficient of perceptron using Equation 4
   for every training image, m do
```

$$c = \begin{cases} 1, & \text{if classified corectly by } \Omega \\ -1, & \text{else} \end{cases}$$

$$\delta_M[m] = \frac{\delta_M[m] * e^{-\rho \cdot c}}{\text{Normalization Factor}}$$

```
   end for
end for
```

2.4 AdaBoost for Object Classification

Figure 3 shows an example of classifying an image with an AdaBoost model containing three ECO features. The figure shows three ECO features in the image. Each ECO feature operates on its own sub-region of the image. As the ECO feature evolves, the size of the sub-region as one of the transformation parameters could change from one transform to the next, just as ECO feature 2 and ECO feature 3 shown in Fig. 2. The number of the transforms included in one ECO feature could also be different for different ECO features.

After ECO features are found, each feature is accompanied by its trained perceptron. So we use AdaBoost to combine their corresponding perceptrons (weak classifiers) to make a stronger classifier. After training, the resulting AdaBoost model consists of a list of perceptrons and coefficients that indicate how much to trust each perceptron. The coefficient for each perceptron, ρ, is calculated using Eq. 4 where δ_w is the error of the perceptron over the training images.

$$\rho = \frac{1}{2} \cdot \ln \frac{1 - \delta_w}{\delta_w} \tag{4}$$

The output of each perceptron is combined according to Eq. 5 where X is the number of perceptrons in the Adaboost model, ρ_x is the coefficient for perceptron x (see Eq. 4), α_x is the output of the perceptron x (see Eq. 2), τ is a threshold value, and c is the final

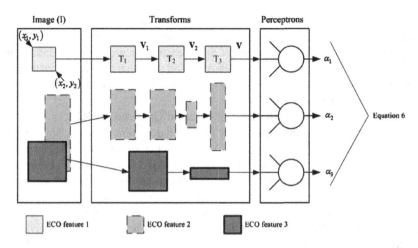

Fig. 3. ECO features and their corresponding perceptrons are combined using Adaboost to classify an image as object or non-object

classification given by the Adaboost model. The threshold τ can be adjusted to vary the tradeoff between false positives and false negatives.

$$c = \begin{cases} 1 & \text{if } \sum_{x=1}^{X} \rho_x \cdot \alpha_x > \tau \\ 0 & \text{else} \end{cases} \tag{5}$$

3   Feature Descriptors

The simplest way to describe an image region is to use an ordered list of pixel intensities to form a feature vector. It's widely used in machine learning area. And the current ECO features use raw pixels to describe image features. Representing an image region in this manner is simple, but has a few drawbacks. It is very sensitive to even small shifts, rotations. If the object shifts slightly in the image, the feature vector may change significantly because each pixel in the image patch changes its location in the feature vector. It is also sensitive to pose, scale and intra-class variations. Therefore this raw-pixel representation relies heavily on good alignment and it will be difficult for object recognition from different viewpoints. Additionally, by using a vector of pixel intensities, the resulting feature descriptor will have a high dimensionality. Its high dimensionality leads to high computational complexity without being robust to some deformations.

Our goal is to use more powerful descriptors for our ECO features. We need a feature descriptor that describes the image region in a way that is invariant to image variations. This kind of feature descriptor, which is a better way of representing the image region than using the raw pixel data, can overcome the limitation of our original ECO features. A natural idea is to describe the distribution of the pixel intensities by histograms to represent image features.

Most of the recently developed feature descriptors using histograms follow the same strategy: they subdivide the image region and compute a histogram of image attributes, either pixel values or gradient orientations, inside each sub-region. Histograms are more robust to image distortions than pixel intensities and they also reserve the spatial information by subdividing the region. In this paper, we experimented on three different feature descriptors in this paper.

The first descriptor was Histograms of Oriented Gradients (HOG) [7]. It's widely used in object recognition tasks especially in human detection. The descriptor is based on the gradient of the image intensity so it's very efficient to compute. The general idea of the Histograms of Oriented Gradients is that local object in an image can be described by the distribution of intensity gradients or edge directions. This descriptor's purpose is to divide the image into small regions called cells, and for each cell a histogram of gradient directions or edge orientations for the pixels within the cell is calculated. The descriptor is obtained by combining all these histograms. The HOG has a few advantages over other descriptors [7]. Since the HOG descriptor operates on localized cells, it is invariant to geometric and photometric transforms. HOG descriptor is one of the most popular descriptors used in object detection research.

The second descriptor we used in this paper was Local Binary Patterns Histograms (LBPH). The basic idea of Local Binary Pattern is to summarize the local structure in an image by comparing each pixel with its neighboring pixels. Following the same idea in HOG descriptor, the LBPH descriptor [8] is to divide the local binary pattern image into several local regions and extract a histogram from each region. The spatially enhanced feature vector is then obtained by concatenating the local histograms of all the local regions. In this way, the spatial information is preserved and embedded in the feature descriptor.

The third descriptor we used was a statistical view of the image region. It contains the Hu set of the invariant moments [9]. This feature descriptor also follows the same idea of HOG and LBPH descriptor. But this descriptor didn't use gradient values of pixel intensities as the first two descriptors. It first divided the image region into several local regions and extracts the Hu set from each of them. Hu moments are scale, rotation and translation invariant. The final descriptor is formed by having all the statistical values in a feature vector.

4   Dataset and Experiments

ECO features were then constructed on the training set and evaluated using the testing set. The classification accuracy, precision and recall were computed. The equations for recall and precision are given in the following equations (Eqs. 6 and 7):

$$\text{precision} = \frac{t_p}{t_p + f_p} \tag{6}$$

$$\text{recall} = \frac{t_p}{t_p + f_n} \tag{7}$$

As discussed in Sect. 2.4, the algorithm can be adjusted to have different thresholds for the tradeoffs between false positives and false negatives. Accepting more true positives will inevitably accept more false positives (or fewer true negatives). Setting the threshold allows the user to determine a desired tradeoff between the two. In our experiments, we chose the thresholds which gave the highest classification accuracy.

Fruit maturity evaluation plays a significant part in food industry. So we collected a dataset for the experiment to evaluate fruit maturity. This dataset contains 500 dry dates and 500 wet dates. Separating dry dates from wet dates is important because excessive drying will cause the date skin to peel and lower the quality and reduce its value.

When using HOG and Hu moments as the feature descriptors to describe the sub-region image, the classification accuracy was exactly the same as the raw pixel representation, approximately 95.9 %. The accuracy was not improved but was also not affected after the dimensionality was reduced significantly from N × M pixels to a few simple histograms. When using LBPH as the descriptor, the classification accuracy was improved from 95.9 % to 97.3 % even with significantly reduced dimensionality. The classification accuracy for the date dataset is given in Table 1.

Table 1. Accuracy comparison using LBPH

	Classification accuracy	False positive	False negative
Original feature	95.9 %	4 of 100 negatives	5 of 122 positives
Using LBPH	97.3 %	1 of 100 negatives	5 of 122 positives

All three selected feature descriptors were able to reduce the dimensionality of ECO features. In our experiments, we computed the gradients for each pixel and divided the input image into 2 × 2 blocks. For each block, we computed a gradient direction histogram over nine directions and then concatenated the histograms to obtain a 36-dimensional feature vector. For a 50 × 50 image sub-region, the dimension of our original ECO features is 2500. Using HOG as its descriptor, the dimension is reduced to 36. For LBPH, after a local binary pattern image for the input image is calculated, each pixel in the local binary pattern image has a value between 0 and 8. Using a histogram to represent each block of the image, the dimension of the image sub-region is reduced from $N \times M$ to $9 \times 4 = 36$. With this new approach, ECO features not only can be trained faster but also becomes scaling, rotation, and translation invariant.

5 Conclusion

In this paper, we have discussed a method to improve our ECO features algorithm. Feature descriptors are used instead of original ECO features, which are just raw pixels of an image sub-regions. It was shown that the performance of our original ECO features has been improved using feature descriptors and our new ECO feature has a number of benefits. First of all, the dimensionality of ECO features are greatly reduced. Additionally, feature descriptors were shown to provide a lot of performance improvement over our original ECO features. For further improvement, we plan to look for more ways to reduce the dimension of ECO features in order to improve the efficiency of the training process.

Acknowledgement. The project was supported by the Small Business Innovation Research program of the U.S. Department of Agriculture, grant number #2014-33610-21951.

References

1. Lillywhite, K., Lee, D.J., Tippetts, B., Archibald, J.: A feature construction method for general object recognition. Pattern Recogn. **46**, 3300–3314 (2013)
2. Mitchell, M.: An introduction to genetic algorithms. MIT press, Cambridge (1998)
3. Mikolajczyk, K., Schmid, C.: A performance evaluation of local descriptors. IEEE Trans. Pattern Anal. Mach. Intell. **27**, 1615–1630 (2005)
4. Shakhnarovich, G., Viola, P., Moghaddam, B.: A unified learning framework for real time face detection and classification. In: Proceedings Fifth IEEE International Conference on Automatic Face and Gesture Recognition, 2002. IEEE (2002)
5. Mäkinen, E., Raisamo, R.: Evaluation of gender classification methods with automatically detected and aligned faces. IEEE Trans. Pattern Anal. Mach. Intell. **30**(3), 541–547 (2008)
6. Viola, P., Jones, M.: Rapid object detection using a boosted cascade of simple features. In: Proceedings of the 2001 IEEE Computer Society Conference on Computer Vision and Pattern Recognition, CVPR 2001, vol. 1, pp. I–511. IEEE (2001)
7. Dalal, N., Triggs, B.: Histograms of oriented gradients for human detection. In: IEEE Computer Society Conference on Computer Vision and Pattern Recognition, 2005. CVPR 2005, vol. 1. IEEE (2005)
8. Ahonen, T., Hadid, A., Pietikäinen, M.: Face recognition with local binary patterns. In: Pajdla, T., Matas, J. (eds.) ECCV 2004. LNCS, vol. 3021, pp. 469–481. Springer, Heidelberg (2004)
9. Hu, M.-K.: Visual pattern recognition by moment invariants. IRE Trans. Inf. Theor. **8**(2), 179–187 (1962)

Ground Extraction from Terrestrial LiDAR Scans Using 2D-3D Neighborhood Graphs

Yassine Belkhouche[1], Prakash Duraisamy[2(✉)], and Bill Buckles[3]

[1] University of Arkansas, Little Rock, AR 72204, USA
yassinemb@gmail.com
[2] Miami University, Oxford, OH 45056, USA
duraisp@miamioh.edu
[3] University of North Texas, Denton, TX 76203, USA
bbuckles@cse.unt.edu

Abstract. We introduce a new method for filtering terrestrial LiDAR data into two categories: Ground points and object points. Our method consists of four steps. First, we propose a graph-based feature, which is obtained by combining 2D and 3D neighborhood graphs. For each point, we assign a number, that is the count of common neighbors in 2D and 3D graphs. This feature allows the discrimination between terrain points and object points as terrain points tend to have the same neighbors in both 2D and 3D graphs, while off-terrain points tend to have less common neighbors between 2D and 3D graphs. In second step, we used c-mean algorithm to quantize the feature space into two clusters, terrain points and object points. The third step consists of repeating the first and the second step using different neighborhood sizes to construct the KNN(k-nearest neighbor) graph. In the final step, we propose a decision-level fusion scheme that combines the results obtained in the third step to achieve higher accuracy. Experiments show the effectiveness of our method.

1 Introduction

Laser instruments are used scan a scene of interest for the purpose of recon-struction and analysis. These scanners collect tremendous amount of 3D point clouds(up one million points per second). The collected point clouds are used in many civilian and military applications. However, these datasets need pre-processing in order to extract the necessary information for the application of interest. Filtering is the process of separating the point clouds into terrain points and object points. Filtering terrestrial LiDAR scans is a key step, it is a very active research area due to the large number of applications that require this preprocessing. This process is a challenging task due to the complexity of scanned scene as well as the scanning spacial resolution. Urban scenes usually contain many complex man-made objects(e.g. buildings, cars, traffic signs) and natural objects(e.g. trees, grass, bushes, humans) which complicate the filter-ing process. The result of the filtering process are two classes: Terrain points

© Springer International Publishing Switzerland 2015
G. Bebis et al. (Eds.): ISVC 2015, Part II, LNCS 9475, pp. 655–663, 2015.
DOI: 10.1007/978-3-319-27863-6_61

and objects points. Terrain points are used for applications such as autonomous robot navigation, DEMs(digital elevation model) generation which are used in many GIS(geographic information system) applications such as flood modeling, change detection, and urban planing. Object points are further classified into different classes such as cars, traffic signs, trees and other objects depending on the application of interest. For autonomous cars navigation, identifying other cars, traffic signs, pedestrians and any kind of obstacles is critical for the safe navigation. Classified points are also used to build 3D city models, which are used for touristic applications, military training, simulation, and planning emergency response.

2   Literature Review

In recent years, there have been a large number of research papers for filtering LiDAR data [1,3,7–9,14–16,18]. However nearly all the papers address the filtering of airborne collected LiDAR data, only few papers tackled the problem of filtering terrestrial laser scans. Douillard et al. [2] introduced an iterative probabilistic method for filtering terrestrial laser scans. The proposed method called GP-INSAC(Gaussian Process Incremental Sample Consensus) assume that most of the points in the dataset belong to the ground, and there are only few outliers. Using this assumption, the GP-INSAC uses a set of seed terrain points to build a model, then it evaluates the remaining points. The inlier are added to the ground points, while the outliers are considered object points. Pirotti et al. [10] presented a method for filtering multi-return terrestrial laser scans. Their method works in natural landscape areas such as forests and hills. They start by identifying ground points candidates using the return number and the corresponding amplitude. The next step is to find an approximation of the ground surface using a progressive morphological filter, where the window size is progressively decreased. To detect the ground points, the authors used the opening operator. Zhou et al. [19] used the assumption that the ground surface is flat, and compute an estimate of the ground plane using the RANSAC algorithm. Points within a small distance to this plane are classified as ground points, the rest are considered object points. Plaza et al. [11] proposed a ground segmentation method for natural environments. The authors used a voxelmap. Initially, voxels are projected on a two dimensional grid, the lowest voxel within each grid cell is considered as ground, then adjacent voxels are classified as either ground or objects using the vertical discontinuity between voxels, isolated area, terrain crest or slope criteria. Clustering-based methods were used to filter terrestrial laser scans, the main idea is to cluster the point clouds then use some criteria to joint the clusters that satisfy terrain characteristics into one cluster, the other clusters are considered object points. Roggero [13] proposed a method for clustering raw LiDAR data for object detection. His algorithm works on raw data, it initially separates terrain and object points using height difference between neighboring points, then he classified object points into either buildings or trees by using the spatial distribution of the points as a discrimination feature.

A LiDAR segmentation method was proposed by Filin et al. [4], Filin method defined a slope adaptive neighborhood system that allows points with similar properties to be assigned to the same cluster. The neighborhood was defined as follows: Using the points within a cylinder centered at a point of interest, only points that are within a distance less than the maximum distance to an estimated local tangent plan are considered as neighbors of the point of interest. This criterion allows the algorithm to connect points on the same surface. A feature vector that consists of the point position and the tangent plan parameters are used by the clustering algorithm to assign points to clusters, which are then classified into ground or object. The method proposed by Rabbani et al. [12] uses smoothness constraint for segmenting LiDAR raw data. The segmentation algorithm proposed by that method uses only an estimation of the surface normals to determine the smoothness of each area. Regions with similar smoothness values are considered as one segment. Sithole et al. [15] introduced a filtering algorithm that uses segmented LiDAR. LiDAR points are segmented into smooth segments, then these segments are classified as ground segments or object segments using geometric relationships with the surrounding segments.

3   Formal Description of the Filtering Problem and Challenges

Given a set of 3D terrestrial point clouds $P_{scan} = \{p_1, p_2, ..., p_N\}$. Our objective is to separate the set P_{scan} into two subsets $T = \{t_1, ..., t_k\}$ and $O = \{O_1, ..., O_r\}$ such that:

- T: The subset consisting of terrain(ground) points.
- O: The subset of consisting of object points.
- $T \cap O = \emptyset$: There are no common points between the two subsets.
- $T \cup O = P_{scan}$: The union of these two subsets will give us the original point clouds.

The most important challenges in filtering LiDAR data are:

- LiDAR points are irregularly spaced (direct image processing tools cannot be used).
- Urban areas are complex scenes that contain several complex structures such as buildings, trees, cars, humans, traffic signs.
- Spacial density of the scans affects the performance of many filtering algorithms. In low density data, there are missing features. these features are used by the filtering algorithms to decide which points are terrain and which points are object, which leads to wrong classification.

4   Contributions

In order to deal with the previous challenges, we proposed a new filtering method that uses graph-based feature and decision level fusion. Our main contributions are:

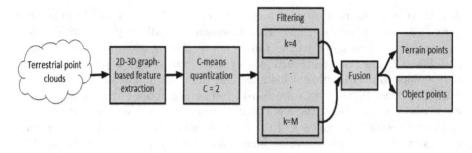

Fig. 1. The filtering process.

- We introduce a 2D-3D KNN graph-based feature that allows the discrimination between terrain points and object points.
- We establish a decision-level fusion technique that combines the decisions obtained using different neighborhood sizes in the graph-based feature extraction to obtain better accuracy.

5   Methodology

The proposed filtering method consists of four steps:

- 2D-3D KNN graph-based feature extraction.
- C-means quantization (since we are classifying the LiDAR data into terrain and object points, we fixed $C = 2$).
- Filtering using the extracted feature with different neighborhood sizes.
- Decision-level fusion of the results obtained from the filtering process.

Figure 1 show the entire filtering process.

5.1   2D-3D KNN Graph-Based Feature Extraction

We generate two k-nearest neighbors graphs, one using two-dimensional(2D) neighborhood system and the second using three-dimensional(3D) neighborhood system. The 2D graph is generated by considering only the (x, y) coordinates of each point(setting the z-coordinate to 0). The 3D graph is constructed by using the (x, y, z) coordinates of each point. After the construction of the two graphs, for each point we find the number of common neighbors r between the 2D and 3D graphs. Each point is associated a feature r_k, where r_k is the number of common neighbors using the neighborhood size k. The idea behind this feature is that terrain points tend to keep most of their neighbors in both the 2D and 3D graphs, while object points tend to have may differences between the 2D and 3D graphs.

5.2   C-Means Quantization

In this paper we used a simple clustering algorithm(C-means). Given a specified number of clusters c, with their corresponding centroids, the C-means algorithm proceeds as follows: It starts by assigning each point to the closest centroid, then the centroids are updated. This process is repeated until there are no changes(the points do not change clusters or the centroids remain the same). In this particle application we observed that the C-means produces noisy results, to deal with this problem, we proceed as follows: after obtaining two clusters, we perform a refinement step in which each point in the KNN graph is assigned to a cluster based on the majority vote of its neighbors. The C-means and the refinement process are described in Algorithm 1.

Data: 3D Point clouds.
Result: Two clusters C_1 and C_2.
$C = 2$;
$Feature = Compute\_2D\_3D\_Feature(data)$;
// C-means Quantization ;
while $(Centoids(k+1) - Centoids(k) > \epsilon)$ **do**
    for $i = 1 : \#\_points$ **do**
        if $(dist(p_i, C_1) \geq dist(p_i, C_2))$ **then**
            Assign p_i to C_1;
        else
            Assign p_i to C_2;
        end
    end
    Centroids(k+1) = Update(Centoids(k));
end
// Refinement step ;
for $(i = 1 : \#\_points)$ **do**
    if $(p_i(nb\_c_1) \geq p_i(nb\_c_2))$ **then**
        Assign p_i to C_1;
    else
        Assign p_i to C_2;
    end
end

Algorithm 1. C-means quantization and refinement.

where $p_i(nb\_c_j)$ is the number of p_i neighbors belonging to cluster C_j.

5.3   Decision-Level Fusion

The proposed filtering algorithm depends on the selected neighborhood size k used for the construction the KNN-graph, in order to remove this dependency we proposed a decision-level fusion algorithm that combines the results obtained at different scales. Our fusion technique works as follows: let $[k_0...k_m]$ be a user defined interval of possible neighborhood sizes, let $T_{k_i} \in [t_0...t_r]$ and $O_{k_i} \in$

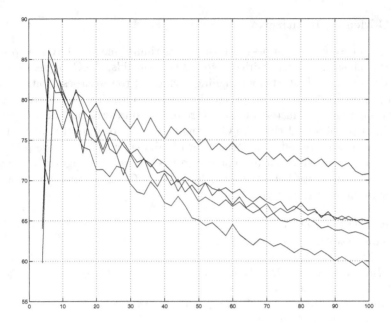

Fig. 2. The number of neighbors k vs. accuracy

$[O_0...O_f]$ be the filtering results using $k = k_i$. The following rules are used to obtain the final results:

- Using a neighborhood size k, each terrain point is labeled $l_k(p_i) = -1$ and object points are labeled $l_k(p_i) = +1$.
- let $L(p_i) = \sum_{k=1}^{m} l_k(p_i)$.
- if $L_{(p_i)} > -m$ then $l(p_i) = +1$.

In order to study the effect on the neighborhood size on the filtering accuracy, we varied the number of neighbors k in the KNN graph and observed the change in the accuracy. Figure 2 shows how the neighborhood size k used to construct the KNN graph affects the accuracy. Varying the value of k from $k = 4$ to $k = 100$ with step of 2, the accuracy start increasing until it reaches it peak around $k = 20$ then it starts declining. In the next section, we show that the fused results always perform better than the individual results.

6   Experimental Results

We used five datasets provided by [5,6] to illustrate the efficiency of the proposed filtering method. These datasets were collected using ground-based Velodyne laser scanner. It was considered as one of the most difficult dataset to filter due to the noisy and sparse point measurements [17]. The characteristics of these datasets are summarized in Table 2. Figure 3 shows a study scene filtered using

Fig. 3. Scene 1: Red points are object, green are terrain (Color figure online).

Table 1. Confusion-matrix

		Classified	
		Terrain	Off-terrain
References	Terrain	a	b
	Off-terrain	c	d

out method, the green points are terrain points and the red points are object points.

The measures used for evaluation are summarized in Table 1.

Where:

- a: is the number of correctly classified terrain points.
- b: is the number of terrain points classified as object points.
- c: is the number of object points classified as terrain points.
- d: is the number of correctly classified object points.

Table 3 shows the accuracy of each dataset. We varied the k value, and show how the fusion method improve the accuracy.

Table 2. Characteristics of study areas

	#Points	Buildings	Trees	Fences	Humans	Cars	Street signs	Others
Area 1	65545	Yes	Yes	Yes	No	No	No	Yes
Area 2	65432	Yes	Yes	Yes	No	No	No	No
Area 3	65858	Yes	Yes	Yes	Yes	Yes	Yes	Yes
Area 4	65398	Yes	Yes	Yes	Yes	Yes	Yes	Yes
Area 5	66044	Yes	Yes	Yes	No	Yes	Yes	Yes

Table 3. Classification Accuracy

	Accuracy							
	$k = 7$	$k = 11$	$k = 19$	$k = 25$	$k = 29$	$k = 35$	$k = 41$	Fused
Dataset 1	77.76	81.48	77.13	75.05	72.45	71.29	69.78	84.70
Dataset 2	78.66	82.28	78.94	77.75	77.80	76.89	75.99	87.02
Dataset 3	78.62	82.21	74.32	73.95	74.77	73.53	72.23	85.62
Dataset 4	72.69	78.05	73.73	71.77	72.10	10.18	71.01	85.04
Dataset 5	79.74	76.51	74.77	71.08	68.97	68.77	68.18	84.68

7   Conclusion

We introduced a new graph-based feature for filtering terrestrial laser scans. This feature is used to discriminate terrain points and object points. To improve the accuracy of our method, we proposed a decision-level fusion scheme that combines the individual results obtained form filtering the point clouds using different neighborhood sizes. Experiments demonstrate that the results of the fusion method performs better than all of the individual results.

References

1. Bartels, M., Wei, H.: Towards DTM generation from LiDAR data in hilly terrain using wavelets. In: Proceedings of 4th International Workshop on Pattern Recognition in Remote Sensing in Conjunction with the 18th International Conference on Pattern Recognition, pp. 33–36 (2006)
2. Douillard, B., Underwood, J., Kuntz, N., Vlaskine, V., Quadros, A., Morton, P., Frenkel, A.: On the segmentation of 3D lidar point clouds. In: IEEE International Conference on Robotic and Autonomation, Shanghai, China, May 2011
3. Elaksher, A.: A framework for generating high quality digital elevation models in urban areas. In: IEEE Southwest Symposium on Image Analysis and Interpretation, pp. 169–172, May 2008
4. Filin, S., Pfeifer, N.: Segmentation of airborne laser scanning data using a slope adaptive neighborhood. ISPRS J. Photogramm. Remote Sens. **60**, 71–81 (2006)
5. Lai, K., Fox, D.: 3D laser scan classification using web data and domain adaptation. In: Robotics: Science and Systems, July 2009
6. Lai, K., Fox, D.: Object recognition in 3D point clouds using web data and domain adaptation. Int. J. Robot. Res. **29**, 1019–1028 (2010)
7. Kraus, K., Pfeifer, N.: Advanced DTM generation from LiDAR data. Int Arch of Photogrammetry and Remote Sensing **34**, 23–30 (2001)
8. Lu, W.: A hybrid conditional random field for estimating the underlying ground surface from airborne lidar data. IEEE Trans. Geosci. Remote Sens. **47**(8), 2913–2922 (2009)
9. Martels, M., Wei, H., Mason, D.: DTM generation from LiDAR using skewness balancing. In: Proceedings of the 18th International Conference Pattern Recognition, pp. 01:566–569 (2006)

10. Pirotti, F., Guarnieri, A., Vettore, A.: Ground filtering and vegetation mapping using multi-return terrestrial laser scanning. ISPRS J. Photogramm. Remote Sens. **76**, 56–63 (2013)
11. Plaza, V., Ababsa, F.E., Garcia-Cerezo, A.J., Gomez-Ruiz, J.: 3D segmentation method for natural environments based on a geometric-featured voxel map. In: IEEE International Conference on Industrial Technology, Sevilla, Spain (2015)
12. Rabbani, T., van den Heuvel, F.A., Vosselman, G.: Segmentation of point clouds using smoothness constraint. In: ISRPRS Commission V Symposium Image Engineering and Vision Metrology, Dresden, September 2006
13. Roggero, M.: Airborne laser scanning: clustering in raw data. Int. Arch. Photogramm. Remote Sens. **34**(3/w4), 227–232 (2001)
14. Sithole, G., Vosselman, G.: Filtering of laser altimetry data using a slope adaptive filter. Int. Arch. Photogramm. Remote Sens. Spat. Inf. Sci. **34**, 203–210 (2001)
15. Sithole, G., Vosselman, G.: Filtering of airborne laser scanner data based on segmented point clouds. In: ISPRS WG III/3, III/4, V/3 Workshop Laser scanning 2005, nschede, September 2005
16. Wack, R., Wimmer, A.: Digital terrain models from airborn laser scanner data -a grid based approach. Int. Arch. of Photogrametry and Remote Sensing, pp. XXXIV:293–296 (2002)
17. Xiong, X., Munoz, D., Bagnell, J., Hebert, M.: 3D scene analysis via sequenced predictions over points and regions. In: IEEE International Conference on Robotics and Automation (ICRA), 2011, pp. 2609–2616, May 2011
18. Zhang, K., Chen, S., Shyu, M., Yan, J., Zhang, C.: A progressive morphological filter for removing nonground measurements from airborne LiDAR data. IEEE Trans. Geosci. Remote Sensing **41**, 872–882 (2003)
19. Zhou, Y., Yu, Y., Lu, G., Du, S.: Super-segments based classification of 3d urban street scenes. Int. J. Adv. Rob. Syst. **9**, 9:248–255 (2012)

Mass Segmentation in Mammograms Based on the Combination of the Spiking Cortical Model (SCM) and the Improved CV Model

Xiaoli Gao, Keju Wang, Yanan Guo, Zhen Yang, and Yide Ma[✉]

School of Information Science and Engineering, Lanzhou University,
Lanzhou 730000, Gansu Province, China
yidema@lzu.edu.cn

Abstract. In this paper, a novel method based on CV model for the mass segmentation is proposed. Firstly, selecting the largest connected region, seeded region growing, and singular value decomposition (SVD) are used to pre-processing. After that apply the Spiking Cortical Model (SCM) on the pre-processed image to locate the lesion. Finally, the mass boundary is accurately segmented by the improved CV model. The validity of the proposed method is evaluated through two well-known digitized datasets (DDSM and MIAS). The performance of the method is evaluated with detection rate and area overlap. The results indicate the proposed scheme could obtain better performance when compared with several existing schemes.

1 Introduction

Breast cancer is the common malignant tumors and remains the leading cause of cancer death among females, accounting for 23 % of the total cancer cases and 14 % of the cancer deaths in the world [1]. Early detection will improve survival rate, unfortunately, only 20 % of the breast cancer patients are diagnosed at the early stage. Mammography is an outstanding method for early detection of breast cancer. While the low contrast and the lesions are blurry, which cause the high misdiagnosis.

To obtain the accurate masses contour plays an important role for mass classification. As a result, various mass detection algorithms were proposed to assist radiologists in the early identification of breast cancer [2]. Segmentation of mammography is an important step of detecting breast cancer. Recently, vast works on active contour have been proposed to image segmentation problems [3–5]. Chan and Vese proposed an active contour (CV) model using level set formulation [6]. The CV model doesn't work well when to extract the mass in mammograms. Meanwhile the CV model performs more excellent characteristics in segmentation of boundaries such as capability of convergence and superior noise robustness and the method is quite sensitive to the initial contour of the object [7]. And the SCM model cannot accurately converge to the boundary of the masses. To this question, we applied the SCM model to the mammogram to obtain the mass location which would be used as the initial contour of improved CV model followed by. The improved CV model [8] proposed energy minimization framework for image segmentation and estimation of

© Springer International Publishing Switzerland 2015
G. Bebis et al. (Eds.): ISVC 2015, Part II, LNCS 9475, pp. 664–671, 2015.
DOI: 10.1007/978-3-319-27863-6_62

bias field to extract the mass in mammograms. The present method can detect the regions of masses in mammograms accurately. The performance of the proposed approach is directly compared to the CV model.

This paper is organized as follows. In Sect. 2, firstly, we introduced the databases we used and present the method of pre-processing. After that brief reviews the level set method, we present our methodology for mass segmentation. Section 3, illustrates some experiments to verify the proposed method. Section 4 gives the conclusions of this paper.

2   Methodology

In this work, segmentation of region of interest (ROI) consists of four steps: 1) Obtaining the Mammogram images; 2) Mammogram pre-processing, to remove the label and pectoral muscle and enhance image; 3) Mass segmentation.

2.1   Image Database

In this work, a set of images selected from Mammography Image Analysis Society (MIAS) database [9] and digital database for screening mammography (DDSM) [10] are used to verify the validity of the method we proposed. The DDSM consists of 2620 cases, available in 43 volumes and each image was about 3000×5000 pixels. The MIAS consists of 322 cases and offered some corresponding information of lesion area such as type, location, severity, central coordinate and radius by experts and each image is 1024×1024 pixels.

2.2   Pre-processing

Due to the low contrast, the pre-processing of mammogram is essential. The pre-processing used to remove the labels, pectoral muscle and enhance the contrast.

2.2.1   Label Removal
The mammogram image usually includes of breast region, label and pectoral muscle. Usually the label and muscle should be removed to reduce the interference. Here we apply the methods of Selecting the largest connected region and seed region growing algorithm [11] to remove the label and muscle respectively.

2.2.2   Image Enhancement
Traditional gray level transformation is not working well since the gray values between mass region, pectoral muscle and gland are similar. As this, the method of singular value decomposition (SVD) [12] was applied to increase the contrast of an image locally. This pre-processing makes the brightness variation between a mass and the surrounding tissue more obvious, facilitating the substantial mass segmentation. The image after pre-processing and mass locate are given below in Fig. 1.

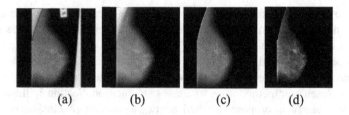

Fig. 1. Pre-processing and localization results (a) Original image (b) Label removed image (c) Muscle removed image (d) Enhanced image

2.3   Segmentation Methods

In order to obtain the accurate contour of the mass, we applied the Spiking Cortical Model (SCM) [13] model to locate the lesions firstly. After that, we applied the improved CV model to obtain the accurate contour.

2.3.1   Mass Candidate Localization

The SCM was applied to locate the mass location because the brightest points in the ROIs are usually found inside the masses and the SCM model is sensitive to the bright point. The initial contour was obtained after input the pre-processing image into the SCM model. The SCM model is show as Fig. 2:

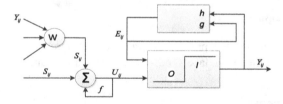

Fig. 2. The SCM model which used to obtain the location of the mass

Where U_{ij} is internal activity, S_{ij} is a stimulus, Y_{ij} is output, E_{ij} is dynamic threshold, W is synaptic weight matrix applied to the linking field, f and g are decay constants, h is threshold magnitude coefficient.

2.3.2   Accurate Segmentation

Since the results of mass location are rough and exist certain gap with the actual boundary, we utilize the improved CV model to obtain accurate segmentation.

2.3.2.1 The CV Model

The CV model assumes to segment the image into two regions: the object and the background. The CV model, which formulates the image segmentation as a problem of minimizing the follow energy function:

$$F^{CV}(\phi, c_1, c_2) = \int_\Omega |I(x) - c_1|^2 H(\phi(x))\, dx + \int_\Omega |I(x) - c_2|(1 - H(\phi(x)))\, dx + v \int_\Omega |\nabla H(\phi(x))|\, dx. \tag{1}$$

Where the first two terms are the date fitting terms. c_1, c_2 are the averages of image I inside and outside counter v respectively. The last term is the regulation term, regulating the length of the level set curve. $H(.)$ is the Heaviside function. $\phi(x)$ is the level set function.

2.3.2.2 Image Model

According to the principle of physical imaging, the observed image I can be modeled as:

$$I = bJ + n. \tag{2}$$

Where J is the true image, b is the component that accounts for the intensity inhomogeneity, here is referred to as a bias field, and n is additive noise.

According to the features of real image and bias field b, we have the following assumptions:

(A1) The bias field b is slowly varying, which implies that each point can be approximated by a constant in a neighborhood of the image domain.

(A2) The true image J can be divided into N disjoint regions $\Omega_1, \dots \Omega_N$, which can be approximated by N different constants $c_1 \cdots c_2$.

2.3.2.3 Energy Minimization of the CV Model

At each point y in the image I, we define a circular neighborhood $O_y \in \{x: |x - y| \le \rho\}$ with a radius ρ, according to the image model in (2) and the assumptions, the image model can be expressed as:

$$I(x) \approx b(y)c_i + n(x) \quad \text{for} \quad x \in O_y \cap \Omega_i. \tag{3}$$

Therefore, the intensities in the set

$$I_y^i = \{I(x): x \in O_y \cap \Omega_i\}. \tag{4}$$

Forms a cluster, $m_i \approx b(y)c_i$ is the cluster center.

Through the property of the local intensity clustering, the intensities in the neighborhood O_y can be divided into N clusters. The K-means clustering algorithm is applied as an iterative process to minimize the clustering criterion for the intensities $I(x)$ in the neighborhood. Through above, the energy function of CV model can be written in a continuous form as follows:

$$F(\phi, c_1, c_2) = \int \sum_{i=1}^{N} \int K(y-x) |I(x) - b(y) c_i|^2 \, dy M_i(\phi(x)) \, dx + vL(\phi) + \mu R_p(\phi). \tag{5}$$

Where $K(y-x)$ is introduced as a nonnegative window function. As the $K(y-x) = 0$, the $x \in O_y$. By minimizing the energy function of the level set with bias fields, we can get the segmented image.

3   Experiment Results and Discussion

In this section, we run a set of experimentations on DDSM and MIAS database and compared the performance of our proposed model. We test the methods on 400 mammograms and each case contains one abnormality. The segmentation results are shown in Fig. 3.

3.1   Experiment Results

In this paper, 400 mammograms were used to be segmented which including 64 mammograms from MIAS database and 336 from DDSM database.

The four images shown as Fig. 3 are selected from MIAS database. The MIAS database has offered the central coordinate and radius of each abnormal region as shown in Fig. 3(a-5)–(d-5), the red curves are the ground truth. As well, we applying the pre-processing method to the mammogram firstly. And then, applying the proposed method to segment the precise contour. The results of the method we proposed are shown in Fig. 3(a-3)–(d-3). The enlarged results are shown as Fig. 3(a-4)–(d-4). The Fig. 3(a-1)–(d-1) and (a-2)–(d-2) are the results of the CV model and the enlarged results respectively. Compared with the typical CV model, our method can completely remove the labels or interference and achieve more robust and accurate results. As we can see, the curves are closer to the ground truth even in blurry region.

The enlarged results of our method shown the margin of the last two images are rough and the others are smooth. The reason is that the severity between the last two lesions and the rest are different, the last two lesions are malignant while others are benign. Our results objectively reflect the pathology characteristics of actual masses to some extent that the malignant masses are always with burrs. This performance is somewhat benefit to the early diagnosis of breast cancer and also indicated the superiority of our method.

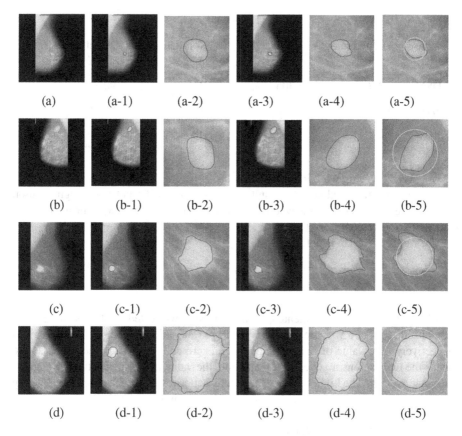

(a)   (a-1)   (a-2)   (a-3)   (a-4)   (a-5)

(b)   (b-1)   (b-2)   (b-3)   (b-4)   (b-5)

(c)   (c-1)   (c-2)   (c-3)   (c-4)   (c-5)

(d)   (d-1)   (d-2)   (d-3)   (d-4)   (d-5)

Fig. 3. The results of MIAS

3.2   Algorithm Perfoemance Analysis

3.2.1   Detection Rate

The mammograms we choosed to test our method are from DDSM and MIAS databases. Our evaluation principle is that the segmented region by the proposed method is completely within the criterion region by the experts. In the case of DDSM database, the criterion region is the outline formed by chain code data, and for MIAS database the criterion region is the circle formed by the center coordinates and the radius. The detection rates of each databases are shown in Table 1. Just as the table shown, 373 images are detected successfully. As the lesions in dense breast images of MIAS are always embedded in the gland and the boundary of the masses are blurry, the mass counters are hard to obtain. Therefore, the detection rate is lower for the MIAS images.

Table 1. Detection rate

Database	Tested images	Detected images	Non-detected images	Detection rate(%)
DDSM	336	305	31	90.77
MIAS	64	57	7	89.06
Total	400	362	48	90.5

3.2.2   Area Overlap Metric

In order to quantify evaluate the results from the three different mass segmentation methods, we employed the area overlap metric. It is defined as the ratio of the intersection of two areas to be compared [14].

The ratio of overlap is defined as in

$$O = \frac{S_{L \cap T}}{S_{L \cup T}}. \tag{6}$$

Where L is the area segmented by the methods we tested, and T is the area of groundtruth. $S_{L \cap T}$ and $S_{L \cup T}$ are the intersection area and union set area of the two regions respectively. The value of O is bound between zero and one. When the value is zero means the two regions no overlap. Meanwhile, the value is one means the two regions are exact overlap. The higher value of O means better result. The average area overlap ratio and the variance of the segmentation results are shown in Table 2.

Table 2. The area overlap ratios

	Mean (%)	Variance (%)
Typical CV model	76.34	7.5498
Proposed method	81.45	4.7958

As show in Table 2, the average area overlap ratio of proposed method is higher than the typical CV model. As well, the variance is lower. In conclusion, our segmented results are much more close to the ground truth.

4   Conclusions

In this paper, a novel segmentation method based on SCM model and CV model was proposed. Through this method, we obtained the accurate counter of the mass. Firstly, we apply a set of methods to pre-processing the mammogram and obtain the location of the mass. Subsequently, we applied the proposed method to segment the accurate counter. As the results and quantitative analysis values shown, the method we proposed has a higher values of different evaluation metrics.

In spite of the higher detection ratio we already obtained, there are still mammograms with masses haven't been segmented. At the same time, the mass segmentation consume a huge of time. Therefore, we need deeper study to segment the mass counter.

Acknowledgements. This work was jointly supported by the National Natural Science Foundation of China (Grant No.61175012), Specialized Research Fund for the Doctoral Program of Higher Education of China (Grant No.20110211110026), the Fundamental Research Funds for the Central Universities of China (Grant No. lzujbky-2013-k06 & -lzujbky-2015-197) and the Central Universities of China under Grant lzujbky-2015-196.

References

1. Jemal, A., Bray, F., Melissa, M., Ferlay, J., Ward, E., Forman, D.: Global cancer statistics. CA Cancer J. Clin. **61**(2), 69–90 (2011)
2. Salmeri, M., Mencattini, A., Rabottino, G., Accattatis, A., Lojacono, R.: Assisted breast cancer diagnosis environment: a tool for dicom mammographic images analysis. In: IEEE International Workshop on Medical Measurements and Applications (MEMEA 2009), pp. 160–165 (2009)
3. Han, X., Xu, C., Prince, J.L.: A topology preserving level set method for geometric deformable models. IEEE Trans. Patt. Anal. Mach. Intell. **25**(6), 755–768 (2003)
4. Caselles, V., Catté, F., Coll, T.: Françoise Dibos.: a geometric model for active contours in image processing. Numer. Math. **66**(1), 1–31 (1993)
5. Malladi, R., Sethian, J.A., Vemuri, B.C.: Shape modeling with front propagation: a level set approach. IEEE Trans. Patt. Anal. Mach. Intell. **17**(2), 158–175 (1995)
6. Chan, T.F., Vese, L.A.: Active contours without edges. IEEE Trans. Imag. Proc. **10**(2), 266–277 (2001)
7. Liu, J., Liu, X., Chen, J., Tang, J.: Mass segmentation in mammograms based on improved level set and watershed algorithm. In: Huang, D.-S., Gan, Y., Gupta, P., Gromiha, M. (eds.) ICIC 2011. LNCS, vol. 6839, pp. 502–508. Springer, Heidelberg (2012)
8. Li, C., Huang, R., Ding, Z., Gatenby, J.C., Metaxas, D.N., Gore, J.C.: A level set method for image segmentation in the presence of intensity inhomogeneities with application to MRI. IEEE Trans. Image Process. **20**(7), 2007–2016 (2011)
9. Suckling, J., Parker, J., Dance, D., Astley, S., Hutt, I., Boggis, C., Ricketts, I., Stamatakis, E., Cerneaz, N., Kok, S., Taylor, P., Betal, D., Savage, J.: The Mammographic Image Analysis Society digital mammogram database. In: International Workshop on Digital Mammography, pp. 211–221 (1994)
10. Heath, M., Bowyer, K., Kopans, D., Moore, R., Kegelmeyer, W.P.: The digital database for screening mammography. In: Proceedings of the Fifth International Workshop on Digital Mammography (IWDM), pp. 212–218 (2001)
11. Mirghasemi, S., Rayudu, R., Zhang, M.: A new image segmentation algorithm based on modified seeded region growing and particle swarm optimization. In: 28th International Conference of Image and Vision Computing (IVCNZ), pp. 382–387 (2013)
12. Bhattacharya, S., Gupta, S., Subramanian, V.K.: Localized image enhancement. Twentieth National Conference on Communications (NCC), pp. 1–6 (2014)
13. Ma, Y.-D., Yuan, J.-X., Zhang, H.-J.: Self-adaptive method using SCM for noise removal in color images. J. Univ. Electron. Sci. Technol. China **41**(5), 754–758 (2012)
14. Zhang, H., Fritts, J., Goldman, S.: Image segmentation evaluation: a survey of unsupervised methods. Comput. Vis. Image Underst. **110**(2), 260–280 (2008)

High Performance and Efficient
Facial Recognition Using Norm
of ICA/Multiwavelet Features

Ahmed Aldhahab[1,2]([✉]), George Atia[1], and Wasfy B. Mikhael[1]

[1] Department of Electrical and Computer Engineering,
University of Central Florida, Orlando, FL, USA
[2] Department of Electrical Engineering, University of Babylon, Hilla, Babylon, Iraq
Aldhahab2012@Knights.ucf.edu, {George.Atia,Wasfy.Mikhael}@ucf.edu

Abstract. In this paper, a supervised facial recognition system is proposed. For feature extraction, a Two-Dimensional Discrete Multiwavelet Transform (2D DMWT) is applied to the training databases to compress the data and extract useful information from the face images. Then, a Two-Dimensional Fast Independent Component Analysis (2D FastICA) is applied to different combinations of poses corresponding to the subimages of the low-low frequency subband of the MWT, and the ℓ_2-norm of the resulting features are computed to obtain discriminating and independent features, while achieving significant dimensionality reduction. The compact features are fed to a Neural Network (NNT) based classifier to identify the unknown images. The proposed techniques are evaluated using three different databases, namely, ORL, YALE, and FERET. The recognition rates are measured using K-fold Cross Validation. The proposed approach is shown to yield significant improvement in storage requirements, computational complexity, as well as recognition rates over existing approaches.

1 Introduction

Facial recognition is an important task of numerous applications in computer vision, pattern recognition, and image processing, which has received renewed attention in recent year due to its wide applicability in security, biomedical diagnosis, control, and personal identification. The efficiency and reliability of different recognition systems depends on several factors, including the computational complexity, storage requirements, and recognition rates. Therefore, extracting efficient and compact features is central to every reliable recognition system.

Numerous algorithms and techniques were proposed and used for facial recognition. Recognition based on Independent Component Analysis (ICA) of wavelet features was proposed in [1]. There, various wavelet functions were used to extract the features by decomposing the face image into 8-levels. Then, the extracted facial features were analysed using 2D ICA and the Euclidean distance is used for accuracy measurements. The algorithm was tested on the ORL

© Springer International Publishing Switzerland 2015
G. Bebis et al. (Eds.): ISVC 2015, Part II, LNCS 9475, pp. 672–681, 2015.
DOI: 10.1007/978-3-319-27863-6_63

database and the recognition rate recorded was 91.5 %. A 2D Principal Component Analysis (2D PCA) approach based on multiresolusion analysis for facial recognition was proposed in [2]. Multi-levels of decompositions of 2D Discrete Wavelet Transform (DWT) are used to extract the facial features. The highest recognition rate achieved using the combination of 2-level 2D DWT/2D PCA was 94.5 % while using 2D PCA alone achieved 90.5 % for the ORL database. The authors in [3] proposed a 2D Discrete Cosine Transform (2D DCT) approach wherein the feature vectors are extracted using 2D DCT and a SVM is used for recognition. The system was tested on the ORL database and a recognition rate of 95 % was achieved. PCA and Daubechies wavelet subbands were presented in [4]. First, PCA was applied into the four levels of Daubechies wavelet subbands to select the features. Then, the City Block distance and Euclidean distance measures were used for accuracy measurements. The system was tested on the ORL database and a recognition rate of 96.87 % was reported. An approach that integrates 2D Stationary Multiwavelet Transform (MWT) and 2D PCA was proposed in [5]. The features were extracted using 2D SMWT, then 2D PCA and the Histogram-Based Method (HBM) were applied to the extracted features. The highest recognition rates achieved were 63 %, 89 %, and 94.5 % for 2D PCA, 2D SMWT, and HBM, respectively. These algorithms were tested on the ORL database.

In this paper, a new approach based on the 2D DMWT and 2D FastICA is proposed. In contrast to prior work, the proposed approach has two key contributions. First, we apply 2D FastICA to 4 different combinations of 2D DMWT features obtained by considering every subimage of the low-low (LL) frequency subband of the MWT from all poses, thereby exploiting the redundancy in these subimages to further reduce the dimensionality. Second, for the combination corresponding to each subimage we incorporate information from different poses for each person, which leads to a rich set of features that account for different illumination, rotations and facial expressions. As such, the features used lead to a notable dimensionality reduction, reduction in computational complexity, and high recognition rates compared to the existing approaches [1,2]. The proposed approach is evaluated using three different databases, namely, ORL, YALE, and FERET, which have different light conditions, angle rotations, and facial expressions. The classification rates of the system are measured using K-fold Cross Validation (CV).

The paper is organized as follows. In Sect. 2, some preliminary background is provided. In Sect. 3, the proposed system is presented. The experimental results and analysis of the proposed approach are presented in Sect. 4. Section 5 provides some concluding remarks.

2 Multiwavelet Transform and ICA

2.1 Discrete MultiWavelet Transform

The Wavelet transform is broadly used in signal processing, image processing, and pattern recognition for Multiresolution Analysis. The Multiwavelet Transform (MWT) naturally extends the traditional scalar case, where only one scaling and wavelet function are used, to the general case of R scaling and wavelet functions. The scaling function $\Phi(t)$ and the wavelet function $\Psi(t)$ are associated with the low pass and high pass filter, respectively [6]. In Multiwavelets with multiplicity R, the multi-scaling function $\Phi(t)$ and wavelet function $\Psi(t)$ can be written in vector notation as

$$\Phi(t) \equiv [\phi_1(t), \phi_2(t), \dots, \phi_R(t)]^T \tag{1}$$
$$\Psi(t) \equiv [\psi_1(t), \psi_2(t), \dots, \psi_R(t)]^T . \tag{2}$$

The case where $R = 1$ corresponds to the traditional scalar wavelet transform. The two-scale equations of Multiwavelets resemble those of scalar wavelet, i.e.,

$$\Phi(t) = \sqrt{2} \sum_{k=-\infty}^{\infty} H_k \cdot \Phi(2t - k) \tag{3}$$

$$\Psi(t) = \sqrt{2} \sum_{k=-\infty}^{\infty} G_k \cdot \Phi(2t - k) . \tag{4}$$

where H_k and G_k are the filter matrices with dimension $R \times R$ for each integer k [7]. In contrast to the scalar wavelet, MWT can simultaneously achieve perfect reconstruction while preserving length (orthogonality), good performance at the boundaries (via linear phase symmetry), and high-order of approximation (vanishing moments). These favorable properties cannot be achieved by the scalar wavelet at the same time, wherefore Multiwavelets offer more degrees of freedom, and potentially superior performance, for image and signal processing applications compared with scalar wavelet [7].

The multiscaling and wavelet functions are often used with multiplicity $R = 2$, in which case they can be written as $\Phi(t) = [\phi_1(t) \ \phi_2(t)]^T$ and $\Psi(t) = [\psi_1(t) \ \psi_2(t)]^T$, respectively [7]. In this paper, we use the well-known GHM filter due to Geronimo, Hardian, and Massopust [8]. The multiscaling and wavelet functions for GHM system have four scaling and wavelet matrices [8]:

$$H_0 = \begin{bmatrix} \frac{3}{5\sqrt{2}} & \frac{2}{5} \\ \frac{-1}{20} & \frac{-3}{10\sqrt{2}} \end{bmatrix}, H_1 = \begin{bmatrix} \frac{3}{2\sqrt{2}} & 0 \\ \frac{9}{20} & \frac{1}{\sqrt{2}} \end{bmatrix}, H_2 = \begin{bmatrix} 0 & 0 \\ \frac{9}{20} & \frac{-3}{10\sqrt{2}} \end{bmatrix}, H_3 = \begin{bmatrix} 0 & 0 \\ \frac{-1}{20} & 0 \end{bmatrix}$$

$$G_0 = \begin{bmatrix} \frac{-1}{20} & \frac{-3}{10\sqrt{2}} \\ \frac{1}{10\sqrt{2}} & \frac{3}{10} \end{bmatrix}, G_1 = \begin{bmatrix} \frac{9}{20} & \frac{-1}{\sqrt{2}} \\ \frac{9}{20} & \frac{1}{\sqrt{2}} \end{bmatrix}, G_2 = \begin{bmatrix} \frac{9}{20} & \frac{-3}{10\sqrt{2}} \\ \frac{9}{10\sqrt{2}} & \frac{-3}{10} \end{bmatrix}, G_3 = \begin{bmatrix} \frac{-1}{20} & 0 \\ \frac{-1}{10\sqrt{2}} & 0 \end{bmatrix}$$

Unlike scalar wavelet, in MWT the low pass H_k and high pass G_k filters are $R \times R$ matrices, which calls for the use of prefiltering. There are several ways to do prefiltering. In this paper, the critically-sampled scheme is used [9].

2.2   Independent Component Analysis (ICA)

ICA has been primarily used for blind source separation (BSS) [10,11]. ICA has also been successfully used for feature extraction in Image Processing, where it can extract independent image bases that are not necessarily orthogonal. While PCA only considers second-order statistics, it is well-known that ICA exploits high order statistics in the data [1]. Moreover, the bases vectors of ICA capture local image characteristics. These are particularly useful properties for facial recognition as important information may be found in the high order relationships between the pixels, and also due to the fact that the local features of human faces are quite robust to facial expressions, illumination and occlusions owing to their non-rigid nature [12].

The idea of ICA is to represent a set of random variables (RVs) using bases functions such that the coefficients of the expansion are statistically independent or nearly independent. Consider observing M random variables $X_1, X_2, \ldots, X_M$ with zero mean, which are assumed to be linear combinations of N mutually independent components $V_1, V_2, \ldots, V_N$. Let the vector $X = [X_1, X_2, \ldots, X_M]^T$ denote an $M \times 1$ vector of observed variables and $V = [V_1, V_2, \ldots, V_N]^T$ an $N \times 1$ vector of mutually independent components. The relation between X and V is expressed as

$$X = AV. \tag{5}$$

where A is a full rank $M \times N$ unknown matrix, called the mixing matrix or the feature matrix [13]. In this paper, X corresponds to the intensities of the pixels of the LL subband of the facial images.

ICA has two main drawbacks. First, it typically requires complex matrix operations [14]. Second, it has slow convergence [10]. To alleviate these two drawbacks, [11] introduced a new technique called FastICA, which is used in this paper. FastICA is computationally more efficient for estimating the ICA components and has a faster convergence rate by using a fixed point iteration algorithm [11]. According to [11,13,15], the mixing matrix A (feature matrix) can be considered as another representation of the face images.

3 Proposed Approach

In this section, the proposed system is presented. The supervised facial recognition system consists of three phases, namely, Preprocessing, Feature Extraction, and Recognition.

3.1 Preprocessing: *This Phase Consists of Two Steps:*

1. The first step aims to convert all dimensions of different image databases used in this paper (see Table 1) to an appropriate one. This dimension is chosen since the algorithm used requires dimensions that are power of two.
2. This step aims to convert all the images into (128×128 *double*) instead of using *uint*8 extension, which is not suitable for the transform used in this paper.

Table 1. The dimensions of the databases

Databases	ORL	YALE	FERET	Proposed
Size	112×92	243×320	384×256	128×128

3.2 Feature Extraction

Feature extraction is an essential component of facial recognition systems. A face image may contain information that is not essential for the recognition purpose. Hence, the following effective tools are applied to get an efficient representation of the images by extracting discriminating and independent features:

1. The 2D DMWT based on MRA is used for:
   (a) Dimensionality reduction.
   (b) Noise reduction.
   (c) Localizing all the useful information in one single band.
2. The FastICA is used for:
   (a) Decorrelating the high order statistics since most of the significant information is contained in the high order statistics of the image [1,2,11–15].
   (b) The ICA features are independent leading to a better representation and hence better identification and recognition rates.
   (c) The ICA features are less sensitive to the facial variations arising from different facial expressions and different poses [1,2,11,13,15].
   (d) Reducing the computational complexity and improving the convergence rate.

First, the 2D DMWT is applied to the different databases. Figure 1 shows an example of applying 2D DMWT to different face images. As shown in Fig. 1, the images are divided into four main subbands with 64×64 dimension and each one is further divided into four 32×32 subimages. From Fig. 1a, b, c, it is easily seen that most of the useful information is localized in the upper left band, which is related to the low-low (LL) frequency band of the DMWT. Thus the LL subband is retained, and all the remaining subbands are eliminated. Therefore, the resulting image matrix is 64×64. The following procedures are applied to the four subimages of the LL subband of DMWT in order to get an efficient representation for the input images.

1. Convert each 32×32 subimage into a 1024×1 vector.
2. Repeat (1) for all subimages of each pose. The resulting features for each pose have dimension 1024×4.
3. For each of the four subimages of the LL subband, combine the 1024×1 subimage vectors of the different poses of each person as shown in Fig. 2-B.

Thus, the combinations shown in Fig. 2-B are defined as:

$$\text{Combination}_i = [S_{pi}] \tag{6}$$

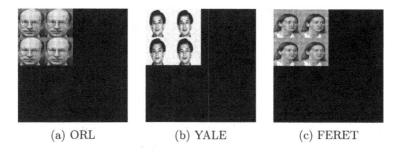

| (a) ORL | (b) YALE | (c) FERET |

Fig. 1. Resultant images after applying 2-D DMWT to the different databases.

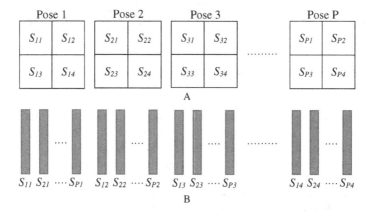

Fig. 2. Figure 2-A represents the subimages of the LL subband of 2D DMWT for P training poses from each person. Figure 2-B is rearranging Fig. 2-A in four combinations.

where $p \in \{1, 2, 3, \ldots, P\}$ and P is the number of poses used in the training mode, $i \in \{1, 2, 3, 4\}$ indexes the four subimages of the LL subband.

It is very useful to extract new features from the original features to reduce the dimensionality of the feature space. To this end, we apply the following four techniques to the *Combinations* above:

A. *Technique 1*
   i. Apply 2D FastICA on each combination as shown in Fig. 3.
   ii. Choose the IF_p column that has the maximum ℓ_2-norm to represent the features of each combination. In this case, the resultant features of each person have dimensions 1024×4 regardless of the number of poses used in both the training and the testing modes.
B. *Technique 2* (shown in Fig. 4)
   i. Apply 2D FastICA on each combination.
   ii. Form a 1024×1 column vector whose entries consist of the ℓ_2-Norm for each row of the resultant features to reduce the dimensionality. The resultant features of each person have dimension 1024×4 regardless of the number of poses used in training mode.

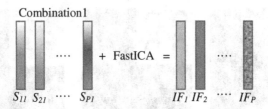

Fig. 3. An example of applying 2D FastICA to the first combination. IF_p represents the resultant Independent features where $p \in \{1, 2, 3, \ldots, P\}$.

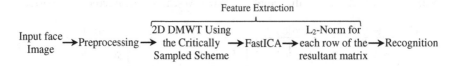

Fig. 4. Proposed Technique 2.

C. *Technique 3*: This is similar to Tech.2, except that we use the mixing matrix A shown in (5) instead of using the resultant FastICA matrix. This is an alternative approach for representing the face images [11, 15].

D. *Technique 4*: Same as Tech.2 without applying step B-i.

4. Repeat [1–3] for all poses of the three different databases.
5. Repeat Tech.1-Tech.4 for all persons in the three different databases. For all techniques, there are 1024×4 features corresponding to each person regardless of the number of poses used in training mode.

3.3   Recognition

A Neural Network (NNT) based on the Back Propagation Training Algorithm (BPTA) is used during the training and the testing phases. Since BPTA is a supervised learning algorithm, it is necessary to choose a desired output for each database used. In this paper, the system is evaluated using three different databases, each with a different number of persons. Therefore, there are 40, 15, 200 desired outputs for the ORL, YALE, and FERET databases, respectively, corresponding to the number of persons in each database. The configuration of the NNT has three layers, namely, an input layer, a hidden layer, and an output layer.

In the training mode, the classifier is configured using the described training features. For testing, the recognition rate is measured as $\frac{D}{L} \times 100\%$, where D is the total number of images correctly matched and L is the total number of images in the database.

In the testing mode, we follow the same steps of the training phase. First, the 2D DMWT is applied to the test image. Second, only the LL subband is retained. Third, each subimage is converted to 1D form. Then, Tech.1-Tech.4 are applied to the 1024×4 resultant features corresponding to the four subimages of the LL subband. For each technique, there is a 1024×1 feature vector for each test image. This vector is fed into the NNT for classification.

4   Experimental Results and Discussion

The experimental results of the proposed techniques are presented in this section. We also compare the results of the proposed approach to some of the state-of-the-art methods. The techniques are tested using three different databases, ORL, YALE, and FERET, which have different facial expressions, light conditions, and rotations. K-fold Cross Validation [16] is used to evaluate our proposed techniques. Different values of K are used in this paper as show in Tables 2, 3, and 4.

4.1   Experimental Results for the ORL Database

The ORL database consists of 40 persons, each with 10 different poses. Therefore, P different poses are used in the training mode, and $10 - P$ poses are used in the testing mode. Table 2 summarizes the results for all different techniques. As shown, even when a small number of poses are used in the training mode, the algorithm achieves high recognition accuracy. As the number of poses increased, the performance is improved.

4.2   Experimental Results for the YALE Database

This database consists of 15 persons, each with 11 different poses. Table 3 summarizes the results of the proposed techniques. As before, our proposed techniques show superior performance.

Table 2. Experimental Results for the ORL Database

K Fold	Training Rates	Recognition Rates for Testing Phase				2D DWT and 2D ICA [1]	2D DWT and 2D PCA [2]
		Tech.1	**Tech.2**	Tech.3	Tech.4		
K=2	100 %	96.75 %	**98.25%**	97.25 %	96.25 %	93.75 %	92.5 %
K=3	100 %	97.5 %	**98.75%**	97.75 %	96.75 %	95.25 %	94.75 %
K=5	100 %	98.25 %	**99.25%**	98.75 %	97.75 %	96.75 %	96.5 %

Table 3. Experimental results for the YALE database

K Fold	Training Rates	Recognition Rates for Testing Phase				2D DWT and 2D ICA [1]	2D DWT and 2D PCA [2]
		Tech.1	Tech.2	Tech.3	Tech.4		
K=2	100 %	96.97 %	**98.18%**	96.97 %	95.15 %	92.73 %	92.12 %
K=3	100 %	97.58 %	**98.79%**	97.58 %	96.36 %	94.55 %	93.94 %
K=5	100 %	98.18 %	**99.39%**	98.78 %	97.58 %	96.36 %	95.15 %

Table 4. Experimental results for the FERET database

| K Fold | Training Rates | Recognition Rates for Testing Phase | | | | 2D DWT and | 2D DWT and |
		Tech.1	Tech.2	Tech.3	Tech.4	2D ICA [1]	2D PCA [2]
K=2	100 %	96.68 %	**98.32%**	97.36 %	95.86 %	92.87 %	92.5 %
K=3	100 %	97.6 %	**98.6%**	98.1 %	96.9 %	95.14 %	94.86 %
K=5	100 %	97.78 %	**99%**	98.46 %	96.96 %	96.14 %	95.68 %

4.3 Experimental Results for the FERET Database

There are 200 persons in this database, each with 11 different poses. Table 4 summarizes the results for the proposed techniques. The results exhibit the same behavior as in the previous databases.

The goal of using FastICA in this paper is to find a set of statistically independent basis images that preserve the structural information of the face images present in the facial expressions, illumination, and rotations. ICA representations [15] are more robust to variations due to different light conditions, changes in hair location, make up, and facial expressions. This efficient representation leads to significant improvement in the recognition rates. As demonstrated in the previous results, Tech. 1, Tech. 2, Tech. 3 outperform Tech. 4, which does not use the FastICA step. Our proposed techniques are shown to outperform the existing method [1,2] in the recognition rates, storage requirements, and computational complexity. The configuration of the Neural Network during the training mode can effect the overall accuracy of the system. Hence, choosing the number of hidden layers, number of neurons in each hidden layer, activation function, training algorithms, and the target performance play an important role in the overall system performance.

5 Conclusion

A facial recognition system based on the integration of Multiresolusion Analysis and FastICA was proposed. The proposed approach leads to notable dimensionality reduction and subsequently less storage requirements, as well as higher recognition rates. These benefits are due to using combinations of different poses for the subimages of the LL subband of the 2D DMWT and the ℓ_2 Norm for 2D FastICA features in the feature extraction step. Therefore, we exploit the redundancy present in these subimages to reduce the dimensions of the used features, which also account for all the variations in one short feature vector per combination. Each person is represented using features of dimension 1024×4 regardless of the number of poses in the training phase. Compared with the existing methods, the proposed approach achieves higher recognition rates with less storage and computation requirements. The proposed techniques were evaluated using three different databases with different light conditions, rotations, and facial expressions. The recognition rates are analyzed using K-fold CV. For example, for K=2, the highest recognition rates achieved were 98.25 %, 98.18 %, and 98.32 % for the ORL, YALE, and FERET databases, respectively.

Acknowledgement. This work was supported in part by NSF grant (CCF - 1320547) and by the Iraqi government scholarship (HCED).

References

1. Kinage, K., Bhirud, S.: Face recognition based on independent component analysis on wavelet subband. In: 3rd IEEE International Conference on Computer Science and Information Technology (ICCSIT). vol. 9, pp. 436–440 (2010)
2. AlEnzi, V., Alfiras, M., Alsaqre, F.: Face recognition algorithm using two dimensional principal component analysis based on discrete wavelet transform. In: Snasel, V., Platos, J., El-Qawasmeh, E. (eds.) ICDIPC 2011, Part I. CCIS, vol. 188, pp. 426–438. Springer, Heidelberg (2011)
3. Hongtao, Y., Jiaqing, Q., Ping, F.: Face recognition with discrete cosine transform. In: Second International Conference on Instrumentation, Measurement, Computer, Communication and Control (IMCCC), pp. 802–805 (2012)
4. Satone, M., Kharate, G.: Face recognition based on pca on wavelet subband. In: IEEE Students' Conference on Electrical, Electronics and Computer Science (SCEECS), pp. 1–4 (2012)
5. Ismaeel, T.Z., Kamil, A.A., Naji, A.K.: Article: human face recognition using stationary multiwavelet transform. Int. J. Comput. Appl. (IJCA) **72**, 23–32 (2013)
6. Zhihua, X., Guodong, L.: Weighted infrared face recognition in multiwavelet domain. In: IEEE International Conference on Imaging Systems and Techniques (IST), pp. 70–74 (2013)
7. Reddy, P.V.N., Prasad, K.: Article: Multiwavelet based texture features for content based image retrieval. Int. J. Comput. Appl. **17**, 39–44 (2011)
8. Geronimo, J.S., Hardin, D.P., Massopust, P.R.: Fractal functions and wavelet expansions based on several scaling functions. J. Approx. Theor. **78**, 373–401 (1994)
9. Strela, V., Heller, P., Strang, G., Topiwala, P., Heil, C.: The application of multiwavelet filterbanks to image processing. IEEE Trans. Image Process. **8**, 548–563 (1999)
10. Bell, A., Sejnowski, T.: An information-maximization approach to blind separation and blind deconvolution. Neural Comput. **7**, 1129–1159 (1995)
11. Hyvarinen, A.: Fast and robust fixed-point algorithms for independent component analysis. IEEE Trans. Neural Netw. **10**, 626–634 (1999)
12. Lihong, Z., Ye, W., Hongfeng, T.: Face recognition based on independent component analysis. In: Control and Decision Conference (CCDC), Chinese, pp. 426–429 (2011)
13. Yuen, P.C., Lai, J.H.: Face representation using independent component analysis. Pattern Recogn. **35**, 1247–1257 (2002)
14. Comon, P.: Independent component analysis, a new concept? Sig. Process. **36**, 287–314 (1994)
15. Bartlett, M., Movellan, J.R., Sejnowski, T.: Face recognition by independent component analysis. IEEE Trans. Neural Netw. **13**, 1450–1464 (2002)
16. Arlot, S., Celisse, A., et al.: A survey of cross-validation procedures for model selection. Stat. Surv. **4**, 40–79 (2010)

Dynamic Hand Gesture Recognition Using Generalized Time Warping and Deep Belief Networks

Cristian A. Torres-Valencia$^{(\boxtimes)}$, Hernán F. García, Germán A. Holguín, Mauricio A. Álvarez, and Álvaro Orozco

Grupo de Investigación En Automática,
Universidad Tecnológica de Pereira, La Julita, Pereira, Colombia
{cristian.torres,hernan.garcia,gahol,malvarez,aaog}@utp.edu.co

Abstract. Body gestures play an important role in human communications, specially hand gestures are the most distinctive features in sign languages. Several works have been proposed in order to recognize hand gestures using static and dynamic approaches. Nevertheless, due to the high variety of signs and the dynamic changes exhibited in different hand motions, a strategy for modeling these dynamic changes in hand signs must be fulfilled. In this work we propose a framework for dynamic hand gesture recognition using a well known method for alignment of time series as the Generalized Time Warping (GTW). Several features are extracted from the aligned sequences of hand gestures based on texture descriptors. Then a methodology for hand motion recognition is carried out based on Convolutional Neural Networks. The obtained results show that the methodology proposed allows an accurate recognition of several hand gestures obtained from the RVL-SLLL American Sign Language Database.

1 Introduction

Human interaction is influenced by body gestures, even for non-deaf people the sign language (SL) plays a big role into the communication process. Several works have been proposed in this field in recent years in order to recognize efficiently different gestures from SL in some languages. As the most structured gesture language, the SL gives a full dictionary of possible words and phrases that vary from each language [1]. The recognition of the SL gestures can be applied into human-computer interfaces (HCI), also as a development tool for technology for deft people and other interesting applications [1].

The recognition of SL could be divided in two main categories, an static sign based recognition and a dynamic sign based recognition. The static approach is focused on the analysis of hand poses that represents specific characters of the alphabet of SL [1]. Several works have been developed in the static based SL recognition, the first step is the segmentation and characterization of the hand region inside the image that is achieved using geometrical descriptors [2],

© Springer International Publishing Switzerland 2015
G. Bebis et al. (Eds.): ISVC 2015, Part II, LNCS 9475, pp. 682–691, 2015.
DOI: 10.1007/978-3-319-27863-6_64

geometrical moments, wavelets [3], texture descriptors, marked color gloves and manually labeled images [1]. From the obtained descriptors, classifiers are trained in order to determine the corresponding static sign represented. Support Vector Machines (SVM) [4] and Neural Networks (NN) [5] are commonly used in this stage for the static recognition.

For the dynamic approach, the idea is to determine the gestures from SL that involve any kind of movement in their representation. This approach has major challenges involved since the analysis of a image sequence and the tracking of the features could present several drawbacks. Single hand and both hand dynamic recognition have been developed, the use of contour features [6], direction histograms [7] is combined in most of the cases with Hidden Markov Models (HMM) [8]. The results obtained in these works is promising but far from optimal and is only applicable to fixed LS since the words or phrases represented vary from each to each language [1].

There are several considerations that should be analyzed when developing a methodology for SL recognition. The variation on the illumination conditions, the possibly unrestricted environment, the size of the set of gestures that is going to be recognized, the invariance in the performance of the gestures are factors that influence the result of the recognition [1]. In order to develop a strategy for SL recognition, we propose the alignment of the realizations of the SL gesture based on the Generalized Time Warping (GTW) [9] that is an extension of the Dynamic Time Warping (DTW) for human motion analysis. After all the sequences are aligned, a feature extraction algorithm known as the Scale Invariant Feature Transform (SIFT) [10] over all the frames in the sequence. Finally a SVM classification algorithm is applied over the set of features in order to determine the SL gesture. The image sequences used in the development of this work are provided from the RVL-SLLL American Sign Database of the Purdue University [11].

This paper is ordered as follows, Sect. 2 presents the description of the methods as well as the database used for the processing of the SL. Section 3 presents the obtained results using the proposed approach and finally the Sect. 4 present the discussion of the results and the future work that could be derived from this paper.

2   Materials and Methods

In this section we describe the methods used for the development of the present work, also a description of the materials is included.

2.1   Database

The image sequences for the hand gesture recognition proposed in this framework are obtained from the RVL-SLL American Sign Language Database [11]. The videos of the database has a range of signed material under controlled an

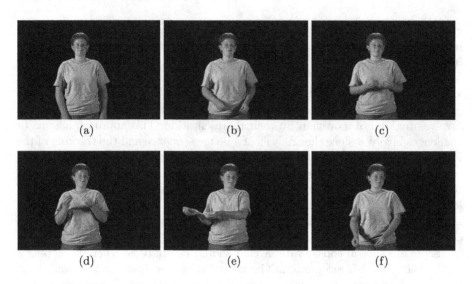

Fig. 1. Image sequence of one realization from the RVL-SLLL database

less-controlled lighting conditions. The database contains handshapes in isolation and in single signs, the American fingerspelling alphabet, numbers, movement in single signs, and examples of short discourse narratives for testing sign recognition in connected linguistic contexts [11]. From the whole database the "Motions" category of signs is used. The motions category contains two handed signs of both hands moving or one moving and the other acting as the basehand. The first ten signs of the motions category are used, being the following signs: PUT (on a shelf), PUT (straight down), PUT (center of table), PUT (left edge of the table), LEFT-GIVE-RIGHT, RIGHT-GIVE-LEFT, I-GIVE-YOU, YOU-GIVE-ME, I-GIVE-LEFT and I-GIVE-RIGHT signs. Some examples of the images from the database re presented in Fig. 1.

2.2   Generalized Time Warping

The analysis of temporal variant series has a wide scope of applications. Dynamic Time Warping has proved it efficiency in alignment of time series and several extensions has been proposed for the alignment of human behavior. Canonical Time Warping (CTW) and Dynamic Manifold Warping (DMW) combines DTW with the addition of canonical correlation analysis and manifold learning to incorporate spatial transformations respectively [9]. Despite the study of those extensions and the possible improvement in the analysis of time series, there are still some limitations related to the basis of DTW. The computational complexity is quadratic in space and time, is focused on the alignment of only two sequences and the temporal alignment of DTW is relies on dynamic programming [9].

Given a set of m time series (i.e. motion capture trajectories), $\{\mathbf{X}_1, \ldots, \mathbf{X}_m\}$, where $\mathbf{X}_i = [\mathbf{x}_1^i, \ldots, \mathbf{x}_{ni}^i] \in \mathbb{R}^{d_i \times n_i}$. In GTW, we want to find a non-linear temporal transformation $\mathbf{W}_i \in \{0, 1\}^{n_i \times l}$ and a low-dimensional embedding $\mathbf{V}_i \in \{0, 1\}^{d \times l}$ for each $\mathbf{X}_i$.

$$J_{gtw}\left(\{\mathbf{W}_i, \mathbf{V}_i\}\right) = \sum_{i=1}^{m} \sum_{j=1}^{m} \frac{1}{2} \left\| \mathbf{V}_i^T \mathbf{X}_i \mathbf{W}_i - \mathbf{V}_j^T \mathbf{X}_j \mathbf{W}_j \right\|_F^2$$

$$+ \left(\sum_{i=1}^{m} \psi\left(\mathbf{W}_i\right) + \phi\left(\mathbf{V}_i\right) \right), \qquad (1)$$

where $\phi(.)$ and $\psi(.)$ are regularization functions, which bias the solution in the space of temporal transformations $\mathbf{W}_i$ and the embedding for spatial transformation $\mathbf{V}_i$, respectively. Ψ and Φ represent the domains for $\mathbf{W}_i$ and $\mathbf{V}_i$ [9]. Generally speaking, optimizing J_{gtw} is a non-convex optimization problem with respect to $\mathbf{W}_i$ and $\mathbf{V}_i$. The procedure is alternated between using a Gauss-Newton for $\mathbf{W}_i$ and optimally computing $\mathbf{V}_i$ using mCCA [9]. These steps monotonically decrease J_{gtw}.

Parametrization of the Temporal Warping: From the temporal warping matrix $\mathbf{W} \in \{0, 1\}^{\mathbf{n} \times l}$ for a single sequence $\mathbf{X} \in \mathbb{R}^{\mathbf{d} \times \mathbf{n}}$. The DP-based approach to optimize $\mathbf{W}$ has a computational cost of $O(nl)$ [9]. In order to reduce computational cost, GTW approximates the warping path $\mathbf{p} \in \{1 : \mathbf{n}\}$, which parametrizes the warping matrix $\mathbf{W}(\mathbf{p})$, as a linear combination of monotonic functions $q \in [1, n]^l$, that is:

$$p \approx \sum_{\bar{c}=1}^{\bar{k}} \bar{a}_{\bar{c}} \bar{q}_{\bar{c}} + \sum_{\dot{c}=1}^{\dot{k}} \dot{a}_{\dot{c}} \dot{q}_{\dot{c}} = \bar{Q}\bar{a} + \dot{Q}\dot{a} = Qa, \qquad (2)$$

where $a = [\bar{a}; \dot{a}] \in \mathbb{R}^k$, $k = \bar{k} + \dot{k}$ is the weight vector and $Q = \left[\bar{Q}, \dot{Q}\right] \in \mathbb{R}^{l \times k}$ is the basis set composed of a constant function $\bar{Q} = [\bar{q}_1, \ldots, \bar{q}_{\bar{k}}] \in [1, n]^{l \times \bar{k}}$ and a monotonically increasing function $\dot{Q} = [\dot{q}_1, \ldots, \dot{q}_{\dot{k}}] \in [1, n]^{l \times \dot{k}}$. As in DTW, the following constraints on the weight a are incorporated, to constrain the warping path $\mathbf{p} = \mathbf{Q}\mathbf{a}$.

Boundary Conditions: The position of the first frame is enforced, $\mathbf{p}_1 = q^{(1)} a \geq 1$, as well as the final frame, $p_l = q^{(l)} a \leq n$, where $q^{(1)} \in \mathbb{R}^{1 \times k}$ and $q^{(l)} \in \mathbb{R}^{1 \times k}$ are the first and last row of the basis matrix $\mathbf{Q} \in \mathbb{R}^{l \times k}$ respectively. In contrast to DTW that imposes tight boundary, GTW relaxes the equality with inequality constrains to allow for a sub-part of X being indexed by $\mathbf{p}$ [9].

Monotonicity: $t_1 \leq t_2 \Rightarrow p_{t_1} \leq p_{t_2}$ are enforced by constraining the sign of weight: $\dot{a} \geq 0$. Notice that constraining the weights is only a sufficient condition to ensure monotonicity but is not necessary [9].

Continuity: To approximate the hard constraint on the step size, the curvature of the warping path is penalized, $\sum_{t=1}^{l} \left\| \nabla q^{(t)} a \right\|_2^2 \approx \| FQa \|_2^2$ where $\mathbf{F} \in \mathbb{R}^{l \times l}$ is the first order differential operator [9].

In summary the warping path is constrained as [9]:

$$\varphi_a(a) = \eta \| FQa \|_2^2, \quad \Psi_a = \{ a | La \leqslant b \}, \tag{3}$$

where $L = \begin{bmatrix} 0_{k \times \bar{k}} & -I_k \\ -\bar{q}^{(1)} & -\dot{q}^{(1)} \\ \bar{q}^{(l)} & \dot{q}^{(l)} \end{bmatrix}$ and $b = \begin{bmatrix} 0_k \\ -1 \\ n \end{bmatrix}$.

Therefore, given a basis set of k monotone functions, all feasible weights belong to a polyhedron in $\mathbb{R}^k$ parametrized by $L \in \mathbb{R}^{(k+2) \times k}$ and $b \in \mathbb{R}^{k+2}$ [9].

2.3   Convolutional Deep Belief Networks

For the purpose of this explanation, all inputs to the algorithm are single-channel time-series data with n_V frames (an n_V dimensional vector) are assumed; but the formulation can be extended to the case of multiple channels [12]. The convolutional restricted Boltzmann machines (CRBMs) are an extension of the "regular" RBM to a convolutional setting, in which the weights between the hidden units and the visible units are shared among all locations in the hidden layer. The CRBM consists of two layers: an input (visible) layer V and a hidden layer H. The hidden units are binary-valued, and the visible units are binary-valued or real-valued [12].

Considering the input layer of an n_V dimensional array of binary units. To construct the hidden layer, consider Kn_W-dimensional filter weights W_K [12]. The hidden layer consists of K "groups" of n_H-dimensional arrays (where $n_H \triangleq n_V - n_W + 1$) with units in group k sharing the weights W^k. There is also a shared bias b_k for each group and a shared bias c for the visible units [12]. The energy function can then be defined as:

$$E(v, h) = -\sum_{k=1}^{K} \sum_{j=1}^{n_H} \sum_{r=1}^{n_W} h_j^k W_r^k v_{j+r-1} - \sum_{k=1}^{K} b_k \sum_{j=1}^{n_H} h_j^k - c \sum_{i=1}^{n_V} v_i. \tag{4}$$

Similarly, the energy function of CRBM with real-valued visible units can be defined as:

$$E(v, h) = \frac{1}{2} \sum_{i}^{n_V} v_i^2 - \sum_{k=1}^{K} \sum_{j=1}^{n_H} \sum_{r=1}^{n_W} h_j^k W_r^k v_{j+r-1} - \sum_{k=1}^{K} b_k \sum_{j=1}^{n_H} h_j^k - c \sum_{i=1}^{n_V} v_i. \tag{5}$$

The joint and conditional probability distributions are defined as follows [12]:

$$P(v, h) = \frac{1}{z} \exp(-E(v, h)) \tag{6}$$

$$P\left(h_j^k = 1 \,|\, v\right) = sigmoid\left(\left(\tilde{W}^k *_v v\right)_j + b_k\right) \qquad (7)$$

$$P\left(v_i = 1 \,|\, h\right) = sigmoid\left(\sum_k \left(W^k *_f h^k\right)_i + c\right) \qquad (8)$$

$$P\left(v_i \,|\, h\right) = Normal\left(\sum_k \left(W^k *_f h^k\right)_i + c, 1\right) \qquad (9)$$

where $*_v$ is a "valid" convolution, $*_f$ is a "full" convolution, and $\tilde{W}_j^k \triangleq W_{n_W - j + 1}^k$. Since all units in one layer are conditionally independent given the other layer, inference in the network can be efficiently performed using block Gibbs sampling. Lee et al. [13] further developed a convolutional RBM with "probabilistic max-pooling", where the maxima over small neighborhoods of hidden units are computed in a probabilistically sound way [12].

For training the convolutional RBMs, contrastive divergence can be used to approximate the gradient effectively. Since a typical CRBM is highly overcomplete, a sparsity penalty term is added to the log-likelihood objective [12]. More specifically, the training objective can be written as

$$minimize_{W,b,c} \quad L_{likelihood}\left(W, b, c\right) + L_{sparsity}\left(W, b, c\right), \qquad (10)$$

where $L_{likelihood}$ is a negative log-likelihood that measures how well the CRBM approximates the input data distribution, and $L_{sparsity}$ is a penalty term that constrains the hidden units to having sparse average activations [12]. This sparsity regularization can be viewed as limiting the "capacity" of the network, and it often results in more easily interpretable feature representations. Once the parameters for all the layers are trained, we stack the CRBMs to form a convolutional deep belief network [12].

2.4 Dynamic Hand Signs Recognition

For every sequence of hand signs, first we compute correspondence descriptors such as SIFT and SURF features (we select the first 100 detected points from both descriptors). Secondly, we align all time-series of hand signs using Generalized time warping in which a spatial transformation of the multimodal data (shape descriptors) is computed using multi-set canonical correlation analysis[1]. Finally, we use two methods to perform the recognition, SVM[2] and DBN[3] both using *one-vs-all* formulation. Figure 2, shows the hand signs recognition scheme.

---

[1] We use the GTW implementation available in http://www.f-zhou.com/ta_code.html.
[2] We use LIBSVM available on https://www.csie.ntu.edu.tw/~cjlin/libsvm/.
[3] We use the Deep Learning toolbox available on http://deeplearning.cs.toronto.edu/codes.

Fig. 2. Overview of the dynamic hand signs recognition scheme. Aligned hand motion sequences are taken as observed points in order to compute some correspondence descriptors. Then, Deep belief networks are used to perform the recognition.

3   Results

In the following sections we show the results for our dynamic hand signs recognition approach using GTW and deep learning. On Sect. 3.1, we test the proposed method for dynamic hand sings alignment from image sequences on the RVL-SLL database. Finally in Sect. 3.2, we test the method on hand signs classification for ten hand motions (see Sect. 2.1).

3.1   Dynamic Hand Signs Alignment

In this experiment we applied GTW to align video sequences of different people performing a similar hand gesture (motion). Each video is encoded using correspondences descriptors such as SURF and SIFT features. Here, we took recordings for five subjects each displaying ten motion gestures. We use as ground-truth alignment uniform time warping (UDTW) using the same descriptors [9]. We use uniform alignment to initialize GTW, using values for λ between $[0.1 - 0.4]$.

Figure 3, shows the results of the dynamic alignment using GTW. Figure 3(d), shows the alignment error for all sequences in database. The results are compared against a common alignment method such as Procrustes dynamic time warping (PDTW). We can notice that PDTW (see Fig. 3(b)) is not able to align the dynamic sequences accurately (space in time) due to the problem of compute correspondence between sequences with different motions. However, Fig. 3(c) shows that GTW warps the hand sequences accurately in both time and space.

3.2   Hand Signs Recognition Results

In order to perform the recognition, first we train a DBN with one input layer of $106 \times N$ inputs units corresponding to the correspondence descriptors of the hand sequences. We use $N = 40$ as the size of the aligned sequences. Besides, for the abstraction and association layers, we use 500-500 and a 10-unit label layer corresponding to the 10 motions classes.

Table 1 shows the results for the hand gesture recognition using DBN. We provide three variants for DBN training (Sigmoid-binary and Siegert are time-stepped model and an event-driven DBN using LIF neurons [14]). The results show that Sigmoid-Binary model brings more accuracy in the recognition process (**91.04 % ± 1.38**), highlighting the benefits of the deep neural training.

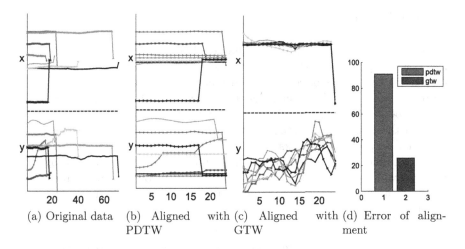

(a) Original data   (b) Aligned with (c) Aligned with (d) Error of align-
PDTW   GTW   ment

Fig. 3. Hand gestures alignment results. (a) Sample of original hand gesture sequences. We align the spatial location of the texture descriptors. (b) Alignment process using PDTW. (c) Alignment process using GTW. (d) Error of alignment (difference between the warping matrix of ground-truth and method).

Table 1. Classification performance on the RVL-SLL database using a deep belief network. We provide accuracy for three neuron models ($Acc \pm std$).

Neuron model	SURF exp.	SIFT exp.
LIF	84.02 ± 3.34	85.14 ± 1.48
Siegert	86.51 ± 2.43	87.88 ± 2.62
Sigmoid-Binary	87.44 ± 4.15	$\mathbf{91.04 \pm 1.38}$

Table 2. Classification performance on the RVL-SLL database using a SVM classifier ($Acc \pm std$).

Kernel	SURF exp.	SIFT exp.
RBF	82.1 ± 5.24	86.24 ± 4.23
Polynomial	83 ± 3.17	84.63 ± 3.45

Also, in order to compare the benefits of the deep neural training, we perform a classification task based on SVM using a RBF and polynomial kernels and one-vs-all validation with 25 repetitions of the experiments. The results in Table 2, show that despite using a SVM classifier, this method do not outperforms the deep neural training over dynamic features of hand gestures. However, the results show that when SIFT are used to compute the correspondence descriptors, the recognition task shows more accurate results.

4   Conclusions

We have presented a method for dynamic hand gesture recognition using generalized time warping and deep belief networks. Our model represent hand gesture sequences as time-series of correspondence descriptors such as SIFT and SURF features. In order to scale all hand gesture sequences both space and time, GTW proves to be a robust way to align time-varying hand sequences. Moreover, the results in the hand gesture recognition process (DBN and SVM experiments) shows that deep belief networks provide a novel and efficient framework to perform a dynamic recognition, due to the importance of unsupervised layer-by-layer pre-training with the Contrastive Divergence algorithm. As future works, we want to analyze more sets of hand gestures in order to capture hierarchical properties of multi-language approaches.

Acknowledgments. C.A. Torres-Valencia is funded by Colciencias under the program: *Formación de alto nivel para la ciencia, la tecnología y la innovación - Convocatoria 647 de 2014*. H.F. García is funded by Colciencias under the program: *Formación de alto nivel para la ciencia, la tecnología y la innovación - Convocatoria 617 de 2014*. All authors thanks the support from the "Maestría en Ingeniería Eléctrica" and "Doctorado en Ingeniería" programs from the Universidad Tecnológica de Pereira.

References

1. Kausar, S., Javed, M.: A survey on sign language recognition. In: Frontiers of Information Technology (FIT 2011), pp. 95–98 (2011)
2. Collumeau, J.F., Leconge, R., Emile, B., Laurent, H.: Hand gesture recognition using a dedicated geometric descriptor. In: 2012 3rd International Conference on Image Processing Theory, Tools and Applications (IPTA), pp. 287–292 (2012)
3. Kiani Sarkaleh, A., Poorahangaryan, F., Zanj, B., Karami, A.: A neural network based system for persian sign language recognition. In: 2009 IEEE International Conference on Signal and Image Processing Applications (ICSIPA), pp. 145–149 (2009)
4. Sinith, M., Kamal, S., Nisha, B., Nayana, S., Surendran, K., Jith, P.: Sign gesture recongnition using support vector machine. In: 2012 International Conference on Advances in Computing and Communications (ICACC), pp. 122–125 (2012)
5. Isaacs, J., Foo, S.: Hand pose estimation for american sign language recognition. In: Proceedings of the Thirty-Sixth Southeastern Symposium on System Theory, pp. 132–136 (2004)
6. Li, H., Greenspan, M.A.: Model-based segmentation and recognition of dynamic gestures in continuous video streams. Pattern Recogn. **44**, 1614–1628 (2011)
7. Nandy, A., Prasad, J.S., Mondal, S., Chakraborty, P., Nandi, G.C.: Recognition of isolated indian sign language gesture in real time. In: Das, V.V., et al. (eds.) BAIP 2010. CCIS, vol. 70, pp. 102–107. Springer, Heidelberg (2010)
8. Ahmed, A., Aly, S.: Appearance-based arabic sign language recognition using hidden markov models. In: 2014 International Conference on Engineering and Technology (ICET), pp. 1–6 (2014)

9. Zhou, F., De la Torre Frade, F.: Generalized time warping for multi-modal alignment of human motion. In: IEEE Conference on Computer Vision and Pattern Recognition (CVPR) (2012)

10. Lowe, D.G.: Distinctive image features from scale-invariant keypoints. Int. J. Comput. Vis. **60**, 91–110 (2004)

11. Martnez, A.M., Wilbur, R.B., Shay, R., Kak, A.C.: Purdue RVL-SLLL ASL database for automatic recognition of american sign language. In: ICMI, pp. 167–172. IEEE Computer Society (2002)

12. Lee, H., Pham, P.T., Largman, Y., Ng, A.Y.: Unsupervised feature learning for audio classification using convolutional deep belief networks. In Bengio, Y., Schuurmans, D., Lafferty, J.D., Williams, C.K.I., Culotta, A., (eds.) NIPS, pp. 1096–1104. Curran Associates, Inc. (2009)

13. Lee, H., Grosse, R., Ranganath, R., Ng, A.Y.: Convolutional deep belief networks for scalable unsupervised learning of hierarchical representations. In: Proceedings of the 26th Annual International Conference on Machine Learning, ICML 2009, pp. 609–616. ACM, New York (2009)

14. O'Connor, P., Neil, D., Liu, S.C., Delbruck, T., Pfeiffer, M.: Real-time classification and sensor fusion with a spiking deep belief network. Front. Neurosci. **7** (2013)

Gaussian Processes for Slice-Based Super-Resolution MR Images

Hernán Darío Vargas Cardona[1]([✉]), Andrés F. López-Lopera[1],
Álvaro A. Orozco[1], Mauricio A. Álvarez[1], Juan Antonio Hernández Tamames[2],
and Norberto Malpica[2]

[1] Faculty of Engineering, Universidad Tecnológica de Pereira, Pereira, Colombia
{hernan.vargas,anfelopera,aaog,malvarez}@utp.edu.co
[2] Laboratorio de Análisis de Imagen Médica Y Biometría,
Universidad Rey Juan Carlos, Móstoles, Spain
{juan.tamames,norberto.malpica}@urjc.es

Abstract. Magnetic resonance imaging (MRI) is a medical technique used in radiology to obtain anatomical images of healthy and pathological tissues. Due to hardware limitations and clinical protocols, MRI data are often acquired with low-resolution. For this reason, the scientific community has been developing super-resolution (SR) methodologies in order to enhance spatial resolution through post-processing of 2D multi-slice images. The enhancement of spatial resolution in magnetic resonance (MR) images improves clinical procedures such as tissue segmentation, registration and disease diagnosis. Several methods to perform SR-MR images have been proposed. However, they present different drawbacks: sensitivity to noise, high computational cost, and complex optimization algorithms. In this paper, we develop a supervised learning methodology to perform SR-MR images using a patch-based Gaussian process regression (GPR) method. We compare our approach with nearest-neighbor interpolation, B-splines and a SR-GPR scheme based on nearest-neighbors. We test our SR-GPR algorithm in MRI-T1 and MRI-T2 studies, evaluating the performance through error metrics and morphological validation (tissue segmentation). Results obtained with our methodology outperform the other alternatives for all validation protocols.

1 Introduction

Magnetic resonance imaging (MRI) is a medical technique used in radiology to obtain anatomical images of healthy and pathological tissues. MRI scanners use strong magnetic fields and radio-waves to form the images of the body using the signal information from protons in water and lipid. In MRI studies, there are two main types of image acquisitions: MRI-T1 and MRI-T2. T1 imaging is based on the exponential recovery of longitudinal magnetization. T2 imaging is based on exponential loss of signal resulting from purely random spin-spin interactions in the transverse of the XY plane [1]. Both T1 and T2 are spin echo studies, and they are needed to adequately evaluate a tissue in MRI protocols.

© Springer International Publishing Switzerland 2015
G. Bebis et al. (Eds.): ISVC 2015, Part II, LNCS 9475, pp. 692–701, 2015.
DOI: 10.1007/978-3-319-27863-6_65

Currently, clinical protocols allow fast acquisitions of a considerable number of slices in different planes. However, spatial resolution is not high enough in many cases for clinical diagnosis [2]. Although 3D Fourier acquisition is commonly the suitable procedure for MRI, where a 3D high-resolution (HR) is required, this option is not available in practice for all desired image contrast mechanisms [3]. For this reason, the scientific community have been developing super-resolution (SR) methodologies to enhance spatial resolution through post-processing of 2D multi-slice images. SR approaches use MRI studies with low-resolution (LR) images for enhancing edge definition in each slice [4], and for improving clinical procedures such as tissue segmentation, registration, target detection, tracking and disease diagnosis.

The MRI resolution problem has been widely studied. There are many algorithms and methods that deal with the issue of LR images due to the available hardware and acquisition methods [5]. The earliest attempts to up-sample resolution in medical images were presented in [6–8], and a method based on inter-slice reconstruction using SR was proposed in [9]. But their performance is reduced in presence of high levels of noise. Also, approaches based on Wiener filter regularization [10], and non-local means filtering [11] have been developed to perform SR images, obtaining acceptable results. Current methods for SR images are based in patch learning and regularization. In [12,13], the authors introduced a dynamic patch scheme applied to MR image reconstruction. The problem with this method resides in the dependence with the number of nearest-neighbors of the patch. In addition, they assume that input LR-MR images are free of noise, despite the intrinsic noise produced by scanners. In [14,15], sparse methodologies with overcomplete dictionaries were introduced, obtaining high quality MR images. However, there are some drawbacks in [14,15] due to the complex optimization algorithms. Finally, an automated method based on random forest algorithm was presented in [16], showing that supervised learning can offer convenient results in SR medical images.

On the other hand, Gaussian processes (GPs) have become the method of choice for non-linear regression. A GP is a collection of random variables, any finite number of which have a joint Gaussian distribution [17]. GPs are practical and flexible models, that offer advantages with respect to the interpretation of model predictions, and provide well-founded framework for learning and model selection [17,18]. Research communities in statistical image processing and supervised learning have showed that the 2D SR image problem can be solved by a Gaussian process regression (GPR) method, where the prediction of the target pixel depends on a set of training samples [19]. In this paper, we introduce a supervised learning methodology based on GPs for super-resolution in MR images. We develop a patch-based GPR algorithm complemented with a filtering stage to enhance edges. We compare our approach with nearest-neighbor interpolation [12], B-splines [20], and a GPR based on nearest-neighbor [19]. We test the SR methods in MRI-T1 and MRI-T2 studies, evaluating the performance employing error metrics and morphological validation (tissue segmentation). Results obtained with our SR-GPR approach outperform to the other alternatives in all validation protocols, concluding that resolution enhancement of MR images with post-processing is a feasible methodology to avoid acquisition time in MRI scanners.

This paper is organized as follows. The materials and methods are described in Sect. 2. This section includes a definition of SR-GPR, and a description of the dataset and the procedure used in this paper. In Sect. 3, we present the outcomes obtained for each SR method and we establish the discussion about the advantages and disadvantages of our proposed methodology. Finally, conclusions are given in Sect. 4.

2   Materials and Methods

2.1   Super-Resolution Gaussian Process Regression (SR-GPR)

A Gaussian Process (GP) is a collection of random variables, any finite number of which have a joint Gaussian distribution [17]. A GP is completely defined by its mean function, $m(\mathbf{x})$, and covariance function, $k(\mathbf{x}, \mathbf{x}')$, such that

$$f(\mathbf{x}) \sim \mathcal{N}\left(m(\mathbf{x}), k(\mathbf{x}, \mathbf{x}')\right),$$

or equivalently,

$$f(\mathbf{x}) \sim \mathcal{GP}\left(m(\mathbf{x}), k(\mathbf{x}, \mathbf{x}')\right),$$

where $f(\mathbf{x})$ is the intensity value of pixel $\mathbf{x}$. In supervised learning, the radial basis kernel (RBF) is commonly employed as covariance function, and it is given by

$$k(\mathbf{x}, \mathbf{x}') = \sigma^2 \exp\left(-\frac{||\mathbf{x} - \mathbf{x}'||^2}{2\theta^2}\right), \tag{1}$$

where θ and σ^2 are the length-scale and the variance hyperparameters, respectively. Be a noisy image with a set of pixels and intensities $\{(\mathbf{x}_i, y_i)\}_{i=1}^n$, where its intensities follow the standard linear regression model $y_i = f_i + \varepsilon$, with $f_i = f(\mathbf{x}_i)$ and Gaussian noise $\varepsilon \sim \mathcal{N}(0, \sigma_n^2)$. The joint distribution of the training intensities, $\mathbf{y}$, and the test intensities, $\mathbf{y}_*$, is given by

$$\begin{bmatrix} \mathbf{y} \\ \mathbf{y}_* \end{bmatrix} \sim \mathcal{N}\left(\begin{bmatrix} \mathbf{0} \\ \mathbf{0}_* \end{bmatrix}, \begin{bmatrix} \mathbf{K}_{\mathbf{y},\mathbf{y}} + \sigma_n^2 \mathbf{I} & \mathbf{K}_{\mathbf{y},\mathbf{y}_*} \\ \mathbf{K}_{\mathbf{y},\mathbf{y}_*}^\top & \mathbf{K}_{\mathbf{y}_*,\mathbf{y}_*} + \sigma_n^2 \mathbf{I}_* \end{bmatrix}\right),$$

where $\mathbf{K}_{\mathbf{y},\mathbf{y}_*}$, $\mathbf{K}_{\mathbf{y},\mathbf{y}}$ and $\mathbf{K}_{\mathbf{y}_*,\mathbf{y}_*}$ denote the matrix covariances between training and test points, training points, and test points, respectively. Here, the conditional distribution for test points is given by [17]

$$\mathbf{y}_* | \mathbf{X}, \mathbf{y}, \mathbf{X}_* \sim \mathcal{N}\left(\bar{\mathbf{y}}_*, \mathrm{cov}(\mathbf{y}_*)\right),$$

with

$$\bar{\mathbf{y}}_* = \mathbf{K}_{\mathbf{y},\mathbf{y}_*}^\top [\mathbf{K}_{\mathbf{y},\mathbf{y}} + \sigma_n^2 \mathbf{I}]^{-1} \mathbf{y},$$
$$\mathrm{cov}(\mathbf{y}_*) = \mathbf{K}_{\mathbf{y}_*,\mathbf{y}_*} + \sigma_n^2 \mathbf{I}_* - \mathbf{K}_{\mathbf{y},\mathbf{y}_*}^\top [\mathbf{K}_{\mathbf{y},\mathbf{y}} + \sigma_n^2 \mathbf{I}]^{-1} \mathbf{K}_{\mathbf{y},\mathbf{y}_*}.$$

In Fig. 1, we show the graphical model for a GPR adapted for single SR images using a 3×3 patch proposed in [19]. Here, the inputs are the intensities of neighbor pixels (predictors), and the output is the intensity of the target pixel we want to predict.

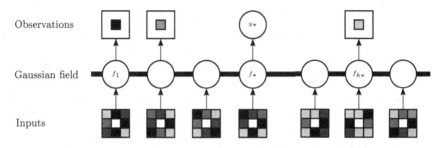

Fig. 1. Graphical model for a GPR applied to a single super-resolution image using a 3×3 patch proposed in [19]. Squares represent observed pixels and circles represent unknown Gaussian fields. The inputs are the intensities of neighbor pixels (predictors), and the output are the intensity of target pixels we want to predict.

2.2 Dataset and Procedure

Dataset: Images of the head were acquired on a General Electric Signa HDxt 3.0T MR scanner using the body coil for excitation, and an 8-channel quadrature brain coil for reception. Subjects were positioned supine with the arms down. The head was positioned in the head-neck coil (MRI). Imaging was performed using an isotropic 3D T1-weighted and 3D T2-weighted Spoiled Gradient Recalled (SPGR) sequence with a repetition time of 10.024 ms, an echo time of 4.56 ms, an inversion time of 600 ms, 1 number of excitations (NEX), an acquisition matrix of 512×512, a resolution of $1 \times 1 \times 1$ mm^3 per voxel, and a flip angle=12. The MRI-T1 and MRI-T2 studies are composed by 160 and 45 slices, respectively. The acquisition was performed using a parallel imaging factor of 2.

Experimental Procedure: First, using the dataset described in Sect. 2.2, we down-sample the T1 and T2 images to obtain 256×256 multi-slice low-resolution (LR) images. Next, we perform the patch-based GPR algorithm to build the 512×512 multi-slice super-resolution (SR) images. We compare against two common methods based on nearest-neighbor (NN) interpolation [12], and B-splines [20] to estimate 512×512 high-resolution (HR) images starting from the

---

Algorithm 1. patch-based SR-GPR approach with filtering.

Require: LR (low-resolution image)
1: GPR-patch training: Maximum likelihood.
2: GPR-patch pixel inference: HR $\leftarrow \mathbf{y}_* | \mathbf{x}, \mathbf{y}, \mathbf{x}_* \sim \mathcal{N}\left(\bar{\mathbf{y}}_*, \text{cov}(\mathbf{y}_*)\right)$
3: 2D filtering stage: SR $\leftarrow$ HR $\otimes H$, with $H = [0 \text{ -1 } 0 \text{ ; -1 5 -1 ; 0 -1 0}]$
4: **return** SR (super-resolution image)

LR studies. Also, we compare with the SR-GPR methodology proposed in [19]. We apply a 2D filtering stage to improve edge enhancement for both methods based on GPR. Next, we compare all the HR and SR images with respect to the ground truth described in Sect. 2.2. As the T1 and T2 studies are 3D images, we evaluate the performance of the methods by computing the average of the mean squared error (MSE) obtained in each slice respect to the ground truth slices. Finally, using the MRI-T1 study, we validate morphologically the SR algorithms through segmentation of different brain tissues such as cerebrospinal fluid (CSF), white matter (WM) and gray matter (GM). We perform the morphological validation using the segmentation module with atlas of the multi-platform software package for visualization and medical image computing (3D-Slicer), available at http://www.slicer.org/ [21]. We average the MSE obtained in each slice per each segmented tissue.

Gaussian Process Regression for Slice-Based MRI Super-Resolution: we develop a patch-based GPR using 12×12 patches with an overlap equal to one sixth. The patch size and the overlap were selected by cross-validation. For the model parameter estimation, we employ the maximum likelihood method, taking a Gaussian function as the likelihood. We train a GP per each patch using a radial basis kernel (RBF) for the covariance function. For obtaining 2D multi-slice SR images, we repeat this procedure for all the slices in the LR-T1 and LR-T2 studies. Finally, we apply a 2D filtering stage to improve edge enhancement. We test several types of filters, and we obtain better results using the Laplacian suppressor filter. The filter stage is not adequate for B-splines and nearest neighbor because the performance worsens. The Algorithm 1 describes the method proposed. In order to preserve the experimental conditions between both SR-GPR methodologies, we set the same patch size and overlap for the GPR approach proposed in [19].

3 Experimental Results and Discussion

3.1 Super-Resolution MR Images Validation

We make a direct comparison between the SR-MR images obtained for each method and the ground truth studies. As we pointed out in Sect. 2.2, we compute the average MSE for the whole up-sampled MRI studies (T1 and T2). Table 1 shows the average MSE results for B-splines, nearest-neighbor (NN),

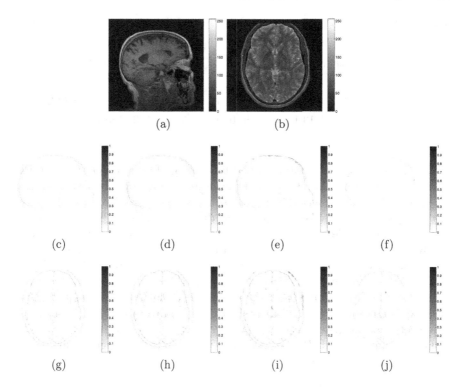

Fig. 2. Ground truth and graphic errors for super-resolution (SR) MRI images validation. The ground truth for T1 slice-100 and T2 slice-20 are showed in (a) and (b), respectively. (c), (d), (e) and (f) correspond to absolute error images of SR-T1 studies for B-spline, nearest-neighbor, SR-GPR1 and SR-GPR2, respectively. Finally, in (g), (h), (i) and (j), the absolute error images for SR-T2 study are showed in the same order than SR-T1 results. The color-bars show the magnitude of the absolute error produced.

the GPR proposed in [19] (SR-GPR1), and our patch-based GPR methodology (SR-GPR2). In Fig. 2, we show graphic errors obtained in SR-MR images validation (T1 and T2). The ground truth for T1 slice-100 and T2 slice-20 are showed in (a) and (b), respectively. Sub-figures (c), (d), (e) and (f) correspond to absolute error images of the SR-T1 image for B-spline, nearest-neighbor (NN), SR-GPR1, and SR-GPR2, respectively. Finally, in sub-figures (g), (h), (i) and (j), the absolute error images for SR-T2 image are showed in the same order used for SR-T1 results.

From Table 1, we can observe that under the same experimental conditions, our methodology for super-resolution (SR-GPR2) outperforms both MRI-T1 and MRI-T2 validation studies. Although the T2 study has a lower signal noise ratio (SNR) than the T1 study, SR-GPR2 works better than the other methods. Regardless of the type of MRI study, our proposed method based on GPR achieves a higher accuracy in super-resolution. The low performance obtained

Table 1. Average MSE results obtained for T1 and T2 respect to the MRI ground truth

MR study	B-spline	NN	SR-GPR1	SR-GPR2
	$\mu \pm \sigma$	$\mu \pm \sigma$	$\mu \pm \sigma$	$\mu \pm \sigma$
T1	27.95 ± 7.282	38.12 ± 10.45	55.89 ± 20.47	$\mathbf{7.806 \pm 1.552}$
T2	48.83 ± 13.14	60.44 ± 16.05	72.99 ± 22.12	$\mathbf{25.65 \pm 7.659}$

by SR-GPR1 is due to the natural smooth behavior of GPs. When the SR-GPR1 estimates new pixels, the GP tends to generate a blurring effect over edges in the up-sampled MR images. For this reason, the low capability to capture high contrast transitions may be a considerable drawback in SR-GPR approaches. In SR-GPR2, this problem is partially solved with the 2D filtering stage. However, for SR-GPR1 this issue remains. From Fig. 2, we observe that high error values are produced by strong changes in edge regions (e.g. skull, lobes and brain structures), concluding that B-splines, NN and SR-GPR1 fail for performing super-resolution in high contrast areas. It is clear that error images for T1 and T2 show a higher accuracy for our proposed method (SR-GPR2). But, there are some prediction mistakes in several pixels due to changes of high contrast that SR-GPR2 cannot avoid. Perhaps, a more robust filtering stage could improve the pixels estimation at edges.

3.2 Morphological Validation

Morphological validation is a key procedure to establish if a SR approach is appropriate or not. Brain tissue segmentation is one of the main applications of MRI-T1 studies. In surgical planning of neuro-degenerative diseases, anatomical segmentation is necessary in order to identify types of tissue in MRI studies. However, LR images present problems such as partial volume effect [14]. This leads to erroneous segmentations because there is not a significant difference among the brain tissue types. Therefore, SR becomes a necessary task for achieving a refined and accurate segmentation. Understanding the importance of tissue segmentation for clinical procedures and diagnosis, we validate morphologically the SR methods over the MRI-T1 study described in Sect. 2.2. For this test, the gold standards are the probabilistic maps of segmentations achieved with the ground truth data using 3D slicer [21]. We segment three types of tissues such as cerebrospinal fluid (CSF), white matter (WM), and gray matter (GM). Similarly, we compute the average MSE in all the tissue segmentations obtained from the SR-T1 results. Table 2 shows the average MSE results for B-splines, nearest-neighbor (NN), SR-GPR1 and SR-GPR2. Also, in Fig. 3, we show some graphic errors obtained for morphological validation. Using the T1 slice-100 from Fig. 2, the gold standards of CSF, WM and GM segmentation are showed in (a), (b) and (c), respectively. Sub-figures (d), (e), (f) and (g) correspond to absolute error images of CSF segmentation for B-splines, nearest-neighbor, SR-GPR1 and SR-GPR2, respectively. Following the same order, from sub-figure (h) to (k) and

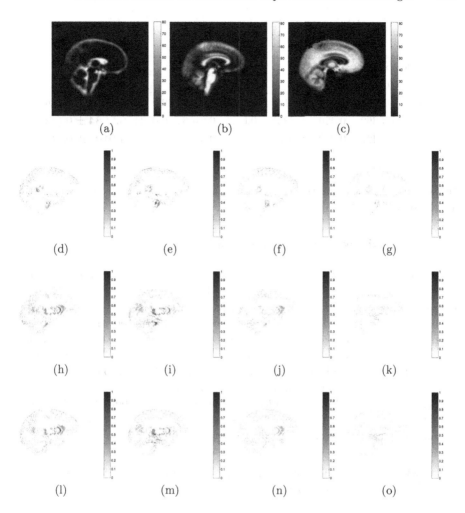

Fig. 3. Ground truth and graphic errors for morphological validation. Using the T1 slice-100 from Fig. 2, the gold standard of CSF, WM and GM segmentation are showed in (a), (b) and (c), respectively. (d), (e), (f) and (g) correspond to absolute error images of CSF segmentation for B-splines, nearest-neighbor, SR-GPR1 and SR-GPR2, respectively. Following the same order, from sub-figure (h) to (k) and from sub-figure (l) to (o), the absolute error images for WM and GM segmentation are showed, respectively. The color-bars show the magnitude of the absolute error produced.

from sub-figure (l) to (o), the absolute error images for WM and GM segmentation are showed, respectively.

From Table 2 and Fig. 2, we can observe that our methodology also outperforms the results in morphological validation. Although SR-GPR1 already outperforms the segmentation provided by the standard techniques, for all the different tissues, SR-GPR2 works better due its ability to represent high contrast

Table 2. Average MSE results obtained respect to the gold standard tissue segmentations.

Tissue	B-splines	NN	SR-GPR1	SR-GPR2
	$\mu \pm \sigma$	$\mu \pm \sigma$	$\mu \pm \sigma$	$\mu \pm \sigma$
CSF	0.0154 ± 0.0142	0.0148 ± 0.0140	0.0079 ± 0.0064	$\mathbf{0.0044 \pm 0.0040}$
WM	0.0362 ± 0.0297	0.0357 ± 0.0322	0.0166 ± 0.0124	$\mathbf{0.0084 \pm 0.0065}$
GM	0.0422 ± 0.0300	0.0406 ± 0.0312	0.0203 ± 0.0132	$\mathbf{0.0105 \pm 0.0071}$

transitions. We conclude then that the patch-based GPR with 2D post-filtering stage provides promising result in SR-MR images.

4   Conclusions and Future Work

In this paper we presented a methodology for super-resolution magnetic resonance (SR-MR) images based on supervised learning. We modeled a Gaussian process regression for 2D multi-slice SR images (SR-GPR2). In order to enhance edge regions, we applied a 2D filtering stage. Results achieved with our proposal outperform to B-spline interpolation, nearest-neighbor, and the GPR method developed in [19]. Our methodology performs better in both SR reconstruction images and morphological validation, making SR-GPR2 a promising approach to enhance spatial resolution in brain MR images.

As future work, the methodology can be extended to perform 3D SR-MRI studies in order to estimate voxels corresponding to new slices. Also, we pretend to improve edge enhancement by using adaptive filters.

Acknowledgments. H.D. Vargas Cardona is funded by Colciencias under the program: *formación de alto nivel para la ciencia, la tecnología y la innovación - Convocatoria 617 de 2013*. This research has been developed under the project: *Estimación de los parámetros de neuro modulación con terapia de estimulación cerebral profunda en pacientes con enfermedad de Parkinson a partir del volumen de tejido activo planeado*, financed by Colciencias with code 1110-657-40687.

References

1. Hollingworth, W., Todd, C., Bell, M., Arafat, Q., Girling, S., Karia, K., Dixon, A.K.: The diagnostic and therapeutic impact of MRI: an observational multi-centre study. Clin. Radiol. **55**, 825–831 (2000)
2. Isaac, J., Kulkarni, J.: Super resolution techniques for medical image processing. In: International Conference on Technologies for Sustainable Development (ICTSD), pp. 1–6 (2015)
3. Hefnawy, A.: An efficient super-resolution approach for obtaining isotropic 3-D imaging using 2-D multi-slice MRI. Egypt. Inf. J. **14**, 117–123 (2013)
4. Malczewski, K.: Inter-K-space motion based strategy for super-resolution in MRI. In: 17th European Signal Processing Conference, pp. 30–34 (2009)

5. Carmi, E., Liu, S., Alon, N., Fiat, A., Fiat, D.: Resolution enhancement in MRI. Magn. Reson. Imaging **24**, 133–154 (2006)
6. Huang, T., Tsai, R.: Multi-frame image restoration and registration. In: Advances in Computer Vision and Image Processing, vol. 1, pp. 317–339 (1984)
7. Kim, S., Bose, N., Valenzuela, H.: Recursive reconstruction of high resolution image from noisy undersampled frames. IEEE Trans. Acoust. Speech Sig. Process. **38**, 1013–1027 (1990)
8. Patti, A., Sezan, M., Teklap, A.: High-resolution image reconstruction from a low-resolution image sequence in the presence of time-varying motion blur. In: Proceedings of ICIP, pp. 343–337 (1994)
9. Greenspan, H., Oz, G., Kiryati, N., Peled, S.: MRI inter-slice reconstruction using super resolution. Magn. Reson. Imaging **20**, 437–446 (2002)
10. Aguena, M., Mascarenhas, N., Anacleto, J., Fels, S.: MRI iterative super resolution with Wiener filter regularization. In: XXVI Conference on Graphics, Patterns and Images, pp. 155–162 (2013)
11. Jafari-Khouzani, K.: MRI upsampling using feature-based nonlocal means approach. IEEE Trans. Med. Imaging **30**, 1969–1985 (2014)
12. Lu, Y., Yang, R.: Super-resolution reconstruction of dynamic MRI by patch learning. In: 12th International Conference on Control, Automation, Robotics and Vision, pp. 1443–1448 (2012)
13. Rousseau, F., Studholme, C.: A supervised patch-based image reconstruction technique: application to brain MRI super-resolution. In: 10th International Symposium on Biomedical Imaging, pp. 346–349 (2013)
14. Rueda, A., Malpica, N., Romero, E.: Single-image super-resolution of brain MR images using overcomplete dictionaries. Med. Image Anal. **17**, 113–132 (2012)
15. Wang, Y., Qiao, J., Jun-bao, L., Ping-Fu, A., Shu-Chuan, C., Rodd, J.: Sparse representation-based MRI super-resolution reconstruction. Meas. **47**, 946–953 (2012)
16. Jog, A., Carass, A., Prince, J.: Improving magnetic resonance resolution with supervised learning. In: IEEE 11th International Symposium on Biomedical Imaging (ISBI), pp. 987–990 (2014)
17. Rasmussen, C.E., Williams, C.K.I.: Gaussian Processes for Machine Learning (Adaptive Computation and Machine Learning). The MIT Press, Cambridge (2005)
18. Murphy, K.P.: Machine Learning: A Probabilistic Perspective (Adaptive Computation And Machine Learning Series). The MIT Press, Cambridge (2012)
19. He, H., Siu, W.C.: Single image super-resolution using Gaussian process regression. In: IEEE Conference on Computer Vision and Pattern Recognition (CVPR), pp. 449–456 (2011)
20. Barmpoutis, A., Vemuri, B., Shepherd, T., Forder, J.: Tensor splines for interpolation and approximation of DT-MRI with applications to segmentation of isolated rat hippocampi. IEEE Trans. Med. Imaging **26**, 1537–1546 (2007)
21. Kikinis, R., Pieper, S.D., Vosburgh, K.G.: 3D slicer: a platform for subject-specific image analysis, visualization, and clinical support. In: Jolesz, F.A. (ed.) Intraoperative Imaging and Image-Guided Therapy, pp. 277–289. Springer, New York (2014)

Congestion-Aware Warehouse Flow Analysis and Optimization

Sawsan AlHalawani[1]([⊠]) and Niloy J. Mitra[1,2]

[1] King Abdullah University of Science and Technology - KAUST,
Thuwal, Saudi Arabia
sawsan.halawani@gmail.com
[2] University College London - UCL, London, UK

Abstract. Generating realistic configurations of urban models is a vital part of the modeling process, especially if these models are used for evaluation and analysis. In this work, we address the problem of assigning objects to their storage locations inside a warehouse which has a great impact on the quality of operations within a warehouse. Existing storage policies aim to improve the efficiency by minimizing travel time or by classifying the items based on some features. We go beyond existing methods as we analyze warehouse layout network in an attempt to understand the factors that affect traffic within the warehouse. We use simulated annealing based sampling to assign items to their storage locations while reducing traffic congestion and enhancing the speed of order picking processes. The proposed method enables a range of applications including efficient storage assignment, warehouse reliability evaluation and traffic congestion estimation.

1 Introduction

With the advancement of technologies, virtual environment applications have a growing need which poses many challenges to model and visualize man-made systems with high realism and accurate behaviour. In computer graphics, the aim of procedural modeling is to create high quality virtual models that closely mimic real world environments. In many cases, the layout of objects within a model plays a vital role in the functionality of the environment. In this paper, we study the layout and positioning of objects in a warehouse.

The placement of objects in a warehouse has a critical impact on the customer service levels and logistics costs. Managers usually aim to achieve the optimum layout which reduces material handling costs, minimizes space requirements, and lowers energy bills. In warehouse design, there are many techniques to approach the assignment of stock to storage locations. *Randomized storage* policy randomly assigns each item to an available location with equal probability. *Dedicated storage* is to assign a precise number of slots to each product which ensures easy tracking but results in wasted space with seasonal demand goods as the assigned slots can not be reused. *Class-based (ABC) storage* policy assigns the most frequently requested items closest to the input and output point where the loading and unloading happens.

G. Bebis et al. (Eds.): ISVC 2015, Part II, LNCS 9475, pp. 702–711, 2015.
DOI: 10.1007/978-3-319-27863-6_66

In this research, we analyze objects layout within a warehouse with the aim of increasing its efficiency. There are many competing factors that we explore. Objects should be placed close to the input and output point so that requests can be quickly processed. However, this can lead to congestions as the density of moving people and carriers near the I/O point will be high and they will start blocking each other. Hence, we investigate to find a good balance between the processing speed and resulted congestion. We aim to harvest the advantages of two well known storage assignment policies. We start with a random assignment which prevents the wasted storage drawback of having seasonal items while providing fast processing of requests which is achieved in class-based storage assignment.

Fig. 1. Starting from a random storage assignment, we solve for congestion minimizing storage assignment while ensuring fast processing of orders. Shelves are depicted in random colors based on the assigned item type. Red, green and blue shelves represent the most, medium and lowest demanded items, while yellow represents empty shelves (Color figure online).

Figure 1 gives an overview of our proposed work. First, we analyze traffic flow within a warehouse and then optimize to find the best storage assignment that enhances the traffic flow. We are mainly interested in two questions: (i) what features are suitable to find a storage assignment policy that reduces traffic congestion ; and (ii) what applications are possible using the proposed method.

2   Related Work

With the growing popularity of virtual words, a great amount of work has been conducted demonstrating various methods for modeling real worlds and generating their configurations. Smelik et al. [1] presented a survey for procedural modeling methods that are useful to generate features of virtual worlds. We discuss some of the most relevant works to our research.

Storage Assignment. The goal of storage assignment policies is to determine which product is to be positioned at which location. This was initially addressed by Hausman et al. [2] who proposed a detailed taxonomy of storage location assignment policies such as randomized, dedicated and class-based storage assignment policies.

Similar to arranging items in a warehouse comes arranging furniture in rooms. Merrell et al. [3] proposed to arrange furniture based on interior design guidelines using a hardware-accelerated Monte Carlo sampler for the density function.

Yu et al. [4] presented a system to automatically arrange a variety of furniture objects and generate realistic indoor scenes.

Layout Structure. Among the most standard layouts are the ones with parallel storage aisles. Some of them add cross aisles in between the parallel aisles. Recently, Gue and Meller [5] introduced the Flying-V and Fishbone designs by relaxing the parallel and orthogonal aisles requirement. Figure 2 shows some examples of different layouts.

Other interesting layouts are building interior layouts. Designing layouts was initially treated as packing problem. Galle et al. [6] implemented an exhaustive algorithm to select the rectangular arrangements satisfying constraints among all possible generations. Wong et al. [7] searched feasible solutions by simulated annealing. Moreover, generating good building layouts was addressed by Bao et al. [8] who encode the spaces and transitions of good layouts in a portal graph and allow the user to explore the plausible layouts. Recently, Liu et al. [9] proposed an exploration method for the interior layouts of precast concrete-based buildings. Moreover, it is not only important to generate a model of the world but to control its details. Vanegas et al. [10] provided a mechanism to interactively edit an urban model using inverse modeling to vary and control the parameters during the modeling process. Additionally, good considerations of deformation analysis and detection should be addressed as investigated by AlHalawani et al. [11].

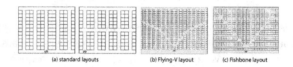

(a) standard layouts   (b) Flying-V layout   (c) Fishbone layout

Fig. 2. Layout (a) shows an example of the standard layouts with parallel aisles [12]. Layouts (b) and (c) show the Flying-V and Fishbone designs [5].

Layout Analysis. Recently, the analysis and optimization of warehouses design have been addressed as in Meller and Gau [13], Meller [14] and Tompkins et al. [15]. Hall [16] analyzed different routing strategies and their impact on order picking efficiency. Similarly, Petersen [17] and [18] studied the impact of different routing policies on the layout by means of simulation.

Analyzing a layout requires a set of features for the evaluation as was addressed by Alhalawani et al. [19] who proposed a set of topological and geometric features to describe the functionality of a street network. Moreover, Gallager [20] proposed a routing algorithm that achieves the total minimum delay in the network. Aslam et al. [21] learn a congestion model based on real data and develop a congestion aware traffic planning system. When shortest paths are congested, passing through them can deteriorate the network efficiency considerably. In Ebrahimi [22], they present an adaptive routing algorithm for on-chip networks that selects a less congested path from a wide range of alternative paths between each pair of source and destination switches. We use a similar concept to evaluate the reliability of traffic within a warehouse by considering possible redundant routes between the I/O point and the pick locations.

3 Warehouse Layout Structure

Typically, order pickers drive through warehouse aisles to retrieve products from their storage locations. Figure 3 (left) shows various aspects of the layout of an order picking area. There are several pick aisles that have racks on both sides in which to store products. Changing from one pick aisle to another is possible through the cross aisles, which are perpendicular to the pick aisles. These cross aisles do not contain pick locations. Adding more cross aisles increases the number of possible routes within a warehouse. The main advantage of having extra cross aisles in a warehouse is the increased number of routing options, which may result in lower travel distances [23].

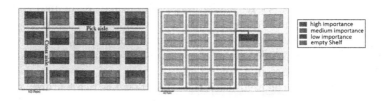

Fig. 3. The left figure gives a top view of a typical order picking area in a warehouse. There are 5 pick aisles and 6 cross aisles. The right figure shows H possible redundant routes between the I/O point and a selected storage shelf s_i (in red). The blue route represents the shortest route (Color figure online).

Each warehouse layout is composed of a set of aisles and a set of storage shelves. To define the aisles, we obtain a set of nodes (i.e., intersections) $\{v_1, v_2, ..., v_N\}$ described by their 2D locations, together with the individual aisles connections $\{e_1, e_2, ..., e_M\}$. Based on the node set and the connectivity of individual aisles, we construct a graph $G = \{V, E\}$ for the entire warehouse layout. The vertex set is defined as $V := \{v_1, v_2, ..., v_N\}$ and the edge set as $E := \{e_{ij}\}$ where $e_{ij} = \overline{v_i v_j}$ denotes an aisle segment.

On each side of a picking aisle, there are storage shelves. We assume to have only one rack of shelves. The set of storage shelves is defined as $S := \{s_1, s_2, ..., s_L\}$ where each node is assigned a type $t \in T$ where T is the set of item types to be stored in the warehouse. Moreover, based on the assigned type, it is associated with a demand $d \in D$ to approximate the importance of the item stored in the location. We assume to have three types; the most, medium and lowest demanded items. In order to access the storage shelves, each node is projected to the nearest aisle from the set E. We also consider a layout that has one input/output point (herein after I/O point) for the loading and unloading of items. We take this point as the bottom left corner in our examples.

4   Methodology

In this section, we describe the analysis of a warehouse in order to find the best items allocation that enhances the traffic flow within a warehouse. Since there are items that are more vital to the business than others, it is common and intuitive to place the most important items closer to the I/O point. However, this will lead to more congestion within the warehouse as more items are concentrated in one region. Therefore, we propose an efficient method that improves warehouse reliability by specifying storage locations and allocating their types.

4.1   Warehouse Flow Analysis

Given a warehouse layout with the empty shelves, we start by randomly assigning item types to some of the storage shelves based on a predefined desired quantity of each type. Then, we enumerate H possible redundant routes from the I/O point to each of the storage locations which are within twice the shortest distance. We also mark the shortest route from the I/O point to the storage location. Therefore, we define the set $R := \{R_1, R_2, ..., R_L\}$ where R_i defines H possible redundant routes for each storage location $\mathbf{s}_i$ such that $R_i := \{\mathbf{r}_1, \mathbf{r}_2, ..., \mathbf{r}_H\}$ where $\mathbf{r}_j$ is a single route composed of a sequence of edges that belong to the aisles edges E. Figure 3 (right) shows a set of possible redundant routes between the I/O point and a selected storage shelf.

Our goal is to place the items such that their flow has two features: (a) items can be accessed and delivered quickly relative to their importance, and to have (b) reliable traffic within the warehouse with the minimum congestion. In order to measure these features, we compute the following terms.

Processing Speed Term. This metric measures how fast it is to reach items with respect to their demand rate. It favors the most demanded items to be closer to the I/O point. We have the storage shelves, their assigned types t and the demand rate for each type d. We compute the processing speed energy term as the sum product of the demand value for type t_j and the average shortest distance to reach all storage shelves of type t_j which is expressed as follows:

$$E_{processing} = \sum_{j;t_j \in T} d_j * \frac{\sum_{i;t_i=t_j} shortest\ distance_i}{n\ elements\ of\ type\ t_j} \tag{1}$$

where d_j is the demand rate for type t_j.

Dispersion Term. In addition to aiming to have the most important items closer to the I/O point, we also want to keep items of the same type close to each other. Dispersion term ensures that the items of the same type are distributed in the same region which is more convenient and has the advantage of faster access for requests with large quantities. In order to achieve this feature, we compute dispersion term based on the distribution of the items of each type as follows:

$$E_{dispersion} = \frac{\sum_{j=1}^{n\ types} \frac{1}{n} \sum_{i=1}^{n\ elements\ of\ type\ t_j} ||z_i - c_j||}{n\ types} \tag{2}$$

where c_j is the centroid of z_i points computed for each type t_j and $||z_i - c_j|| = \sqrt{(x_i - c_1)^2 + (y_i - c_2)^2}$

Congestion Term. We use this term to ensure having the minimum congestion within the warehouse. In order to evaluate the reliability of traffic, we consider the layout graph as a compound system of parallel and serial components as described in [24]. Each individual route along the aisles between the I/O point and the storage shelf s_i is composed of a series of edges while the multiple redundant routes are the parallel components. First, we start by computing edge probability for each edge $e_k \in E$ in the graph which is based on the number of times an edge is to be used by all redundant paths in R denoted as K such that:

$$p\_edge_k = \frac{1}{K} \tag{3}$$

Then, using the series component probability of each aisle being composed of many edges connected in series, we find the probability of each redundant route:

$$p\_route_j = \prod_{k \in r_j} p_k \tag{4}$$

By considering the redundant routes $R_i := \{r_1, r_2, ..., r_H\}$ for each storage shelf s_i as a parallel system, we can find congestion rate as follows:

$$E_{congestion} = 1 - \prod_{r_j \in R_i} (1 - p\_route_j) \tag{5}$$

4.2   Storage Assignment Optimization

We aim to have an improved traffic flow within a warehouse by efficiently allocating the items in their storage places. We use the following energy to evaluate the traffic flow efficiency of a warehouse as follows:

$$E = \lambda(E_{processing} + E_{dispersion}) + (1 - \lambda)E_{congestion} \tag{6}$$

where λ determines the relative contributions among the terms.

In order to find the best items allocation, we minimize the energy given in Eq. 6 using a simulated annealing based sampling (SA) [25]. Initially, we start with random allocation and random type assignment. We also set $E \leftarrow \infty$ and $T = 500$. In each SA step, we randomly select two items and swap their types. We accept the new solution with energy $E(S)$ if $E_{new} \leq E(S)$; else we accept the new solution with probability of $exp(-(E_{new} - E(S))/t)$, where t is the temperature; else we reject the new layout and retain the old one. If we accept a solution, we set $E \leftarrow E_{new}$. In the annealing schedule, we reduce temperature t and continue with the iterations. We stop if either the maximum number of steps $(500 - 1000$ in our tests) has been reached, or when $E < threshold$. An example of the results with the intermediate steps is shown in Fig. 1.

5 Evaluation and Applications

We evaluated our method on different sizes and layouts of warehouses. The results show that our framework is simple and yet effective for the analysis of warehouse layouts. It can be used to efficiently generate storage assignment which maximizes the reliability and productivity of a warehouse.

Processing Speed Analysis. Our processing speed term decreases as the most demanded items are nearer to the I/O point while the least demanded items are further which implies a faster order picking process. Therefore, it is useful to use this term to evaluate the processing speed for a given storage assignment policy. Figure 4 shows different storage assignments used in our evaluation and Table 1 shows the values to compare the efficiency of these assignments.

Table 1. A comparison between the distances to reach different types and the processing speed metric for various storage assignment policies (see Fig. 4 for a visualization of the assignments). The average distance to reach the most demanded items of type $t = 1$ (shown in red in Fig. 1) should be the least distance. Clearly, class-based and our assignment has the least distance to reach items of type $t = 1$. Our proposed assignment produces the lowest score which implies the fastest processing.

Assignment type	Random	Class-based	Opposite class-based	Our assignment
average distance for $t = 1$	56.25	40.9	78.6	40.6
average distance for $t = 2$	74.7	66.4	54.7	54.6
average distance for $t = 3$	71.4	104.6	29.4	86.5
processing speed metric	173.6	150.2	204.3	132.5

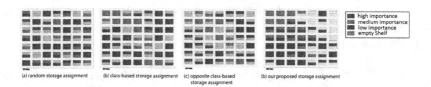

(a) random storage assignment   (b) class-based storage assignment   (c) opposite class-based storage assignment   (b) our proposed storage assignment

high importance
medium importance
low importance
empty Shelf

Fig. 4. Different storage assignment policies used to allocate items in a 8x8 warehouse. Our assignment optimizes the allocation of items to achieve the fastest processing speed. As shown in Table 1, our assignment has the fastest processing speed value.

Storage Assignment Analysis. We compare our storage assignment with some of the well known policies. The results in Fig. 5 show that our assignment reduces congestion level in a warehouse. The random and class-based storage have comparable congestion rates as they both depend on the randomness in their assignment without considering the traffic flow within a warehouse.

Traffic Flow Analysis. Our framework is able to analyze congestion levels based on the value computed using Eq. 3. In our simulator, we render the aisles

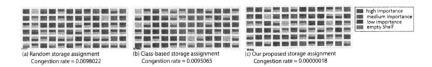

(a) Random storage assignment
Congestion rate = 0.0098022

(b) Class-based storage assignment
Congestion rate = 0.0095065

(c) Our proposed storage assignment
Congestion rate = 0.00000018

Fig. 5. An example showing a comparison between the two standard storage methods and our proposed method. We use 7x13 warehouse structure.

with different colors based on the link probability at each aisle edge. Figure 6 shows that our storage assignment reduces the number of congested aisle edges as layout (c) has the least number of red edges. Moreover, the overall congestion score is lower than the two common storage assignments.

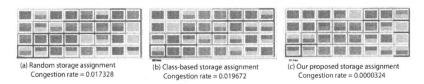

(a) Random storage assignment
Congestion rate = 0.017328

(b) Class-based storage assignment
Congestion rate = 0.019672

(c) Our proposed storage assignment
Congestion rate = 0.0000324

Fig. 6. Visualization of the traffic flow for different storage assignment policies. Red, green and yellow edges denote high, medium and low congestion levels, respectively. Our result (layout c) has the least number of red (most congested) edges (Color figure online).

Cross Aisles Evaluation. One factor affecting the efficiency of processes in a warehouse is the number of cross aisles in its layout structure. Adding more cross aisles increases the number of possible different routes and reduces the probability of congestion in the aisles which can be evaluated using Eq. 5. Figure 7 shows a demonstration of these results.

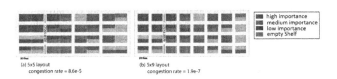

(a) 5x5 layout
congestion rate = 8.6e-5

(b) 5x9 layout
congestion rate = 1.9e-7

Fig. 7. Evaluating congestion score for a warehouse (a) with 5 cross aisles and a warehouse (b) with 9 cross aisles shows the decrease in congestion after adding more aisles.

Congestion-Aware Layout Synthesis. We demonstrated earlier that congestion level decreases as we add more cross aisles. Therefore, we use our proposed congestion score to edit a layout and improve its reliability. We start from a layout that has many picking aisles and only cross aisles at the borders. The items are assigned to their storage positions previously using our method. Then, we randomly add edges which represent parts of cross aisles. We evaluate congestion rate and minimize it to find the best layout which increases the warehouse reliability. Figure 8 shows the result of this synthesis.

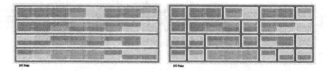

Fig. 8. Starting from the (left) layout with a congestion score equals to 0.7092 and randomly adding cross aisles edges to reduce congestion score to 0.3139 (layout on the right). Original aisles are shown in orange while the added ones are shown in red (Color figure online).

6   Conclusion

We presented an algorithm for warehouse storage assignment policy that aims to assign the items to their storage locations while lowering congestion of moving vehicles within a warehouse. We evaluated our framework on many warehouse layouts to generate a storage assignment that minimizes congestion, while current storage assignment methods do not consider congestion in their policies.

The proposed method degenerates to a class-based storage assignment when we have a full warehouse without any empty storage locations since the reliability score will be the same in every optimization step. In the future, we would like to investigate coupling reliability estimate with the demand value to overcome this issue. Moreover, we do not consider picking multiple items in a single trip which has a great impact on the warehouse performance which we plan to investigate later. The proposed scores can be used to evaluate different warehouse layout structures as well as varying many warehouse parameters such as the number of cross aisles or the placement of the I/O point. This will lead to novel warehouse layout structure synthesis possibilities.

Acknowledgments. We thank the anonymous reviewers for their useful suggestions and Dong-Ming Yan for his valuable assistance in preparing the simulation framework. This work was partly supported by an Anita Borg Google PhD scholarship award.

References

1. Smelik, R.M., Tutenel, T., Bidarra, R., Benes, B.: A survey on procedural modelling for virtual worlds. Comput. Graph. Forum **33**, 31–50 (2014)
2. Hausman, W., Schwarz, L.B., Graves, S.C.: Optimal storage assignment in automatic warehouse systems. Manage. Sci. **22**, 629–638 (1976)
3. Merrell, P., Schkufza, E., Li, Z., Agrawala, M., Koltun, V.: Interactive furniture layout using interior design guidelines. ACM Trans. Graph. **30**, 87:1–87:10 (2011)
4. Yu, L.F., Yeung, S.K., Tang, C.K., Terzopoulos, D., Chan, T.F., Osher, S.J.: Make it home: automatic optimization of furniture arrangement. ACM Trans. Graph. **30**, 86:1–86:12 (2011)
5. Gue, K.R., Meller, R.D.: Aisle configurations for unit-load warehouses. IIE Trans. **41**, 171–182 (2009)

6. Galle, P.: An algorithm for exhaustive generation of building floor plans. Commun. ACM **24**, 813–825 (1981)

7. Wong, D.F., Liu, C.L.: A new algorithm for floorplan design. In: Proceedings of the 23rd ACM/IEEE Design Automation Conference. DAC 1986, Piscataway, NJ, USA, pp. 101–107. IEEE Press (1986)

8. Bao, F., Yan, D.M., Mitra, N.J., Wonka, P.: Generating and exploring good building layouts. ACM Trans. Graph. **32**, 1 (2013)

9. Liu, H., Yang, Y.L., AlHalawani, S., Mitra, N.J.: Constraint-aware interior layout exploration for precast concrete-based buildings. Vis. Comput. **29**, 663–673 (2013). (CGI Special Issue)

10. Vanegas, C.A., Garcia-Dorado, I., Aliaga, D.G., Benes, B., Waddell, P.: Inverse design of urban procedural models. ACM Trans. Graph. **31**, 168:1–168:11 (2012)

11. AlHalawani, S., Yang, Y.L., Liu, H., Mitra, N.J.: Interactive facades: analysis and synthesis of semi-regular facades. Comput. Graph. Forum (Eurographics) **32**, 215–224 (2013)

12. Tsige, M.T.: Improving order-picking efficiency via storage assignment strategies (2013)

13. Meller, R.D., Gau, K.Y.: The facility layout problem: Recent and emerging trends and perspectives. J. Manuf. Syst. **15**, 351–366 (1996)

14. Meller, R.: Optimal order-to-lane assignments in an order accumulation/sortation system. IIE Trans. **29**, 293–301 (1997)

15. Tanchoco, J.M.A., Tompkins, J.A., White, J.A., Bozer, Y.A.: Facilities Planning. Wiley, New York (2002)

16. Hall, R.W.: Distance approximations for routing manual pickers in a warehouse. IIE Trans. **25**, 76–87 (1993)

17. Petersen, C.G.: An evaluation of order picking routeing policies. Int. J. Oper. Prod. Manage. **17**, 1098–1111 (1997)

18. Petersen, C.G.: The impact of routing and storage policies on warehouse efficiency. Int. J. Oper. Prod. Manage. **19**, 1053–1064 (1999)

19. AlHalawani, S., Yang, Y.L., Wonka, P., Mitra, N.J.: What makes london work like London? Comput. Graph. Forum **33**, 157–165 (2014)

20. Gallager, R.: A minimum delay routing algorithm using distributed computation. IEEE Trans. Commun. **25**, 73–85 (1977)

21. Aslam, J., Lim, S., Rus, D.: Congestion-aware traffic routing system using sensor data. 2012 15th International IEEE Conference on Intelligent Transportation Systems, pp. 1006–1013 (2012)

22. Ebrahimi, M., Daneshtalab, M., Farahnakian, F., Plosila, J., Liljeberg, P., Palesi, M., Tenhunen, H.: Haraq: Congestion-aware learning model for highly adaptive routing algorithm in on-chip networks. In: Proceedings of the 2012 6th IEEE/ACM International Symposium on Networks-on-Chip, NoCS 2012, pp. 19–26 (2012)

23. Vaughan, T.S., Petersen, G.C.: The effect of warehouse cross aisles on order picking efficiency. Int. J. Prod. Res. **37**, 881–897 (1999)

24. Myers, A.: Complex System Reliability, 2nd edn. Springer, London (2010)

25. Kirkpatrick, S., Gelatt, C.D., Vecchi, M.P.: Optimization by simulated annealing. Sci. **220**, 617–680 (1983)

Building of Readable Decision Trees
for Automated Melanoma Discrimination

Keiichi Ohki[1], M. Emre Celebi[2], Gerald Schaefer[3], and Hitoshi Iyatomi[4(✉)]

[1] Graduate School of Science and Engineering, Hosei University, Saitama, Japan
[2] Department of Computer Science,
Louisiana State University in Shreveport, Shreveport, USA
[3] Department of Computer Science, Loughborough University, Loughborough, UK
[4] Graduate School of Science and Engineering, Hosei University, Tokyo, Japan
iyatomi@hosei.ac.jp

Abstract. Even expert dermatologists cannot easily diagnose a melanoma, because its appearance is often similar to that of a nevus, in particular in its early stage. For this reason, studies of automated melanoma discrimination using image analysis have been conducted. However, no systematic studies exist that offer grounds for the discrimination result in a readable form. In this paper, we propose an automated melanoma discrimination system that it is capable of providing not only the discrimination results but also their grounds by means of utilizing a Random Forest (RF) technique. Our system was constructed based on a total of 1,148 dermoscopy images (168 melanomas and 980 nevi) and uses only their color features in order to ensure the readability of the grounds for the discrimination results. By virtue of our efficient feature selection procedure, our system provides accurate discrimination results (a sensitivity of 79.8 % and a specificity of 80.7 % with 10-fold cross-validation) under human-oriented limitations and presents the grounds for the results in an intelligible format.

1 Introduction

Melanoma is a fatal skin cancer and among the cancer types that very frequently become metastatic. The five-year survival rate of patients diagnosed with advanced melanoma is reported to be about 7 % [1]. Therefore, early detection of a melanoma is crucial; however, even experts cannot easily diagnose it, in particular in its early stage. Dermoscopy [2], a non-invasive skin imaging technique, and associated diagnosis criteria, such as the ABCD rules [3] and 7-point check list [4], were developed to improve diagnostic accuracy. However, dermoscopic diagnosis is often subjective and is therefore associated with poor reproducibility and low accuracy, in particular when performed by inexperienced dermatologists. Argentiano et al. reported that the accuracy of melanoma diagnosis performed by experts using the criteria mentioned above was about 75–84 % and both the inter-observer and intra-observer diagnosis agreement were large [5]. If an accurate, easy, and objective diagnosis method could be established,

© Springer International Publishing Switzerland 2015
G. Bebis et al. (Eds.): ISVC 2015, Part II, LNCS 9475, pp. 712–721, 2015.
DOI: 10.1007/978-3-319-27863-6_67

it would facilitate early detection of melanomas and reduce melanoma-related deaths. To address this issue, studies on automated melanoma diagnosis have been conducted [6,10]. Rubegni et al. [6] trained an artificial neural network (ANN) using 200 melanomas and 350 nevi and 48 extracted parameters. Their ANN achieved a sensitivity (SE: melanoma detection accuracy) of 94.3 % and a specificity (SP: nevus detection accuracy) of 93.8 % on a test dataset consisting of 17 melanomas and 21 nevi. Celebi et al. [7] developed a support vector machine classification model using 88 melanomas and 479 nevi. Their system achieved an SE of 93.3 % and SP of 92.3 % under Monte Carlo cross-validation. In 2004, we opened our Internet-based melanoma screening system for dermatologists [8,9]. Our current system (http://dermoscopy.k.hosei.ac.jp) comprises approximately 1,500 dermoscopy images and achieved both an SE and SP of 86 % for tumors in normal body parts [9], and an SE of 91 % and SP 93 % for tumors in acral volar regions [10] under leave-one-out cross-validation. Our system provides discrimination results as well as a malignancy score. Although several conventional systems attained a discrimination performance equivalent or superior to that of expert dermatologists in numeric terms, most of these automated systems used a limited type of tumor image, the condition of which was good, i.e., the brightness and color were appropriate and many artifacts, such as hairs, bubbles, and the border of the scope, were not included. Accordingly, taking the practical use of these automated systems into consideration, it is considered desirable that the system can offer the grounds for the discrimination results, i.e., the reason(s) why the system yields the results, in order to ensure reliability. From this viewpoint, previously we proposed a quantification method for each clinical finding defined in the ABCD rule [3] and 7-point checklist [4], which attained a good estimation performance [11]. However, this research was conducted based on only 105 dermoscopy images because of the difficulty in obtaining a reliable gold standard (15 items for each case; diagnosis by multiple experts was required), and therefore, further investigations were required in order to attain robustness. In this paper, we propose an automated melanoma discrimination system utilizing a Random Forest (RF) technique that is capable of providing discrimination and its associated grounds [12]. RF is a machine-learning algorithm that aggregates several decision "trees" as weak learners to form a highly accurate classifier "forest." The discrimination result of each decision tree is obtained by following a path from the root node to the leaf node and the majority vote of the nodes determines the final decision of the RF. In this study, the grounds for the discrimination results is expressed in the form of the built tree, which has important image features. So that the grounds for the discrimination results are readable, the image features used in the RF classifier should be intelligible and the number of tree layers limited. To achieve this, we have to seek important and readable image features for classifying melanomas that attain a certain classification accuracy under these limitations. In this paper, we also propose an effective feature selection method for constructing a sophisticated RF classifier.

2   Random Forests

RF is an accurate classification algorithm that aggregates a multitude of decision trees as weak learners. Each tree is built based on a limited number of randomly selected samples and their features, allowing them to overlap. This strategy results in a lower correlation among each tree and accordingly, the formed "forest" is expected to be resistant to over-fitting and is capable of producing an accurate classification.

In each node in a decision tree, the feature used for dividing data was determined by means of its entropy. When we assume that data are divided into N groups at the branch node k, the entropy of the k-th node $H(k, p)$ is expressed as

$$H(k, p) = -\sum_{i=1}^{N} p(C_i|p) \log p(C_i|p), \tag{1}$$

where $p(C_i|p)$ represents the ratio according to which given data are divided into class $C_i (i = 1, 2, ..., N)$ using feature p. Here, an expected entropy reduction at the node k is

$$\eta(k, p) = H(k, p) - \sum_{j=1}^{M(k)} H(k_j, p), \tag{2}$$

where $k_j (j = 1, 2, ..., M(k))$ is the child node of the node k. This amount, $\eta(k, p)$, represents the effect of the data partitioning by feature p at the node k. In the k-th node, the applicable feature $p^*(k)$ is selected from $\forall$ p such that the corresponding $\eta(k, p)$ is a maximum.

$$p^*(k) = \underset{p}{\operatorname{argmax}}\, \eta(k, p). \tag{3}$$

Note that the applicable number of feature candidates for each node, i.e., the parameter pool, is suggested to be the square root of that of the all features in order to ensure independence among the trees [12].

3   Melanoma Discrimination System Capable of Offering the Grounds

Our proposed system is composed of the following five phases.

- Tumor area extraction
- Calculation of the image features
- Selection of the important features with RF
- Training of RF
- Presentation of the classification results and decision tree.

The first to the fourth phases are for constructing the RF classifier and the final phase yields the discrimination results.

In pattern recognition problems, the design and selection of important features for recognition constitute one of the most important processes. RF plants a multitude of highly independent decision trees by sampling the training data and their features randomly. Here, when the training data contain considerable noise and/or the correlations among their features are high, the overall performance of the RF is decreased. Accordingly, we also propose a novel feature selection method using RF, the effectiveness of which we evaluated. The details are described in the following sections.

3.1 Tumor Area Extraction

The color properties in the peripheral area of a tumor are known to be of particular importance for classifying melanomas. Therefore, most studies on automated melanoma classification perform tumor area segmentation as an important preprocessing step. For this reason, many methods for automated tumor segmentation from dermoscopy images were proposed [9,12–15]. In this study, we applied our former method [9], the performance of which was evaluated with several state-of-the-art methods and confirmed [15]. Because of the space limitation, please refer to our original article for details.

3.2 Calculation of the Image Features

We designed and used a total of 120 color image features extracted from tumors in this study based on the consideration of the ABCD rule [3]. We named them "base-parameters." We emphasize here that we did not use image features that we could not easily understand, such as textures, frequency components, or coefficients calculated by known robust features, i.e., SIFT, SURF etc., in order to attach great importance to the readability of the "grounds" of the discrimination results. For calculating the base-parameters, we focused on the tumor body $(T)$and the periphery of the tumor (P), and the difference between the tumor and its surrounding skin $(S\text{-}T)$ and that between the tumor and its periphery $(T\text{-}P)$. In each part of the tumor, we calculated the mean (μ), minimum (min), maximum (max), standard deviation (σ), and skewness (skew) values for each RGB and HSV color channel, i.e., $5 \times 6 = 30$ features in each part. Note that the peripheral part of the tumor was defined as the region inside the tumor's border that has an area equal to 30 % of the tumor area as determined by a recursive dilation process applied to the outer border, working inward from the border of the extracted tumor [9].

3.3 Selection of the Important Features with RF

In this phase, we determined a set of important features for classifying melanomas, called "important-parameters," as a subset of the "base-parameters." In classification or machine learning problems, feature selection is a requisite process for achieving robust and reliable systems. For example, the incremental stepwise method [16] has been widely used because of the theoretical evidence of its effectiveness; however, it does not always lead to an appropriate solution in the case of non-linear systems, in particular those including iterative processes. Accordingly, we propose a new novel feature selection method that uses the mechanism of RF. First, we trained RF with the base-parameters and calculated the effectiveness of the parameter p, $\eta(k, p)$, in each decision tree. Then, we calculated the overall influence of the parameter p by using

$$\alpha(p) = \frac{\sum_k \eta(k, p)}{\sum_k \sum_p \eta(k, p)}. \tag{4}$$

It should be noted here that $\sum_p \alpha(p) = 1$ because of its normalized definition. In this study, we selected the number of decision trees and their depth to be 100 and 10, respectively. We selected features having $\alpha(p)$ equal to or larger than 0.01 as the important-parameters based on our preliminary experiments.

3.4   Training of RF

We trained the RF with the important-parameters. In the k-th node, given data are divided into two groups by means of the selected parameter $p^*(k)$ and associated threshold $\theta^*(k)$. The threshold $\theta^*(k)$ is determined by (1) generating several random values $\theta(k)$ as the candidates of the threshold between the maximum and minimum value in terms of $p^*(k)$ of the incoming data, (2) calculating the amount of entropy reduction for each $\theta(k)$, and (3) selecting $\theta^*(k)$ with the largest entropy reduction. In addition, it is known that if the training dataset has a class bias, the performance of the built RF will be also biased. In this study, we introduced a reverse biased sampling and weighting technique to normalize the class bias in the training dataset [12].

3.5   Exhibition of the Classification Result and Decision Tree

Melanoma classification is performed using the built RF classifier. Note that the final decision was made by majority voting of each decision tree. We selected one decision tree with the highest average of $\alpha(p)$ among all trees yielding the majority result. The system provided the path from the root node to the leaf node, i.e., the final discrimination result, as the grounds for the result.

4   Experiment

In this study, we used a total of 1,148 dermoscopy images having an established diagnosis (168 melanomas and 980 nevi) from Naples, Graz, and Keio Universities. We obtained the important-parameters using the method described above. In order to confirm the effectiveness of our method, in particular its feature selection process, described in Sect. 3.3, we conducted three comparative experiments.

- RF trained with the base-parameters (performance baseline)
- RF trained with parameters selected using the stepwise method
- RF trained with the important-parameters (proposed method).

We used sensitivity and specificity as evaluation criteria and the classification performance of the RF was evaluated using a 10-fold cross-validation strategy.

5 Results

A total of 24 color features were selected as the important-parameters. Their top five features are summarized in Table 1. We determined that the number of decision trees and their depth were 50 and 8, respectively, according to the preliminary experiments. The final classification results for melanomas are summarized in Table 2. The proposed method (Experiment 3) attained the best performance values: SE = 79.8 % and SP = 80.7 %. Figure 1 shows an example of a test dermoscopy image (nevus) and its discrimination path in the built decision tree. Figure 2 shows its partial enlargement. In a rectangular (non-terminal) node in the figure, the upper line represents the number of data incoming to the node and the lower line represents the conditions for data division, i.e., inaction with the image feature $p^*(k)$ and the corresponding threshold $\theta^*(k)$). If the condition is satisfied, the algorithm follows the left link; otherwise, the right. The nodes having rounded corners in the lower part of figure represent the final discrimination result; gray and white indicate malignant (M) and benign (N) cases, respectively. The upper line indicates the number of data incoming to the node and the lower indicates those classified in each category. Note here that this number is compensated by the reciprocal ratio of malignant and benign

Table 1. Selected important-parameters (top 5)

Rank	Feature	Description	$\alpha(p)$ (%)
1	σ_G^P	Standard deviation of green in the periphery	2.3
2	μ_B^P	Average of blue in the periphery	2.2
3	σ_R^T	Standard deviation of red in the tumor	2.0
4	σ_V^{T-P}	Standard deviation of intensity between the tumor and its periphery	1.8
5	σ_G^T	Standard deviation of green in the tumor	1.6

Table 2. Comparison of melanoma discrimination performance

Experiments	Feature selection	No. of features	SE (%)	SP (%)
Experiment 1	no	120	69.9	70.5
Experiment 2	stepwise [16]	14	70.2	68.8
Experiment 3	RF (proposed)	24	79.8	80.7

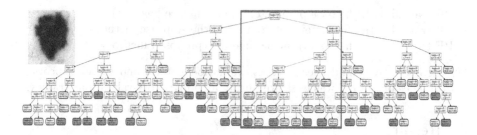

Fig. 1. Discrimination of a nevus with the built decision tree.

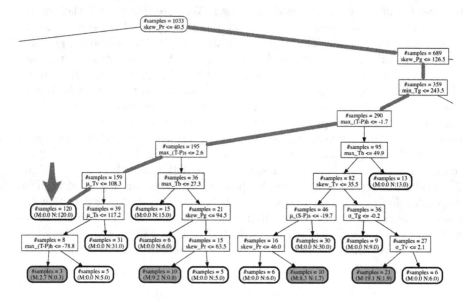

Fig. 2. Enlarged view of Fig. 1.

Table 3. Selected important-parameters (top seven)

Rank	Feature	Description	$\alpha(p)$ (%)
1	σ_V^T	Standard deviation of intensity in the tumor	21.9
2	$\max_B^P$	Maximum of blue in the periphery	11.5
3	skew_V^{T-P}	Difference of skew in intensity between the tumor and its periphery	10.7
4	$\min_G^T$	Minimum of green in the tumor	8.2
5	μ_B^T	Average of blue in the tumor	7.6
6	skew_V^T	Skew of intensity in the tumor	6.7
7	σ_G^T	Difference of standard deviation in green between the tumor and its periphery	4.2

cases, and each tree was trained with 90 % of the training data because of the 10 fold cross-validation strategy. The image features with larger importance $\alpha(p)$ in Experiment 3 are summarized in Table 3.

6   Discussion

We confirmed that the RF classifier with the proposed feature selection method (Experiment 3) attained the best classification performance of all the methods. When the incremental stepwise input selection was applied (Experiment 2), the classification performance was worse than that of Experiment 1. The incremental stepwise method is designed for selecting appropriate feature sets for single linear models with a statistical t-test. We considered this one of the best solutions for a single linear model, but the limited number of features largely reduced the degree of freedom of the parameter pools and accordingly caused the performance degradation of RF. RF requires a certain degree of freedom in a parameter pool; the proposed feature selection method provided this. In Table 3, it can be seen that the most significant feature was the standard deviation of the intensity in the tumor body (σ_V^T). We consider this represents the polychromism of the tumor caused by the variation in the depth of melanin from the epidermis. This is consistent with the diagnostic criteria used in clinical practice. The second and the following features represent the blue, intensity, or green (highly correlated with the intensity) property of the tumor or its periphery. The results are also considered to show that our proposed feature selection method selected image features that are important for classifying melanomas appropriately. In the RF algorithm, test data are inputted to all the built decision trees and in each tree inputted data items remain sorted by the condition described in each node until they reach the terminal. The proposed system shows the process of a typical tree as the grounds for the result. Again, because we attached the highest importance to building easy-to-read "grounds for the result," and to suppressing over-learning, we limited the number of layers of the tree, and the base-parameters comprise only color features. Because of these human-oriented limitations, on the one hand, the numerical classification performance of our system gave a slight advantage to other state-of-the-art methods that focus on only the classification performance, while on the other, its performance is comparable to that of expert dermatologists. However, our current system provides the grounds of the discrimination expressed as a tree with numerical values. We are currently investigating a method for converting the numerical value to a more intelligible linguistic expression. We will report the results of this endeavor and upload this function on our Internet melanoma classification system in the near feature. However, our method successfully attains a performance comparable to that of expert dermatologists.

7   Conclusion

In this paper, we proposed an automated melanoma discrimination system capable of offering both the discrimination results and their grounds by means of utilizing a Random Forest (RF) technique. Owing to the introduction of our feature selection method for RF, our system is capable of providing not only a good classification performance, but also the grounds for the discrimination results in an intelligible format.Our system provides the grounds of discrimination in the tree expressions with numerical values. We are currently investigating to convert this into more intelligible linguistic expressions, and will incorporate this functionality into our Internet melanoma classification system in the near feature.

Acknowledgement. This research was partially supported by the Ministry of Education, Culture, Science and Technology of Japan (Grant in-Aid for Fundamental research program (C), 26461666, 2014-2017).

References

1. National Cancer Institute: SEER Cancer Statistics Review 1975–2012. NCI, Bethesda, MD (2015)
2. Soyer, H.P., Smolle, J., Kerl, H., Stettner, H.: Early diagnosis of malignant melanoma by surface microscopy. The Lancet **2**(8562), 803 (1987)
3. Stolz, W., Riemann, A., Cognetta, A.B., Pillet, L., Abmayr, W., Holzel, D., et al.: ABCD rule of dermoscopy: a new practical method for early recognition of malignant melanoma. Eur. J. Dermatol. **4**(7), 521–527 (1993)
4. Argenziano, G., Fabbrocini, G., Carli, P., Giorgi, V.D., et al.: Epiluminescence microscopy for the diagnosis of doubtful melanocytic skin lesions comparison of the ABCD rule of dermatoscopy and a new 7-point checklist based on pattern analysis. Arch. Dermatol. **134**(12), 1563–1570 (1998)
5. Stolz, W., Falco, O.B., Bliek, P., Kandthaler, M., Burgdorf, W.H.C., Cognetta, A.B.: Color Atlas of Dermatoscopy, 2nd enlarged and completely revised edn. Blackwell Publishing, Berlin (2002)
6. Rubegni, P., Cecenini, G., Burroni, M., Perotti, R., Del'Eva, G., Sbano, P., et al.: Automated diagnosis of pigmented skin lesions. Int. J. Cancer **101**(6), 576–580 (2002)
7. Celebi, M.E., Kingravi, H.A., Uddin, B., Iyatomi, H., et al.: A methodological approach to the classification of dermoscopy images. Comput. Med. Imaging Graph. **31**(6), 362–373 (2007)
8. Oka, H., Hashimoto, M., Iyatomi, H., Argenziano, G., Soyer, H.P., Tanaka, M.: Internet-based program for automatic discrimination of dermoscopic images between melanomas and Clark nevi. British J. Dermatol. **150**(5), 1041 (2004)
9. Iyatomi, H., Oka, H., Saito, M., Miyake, A., Kimoto, M., Yamagami, J., et al.: Quantitative assessment of tumor extraction from dermoscopy images and evaluation of computer-based extraction methods for automatic melanoma diagnostic system. Melanoma Res. **16**(2), 183–190 (2006)
10. Iyatomi, H., Oka, H., Celebi, M.E., Ogawa, K., Argenziano, G., Soyer, H.P., Koga, H., Saida, T., Ohara, K., Tanaka, M.: Computer-based classification of dermoscopy images of melanocytic lesions on acral volar skin. J. Invest. Dermatol. **128**(8), 2049–2054 (2008)
11. Iyatomi, H., Oka, H., Celebi, M.E., Tanaka, M., Ogawa, K.: Parameterization of dermoscopic findings for the Internet-based melanoma diagnostic system. In: Proceedings of the IEEE CIISP 2007, pp. 183–193 (2007)

12. Breiman, L.: Random forests. Mach. Learn. **45**(1), 5–32 (2001)
13. Joel, G., Philippe, S.S., et al.: Validation of segmentation techniques for digital dermoscopy. Skin Res. Technol. **8**(4), 240–249 (2002)
14. Celebi, M.E., Iyatomi, H., Schaefer, G., Stoecker, W.V.: Lesion border detection in dermoscopy images. Comput. Med. Imaging Graph. **33**(2), 148–153 (2009)
15. Norton, K.A., Iyatomi, H., Celebi, M.E., Ishizaki, S., et al.: Three-phase general border detection method for dermoscopy images using non-uniform illumination correction. Skin Res. Technol. **18**(3), 290–300 (2012)
16. Hocking, R.R.: The analysis and selection of variables in linear regression. Biometrics **32**(1), 1–49 (1976)

A Novel Infrastructure for Supporting Display Ecologies

Christian Eichner[✉], Martin Nyolt, and Heidrun Schumann

University of Rostock, Rostock, Germany
{christian.eichner,martin.nyolt,heidrun.schumann}@uni-rostock.de

Abstract. We introduce a novel approach for display ecologies that aims to support users in presentation and discussion scenarios by applying assistance from Smart Meeting Rooms (SMR). We present an infrastructure that allows multiple users to easily integrate their mobile devices into the device ensemble of a SMR and to utilize its large displays to show contents like slides, pictures and other data visualizations. With a tailored editor, multiple users can easily share contents and interactively coordinate the display of information. The content is automatically distributed to the displays of the SMR based on user-defined spatial and temporal links between contents as well as on semantic networks. Further, intention recognition is used to automatically adapt the representation of contents with regard to the current situation. In this way we provide a user-driven smart steering that supports users by automatically reducing their effort to configure and to work with display ecologies.

1 Introduction

The increasing spread of mobile devices such as personal laptops, smart phones or tablets, and the increasing number of large public displays provide new ways to distribute and access visual representations. When multiple such devices are collaboratively combined in order to create, share, and display several views they form a so-called display ecology. Such environments provide the technical basis for distributing different types of contents across multiple displays, allowing users to share their slides, documents and images. In this way, presentation and discussion scenarios are well supported. This can be very helpful in various domains. However, in a recent survey [1] on supporting visual analysis through display ecologies four inherent challenges are described that can be summarized as follows:

- *C1*: The user must combine multiple displays into one holistic space.
- *C2*: The user must transfer different data and views for coordinating information across displays.
- *C3*: The user must link certain views across different displays.
- *C4*: The user must add and remove displays as well as particular views on demand.

© Springer International Publishing Switzerland 2015
G. Bebis et al. (Eds.): ISVC 2015, Part II, LNCS 9475, pp. 722–732, 2015.
DOI: 10.1007/978-3-319-27863-6_68

Our aim is to address these challenges. We want to reduce the burden of the user by introducing a novel infrastructure that enables a **smart steering** of dynamic display ecologies. This means, we want that users only need to specify layout constraints, whereas display combination, information distribution, and dynamic layouts are computed automatically.

The background for our work are presentation and discussion scenarios in a smart meeting room (SMR). Smart meeting rooms are environments in which heterogeneous display devices (projectors, stationary and mobile displays) provide abundant space for representing views. Tracking devices and sensors deliver several types of information such as user positions or view directions. Associated software tools provide customized user assistance by (i) evaluating sensor data to recognize certain situations and reason about the users' intentions, (ii) integrating additional displays and other devices into the existing device ensemble, and (iii) transferring data between multiple devices on demand. However, in order to fully take advantage of these capabilities in terms of using a SMR as a display ecology further investigations are required.

In this paper we focus on supporting presentation and discussion scenarios. Our approach is built upon the infrastructure of a SMR and consists of three components:

– An editor enables a user-driven coordination of information. So-called spatial and temporal links between different contents can be added interactively. In this way spatial and temporal constraints are defined to steer how contents should be arranged on the available displays and how contents should be replaced over time.
– A semantic description includes one or more semantic networks. Each network describes a particular topic. The nodes of such a network represent individual parts of the content. Related nodes are connected via so-called semantic links. Semantic links are either defined by the user or extracted automatically. They will be used to arrange views or to support search for additional contents during a discussion.
– A support component provides methods for recognizing a user's intention, assigning views to different displays and generating layouts. These methodes enable a high degree of customized user assistance.

The three described components enable a smart configuration of display ecologies in a smart environment. In the following we will discuss them in more detail.

The remainder of the paper is organized as follows. Section 2 gives an overview of the background and related work. Section 3 describes our setup and discusses the problems to be solved. The editor component as well as the semantic description are presented in Sect. 4. Since we focus on user assistance two separate sections explain the automatic computations: Sect. 5 describes the distribution and layout of views and Sect. 6 discusses the use of intention recognition to support dynamically changing discussions. We conclude the paper with an overview of the implementation in Sect. 7 and a description of possible applications in Sect. 8.

2   Background and Related Work

A key property of smart environments is the provided assistance [2,3]. Examples are smart homes [4], applications in health-care [5], assistive systems that focus on lecture scenarios [6,7] and in particular smart meeting rooms [8,9], which record meetings for archiving, organizing and automatically summarizing the contents. Typically, audio-visual recordings are applied, although sometimes further context information such as activities or attention are tracked as well. However, the available assistance in SMRs has not yet been leveraged to foster display ecologies for visualization, presentation, and discussion.

To incorporate personal contents from mobile devices in SMRs, systems were developed to easily share windows or certain views from mobile personal devices and to display these views on the screens of the environment [10–12]. In such systems the large displays of the environment serve as a kind of shared extended desktop of the personal devices. Views can be exchanged using a local network or the Internet [12]. In addition, local inputs might be transmitted to the displays to enable multiple users to operate a single application [13,14]. These approaches can greatly support collaborative work and allow users to jointly create new contents. However, these systems still rely on much manual user interaction to define where contents are to be presented in the environment (which display) and on the displays (which position). Window managers and automatically generated layouts can be applied to reduce user interaction [15–17]. These approaches address collaborative data exploration and visual analysis by supporting the user in determining where manually selected contents should be displayed. However, they lack assistance for quickly accessing contents during a meeting.

Chung et al. provide a comprehensive overview on different approaches for display ecologies [1]. They describe challenges, and design considerations, and summarize the fundamental approaches based on the tasks at hand. This overview makes clear that only a few approaches focus on supporting users by automatic configuration. Software that is designed for advanced, collaborative work in display ecologies lack the capabilities to manage dynamic device ensembles and to incorporate support from assistance systems and situation recognition.

We conclude that current display ecologies do not utilize the advanced capabilities of SMRs. On the other hand SMRs mainly support dynamic hardware configurations and interactions, rather than addressing a sophisticated presentation of information. We aim at bridging this gap by (1) providing a way to easily integrate users' mobile devices and contents into an existing display ecology, (2) providing constraints on how to use them, and (3) applying automatic assistance to display these contents with respect to the current situation.

3   Setup

Our Smart Meeting Room (SMR) is a smart environment with many different devices and a complex infrastructure. Up to seven projection screens can be used in parallel. A video matrix crossbar connects up to nine user VGA inputs and seven VGA signals from a central server to the projectors. In practice it has been shown that it can be still a problem to effortlessly connect user devices to projectors. Most common problems are the mix in standards and available connectors and adapters (VGA, HDMI, DVI, DP) and technical incompatibilities. Therefore, three miniature PCs are directly connected to the seven projectors. They act as an interface between personal devices and projectors. In this way, views from different personal devices can be shown on the displays of the SMR immediately and simultaneously. The personal devices are connected to the infrastructure via Ethernet and Wi-Fi. The sensors and actuators of our SMR provide, for example, information about user position and enable control of the screens and projectors.

The infrastructure of the SMR is controlled by a middle-ware [18] that implements several paradigms for coupling the components and providing control of the environment. Direct message exchange, event dispatching and publish/subscribe provide flexible communication between devices, services and applications. Applications are deployed from a web server using Java Web Start technology. This way, users can just enter the room, open their web browser to connect with the infrastructure, and start their applications to generate views to be shown in the SMR. Figure 1 gives an overview of the setting we are using. Figure 2 shows a photograph of the SMR.

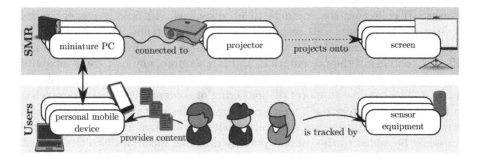

Fig. 1. Overview of the display ecology with the devices of our SmartLab and the personal devices of the users.

In summary, our SMR realizes the combination of displays and personal devices into one holistic space (*C1*), the update of the device ensemble after adding or removing displays or personal devices (*C4*), and the transfer of information (*C2*). So, three of four challenges summarized in Sect. 1 are already addressed. However, the semantic coordination and linking of views of multiple users on multiple displays is still an unsolved problem. The users need a way to steer which content should be displayed and how it should be laid out to generate meaningful arrangements. An appropriate interface is needed that supports the organization of contents by defining different types of constraints. We will discuss this issue in the following section.

Fig. 2. Photograph of the smart environment in a meeting situation.

4   Interactive Coordination of Information

To collaboratively utilize a display ecology users should be enabled to steer *which* content parts should be displayed and *how* these content parts will be arranged on the displays. To address this issue, we have developed an editor that supports 3 tasks: (i) collecting the contents to be displayed, (ii) linking content parts to steer the layout and (iii) searching for related contents. The editor can be executed on any personal device. In this way, users get individual access to the different contents of the presentation, and furthermore, any user can contribute to the presentation with own contents. The user interface of the editor is shown in Fig. 3. The top row supports the gathering of contents. It shows an overview on the so-called content pool, which contains all files that can be published. All provided files of all users are symbolized with small thumbnails. New contents can easily be added to the content pool by dragging local files into the content pool. These new contents are then automatically transmitted to other devices of the SMR via the middle-ware. The content is automatically grouped with regard to the contributing users. Tabs are used to group contents by user.

The bottom of the editor UI gives an overview of the presentation graph. It supports the linking of individual contents to steer the display of the presentation. For this purpose, contents are interactively selected from the content pool and dragged into the presentation graph. Relevance values can be assigned to emphasize information of particular interest, which will then be prioritized by assigning more display space to them. To steer the final layout, the individuals contents need to be set in relation. A user can easily define spatial and temporal links by interactively drawing a line between thumbnails. A spatial link specifies a spatial constraint: Connected contents are to be displayed in spatial proximity. A temporal link represents a temporal constraint: Such contents are linked in the chronology of the presentation and should therefore be displayed at roughly the same position. In other words, successive content is shown in a stable location.

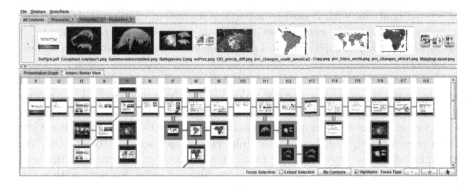

Fig. 3. The editor UI provides content sharing via the content pool (top) and allows modification of the presentation graph (bottom).

The editor interface supports the generation of presentations. However, during a discussion, often further information that belongs to the topic could be required. To support the searching for related contents, a semantic description is provided. The semantic description contains one or more semantic networks each belonging to a particular topic.

Fig. 4. The semantic network allows quick access to related content. Left: An overview allows to explore the structure of the semantic networks. An integrated lens provides a detailed view of a content. Right: An attribute based layout allows to search for particular contents and to explore relations with respect to specific attributes.

A semantic network describes the additional connections between slides, pictures and other contents. Contents are connected, if they are about the same subject, if one content complements another content, or if there exist other relationships, e.g. based on the document structure. The semantic description contains every content that was shared by any user, regardless whether it is currently displayed or not. A content may be used more than once in the course of a presentation, but the semantic description contains it only once. The semantic description can be used in two ways: (i) Semantic links can be used to search for related contents with regard to a particular content of interest. (ii) Semantic networks as a whole can be used to get an overview of all related contents regarding a particular topic. Figure 4 shows the visual representation of a semantic network.

The semantic networks can be created or provided in several ways. First, existing semantic networks can be uploaded by the users. Second, users can interactively define semantic links between content parts either by drawing lines in the visual representation of the semantic network or by using the editor view of the presentation. Third, semantic links can be extracted automatically. We add links based on keywords, the table of contents, the structure of the text (e.g., in sections and subsections), the embedding of other contents within the document hierarchy (e.g., pictures in a PDF), hyperlinks, references to the cited literature and so forth.

In summary, semantic networks serve as a basis to link views automatically across several displays. Spatial and temporal links, on the other hand, steer the configuration of layouts. This will be described in the following section.

5   Automatic Content Distribution and Layout Generation

Our aim is to reduce the burden of the user. To this end, we provide an automatic computation of the display. This includes two tasks. First the individual contents need to be distributed, i.e. they need to be assigned to the displays. Second, all contents that will be shown on the same display need to be arranged. The automatic configuration is based on user-driven constraints, i.e. the defined temporal and spatial links as well as the assigned relevance values. Based on our previous approaches [19,20] the configuration is determined by maximizing a quality function Q. It is defined as the weighted sum of the spatial proximity quality Q_s, the temporal quality Q_t and the display quality Q_d.

$$Max : Q = \alpha \cdot Q_s + \beta \cdot Q_t + \gamma \cdot Q_d$$

Here, the spatial proximity quality Q_s is high if positions of views that are connected by a spatial link are close to each other. This ensures that the calculated layout places these contents next to each other. The temporal quality Q_t rates how much the content distribution and layouts change when new contents are added or removed. If contents are moving too much on the displays, users could lose the focus on specific content of interest. Therefore, the temporal quality is high if temporally linked contents roughly stay at the same displays and positions of a layout. Finally the display quality Q_d quantifies the visibility of each view. A good layout should place contents on displays, where (most) users are able to see them well. To accomplish this, we consider the display size, the position of the displays in the room, the position and viewing direction of the audience and the size of a view in the generated layouts.

By adjusting the weights α, β, and γ, users can steer the influence of spatial, temporal and display quality on the layout generation. Thus they can decide whether the automatic layout calculation should produce layouts that tend to place related contents side by side, layouts that change as little as possible when views are added or removed, or layouts that ensure a good visibility of views. Adaptations of the weights or of the underlying constraints are automatically recognized and result in an immediate update of the content distribution. Layout calculations and updates are realized at interactive rates ($\sim$30 Hz). In this way users get fast feedback to their adaptations.

6   Intention Recognition and Discussion Support

In dynamic discussions, it is not uncommon that users have to fequently change visible content compositions by adding or removing contents ($C4$) and adapt the layouts on the displays for certain tasks ($C3$: link and compare visualizations). For this they have to interact with various devices and operate the editor, which allows for a flexible configuration. We aim to reduce the manual effort required to do these adaptations. We support various tasks with customized user assistance based on the capability of our SMR. For example we support the following typical tasks:

- *Navigation to a particular content* is supported by automatically:
  - Showing a temporal outline that can be used to bring up previously shown content compositions (see Number 2 in Fig. 5(b)).
  - Showing the semantic network and allowing interactive exploration (see Fig. 4).
- *Search for related content* is supported by automatically:
  - Applying highlights to currently shown contents (see Fig. 5(a)).
  - Recommending semantically related content from the audience or other presenters (see Number 3 in Fig. 5(b)).
- *Comparison of contents* is supported by an adaptation of the layout to show semantically linked contents side by side.

To provide the automatic support it is necessary to get a more detailed insight about the situation and the tasks of the users. This requires to reason about the state and behavior of the users in the context of the current situation. Most actions and states are not observable directly, for example the SMR provides no sensor to distinguish between search and comparison behavior. Therefore, we need to estimate the state and actions from all available observations. The observations are noisy and user behavior may *likely* belong to a particular state. Our approach is to build a probabilistic human behavior model that reflects the states, actions, and observations. Using a Bayesian filter suited for such models, which relates observations to the situation, we can infer the current tasks of the users. As is typical for probabilistic inference, each task is associated with some probability. Our support will only be executed automatically if the task can be recognized with sufficient certainty (i.e. with probability larger than some user-defined threshold). A detailed description of our inference algorithm is provided in [21].

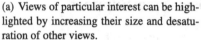

(a) Views of particular interest can be high-lighted by increasing their size and desaturation of other views.

(b) In addition to the currently shown views (1) a temporal outline (2) can be shown to support temporal navigation. Semantically related contents (3) are proposed to support search for contents.

Fig. 5. Examples for automatic adaptation to support typical user tasks.

7 Implementation

We implemented the presentation tool "multipresenter"[1] which incorporates our solutions. Figure 6 shows its components and integration into the setup of our SMR (see Sect. 3). The editor runs on the mobile devices and allows users to share their contents and provide semantic networks. Each instance of the editor connects to the infrastructure of the SMR and is automatically integrated into its middleware [18]. When users provide contents and semantic information via the editor, the data is transmitted by the middleware using network communication. In this way, all components in the middleware (i.e. editor instances on other mobile devices or the support components) can access these data. The layout component takes the user-provided constraints and automatically generates a suitable content distribution and layouts for the displays. Based on these gen-

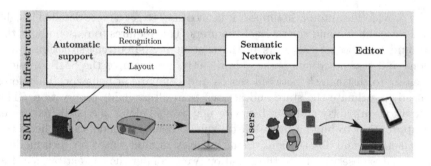

Fig. 6. Overview of the complete infrastructure and its integration into our Smart Meeting Room (SMR).

---

[1] A video that demonstrates the multipresenter is available at http://youtu.be/A7XQzq4zavk.

erated layouts, contents are distributed to the miniature PCs of the SMR and displayed by the connected projector. The situation recognition considers the sensor data (e.g., from the positions of the users) to infer the current situation. If it is confident about a certain situation, it may adapt the display of contents (e.g., by changing the layout generation) in order to support the users in their current task.

8   Conclusion

In this paper we presented an approach that makes use of the advanced capabilities of a SMR. It allows an easy interactive configuration of a multi-display environment and further supports users with automatic generated view layouts. By using intention recognition, we can infer the tasks of the users and apply customized assistance. In this way, we address the challenges *C1-C4* of display ecologies.

We have tested our approach in two settings. First, we applied it to a discussion scenario on the impact of climate change [20]. Second, we used our system in a teaching scenario, which was part of a project at our university to improve learning and teaching conditions. In both cases we got positive feedback from domain experts as well as from non-technical experts, which found our tool "very helpful" and "easy to use".

Acknowledgements. This research is supported by the German Research Foundation (DFG) within the research training group GRK 1424 MuSAMA (Multimodal Smart Appliance Ensembles for Mobile Applications).

References

1. Chung, H., North, C., Joshi, S., Chen, J.: Four considerations for supporting visual analysis in display ecologies. In: Proceedings of Visual Analytics Science and Technology (VAST). IEEE (2015)
2. Augusto, J., Nakashima, H., Aghajan, H.: Ambient intelligence and smart environments: a state of the art. In: Nakashima, H., Aghajan, H., Augusto, J.C. (eds.) Handbook of Ambient Intelligence and Smart Environments, pp. 3–31. Springer, New York (2010)
3. Cook, D.J., Das, S.K.: How smart are our environments? An updated look at the state of the art. Pervasive Mob. Comput. **3**, 53–73 (2007)
4. Rashidi, P., Mihailidis, A.: A survey on ambient-assisted living tools for older adults. IEEE J. Biomed. Health Inf. **17**, 579–590 (2013)
5. Hoey, J., Boutilier, C., Poupart, P., Olivier, P., Monk, A., Mihailidis, A.: People, sensors, decisions: customizable and adaptive technologies for assistance in healthcare. ACM Trans. Interact. Intell. Syst. **2**, 20:1–20:36 (2013)
6. Cao, L., Chen, W., Zhang, X., Huang, K.: A smart meeting management system with video based seat detection. In: Proceedings of International Conference on Internet Multimedia Computing and Service, ICIMCS 2014, pp. 232:232–232:236. ACM (2014)

7. Vetterick, J., Garbe, M., Dähn, A., Cap, C.H.: Classroom response systems in the wild: technical and non-technical observations. Int. J. Interact. Mob. Technol. (i-JIM) **8**, 21–25 (2014)

8. Budkov, V.Y., Ronzhin, A.L., Glazkov, S.V., Ronzhin, A.L.: Event-driven content management system for smart meeting room. In: Balandin, S., Koucheryavy, Y., Hu, H. (eds.) NEW2AN 2011 and ruSMART 2011. LNCS, vol. 6869, pp. 550–560. Springer, Heidelberg (2011)

9. Yu, Z., Nakamura, Y.: Smart meeting systems: a survey of state-of-the-art and open issues. ACM Comput. Surv. **42**, 8 (2010)

10. Tan, D.S., Meyers, B., Czerwinski, M.: WinCuts: Manipulating arbitrary window regions for more effective use of screen space. In: CHI, pp. 1525–1528. ACM (2004)

11. Biehl, J., Baker, W., Bailey, B., Tan, D., Inkpen, K., Czerwinski, M.: Impromptu: a new interaction framework for supporting collaboration in multiple display environments and its field evaluation for co-located software development. In: Proceedings of CHI. ACM (2008)

12. Yang, J., Wigdor, D.: Panelrama: enabling easy specification of cross-device web applications. In: Proceedings of CHI, pp. 2783–2792. ACM (2014)

13. Badam, S.K., Elmqvist, N.: Polychrome: a cross-device framework for collaborative web visualization. In: Proceedings of ITS 2014, pp. 109–118. ACM (2014)

14. Radloff, A., Lehmann, A., Staadt, O., Schumann, H.: Smart interaction management: an interaction approach for smart meeting rooms. In: IE, pp. 228–235 (2012)

15. Waldner, M., Steinberger, M., Grasset, R., Schmalstieg, D.: Importance-driven compositing window management. In: Proceedings of CHI, pp. 959–968. ACM (2011)

16. Radloff, A., Luboschik, M., Schumann, H.: Smart views in smart environments. In: Dickmann, L., Volkmann, G., Malaka, R., Boll, S., Krüger, A., Olivier, P. (eds.) SG 2011. LNCS, vol. 6815, pp. 1–12. Springer, Heidelberg (2011)

17. Heider, T., Kirste, T.: Automatic vs. manual multi-display configuration: a study of user performance in a semi-cooperative task setting. In: Proceedings of BCS-HCI, pp. 43–46 (2007)

18. Bader, S., Kirste, T.: An overview of the helferlein-system. In: Proceedings of MMS (2013)

19. Radloff, A., Luboschik, M., Sips, M., Schumann, H.: Supporting display scalability by redundant mapping. In: Bebis, G., Boyle, R., Parvin, B., Koracin, D., Wang, S., Kyungnam, K., Benes, B., Moreland, K., Borst, C., Di Verdi, S., Yi-Jen, C., Ming, J. (eds.) ISVC 2011, Part I. LNCS, vol. 6938, pp. 472–483. Springer, Heidelberg (2011)

20. Eichner, C., Nocke, T., Schulz, H.J., Schumann, H.: Interactive presentation of geo-spatial climate data in multi-display environments. ISPRS Int. J. Geo-Inf. **4**, 493 (2015)

21. Nyolt, M., Krüger, F., Yordanova, K., Hein, A., Kirste, T.: Marginal filtering in large state spaces. Int. J. Approx. Reason. **61**, 16–32 (2015)

Visualizing Software Metrics in a Software System Hierarchy

Michael Burch[(⊠)]

VISUS, University of Stuttgart, Stuttgart, Germany
michael.burch@visus.uni-stuttgart.de

Abstract. Various software metrics can be derived from a software system to measure inherent quantitative properties such a system can have or not. The general problem with these metrics is the fact that many of them may exist with varying values making an exploration of the raw metric data a challenging task. As another data dimension we have to deal with the hierarchical organization of the software system since we are also interested in software metric correlations or anomalies on different hierarchy levels. In this paper we introduce a visualization concept which shows the hierarchical organization of the software system on the one hand, but also the list of software metrics attached to each hierarchy level on the other hand. This interactive technique exploits the strengths of the human visual system that allow fast pattern recognition to derive similar or different metric patterns in the software hierarchy. The provided visualization technique targets the rapid finding of insights and knowledge in the typically vast amounts of multivariate and hierarchical software metric data. We illustrate the usefulness of our approach in a case study investigating more than 70 software metrics in the Eclipse open source software project.

1 Introduction

When analyzing or visualizing a software system [1,2] we may compute software metrics [3] on which we base our analyses. A software metric describes the quantitative values of a specific property in any of the software entities, i.e., either in the complete system, in the packages, subpackages, directories, subdirectories, files, classes, or in even finer levels of hierarchical granularity.

There is an endless list of software metric candidates like the number of lines of code (LOC), the number of variables, or the number of commented lines which can (at least in our case) be easily computed ending up in a multivariate dataset in which the rows are hierarchically organized. A row in a multivariate data table expresses the software metric values measured for a certain software entity. The columns on the other hand express the values for a specific software metric for each of the measured software entities.

Analyzing and visualizing such multivariate data has been researched before [4,5] but in our case we have to deal with an additional hierarchical organization of the table rows. Applying standard visualization techniques such as parallel coordinates, scatter plot matrices, or glyph-based approaches either cannot

© Springer International Publishing Switzerland 2015
G. Bebis et al. (Eds.): ISVC 2015, Part II, LNCS 9475, pp. 733–744, 2015.
DOI: 10.1007/978-3-319-27863-6_69

support an additional exploration of the hierarchical structure, suffer from over-plotted data points, or do not scale to many software metrics.

In this paper we introduce a concept based on orthogonal node-link tree diagrams in a top-down layout for displaying the software hierarchy. The soft-ware metrics are visually encoded as stacked color coded bars with optionally displayable varying thicknesses in the same stacking order for each software entity. This visualization concept allows to visually scale to many hierarchy elements side-by-side clearly revealing the hierarchical structure while simulta-neously showing the visual patterns of the software metrics for each entity in a clutter-free representation without overplotted points or polylines as in scatter-plot matrices or parallel coordinates plots.

Our visualization tool supports interaction techniques such as filtering in the hierarchy as well as in the number of software metrics and in their values, changing of metric stacking order, or traversing of the hierarchy. We illustrate the usefulness of our technique by applying it to software metric data computed from the Eclipse open source software project which contains 73 software metrics in a software hierarchy composed of many entities.

2   Related Work

Multivariate data has been researched a lot in the past and the most prominent visualization technique is the parallel coordinates plot introduced by Inselberg and Dimsdale [5]. The original parallel coordinates plot has been varied and enhanced in recently designed techniques, for example by bundled versions [6] to visually emphasize correlations which have the negative impact of changing the straight lines into curved ones generating different visual patterns. Heinrich et al. [7] introduce a density field-based continuous version of these line-based diagrams to reflect hot spots and to better unhide visual patterns. Interaction techniques have been added to explore the data for correlations such as brushing and linking [8,9] or axes reordering [4]. But such diagrams are not useful for a data scenario consisting of many hierarchically organized software entities for which lots of software metrics are computed.

Multivariate data can easily be regarded by modeling it as a tabular scheme consisting of rows and columns. Each row expresses an object (or software entity in this specific domain) whereas the table columns are the measurements, i.e., the values of the software metrics, for each of the software entities. In a parallel coordi-nates plot the individual rows are visually encoded as polylines with line segments connecting the points on parallel equidistant axes reflecting the corresponding quantitative metric values. This approach allows to explore a multivariate dataset for correlations but, on the negative side, it does not scale to many software met-rics nor does it allow to display the hierarchical organization of the table rows which are in our case the software entities. The hierarchical organization is very crucial in software development and analyzing correlations and anomalies on dif-ferent levels of hierarchical granularity is important to detect possible software design flaws which only happen in a certain development area, i.e., in a subhier-archy of a software system for which a specific developer might be responsible.

There are also scatterplot matrices, a combination of individual scatterplots organized in a matrix structure, that benefit from showing all software metric comparisons in each of the single scatterplots but which also do not scale for many software metrics. Moreover, only color coding might be a useful concept to visualize the hierarchical structure of the entities since points instead of lines are used, but as a negative consequence, only a limited number of colors can be reliably perceived to make comparisons due to the limitations of the human's visual system [10]. Consequently, scatterplot matrices are also not applicable to the vast amounts of multivariate and hierarchical software metric data.

Another drawback of parallel coordinates plots and scatterplot matrices is the similarity of values, i.e., software metrics in our case. The design of the diagram would map them to unique lines or points overlaying each other and consequently making them hard to visually compare and analyze. A similar problem is the structure of the multivariate data which might produce densely packed regions in a parallel coordinates plot or in a scatterplot matrix which produces vast amounts of visual clutter [11]. Moreover, in a parallel coordinates plot individual polylines cannot be tracked and followed by the eye making comparison tasks difficult to solve.

This is different in glyph-based techniques which map each list of metric values to an individual non-overlapping glyph. Such glyph-based designs are for example software feathers [12], leaf glyphs [13], star plots [14], or Chernoff faces [15]. In these techniques we can positively mention that the hierarchical organization can easily be achieved by just mapping each glyph to a standard hierarchy visualization such as a node-link diagram [16], a treemap [17], an icicle [18], an indented plot [19,20], or a fractal approach such as a generalized Pythagoras tree [21]. Although the hierarchical structure can easily be detected depending on the strength of the visual metaphor for information hierarchies this visual design would render the glyphs to very small display regions making them unreadable due to the fact that our technique should be visually scalable in both dimensions of the multivariate data, the rows and the columns, i.e., the software entities and the software metrics.

Our novel technique scales to many software metrics and software entities where the hierarchical structure is still visible and derivable. Moreover, visual patterns such as correlation patterns or anomalies can still be recognized.

3   Data Model and Transformations

Before we start designing an interactive visualization technique for multivariate and hierarchical data we come up with a data model and transformations of the data in order to achieve better readability prerequisites. To reach this goal we allow permutations of the metrics in the metric lists in all of them simultaneously to achieve mental map preservation [22] in the visually encoded data. Moreover, we describe a strategy that takes the Optimal Linear Arrangement Problem (OLAP) [23] into account to reduce the amount of differences in neighbored metric lists. This has to be done recursively in each of the subhierarchies resulting in a traversed software hierarchy.

3.1 Multivariate and Hierarchical Data

We model multivariate data as a matrix A with entries $v_{i,j} \in \mathbb{R}_0^+$ where $1 \leq i \leq m$ and $1 \leq j \leq n$. Each entry in the matrix expresses a value $v_{i,j}$ of the software metric j of software entity i. Software metrics are measured properties of a quantitative nature but those might also be extended to express categorical, numerical, or any qualitative textual measure. Those only demand for a different kind of visual encoding and are not further treated in particular in this work.

The $m \in \mathbb{N}$ rows are hierarchically organized, i.e., each row r_i, $1 \leq i \leq m$ builds an element of a hierarchy which can be modeled as

$$H = (V, E)$$

in which the vertices

$$V := \{r_1, \ldots, r_m\}$$

are the rows r_i from the matrix and the edges

$$E \subset V \times V$$

model the parent-child relationships in the hierarchy H.

The columns c_j, $1 \leq j \leq n$ build the basis for traversing the subhierarchies, i.e., the m rows r_i, by taking into account the Optimal Linear Arrangement Problem (OLAP) [23]. We hereby compute a distance value based on the software metrics multiplied by a factor expressing this linear hierarchical distance for a certain hierarchy traversal, i.e., table row permutation.

3.2 Metric Ordering and Arrangement

Let $v_{i,j} \in \mathbb{R}_0^+$ be the metric values from the tabular scheme. For the visualization design we allow several stacking strategies which are dependent on the values of the individual metric values. To achieve a suitable order criterion for the values in the metric lists we first compute the sum of all metric lists

$$S_{sum}(j) := \sum_{i=1}^{m} v_{i,j} .$$

This strategy results in a single aggregated metric value list which can then be used to apply an ordering strategy, i.e., by either starting with the lowest values or the largest ones on top, i.e., in an increasing or decreasing order. It may be noted that the ordering should be the same for each individual list attached to each hierarchy element due to the fact that the mental map should be preserved to allow visual comparisons among the many values.

We do not have to sum up the values for the ordering criterion but we can also use the averages for each individual metric which gives a list of average metric values. This can be achieved by applying

$$S_{ave}(j) := \frac{\sum_{i=1}^{m} v_{i,j}}{m} .$$

Additionally, the maximum or minimum values can be used as a stacking order criterion, i.e., either

$$S_{max}(j) := \max_i v_{i,j}$$

or

$$S_{min}(j) := \min_i v_{i,j}\,.$$

3.3 Traversal of Hierarchy

The stacking order plays a crucial role to visually explore the quantitative metric data. But we might also be interested in traversing the subhierarchies in a way to reduce the metric differences between neighbored hierarchy elements. This problem is referred to as the Optimal Linear Arrangement Problem (OLAP) or the MINLA problem which is an NP-hard optimization problem, i.e., we cannot expect to generate an optimal solution with an efficient runtime complexity. To obtain a suitable hierarchy traversal we come up with a heuristics solution as also described in the work of van den Elzen et al. dealing with dynamic graphs [24].

We take into account the single subhierarchies and map them to a linear list of natural numbered identifiers describable by

$$H_{sub} := \{s_1, \ldots, s_k\}\,.$$

Each s_i is attached by a list of metric values $v_{i,j}$. We compute the overlap Δ of the subhierarchies with respect to the subhierarchy element positions H_{sub} (which are a permutation σ of the hierarchy elements of the s_i) and the differences of the metric lists by using the formula

$$\Delta(\sigma) := \sum_{i=1}^{k}\sum_{l=i}^{k} \mid \sigma(i) - \sigma(l) \mid \cdot \sum_{j=1}^{n} \mid v_{\sigma(i),j} - v_{\sigma(l),j} \mid\,.$$

The OLA problem follows the goal to reduce the amount of overlap expressed by the value of Δ for a certain permutation σ of the hierarchy elements. This means we are looking for a σ which fulfills $min\{\Delta(\sigma)\}$. In the best case this overlap can be reduced to a minimum but typically, we only achieve a local minimum solution which gives a good solution to this problem. If a subhierarchy does not contain that many elements we can compute the Δ-values for all of the permutations which consequently results in the optimal solution for the OLA problem.

4 Visual Design

Since we are dealing with multivariate quantitative data in which the table rows are hierarchically organized due to the fact that they model the software entities we have to integrate two different aspects in our visual design. On the one hand the software metric lists have to be visually encoded allowing to derive different and common pattern structures and on the other hand we need a visual representation of the hierarchical organization of a software system.

4.1   Hierarchy Representation

There are various ways to visualize an information hierarchy [25]. We find a node-link metaphor in a top-to-bottom fashion best suitable to represent a software hierarchy. The node-link metaphor easily reflects the hierarchical organization and additional values can be attached to the individual links which uniquely represent a hierarchy element. To effectively support metric comparison tasks we use an orthogonal tree node-link layout [26] since there the single links run in parallel on different hierarchy layers depending on the hierarchical granularity. A similar concept has already been applied by Burch and Weiskopf [27] for visualizing evolving water levels in a river hierarchy. But there subhierarchies cannot be traversed and the stacking order cannot be changed due to its dependency on time which is in focus of this work.

4.2   Visual Encoding of Software Metrics

We use the links of the hierarchy representation to stack the software metrics by using color coding for the quantities. This has the benefit of giving each software metric the same vertical space, i.e., it is visually treated similarly except for the fact that a different color coding is applied depending on the value of the corresponding software metric. Using the same height for each metric only varies between the color codings not in the shapes supporting mental map preservation and allows easily comparable and identifiable visual patterns which can be remapped to the data to use the visualization as an overview representation with the goal to serve as a visual analysis tool. Interaction techniques can then be applied to look for details on demand or to change the views, stacking orders, or apply a hierarchy traversal.

Figure 1(a) illustrates how the software metrics are stacked and how they are attached to an individual link which represents a parent-child relationship from the hierarchy but which also uniquely represents an individual software entity. In the figure three metrics are color coded for three software entities which are labeled 1, 2, and 3. By the vertical alignment of the colored metric boxes we can easily see if there are similar patterns or if the metric values are different for each of the hierarchy elements. Due to its overlapping-free representation the display does not suffer from visual clutter and it visually scales to many entities if it is scaled-down to pixel sizes.

The visual appearance of the stacked metric values can also be changed on user's demand. There are different options to display the quantities such as color only, thickness only, or color and thickness (Fig. 1(b)). Color and thickness in combination uses two visual features to represent a quantity which is beneficial in this technique since the visual differences are made more apparent by the varying shapes and colors.

Our visualization tool supports several interaction techniques to manipulate the views and to navigate and browse in the data. These interactions take into account either the metric values, the hierarchical organization of the software system, or more general interactions like filtering or details on demand.

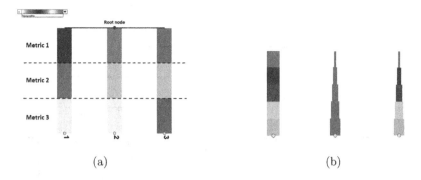

(a)   (b)

Fig. 1. A link from the tree diagram is used to stack the software metrics. Each metric value is mapped to the same vertical space and color coding is used to visually encode the metric value (a). The quantities can either be represented as color codings, as horizontal thicknesses, or as a hybrid of both (b) (Color figure online).

To reliably use a visualization technique an analyst should be able to efficiently identify visual patterns which should be remapped to the displayed data in order to use the visualization as a data analysis and exploration tool. In our dataset scenario we can identify several visual patterns concerning either the hierarchical organization, the metric values for an individual software entity, the metric values compared between different software entities, or which is the biggest benefit, the metric values compared for software entities on different levels of hierarchical granularities.

5   Case Study

To illustrate the usefulness of our visualization technique we computed 73 software metrics from the Eclipse open source software project. The metrics that we considered in this case study can be classified into different groups: Documentable items, types, methods, ratios, words, Kincaid, flesh index, and SMOG-grading.

Figure 2 shows a small part of the hierarchically organized Eclipse software project with link-attached software metric lists containing all 73 software metrics from above in the same order for each hierarchy element. We can directly observe that there are similar pattern structures in all of the software elements. The displayed hierarchy part consists of seven layers and the hierarchy is more or less balanced. We applied a logarithmic color coding since otherwise we will not be able to see many differences in the metric values. The values for the *documentable items* are much higher than all the others. This can also be observed by searching for yellow or green colored boxes which are always close to the top of each orthogonal link which reflect the visual appearances of the *documentable items* metrics.

There are many ways to dig deeper in the data. As a next step we decide to collapse the hierarchy apart from the *ant* subdirectory. By doing this we obtain the representation given in Fig. 3. This leaves more space to display the data in a larger

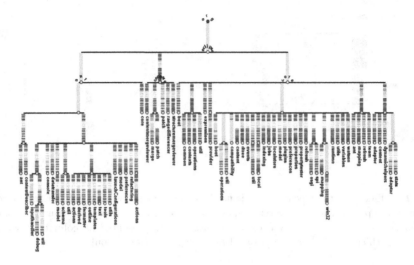

Fig. 2. A small portion of the Eclipse project hierarchy with 73 attached software metrics. We can easily observe the hierarchical structure and simultaneously the software metric lists attached to the orthogonal links. Color coding can be used to explore and compare the data for patterns and correlations (Color figure online).

display area. Due to this won space we decide to also apply the thickness encoding together with the color coding which gives us more apparent visual features for the metric lists. Now we can directly see that the *documentable items* metrics are much larger than most of the others (remember that we applied a logarithm).

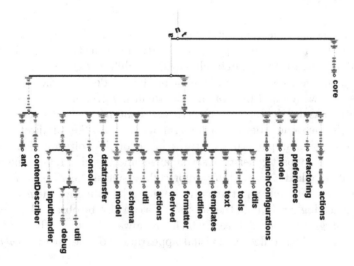

Fig. 3. Collapsing the hierarchy to only inspect the *ant* subdirectory (Color figure online).

In a next step we decide to filter out all other software metrics to only see the *documentable items* metrics in more detail. Figure 4 illustrates this scenario for only 6 metric values in the collapsed *ant* subdirectory. Again we use color coding and thickness as visual features to display the software metrics. Now we can see that there is a decreasing behavior of the metric values which is a natural phenomenon due to the fact that the very first metric counts all documentable items which means that these must be the largest values for all of the hierarchy elements. This phenomenon can mostly be seen in the green or yellow color codings for the topmost boxes at each link.

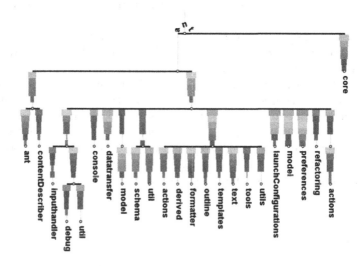

Fig. 4. Filtering for the *documentable items* in the *ant* subdirectory (Color figure online).

There are many more insights that might be derived from the visualized data. Applying filtering for the metric values can help to find only the larger metrics for example. This step can then be used to further extract important hierarchy elements which need a deeper investigation.

6   Conclusion and Future Work

In this paper we introduced an approach for visualizing lists of quantitative software metrics in a software system hierarchy. We base our concept on orthogonal node-link tree diagrams since those allow a visual comparison of the individual values on different levels of hierarchical granularity. The stacking order remains fixed and aligned in each of the hierarchy levels and for each of the software entities. The perceptual abilities of the human's visual system to recognize and compare visual patterns support to see correlations and anomalies in the color coded metric stripes. Traditional multivariate data visualization techniques like parallel coordinates plots, scatterplot matrices, or glyph-based approaches such as star plots or Chernoff faces cannot be applied effectively since they either do not visually scale to thousands of software entities or they do not reflect the hierarchical organization of the software entities. In our visualization tool interaction techniques are applicable to have an impact on the stacking order of the software metrics and to traverse the software subhierarchies in order to reduce the amount of visual differences in the stripes by taking into account the Optimal Linear Arrangement Problem (OLAP). We illustrated the usefulness of our technique in a case study investigating more than 70 software metrics in the Eclipse open source software project which contains many hierarchically structured software entities.

For future work we plan to evaluate the visualization technique in a comparative user study. Standard visualization techniques for multivariate data should be compared. Eye tracking [28–30] might be an option to explore visual attention paid to each of the displayed visualizations in order to analyze the eye movement data for visual task solution strategies applied by the study participants [31]. Moreover, since we are focusing on software visualization and software engineering we should test our visualization with different software projects and also with expert users from the field of software engineering. More interaction techniques should be integrated, for example, to analyze the data on a finer granular level such as the source code level. This is important to finally understand problems which have been detected in the software metric visualization.

References

1. Diehl, S.: Software Visualization - Visualizing the Structure, Behaviour, and Evolution of Software. Springer, Heidelberg (2007)
2. Burch, M., Diehl, S., Weißgerber, P.: EPOSee - A tool for visualizing software evolution. In: Proceedings of the 3rd International Workshop on Visualizing Software for Understanding and Analysis, VISSOFT, pp. 127–128 (2005)
3. Fenton, N.: Software Metrics A rigorous and practical approach. Chapman & Hall/CRC Innovations in Software Engineering and Software Development Series, 3rd edn. CRC Press, Boca Raton (2014)
4. Heinrich, J., Weiskopf, D.: Parallel coordinates for multidimensional data visualization: Basic concepts. Comput. Sci. Eng. **17**, 70–76 (2015)
5. Inselberg, A., Dimsdale, B.: Parallel coordinates: a tool for visualizing multidimensional geometry. In: IEEE Visualization, pp. 361–378 (1990)
6. Heinrich, J., Luo, Y., Kirkpatrick, A.E., Weiskopf, D.: Evaluation of a bundling technique for parallel coordinates. In: Proceedings of the International Conference on Computer Graphics Theory and Applications, pp. 594–602 (2012)
7. Heinrich, J., Bachthaler, S., Weiskopf, D.: Progressive splatting of continuous scatterplots and parallel coordinates. Comput. Graph. Forum **30**, 653–662 (2011)
8. Cleveland, W., McGill, M.: Dynamic Graphics for Statistics. Wadsworth, Belmont (1988)
9. Ward, M.O.: Xmdvtool: Integrating multiple methods for visualizing multivariate data. In: Proceedings of IEEE Visualization, pp. 326–333 (1994)

10. Ware, C.: Visual Thinking: for Design. Morgan Kaufmann Series in Interactive Technologies, San Francisco (2008). Paperback

11. Rosenholtz, R., Li, Y., Mansfield, J., Jin, Z.: Feature congestion: a measure of display clutter. In: CHI, pp. 761–770 (2005)

12. Beck, F.: Software feathers - figurative visualization of software metrics. In: Proceedings of the 5th International Conference on Information Visualization Theory and Applications, pp. 5–16 (2014)

13. Fuchs, J., Jäckle, D., Weiler, N., Schreck, T.: Leaf glyph - visualizing multi-dimensional data with environmental cues. In: Proceedings of the 5th International Conference on Information Visualization Theory and Applications, pp. 195–206 (2015)

14. Chambers, J., Cleveland, W.S., Kleiner, B., Tukey, P.A.: Graphical Methods for Data Analysis. Wadsworth, Belmont (1983)

15. Chernoff, H.: The use of faces to represent points in k-dimensional space graphically. J. Am. Stat. Assoc. (American Statistical Association) **68**, 361–368 (1973)

16. Reingold, E.M., Tilford, J.S.: Tidier drawings of trees. IEEE Trans. Software Eng. **7**, 223–228 (1981)

17. Shneiderman, B.: Tree visualization with tree-maps: 2-D space-filling approach. ACM Trans. Graphic. **11**, 92–99 (1992)

18. Kruskal, J., Landwehr, J.: Icicle plots: better displays for hierarchical clustering. Am. Stat. **37**, 162–168 (1983)

19. Burch, M., Raschke, M., Weiskopf, D.: Indented pixel tree plots. In: Bebis, G., Boyle, R., Parvin, B., Koracin, D., Chung, R., Hammoud, R., Hussain, M., Kar-Han, T., Crawfis, R., Thalmann, D., Kao, D., Avila, L. (eds.) ISVC 2010, Part I. LNCS, vol. 6453, pp. 338–349. Springer, Heidelberg (2010)

20. Burch, M., Schmauder, H., Weiskopf, D.: Indented pixel tree browser for exploring huge hierarchies. In: Bebis, G., Boyle, R., Parvin, B., Koracin, D., Wang, S., Kyungnam, K., Benes, B., Moreland, K., Borst, C., Di Verdi, S., Yi-Jen, C., Ming, J. (eds.) ISVC 2011, Part I. LNCS, vol. 6938, pp. 301–312. Springer, Heidelberg (2011)

21. Beck, F., Burch, M., Munz, T., Silvestro, L.D., Weiskopf, D.: Generalized Pythagoras trees for visualizing hierarchies. In: Proceedings of the 5th International Conference on Information Visualization Theory and Applications, pp. 17–28 (2014)

22. Archambault, D., Purchase, H.C.: The "map" in the mental map: experimental results in dynamic graph drawing. Int. J. Hum Comput Stud. **71**, 1044–1055 (2013)

23. Garey, M.R., Johnson, D.S.: Computers and Intractability: A Guide to the Theory of NP-Completeness. W.H. Freeman, New York (1979)

24. van den Elzen, S., Holten, D., Blaas, J., van Wijk, J.J.: Reordering massive sequence views: enabling temporal and structural analysis of dynamic networks. In: Proceedings of the IEEE Pacific Visualization Symposium, pp. 33–40 (2013)

25. Schulz, H.: Treevis.net: a tree visualization reference. IEEE Comput. Graphic. Appl. **31**, 11–15 (2011)

26. Burch, M., Konevtsova, N., Heinrich, J., Höferlin, M., Weiskopf, D.: Evaluation of traditional, orthogonal, and radial tree diagrams by an eye tracking study. IEEE Trans. Vis. Comput. Graphic. **17**, 2440–2448 (2011)

27. Burch, M., Weiskopf, D.: Visualizing dynamic quantitative data in hierarchies - TimeEdgeTrees: Attaching dynamic weights to tree edges. In: Proceedings of the International Conference on Information Visualization Theory and Applications, pp. 177–186 (2011)

28. Blascheck, T., Kurzhals, K., Raschke, M., Burch, M., Weiskopf, D., Ertl, T.: State-of-the-Art of Visualization for Eye Tracking Data. In: EuroVis - STARs, pp. 63–82 (2014)
29. Blascheck, T., Burch, M., Raschke, M., Weiskopf, D.: Challenges and perspectives in big eye-movement data visual analytics. In: Proceedings of the 1st International Symposium on Big Data Visual Analytics (2015)
30. Kurzhals, K., Fisher, B.D., Burch, M., Weiskopf, D.: Evaluating visual analytics with eye tracking. In: Proceedings of the Fifth Workshop on Beyond Time and Errors: Novel Evaluation Methods for Visualization, BELIV, pp. 61–69 (2014)
31. Burch, M., Andrienko, G.L., Andrienko, N.V., Höferlin, M., Raschke, M., Weiskopf, D.: Visual task solution strategies in tree diagrams. In: Proceedings of IEEE Pacific Visualization Symposium, pp. 169–176 (2013)

Region Growing Selection Technique for Dense Volume Visualization

Lionel B. Sakou[1], Daniel Wilches[1]([✉]), and Amy Banic[1,2]

[1] Department Computer Science, University of Wyoming, Laramie, USA
lionel.sakou@gmail.com, dwilches@uwyo.edu
[2] Idaho National Laboratory, Idaho Falls, USA
abanic@cs.uwyo.edu

Abstract. Selection is a fundamental task in volume visualization as it is often the first step for manipulation and analysis tasks. The presented work describes and investigates a novel 3-Dimensional (3D) selection technique for dense clouds of points. This technique solves issues with current selection techniques employed in such applications by allowing users to select similar regions of datasets without requiring prior knowledge about the structures within the data, thus bypassing occlusion and high density. We designed a prototype and experimented on large dense volumetric datasets. The preliminary results of our performance evaluation and the user-simulated test show encouraging results and indicate in which environments this technique could have high potential.

1 Introduction and Motivation

Visualization, specifically volume visualization, allows users to distinctively observe and validate models' features through more precise pattern detection, allowing various types of information to be extracted. Visualization also enhances the learning of systems and has become part of the mainstream analytical tools in the future for humans in science, engineering, medicine, education, and other disciplines [1].

Selecting regions in volume visualizations may be difficult in environments where data rendering is coarse or high density. Our research presents a 3D selection technique that interactively suggests regions from a dense multi-variate volumetric dataset in real-time based on the user's explicit and implicit interest on the dataset. One of the innovative parts of this work lies in how the implicit interest is determined. For this, our technique uses all the attributes (the different variables in the data) of the user's initial selected volume to deduce the user's interest. Subsets of the cloud exhibiting the inferred user's interest are then visually grouped by iteratively applying a region growing algorithm. This approach is user-centered: the technique allows the user to interact with the visualization system and analyses the current selection to suggest similar regions using the results from this analysis. Thus incorporating a user-system symbiotic interface in which the human's visual processing power and the computer's analysis speed are combined to get the desired results.

© Springer International Publishing Switzerland 2015
G. Bebis et al. (Eds.): ISVC 2015, Part II, LNCS 9475, pp. 745–754, 2015.
DOI: 10.1007/978-3-319-27863-6_70

Our method can be applied for efficient selection in object datasets that arise in several visualization domains including medical visualization, scientific visualization, sculpting systems, etc.

In this paper we present the system tests we used to evaluate our technique in large datasets, we present the results of our evaluation and in which scenarios it's used is recommended or not recommended.

2   Related Work

Bowman et al. [2] give a comprehensive review of the state-of-the art of inter-action techniques referred to as the four "basic" interaction tasks for 3D visu-alizations and virtual reality (VR) [3]: navigation, selection, manipulation, and system control. They examine a number of well adopted approaches to object selection in 3D visualizations.

Ray-casting is the most utilized technique as it is fast and easily under-standable [4,5]. In this technique a ray is extended from the users' tool and intersected against objects in the scene. The intersected objects become then selected. While such techniques are generally efficient for selecting single and large objects, they are less well suited for selecting small or occluded objects in cluttered environments.

Another approach is to use elements such as brushes to select sets of pixels of interest. In Brushing, users localize subsets of the data using range selections. A user can add more to the selection simply by moving the brush. While this selection metaphor is very intuitive, the user has to spend a lot of time adjusting the brushes, especially in large multivariate environments. Selection by expansion like our proposed technique provides a faster time for selection tasks due to less action required from the user.

Other selection techniques are based on the user defining a temporary lasso [6] or other selection shapes [7] for multiple objects selection. All objects that lie within the lasso boundary are selected. The evaluations of these techniques show that they are effective in dense particle clouds. However, these approaches are not always straightforward because in some cases precise drawn strokes are required, and in other they may include unwanted elements along with the selection.

Our literature survey suggests that few research has been done on user-guided or automatic selection techniques. And that even these approaches do not con-sider the hidden attributes of virtual objects in multivariate datasets [8,9]. Both the limitations of these techniques and the lingering issues related with selection techniques mentioned above provide inspiration for the dense environment 3D selection technique we propose and evaluate.

Our selection technique builds upon concepts used by Huang and Ma [10] for region growing but applied to multivariate datasets, and by Zhou and Hansen [11] for multivariate data exploration but applied as a more automatic technique where users don't need to define a selection region by hand. Our selection tech-nique utilizes the user's initial selection to interactively suggest regions of a large volumetric dataset to the user. The similarity of hidden attributes and visual fea-tures is used to expand a selection by visually grouping all similar subsets of the dataset. This helps solve selection issues in occluded or dense environments by allowing users to locate and select ROIs quickly and more conveniently without requiring prior knowledge about the structures within the data.

3 Technique Design

3.1 Human-Computer Symbiosis Framework

In our method, we aim to leverage the human visual capabilities to support better visualization of large volumetric data while assisting with physical limitations. Our technique should therefore support a complementary relationship between the visualization system and the user. This idea is related to the notion of human-computer symbiosis [12], which assigns low-level tasks to the computing system in order to support the user in high-level tasks.

3.2 Design Overview

In our technique we assume that the initial selection made by the users contains a group of objects which is representative of the users' interests. This initial selection is done by overlapping a predefined shape (a box) on the data. The data that lies inside this box is regarded as the initial selection. Then, an algorithm then expands the selection from to all similar objects (according to some constraints defined by the user) is made. The user is able to adjust the parameters of this expansion algorithm, and to manually add or remove objects from the selected or suggested volumes. Figure 1 depicts a generalization of this iterative approach, which can be described as an iterative user-application symbiotic selection process for real-time visualization. Specifically, visualization is enhanced by automatically suggesting potential ROIs to the users and giving them a chance to refine them. The technique takes advantage of the users' visual aptitude to decide when and how to expand the selection, while providing them with high-speed computer power, forming then a partnership between users and computer.

3.3 Computing User's Explicit and Implicit Interest

In this technique we assume the explicit interest of the user is based on human perception. So further insight into the visual aspects of its initial selection will tell what this interest is. The explicit interest is then calculated as a certain range of values around the median of the visual attribute. Any object which shares a similar enough value for this attribute could be of the interest of the user.

However, multivariate 3D datasets contain more data that can be visually represented. These extra attributes, hidden to the user, constitute the implicit interest of our selection technique. i.e. if a user selects several datapoints that share similarities in a non-visual attribute then we can assume we have an

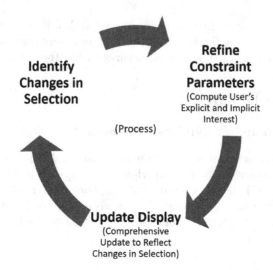

Fig. 1. A generalization of the iterative process that shapes our selection technique. It depicts a user-application symbiosis framework.

implicit pattern in the data. In order to decide which points are similar to our selection we also calculate the medians of the non-visual attributes and define a certain range around them. Objects with attributes laying in those ranges may be considered similar as well.

We note that an optional feature of this technique is to allow the user to adjust a set of thresholds that affect the aforementioned ranges. This helps establishing the relevance of hidden attributes for selection expansion. Increasing a threshold value would have the effect of raising the requirements of the particular attribute in cluster membership. Decreasing the threshold value would have the opposite effect. At the minimum value of 0, attributes are irrelevant in the process, only the visual features are relevant.

3.4   Suggesting Selection Expansion

Expanding the selection consists of using the inferred explicit and implicit interests to group subsets similar to the initial selection. Figure 2 depicts a generalization of the expansion process. Grouping begins by identifying all objects in the dataset that display the explicit and implicit interests. The next step is to identify clusters containing at least some minimum of similar objects, which are then added to the initial selection.

Some of the clustering algorithms [13,14] are based on knowing the expected spatial layout of the objects to be clustered. As we do not want to introduce such constraint in our technique, our technique identifies all similar subsets of any shape and size. Our selected clustering technique is an adaptation to the density-based algorithm DBSCAN [15] presented by Gaonkar and Sawant [16].

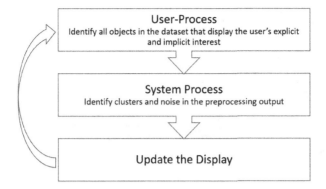

Fig. 2. A generalization of the steps employed to group suggested regions.

This algorithm takes as input a set and identifies subsets of high and low densities around it. This adaptation automates the choosing of some parameters the original DBSCAN relies upon, so clusters are identified automatically with arbitrary shape and size.

4   Prototype Implementation

In order to explore the practicality and the potential benefits of our selection technique we implemented an experimentation system. This system includes a volume explorer and a visualizer, and was designed for users with basic knowledge of volume visualization and interaction techniques. At present, the system consists of a volume renderer, two 2D renderers for specifying the initial volume box, and buttons to add or subtract from the selection the data bounded by the box. Figure 3 shows the user interface with the volume renderer on the left and the two 2D renderers on the right. The top-right renderer displays a 2D representation of the dataset from the X-Y axis perspective while the bottom-right renderer displays the data from the X-Z perspective. They are both used to perform rectangular selections which in turn are used to define a 3D box in the volume renderer. We use a basic 3D box lasso selection scheme as defined by Ulinski et al. [7].

Upon defining the 3D box, the user uses the "Add" button to add the points lying in the cube to the user's current selection, which is displayed on the main renderer using a different style. The data volume can be rotated and translated through mouse control to allow better exploration.

After the user's initial selection, the selection is interactively expanded to all similar points. Adding or subtracting from the current selection updates the expansion. At any time, users can specify attributes weight to refine the expansion. An Attribute weight, by default set to 1, specifies the importance of the attribute in the expansion process.

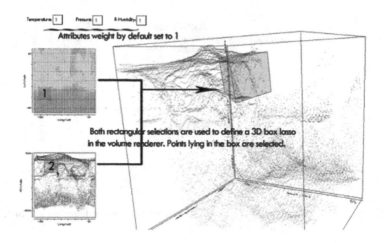

Fig. 3. Selection using the 3D-box lasso style.

4.1   Technique Implementation

Currently, the system runs on an i7-2670QM CPU 2.20 Ghz machine with 12.0 GB of RAM. The spatial rendering is done with local rendering parameters running under 64-bit Microsoft Windows 7. The system contains a NVIDIA GeForce GTX 560M video card with 8948 MB of RAM. Java was chosen as the language for development. We used NetCDF datasets provided by the Unidata Program Center.

Two main Java libraries were used in the development of our experiment interface: (1) The NetCDF-Java library, a Java framework to read and write scientific data formats similar to NetCDF files, and (2) VisAD, a Java class library for interactive and collaborative visualization and analysis of NetCDF data, 2-Dimensional (2D) as well as 3D.

5   System Evaluation

We applied our selection technique to NetCDF datasets ranging from 80000 to 380000 points with inherent visual and hidden attributes similarities. The number of attributes ranged from 3 to 10 of the following: temperature, pressure, pressure-vertical-velocity, relative-humidity, u-component of wind, v-component of wind, density, precipitation, geo-potential height, and visibility at each point. The visual attribute, color, denotes the altitude value at a point. Our choice of the size of the datasets and the number of attributes aimed at simulating real large dense volumetric datasets.

The first part of our experiment consisted on loading a dataset in our application, performing an initial selection, specifying attribute weights and evaluating the suggested regions. We used the Adjusted Rand Index (ARI) [17] to measure the similarity between two data clusterings: the expected and the actual result.

This performance measure calculates the fraction of correctly classified and misclassified elements, which is necessary to ensure the suggested regions accurately reflect our deduction of the user's implicit and explicit interest as defined before. We evaluated how close the suggested clusters were to the expected clusters. An ARI of 1 means that all similar points are correctly clustered and that all expected clusters were found.

For the second part of the experiment, we simulated the users behavior for evaluating the accuracy of our selection technique in deducting the users' interests. This was done to ensure that all aspects of the users' interaction was evaluated. Three tests were conducted, one with three attributes, one with seven and one with 10. For each test we ran five trials, each with four dataset sizes: 80000 points, 120000 points, 260000 points and 380000 points.

Accuracy and error were computed by analyzing the number of points in the expansion that were of interest to the user and the number of points that were of interest to the user but were not suggested. Points were considered errors if they were suggested to the user but were not of interest (Type 1 Error) or if they were not suggested but were of the users' interest (Type 2 Error).

As this techniques aims to be considered interactive we also recorded system response times: (1) time to identify clusters within suggested points, and (2) overall completion time starting with a selection and finishing with the update of the display.

6   Results and Discussion

Table 1 shows the average ARI over 5 runs for each dataset size for the specified number of attributes. The visual grouping at the base of our implemented selection technique performs as we expected as it expands the initial seed to similar clusters as defined by the inferred user's implicit and explicit interest with no significant errors.

In the 380.000 point dataset with 10 attributes, the average ARI over 5 trials was ~0.97, indicating that the suggested clusters in the expansion are about 97 % of what we expected (lowest $\sigma = 0.105$, highest $\sigma = 0.556$). We believe the slight variation in ARI values is attributed to the proficiency of the DBSCAN algorithm in determining density around points. The average distribution of points in the dataset also played a part in effectively estimating density regions, which also shows that the quality of our approach depends significantly on the quality of the clustering algorithm.

The accuracy and error results obtained for our implementation are encouraging (Table 2). Accuracy tests yielded on average a highest accuracy of 95.8 % (mean $\sigma = 0.062$) and a lowest accuracy of 78.3 % (mean $\sigma = 0.183$). On average 87.05 % of points suggested in the selection expansion were of interest to the user. We recorded a 10.43 % average error rate over all tests. We did not see any specific trend in accuracy or error but we noticed that the suggested regions were more accurate when the user's initial selection included points of high density.

Average completion time results are summarized in Tables 2 and 3. Average processing times show that the number of attributes did not have a significant

Table 1. Adjusted rand indices.

Datasets Size	N. Attributes	Avg. ARI	Std. Deviation
80K	3	0.96441	0.123
	7	0.96701	0.180
	10	0.98448	0.135
120K	3	0.96291	0.328
	7	0.98230	0.114
	10	0.98928	0.105
260K	3	0.95111	0.252
	7	0.98376	0.308
	10	0.93291	0.556
380K	3	0.94440	0.120
	7	0.95231	0.281
	10	0.97262	0.316

effect on datasets of size below 120000. Beyond 120000, response times were approximately in the order of 5s (the highest was ~ 7s). These results show that the system presents high latency for presenting the results, which makes it unsuitable for an interactive technique. However, this technique can be used in environments where accuracy is more important than speed, or where the amount of information is of less than 120000 points.

Table 2. Accuracy and error results. (SD: standard deviation)

Dataset Size	Num. Attrs	Accuracy		Error Type 1		Error Type 2	
		Mean	SD	Mean	SD	Mean	SD
80K	3	80 %	0.108	4.80 %	0.118	7.27 %	0.137
	7	95.8 %	0.062	1.28 %	0.050	5.05 %	0.102
	10	87 %	0.129	6.03 %	0.102	4.81 %	0.148
120K	3	90.5 %	0.147	2.21 %	0.084	7.83 %	0.228
	7	86.3 %	0.109	2.08 %	0.078	6.10 %	0.092
	10	83 %	0.283	4.13 %	0.103	7.58 %	0.134
260K	3	90.2 %	0.151	4.28 %	0.184	6.47 %	0.204
	7	92 %	0.163	2.86 %	0.083	4.72 %	0.120
	10	86.5 %	0.182	1.25 %	0.042	6.71 %	0.218
380K	3	78.3 %	0.183	4.12 %	0.120	14.02 %	0.292
	7	90 %	0.157	4.30 %	0.108	3.68 %	0.092
	10	85 %	0.210	3.58 %	0.153	7.7 %	0.216

Table 3. Completion times results in seconds. (Mean CLT: mean clustering time, Mean CT: mean completion time, SD: standard deviation)

Dataset Size	Num. Att	Mean CLT	Mean CL	SD CLT	SD CL
80K	3	2.153	3.204	0.13	0.08
	7	1.094	2.018	0.03	0.02
	10	2.192	3.512	0.06	0.08
120K	3	2.826	3.493	0.05	0.03
	7	1.493	2.202	0.03	0.03
	10	2.006	3.184	0.10	0.06
260K	3	3.031	4.421	0.12	0.14
	7	4.156	5.127	0.08	0.03
	10	4.621	5.322	0.21	0.12
380K	3	4.866	6.763	0.08	0.02
	7	5.230	7.541	0.05	0.03
	10	4.814	6.326	0.03	0.03

7 Summary and Conclusion

We designed 3D selection technique that suggests regions of a volumetric dataset based on the user's explicit and implicit interests on the dataset. We use the conventional gestalt laws of human visual perception [18] and hidden objects attributes to deduce the user's interest on a objects cloud. We evaluated this technique with dataset of varying size and found the latency times for bigger datasets may not be enough for incorporating this technique in an interactive environment. Nonetheless, from the response times for smaller datasets we demonstrated this approach may be suitable for small datasets, or in environments when accuracy is more important than response time. This approach is a step towards a more user-centered, relational approach to 3D selection.

Our contribution will be valuable for future designers of volumetric display applications, as object selection will be a fundamental technique for any such application. While our research focused on selection techniques for volumetric displays, the technique we have designed could also be implemented in large VR environments.

References

1. Ward, M., Grinstein, G., Keim, D.: Interactive Data Visualization: Foundations, Techniques, and Applications. A.K. Peters Ltd, Natick (2010)
2. Bowman, D.A., Kruijff, E., LaViola, J.J., Poupyrev, I.: 3D User Interfaces: Theory and Practice. Addison Wesley Longman Publishing Co. Inc, Redwood City (2004)
3. Bowman, D.A., Kruijff, E., LaViola, J.J., Poupyrev, I.: An introduction to 3-D user interface design. In: Presence: Teleoperations and Virtual Environments, vol. 10, pp. 96–108 (2001)

4. Grossman, T., Balakrishnan, R.: The design and evaluation of selection techniques for 3D volumetric displays. In: Proceedings of the 19th Annual ACM Symposium on User Interface Software and Technology UIST 2006, pp. 3–12. ACM, New York (2006)
5. Poupyrev, I., Weghorst, S., Billinghurst, M., Ichikawa, T.: Egocentric object manipulation in virtual environments: Empirical evaluation of interaction techniques (1998)
6. Yu, L., Efstathiou, K., Isenberg, P., Isenberg, T.: Efficient structure-aware selection techniques for 3D point cloud visualizations with 2DOF input. IEEE Trans. Visual. Comput. Graph. **18**, 2245–2254 (2012)
7. Ulinski, A., Zanbaka, C., Wartell, Z., Goolkasian, P., Hodges, L.: Two handed selection techniques for volumetric data. In: 2007 IEEE Symposium on 3D User Interfaces, 3DUI 2007 (2007)
8. Jang, J., Rossignac, J.R.: Multiple object selection in pattern hierarchies (2007)
9. Kamat, V.R.: Enabling 3D visualization of simulated construction operations. Ph.D. thesis (2000)
10. Huang, R., Ma, K.L.: RGVis: region growing based techniques for volume visualization. In: 2003 Proceedings of the 11th Pacific Conference on Computer Graphics and Applications, pp. 355–363 (2003)
11. Zhou, L., Hansen, C.: Transfer function design based on user selected samples for intuitive multivariate volume exploration. In: 2013 IEEE Pacific Visualization Symposium (PacificVis), pp. 73–80 (2013)
12. Licklider, J.C.R.: Man-computer symbiosis. In: Transactions on Human Factors in Electronics, vol. 1, pp. 4–11 (1960)
13. Steinbach, M., Ertz, L., Kumar, V.: The challenges of clustering high dimensional data. In: Wille, L. (ed.) New Directions in Statistical Physics, pp. 273–309. Springer, Heidelberg (2004)
14. Jain, A.K., Dubes, R.C.: Algorithms for Clustering Data. Prentice-Hall Inc, Upper Saddle River (1988)
15. Ester, M., Kriegel, H.P., Sander, J., Xu, X.: A density-based algorithm for discovering clusters in large spatial databases with noise, pp.226–231. AAAI Press (1996)
16. Gaonkar, N., Sawant, K.: AutoEpsDBSCAN: Dbscan with EPS automatic for large dataset. IJACTE **2**, 11–16 (2013)
17. Steinley, D.: Properties of the hubert-arable adjusted rand index. In: Psychological methods, vol. 9, p. 386 (2004)
18. Wertheimer, M.: Untersuchungen zur lehre von der gestalt. II. Psychologische Forsch. **4**, 301–350 (1923)

Computing Voronoi Diagrams of Line Segments in $\mathbb{R}^K$ in $O(n \log n)$ Time

Jeffrey W. Holcomb[1,2(✉)] and Jorge A. Cobb[2]

[1] Holcomb Technologies, Irving, TX 75062, USA
JHolcomb@holcombtechnologies.com
[2] Department of Computer Engineering, University of Texas at Dallas,
Richardson, TX 75080, USA
cobb@utdallas.edu

Abstract. The theoretical bounds on the time required to compute a Voronoi diagram of line segments in 3D are the lower bound of $\Omega(n^2)$ and the upper bound of $O(n^{3+\varepsilon})$. We present a method here for computing Voronoi diagrams of line segments in $O(2a_k n \log 2n + 2b_k n \log 2n + 14n + 12c_1 n)$ for k-dimensional space. We also present a modification to the Bowyer-Watson method to bring its runtime down to a tight $O(n \log n)$.

1 Introduction

Voronoi diagrams are important geometrical figures with industrial applications ranging from air traffic control [1] to image processing [2, 3]. The theoretical bounds on the time required to compute a Voronoi diagram of line segments in 3D are a lower bound of $\Omega(n^2)$ [4] and an upper bound of $O(n^{3+\varepsilon})$ [5]. We present a novel method for the derivation of Voronoi diagrams of line segments in $\mathbb{R}^k$ in $O(2a_k n \log 2n + 2b_k n \log 2n + 14n + 12c_1 n) \approx O\ (n \log n)$ time.

Our new Voronoi diagram method relies heavily on the Bowyer-Watson method for its implementation. Because of this dependence, we also developed a modification for the Bowyer-Watson algorithm that brings its runtime to $O(a_k n \log n + b_k n \log n)$ for all cases, or essentially a tight $O(n \log n)$.

We have structured our paper as follows. The next section is *Background*: where we present general material about Voronoi diagrams, material on the Bowyer-Watson method, and a brief survey of material on previous investigations into computing Voronoi diagrams of line segments in 3D. Our third section is *Research*, and is where we present our research into tightening the bounds for the Bowyer-Watson method and computing Voronoi diagrams of line segments in 3D. The fourth section is *Conclusion*, and is where summarize our research findings.

2 Background

"in modum cujufdam vorticis, in cujus centro eft Sol"

— Rene Descartes, 1644 [6]

© Springer International Publishing Switzerland 2015
G. Bebis et al. (Eds.): ISVC 2015, Part II, LNCS 9475, pp. 755–766, 2015.
DOI: 10.1007/978-3-319-27863-6_71

Voronoi diagrams are geometric entities defined over a set G of generators g_i. These generators are utilized to partition a data set, data space, metric space, or etc. into, or according to, Voronoi cells C; where, a Voronoi cell C_i is defined as the region of space such that every point p_i in C_i is closer to the generator g_i for C_i than any other generator $g_j \in G$. The Voronoi diagram VD is then defined as the collection of such cells C_i and their associated generators g_i.

The very first Voronoi diagrams where presented in Descartes' Principles of Philosophy. Interestingly, the Voronoi diagrams presented by Descartes were just a graphical rendering utilized to describe the properties of his vortex theory: one of the earliest theories describing relativistic motion, and a distant ancestor to Einstein's theory of relativity. Unfortunately for Descartes, he never formalized how to mathematically construct his diagrams; so, the name for these diagrams was passed to Voronoi instead of Descartes (Fig. 1).

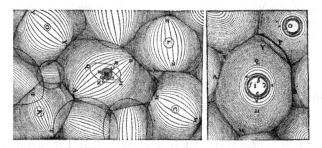

Fig. 1. Two Voronoi diagrams taken from Rene Descartes Principles of Philosophy.

Dirichlet was the first one to use Voronoi diagrams mathematically in 1850 during his investigation of quadratic forms [7], but did not go as far as formally defining the diagrams that he was using. Later, in 1908, Georgy Voronoi finally formally defined the Voronoi diagram while performing his own investigation into the properties of quadratic numbers [8–10].

2.1   Voronoi Diagrams Created from Point Generators

The easiest, and most well studied, form of a Voronoi diagram is that produced by a set of point-generators. The first method for generating a Voronoi diagram for a set of point-generators in $\mathbb{R}^k$ is the Bowyer-Watson algorithm [11, 12]. The pseudo code for the Bowyer-Watson algorithm is listed below.

```
Bowyer-Watson(point_set)
1.    Bowyer Table BT ← Ø;
2.    BP ← compute bounding polytope for point_set
3.    BT.Add(BP);
4.    for each point in point_set
5.       temp_polytope ← Ø;
6.       to_delete ← Ø;
7.       for each simplex in BT;
8.          if point is inside circumcircle of simplex;
9.             to_delete.Add(simplex);
10.      for each simplex in to_delete;
11.         for each polytope in simplex;
12.            if polytope ∉ (to_delete - simplex);
13.               temp_polytope.Add(polytope);
14.      for each simplex in to_delete;
15.         remove entry corresponding to simplex from BT;
16.      for each polytope in temp_polytope;
17.         new_simplex ← form a new simplex by connecting
each vertex in polytope to point;
18.         BT.Add(new_simplex);
19.   for each simplex in BT;
20.      if simplex contains a vertex from BP;
21.         remove simplex from BT;
22.   return BT;
```

The average runtime for the Bowyer-Watson algorithm is $O(a_k n \log n + b_k n)$; however, it has a worst case run time of $O(n^2)$ for special cases involving multiple generators, more than $k + 1$, sharing the same Voronoi vertex, see Fig. 2a. In Sect. 3.1 through Sect. 3.3 we present work showing how to eliminate this $O(n^2)$ runtime for the worst case scenario: thereby making the runtime for the Bowyer-Watson algorithm a tight $O(n \log n)$.

2.2 Voronoi Diagrams Created from Line Segment Generators

Despite a well-developed understanding of Voronoi diagrams of sets of point generators, not much is known about more complex generators such as line segments. Several methods have been developed for the construction of Voronoi diagrams of line segments in 2D [13–16], but virtually nothing is known of Voronoi diagrams of line segments in 3D. [13, 17] were able to compute Voronoi diagrams of lines in 3D, but not line segments. [4] proved that the lower bound for the computation of a Voronoi diagram of line segments in 3D is $\Omega(n^2)$, and [5] showed that the upper bound is $O(n^{3+\varepsilon})$. [18] was able to compute a Voronoi Diagram of line segments in 3D, but the runtime is $O(nk)$ for n line segments and k points: meaning that it performs well on discrete sets but has potentially unbounded runtime with respect to the task of partitioning a data space. In Sect. 3.4 we present a method for computing the Voronoi diagram of line segment generators in $O(n \log n)$: which even beats the theoretical lower limit.

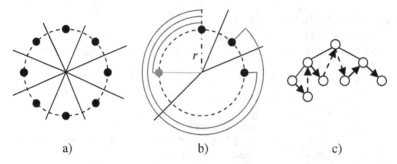

Fig. 2. (a) and (b) he show the worst case scenario for the Bowyer-Watson algorithm. (a) is an example Voronoi diagram for a set of nine point-generators, where all nine point generators lie on the circumcircle for the same Voronoi vertex. (b) is an example Voronoi diagram for a set of three point generators and a fourth, new point generator that falls on the circumcircle of the original generators; and where r is a reference line from the circumcenter to an arbitrary position on the circumcircle. The blue lines in (b) indicate radial distance. (c) shows a combined red-black tree and sorted linked list. The red-black tree edges are in solid lines. The sorted linked list is completed using the dashed lines and the lines with arrows; where the arrows point in the direction of increasing key value (Color figure online).

2.3   Line Segment Voronoi Diagram Cell Surfaces

Part of the difficulty of computing Voronoi diagrams of line segments is that the cellular bounds are no longer regular. For point generators we only have to deal with lines in 2D and planes in 3D and higher dimensions. For line segment generators, and other more complex generators, we now have to concern ourselves with curves and curved surfaces. To this end, [18] defined the four modes of interaction for line segments in higher dimensional space, see Fig. 3, and their associated equations.

The first case for two interacting line segments is that where there does not exist an endpoint e for L1 such that e is closer to L1 than any other point in L1 for all points in L2, and *vice versa*. The equation that defines the surface of separation for the first case is:

$$f(s) = \left(s - \text{proj}_{L_4} s\right) \cdot \left(s - \text{proj}_{L_4} s\right) - \left(s - \text{proj}_{L_3} s\right) \cdot \left(s - \text{proj}_{L_3} s\right) = 0 \qquad (1)$$

The second case is where there exists and endpoint e in L1 such that e is closer to all of the points in L2 than any other point in L1, or *vice versa*. The equation that defines the surface of separation for the second case is:

$$f(s) = \left(s - \text{proj}_{L2} s\right) \cdot \left(s - \text{proj}_{L2} s\right) - \left(s - p_{2.2}\right) \cdot \left(s - p_{2.2}\right) = 0 \qquad (2)$$

The third case is where one of the endpoints of L1 is identical to one of the endpoints of L2. The equation that defines the surface of separation for the third case is:

$$f(s) = \left(L_{4,dir} \times L_{1,dir}\right) \times \left(\left(L_{4,dir} + L_{1,dir}\right) - p_{1,2}\right) \cdot \left(\text{proj}_{Hp1,2} s - p_{1,2}\right) = 0 \qquad (3)$$

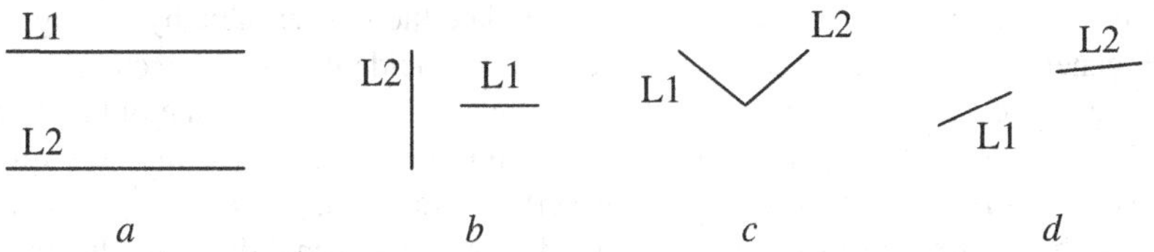

Fig. 3. The four cases for line segment interaction. (*a*) is the case where $\neg\exists e \in L1$: $\forall p \in L1$ $\|p' - e\| < \|p' - p\|$ $\forall p' \in L2$ and *vice versa*. (*b*) is the case where $\exists e \in L1$: $\forall p \in L1$ $\|p' - e\| < \|p' - p\|$ $\forall p' \in L2$, but not *vice versa*. (*c*) is the case where $\exists e \in L1$, $\exists e' \in L2$: $e = e'$. (*d*) is the case where $\exists e_1 \in L1$: $\forall p \in L1$ $\|p' - e_1\| < \|p' - p\|$ $\forall p' \in L2$ and *vice versa*.

The fourth, and final, case is where one of the endpoints e_1 in L1 is closer to all of the points in L2 than any other point in L1 and *vice versa*. The equation that defines the surface of separation for the fourth case is:

$$f(s) = (s - p_{4.2}) \cdot (s - p_{4,2}) - (s - p_{7.1}) \cdot (s - p_{7.1}) = 0 \qquad (4)$$

3 Research

In this section we present the research that we performed with respect to Voronoi diagrams. In the first subsection, we present a novel red-black tree/sorted linked list hybrid data structure that is capable of leveraging both the quick insertion properties of red-black trees and the right/left query properties of a linked list. In the second sub-section we present our modification to the Bowyer-Watson algorithm that brings the runtime for the Bowyer-Watson algorithm down to a tight $O(n \log n)$. In the third subsection we discuss the runtime for our modified Bowyer-Watson method. In the fourth subsection we present a method for computing the Voronoi diagram of a set of line segment generators in $\mathbb{R}^k$. In the fifth subsection we discuss the runtime for our new method for computing Voronoi diagrams for line segment generators in $\mathbb{R}^k$.

3.1   Motivation for and Properties of a Hybrid Data Structure Built from a Red-Black Tree and a Sorted Linked List

As discussed in Sect. 2.1, the Bowyer-Watson algorithm is a method for computing the Voronoi diagram of a set of point-generators in $\mathbb{R}^k$. The runtime for the Bowyer-Watson algorithm is normally $O(n \log n)$ with the exception of the special case presented in Fig. 2a; which has a runtime of $O(n^2)$. Here we present a slight modification to the Bowyer-Watson method that brings the runtime down to a tight $O(n \log n)$.

There are two key observations that enable our improvement of the Bowyer-Watson method. The first is that the participation of the new generator in the circumcircle of the previous generators can be detected by measuring the distance of the new generator to the Voronoi vertex associated with the original set of generators, see Fig. 2b. The second observation is that the set of generators that lie along a circumcenter for a Voronoi vertex v can be ordered according to their arc length with respect to a reference position r, see Fig. 2b.

With these two new observations we first update the Bower table by adding a new sorted generator index that stores the generators incident upon the current vertex according to their radial distance from the reference r. This changes the old table entry from the triple < vertex, forming points, neighboring vertices > to the new quadruple < vertex, forming points, neighboring vertices, radial index > ; where, the radial index is defined as a set of k - 1 trees, for a k-dimension Voronoi diagram, defined over a combination of a red-black tree and a sorted linked list, see Fig. 2c, that uses a node defined over the 6-tuple < *parent*, *rb_left*, *ll_left*, *radial*, *rb_right*, *ll_right*, *vertexID* > ; where the *ll_left* and *ll_right* values are specific to the sorted linked list, and the *parent*, *rb_left*, and *rb_right* values are specific to the red-black tree. Using a combined red-black tree and sorted linked list enables us to quickly add a new generator to our index while still being able to quickly retrieve neighbor generator information from the sorted index.

As a standard algorithm, a version of the original red-black tree method can be found in [19]. To achieve the new hybrid red-black tree/sorted linked list hybrid we only need to modify the RB-INSERT function from [19]. The new RB-INSERT function is as follows:

```
RB-Insert (z)
  y ← nil[this];
  x ← root[this];
  l ← nil[this];
  r ← nil[this];
  while x ≠ nil[this];
    y ← x;
    if radial[z] < radial[x];
      x ← rb_left[x];
      l ← y;
    else;
      x ← rb_right[x];
      r ← y;
  parent[z] ← y;
  if y = nil[this];
    root[this] ← z;
  else if radial[z] < radial[y];
    rb_left[y] ← z;
    ll_left[y] ← z;
    ll_left[z] ← l;
    ll_right[z] ← y;
    r ← y;
  else;
```

```
  rb_right[y] ← z;
  ll_right[y] ← z;
  ll_left[z] ← y;
  ll_right[z] ← r;
  l ← y;
rb_left[z] ← nil[this];
rb_right[z] ← nil[this];
color[z] ← red;
RB-Insert-Fixup(this, z);
return <l.vertexID, r.vertexID>;
```

As can be seen from the pseudo code for the new RB-INSERT function, we can maintain the sorted state of the linked list structure embedded in the red-black tree by simply adding a few lines to the RB-INSERT function. This new hybrid data structure allows us to return the two neighboring generators along the circumference of the circumcircle for the simplex affected by the new generator in $O(\lg m)$ time: where m is the total number of generators that lie on the circumcircle of the current Voronoi vertex.

3.2   The Modified Bowyer-Watson Algorithm

To maintain a tight $O(n \lg n)$ runtime for the Bowyer-Watson Algorithm we modify the original method to detect and accommodate the worst-case scenario. To detect the worst-case scenario we simply add a test for equal distance to the Bowyer-Watson method in addition to the original less than/less than or equal to test. To accommodate the worst-case scenario we utilize the red-black tree/sorted linked list hybrid data structure discussed in Sect. 3.1 to implement the following partial method:

```
WorstCaseFix(Bower Table BT, generator g)
  for each simplex in BT;
    if ||circumcenter - point||
                    = circumcircle.radius;
      if simplex.radial_index.Size = 0;
        for each point in simplex.forming_points;
          simplex.radial_index.Insert(point);
      simplex.forming_points.Add(point);
      polytope ← ∅;
      for each dimension;
        pair ← simplex.radial_index.Insert(point);
        polytope.Add(edge from point to pair.first);
        polytope.Add(edge from point to pair.second);
        BT.Delete(edge from pair.first to pair.second);
      temp_polytope.Add(polytope);
    else if point is inside circumcircle of simplex;
      to_delete.Add(simplex);
```

This partial method is inserted into the original Bowyer-Watson method as a replacement for lines 7 through 9 of the original Bowyer-Watson method.

3.3 Strict $O(n \log n)$ Runtime for the Modified Bowyer-Watson Algorithm

If we look at the runtime for the modified Bowyer-Watson algorithm we see that there is no noticeable change in runtime with respect to the normal case. This means that if none of our generators fall into the worst-case category then we have the original runtime of $O(a_k n \log n + b_k n)$.

For the worst-case scenario, we no longer delete the Voronoi vertex that participates in the worst-case: thereby removing the extra time cost associated with completely rebuilding the Voronoi diagram for every generator that participates in the common Voronoi vertex. Instead, we only delete the edge connecting the two neighboring generators, and insert two additional edges connecting the new generator to its two neighbors; where this delete and insert has the constant cost of $O(3)$.

Computing the identity of the two neighbors for the new generator, as well as inserting the new generator into the radial index, can be done in $O(\lg m)$; where m is the number of generators currently in the index. If all n generators have to be added to the radial index, the absolute worst case scenario, then we will have a runtime of $O(n \lg n)$. If only a subset of m generators need to be added to some radial index then we will have a runtime of $O(m \lg m)$, for some $m \le n$. Since we need a separate radial index for each dimension, we have a final runtime for the partial code of Sect. 3.2 of $O(km \lg m)$ for k-dimensional space.

Inserting this partial code block into the Bowyer-Watson algorithm, if all of the generators lie along the circumcircle of a single Voronoi vertex then the runtime is dictated by the pseudo code from Sect. 3.2, and therefore has a runtime of $O(b_k n \lg n)$; where the k-term is absorbed by b_k.

Since the $O(km \lg m)$ runtime only comes from trying to insert a new generator into the current Voronoi diagram it only affects the second term of the original runtime for the Bowyer-Watson algorithm. This changes the original runtime from $O(a_k n \log n + b_k n)$ to $O(a_k n \log n + b_k n \lg m)$; where the first term is still associated with the amount of time to sort G, and the second term is associated with the amount of time to insert all n generators from G into the new Voronoi diagram. Since $m \le n$ we can simplify the runtime for the modified Bowyer-Watson method to $O(a_k n \log n + b_k n \lg n)$.

We can further reduce this runtime to $O(a_k n \log n + b_k n \log n)$ by using a tree with a branching degree of ten instead of two for our red-black tree. Such a degree ten tree will produce a tight $O(n \log n)$ runtime, but only because we can define a constant time search for each level of the tree. However, using a degree ten tree instead of degree two may not increase performance in real world applications.

3.4   Computing the Voronoi Diagram of for a Set of Line Segment Generators in $\mathbb{R}^K$

The hardest part about computing the Voronoi diagram for any complex geometric form of generator is determining the geometrical and spatial interrelationships between the generators. To solve this problem for line segment generators we begin by using the Bowyer-Watson algorithm to compute the Voronoi diagram of the set of endpoints for our line segments, see Fig. 4.

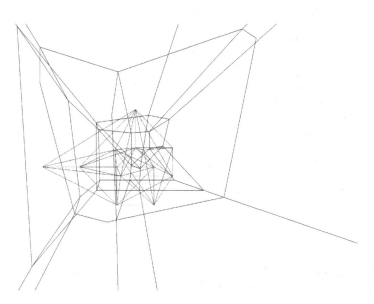

Fig. 4. The Voronoi diagram and Delaunay tessellation of the endpoints for a set of line segment generators.

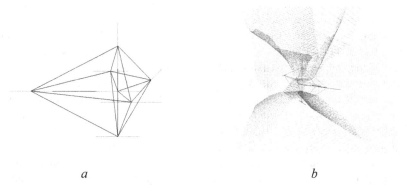

a   *b*

Fig. 5. (*a*) The Delaunay tessellation of a set of line segment generators. (*b*) The Voronoi diagram for the same set of line segment generators as in (*a*).

Next, we scan the edge list for the Delaunay tessellation computed from the Bowyer-Watson algorithm for inter-line segment edges. If two line segments are found to share an edge in this first Delaunay tessellation then they keep the edge in the final tessellation; however, each line segment pair is only allowed to keep one edge. Figure 5 shows such a final tessellation where the edges belonging to the Delaunay tessellation of the line segment generators connect to the center point of the line segment.

Finally, we utilize the Delaunay tessellation of the line segment generators to generate the Voronoi diagram of the line segment generators. The pseudo code for computing the Voronoi diagram of a set of line segment generators is:

```
Voronoi diagram (line_segments)
  line_segment_adjaceny_list ← Ø;
  Voronoi_diagram ← Ø;
  Endpoints ← line_segments.GetEndPoints();
  Endpoints.Sort();
  Bowyer_Table ← Bowyer_Watson_Algorithm(Endpoints);
  for each edge e in Bowyer_Table;
    line_segment_adjaceny_list.Add_Unique(e);
  for each edge e in line_segment_adjaceny_list
    surface ← compute surface of separation for e using
      equations 1 through 4;
    Voronoi_diagram.Add(surface);
  return Voronoi_diagram.
```

3.5 $O(n \log n)$ Runtime Complexity for the Derivation of the Voronoi Diagram of Line Segments in $\mathbb{R}^K$

The runtime of the method presented in Sect. 3.4 is closely tied to the runtime of the Bowyer-Watson algorithm; which [12] showed to be $O(akn \log n + b_k)$. Modifying the Bowyer-Watson algorithm to better handle the worst case scenario, as shown in Sect. 3.2, brings this runtime to a tight $O(a_k n \log n + b_k n \log n)$ We can extract the set of endpoints in $O(2n)$, where n is the number of line segment generators. Similarly, the Bowyer-Watson algorithm can be computed for the set of endpoints in $O(2a_k n \log 2n + 2b_k n \log 2n)$. We can reduce the Delaunay tessellation of the set of endpoints to a Delaunay tessellation of the set of line segments in $O(|E|)$, where $|E|$ is the size of the edge list for the Delaunay tessellation of the endpoints for our line segment generators. Similarly, we can compute the Voronoi diagram for our line segments from the Delaunay tessellation of the line segments in $O(c_1|E|)$; where c_1 is the ratio of edges in the Delaunay tessellation of endpoints to the Delaunay tessellation of line segments. This gives us a total runtime of $O(2n + 2a_k n \log 2n + 2b_k n \log 2n + |E| + c_1|E|)$. We know from Kepler's Conjecture [20, 21] that the tightest packing of spheres, or average upper bound on the number of line segments that a central line segment can interact with, is twelve neighbors. That is, the highest average edge degree that we can ever have for line segments that are not on the edge of our Voronoi diagram is an edge degree of twelve: any other edge degree would lead to a lower overall average edge degree. For example, if we consider the worst case for the unmodified Bowyer-Watson algorithm and construct a spherical mono-layer of n line segments then each mono-layer line segment could have at most eight neighbors. If we then add a new line segment to the center of this line segment mono-layer then the central line segment will have every single other line segment as its neighbor, but its neighbors will only be a degree nine node in their associated Delaunay tessellation. This will produce an average node degree determined by the equation

$$\lim_{n \to \infty} \frac{n + 9n}{n + 1} < 12. \tag{5}$$

As such, we can restate the final runtime as $O(14n + 12c_1n + 2a_kn \log 2n + 2b_kn \log 2n) \approx O(n \log n)$.

While it is possible to achieve a higher average node degree for varying sized line segments, the node degree contribution to the equations is still $d \times n$; where d is the highest average node degree for the given distribution of line segment generator sizes.

4 Conclusion

We presented here a modification to the Bowyer-Watson method that enables the computation of Voronoi diagrams for sets of point generators in a tight $O(n \log n)$ time: thereby removing the $O(n2)$ upper-bound associated with the special case for the Bowyer-Watson method that involves $k + 2$ or more generators sharing the same Voronoi vertex. Having resolved the issue associated with such trouble-maker Voronoi vertices, we next presented a novel method for the derivation of a Voronoi diagram in $\mathbb{R}^k$ for a set of line segment generators; where, our novel method runs in $O(n \log n)$ time.

It should be noted that our novel method is associated with only a moderate increase in the spatial requirements for the Bowyer-Watson method. Specifically, the increase in spatial requirements for the modified Bowyer table is $O(kn)$ for the reference to the radial indices. Each radial index only requires at most $O(n)$, but if all nodes participate in the radial index then we will only have one radial index. Also, as more and more generators participate in a single radial index it becomes impossible for any given generator to be involved in more than two radial indices at once. As such, as the radial indices become large, the number of radial indices becomes small. Finally, the line segment adjacency list and surface list only requires $O(12n)$ space each. This gives us a total increase in spatial cost of $O((3 k + r + 14)n)$; where r is the number of radial indices required.

References

1. Delahaye, D., Puechmorel, S.: 3D airspace sectoring by evolutionary computation. In: Proceedings of the 8th Annual Conference on Genetic and Evolutionary Computation, pp. 1637–1644 (2006)

2. Cheddad, A., Mohamad, D., Manaf, A.: Exploiting voronoi diagram properties in face segmentation and feature extraction. Pattern Recogn. **41**, 3842–3859 (2008)

3. Sabha, M., Dutré, P.: Feature-based texture synthesis and editing using voronoi diagrams. In: Sixth International Symposium on Voronoi Diagrams, pp. 165–170 (2009)

4. Agarwal, P.D., Shwarzkopf, O., Harir, M.: The Overlay of Lower Envelopes and its Applications. Disc. Comput. Geom. **15**, 1–13 (1996)

5. Sharir, M.: Almost tight upper bounds for lower envelopes in higher dimensions. Disc. Comput. Geom. **12**, 327–345 (1994)

6. Descartes, R.: Principia Philosophiæ, Amsterdam (1644)

7. Dirichelt, P.: Über die Reduction der positiven quadratischen Formen mit drei unbestimmten ganzen Zahlen. Journal Für Die Reine Und Angewandte Mathematik, Berlin **40**, 209–227 (1850)

8. Voronoi, G.: Nouvelles applications des paramètres continus à la théorie des formes quadratiqes, Premier Mémoire, Sur quelques propriétés des formes quadratiques positives parafites. Journal Für die reine und angewandte, Mathematik, Berlin **133**, 97–102 (1908)

9. Voronoi, G.: Nouvelles applications des paramètres continus à la théorie des formes quadratiqes, Deuxième Mémoire, Recherches sur les parallélloèdres primitifs. Journal Für die reine und angewandte, Mathematik, Berlin **134**, 198–287 (1908)

10. Voronoi, G.: Nouvelles applications des paramètres continus à théorie des formes quadratiqes, Deuxième Mémoire, Recherches sur les parallélloèdres primitifs. Journal Für die reine und angewandte, Mathematik, Berlin **136**, 67–182 (1909)

11. Watson, D.F.: Computing the n-dimensional delaunay tessellation with application to voronoi polytopes. Comput. J. **24**(2), 167–172 (1981)

12. Bowyer, A.: Computing Dirichlet Tessellations. Comput. J. **24**(2), 162–166 (1981)

13. Boada, I., Coll, N., Madern, N., Sellarès, J.: Approximations of 2D and 3D Generalized Voronoi Diagrams. International Journal of Computer Mathematics **85**(7), 1003–1022 (2008)

14. Held, M.: VRONI: an engineering approach to the reliable and efficient computation of voronoi diagrams of points and line segments. Comput. Geom. **18**(2), 95–123 (2001)

15. Held, M., Huber, S.: Topology-oriented incremental computation of voronoi diagrams of circular arcs and straight-line segments. Comput. Aided Des. **41**, 327–338 (2009)

16. Gold, C., Remmele, P., and Roos, T.: Voronoi diagrams of line segments made easy. In: Canadian Conference on Computational Geometry (1995)

17. Hemmer, M., Setter, O., Halperin, D.: Constructing the exact voronoi diagram of arbitrary lines in three-dimensional space with fast point-location. In: 18th Annual European Symposium, pp. 6–8 (2010)

18. Holcomb, J.W., Cobb, J.A.: Voronoi diagrams of line segments in 3D, with application to automatic rigging. In: Bebis, G., Boyle, R., Parvin, B., Koracin, D., McMahan, R., Jerald, J., Zhang, H., Drucker, S.M., Kambhamettu, C., El Choubassi, M., Deng, Z., Carlson, M. (eds.) ISVC 2014, Part I. LNCS, vol. 8887, pp. 75–86. Springer, Heidelberg (2014)

19. Cormen, T., Leiserson, C., Rivest, R., Stein, C.: Introduction to Algorithms. The MIT Press, Cambridge (2001)

20. Kepler, J.: Strena seu de Nive Sexangula, 1611

21. Hales, T.: A proof of the kepler conjecture. Ann. Math. **162**, 1065–1185 (2005)

Visualizing Aldo Giorgini's Ideal Flow

Esteban Garcia Bravo[✉] and Tim McGraw

Purdue University, West Lafayette, USA
garciao@purdue.edu

Abstract. This paper offers a more detailed analysis of the mathematical methods that Aldo Giorgini used and aims to serve as a supplement to the biographical sketch of Giorgini's life, *Cybernethisms*. Giorgini's career as a civil engineer led him to become a pioneer of computer art, as he repurposed the frameworks for turbulence visualization into works of art. Specifically, we studied Giorgini's original materials to gain insight into the mathematical methods that Giorgini used in his visualizations of fluid dynamics. Given today's computer memory and specialized frameworks, the limitations that Giorgini worked with in his time now seem foreign. We wanted to further understand Giorgini's use of technology and make it more accessible by creating an interactive web tool named Ideal Flow, that allows users to see fluid dynamics variations in real time. Using WebGL, we were able to implement Giorgini's frameworks using a modern tool and thus fully enable new viewers to celebrate Giorgini's contributions. Future generations now have the opportunity to familiarize themselves with Giorgini's unique approach to art and science, as well as computer art history in general.

1 Introduction

This paper is the result of a multi-year study of the life of a relatively unknown computer arts pioneer named Aldo Giorgini (1934–1994). Prior research investigating the contribution of this individual to the artistic and scientific community has been explored through a biographical sketch, *Cybernethisms* [1]. This book analyzes primary source materials found at Giorgini's estate in West Lafayette, Indiana and unearths lost art, manuscripts, computer code and documentation of software developments. However, *Cybernethisms* did not entirely analyze the mathematical methods that Giorgini used to design his artworks. In this paper we are interested in the mathematical equations that Giorgini created to visualize the complexity of fluid dynamics. These historical visualizations show deep understanding of natural phenomena through computer simulations, while being representative of Giorgini's own artistic voice. An example of this body of work can be seen in Fig. 1 titled *Ideal Flow*.

Giorgini's algorithms, developed mainly during 1975 to 1978, on fluid dynamics had not been explored before. Rather than focusing on his entire life, we decided to focus on a specific period concerning fluid simulation, which constitutes his most extensive and obscure work. This gap allowed us to look again at this documentation (now hosted at the Virginia Kelly Karnes Archives and Special Collections Research Center at Purdue University) to revisit and reinterpret Giorgini's visually engaging art-science practice through a modern-day visualization tool.

© Springer International Publishing Switzerland 2015
G. Bebis et al. (Eds.): ISVC 2015, Part II, LNCS 9475, pp. 767–775, 2015.
DOI: 10.1007/978-3-319-27863-6_72

Fig. 1. Ideal Flow, 1976 Acrylic on Mylar. 36" × 36" ©Aldo Giorgini.

Simulating Nature. Starting in 1966 as a research fellow at the National Center for Atmospheric Research (NCAR) in Boulder, Colorado, Aldo Giorgini began his first computer simulations. Giorgini was a recent PhD graduate in Civil Engineering from UC Fort Collins when he applied statistical methods to visualize meteorological data. Giorgini's area of study was turbulence simulation. After the NCAR, Giorgini became an Assistant Professor of Civil Engineering at Purdue University in 1967. At Purdue, Giorgini had access to the university's mainframe computer and a plotter, resuming his research on computational simulation and visualization. Our sample of Giorgini's computer code dates back to 1971 from a report titled "Numerical Simulation of the Navier-Stokes equations in Fourier space" published by the School of Civil Engineering at Purdue [2].

The report was co-authored by Giorgini's first PhD student named J.R. Travis. The nearly 300-page report is a comprehensive catalog of different type visualizations, theory, computer code and mathematical models. The illustrations in the report were more commonly plotted graphs displaying two-dimensional datasets like in a pareto chart or three-dimensional datasets represented in a perspective view (Fig. 2). These are only two among many other illustrations that also inspired Giorgini to create new art forms.

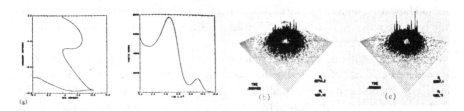

Fig. 2. A sample of illustrations from pages 94 and 100 of the *Numerical Simulation of the Navier-Stokes equations in Fourier space* report from 1971.

Art Appreciation. Aldo Giorgini was originally Italian, born in Voghera in the province of Pavia in 1934. His family moved to Eritrea in 1938 during Mussolini's dictatorship. Upon his return to Voghera, already in his teenage years, Giorgini worked with Ambrogio Casati, one of the futurist painters, restoring frescoes damaged during WWII. Voghera had been heavily bombed by the allies and some of the realistic depictions from a renaissance tradition had to be completely repainted. This experience allowed Giorgini to discover the world of visual representation through lines, shapes and color, a passion that would never leave him. Giorgini's call for artistic expression spontaneously resurfaced in his adult years while doing research in hydromechanics. At the same time that Giorgini was developing the mathematical systems, he also had an art studio at the basement of his home, where he started experimenting mixing oil paints, water-based paints and pigments together. The result was a series of small paintings on canvas paper that he called "Chastiques" (Fig. 3).

Fig. 3. An example of *Chastique* made in 1972 in collaboration with Dan Cook

Giorgini called this "Chastiques" alluding to the term "stochastic" to define a composition that was randomly determined. There is a subtle connection between the paintings and the visualizations he was making for hydromechanics research, as they were both ways to observe and visualize unpredictability. Stochastic systems use probability theory to determine patterns in a seemingly chaotic dataset. The mathematical foundations for the computer code allowed Giorgini to make his first computer simulations of fluid flow. This research on hydraulics awakened Giorgini's interest in art. Most specifically, the illustration on page 136 from the 1971 report became Giorgini's first computer art design entitled "Don't go Bananas" (Fig. 4). Giorgini explained a few years later that he "started 'playing around' with some of the computer drawings that were made as illustration of the research done. From here to the purposeful use of the computer as an art tool the pace was very short" [3].

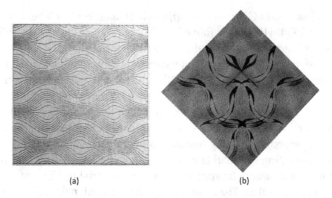

(a)   (b)

Fig. 4. (a) A design in "Numerical simulation of the Navier-Stokes equations in Fourier space" from 1971 p. 136 and (b) "Don't go Bananas" from 1973, India ink on Mylar 36" × 36"

Software Art Documentation. In his lifetime, Giorgini published 78 "technical reports" at the School of Civil Engineering. The topics were mainly on computational hydromechanics. Unfortunately, out of the 78, only six that are listed in his professional resumé [4] are computer art projects, and, unfortunately, only one of them has been found. Giorgini's most well known works were created using the frameworks called "Surfaces," "Fields" and "Photo". There are also many manuscripts and art statements that describe his intent with each framework. The artifacts produced with these programs have been acquired by various museums such as the Indianapolis Museum of Art and the Carnegie Museum of Art, and have also been referenced in several books on computer art. According to the professional resumé, two of the lost art reports were called "Project" and "Light" and dated from 1975 to 1978. These were Giorgini's most prolific works and unfortunately, there is no documentation or any art statements that help explain them in quantity.

During this underexamined period Giorgini produced close to 300 original paintings characterized to be entirely water inspired. "Project" and "Light" focused extensively on the ideal flow of fluids around circular cylinders. Giorgini experimented with parameters such as time and viscosity to create multiple iterations of his work. An example of these iterations can be seen in Fig. 5 from the series "Perturbation of a Connector Flow". Other works in the series were iterative such as "Cylinder Ozeen", "Flow around two cylinders", "Ideal Flow", "Ozeen Half/Half", "Ozeen Relative", "Ozeen Vorticity" and "Light". Each of these had at least ten versions that were hand-painted as separate artworks. The process consisted of designing an algorithm, then using a CalComp pen plotter to draw the lines and finally painting with India ink or acrylic the characteristic black and white patterns. The painted works were finally reproduced using an electrostatic printer on paper or mylar.

One of the most enigmatic series is "Light" because of the vast number of paintings found contrasts with a lack of documentation on this particular creative process. In contrast to the other series, "Light" had close to 60 iterations. Light was a temporary name that Giorgini used for a project that became a mural installed at the Potter Engineering Library at Purdue in 1977 (Fig. 6). The finalized pieces were installed and

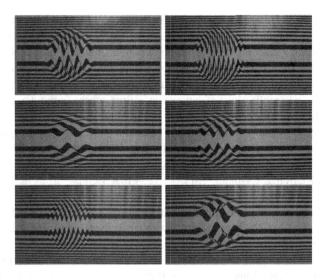

Fig. 5. Six iterations of "Perturbation of a Connector Flow" Each is a 58" × 38" electrostatic print on paper.

renamed as "Fiat Lux" and "Sculptural Forms". "Sculptural forms" is an assemblage of 6 frames while "Fiat Lux" is a single 144" by 87" piece.

Fig. 6. Aldo Giorgini posing next to "Fiat Lux" at the Potter Engineering Center circa 1977.

The principles of these large-scale installations were the principle of the cylinders placed on the space that created a rippled effect due to factors in fluid mechanics, like for example density, vorticity and turbulence. We consider that Giorgini's strengths as a Civil Engineer allowed him to make very complex forms. In "Fiat Lux" Giorgini placed several cylinders on the geometric space. This is the only one of this kind, because the majority of artworks from that period placed only one cylinder.

In the early 1980s Giorgini's art practice diminished gradually, giving more focus to scholarship in computational hydromechanics. His courses were on related topics such as "Stochastic Concepts" and "Computational Visualization of Fluid Flow". His papers became increasingly more about simulating fluid phenomena and less about artistic. The creative exploration of forms during the decade of the 1970s gave Giorgini a lot of experience creating visual outputs in a time where computer displays were unavailable. Giorgini did not have a display terminal until 1979.

One point on how his art influenced his scientific endeavors can be appreciated on the many illustrations that were included in his papers and other academic publications. One example is in the manuscript "Influence of Perturbation Amplitude on Vortex Shedding Past a Circular Cylinder" [5] from 1984 that used pictorials that use the same graphics foundation as "Fiat Lux".

After this analysis of primary source materials, we were intrigued by Giorgini's visualization methods. However, we still wanted to learn more about how the fluid mechanics factors and the cylinder affected the visual composition. Our intent was to further understand Giorgini's use of technology by creating an interactive tool that allowed users to see iterations in real time. We called this implementation and website Ideal Flow, taking the name from the series that inspired this research. In his time, Giorgini had to do a lot of work to produce a single image using a mainframe computer. Visualization was the result of a long process that included creating a mathematical model, transferring into code via punched cards and finally plotted. This contrasts with today's increased computer memory and more specialized visualization frameworks.

Implementation. Ideal flow was implemented in WebGL [6] which runs in most modern web browsers without requiring any plugins. In order to achieve good performance at the expense of physical accuracy we used an Eulerian approach to solving the Navier-Stokes equations on the graphics processing unit (GPU).

Jos Stam [7] described a sequence of operations that result in a stable fluid simulation which captures the turbulent swirling motion of real fluids. The simulation proceeds by discretizing the domain of the fluid into a rectangular grid and numerically representing the physical parameters, such as density, velocity, temperature and pressure, as 2D arrays. Harris [8] presented a GPU implementation of Stam's solution using textures to store the physical parameters and fragment shaders to perform the numerical simulation. This permits the parallel processing power of the GPU to be exploited for fast simulation while keeping the fluid parameters on the GPU for fast rendering. Rideout [9] described an extension of Harris' fluid implementation which handled arbitrary obstacles in the flow which were also discretized and stored as textures.

Ideal flow uses a solver based on Rideout's approach with a few modifications. Since Rideout's OpenGL implementation is geared specifically to simulating smoke, we needed to add the viscous diffusion simulation step described by Harris so that we could represent liquid flows. We also needed to modify the implementation for WebGL, which doesn't support the same set of shader instructions as those used in Rideout's OpenGL implementation. WebGL also does not guarantee support for filtering of floating point textures. To provide a uniform experience for users of all browsers and video drivers we emulated texture filtering in the fragment shader by averaging multiple texture reads.

For rendering smoke and clouds, simply alpha blending the density texture over the scene gives a plausible visualization which fades to transparent as suspended particles or water droplets disperse. However, to match the monochromatic style of many of Giorgini's works we opted for a less physically-based rendering equation given by

$$\text{color } = \text{ C smoothStep}\,(0.2,\,0.8,\,\sin\,(v/s)) \tag{1}$$

where v is the input field value (temperature, density or pressure as selected by the user), s is a scale factor which determines the frequency of stripes and C is a base color (white in our implementation). The smoothstep function (Fig. 7) is an OpenGL shading language (glsl) function which rises smoothly from zero to one over the interval [edge1, edge2]. For all values of x less than edge1, smoothstep (edge1, edge2, x) returns 0 and all values of x greater than edge2 it returns 1. Applying the smoothstep function to a sinusoid results in a pattern of alternating black and white stripes. An example comparing fluid density with a simple linear mapping to the mapping from Eq. 1 is shown in Fig. 8.

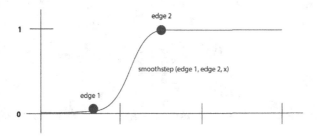

Fig. 7. The smoothstep function defines a smooth transition from zero to one defined by the value of edge 1 and edge 2

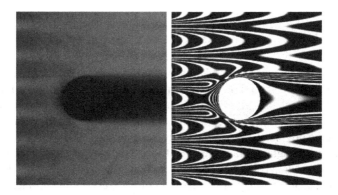

Fig. 8. Linear mapping of density to color (left) and our Giorgini-styled rendering equation (right).

The WebGL simulation was rendered into an html5 canvas element, and additional sliders were added to the document so that simulation parameters such as time step, dissipation, buoyancy and ambient temperature can be dynamically changed by the user. The cylinder within the flow can be scaled and translated with the mouse. This website is available at http://www.snebtor.org/idealflow (Fig. 9).

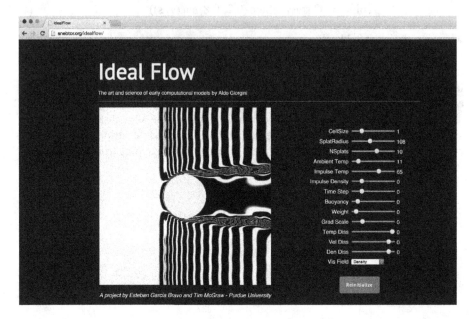

Fig. 9. Ideal Flow website

2 Conclusions

The report titled "Numerical Simulation of the Navier-Stokes equations in Fourier Space" in 1971 marks the origin of Giorgini's art, but it also laid out a foundational mathematical method that allowed him to simulate uncertainty in fluid dynamics. Through the years, Aldo Giorgini gained mastery in visualization through artistic experimentation at the origins of computational hydromechanics. Today, modern computers have more robust imaging libraries and hardware, but some of the mathematical equations for fluid simulation continue to be the same that were used by Giorgini. Researchers in fields of media archaeology or preservation of early computing artifacts have often found questions on how to preserve these materials due to the ephemeral nature of code-based practices. There is not an easy answer, but our focus was on making Giorgini's forgotten work visible through an accessible, modern day interactive tool. We found WebGL as a feasible way to display historic computer visualizations for new generations.

References

1. Bravo, E.G.: Cybernethisms: Aldo Giorgini's Computer Art Legacy. Purdue University Press, New York (2015)
2. Travis, J.R., Giorgini, A.: Numerical simulation of the Navier-Stokes equations in Fourier space. Virginia Kelly Karnes Archives and Special Collections Research Center at Purdue University (1971)
3. Giorgini, A.: Aesthetics in Technology, Page 10, Virginia Kelly Karnes Archives and Special Collections Research Center at Purdue University (1975)
4. Giorgini, A.: Professional Resumé. Virginia Kelly Karnes Archives and Special Collections Research Center at Purdue University (1984)
5. Giorgini, A., Alfonsi, G.: Influence of Perturbation Amplitude on Vortex Shedding Past a Circular Cylinder. Virginia Kelly Karnes Archives and Special Collections Research Center at Purdue University (1984)
6. Matsuda, K., Lea, R.: WebGL Programming Guide: Interactive 3D Graphics Programming with WebGL. Addison-Wesley, Upper Saddle River (2013)
7. Stam, J.: Stable fluids. In: Proceedings of the 26th Annual Conference on Computer Graphics and Interactive Techniques. ACM Press/Addison-Wesley Publishing Co. (1999)
8. Harris, M.: Fast fluid dynamics simulation on the GPU. GPU Gems 1, 637–665 (2004)
9. Rideout, P.: Simple Fluid Simulation. The Little Grasshopper Graphics Programming Tips (2010). http://prideout.net/blog/?p=58

Restoration of Blurred-Noisy Images Through the Concept of Bilevel Programming

Jessica Soo Mee Wong[(✉)] and Chee Seng Chan

Center of Image and Signal Processing,
Faculty of Computer Science and Information Technology, University of Malaya,
50603 Kuala Lumpur, Malaysia
jessica@angkasa.gov.my, cs.chan@um.edu.my

Abstract. Finding a compromise between regularity to remove noise and preserving image fidelity for natural images is unarguably a non-trivial problem. This paper proposes a new image restoration algorithm that executes an optimal tradeoff between sharpness and noise to warrant an acceptable result of image restoration based on bilevel programming. The algorithm demands an objective functions to perform denoising on the degraded image for the lower-level problem using the curvelet-based denoising method, while the upper-level problem with ultimate objective function that is to obtain restored image by performing deblurring to the denoised image using an improved Wiener filter. Experiments were conducted for synthetically blurred and noisy images. The experimental result shows that the algorithm successfully restores image detail. Numerical measurements of the image quality reveal that the algorithm is comparable with other state-of-the-art methods and has the advantage for image contrast and preserving edge details.

Keywords: Image restoration · Image fidelity · Regularization · Curvelet transform · Wiener filter · Bilevel programming

1 Introduction

It is inevitable that image degradations will occur in the imaging chain due to environmental factors and the physical limitations of the imaging system. Two distinct processes results from the degradation: the deterministic blur and the random noise; such degenerative effects make a great influence on identification and extraction of the useful information residing in the images. Therefore, restoration of the degraded images is unarguably a necessity step. It aims to reconstruct the original high-quality image f from its degraded observed version g, which is a typical ill-posed linear inverse problem that generally formulated as:

$$g(x, y) = h(x, y) * f(x, y) + \eta(x, y). \tag{1}$$

where h is a matrix representing a non-invertible linear degradation blur, also known as the Point Spread Function (PSF). "*" is the convolution operation and η is usually taken to be *additive white Gaussian noise (AWGN)*. Image restoration problems have been the

© Springer International Publishing Switzerland 2015
G. Bebis et al. (Eds.): ISVC 2015, Part II, LNCS 9475, pp. 776–786, 2015.
DOI: 10.1007/978-3-319-27863-6_73

subject of intensive research efforts among scholars and researcher fellows over the decades [1–14]. Based on the review, one limitation of standard restoration methods used in the literature is that no attempt is made to explicitly reduce the amount of noise in the degraded image before performing image deblurring. Instead, noise is removed implicitly by the regularizer, at the cost of image fidelity. In practice, no regularizer can perfectly model the characteristics of the original image, therefore, a compromise must be found between regularization (to remove noise) and preserving data fidelity. To rectify this problem, restoration methods have been developed to decouple denoising and deblurring into separate steps, thus becoming a two-step or hybrid algorithm. For example, in [8], the authors presented a two-step image restoration method using a Bayes least squares Gaussian scale mixture (BLS-GSM) formulation. In the first step, BLS-GSM was used for global blur image compensation, and another deblurring step was used in the second step for denoising, in an overcomplete pyramid. In [9], neutron radiography images were restored using a dual-tree complex wavelet transform (DT-CWT) for noise suppression and a Lucy-Richardson (LR) algorithm for deconvolution. In [10], the authors used a wavelet- based framework to decompose images into cartoon, texture and noise components and the approximate bounded variations approach in the next step to perform image denoising and deblurring simultaneously. Mahbubur Rahman *et al.* [11] proposed a hybrid-type image restoration algorithm that takes the advantage of the diagonalization property of the discrete Fourier transform (DFT) for the deconvolution process and space-frequency localization property of the discrete wavelet transform (DWT) for denoising process. In [12], another hybrid restoration approach that employs both Fourier- and wavelet-based techniques for image restoration was presented. The hybrid technique is commonly referred to as ForWarD; Fourier-Wavelet Regularized Deconvolution. Wen *et al.* [13] studied other deblurring schemes in ForWarD, and a two-step algorithm based on the ForWarD was developed using a combination of constrained least-squares restoration and adaptive regularized least-squares restoration. Most recently, Yang et al. [14] proposed a two-step algorithm to restore degraded image by incorporating a curvelet-based empirical Wiener filter with a spatial-based joint non-local means filter. The method restores the image in the frequency domain to obtain a noisy result with minimal loss of image components, followed by an empirical Wiener filter in the curvelet domain to attenuate the leaked noise, then a joint non-local means filter by using the curvelet deblurring result to suppress the leaked noise.

One of the benefits of decoupling the deblurring and denoising stages is that different mathematical techniques can be applied to each problem. Decoupling allows amplified noise, leaked noise or artifacts caused by deblurring, to be handled in a different domain, thus effectively increase the signal to noise of the final image. Bi-level programming (BLP) is a branch of optimization where one problem is embedded (nested) within another. The outer optimization task is commonly referred to as the upper-level optimization task, and the inner optimization task is commonly referred to as the lower-level optimization task. The lower-level problem appears as a constraint, such that only an optimal solution to the lower-level optimization is a feasible candidate to the upper-level optimization problem. The idea of decoupling the deblurring and denoising stages can be designed as a nested problem, where both can carry one or more objective function to solve the degradation issues and ultimately achieve the restoration objective. BLP represents an interesting and rich field of mathematical programming and although some important results have already been obtained [15–20], it is still a fertile area for research, especially in the image restoration problem. To the best of our knowledge, image restoration techniques based on this approach has not yet received a broad attention in the literature. Only a few articles related to this class of problems in the literature [21–25] was found, and the studies were concentrated in parameter learning for variational image denoising models. BLP has an advantage as compared to the conventional iterative method that frequently use in image restoration technique, is that it has the ability optimize many parameters simultaneously. This paper proposed a new image restoration algorithm that executes an optimal tradeoff between sharpness and noise to warrant an acceptable result of image restoration based on the concept of bilevel programming (BLP).

The remainder of the paper is organized as follows. Section 2 elaborates the description of Bilevel optimization problem. Section 3 proposes a new image restoration technique containing nested denoising method in a deblurring method formed by the concept BLP, and gives the implementation details of the restoration technique. Preliminary experimental results are reported in Sect. 4. In Sect. 5, the summary of this paper is presented.

2 Description of Bilevel Optimization Problem

The first significant work on BLP was presented by Heinrich Von Stackelberg in his 1934 monograph on the market economy [26, 27]. The dynamic program that Stackelberg developed, known as a Stackelberg game, lead to a new area of mathematical programming known as "Game Theory". The original formulation for BLP only appeared in 1973, in a paper authored by Bracken and McGil [28], although it was Candler and Norton [29] that first used the designation bilevel programming. Bilevel programming have since been applied to a variety of application, and among the recent one including, Transportation - Network design problem [15, 16]; Management– Coordination of supply demand [17, 18]; Planning - credit allocation [19]; Engineering design- Optimal design problem [20]. Important surveys of BLP include those by Kolstad [30], Colson et al. [31] and Hongli et al. [32], among others.

A bilevel optimization problem contains two levels of optimization tasks such that the optimal solution of the lower-level problem may implicitly determine the feasible space of the upper-level optimization problem. There are two types of variables in these problems; namely, the upper-level variables $x_u \in X_U \subset R^n$, and the lower-level variables $x_l \in X_L \subset R^m$. The lower-level problem is solved with respect to the lower-level variables while the upper-level variables act as parameters to the optimization problem. If F and f are vector value functions ($F: R^n \times R^m \rightarrow R^p$ and $f: R^n \times R^m \rightarrow R^q$), then an optimistic bilevel optimization problem can be described as follows:

$$
\min_{x_u \in X_U, x_l \in X_L} F(x_u, x_l) \text{ subject to} \begin{cases} G(x_u) \leq 0, H(x_u) = 0 \\ x_l \text{ solves} \begin{cases} \min_{x_l \in X_L} f(x_u, x_l) \\ s.t. g(x_u, x_l) \leq 0, h(x_u, x_l) = 0 \end{cases} \end{cases} \tag{2}
$$

where $F(x_u, x_l)$ is the upper-level objective function, and $G(x_u)$ and $H(x_u)$ are the upper-level inequality and equality constraints respectively, while $f(x_u, x_l)$ is the lower-level objective function, and $g(x_u, x_l)$ and $h(x_u, x_l)$ are lower-level inequality and equality constraints respectively. It should be noted that the lower-level optimization problem is optimized only with respect to the variables x_l, and the variable vector x_u is kept fixed. The objective function and constraints of the upper-level optimization problem are not only related to the variables x_l but also the optimal solution of the lower-level problem.

3 The Image Restoration Problem

Consider the linear image degradation model in (1), the process described by this model can be decoupled into two stages: in the first stage the original image $f \in R^n$ is subject to a blur $h \in R^{m \times n}$ due to a number of reasons, such as motion, defocusing and environmental factors, while in the second stage the blurred image is corrupted by noise $\eta \in R^m$ originates in the imaging-formation process, the transmission process, or a combination of both to produce the degraded image $g \in R^m$. So hypothetically, the degradation process can be reversed in separate stages.

The transformation of (1) in a Fourier domain leads to:-

$$G(u, v) = H(u, v) \cdot F(u, v) + N(u, v). \tag{3}$$

where G, F, H, and N are the Fourier transforms of the corresponding lowercase variables. The pair (u, v) represents the location in the spatial frequency domain, H is the degradation function based on PSF estimation and $\bullet$ indicates multiplication. Therefore, reconstruction of ideal image can be expressed as:-

$$
\begin{aligned}
F(u, v) &= \frac{G(u, v)}{H(u, v)} - \frac{N(u, v)}{H(u, v)} \\
&= [G(u, v) - N(u, v)] \cdot \frac{1}{H(u, v)}
\end{aligned} \tag{4}
$$

which can be decoupled into denoising and deblurring stage. Here $[G(u, v) - N(u, v)]$ denotes the spectral image of image denoising for lower-level problem. Points of sharp variations and discontinuities are two important properties for analyzing an image or signal. The Curvelet Transform [33] is based in an anisotropic notion of scale and high directional sensitivity in multiple directions; therefore it can deal with lines or superplane singularities effectively. Due to this consideration, in the lower-level problem, curvelet multiscale transform is employed to decompose degraded image into a different scale to remove image noise. Ideally the denoised image supposedly contains no noise, so $F(u, v)$ can be obtained using simple inverse filtering $\frac{1}{H(u,v)}$. However, in the ill-posed restoration problem, leaked noise, will always exist. In order to make it a well-posed problem, it requires a deblurring method that able to introduce additional information about the image using an explicit regularization term. The Wiener filter is a favor to many because of its simplicity and speed. Besides, it is known to be able to execute an optimal trade-off between inverse filtering and noise smoothing. For this work, the improved wiener filter proposed by [34], is used for image deblurring in the upper-level problem.

3.1   The Optimistic Formulation of BLP for Image Restoration

The general bilevel Optimization for image restoration can be defined as follows:

$$\min_{\alpha \geq 0} B(\hat{f}(\alpha), \hat{g}) \ subject \ to \begin{cases} C(\hat{f}) > C(\hat{g}), D(\hat{f}, \hat{g}) \geq 0 \\ \hat{g} \in \ \text{argmin} \begin{cases} b(\hat{f}, \hat{g}) \\ s.t. \ c(\hat{g}, g) > 0 \end{cases} \end{cases} \tag{5}$$

Where $\hat{f} \in R^n, \hat{g} \in R^m$ and $g \in R^m$ represent the restored image, the denoised image and the degraded input image, respectively. The lower-level objective function $b(\hat{f}, \hat{g})$ performs denoising on the degraded image g while the upper-level objective function $B(\hat{f}(\alpha), \hat{g})$ deblurs $\hat{g}$ to obtain the restored image $\hat{f}$. The $C(\hat{f}) > C(\hat{g}), D(\hat{f}, \hat{g}) \geq 0$ and $c(\hat{g}, g) \geq 0$ are the inequality constraint of fitness value for the upper-level problem and lower-level problem, respectively. The fitness function for inequality constraint C is formulated based on one of the most widely used quality metric, which is the improvement of signal-to-noise (ISNR), while fitness function for inequality constraint D is formulated using Structural Similarity (SSIM) Index quality assessment [35]. The aim of the bilevel problem of (5) is to find a parameter vector that able to obtain an estimate $\hat{f}$, that is as close as possible to the original input image f from solution $\hat{g}$ such that $B(\hat{f}(\alpha), \hat{g})$ attains a minimum $\left\| \hat{f} - f \right\|_2^2$.

Upper-level Problem. The upper-level objection function that corresponds to a parameter optimization to deblur the denoised estimate $\hat{g}$ is defined as follows:-

$$B(\hat{f}(\alpha), \hat{g}) = \min_{\alpha \geq 0} \|Hf - \hat{g}\|_2^2 + W(\hat{f}(\alpha))$$
$$ISNR(\hat{f}) > ISNR(\hat{g}), \ SSIM(\hat{f}, \hat{g}) \geq 0.7 \tag{6}$$

where W is the improved wiener filter by [34]. The upper-level optimization tasks in (6) have conflicting objectives; to perform deblurring to sharpen the image while suppressing residual noise by smoothing the image. Therefore, the optimal decision on the tradeoff between sharpness and noise is determined in accordance with (6). The SSM index is a decimal value comprised between -1 and 1, the closer SSIM value to 1 the higher is the quality of the measured image. For this work, the SSIM value is set to 0.7 to warrant a restored image with acceptable quality.

Lower-level Problem. The objective function for the lower-level problem is to perform denoising on the degraded image. The objective function is defined as follows:-

$$b(\hat{f}, \hat{g}) = \min_{\hat{g} \geq 0} \|\hat{g} - g\|_2^2 + \sum_{j,l} (\langle g, \Psi_{j,i,k} \rangle)$$

$$\text{subject to } c(\hat{g}, g) = \xi - \|\hat{g} - g\|_2^2 \geq 0$$

(8)

where $\Psi_{j,i,k}$ denotes the curvelet transform and ξ is the noise power bound. The noise bound is the trace of the noise covariance. With underlying assumption that the noise is AWGN with variance σ^2, then the noise power bound can be approximated as

$$\xi = n\sigma_v^2$$

(9)

where n is the length of noise vector v. The curvelet coefficients of $\hat{g}$ are obtained for each $\langle g, \Psi_{j,i,k} \rangle$ using the wrapping-based curvelet transform as it is simpler to understand and fast. Details about the architecture of curvelet transform via wrapping can be found in [36]. The traditional Curvelet transform denoising method has side effects of producing pseudo-gibbs effects where it would appear slightly "scratches" and "ringing" in the reconstructed image. To reduce these side effects, an improved curvelet-based denoising method proposed by [37] is adopted in this work.

4   Experimental Results

In this section, image restoration results obtained by the proposed algorithm are presented for quality assessment. The quality assessments are focused on the effectiveness of proposed method, the efficiency of the algorithm is not reported in this paper. For this preliminary work, the canonical images of Lena and Barbara are chosen for experiments; these images are the classic amalgam of structural details and texture in natural images, and both have multiscale edge features as well in the form of outline of objects in the background that are ideal for deblurring and denoising visual quality assessment. For the experiments, these images are convolved with five different combinations of known blur kernel and different amounts of AWGN as tabulated in Table 1.

Table 1. Description of the observation parameters for the five experiments

Exp #	Blur Kernel, h	Noise Variance, δ^2
Exp 1	$h_{i,j} = (1 + i^2 + j^2)^{-1}$, for $i, j = -7,...,7$ (Barbara)	8
Exp 2	9×9 uniform kernel (Barbara)	0.308
Exp 3	$[1,4,6,4,1]^{\mathrm{T}}[1,4,6,4,1]/256$ (Barbara)	49
Exp 4	Gaussian PSF (25×25) with standard deviation 1.6 (Lena)	4
Exp 5	Gaussian PSF (35×35) with standard deviation 2.6 (Lena)	4

The combinations of observation parameter shown in Table 1 were previously used by [6, 14]. In [14], the authors compared their experiments in term of ISNR with some state-of-art method, namely the FTVd [3], ShearDec [4], CGMK [5], L0-Abs [6] and ForWaRD [12]. Due to the difficulty of access the codes of these methods, the ISNR results and visual results published in [14] are used as the benchmark for performance comparison. Although ISNR can measure the improvement of intensity between two images, it is well-known that it may fail to describe the visual perception quality of the image. The SSIM index proposed in [35] is one of the most commonly used measures for image visual quality assessment. Compared with ISNR, SSIM can better reflect the structure similarity between the target image and the reference image. Therefore, this quality measurement metric is included for assessment of restoration results of the proposed method.

Table 2 presents the comparison of ISNR results from the proposed method with the ISNR results of other state-of-the-art methods published by [14]. The results demonstrate that the proposed method outperforms the other methods in three of the test cases. It achieved the highest ISNR in test case Exp 2, Exp 4 and Exp 5, and the second and third highest ISNR in Exp 1 and Exp 3, respectively. The proposed method is capable of acquiring a high SSIM value for each test case, which indicates the capability of the proposed method in reconstructing the luminance, contrast and structure details of degraded images.

Table 2. Performance comparison of the proposed method with previous state-of-the-art methods in ISNR, in dBs [14]. Best results are highlighted in **bold** font. Bracketed results are the SSIM value of the proposed method.

Methods	Exp 1	Exp 2	Exp 3	Exp 4	Exp 5
ForWaRD	1.87	4.02	0.94	3.74	2.34
FTVd	1.67	4.63	0.67	3.63	2.61
ShearDec	2.54	4.57	1.72	3.90	2.73
L0-AbS	1.68	3.81	0.78	4.11	2.87
CGMK	1.34	3.55	0.44	3.93	2.52
Method [14]	**3.97**	5.01	**2.27**	4.27	3.08
Proposed method	3.33	**5.08**	1.66	**4.56**	**3.26**
	(0.79)	(0.85)	(0.82)	(0.88)	(0.83)

Some visual results of the restored images obtained by the stipulated methods are presented in Figs. 1 and 2. For test case Exp 3, the ISNR result of the proposed method from Table 2 obviously shown that the proposed method is outperformed by Method [14] and ShearDec [4]. However, by visual observation and its quantitative measurement of SSIM of 0.82, it is can be noticed that the restored image in Fig. 1(h) preserves more textural features than other methods.

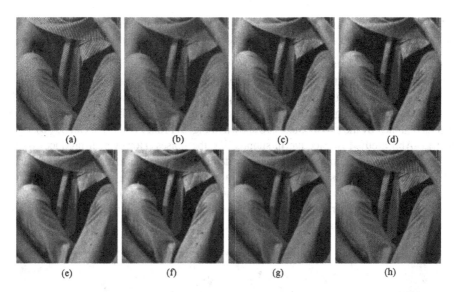

Fig. 1. Visual quality comparison of restoration results from test case Exp 3. (a) original image; (b) blurred and noisy image; (c) FTVd result, ISNR = 0.67; (d) ShearDec result, ISNR = 1.72; (e) L0-AbS result, ISNR = 0.78; (f) CGMK result, ISNR = 0.44; (g) Method of [14], ISNR = 2.27; (h) the proposed method, ISNR 1.66, SSIM 0.82

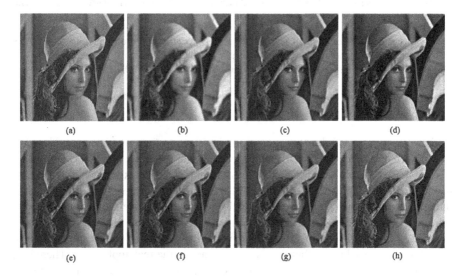

Fig. 2. Visual quality comparison of restoration results from test case Exp 5. (a) original image; (b) blurred and noisy image; (c) ForWaRD, ISNR = 2.34; (d) FTVd result, ISNR = 2.61; (e) ShearDec result, ISNR = 2.73; (e) L0-AbS result, ISNR = 2.87; (g) Method of [14], ISNR = 3.08; (h) the proposed method, ISNR 3.26, SSIM 0.83

An additional set of experiments is conducted to test the robustness of proposed algorithm in handling noise and blur in decoupling manners using BLP. In this experiment, the original image of Lena is convolved with gaussian PSF with spatial 2nd order moment $\sigma_b^2 = \sigma_x^2 + \sigma_y^2$ (with $\sigma_x = \sigma_y = 2.6$) and noise variance of 225. The degraded image is experimented and compared to the L0-Abs Method [6] which Matlab® code is available for download[1]. The computed ISNR results show that the L0-AbS method yields a value of 5.39 which is better than 4.77 obtained by the proposed method. However, in term of SSIM, the proposed shows a good result of 0.76 as compared to 0.46 of L0-AbS method. The results obtained by the two methods are shown in Fig. 3. From Fig. 3(c), it can be observed that Lo-AbS method is effective in suppressing the noises. However, it produces over-smoothed results and eliminates much image details, which results in a poorer SSIM value. The proposed method is very competitive in recovering the image structures. Unfortunately, it cannot escape from generating some annoying artifacts in the smooth regions. Nevertheless, by comparing with these two methods, the proposed method provides more visually pleasant results as shown in Fig. 3(d).

Fig. 3. Visual quality comparison of restoration between the proposed method and L0-AbS method in suppressing noise. (a) original image; (b) blurred and noisy image; (c) L0-Abs, ISNR = 5.39, SSIM 0.46; (d) the proposed method, ISNR 4.77, SSIM 0.76

5   Conclusions

In this work, the use of bilevel programming in image restoration is being explored to find a new image restoration algorithm that executes an optimal tradeoff between sharpness and noise to warrant an acceptable restoration result. The experimental results are encouraging and show the effectiveness of the algorithm in restoring image details. Numerical measurements of the image quality reveal that the algorithm is comparable with some of the state-of-the-art methods and has the advantage for image contrast and preserving edge details. For this preliminary work, the blur kernel and the noise variance are assumed exactly known, but this is not the case in the real application. Therefore, future works include improvement in the lower-level problem, experiments on the effectiveness and reliability of the algorithm in restoring real images, specifically remotely sensed images, and also investigation on efficiency and robustness of the algorithm that maximizes image fidelity within the computational limit.

---

[1] http://www.csee.wvu.edu/~xinl/source.html .

References

1. Cannon, M.: Blind deconvolution of spatially invariant image blurs with phase. IEEE Trans. Acoust. Speech Signal Process. **24**(1), 58–63 (1976)
2. Ayers, G.R., Dainty, J.C.: Iterative blind deconvolution method and its applications. Opt. Lett. **13**(7), 547–549 (1988)
3. Pan, H., Blu, T.: An iterative linear expansion of thresholds for ℓ_1-based image restoration. IEEE Trans. Image Process. **22**(9), 289–3715 (2013)
4. Patel, V.M., Easley, G.R., Healy, D.M.: Shearlet-based deconvolution. IEEE Trans. Image Process. **18**(12), 2673–26858 (2009)
5. Chantas, G., Galatsanos, N.P., Molina, R., Katsaggelos, A.K.: Variational bayesian image restoration with a product of spatially weighted total variation image priors. IEEE Trans. Image Process. **19**(2), 351–362 (2010)
6. Portilla, J.: Image restoration through L0 analysis-based sparse optimization in tight frames. In Proceedings of 16th IEEE International Conference on Image Processing (ICIP) 2009, pp. 3909–3912 (2009)
7. Zhang, J., Zhao, D., Xiong, R., Ma, S., Gao, W.: Image restoration using joint statistical modeling in space-transform domain. IEEE Trans. Circuits Syst. Video Technol. **24**(6), 915–928 (2014)
8. Guerrero-Colon, J., Portilla, J.: Deblurring-by-denoising using spatially adaptive gaussian scale mixtures in overcomplete pyramids. In: Proceedings of IEEE International Conference on Image Processing (ICIP) 2006, pp. 625–628 (2006)
9. Jin, W.: Image restoration in neutron radiography using complex-wavelet denoising and lucy-richardson deconvolution. In: Proceedings of 8th IEEE Conference on Signal Processing (ICSP) 2006, pp. 16–20 (2006)
10. Daubechies, I., Teschke, G.: Variational image restoration by means of wavelets: simultaneous decomposition, deblurring and denoising. Appl. Comput. Harmonic Anal. **19**(1), 1–16 (2005)
11. Mahbubur Rahman, S.M., Omair Ahmad, M., Swamy, M.N.S.: Improved image restoration using wavelet-based denoising and fourier-based deconvolution. In: Proceedings of the 51st IEEE International Midwest Symposium on Circuits and Systems, pp. 249–252 (2008)
12. Neelamani, R., Choi, H., Baraniuk, R.: ForWaRD: fourier-wavelet regularized deconvolution for ill-conditioned systems. IEEE Trans. Signal Process. **52**(2), 418–433 (2004)
13. Wen, Y.W., Ching, W.K., Ng, M.K., Liu, H.: A hybrid algorithm for spatial and wavelet domain image restoration. In: Proceedings of SPIE - The International Society for Optical Engineering. pp. 59605 V–59605V–8 (2005)
14. Yang, H., Zhang, Z.B., Wu, D.Y., Huang, H.Y.: Image Deblurring using Empirical Wiener Filter in the Curvelet Domain and Joint Non-Local Means Filter in The Spatial Domain. Imaging Sci. J. **62**(3), 178–185 (2014)
15. Kim, J.R., Jo, J.B., Yang, H.-K.: A solution for bi-level network design problem through nash genetic algorithm. In: Szczuka, M.S., Howard, D., Ślęzak, D., Kim, H.-k., Kim, T.-h., Ko, I.-s., Lee, G., Sloot, P.M. (eds.) ICHIT 2006. LNCS (LNAI), vol. 4413, pp. 269–280. Springer, Heidelberg (2007)
16. Huo, Y., Chen, J., Chen, L.J.: Reasonable scale of integrated transportation network based on bilevel programming. J. Chongqing Jiaotong Univ. (natural science) **29**(5), 791–795 (2010)
17. Shee, D., Tang, T., Tzeng, G.: Modeling the supply-demand interaction in electronic commerce: a bi-level programming approach. J. Electron. Commer. Res. **1**(2), 79–93 (2000)

18. Huang, W.-Q., Li, P.-X., Zhang, R.-H.: The application of bi-level programming in airline revenue management. J. Comput. Eng. Appl. **7**, 188–192 (2006)
19. Bard, J.F., Plummer, J., Sourie, J.C.: A bilevel programming approach to determining tax credits for biofuel production. Eur. J. Oper. Res. **120**, 30–46 (2000)
20. Dong, H., Guarneri, P., Fadel, G.L.: Bi-level approach to vehicle component layout with shape morphing. J. Mech. Des. **133**, 1–8 (2011)
21. Samuel K.G.G., Tappen, M.: Learning optimized map estimates in continuously-valued MRF models. In: Proceedings of the IEEE Conference on Computer Vision and Pattern Recognition (CVPR) 2009, pp. 477–484 (2009)
22. Tappen, M.F.: Utilizing variational optimization to learn markov random fields. In: Proceedings of the IEEE Conference on Computer Vision and Pattern Recognition (CVPR) 2007, pp. 1–8 (2007)
23. Tappen, M.F., Liu, C., Adelson, E.H., Freeman, W.T.: Learning gaussian conditional random fields for low-level vision. In: Proceedings of the IEEE Conference on Computer Vision and Pattern Recognition (CVPR) 2007, pp. 1–8 (2007)
24. Fehrenbach, J., Nikolova, M., Steidl, G., Weiss, P.: Bilevel image denoising using gaussianity tests. In: Aujol, J.-F., Nikolova, M., Papadakis, N. (eds.) SSVM 2015. LNCS, vol. 9087, pp. 117–128. Springer, Heidelberg (2015)
25. Kunisch, K., Pock, T.: A bilevel optimization approach for parameter learning in variational models. SIAM J. Imaging Sci. **6**(2), 938–983 (2013)
26. Stackelberg, H.V.: Market Structure and Equilibrium. Springer, New York (1934)
27. Stackelberg, H.V., Peacock, A.: The theory of the market economy. Oxford University Press, London (1952)
28. Bracken, J., McGill, J.: Mathematical programs with optimization problems in the constraints. Oper. Res. **21**, 37–44 (1973)
29. Candler, W., Norton, R.: Multilevel programming. Technical report 20, World Bank Development Research Center, Washington D.C. (1977)
30. Kolstad, C.D.: A Review of Literature on Bi-Level Mathematical Programming. Technical report LA-10284-MS, Los Alomos National Laboratory, New Mexico (1985)
31. Colson, B., Marcotte, P., Savard, G.: Bilevel programming: a survey. 4OR Quart. J. Oper. Res. Springer-Verlag **3**(2), 87–107 (2005)
32. Hongli, G., Juntao, L., Hong, G.: A survey of bilevel programming model and algorithm. In: Proceedings of IEEE International Symposium on Computational Intelligence and Design, pp.199–203 (2011)
33. Candès, E.J., Donoho, D.L.: Curvelets - A Surprisingly Effective Nonadaptive Representation for Objects with Edges. Curves and Surfaces, pp. 105–120. Vanderbilt University Press, Nashville, Tennessee (2000)
34. Wong, S.M.J.: Modulation transfer function compensation through a modified wiener filter for spatial image quality improvement. ISI Book Comput. Simul. Mod. Sci. **5**, 177–182 (2010)
35. Wang, Z., Bovik, A.C., Sheikh, H.R., Simoncelli, E.P.: Image quality assessment: from error visibility to structural similarity. IEEE Trans. Image Process. **13**(4), 600–612 (2004)
36. Candès, E., Demanet, L., Donoho, D., Ying, L.: Fast discrete curvelet transforms. SIAM J. Multiscale Model. Simul. **5**(3), 861–899 (2006)
37. Jiang Tao, Z.X.: Research and application of image denoising method based on curvelet transform. Int. Arch. Photogrammetry Remote Sens. Spat. Inf. Sci. (ISPRS Archives) **XXXVIII**, 363–368 (2008)

Free-Form Tetrahedron Deformation

Ben Kenwright$^{(\boxtimes)}$

Edinburgh Napier University, Edinburgh, UK
b.kenwright@napier.ac.uk

Abstract. Deformation mechanics in combination with artistic control allows the creation of remarkably fluid and life-like 3-dimensional models. Slightly deforming and distorting a graphical mesh injects vibrant harmonious characteristics that would otherwise be lacking. Having said that, the deformation of high poly complex shapes is a challenging and important problem (e.g., a solution that is computationally fast, exploits parallel architecture, such as, the graphical processing unit, is controllable, and produces aesthetically pleasing results). We present a solution that addresses these problems by combining a tetrahedron interpolation method with an automated tetrahedronization partitioning algorithm. For this paper, we focus on 3-dimensional tetrahedron meshes, while our technique is applicable to both 3-dimensional (tetrahedron) and 2-dimensional (triangulated planar) meshes. With this in mind, we compare and review free-form deformation techniques over the past few years. We also show experimental results to demonstrate our algorithm's advantages and simplicity compared to other more esoteric approaches.

Keywords: Deformation · Convex hulls · 3d · Tetrahedrons · Free-form · Convex · Video games · Real-time computer generated · Interactive

1 Introduction

Deformation. The deformation of a structure enables the emphasis of fluid life-like features. An inspiring example is shown in Fig. 1, which illustrates a 'stone' sculpture that captures flesh-like characteristics through the deformation of two bodies touching. The free-form deformation (FFD) techniques reduce the manual work by embedding a complex shape in a relatively simple control lattice. The fundamental FFD methods adopt spline functions to achieve smooth interpolation results [1–3]. While high-quality results are achievable, the evaluation procedure requires the information of a relative large number of neighbouring control points or even the whole control mesh, which can be computationally expensive. A local solution based upon the barycentric interpolation is possibly the simplest and most widely adopted solution in many graphics-based tools. The barycentric weighting in 3-dimensions involves only four vertex values of a tetrahedron. Since each graphical vertex depends on only four local control points, this enables an extremely fast computation that is ideally suited to massively parallel architectures. Due to the method only depending on local points

© Springer International Publishing Switzerland 2015
G. Bebis et al. (Eds.): ISVC 2015, Part II, LNCS 9475, pp. 787–796, 2015.
DOI: 10.1007/978-3-319-27863-6_74

Fig. 1. Artistic Control - Inspiring example of deformation in this life-like marble statue by the 17th-century sculptor Gian Lorenzo Bernini (The Rape of Proserpina) [4]. While the statue is stone, it captures the living essence of flesh due to the ripples in the skin as the fingers slightly dig into the body.

(i.e., within a tetrahedron and not neighbouring points), the deformation may produce undesirable artifacts at the boundaries with sparse density. Our method reduces this by dynamically increasing the control mesh count resolution at targeted areas (i.e., adding/removing control points to cause the encapsulating mesh to be re-tetrahedronized and reduce the attenuation).

High Poly Graphical Meshes. The trend for realism and the advancement of computational power, means graphical models, have become more detailed and complex, e.g., it is not uncommon for real-time graphical meshes to have hundreds of thousands of vertices. The process of deforming the vertices within a mesh is the domain of soft-body mechanics. Soft-body mechanics in combination with traditional animation techniques greatly enhance the realism of a scene/virtual world. *Until recently, this process was simply too expensive to consider in real-time environments.* However, with the advancement of computational power, such as, the graphical processing unit (GPU) and massively parallel processing, soft-object animations are becoming a viable real-time solution [5,6]. The deformation of an object occurs by moving the vertices of the graphical mesh. Typically, due to the high poly count - groups of vertices are attached to control points that artists can move using either pre-recorded animations or parametric curves to attain the desired effect. A crucial factor is that the mesh object has to have a sufficient number of vertices/face. As if the polygon resolution is low, the deformations give rise to a degradation in silhouette edge aliasing.

Contribution. The key contributions of this paper are: (1) vector representation of the tetrahedron free-form deformation, (2) non-linear formation and distribution of tetrahedron regions for controlled deformation (i.e., modified on-the-fly), (3) automatic tetrahedronization of the encapsulating control mesh using the control points to target areas of fidelity as needed to reduce boundary artifacts.

2   Related Work

Control Mesh. The creation and control of deformable meshes is an important component in multiple research areas, such as, engineering material analysis and safety testing. However, we focus on interactive solutions for real-time systems, such as, art programs and virtual worlds, since it allows the user to visually experience a more life-like and engaging process. We review a number of important papers and concepts for solving deformation problems in different contexts (e.g., character animation and material analysis) in this section. Figure 2, presents a visual time-line of key publications in the area, starting with the initial free-form deformation work by [1]. Theoretically speaking, any deformation can be achieved by manually manipulating individual vertices, which is, however, impractical due to the complexity of meshes in practice. Hence the embedded shape is deformed via interpolation.

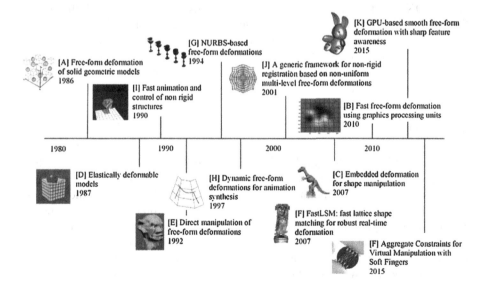

Fig. 2. Timeline - We show a brief visual overview of deformation techniques and their applications during the past few decades. [A] [1], [B] [6], [C] [7], [D] [8], [E] [9], [F] [10], [G] [11], [H] [12], [I] [13], [J] [14], [K] [5], [L] [15]

Free-Form Deformation (FFD). Many algorithms have been presented around the free-form deformation concept. The **free-form deformation technique, commonly known as FFD** has been around for some time, with it first documented in a SIGGRAPH paper by Sederburg and Parry in 1986 [1]. This FFD technique is used in numerous commercial modelling and animation packages, and forms the groundwork of our work. The technique is intuitive and straightforward to implement and forms the basis of a number of other techniques (e.g., hierarchy ffd [12]). A number of important deformation techniques have been published over the years based on the FFD concept, such as, lattice structure [1,16], surfaces [17], line feature [18], control curves [3,19], and points [9]. Additional work has focused on computational speed-ups [9,20] providing real-time solution. Then again, there are commercial solutions available within packages (e.g., for video games and editing tools), however, these deformation techniques are typically propriety owned and not shared publically. Crucially, the deformation animation solution should not consume all the system resources. For real-time applications, only a small amount of the overall computational resource is allocated to the animation, since the application needs to run a number of components, like graphics, artificial intelligence, networking, and game-play features.

Our Work. The tetrahedron free-form deformation method was presented by Kenwright and Lane [21] who solved a system of equations by inverting a 3x3 matrix. However, we use a compact vector formulation, as with advancements in parallel processing architecture, the calculation of operations, such as, the dot and cross product may be calculated in parallel to improve through-put (e.g., Single Instruction Multiple Data (SIMD) - MMX, AltiVec, 3DNow, SSE, and NEON just to name a few). Our method also builds upon Sederburg and Parry's [1] formulation, as in their original work, it does not explicitly emphasis the concept can be applied directly to 'unstructured' grids containing tetrahedral cells. We generate a tetrahedron lattice structure algorithmically on-the-fly using a Delauney algorithm, enabling us to add and remove control points dynamically without having to manually reconfigure the control mesh.

3   Method

Overview. We take the graphical mesh and add control points to areas that require deformation. Control points can be distributed non-linearly across the mesh with more control points being added to areas that require finer detail. Triangulate the control points into an array of tetrahedron elements that encapsulate the graphical mesh. The vertices of the graphical mesh are associated with the tetrahedron that encloses them. Each vertex is attached to the control points for the tetrahedron, and the initial deformation weights are calculated. As the control points are translated the associated graphical mesh is updated to reflect the deformation.

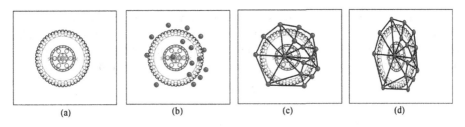

(a)   (b)   (c)   (d)

Fig. 3. Steps - Algorithm steps. (a) original graphical mesh, (b) randomly scatter control points ability to add more control points to areas that require finer detail, (c) triangulate the control points into tetrahedron volumes (using Delauney triangulation algorithm), and (d) calculate weights for graphical vertices and their associated tetrahedrons.

$$U = B - A$$
$$S = C - A \qquad (1)$$
$$T = D - A$$

Tetrahedron Free-Form Deformation (FFD). The free-form deformation concept is typically represented as a uniform grid. However, we are also able to apply the FFD principle to other geometric volumes, such as, tetrahedrons. Defining the four corners of the tetrahedron (ABCD) and the point within (X) as shown in Fig. 4. The interpolation function for a tetrahedron cell is linear, with the lattice space defined by Eq. 2 below:

$$X(s, t, u) = A + sS + tT + uU \qquad (2)$$

where A is the origin of the local coordinate system and S, T, and U lie along the edges of the FFD tetrahedron. Note that for any point interior to the lattice $0 < s < 1$, $0 < t < 1$, and $0 < u < 1$. We calculate s, t and u using Eq. 3 with A set as one of the corner points of the tetrahedron.

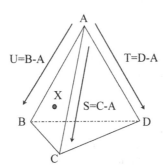

Fig. 4. Tetrahedron - A tetrahedron is a polyhedron composed of four triangular faces, three of which meet at each corner or vertex. It has six edges and four vertices and provides a simple non-ambiguous representation of a geometric region.

The natural coordinates for s, t and u are taken to vary from 0 to 1 in the non-dimensional cell. With reference to Fig. 4, Eq. 3 calculates the weight coordinates. Our vector formulation also correlates with the work of Kenwright and Lane [21] who solved a system of equations by inverting a 3x3 matrix.

$$s = \frac{T \times U(X - A)}{T \times U \cdot S},$$

$$t = \frac{S \times U(X - A)}{S \times U \cdot T},$$

$$u = \frac{S \times T(X - A)}{S \times T \cdot U},$$

$$X = A + sS + tT + uU$$

(3)

where s, t, and u are the scalar weights for the point within the tetrahedron. The scalar weights are calculated initially at the start (i.e., before any deformation takes place). As the control points (corner vertices of the tetrahedron) move, the morphed vertex X is recalculated.

Triangulation (Tetrahedron-ization). The tetrahedron possesses a number of interesting and unique properties that make it ideally suited to geometric topology (and is sometimes called a 'triangular pyramid'). While the tetrahedron is one of the five platonic solids (e.g., cube and dodecahedron), the tetrahedron is the only platonic solid with 'no' parallel faces. The tetrahedron is also one of the simplest 3D shapes for representing a volume (i.e., three triangles - any fewer and we have a plane). We set the tetrahedron corners as the control points for the lattice structure. The lattice structure engulfs the graphical mesh and is made up of tetrahedron elements. We construct the tetrahedron elements using a Delaunay triangulation algorithm [22]. The steps for the generation the tetrahedron structure from the control points is given below in Algorithm 1.

Algorithm 1. Delaunay tetrahedronize algorithm.

Data: In: array control points
Result: Out: array tetrahdrons

1 initialization;
2 array of control points;
3 construct super tetrahedron that encapsulates all the control points and add it to the final list;
4 **for** *each control points i* **do**
5   if a control point i is inside a tetrahedrons circumcenter remove it from the list and store it;
6   find all the non shared faces for the removed tetrahedrons;
7   control point i and the non shared faces create new tetrahedrons and add them back to the tetrahedron list ;
8 **end**
9 remove any tetrahedrons that contain references to the super tetrahedron from the final list;
10 return final list of tetrahedrons ;

Collision Detection. Tetrahedron elements form the foundation of a number of high-performance collision detection algorithms [23,24]. The collision detection algorithm is able to provide contact information (e.g., contact position, penetration depth, and separating normal). The contact information enables us to add additional control points to sparse regions to aid the fidelity of the deformation.

4   Experimental Results

Our approach does not require any off-line processing or complex solvers. The method is well suited to massively parallel architecture, such as, the graphical processing unit (GPU). The simulations with all the test models were implemented on a desktop machine with 3.2 GHz Intel i7 CPU and NVIDIA GeForce GTX 480 GPU. The number of control points allowed greater fine detailed influence over the deformation. Additionally, reducing the regional influence (i.e., from global to local) affects the computational cost and visual appearance.

Listing 1.1. Example implementation to show to calculate the three tetrahedron weights, given the tetrahedron corners and the vertex position (within) the tetrahedron volume.

```
1   void CalculateTetrahedronWeight(
        const vector<Vector3>& tetrahedronCorners,
        const Vector3& point,
        float& s, float& t, float& u)
    {
6   DBG_ASSERT(tetrahedronCorners.size()==4);

        const Vector3& A = tetrahedronCorners[0];
        const Vector3& B = tetrahedronCorners[1];
        const Vector3& C = tetrahedronCorners[2];
11      const Vector3& D = tetrahedronCorners[3];

        const Vector3& X = point;

        Vector3 S = B-A;
16      Vector3 T = C-A;
        Vector3 U = D-A;

        s = Dot( Cross(T, U), X-A ) / Dot( Cross( T, U ), S );
        t = Dot( Cross(S, U), X-A ) / Dot( Cross( S, U ), T );
21      u = Dot( Cross(S, T), X-A ) / Dot( Cross( S, T ), U );

        DBG_ASSERT(s>=0 && s<=1.0f);
        DBG_ASSERT(t>=0 && t<=1.0f);
        DBG_ASSERT(u>=0 && u<=1.0f);
26  }// CalculateTetrahedronWeight(..)
```

Scalability. We endeavoured to automate trivial parts of the deformation process rather than depending on artist intervention for modelling the underlying low-poly control mesh. This enables us to adapt the detail of the transformation for different target audiences (i.e., reduce the model complexity to more coarser representations for environments with limited resources, such as, memory and processing power).

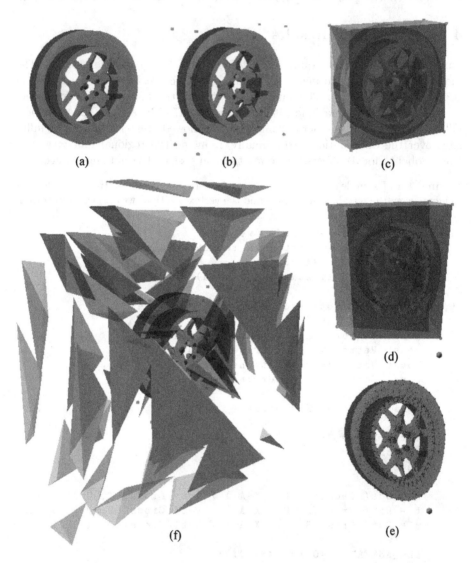

Fig. 5. Simulation - Graphical mesh 12270 vertices, 23 control points, 27 tetrahedrons. (a) original mesh, (b) placement of control points, (c) tetrahedronize control points, (d) assign graphical vertices to tetrahedrons (coupled deformation), (e) graphical mesh is deformed based on the tetrahedron/control points, and (f) exploded view of the generated tetrahedron structure.

Limitations. The method only considers the single tetrahedron case based on a weighted barycentric coordinate interpolation. This can cause discontinuities and artifacts at boundaries. This problem can be addressed by encapsulating multiple levels (i.e., hierarchy of tetrahedrons and the weighted average of neighbouring cells). This problem was also addressed by Huang [2] by formulating modified global representation.

5   Conclusion/Discussion

In conclusion, we have presented a straightforward non-linear method for creating soft-body deformations. The solution is able to create a non-linear free-form deformation distribution (i.e., compared to the common grid-distribution). The control points can be distributed across the shape surface to focus additional level of detail in areas that require it. Additional work would be the automation of the control point placement. For example, automatically creating a reduced poly convex hull and placing control points based on the distribution of graphical vertices (i.e., concentrate greater fidelity in areas that allow it).

Acknowledgements. A special thank you to reviewers for taking time to out of their busy schedules to provide insightful comments and suggestions to help to improve the quality of this paper.

References

1. Sederberg, T.W., Parry, S.R.: Free-form deformation of solid geometric models. In: ACM SIGGRAPH Computer Graphics. vol. 20, pp. 151–160. ACM (1986)
2. Huang, J., Chen, L., Liu, X., Bao, H.: Efficient mesh deformation using tetrahedron control mesh. In: Proceedings of the 2008 ACM Symposium on Solid and Physical Modeling, pp. 241–247. ACM (2008)
3. Kho, Y., Garland, M.: Sketching mesh deformations. In: ACM Siggraph 2007 Courses vol. 41, ACM (2007)
4. Avery, C.: Bernini: genius of the Baroque. Thames & Hudson, London (2006)
5. Cui, Y., Feng, J.: Gpu-based smooth free-form deformation with sharp feature awareness. Comput. Aided Geom. Des. **35**, 69–81 (2015)
6. Modat, M., Ridgway, G.R., Taylor, Z.A., Lehmann, M., Barnes, J., Hawkes, D.J., Fox, N.C., Ourselin, S.: Fast free-form deformation using graphics processing units. Comput. Methods Programs Biomed. **98**, 278–284 (2010)
7. Sumner, R.W., Schmid, J., Pauly, M.: Embedded deformation for shape manipulation. ACM Trans. Graph. (TOG) **26**(80), 1–57 (2007)
8. Terzopoulos, D., Platt, J., Barr, A., Fleischer, K.: Elastically deformable models. In: ACM Siggraph Computer Graphics. vol. 21, pp. 205–214. ACM (1987)
9. Hsu, W.M., Hughes, J.F., Kaufman, H.: Direct manipulation of free-form deformations. In: ACM Siggraph Computer Graphics. vol. 26, pp. 177–184. ACM (1992)
10. Rivers, A.R., James, D.L.: Fastlsm: fast lattice shape matching for robust real-time deformation. ACM Trans. Graph. (TOG) **26**(82), 82:1–82:6 (2007)
11. Lamousin, H.J., Waggenspack Jr, W.N.: Nurbs-based free-form deformations. IEEE Comput. Graph. Appl. **14**, 59–65 (1994)

12. Faloutsos, P., Van De Panne, M., Terzopoulos, D.: Dynamic free-form deformations for animation synthesis. IEEE Trans. Vis. Comput. Graph. **3**, 201–214 (1997)
13. Witkin, A., Welch, W.: Fast animation and control of nonrigid structures. In: Proceedings of the 17th Annual Conference on Computer Graphics and Interactive Techniques, pp. 243–252. ACM (1990)
14. Schnabel, J.A., Rueckert, D., Quist, M., Blackall, J.M., Castellano-Smith, A.D., Hartkens, T., Penney, G.P., Hall, W.A., Liu, H., Truwit, C.L., Gerritsen, F.A., Hill, D.L.G., Hawkes, D.J.: A generic framework for non-rigid registration based on non-uniform multi-level free-form deformations. In: Niessen, W.J., Viergever, M.A. (eds.) MICCAI 2001. LNCS, vol. 2208, pp. 573–581. Springer, Heidelberg (2001)
15. Talvas, A., Marchal, M., Duriez, C., Otaduy, M., et al.: Aggregate constraints for virtual manipulation with soft fingers. IEEE Trans. Vis. Comput. Graph. **21**, 452–461 (2015)
16. MacCracken, R., Joy, K.I.: Free-form deformations with lattices of arbitrary topology. In: Proceedings of the 23rd Annual Conference on Computer Graphics and Interactive Techniques, pp. 181–188. ACM (1996)
17. Feng, J., Ma, L., Peng, Q.: A new free-form deformation through the control of parametric surfaces. Comput. Graph. **20**, 531–539 (1996)
18. Beier, T., Neely, S.: Feature-based image metamorphosis. In: ACM SIGGRAPH Computer Graphics. vol. 26, pp. 35–42. ACM (1992)
19. Barr, A.H.: Global and local deformations of solid primitives. In: ACM Siggraph Computer Graphics. vol. 18, pp. 21–30. ACM (1984)
20. Montagnat, J., Delingette, H., Ayache, N.: A review of deformable surfaces: topology, geometry and deformation. Image Vis. Comput. **19**, 1023–1040 (2001)
21. Kenwright, D.N., Lane, D.A.: Optimization of time-dependent particle tracing using tetrahedral decomposition. In: Proceedings of the 6th Conference on Visualization 1995, pp. 321. IEEE Computer Society (1995)
22. George, P.L., Borouchaki, H.: Delaunay triangulation and meshing. Hermes, Paris (1998)
23. Bergen, G.V.D.: A fast and robust gjk implementation for collision detection of convex objects. J. Graph. Tools **4**, 7–25 (1999)
24. Jiménez, P., Thomas, F., Torras, C.: 3d collision detection: a survey. Comput. Graph. **25**, 269–285 (2001)

Innovative Virtual Reality Application for Road Safety Education of Children in Urban Areas

Taha Ridene[1(⊠)], Laure Leroy[2], and Safwan Chendeb[2]

[1] U2IS, Ensta ParisTech, Palaiseau, France
taha.ridene@ensta-paristech.fr
[2] Paris 8 University, Saint-Denis, France
{laure.leroy,safwan.chendeb}@citu.fr

Abstract. In order to make children develop good safety habits on the streets, it is very important to educate them on this subject at an early age. The technological advancements allow the creation of applications for training assistance and support. Virtual Reality and Augmented Reality are some of the best suitable scientific domains for successful training applications. In this paper, we present an innovative application for child risk prevention and education in urban are as; this one is based on a collaborative Research & Technologies platform which refers to the dynamic simulation of a city containing artificial intelligence and behavior modeling for pedestrians, crowds, vehicles and traffic in 3D visual and audio environment. We propose an interactive scenario for child risk education and prevention. We experiment it in an autonomous city (Paris) represented in a virtual environment and including artificial intelligence. This scenario takes into account the social implication and the relation between real and virtual actors.

1 Introduction

Since 2008, half of the world's population has been concentrated in urban areas and no reverse trend is in sight. 30 cities may at present claim a population well in excess of 10 million inhabitants. The populations of the cities of Mumbai (formerly named Bombay), Seoul, Mexico City and New York are estimated at more than 20 million Seoul, with Tokyo leading the way with a staggering count of over 37 million residents [1].

One of the major challenges facing megacities has its roots in the safety of pedestrians. For example, in France and in 1995 alone, 5819 children aged fewer than fifteen fell victims to urban accidents, resulting in a fatal outcome in 91 instances. A number of elaborate studies have been conducted to find ways of coping with such safety hazards and prevent them. We can for instance highlight works conducted in *ADAS* (Advanced Driver Assistance Systems) areas. But reactive systems are not sufficient on their own to eradicate those issues and emphasis should proactively be placed on prevention instead. Under this assumption, we propose, within the framework of the *TerraDynamica* project, to apply Virtual Reality techniques to prevention of urban risks involving particularly children.

© Springer International Publishing Switzerland 2015
G. Bebis et al. (Eds.): ISVC 2015, Part II, LNCS 9475, pp. 797–808, 2015.
DOI: 10.1007/978-3-319-27863-6_75

The main contributions to this paper can be identified as follows: (1) firstly, we explain in what way we think our system should be viewed as innovative compared to similar existing platforms; (2) secondly, we present an original scenario of risk prevention based on a modular collaborative architecture, using an effective way of promoting the transfer, into a virtual reality environment, of the knowledge acquired by the child; (3) lastly, we show how to elaborate a testing protocol in a real use case applied to this risk prevention scheme.

This paper is organized as follows. Section 2 elaborates on the virtual reality learning concepts; Sect. 3 presents the strengths of our Platform; Sect. 4 reveals the particulars of our scenario; Sect. 5 presents the overall architecture; Sect. 6 presents our testing protocol and a practical use case. We finally put forward our conclusions and prospects for future developments.

2   Context and Related Studies

2.1   Knowledge Transfer in a Virtual Real World

One submission is that immersion in a realistic environment fosters the learning process [2]. A number of studies have attempted to assess the effectiveness of immersion learning, as opposed to learning in a 2D environment. They have established that a combination of various sensorial methods facilitates the comprehension of a problem [3]. Other studies state that physical immersion encourages learning (and should probably even be looked upon as a prerequisite) [4,5]. The hypothesis that an immersive environment is closer to reality than a nonimmersive one and as such permits a easier knowledge transfer [6] is also advocated by Adams [7], even though some studies seem to indicate that an empirically felt improvement should probably not be taken for granted [8–10]. He also states that the motivation of the learning subject was improved. But it also seems that interactions can form a potential source of distraction to learners in the early stages, and all the more so if the subjects possess a limited intuitive handling capacity [9].

2.2   Road Risk Awareness for Children

There is a road risk awareness campaign in France called "Permis Piéton" (pedestrian license)[1] which aims is to foster the children's alertness to dangers on the streets. This is taught in French classes, and a number of little games are available on internet[2]. None of those games, however, belongs to the immersive category. Meir and Simpson showed that children are less aware of the hazard of crossing a street due to their low level of experience [11,12]. We think as Charron that training can increase their capacity to avoid risk [13], as Schwebel, McComas and Thomson showed it [14–17].

---

[1] "Permis Piéton" (pedestrian License) campaign: http://www.permispieton.com

[2] "Permis Piéton" (pedestrian License) game: http://www.permispieton.com/jeu4.php

2.3   Awareness in Virtual Reality

Before identifying ways of making children alert to road traffic hazards in virtual reality, we have to consider whether and to what extent what they have "learned" and felt in the real world can be replicated in a virtual one. Obviously, if we accept that strong emotions such as exposure to danger play a significant part in the learning process, these elements should also be taken into account in the virtual reality environment.

2.4   Experience of Fear

A study conducted by Maano (Virtual Reality Centre for the Mediterranean) stated that mentally sound subjects (i.e. not predisposed to phobias) will walk through a virtual corridor much faster when it is ablaze (this is simulated by virtual flames and smoke) than when no danger is perceived. This would tend to suggest that a virtual situation comprising a potential danger can also be a source of real stress [18]. It follows that, since a subject retains a lesson better under the influence of acute danger, the same process should very likely take place in comparable virtual circumstances.

2.5   Emotive Avatar Influence

Imitative learning is also achievable in Virtual Reality. Lets suppose for instance that we have an avatar suggesting behaviors to be mimicked by students. There exists a relationship between the manner in which an avatar is identified or perceived and the behavior of the subject who looks at or interacts with it. For example, in a study of Fox, subjects are immerged in front of an avatar which consists of 3D photos of themselves eating. The avatar grows bigger when eating chocolate, loses weight when eating carrots or else remains the same size irrespective of what he eats. After immersion, the subject was left to his own devices in front of a bowl of carrot and another bowl of chocolate; then, we check what he favors to eat under the influence of the avatars behavior. Female subjects were found to consume less candy when avatars changed size while male ones did consume a lot of it [8]. This would tend to suggest that, although imitative learning works well in pair with emotional involvement, it is probably best to choose an avatar to which people can readily identify themselves [19].

2.6   Milgram's Experiment in Virtual Reality

The Milgram experiment analyses obedience patterns to authority figures. It consisted of requiring the subject to inflict increasingly severe electric shocks to another subject (electric shocks were actually fake and the receiving subject played by an actor) under control of a fake physician [18]. This experiment was replicated in virtual reality. This time, an avatar was the electrocuted subject. This avatar can be set to express pain or not. It was noted by Slater that the subjects reactions were fundamentally identical as those already observed by Milgram [20]. We can therefore assume that a similar line of behavior might be expected from the subjects in this instance.

2.7   Existing Platforms

Various lines of work were conducted in relation to the pedestrian presentation in Virtual Reality, and they are related to quite a number of applications that would fit in the preventive measures category. Agent-street research project 28 includes a city evacuation drill scenario in second life[3]. Legion[4] coupled to Aimsun[5] proposes a platform for pedestrian simulation, and it can be used by various tools to analyze and study pedestrian behavior. Urban Analytic Framework[6] comprises a module for Agent safety analysis. NOMAD and SimPed [7] are two tools proposed by the Transport & Planning Department of Delft University of Technology for a microscopic simulation of pedestrian comportment. The *CATT Lab*[8] works in combination with the I95 Corridor Coalition and Forterra Systems Inc. to create an intensive training program that uses three-dimensional, multi-player computer gaming simulation technology and distance-based learning technologies. The program will present typical incident situations and allow the participants to play out their normal roles in what is essentially a highly structured and recorded video game.

3   Proposed Architecture

3.1   General Principle

We produced a platform for developing and integrating virtual reality applications. Our application is very specific in that is cooperatively and modularly designed. Figure 1 shows the building blocks of this application.

3.2   Architecture Principle

Our platform consists in segregating the entire system into three main modules, and to produce, in respect of each module, an abstract interface which is accessible by the other parts of the entire system. This sharing scheme allows managing the system according to a well-defined authorization hierarchy, both in terms of visibility and accessibility.

3.3   Modules

Three main internal modules are defined as follows:

- The MAC module which manages the following actors: pedestrians, cars, demand-triggered traffic lights, obstacles...

---

[3] Second life: http://secondlife.com/

[4] Legion: http://legion.com

[5] Aimsun: http://www.aimsun.com/wp/

[6] Quadstone Paramics: http://www.pedestriansimulation.com/highlights_analytics.php

[7] www.pedestrians.tudelft.nl

[8] CATT Lab: http://www.cattlab.umd.edu/

- The Permits Module which manages the following actors: children, the protected mates, Non-voice Interactions, Stereoscopy Management.
- The ACA Module which manages Voice Command, the Virtual Assistant, and the business logic.

One external module interacts with the ACA module, which is the Davi[9] Platform. This platform includes a conversational processing and a voice recognition tool.

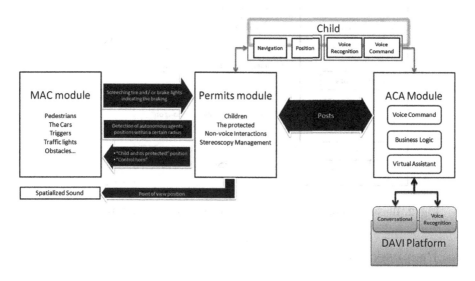

Fig. 1. Illustration of the layered model which is based on the whole platform architecture for producing application. Three main modules are presented: MAC module; permits module; and ACA module which interact with Davi platform.

3.4 Interaction Between Modules

- Mac ⟶ Permits module: (1) Screeching tire and/or brake lights suggesting braking; (2) The Permits app scans the positions of autonomous agents within a defined radius.
- Permits ⟶ Mac module: (1) Child and its protected mate position; (2) Sounding horn point of view position
- Permits ⟷ ACA module: Posts: the posts are detailed in Fig. 2.
- Child ⟷ Permits Module: Control and response
- ACA ⟷ Child: Control and response.
- ACA ⟷ Davi platform: voice command.

---

[9] Davi: http://www.davi.fr/

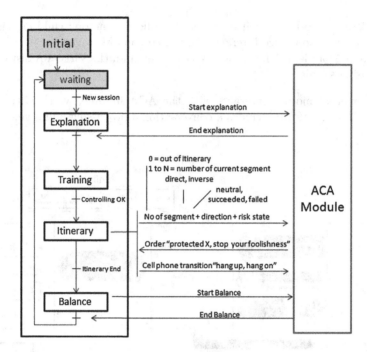

Fig. 2. Illustration of the posts between permits module and ACA module.

4 Proposed Scenario

The learner's mission consists of moving around the city (3D environment) while avoiding the pitfalls and dangers of the street. He acts as a guide for a group of friends and must ensure their safety as well as his own [21]. The path that children will follow runs across the Place de la République in Paris. This place is currently undergoing renovation works, but, the old Place has been replicated in the virtual environment (see Fig. 3 (left and middle)).

Fig. 3. Place de la République in Paris (left) Real Old (France). (middle) Virtual environment. (right) Mapping of path and risks locations. The black line corresponds to the path of the child, red lines to the location of the risk (Color figure online).

4.1   Characters

Protected Friends. Protected friends are small cartoon characters. They occasionally behave recklessly on the street, and they would not be above crossing the street without pausing to look for cars. As we could read in Sect. 2, subjects are often interacting with an avatar. We assume that children can be more focused on the safety of their little mates than of their own. They also give a sense of empowerment to the subject.

– They gravitate in an area facing the player;
– They are always visible to the player;
– They respond to commands from the player;
– They act differently depending on the actual coordinates on the path.

Autonomous Conversational Agent. The virtual teacher is managed by technologies conversational agent. She can:

– Guide the child on his path,
– Correct errors and redefine the rules of good conduct,
– Answer any questions that might be asked by the learner.

As explained in Sect. 2, a subject is more reactive to an emotion conveying avatar and accordingly we wish ours to appear as enticing and expressive as possible.

4.2   Autonomous City and Artificial Intelligence

The behavior of the different agents in the city is not predefined by a script. These are granted their own power of decision and follow motivations of their own. In this way and in contrast to many comparable applications, their behavior is not predetermined by a script. This means that all the moves and shifts of trajectory by the pedestrians and cars in our simulation are not subject to any form of active monitoring on our part. Just like in the real world, life in our virtual city does not fit in a preexisting mould and is therefore no less unpredictable, and that is definitely the way things should be.

Agents are completely autonomous in the decisions they chose to make, and the latter spring from motivations which are theirs and theirs alone. They are thus in a position to make up their own mind about what they should do next on the path to their destination [22–24] and to alter their trajectory at a very short notice to take incoming pedestrians or cars into account [24].

4.3   Proposed Risks

We will confront the child with risks commonly present on the streets. These risks are scattered around the "Place de la République". An enumeration of the potential risks is provided for in Fig. 3 (right). A list of risks is introduced in Fig. 4.

Risk 1	Risk 2	Risk 3	Risk 4	Risk 5	Risk 6	Risk 7	Risk 8	Risk 9

Fig. 4. Summary of presented risk. Risk 1, observing the red light at the zebra crossing. Risk 2, pedestrian crossing without traffic light. Risk 3, obstructing obstacle. Risk 4, driveway. Risk 5, do not walk on the sidewalk. Risk 6, crowd coming from opposite direction. Risk 7, non obstructing obstacle. Risk 8, Emergency vehicles. Risk 9, hidden vehicle.

4.4 Interaction

Shifts in Trajectory. Major shifts in trajectory (i.e. when the child is moving along the pavement) are produced by means of a joystick. Minor shifts are in contrast derived from the very physical movement of the child. As the latter is being tracked by ARTracking[10], his moves on the ground screen (3×4 m) are continuously translated into coordinates to that effect.

Call Teacher. The child has a fake phone which is tracked. When he puts it to his ear, the virtual teacher is summoned on the phone. The child speaks into a microphone, and the interaction needed is obtained by having its speech processed by voice recognition techniques.

Call Protected Friends. The child can call back protected friend through basic vocal commands spoken into his microphone.

5 Experimentation

5.1 The SAS

Our CAVE (see Fig. 5) is an immersive room with a wall screen and a floor screen. The subject is standing on the floor screen. The two screens are in INFITEC[11] stereoscopic vision, which allows depth perception. We have also an ARTracking device which lets us know where the head and hands of the subjects are. The floor screen is 3*4 m and the wall screen is 4*3 m. We might wish to add other tracking systems (Kinect[12] or other) without having to alter our entire software configuration. To that effect we need a generic system which permits the use of various types of hardware without having to reinvent the wheel every time. We might also wish to alter some of the cognitive interactions (gestures or other) without having to redesign the hardware or the application. In this regard, an

---

[10] Artracking: http://www.artracking.com/home/

[11] Infitec: http://www.infitec.net/index.php/en/

[12] Kinect: http://www.microsoft.com/enus/kinectforwindows/

Fig. 5. Le SAS: an immersive room with 2 screens: wall; floor.

interaction server will be perfectly adapted. We don't use an HMD like Simpson [12] because, when there is classroom, it is interesting that other students can see and comment what the subject do. We do not use a large screen, because we want to preserve the peripheral vision which is important for perception.

5.2 Experiment

Protocol. We have submitted our application to the public on the "Futur en Seine"[13] event. It has been tested by 40 children to ascertain its stability and robustness. We wanted to study the children response to the application, both in terms of subjective and overall perception.

The experiment has been done in the SAS (see Fig. 6) with interactions described in Sect. 4.4. We have also study the efficiency of our interactions in functional and cognitive point of view.

Details of the Experiment. 40 children have tested our application; they are between 8 and 15 years old and most of them have followed the "permis

Fig. 6. Children in front of Permis Pieton.

---

[13] Futur en Seine: http://www.futurenseine.fr/fens2013/en/

pieton" program. They were equipped with interfaces Immersion to get into the SAS (joysticks, glasses with tracking). How to manipulate the interface was explained to them orally. In addition, the experiment had began with an explanation of the conversational context and a small training session for the child to understand how to manipulate interfaces and recall its "protected". We let the child go through the scenario of the application before asking a few questions. The questions were relatively short, as they address a general public who may not want to take too much time.

- Do you think the behavior of pedestrians is consistent?
- Do you think the behavior of the car is consistent?
- Do you think the city is realistic?
- Would you repeat that kind of simulation?
- What do you prefer in this kind of simulation?
- What would you change?

We have also asked some question to the parent and teacher:

- Do you think this kind of incentive is interesting?
- Do you think the behavior of pedestrians is consistent?
- Do you think the behavior of cars is consistent?
- Do you think the city is realistic?

5.3   Results

Feedback of Children. The children had no difficulty navigating in the virtual city. They found the behavior of pedestrians very consistent, some of them (8 %) even tried to talk to them, others were afraid of the virtual crowd who advanced on them, these behaviors indicate a good acceptance of pedestrians.

They also found the behavior of the car consistent; some of them playfully crossed the streets gradually to "annoy drivers because they can not do it on real streets". The kids were impressed with the car which came out of the garage. They also believed the city was very realistic, some children living very close to the "Place de la République" told us where they lived, many found their way with shops they know in the real place. In general, the perception of "pedestrian license" was very enthusiastic, a child did not want to leave the SAS (the parents had to drag him outside) during passages class, many children wanted to interact with the "protected". It seems that many of them related to "Foxy" (one of protected who made many mistakes) and yelled at him to come back when it does not respect the rules of the road or when he was a bit slow to catch up with them. However, they felt that the lights were not visible enough in the immersive room to know when they could go through or not.

Feedback of Adults. The return of the adults was also very positive. They thought that the idea was very interesting and many have asked if schools came to test the device. They thought the behavior of pedestrians and cars coherent, although "too wise for Parisians". The city was found to be very realistic; many have recognized the Republic Square without being told about it.

6   Conclusion and Outgoing

We presented in this paper an innovative virtual reality platform used for traffic accident prevention applications. In this term, we designed a detailed specific scenario for child risk prevention in urban areas. This scenario is tested in virtual reality environment and was shown to the public testing in Futur en Seine event.

We have seen that it is very well perceived by children and adults and that it can interest kids. But we also wish to study the perception, by the children, of the speed of our virtual cars. We are fully aware that speed assessment taking place in the real world cannot be a perfect match to its proposed substitute in the virtual one, and therefore, we find it desirable to try to put a figure on the time of reaction exhibited by children when responding to virtually moving objects.

The experiment will be done in the SAS with interactions described in Sect. 4.4. We will also study the efficiency of our interactions in functional and cognitive point of view. After that, we want to study the efficiency of this kind of virtual learning on the real behavior of the children in the streets.

Our platform is modular and extensible, which allows testing other kinds of scenarios. Those modules can be integrated as application part of VRSIX (Virtual Reality Server of Interaction eXtensible).

References

1. Lavaud, J.: Enfants victimes d'accidents de la circulation. Évolution **375**, 2885 (1995)
2. Burkhardt, J.M., Lourdeaux, D., Mellet-dHuart, D.: La réalité virtuelle pour lapprentissage humain. Le traité de la réalité virtuelle, vol. 4 (2006)
3. Dede, C., Salzman, M.C., Loftin, R.B.: Maxwellworld: learning complex scientific concepts via immersion in virtual reality. In: Proceedings of the 1996 International Conference on Learning Sciences, International Society of the Learning Sciences, pp. 22–29 (1996)
4. Malik, E., Martin, B., Pecci, I., Vivian, R.: Le retour de force: une aide à lapprentissage des mathématiques pour les enfants déficients visuels. In: Proceedings of JIM, pp. 126–135 (2001)
5. Winn, W.: The impact of three-dimensional immersive virtual environments on modern pedagogy. University of Washington, Human Interface Technology Laboratory, Seattle, WA (1997)
6. Stanney, K., Salvendy, G.: Aftereffects and sense of presence in virtual environments: formulation of a research and development agenda. Int. J. Hum. Comput. Interact. **10**, 135–187 (1998)
7. Adams, N., Lang, L.: Vr improves motorola training program-vr helps one company administer its advanced training course on operating robotic assembly lines to more sites. AI Expert **10**, 13–14 (1995)
8. Byrne, C.M.: Water on tap: the use of virtual reality as an educational tool. Ph.D. thesis, University of Washington, Washington DC (1996)
9. Gay, E., Greschler, D.: Is virtual reality a good teaching tool. Virtual Reality Spec. Rep. **1**, 51–59 (1994)

10. Merickel, M.L.: The relationship between perceived realism and the cognitive abilities of children. J. Res. Comput. Educ. **26**, 371–381 (1994)
11. Meir, A., Parmet, Y., Oron-Gilad, T.: Towards understanding child-pedestrians hazard perception abilities in a mixed reality dynamic environment. Transp. Res. Part F: Traffic Psychol. Behav. **20**, 90–107 (2013)
12. Simpson, G., Johnston, L., Richardson, M.: An investigation of road crossing in a virtual environment. Accid. Anal. Prev. **35**, 787–796 (2003)
13. Charron, C., Festoc, A., Guéguen, N.: Do child pedestrians deliberately take risks when they are in a hurry? an experimental study on a simulator. Transp. Res. Part F Traffic Psychol. Behav. **15**, 635–643 (2012)
14. Schwebel, D.C., McClure, L.A.: Using virtual reality to train children in safe street-crossing skills. Inj. Prev. **16**, e1–e1 (2010)
15. Schwebel, D.C., Gaines, J., Severson, J.: Validation of virtual reality as a tool to understand and prevent child pedestrian injury. Accid. Anal. Prev. **40**, 1394–1400 (2008)
16. McComas, J., MacKay, M., Pivik, J.: Effectiveness of virtual reality for teaching pedestrian safety. CyberPsychol. Behav. **5**, 185–190 (2002)
17. Thomson, J.A., Tolmie, A.K., Foot, H.C., Whelan, K.M., Sarvary, P., Morrison, S.: Influence of virtual reality training on the roadside crossing judgments of child pedestrians. J. Exp. Psychol. Appl. **11**, 175 (2005)
18. Maïano, C., Therme, P., Mestre, D.: Affective, anxiety and behavioral effects of an aversive stimulation during a simulated navigation task within a virtual environment: a pilot study. Comput. Hum. Behav. **27**, 169–175 (2011)
19. Hoorn, J.F., Konijn, E.A.: Perceiving and experiencing fictional characters: an integrative account1. Japan. Psychol. Res. **45**, 250–268 (2003)
20. Milgram, S.: Behavioral study of obedience. J. Abnorm. Soc. Psychol. **67**, 371 (1963)
21. Frasson, C., Mengelle, T., Aïmeur, E., Gouardères, G.: An actor-based architecture for intelligent tutoring systems. In: Lesgold, A.M., Frasson, C., Gauthier, G. (eds.) ITS 1996. LNCS, vol. 1086, pp. 57–65. Springer, Heidelberg (1996)
22. Bourgois, L., Auberlet, J.M.: Pedestrian agent based model suited to heterogeneous interactions overseen by perception. In: Weidmann, U., Kirsche, U., Schreckenberg, M. (eds.) Pedestrian and Evacuation Dynamics 2012, pp. 847–859. Springer, Heidelberg (2014)
23. Bourgois, L., Saunier, J., Auberlet, J.M.: Towards contextual goal-oriented perception for pedestrian simulation. In: ICAART 2012: 4th International Conference on Agents and Artificial Intelligence, p. 6 (2012)
24. Campano, S., Sabouret, N., De Sevin, E., Corruble, V.: The resource approach to emotion. In: Proceedings of the 11th International Conference on Autonomous Agents and Multiagent Systems, International Foundation for Autonomous Agents and Multiagent Systems, vol. 3, pp. 1191–1192 (2012)

Vision-Based Vehicle Counting with High Accuracy for Highways with Perspective View

Mohammad Shokrolah Shirazi[✉] and Brendan Morris

University of Nevada, Las Vegas, USA
shirazi@unlv.nevada.edu, brendan.morris@unlv.edu

Abstract. Vehicle detection by motion is still a common method used in vision-based tracking systems due to vehicles' continuous motion on highways. However, counting accuracy is affected for highways with perspective view due to long-time merging (i.e. blob merging or occlusion) events. In this work, a new way of vehicle counting with high accuracy using two appearance-based classifiers is proposed to detect merging situations and handle vehicle counts. Experimental results on three Las Vegas highways with differing perspective views and congestion difficulties show improvement in counting and general applicability of the proposed method. Moreover, tracking and counting results of a highly cluttered highway indicates greater counting improvement (89 % to 94 %) for highly congested situations.

1 Introduction

Traffic-count data is a vital component in transportation since it provides useful information for highway management, traffic safety, and signal optimization. Although vehicle counts can be obtained manually, automatic counting with high accuracy is desirable for scalability, availability, and simplicity [1,2]. One method to obtain automatic counts is using computer vision-based systems.

Nowadays, computer vision-based traffic monitoring systems have become more popular with practical applications in comparison with traditional loop detectors since they provide a wide range of useful information like vehicle trajectory, classification and speed with minimal installation/maintenance disruption and cost. However, the accuracy of these vision systems requires tackling challenging issues like the low quality of videos, shadows, lighting variations and merging. The merging problem is the joining of several vehicles into one foreground blob due to their geometric closeness and 2D image projection. Merging is amplified when there is highly cluttered video (dense traffic) or when camera viewing angle does not provide good perspective leading to corrupted vehicle trajectory and vehicle count [3]. Three examples of different highways with perspective problem are shown in Fig. 1. There are two different scenarios for merging events:

1. Objects are initially separated but become occluded or connected during tracking.

© Springer International Publishing Switzerland 2015
G. Bebis et al. (Eds.): ISVC 2015, Part II, LNCS 9475, pp. 809–818, 2015.
DOI: 10.1007/978-3-319-27863-6_76

2. Objects are always viewed together or occluded at the appearance and exit of a scene. This is known as a combined event. This problem is shown in Fig. 2 for two vehicles.

Merging detection and handling is an active research area and many methods have been proposed to solve it, but they mostly consider the first scenario and they do not address vehicle counts [4–6]. Prominent early approaches designed an additional step for occlusion reasoning by examining the predicted overlap area between active contours that are obtained by Kalman filter tracking [5]. In [6], blob merging is detected by creating bipartite graph matching between moving objects (blobs) of the current and previous frames.

(a) I15/STARR   (b) I15/SLOAN   (c) 95/S RANCHO

Fig. 1. Three different highways with perspective problem

Another way to handle vehicle merging is to directly using more complicated tracking methods [7] than using background subtraction. Particle filters [7] have been used as a robust way of tracking objects represented by their probability density functions. However, they are not successful for the combined event [7] since particles needs to be initialized to sample each separated object. In addition, sampling multiple objects by particles limits the real-time applicability.

In this paper, a new way of counting vehicles is presented for highways to tackle the low counting accuracy due to merging events. The merging event is addressed by simple and general appearance-based classifiers to learn the complex nature of merging samples. Two appearance-based classifiers are trained to detect merging events and predict the number of merging vehicles. A vehicle tracking system is developed and benefits using classifiers to provide high vehicle counting accuracy. The evaluation is carried out on variety of highway cameras at different views to highlight the generality of the framework.

The rest of the paper is organized as follows. Different appearance-based feature extraction techniques are described in Sect. 2. The learning algorithm is presented in Sects. 3 and 4 describes the vehicle tracking system. Section 5 presents experimental results for different traffic congestion scenarios and finally, Sect. 6 concludes the paper.

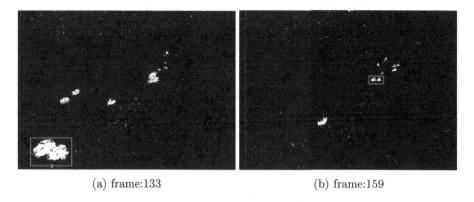

(a) frame:133   (b) frame:159

Fig. 2. An example of combined event (a) Two vehicles, detected as one, enter the scene, (b) Two vehicles travel together and exit the tracking area

2   Appearance-Based Feature Extraction

2.1   Wavelet Features

Wavelets are essentially the multi-resolution function approximation methods that provide a hierarchical decomposition of a signal or image. Wavelet features are used in our study since they encode edge information from multiple resolution levels. Wavelet decomposition coefficients are used in this study since they have been used for vehicle detection in [8]. The coefficients in the HH sub-band (high-pass rows and columns of an image) of the first level are discarded since they encode mostly the fine details and noises [8].

2.2   Gabor Features

The Gabor filter is a linear filter widely used for vehicle detection and classification [9]. The Gabor filter works like a local band pass filter and its frequency and orientation representation is similar to the human visual system. Given an input image, each sub-image include merging and single vehicles are scaled to 32×32 image pixels and then convolved with a Gabor filter banks (i.e. 2 scales, four orientations).

The magnitude of the Gabor filter responses are collected after convolution of each sub-image sample with Gabor filter banks. They are represented by two moments: the mean $\mu_{i,j}$ and the standard deviation $\sigma_{i,j}$ which shows jth block convolved with ith filter. This implies that only the statistical properties of pixels group are considered to create the feature vector. Since 8 filters are used for 9 blocks, $9 \times 8 \times 2$ dimensional feature vector is obtained after the filtering process and collecting μ and σ values of each block. 9 blocks with 50 % overlap is shown for 32×32 sub-image in Fig. 3a.

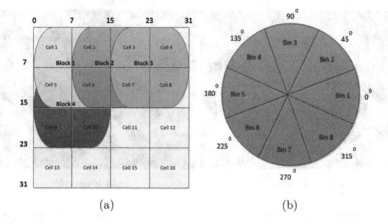

Fig. 3. (a) Dividing 32×32 sub-image into 16 cells and 9 overlapping blocks for Gabor, EOH and HOG, (b) 8 histogram bins used to quantize the gradient orientation of EOH

2.3 Edge Oriented Histogram (EOH)

Edge oriented histogram features [10] are used in this study since they have been widely used for pedestrian detection in intelligent transportation system applications. An EOH feature is obtained by computing the gradient orientation of each pixel in an image region. The orientations are then quantized into a predefined number of ranges shown in Fig. 3b and the histograms are computed. The detailed steps are as follows. First, the image sample is scaled to 32×32 pixels and smoothed with a Gaussian filter. Then horizontal and vertical gradients of pixel values are calculated and the gradient orientation θ is calculated for each pixel. Finally, the orientations are accumulated into eight histogram bins and eight histogram values of each block are concatenated to form a 9×8 dimensional feature vector.

2.4 Histogram of Oriented Gradients (HOG)

Object recognition using HOG features with SVM is quite popular for vehicles and pedestrians. This feature counts the occurrences of gradient orientation computed on a dense grid of uniformly spaced cells to characterize edge-like appearance [11,12]. The histogram of gradients for 9 blocks are calculated in this study similar to steps explained for EOH. However, the histogram of magnitudes is created. Since 9 bins are used, feature vector will have $3 \times 3 \times 9 \times 4$ elements.

3 Support Vector Machines

Two classifiers are trained by SVM in this study to detect a merging event and estimate number of vehicles that are involved in the event. SVMs are primarily binary classifiers with the supervised learning approach to learn linear or nonlinear decisions [13]. Given a set of points, which belong to either of two classes, SVM finds the hyper-plane leaving the largest possible fraction of the same class points on the same side, while maximizing the distance of each class from the hyper-plane. In this paper, we have used Radial Basis Kernel (RBF) since experiments showed better results for our test set.

4   Vehicle Tracking System

The structure of the proposed system is shown in Fig. 4. The vehicle detection and tracking framework is briefly presented [3] and incorporated classifiers for handling vehicle counts are described with more details.

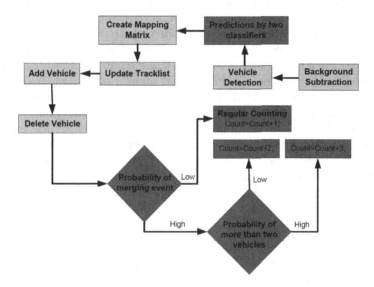

Fig. 4. Vehicle tracking system, count handling parts are shown with orange color (Color figure online)

4.1   Vehicle Detection and Tracking

Vehicle detection is performed using a standard adaptive background subtraction modeling technique. The image background is modeled using Gaussian Mixture Models [14] to address lighting changes and shadow. Moving objects (vehicles) are detected as pixels that do not fit any of the k background Gaussian models.

The dynamic model and appearance constraints are used to find a match and update a track. Vehicle dynamics are modeled using a constant velocity Kalman filter where the state matrix consists of the bounding box and velocity. The predicted location of a track and the centroid of detection must be within a small error to match the trajectory dynamic model. Moreover, the appearance of detection must match that of a track for association using shape measurements. Tracks are maintained using a greedy match algorithm that associates the detection that is closest both in dynamics and appearance [3]. When detection does not match the existing tracks in the track list, a new track is created. If an existing track does not find detection for 3 consecutive frames, it is marked for deletion.

4.2 Merging Detection and Vehicle Count Estimation

Detected vehicles after background subtraction are resized to 32×32 pixels and features are extracted and given to the SVM classifiers for prediction. The first classifier predicts the merging event and second one determines if there are two or more vehicles involve in the merging event. Positive and negative predictions of each track are kept during tracking for final reasoning of vehicle counts.

The probability of the merging event for each track is estimated by $P_1 = \frac{o}{t}$ where o is the number of positive predictions of merging event by the first classifier and t is a track lifetime based on frame numbers. Another probability is computed by $P_2 = \frac{k}{o}$ where k indicates the number of positive predictions for more than two vehicles involved in merging events. This is computed when the computed probability value, P_1 is higher than the threshold value δ. When a track is deleted, tracking system usually counts it by 1. The vehicle count data increases by 2 if the probability of merging event, P_1 is high but the probability of more than two vehicles P_2 is low. It increases by 3 if both calculated probabilities are high. This helps to reduce the effect of false positives (i.e. wrong predictions) at each frame by accumulating all predictions of a track.

5   Experimental Results

Experimental results include two steps. Firstly, different appearance based feature extraction techniques were implemented and compared to incorporate the highest performance classifiers in the tracking system. Finally, the full vision-based vehicle tracking and counting system was implemented by C++ along with OpenCV 2.3. The SVM classifiers were trained using the open-source LIBSVM [13] package.

5.1   Classifiers

1707 and 2595 sub-image samples of merging and single vehicles were collected from 11 highway videos in Las Vegas [15] to cover a wide variety of merging cases with different vehicles' direction and scaling (listed in Table 1). This helps to produce the general classifiers to work on different highways with different view and calibration settings. Figure 5 shows some sub-image samples used to train the vehicle merging detector classifier.

Training sub-image samples were resized to 32×32 pixels and HOG, EOH, Gabor and Wavelet features were extracted to train classifiers. 4 separated SVM

(a)   (b)

Fig. 5. Sub-image samples to train merging detector classifier: (a) Positive samples, (b) Negative samples

Table 1. Highway names and corresponding number of samples used for training merging detector classifier

# Camera [15]	Highway	# Positives	# Negatives
106	I-15/DESERT INN NORTH	516	673
109	I-15/TROPICANA NORTH	145	219
112	I15/SUNSET SOUTH	145	101
119	I15/SILVERADO N SOUTH	98	410
129	I-15/JEAN N MM16 SOUTH	145	362
131	I15/JEAN MM13 EAST	51	100
202	95/S RANCHO EAST	212	246
209	95/TORREY PINES WEST	30	60
408	I-515/WYOMING SOUTH	247	245
410	I-515/DESERT INN SOUTH	29	70
412	I-515/TROPICANA SOUTH	89	109

classifiers were trained for each feature set using RBF kernel. The ROC plot of merging detector classifier is shown in Fig. 6a using 75 % and 25 % of train and test dataset. HOG has the highest area under curve (AUC), indicating better performance in comparison with other feature extraction techniques.

The proposed system uses the vehicle count classifier to estimate vehicle count when high probability of merging happens during a track life-time. This helps to directly handle the combined event (second merging scenario) and correctly estimate the data count. 433 and 1274 positive and negative samples of two vehicles and more than two vehicles were arranged to train the vehicle count classifier (See Fig. 7). The ROC plot of the counting classifier (see Fig. 6b) shows the better performance of HOG feature.

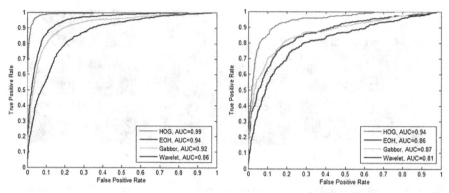

(a) ROC plot of merging detector classifier   (b) ROC plot for vehicle count classifier

Fig. 6. Comparison of appearance-based classifiers for vehicle merging handling

(a)   (b)

Fig. 7. Sub-image samples to train vehicle count classifier (a) Positive samples, (b) Negative samples

5.2   System Evaluation

The proposed system was evaluated for 3 different Las Vegas highways with differing perspective views and vehicles merging difficulty. New highway videos were selected with different congestion level which were not used in the training steps. Since no samples have been chosen from these videos to train SVM classifiers, the generality of the method is evaluated.

Highway videos were captured at 8 frames per second and they were evaluated for 1080 frames in total. Evaluation results of three different highways are shown in Fig. 8. Plots show vehicle counts for 9 intervals of 15 s (120 frames) showing the performance of the proposed system with high resolution under different traffic congestion.

Figure 8a shows large gap between pure tracker (Tracker without SVM classifiers) and the ground truth, indicating poor performance. However, the counting performance gets close to the ground truth by incorporating SVM classifiers. The evaluation of the second highway (Fig. 8b) shows the similar pattern in counting using tracker with SVM classifiers (Tracker Plus) and the ground truth.

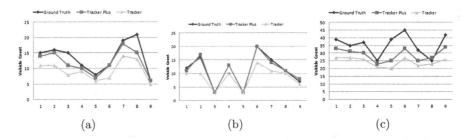

Fig. 8. Performance comparison of proposed system, manual counting and pure tracker with no vehicle count handling: (a) 128-I15/ Sloan S MM 21, (b) 132-I15/ JEAN S MM 9, (c) 102-I15/ Bonneville South

Since the third highway video is noisy and highly cluttered, the tracker results in poor counting accuracy. However, more accurate counting results are provided by incorporating SVM classifiers (Fig. 8c). Counting error is calculated as an absolute difference of ground truth (e.g., manual count) and automatic count [3]. Average of counting error (i.e. $E(ce)$) and average of vehicle counts (i.e. $E(vc)$) are shown in Table 2.

Table 2. Average errors and average vehicle counts using Tracker and Tracker + Classifiers

Highway	Manual count	E(vc)		E(ce)	
		Tracker	Tracker*	Tracker	Tracker*
I15/JEAN S MM 9	10.77	8.55	11.11	2.22	0.77
I15/Sloan S MM 21	13.55	9.33	11.88	4.22	1.66
I15/Bonneville	35.44	24.33	29	11.11	6.88

The average counting error for tracker plus (tracker+ SVM classifiers) is significantly lower and it is closer to zero (see Table 2). The counting accuracy improves from 89 % to 94 % for third highway, indicating greater improvement and the benefit of the proposed system for highly congested highways.

6   Concluding Remarks and Future Work

In this paper, appearance based techniques along with support vector machines are introduced to estimate vehicle counts of highways with perspective view. Two SVM classifiers are trained with positive and negative samples from real test sub-image samples in order to detect merging events including combined events and estimate number of involved vehicles. System evaluation showed the high efficiency of using appearance based classifiers to improve the traffic count; however, some other important features such as local binary pattern and integral channel features should be evaluated in future to provide the best choice for appearance-based classifiers.

References

1. Shirazi, M.S., Morris, B.: A typical video-based framework for counting, behavior and safety analysis at intersections. In: Intelligent Vehicles Symposium (IV), pp. 1264–1269. IEEE (2015)
2. Shirazi, M.S., Morris, B.: Observing behaviors at intersections: a review of recent studies and developments. In: Intelligent Vehicles Symposium (IV), pp. 1258–1263. IEEE (2015)
3. Shirazi, M.S., Morris, B.: Vision-based turning movement counting at intersections by cooperating zone and trajectory comparison modules. In: Proceedings of 17th International IEEE Conference on Intelligent Transportation Systems, Qingdao, China, pp. 3100–3105 (2014)
4. Bousetouane, F., Vandewiele, F., Motamed, C.: Occlusion management in distributed multi-object tracking for visual-surveillance. Pattern Recogn. Image Anal. **25**, 295–300 (2015)
5. Lipton, A.J., Fujiyoshi, H., Patil, R.S.: Moving target classification and tracking from real-time video. In: Proceedings of Fourth IEEE Workshop Applications of Computer Vision, Princeton, New Jersey, pp. 8–14 (1998)
6. Fang, W., Zhao, Y., Yuan, Y., Liu, K.: Real-time multiple vehicles tracking with occlusion handling. In: 2011 Sixth International Conference on Image and Graphics (ICIG), San Juan, Puerto Rico, pp. 667–672 (2011)
7. Koller-Meier, E.B., Ade, F.: Tracking multiple objects using the condensation algorithm. J. Robot. Auton. Syst. **34**, 93–105 (2001)
8. Sun, Z., Bebis, G., Miller, R.: Monocular precrash vehicle detection: features and classifiers. IEEE Trans. Image Process. **15**, 2019–2034 (2006)
9. Sun, Z., Bebis, G., Miller, R.: On road vehicle detection using gabor filters and support vector machines. In: Proceeding of the IEEE International Conference in Digital Signal Processing, pp. 1019–1022 (2002)
10. Teoh, S.S., Braunl, T.: Symmetry-based monocular vehicle detection system. J. Mach. Vis. Appl. **23**, 831–842 (2012)
11. Dalal, N., Triggs, B.: Histograms of oriented gradients for human detection. In: Proceeding of IEEE Conference on Computer Vision and Pattern Recognition (2005)
12. Shirazi, M.S., Morris, B.: Contextual combination of appearance and motion for intersection videos with vehicles and pedestrians. In: Bebis, G., Boyle, R., Parvin, B., Koracin, D., McMahan, R., Jerald, J., Zhang, H., Drucker, S.M., Kambhamettu, C., El Choubassi, M., Deng, Z., Carlson, M. (eds.) ISVC 2014, Part I. LNCS, vol. 8887, pp. 708–717. Springer, Heidelberg (2014)
13. Chang, C.C., Lin, C.J.: LIBSVM: a library for support vector machines. ACM Trans. Intell. Syst. Technol. **2**, 1–27 (2011)
14. Stauffer, C., Grimson, W.E.L.: Adaptive background mixture models for real-time tracking. In: Proceedings of IEEE Conference on Computer Vision and Pattern Recognition, pp. 246–252 (1999)
15. Rtc/fast live freeway traffic cams. (http://www.rtcsnv.com/fast-cam/)

Automatic Motion Classification for Advanced Driver Assistance Systems

Alok Desai, Dah-Jye Lee[✉], and Shreeya Mody

Department of Electrical and Computer Engineering,
Brigham Young University, Provo, UT 84602, USA
djilee@byu.edu

Abstract. Many computer vision applications need motion detection and analysis. In this research, a newly developed feature descriptor is used to find sparse motion vectors. Based on the resulting sparse motion field the camera motion is detected and analyzed. Statistical analysis is performed, based on polar representation of motion vectors. Direction of motion is classified, based on the statistical analysis results. The motion field further is used for depth analysis. This proposed method is evaluated with two video sequences under image deformation: illumination change, blurring and camera movement (i.e. viewpoint change). These video sequences are captured from a moving camera (moving/driving car) with moving objects.

1 Introduction

Motion detection and analysis is an important step for solving computer vision problems such as visual odometry [1], depth from motion [2], structure from motion [3], navigation, and many others. These are common problems in applications like video surveillance [4], robot navigation [5] and advanced driver assistance systems (ADAS) [6].

An ADAS is designed to include safety features to avoid accidents by altering the driver to potential problems, or taking over control of the vehicle to enhance or improve driving safety. These features may include automatic lighting, cruising, and braking control. They can also be used to alert driver to other cars or dangers, keep the driver in the correct lane. In this research, we develop a feature based motion classification and analysis algorithm for autonomous driving or driver assistance systems.

A vehicle may be driven in urban, suburban, or rural environments and with a variety of road types, speeds, daylight conditions, and seasons. Driving situation is quite unpredictable for ADAS applications especially when other moving vehicles and pedestrians are involved [7]. Motion analysis accuracy in different driving situations is critical for an ADAS.

Visual motion is a process of extracting spatial and temporal changes in an image sequence. It is assumed that the relative motion between the camera and the scene causes these changes in the image. Motion analysis can be categorized into four cases,

- Still camera, single moving object, and constant background
- Still camera, multiple moving objects, and constant background

© Springer International Publishing Switzerland 2015
G. Bebis et al. (Eds.): ISVC 2015, Part II, LNCS 9475, pp. 819–829, 2015.
DOI: 10.1007/978-3-319-27863-6_77

- Moving camera, relatively constant scene
- Moving camera, multiple moving objects.

Motion estimation methods can be grouped into two categories. One uses feature matching techniques to find corresponding features between two frames to obtain a sparse motion field. The other one uses differential techniques based on spatial and temporal variations of the image brightness to obtain a dense motion field. There is usually large camera or object movement or long baseline between frames for ADAS applications. Differential techniques are not suitable for ADAS because they only work well for very small movement (short baseline). In this research, we use a newly developed SYnthetic BAsis (SYBA) descriptor [8] for the most challenging case, a moving camera with multiple moving objects.

The feature matching process involves three steps: feature detection, feature description, and matching high quality feature points. Feature description and matching are challenging and time-consuming processes. Many feature descriptors are available, but most of them are very computationally complex, and not suitable for real-time applications. An efficient feature descriptor should avoid using many floating point computations, including square root, division, and exponential operations, to be able to operate at real-time speeds.

Descriptors like Scale Invariant Feature Transform (SIFT) [9] and Speeded-UP Robust Features (SURF) [10] are well-known, but they have limitations. SIFT is an orientation and magnitudes-of-intensity gradient-based feature descriptor. SIFT works well with many image deformations including rotation and scaling. Its major drawbacks are complexity, computation time, and storage requirements. For these reasons, it is not suitable for many resource-limited platforms and real-time applications. SURF computes descriptors using integral images and 2-D Haar Wavelet transform. Its main disadvantage is that it requires 256 bytes to encode 64 floating-point values. Ke and Sukthankar developed a descriptor and applied Principal Components Analysis (PCA) to a normalized image gradient patch [11]. PCA-SIFT performs better than the SIFT descriptor on artificially generated data. At the same time, it has the benefit of reducing high frequency noise in the descriptors. The drawback is that it is not tuned to obtain a sub-space that will be discriminative for matching [12]. In order to perform well, Low-dimensional Embedding (LDE) requires labeled training data, which are difficult to obtain.

The main objective of this work is to analyze the motion of the vehicle for ADAS. We use our newly developed SYBA descriptor for depth and movement estimation. Based on two frames, the camera motion is estimated for four different situations, (1) Pan left, (2) Pan right, (3) Sensor receding, and (4) Sensor approaching. Experiments are performed on images captured from car, with moving camera and moving objects to show the effectiveness of our algorithm.

This paper is organized as follows; we describe the SYBA descriptor algorithm and matching strategy in Sect. 2. Section 3 discusses the motion classification techniques. To validate our proposed algorithm we performed experiments on two sets of video sequences captured from a moving camera. The results are presented in Sect. 4 and followed by a summary and ideas for future work in Sect. 5.

2   SYBA Descriptor and Matching

Our SYBA descriptor is inspired by recent work in compressed sensing [13]. This work, which introduces and uses a new structure called a synthetic basis function, is perfectly applicable as a feature description. The basic structure and the use of synthetic basis functions can be easily explained using the popular Battleship game as an analogy. The synthetic basis functions (random yes and no patterns) are used as guesses in this game. The maximum number of different random patterns (guesses) that is required to locate all ships using synthetic basis functions [13] is expressed as.

$$M = \lceil K \ln N / K \rceil \tag{1}$$

where N is the number of squares on the game board $(n \times n)$ and K is the number of queried battleship locations. M is the number of random patterns required to locate all ships and is maximum when $K = N/2$. If K changes, then the number of random patterns that is required to locate all of the ships, may change as well. M random patterns are adequate to locate all possible locations of ships on patch size $n \times n$.

An innovative idea for feature description using synthetic basis functions to generate a set of intensity similarities called SYthetic BAsis (SYBA) descriptor is developed. It uses a number of synthetic basis images (SBIs) to compute the similarity between a small image region surrounding a detected feature point, called a feature region image (FRI), and the randomly generated synthetic basis images (SBIs). These similarity measures are used as a feature descriptor.

2.1   SYBA Descriptor Algorithm

The objective of the SYBA descriptor is to describe a small region surrounding a feature point in a unique way so that feature regions have the same descriptor elements and are matched together. The flow-diagram is illustrated in Fig. 1. This algorithm needs to be implemented in a target system.

First, the algorithm detects feature points from an image to create a feature list. Any feature detector can be used for this purpose. For each feature on the feature list, its feature region is cropped and saved as a 30×30 feature region image (FRI). Then the average intensity is calculated and used in thresholding to generate a binary FRI.

The algorithm calculates the average intensity (g) of an FRI as,

$$g = \frac{\sum_{x,y} I(x, y)}{p} \tag{2}$$

where p is the number of pixels in the FRI (900 in this case) and $I(x, y)$ is the intensity value at location (x, y). To generate a binary FRI, the average intensity g is used as a threshold value. If pixel intensity at the location (x, y) is brighter than g, the binary FRI is set to one. If pixel intensity at the location (x, y) is darker than g, the binary FRI is set to zero. A binary FRI keeps the spatial and the structural information. This information is used to describe the feature region.

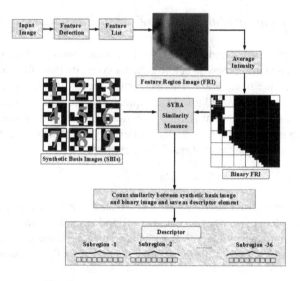

Fig. 1. The SYBA descriptor flow-diagram

The next step of the SYBA descriptor algorithm is to measure the similarity between the binary FRI and each of the 9 SBIs (according to Eq. (1)). This step is called SYBA similarity measure (SSM). The outcome of SSM represents a unique feature descriptor. Once the descriptor has been calculated for the features on the feature list, matching between two feature points is done by calculating the $L1$ norm of the two feature descriptors in different images. A Two-pass matching strategy is developed to find matching feature pairs as shown in Fig. 1 and explained in Sect. 2.2.

The SYBA descriptor size is easily adjusted to changes in SBI and FRI size. For the experiments, SYBA with a 5 × 5 size was implemented and named SYBA 5 × 5. According to (1), the maximum number of SBIs required for SYBA 5 × 5 is 9 if half of the pixels ($N = 25 \ and \ K = 13$) are black. Figure 2 shows a detailed example of the SSM calculation for SYBA 5 × 5. An example of 9 5 × 5 SBIs is shown in Fig. 2(a). The first step of the SSM calculation is to divide the 30 × 30 binary FRI into 36 equal-sized 5 × 5 subregions (Fig. 2(b). Then each subregion of the binary FRI is compared with each of the 9 SBIs (Fig. 2(a)), which are generated using (1). The last step of the SSM process is to count the number of times the FRI subregion and SBI both contain a black pixel at the same pixel location. The same process is carried out for feature matching in Sect. 4.

Our experiment proves that only 13 black pixels (according to (1)) in the SBI are needed for unique comparisons between a 5 × 5 subregion and a 5 × 5 SBI. Use the highlighted 5 × 5 subregion (shown in Fig. 2(b)), and the first SBI (see Fig. 2(a)) to explain how the pixel-by-pixel operation is performed between each subregion of FRI and SBIs. There are 7 black corresponding pixels at the same locations between the FRI and the first SBI. The comparison result is shown in Fig. 2(c). This number is used to describe the descriptor element. The same subregion compared with the second SBI has 7 black pixels at the same location (shown in Fig. 2 (d)).

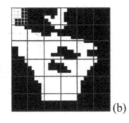

Fig. 2. (a) Nine 5×5 Synthetic basis images (SBIs), (b) 36 5×5 subregions of a feature region image (FRI), (c) Similarity count between the first SBI and the highlighted 5×5 subregion, and (d) Similarity count between the second SBI and the highlighted 5×5 subregion.

By experiment we confirmed that only 13 black pixels (according to Eq. (1)) in the SBI are required for unique comparisons. When subregions are compared with all 9 SBIs, each subregion will yield 9 numbers ranging from 0 to 13 which are used as a feature descriptor. Therefore, a 30×30 FRI with 36 5×5 subregions will require a feature descriptor size of 36 (sub-regions) $\times$ 9 (SBIs) $\times$ 4 bits (0 ~ 13) = 1,296 bits.

The number of operations required to calculate SYBA 5×5 is as follows. The SYBA 5×5 descriptor with 30×30 FRI size and 36 5×5 subregions required 324 (9×36) comparisons between SBIs and FRI. The number of summation operations required to calculate the SYBA descriptor is 324. The total number of operations including summation and comparison that are required to calculate the SYBA descriptor is 648.

2.2 Matching Features

SYBA descriptor is used to find the best feature point matching pairs in two image frames. To find matching pairs, common comparison metrics such as Euclidean or Mahalanobis distance are not used due to their computational complexity. The L1 norm that is calculated as the absolute difference between two points is used instead.

To understand the matching process, consider an example in which the first frame has n feature points and the second frame has m feature points. In matching n feature points to m feature points, the best matching pairs should have the minimum sum of the absolute differences (d_{ij}) between two descriptors,

$$d_{ij} = \sum_{k=1}^{w} |x_k - y_k|, \quad i = 1 \, to \, n, \, j = 1 \, to \, m \tag{3}$$

where x_k and y_k are the regional counts for the feature points in the first and second image, respectively and w represents the descriptor length. The value of w is the product of the number of subregions and the number of SBIs and is $36 \times 9 = 324$ for SYBA 5×5. We select each feature point's descriptor in the first frame and match it to all feature point descriptors in the second frame to find its best match. An example of SYBA descriptor calculation is shown in Fig. 3. Each row represents the feature descriptor of a feature point. The absolute difference d value in this example is 2.

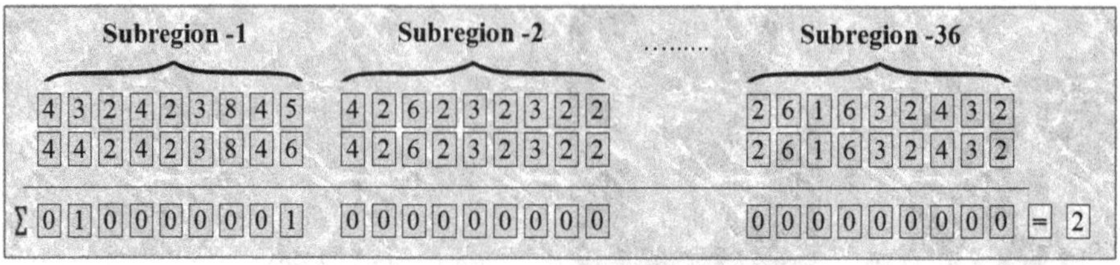

Subregion -1	Subregion -2	....	Subregion -36
4 3 2 4 2 3 8 4 5	4 2 6 2 3 2 3 2 2		2 6 1 6 3 2 4 3 2
4 4 2 4 2 3 8 4 6	4 2 6 2 3 2 3 2 2		2 6 1 6 3 2 4 3 2
Σ 0 1 0 0 0 0 0 0 1	0 0 0 0 0 0 0 0 0		0 0 0 0 0 0 0 0 0 = 2

Fig. 3. The SYBA descriptor calculation. Each row represents a feature descriptor.

Once the d value is calculated between descriptors, a two-pass matching strategy is used as follows. The first pass of this matching strategy is to find the minimum distance d between one feature in the first image and all features in the second image. The feature with the smallest unique d in the second image is considered a match to this feature in the first image. Otherwise, the feature is ignored. The second pass is to confirm the matched feature in the second image also has the unique shortest distance d to its match in the first image. The feature that has the smallest d in the first image is considered a match to this feature in the second image. Otherwise, the feature is ignored. Our method ensures a unique one-to-one match and eliminates any possible ambiguity. The result of this two-pass matching strategy is considered to be matched feature pairs.

3   Motion Classification Techniques

Motion vectors between two frames are calculated using SYBA feature matching. Motion vectors are represented in polar coordinates (magnitude and orientation) for vehicle movement classification. Depth analysis is performed by segmenting the motion field into different regions based on the motion vector length. We show that polar coordinates and motion field segmentation are advantageous for vehicle movement classification.

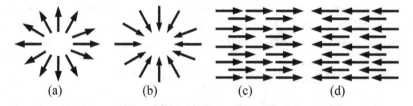

Fig. 4. Four types of camera motion (a) camera moves forward, (b) camera moves backward, (c) camera pans to left, and (d) camera pans to right.

Four types of camera motion are very common in general driving scenarios. The motion fields representing these motions are shown in Fig. 4. A motion field is a result of the vehicle's movement such as forward, reverse or turn. When the car moves forward, the camera imaging sensor approaches the scene (i.e. input image zooming in) and the motion vectors diverge from the vanishing point. Similarly, when a car moves backward, the camera recedes from the scene and the motion vectors converge to the vanishing point. When the car makes a left or right turn, all motion vectors are close to parallel, as shown in Fig. 3 (c, d).

4 Experiment Results

Our motion classification and depth analysis methods were tested using two image sequences from the KITTI dataset [14] and a new sequence captured in Provo, Utah. All sequences were captured in different traffic scenarios and at different locations. As a result, these image sequences have different light conditions, shadow presence, and different numbers of cars, pedestrians, cyclists, bikers, and high slopes.

As explained earlier, SYBA feature matching was used to obtain the magnitude and orientation of each motion vector between the corresponding feature points. The magnitude and orientation were calculated as follows. A feature point location (x_1, y_1) in the previous frame was matched to a feature point location (x_2, y_2) in the current frame. The magnitude (l) is the distance between the two matching points. The orientation (θ) is the angle between the motion vector and the vertical axis on the image with 0 degree pointing up and the positive angle increasing in the clockwise direction and negative angle increasing in the counter clockwise direction.

Figure 5(a) shows a typical motion field that represents the orientation in degrees. Figure 5(b) is the zoomed-in detail of the highlighted red rectangle region shown in Fig. 5(a). The green crosses in Fig. 5(a) indicate the feature point locations in the previous frame and the red circles indicate the matching feature in the current frame. The yellow line connecting a green cross to its corresponding red circle represents the length of the motion vector. The orientation of most of the motion vectors shown in Fig. 5(a) is around −90 degrees according to our definition. This motion field represents a right turn motion, which has most of the motion vectors pointing to the left and approximately −90 degrees from vertical.

(a)   (b)

Fig. 5. (a) An example of a motion vector with orientation (original size). (b) Red rectangle is cropped and zoomed in to highlight the orientation details (Color figure online).

Our motion classification algorithm is based on a statistical analysis of motion vector orientation distribution. We divide the motion field into left and right halves.

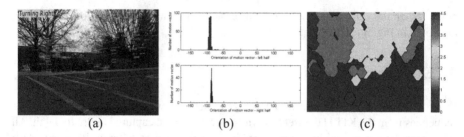

(a)   (b)   (c)

Fig. 6. Experiments on a newly captured sequence. (a) Feature points from the previous frame (green crosses) are matched to feature points in the current frame (red circles). (b) Both left and right histograms peak at around −90 degrees indicating a right turn motion. (c) Motion vectors near the camera have the largest movement (red regions). The furthest regions (least movement) are highlighted in blue (Color figure online).

A 1D histogram of the motion vector orientation is constructed for each half. For a right turn motion as shown in Fig. 6(a), the majority of motion vector orientations in both left and right halves are close to −90 degrees. Figure 6(b) shows that the left and right histograms peak at around −90 degrees.

Depth analysis is performed by segmenting the motion field into different regions based on the length of motion vectors. Figure 6(c) shows the motion vector length segmentation result. Motion vectors near the camera having the largest movement are segmented and shown in red. The furthest regions (least movement) are highlighted in blue. This depth analysis provides a rough estimate of the 3D scene and can be used for time to impact or obstacle detection.

Figures 7, 8 and 9 show results from the same sequence for a left turn motion, approaching and receding sensor. For the left turn motion (Fig. 7), both left and right histograms peak at approximately 90 degrees. For the sensor approaching motion (Fig. 8), the majority of motion vectors in the left half of image have a negative motion vector orientation and the majority of motion vectors in the right half of image have a positive motion vector orientation. It is the opposite for the sensor receding motion (Fig. 9), positive motion vector orientation in the left half and negative motion vector orientation in the right half of the image.

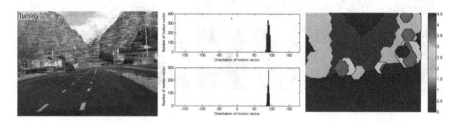

Fig. 7. Motion classification and depth analysis results of a left turn motion.

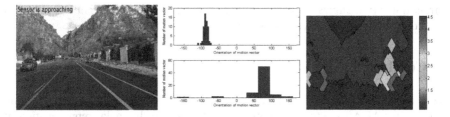

Fig. 8. Motion classification and depth analysis results of a camera approaching motion.

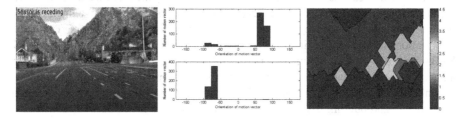

Fig. 9. Motion classification and depth analysis results of a camera receding motion.

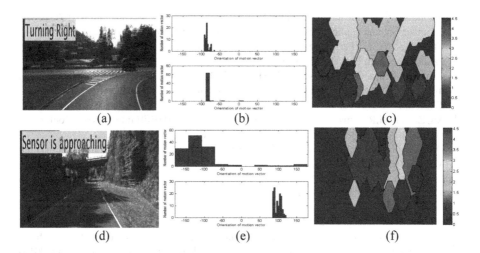

Fig. 10. Experiment on the *KITTI* dataset caes-1. (a, b) Feature points from the previous frame (green crosses) are matched to feature points in the current frame (red circles). Based on motion vector orientation, the direction of the camera is classified as right turn (c) and approaching (d). (e, f) Motion vectors near the camera have the largest movement (red regions). The furthest regions (least movement) are highlighted in blue (Color figure online).

A KITTI sequence was tested in the same manner. Feature matching results using the SYBA descriptor is shown in Fig. 10(a) and (d). As shown in the figures, feature points from the previous frame (green crosses) are matched to feature points in the current frame (red circles). As explained previously, the motion direction is classified based on the statistical analysis of motion vector orientation.

5   Conclusion

In this paper, we have shown that the SYBA descriptor provides high feature matching accuracy for ADAS - automatic motion classification so that polar representation of motion vectors can be calculated. Once polar representation of motion vectors are calculated, statistical analysis can be performed to classify the motion and estimate the depth information for the 3D scene. Depth analysis is performed by segmenting the motion field into different regions based on the length of motion vectors. The depth information can be used for time-to-impact calculation or obstacle detection in ADAS. SYBA is able to create a feature descriptor without using complex floating-point operations. This advantage makes it an excellent candidate for hardware implementation in an embedded vision sensor. Future work involves implementing the SYBA descriptor on hardware and building a small embedded vision sensor.

References

1. Fraundorfer, F., Scaramuzza, D.: Visual odometry: Part II: matching, robustness, optimization, and applications. Robot. Autom. Mag., IEEE **19**(2), 78–90 (2012)
2. Chen, L., Wei, H., Ferryman, J.: A survey of human motion analysis using depth imagery. Pattern Recognit. Lett. **34**(15), 1995–2006 (2013)
3. Civera, J., Grasa, O.G., Davison, A.J., Montiel, J.M.: 1-point RANSAC for extended Kalman filtering: application to real-time structure from motion and visual odometry. J. Field Robot **27**(5), 609–631 (2010)
4. Xu, Z., Wu, H.R.: Smart video surveillance system. In: 2010 IEEE International Conference on Industrial Technology (ICIT), pp. 285–290 (2010)
5. Desouza, G., Kak, A.: Vision for mobile robot navigation: a survey. IEEE Trans. Pattern Anal. Mach. Intell. **24**(2), 237–267 (2002)
6. Onkarappa, N., Domingo Sappa, A.: Speed and texture: an empirical study on optical-flow accuracy in ADAS scenarios. IEEE Trans. Intell. Transp. Syst. **15**(1), 136–147 (2014)
7. Onkarappa, N.: Optical flow in driver assistance systems. Ph.D. dissertation, Universitat Autònoma de Barcelona (2013)
8. Desai, A., Lee, D.J., Ventura, D.: An efficient feature descriptor based on synthetic basis functions and uniqueness matching strategy. Comput. Vis. Image Underst. **142**, 37–49 (2016)
9. Lowe, D.G.: Distinctive image features from scale-invariant keypoints. Int. J. Comput. Vis. **60**, 91–110 (2004)
10. Bay, H., Tuytelaars, T., Van Gool, L.: SURF: speeded up robust features. In: Leonardis, A., Bischof, H., Pinz, A. (eds.) ECCV 2006, Part I. LNCS, vol. 3951, pp. 404–417. Springer, Heidelberg (2006)
11. Ke, Y., Sukthankar, R.: PCA-SIFT: a more distinctive representation for local image descriptors. In: Proceedings of the 2004 In: IEEE Computer Society Conference on Computer Vision and Pattern Recognition, CVPR 2004, vol. 2, pp. II-506–II-513 (2004)
12. Hua, G., Brown, M., Winder, S.: Discriminant embedding for local image descriptors. In: IEEE 11th International Conference on Computer Vision, pp. 1–8 (2007)

13. Anderson, H.: Both lazy and efficient:Compressed sensing and applications. Technical report, (Sandia National Laboratories), Report number: 2013–7521P (2013)
14. Geiger, A., Lenz, P., Urtasun, R.: Are we ready for autonomous driving? the KITTI vision benchmark suite. In 2012 IEEE Conference on Computer Vision and Pattern Recognition (CVPR), pp. 3354–3361 (2012)

Shared Autonomy Perception and Manipulation of Physical Device Controls

Matthew Rueben[✉] and William D. Smart

School of Mechanical, Industrial and Manufacturing Engineering,
Oregon State University, Corvallis, OR 97331, USA
{ruebenm,bill.smart}@oregonstate.edu

Abstract. As robots begin to enter our homes and workplaces, they
will have to deal with the devices and appliances that are already there.
These devices are invariably designed with human perception and manipulation abilities in mind. Unfortunately, this often makes them hard for
robots to interact with autonomously. Control elements, such as buttons,
switches, and knobs, are often hard to identify with current sensors. Even
when they are found, it is often not clear how they should be manipulated to achieve a specific goal without extensive background knowledge
of the specific device and task.

In this paper, we describe a shared-autonomy approach to the identification and operation of these device controls. A human operator provides assistance with perception and high-level planning, while the robot
takes care of the low-level actions that depend on closed-loop sensor feedback. We demonstrate our approach by controlling a consumer electronics
device, despite not being able to autonomously sense some of its control
elements, and give the results of some initial evaluations suggesting that
the shared autonomy interface is both easier to use and more efficient
than a direct teleoperation interface.

1 Introduction

As robots begin to enter our homes and workplaces, they will have to deal with
the devices and appliances that are already there. For example, fetching a cup of
coffee involves interacting with the coffee machine, the cupboard, and possibly
the refrigerator and other doors in the environment. All of these devices (we
count doors as devices for the purposes of this paper) were designed for humans
to operate, with affordances tailored to human perception and manipulation
abilities. As robots become more and more common in our homes and offices, we
can expect consumer devices of the future to have more robot-friendly controls;
either physical controls that are easy to autonomously identify and manipulate,
or a wireless interface to the device that the robot can directly access. At present,
these devices do not exist, and the first wave of personal robots will have to
interact with devices built for humans.

Devices that are easy for humans to operate often cause problems for
robots [1]. In teleoperation settings, the lack of tactile feedback often makes

© Springer International Publishing Switzerland 2015
G. Bebis et al. (Eds.): ISVC 2015, Part II, LNCS 9475, pp. 830–841, 2015.
DOI: 10.1007/978-3-319-27863-6_78

manipulation of buttons and switches awkward and clumsy [2]. Similarly, limitations on perception often make it hard for the remote operator to clearly see what is shown on small displays on electrical devices. In the autonomous setting, perception of small buttons and switches is often difficult due to sensor limitations and poor lighting conditions.

In this paper, we present a shared autonomy approach to the operation of physical device controls. A human operator gives high-level guidance, helps identify controls and their locations, and sequences the actions of the robot. Autonomous software on our robot performs the lower level actions that require closed-loop control, and estimates the exact positions and parameters of controls. We describe the overall system, and then give the results of our initial evaluations, which suggest that the system is effective in operating the controls on a physical device.

2 Related Work

Goodfellow et al. [3] sketch out three general ways in which human operators might provide assistance to a mostly-autonomous robot: helping with navigation, perception, and manipulation. Our system falls into both the perception and manipulation categories that they define. Our system is also similar in spirit to Fong, Thorpe, and Baur's notion of collaborative control [4], where the robot asks (often quite complex) questions of a human operator to help overcome the shortcomings of its autonomy.

Our approach is also an example of what Atherton and Goodrich call "perception by proxy" [5], where a human helps with perception problems that the robot cannot solve autonomously. Pitzer et al. [6] report on a system that uses human perceptual assistance in a similar way to ours to pick up objects that the robot can't find on its own. This task, however, is completely scripted, and the robot only receives help with perception. In our work, we allow the user to also have control over the actions executed to complete the task; this is to provide for the general case in which the robot does not always know what needs to be done. Nguyen et al. [7], on the other hand, show how novel behaviors can be constructed for a robot by a human assistant. The human gives no help with perception, but constructs hierarchical finite state machines that define tasks that are subsequently performed autonomously by the robot. Srinivasa et al. [8] and Beetz et al. [9] both describe relatively complete systems (i.e. with navigation, perception, and manipulation capabilities), the former focusing on manipulating objects and doors while the latter performs a high-level task: making breakfast. Neither treats household devices nor the shared autonomy approach to the extent that this paper does.

3 Finding and Manipulating Device Controls

Controls on devices engineered for humans are designed with human perception and manipulation abilities in mind. Switches and knobs are designed for a typical human hand to operate, with all the related assumptions about dexterity. Displays are designed to be readable for someone with (often slightly less than) normal vision. While this makes life relatively easy for humans, it does not necessarily make it so for robots with different perceptual and manipulative abilities.

Many controls are difficult for a robot to percieve, especially in the general case where the specific control has never been seen before. Accurate visual identification and pose estimation of, for example, a light switch is hard because of the wide variety of switch designs. The same is true for other switches, knobs, sliders, buttons, and the like. Some progress has been made in the identification of power outlets [10], but this is predicated on having a previously-known template to match.

Adding depth information does not help much, since many of the controls we want to manipulate are small, and often close to the noise threshold of currently-available depth sensors typically installed on a mobile robot. This makes it extremely difficult to segment the controls from the other parts of the device.

To make matters worse, many devices have a design aesthetic that makes the controls harder to sense effectively. Small matte black buttons on a matte black device, a popular choice in consumer electronics for decades (illustrated in Fig. 1), makes it almost impossible to autonomously detect the controls.

Fig. 1. The Yamaha RX-V390 stereo receiver, a typical consumer electronic device, showing matte black controls on a matte black background.

Teleoperation can be used to solve the perception problem by simply relaying the sensor information, generally in the form of camera or RGB-D images, to a human operator. The operator then identifies the appropriate controls visually and teleoperates the robot to manipulate them. However, this can be problematic for three reasons. First, when the robot is manipulating a control, its hand often occludes the control, making effective teleoperaton difficult. This can be alleviated to some extent by adding additional sensors such as cameras in the robots forearm, but we can never guarantee that this will solve the problem in all cases. Second, the lack of haptic feedback in most teleoperation systems makes the actual manipulation of the control tricky. It is hard to know when sufficient contact is being made, especially on touchpads or when gripping knobs, which often results in overly-violent interactions. Finally, although the device controls might be easy to see for a human standing in front of the device, they are often more difficult to pick out of a video stream captured by the robot's cameras.

Autonomous control, on the other hand, can solve some of the manipulation problems, since closed-loop controllers can be written that effectively manipulate controls once their size, shape, type, and location are known. However, autonomously perceiving controls to determine these required attributes remains an open problem.

In the next section, we describe our approach to blending human perception with autonomous manipulation to overcome these problems and create an effective system for manipulating device controls.

4   Approach

Regardless of the specifics of the controls to be manipulated by the robot, our approach follows the same general steps. We outline these steps below, and then go on to discuss our specific implementation in the next section. In each of these steps, we use the human to give information that assists the autonomous perception and manipulation capabilities of the robot.

4.1   Identify and Localize the Device

For the work reported in this paper, we assume that all of the controls to be manipulated lie in the same plane of a device or surface. The human first ensures that the robot is in front of the device, either by directly teleoperating it, or by using autonomous navigation routines. They then specify the general area of the controls on a sensor image displayed from the robot, and the robot then goes on to form a model of the surface on which these controls lie. In this paper, this is a simple planar model, estimated from the robot's sensor data.

The location of the device is stored in a global map of the robot's environment, along with a semantic name. Assuming the robot is well-localized, this allows the human to command the robot to return to the device without directly driving it there, by simply supplying its name.

4.2   Select the Control Type

Once the surface of the device is identified and estimated, the human selects one of a fixed set of control types to identify. Our system currently supports buttons, toggle switches, and knobs. Specifying the type of the control will allow the robot to more easily detect and localize the control in the next step.

4.3   Identify and Localize the Control

Once the control type is selected, the human then identifies the approximate location of the control. Using this location and the model of the front of the device as a starting point, the robot then more accurately localizes the control, and stores this location in its internal model of the device.

Localization is done in a control-specific manner, based on the physical properties of the control. Knobs, for example, are generally cylindrical with an axis normal to the device surface.

Once the control is localized, its position is stored, along with a name supplied by the human, in the model of the device that we maintain. Again, this name allows the user to interact with the control semantically after it is initially identified.

4.4   Perform the Manipulation

Once all of the controls are identified and localized, the robot can manipulate them. As before, the human gives the high-level direction, specifying the control and, if necessary, the parameters of the manipulation (for example, the position to turn a knob to). The robot then performs the manipulations using closed-loop, control-specific routines that take advantage of sensor data that would be unavailable in a traditional teleoperation setting.

If there are several manipulations, they can be performed either one at a time or all at once. Operations can be named; flipping a particular switch up might, for example, be labeled "power on", once again allowing the user to interact with the device at a higher level in the future by simply referring to this semantic name. Sequences of manipulations can also be grouped together and named.

4.5   Simplifying Assumptions

Our current approach makes two main simplifying assumptions, both of which can be relaxed with some additional work. The first is that the device controls lie on a plane. This is true for many controls, both on consumer devices and in buildings (light switches, door knobs, drawer handles, and the like). This assumption makes the specification and estimation of the device straightforward, as we describe in Sect. 5.3. However, this assumption can be relaxed by allowing the user to specify a sequence of planes, or by fitting a more complex model to the device. While both of these involve more complex implementations, they do not change the fundamental approach.

Our second simplifying assumption is that there are a limited number of control types that we need to manipulate. This allows us to create special-purpose detectors, localizers, and manipulation routines for each type. We chose our initial set of controls to be representative of the types of controls likely to be encountered by a robot in a home setting, as well as to illustrate our approach. It is by no means an exhaustive set. While we believe that the set of control-specific implementations needed in practice is finite, the actual size of this set is an open question. It may, however, be possible to incorporate mechanisms that will allow the human to teach the robot how to identify, localize, and manipulate novel controls, as we discuss in Sect. 6.

5   Implementation

Our implementation was performed on a Willow Garage PR2 robot using the Robot Operating System infrastructure [11], the Point Cloud Library [12], and OpenCV [13]. All of our software is freely available at our open-source reposi- tory.[1] Visual sensing was done using a Microsoft Kinect sensor mounted on the head of the PR2. The implementation was validated with a Yamaha RX-V390 stereo receiver (a representative piece of consumer electronics, shown in Fig. 1) and a typical household light switch panel (Fig. 2).

Fig. 2. The ready position of the PR2 end-effector (left), and the three manipulations that we perform: buttons, switches, and knobs.

5.1   Identifying and Localizing the Device

We assume that the robot starts in front of the device, with a clear view of it, and able to physically reach all of the relevant controls. The human operator can teleoperate the robot to this position using standard tools in ROS. Since all device locations are recorded as coordinate frames, the robot can autonomously return to them when necessary. Accumulated localization errors will likely pre- vent the robot from returning to the exact position, but this can be compensated for by re-estimating the device model and performing the appropriate odometry corrections when the device is reacquired.

The human is presented with a 2d image of the front of the device from the Kinect sensor. Using a mouse, the human draws a rectangle on this image, corresponding to a section of the plane of the device containing the controls (the front of the stereo receiver, or the wall in which the light switches are embedded). The corners of this region define a frustum in 3d space, and we filter the 3d point cloud from the Kinect sensor to reject all points not in this frustum. We then estimate the most likely plane defined by the non-rejected points using the RANSAC algorithm [14]. This results in a planar model for the part of the device defined by the human.

---

[1] http://github.com/OSUrobotics

We place a coordinate frame (represented by a ROS TF frame) on the detected plane of the device, with the z-axis normal to the plane. We store the location of this frame in the global coordinate frame (the ROS `map` frame), and allow the human to enter a name for the frame. This will allow the robot to autonomously return to the device in the future using the built-in ROS navigation system.

5.2   Selecting the Control Type

In the current implementation, we allow the human operator to select from only three types of controls: buttons, toggle switches, and cylindrical knobs. The detection, localization, and manipulation routines for each control type are specialized to that control. Buttons are pressed, switches are flipped up or down, and knobs are grasped and turned. The operator selects the control type from this set of known control types.

5.3   Identifying and Localizing the Control

Controls are identified by the human operator by mouse clicks on the 2d image of the device. These points are projected into 3d and then passed to the control-specific localization routines. Each of the localization routines results in the placement of a local coordinate frame for the control; this frame is later used for manipulation.

Localizing Buttons. In this paper, we make the assumption that the buttons the robot will interact with are hard to directly identify, even with human guidance. We found this to be the case with several test objects, such as the stereo receiver shown in Fig. 1. Accordingly, we rely on the human operator to localize each button with a mouse click in the 2d image that they are presented with.

We calculate the ray defined by the selected 2d point and intersect it with the plane of the device. We place a coordinate frame, corresponding to the button, at this intersection point, with the x-axis parallel to the floor and the z-axis normal to the device plane. We publish this coordinate frame as a ROS TF frame, with the device frame as the parent. This allows us to define the position of the button in the local coordinate frame of the device.

Localizing Switches. We localize switches similarly to buttons, by relying on human guidance. While toggle switches have an element that sticks out from the plane of the device, this is generally too small to be reliably detected from point cloud data. Although standard light switches can be reliably detected using template-matching techniques in images, this is harder for arbitrary toggle switches, and we leave it for future work.

As with buttons, the localization of switches results in a local coordinate frame for the switch, where the x-axis is parallel to the ground plane, the switch travels along the y-axis, and the z-axis is normal to the surface of the device.

Localizing Knobs. For this work, we assume that knobs are cylindrical and that they protrude from the device plane enough to be detected by the Kinect sensor. The human user draws a rectangle around the knob in the 2d camera image and, once again, we project the corners of this rectangle into 3d, defining a frustum. We reject all points outside of this frustum, as well as all points that lie within 1 cm of the estimated device plane. This leaves points that mostly correspond to the knob itself, along with a small number from the device plane that escaped rejection because of measurement noise.

We estimate the parameters of the knob by applying the RANSAC algorithm with a cylindrical model, or by using a heuristic approach. The heuristic approach, which we found to work more robustly than RANSAC did for our test devices, proceeds as follows.

The front surface of the knob is assumed to be a plane parallel to the device plane, passing through the point in the frustrum that is furthest in front of the device plane. The diameter of the knob is estimated by calculating the spread of the non-rejected points along the x-axis. The axis of rotation of the knob can then be estimated; it is normal to the device plane and passes through the mid-point of the knob. The x-coordinate of the mid-point is the middle of the calculated x-axis range. The y-coordinate is one half of the estimated diameter down from the point in the frustum with the largest y-coordinate.

Once again, we end up with a coordinate frame for the knob, with the z-axis parallel to the device plane normal, and a position relative to the device coordinate frame.

5.4 Performing the Manipulation(s)

Once we have identified and localized each of the controls and attached a coordinate frame to each of them, we can manipulate them. For each control, we use a Cartesian controller to move the robot's end-effector either parallel or perpendicular to the device plane. We assume that there are no transient obstacles between the robot and the device and, hence, perform no obstacle avoidance as we move.

Each of the manipulations involves moving the end effector to a ready position, close to the z-axis of the control to be manipulated, oriented normally to the surface. The robot then performs the control-specific manipulation, and returns to this ready position. The robot's fingertip pressure sensor arrays are used to ensure that only safe forces are applied to the device.

Manipulating Buttons. Buttons are the most straightforward control to manipulate. The end-effector is brought to a ready position on the z-axis of the button coordinate frame. It then moves towards the device plane along this axis until the pressure sensors in the gripper fingers detect a firm contact. It then retreats along the z-axis, back to the ready position.

Manipulating Switches. For switches, the end-effector is once again brought to a ready position on the control z-axis, with the wrist rotated so that the gripper fingers move along the y-axis of the control (see Fig. 2) and the gripper is opened. The end-effector is moved in until the fingertip pressure sensors register a contact with the front of the device. The end-effector is then withdrawn slightly and moved either up or down (depending on the direction selected by the human operator), parallel to the device plane. Sensed contact with the finger indicates that it has touched the toggle of the switch, and that the switching manipulation is therefore likely to have been successful.

Manipulating Knobs. The most complex manipulation that we implement is turning knobs. The end-effector is brought to the same ready position as for switches, with the gripper fingers opened far enough to accommodate the knob. The end-effector is moved in along the z-axis until contact is detected, and then withdrawn slightly. To rectify any position error in the end-effector, a ROS package called pr2_gripper_sensor_action is used to gently grasp the knob several times, moving the end-effector parallel to the device until the grasp is centered on the fingertip pressure sensor arrays. The fingers are then closed until a firm contact (of 5N) with the knob is sensed.

Next, a graphical interface, shown in Fig. 3, is presented to the human operator, allowing them to select the final position of the knob. Once this position is selected and confirmed, the robot executes the turn. The human can inspect the knob position in the interface using images from the PR2's forearm camera. Once the human operator is satisfied with the knob position, the interface is closed, the grippers are opened, and the end-effector returns to the ready position.

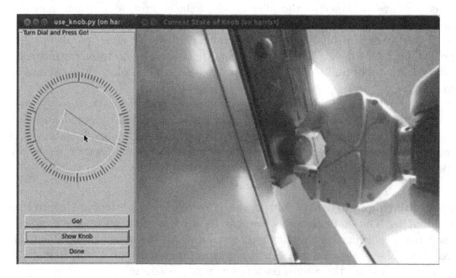

Fig. 3. The graphical interface used to select knob positions. The image on the right is from a camera in the forearm of the robot, which typically affords the best view of manipulations.

6   Discussion

In this paper, we have presented a shared-autonomy approach combining high-level human input and low-level autonomous control to manipulate the controls on everyday devices designed for human use. Our current implementation is capable of working with buttons, switches, and knobs, and can be easily extended to deal with other control types. In our preliminary evaluations, our interface compares favorably with a robust direct teleoperation interface from the literature. Our two main goals for future work are to perform a formal user study and to improve our system as discussed below.

While our current implementation works well in practice, it does have a number of shortcomings. Since we make extensive use of coordinate transforms between the end-effector and the robot's sensors, our technique is heavily dependent on an accurately-calibrated robot. In our laboratory setting, we observe end-effector error of up to 2 cm. We were able to ignore this error for our buttons and switches with only occassional misses, and used the features described in Sect. 5.4 to robustly grasp knobs.

It can be hard to see controls as they are being manipulated, since they are often occluded by the end-effector. This presents a particular problem when turning knobs. While the human can effectively turn the knob, it is often hard for them to know *where* to turn the knob to, since they cannot see it. The solution to this is either to try to position the forearm camera in such a way as to make the knob more visible, or to replicate the markings around the knob in the graphical interface presented to the operator. In addition to providing the human with a better view of the controls, it will also help to do the same for the robot. Measurement error is not uniform across the field of view of the Kinect, and ensuring that objects are centered will potentially lead to better position estimation.

For buttons and switches, we currently trust the human operator to locate them accurately enough for the robot to perform the manipulation. Part of the motivation for this is that these controls, at least in the devices that we have been working with, are hard to identify with RGB-D sensors due to their color and size (see Fig. 1). This will not be true for all controls, however, and we plan to integrate automatic recognition and localization like that described by [10] for those controls that admit it. The human operator will most likely have to give some initial assistance to the robot by more accurately identifying the control (perhaps drawing a bounding box around it). We will then attempt to learn a template, which we can subsequently use to re-identify the control and more precisely localize it.

Our current system allows the human to chain together simple sequences of manipulations. However, we would like to build (and semantically name) more complex sets of operations, in the spirit of ROSCo [7]. Once these sets of operations are defined, it will allow the operator to initiate them by simply referring to their label ("make popcorn", for example).

Finally, we want to address our reliance on precise calibration by implementing a visual servoing scheme for the end-effector. If we can reliably recognize and localize the control in an RGB-D image, we can use this camera to visually servo the end-effector to the correct position, overcoming small calibration errors. Since some of the controls that we want to manipulate are small compared to the size of the gripper fingers, this will be an important part of increasing the robustness of our approach.

References

1. Ciocarlie, M., Hsiao, K., Leeper, A., Gossow, D.: Mobile manipulation through an assistive home robot. In: Proceedings of the IEEE/RSJ International Conference on Robots and Systems (IROS), Vilamoura, Portugal, pp. 5313–5320 (2012)
2. Sankaran, B., Pitzer, B., Osentoski, S.: Failure recovery with shared autonomy. In: 2012 IEEE/RSJ International Conference on Intelligent Robots and Systems (IROS), pp. 349–355. IEEE (2012)
3. Goodfellow, I.J., Koenig, N., Muja, M., Pantofaru, C., Sorokin, A., Takayama, L.: Help me help you: interfaces for personal robots. In: Proceedings of the 5th ACM/IEEE International Conference on Human-Robot Interaction (HRI), Osaka, Japan, pp. 187–188 (2010)
4. Fong, T., Thorpe, C., Baur, C.: Robot, asker of questions. Robot. Auton. Syst. **42**, 235–243 (2003)
5. Atherton, A.J., Goodrich, M.A.: Perception by proxy: humans helping robots to see in a manipulation task. In: Proceedings of the 6th ACM/IEEE International Conference on Human-Robot Interaction (HRI), Lausanne, Switzerland, pp. 109–110 (2011)
6. Pitzer, B., Styer, M., Bersch, C., DuHadway, C., Becker, J.: Towards perceptual shared autonomy for robotic mobile manipulation. In: Proceedings of the IEEE International Conference on Robotics and Automation (ICRA), Shanghai, China, pp. 6245–6251 (2011)
7. Nguyen, H., Ciocarlie, M., Hsiao, K., Kemp, C.C.: ROS commander (ROSCo): behavior creation for home robots. In: Proceedings of the IEEE International Conference on Robotics and Automation (ICRA), Karlsruhe, Germany (2013)
8. Srinivasa, S.S., Ferguson, D., Helfrich, C.J., Berenson, D., Collet, A., Diankov, R., Gallagher, G., Hollinger, G., Kuffner, J., Weghe, M.V.: HERB: a home exploring robotic butler. Auton. Robots **28**, 520 (2010)
9. Beetz, M., Klank, U., Kresse, I., Maldonado, A., Mosenlechner, L., Pangercic, D., Ruhr, T., Tenorth, M.: Robotic roommates making pancakes. In: 2011 11th IEEE-RAS International Conference on Humanoid Robots (Humanoids), pp. 529–536 (2011)
10. Eruhimov, V., Meeussen, W.: Outlet detection and pose estimation for robot continuous operation. In: Proceedings of the IEEE/RSJ International Conference on Robots and Systems (IROS), San Francisco, CA, pp. 2941–2946 (2011)
11. Quigley, M., Conley, K., Gerkey, B., Faust, J., Foote, T., Leibs, J., Berger, E., Wheeler, R., Ng, A.: ROS: An open-source robot operating system. In: Proceedings of the ICRA 2009 Workshop on Open-Source Robotics (2009)

12. Rusu, R.B., Cousins, S.: 3D is here: point cloud library (PCL). In: Proceedings of the IEEE International Conference on Robotics and Automation (ICRA), Shanghai, China, pp. 1–4 (2011)
13. Bradski, G.: The OpenCV library. Dr. Dobb's J. Softw. Tools **25**, 120–126 (2000)
14. Fischler, M.A., Bolles, R.C.: Random sample consensus: a paradigm for model fitting with applications to image analysis and automated cartography. Commun. ACM **24**, 381–395 (1981)

Condition Monitoring for Image-Based Visual Servoing Using Kalman Filter

Mien Van, Denglu Wu, Shuzi Sam Ge, and Hongliang Ren[(✉)]

Advanced Robotic Centre, Faculty of Engineering,
National University of Singapore, Singapore, Singapore
ren@nus.edu.sg

Abstract. In image-based visual servoing (IBVS), the control law is based on the error between the current and desired features on the image plane. The visual servoing system is working well only when all the designed features are correctly extracted. To monitor the quality of feature extraction, in this paper, a condition monitoring scheme is developed. First, the failure scenarios of the visual servoing system caused by incorrect feature extraction are reviewed. Second, we propose a residual generator, which can be used to detect if a failure occurs, based on the Kalman filter. Finally, simulation results are given to verify the effectiveness of the proposed method.

1 Introduction

Visual servoing, has been applied extensively in robotics to enhance sensing capability. The goal of this task is to calculate the control input that was applied to the robot system so that the predefined image features can converge to the desired static reference features. Generally, the visual servoing can be classified into three categories: (1) position based visual servoing (PBVS) [1], where the control input is designed based on the feedback of 3D data such as the robotic system pose; (2) image-based visual servoing (IBVS) [2], where the control input is designed based on the feedback of 2D data defined in the image plane, and (3) hybrid visual servoing [3], where both 2D and 3D data is combined as the feedback. Among them, IBVS is widely applied due to it's easy in implementation and robustness with calibration error and measurement noise [4–6]. In the IBVS, the control law is based on the different between the current and desired features on the image plane, which can be static for positioning problems or dynamic tracking problems. Because the control law is determined based on the image plane data, the system is working well only when all the designed features are correctly extracted. Toward this research direction, reliable feature extraction methods have been developed in computer vision, such as SIFT, SURF, BRIEF, KANSAC, etc. [7]. In addition, reliable feature tracking methods has also been developed to enhance the robustness of feature tracking based on kalman filter or Particle filter [15], etc. However, the extraction of the designed image features is not always obtained correctly; some features will be appeared in or disappeared from the image during visual servoing [8]. Generally, there are two reasons making the appearance/disappearance of image features during visual servoing: (1) when the camera is moving, some parts of the

G. Bebis et al. (Eds.): ISVC 2015, Part II, LNCS 9475, pp. 842–850, 2015.
DOI: 10.1007/978-3-319-27863-6_79

object which contain the designed features, is out of the field of view (FOV) of camera, and (2) due to the environment noise or obstacles, some designed features will be disappeared (occlusion), or some undesired features will be appeared. The failure scenarios of the visual servoing due to the appearance/disappearance of the image feature are illustrated in Fig. 1. Figure 1(a) illustrates visual servoing in normal operation; the system extracts the designed image feature points correctly. Figure 1(b) illustrates a faulty scenario when a feature point is disappearing. Figure 1(c) illustrates a faulty scenario when the vision system extracts an undesired feature point (feature point 5), which has the similar property with a designed feature (feature point 4), while the designed feature point 4 is occluded. In this situation, the visual servoing system will misunderstand that the feature 5 is a true designed image feature point instead of the feature point 4. It is easy to recognize that the failure due to the disappearance of a feature point can be easily detected by visual servoing control system based on the prior knowledge about the number of designed features. However, the faulty situation illustrated in Fig. 1(c) is a difficult term, the classical visual servoing control system has a difficult to detect this kind of fault. The fault due to appearance/disappearance of a feature can be generally described as a bias sensor fault shown in Fig. 1(d); the vision system reads a wrong displacement of the designed image feature.

In order to avoid the designed image features moving out of the FOV of camera during visual servoing, several approaches have been developed for both PBVS and IBVS. In [8], a continuous control has been developed by weighting the displacement of the image features. In [9], a controller based on a switching among PBVS and backward motion has been developed. In [10], a randomized kinodynamic path planning algorithm has been developed to maintain three problems: (1) avoiding visual occlusion of target features caused by the workspace obstacles, robot's body, or the target itself; (2) avoiding collision with obstacles, and (3) joint limits. Other approaches to deal with the image feature occlusion problem during visual servoing are to combine multi sensors as in [11, 12]. The above developed approaches can be applied to detect the system failure due to the disappearance of a designed image feature. However, for the faulty scenarios mentioned on Fig. 1(c) and (d), there are no reports to detect this kind of faults.

In this paper, unlike the previous approaches which focus on increasing visual tracking performance, a condition monitoring scheme is developed for the first time to monitor whether the designed image features are correctly extracted or not. Detection and identification tasks are activated by approximating parameters related to potential failures based on the Kalman filter algorithm. The Kalman filter algorithm is used here because it has a good capability to approximate the Gaussian noise system parameters [13, 14]. We note that our work is preliminary, and as such we have not directly compared it with other work; we leave such such comparison for future works.

The paper is organized as follows. In Sect. 2, image-based visual servoing system and its model are derived. In Sect. 3, the proposed fault detection and isolation is described. Simulation tests are performed to validate the theoretical development, in Sect. 4. Finally, the conclusion is given in Sect. 5.

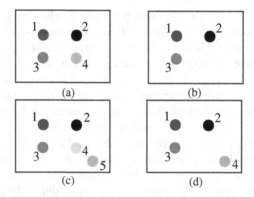

Fig. 1. Illustrated feature extraction scenarios: (a) normal extracted feature; (b) one designed feature disappears; (c) one designed feature is occluded while an undesired feature is appeared and (d) physical change of feature

2 Image Based Visual Servoing and Its Model

In this section, the visual servoing dynamic system is presented based on a pinhole camera. Without loss of generality, this paper considers points as the object features to perform the visual servoing task, being the approach extensible to any other kind of features.

Denote $[X_c, Y_c, Z_c]$ be the axes of the camera frame C attached at the center of the camera O_c, as depicted in Fig. 2. The coordinates of the image frame I are given by $[u, v]^T$ with O_I denoting the center of the image. Noted that the Z_c axis of the camera frame is perpendicular to the image plane transversing O_I. In this way, given a set of n fixed 3D points $P_i = [x_i, y_i, z_i]^T$, $i = 1, \ldots, n$ expressed in the camera frame, the corresponding 2D image feature $s_i = [u_i, v_i]^T$, $i = 1, \ldots, n$ are given as follows [5]:

$$
s_i = \begin{bmatrix} u_i \\ v_i \end{bmatrix} = \frac{\lambda}{z_i} \begin{bmatrix} x_i \\ y_i \end{bmatrix} \tag{1}
$$

where λ is the focal length of the camera. Thus, the effect of the camera motion of the feature coordinates at the image plane is given by:

$$
\dot{s}_i = L_i(s_i, z_i)V_c, \quad i = 1, \ldots, n \tag{2}
$$

where

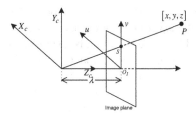

Fig. 2. The geometric model of a pinhole camera

$$
L_i(s_i, z_i) = \begin{bmatrix} -\dfrac{\lambda}{z_i} & 0 & \dfrac{\lambda u_i}{z_i} & u_i v_i & -\left(1+u_i^2\right) & v_i \\[3mm] 0 & -\dfrac{\lambda}{z_i} & \dfrac{\lambda v_i}{z_i} & \left(1+v_i^2\right) & -u_i v_i & -u_i \end{bmatrix}
$$

is the interaction matrix, and $V_c = (v_c, \omega_c)$ denotes the spatial velocity of the camera, with v_c the instantaneous linear velocity of the origin of the camera frame and ω_c the instantaneous angular velocity of the camera frame.

Let $W_b = [X_b, \theta_b]^T = [x_a, y_b, z_b, \alpha_b, \beta_b, \gamma_b] \in \Re^{6 \times 1}$ denotes the pose of the end-effector in the robot base frame (we choose Euler angles to represent the orientation of the end-effector), and a vector $s = [u_1, v_1, \ldots, u_n, v_n] \in \Re^{2n \times 1}$ which contains n image points. Then, the relation between velocity of $\dot{s}$ and $\dot{W}_b$ is given by

$$
\dot{s} = J_x \dot{W}_b, \tag{3}
$$

where $J_x = [L_1, \ldots, L_n] \; .T \in \Re^{2n \times 6}$ is known as the image Jacobian, and $T \in \Re^{6 \times 6}$ is the mapping to transform velocities expressed in the camera frame to the robot base frame.

By using a local model based on the interaction matrix, the system (3) can be represented in the state space form as [15]:

$$
\begin{aligned}
x_{k+1} &= Ax_k + B_k u_k + \upsilon_k, \quad \omega \; \square \; N(0, Q_k) \\
y_k &= Cx_k + \zeta_k, \qquad \qquad \zeta_k \; \square \; N(0, R_k)
\end{aligned} \tag{4}
$$

where $x = (s_1, \ldots, s_n)^T$ is the system state vector, $u = \dot{W}_b$ represents the control input, $N(\mu, \sum)$ refers to Gaussian noise with expected value μ and covariance $\sum$, and

$$
A = I_{(2 \times n, 2 \times n)}, B = J_{x(2 \times n, 6)} \tag{5}
$$

$$
C = I_{(2 \times n, 2 \times n)} \tag{6}
$$

Control actions are computed based on the standard IBVS algorithm [5].

When one of the fault types illustrated in Fig. 1 occurs, the visual servoing control system cannot determine the exact location displacements of the image feature points. That is, the actual extracted image features can be expressed as $\bar{s}(t) = s(t) + \Delta s(t)$, $s(t)$ being the true but unknown image feature output (i.e. the image feature displacements), while $\Delta s(t)$ be the vector of the fault signals acting on it.

3 Fault Detection

To detect the fault, parameters related to potential failures are estimated. Since the presented state space model in (4) contains noises, the Kalman filter is developed in this paper to handle the noises. The estimation of the state vector based on Kalman filter is updated by using the two group equations.

(1) *Time Update*: Compute a priori quantities for iteration $k+1$

$$\hat{x}_{k+1}^- = A_k \hat{x}_k + B_k u_k$$
$$P_{k+1}^- = A_k P_k A_k^T + Q \tag{7}$$

where P_k and P_{k+1}^- are the estimation error covariance matrix at time k and error covariance matrix at time $k+1$, respectively. Both provide a quantitative evaluation of the quality of this estimation and of this prediction.

(2) *Measurement Update*: Try to improve the prediction $\hat{x}_{k+1}$ based on the measurement available at time $k+1$

$$K_{k+1} = P_{k+1}^- C^T \left(C P_{k+1}^- C^T + R\right)^{-1}$$
$$\hat{x}_{k+1} = \hat{x}_{k+1}^- + K_{k+1}\left(y - C\hat{x}_{k+1}^-\right) \tag{8}$$
$$P_{k+1} = P_{k+1}^- - K_{k+1} C P_{k+1}^-$$

Thereafter, the output estimate is calculated as

$$\hat{y} = C\hat{x} \tag{9}$$

When a fault occurs, the state estimate is used to generate a failure decision. The purpose of such a decision is to provide a reliable indication of whether or not a failure exists. This is achieved by using the Kalman filter error e, which is defined as the different between the plant measurement and the Kalman output estimate, i.e.,

$$e = y - \hat{y} \tag{10}$$

The state condition of a feature point is represented by two state variables u and v. To easy in detection a failure of a feature point, we gather the information of two state variables u and v to the state variable s by introducing the following root mean square estimation error:

$$e_{s_i} = \sqrt{e_{u_i}^2 + e_{v_i}^2} \tag{11}$$

where e_{u_i} and e_{v_i} represent the Kalman estimation error of the state variables u_i and v_i of the feature point i, respectively, and e_{s_i} is now used to represent the Kalman estimation error of the feature point i.

In a normal operation, the observer state $\hat{x}$ approximates the true state x with a small error e_s due to noises, i.e., $e_{s_i} \leq T_i$. However, if a failure occurs, the value of e_{s_i} is likely to become a larger one. Hence, the decision is made if the decision function e_{s_i} overshoots the predetermined threshold T_i, i.e., $e_{s_i} > T_i$.

4   Simulation Results

In order to verify the developed fault diagnosis scheme for visual servoing, we have simulated an image-based visual servoing whose objective is to position a perspective free-flying camera with respect to the target, which is composed of four points. The size of the image is 1000×1000 pixel. The coordinate of the four points in the initial image and desired image are shown by dashed and dot-dashed lines in the image space in Fig. 3(a). To simulate real vision scenario, Gaussian white noise with a standard deviation $\sigma = 0.5$ pixel is added on the image data. For Kalman filter, using a trial-and-error procedure, the parameters are chosen as $Q = \text{diag}(10,10,10,10)$ and $R = \text{diag}(1,1,1,1)$.

4.1   Visual Servoing in Normal Operation

The tracking performance of the image-based visual servoing system is shown in Fig. 3. The result shows that the system provides a good tracking performance. The Kalman filter estimation errors, shown in Fig. 4, indicate that the errors of all the feature points are close to zero after few iterations when the system in normal operation.

We assume that the fault can only occur after the Kalman filter estimation error converges to zero (after iteration 5 in Fig. 4). Then, from analysis in Sect. 4.A, the thresholds are selected as shown in Fig. 4 (red line). A fault is detected and isolated whenever the residual exceeds its corresponding selected threshold.

4.2   Visual Servoing in Faulty Scenarios

In this case, we assumed that the feature points 3 and 4 are incorrectly extracted by visual servoing control system at the iteration 25. It means we change the displacement location of the features 3 and 4 by adding a fault term $\Delta s = [0,0,0,0,-100,-50,$ $100,50]$. This kind of failure, illustrated in Fig. 1(d), can be used to represent both the appearance (illustrated in Fig. 1(c)) and disappearance (illustrated in Fig. 1(b)) of the features 3 and 4 in general. For example, when the feature points 3 and 4 are occluded, the fault function can be described as $\Delta s = [0,0,0,0,-u_3,-v_3,-u_4,-v_4]$.

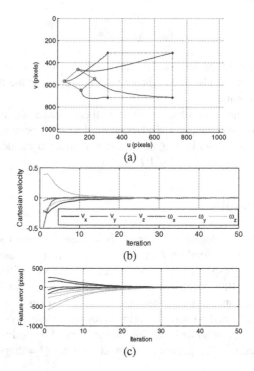

Fig. 3. Visual servoing tracking performance in normal operation. (a) Image space, (b) control inputs, (c) Image error.

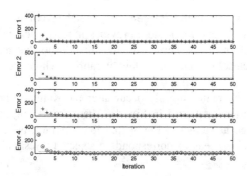

Fig. 4. Kalman filter estimation error when the system in normal operation and the selected threshold values.

The responses of the tracking performances of the visual servoing system under the effect of fault are shown in Fig. 5. The figure indicates that the fault decreases the tracking performance of the whole visual servoing system significantly. Due to the effect of the failure of the image feature point, the velocity control input is

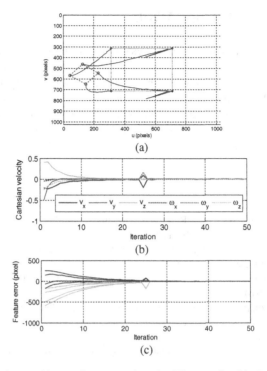

Fig. 5. Visual servoing tracking performance when the failures existed in feature points 3 and 4 at 25 iterations. (a) Image space, (b) control inputs, (c) Image error.

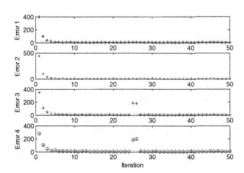

Fig. 6. Kalman filter state estimation error when the failures existed in the feature points 3 and 4.

discontinuous at the iteration 25 as shown in Fig. 5(b). The corresponding Kalman filter estimation errors are shown in Fig. 6, which indicate that the faults has been occurred in the feature points 3 and 4 at the iteration 25. Thus, the fault has been correctly detected and isolated.

5   Conclusions and Future Works

The failure scenarios of visual servoing system due to the incorrect feature extraction tasks are first reviewed in this paper. Then, a condition monitoring scheme based on the Kalman filter algorithm is proposed. The simulation results verify that the proposed scheme has a good capability to identify the faults. After a fault is detected and isolated, the controller of the visual servoing need to be reconfigured to decrease the effects of fault in the system. This task will be addressed in future work.

References

1. Wilson, W.J., Hulls, C.C.W., Bell, G.S.: Relative end-effector control using certesian position based visual servoing. IEEE Trans. Robot. Autom. **15**, 684–696 (1996)
2. Keshmiri, M., Xie, W.F., Mohebbi, A.: Augmented image-based visual servoing of a manipulator using acceleration command. IEEE Trans. Ind. Electron. **61**, 5444–5452 (2014)
3. Malis, E., Chaumette, F., Boudet, S.: 2(1/2)d visual servoing. IEEE Trans. Robot. Autom. **15**, 684–696 (1996)
4. Malis, E., Mezouar, Y., Rives, P.: Robustness of image-based visual servoing with a calibrated camera in the presence of uncertainties in the three-dimensional structure. IEEE Trans. Robot. **26**, 112–120 (2010)
5. Chaumette, F., Hutchinson, S.: Visual servoing control, part I: basic approaches. IEEE Robot. Autom. Mag. **13**, 82–90 (2006)
6. Sharifi, F.J., Deng, L., Wilson, W.J.: Comparison of basic visual servoing methods. IEEE Trans. Mechatron. **16**, 967–983 (2011)
7. Corke, P.I., Hutchinson, S.A.: A new portioned approach to image-based visual servoing control. IEEE Trans. Robot. Autom. **17**, 507–515 (2001)
8. Aracil, N.G., Malis, E., Santonja, R.A., Vidal, C.P.: Continuous visual servoing despite the changes of visibility in image features. IEEE Trans. Robot. **21**, 1214–1220 (2005)
9. Chesi, G., Hashimoto, K., Prattichizzo, D., Vicino, A.: Keeping features in the field of view in Eye-In-Hand visual servoing: A switch approach. IEEE Trans. Robot. **20**, 908–913 (2004)
10. Kazemi, M., Gupta, K.K., Mehrandezh, M.: Randomized kinodynamic planning for robust visual servoing. IEEE Trans. Robot. **29**, 1197–1211 (2013)
11. Folio, D., Cadenat, V.: A sensor based controller able to treat total image loss and to guarantee non-collision during a vision-based navigation task. In: IEEE/RSJ International Conference on Intelligent Robots and Systems (IROS), pp. 3052–3057 (2008)
12. Lippiello, V., Siciliano, B., Villani, L.: Position-based visual servoing in industrial multirobot cells using a hybrid camera configuration. IEEE Trans. Robot. **23**, 73–86 (2007)
13. Huang, S., Tan, K.K., Lee, T.H.: Fault diagnosis and Fault tolerant control in linear drives using the kalman filter. IEEE Trans. Ind. Electron. **59**, 4285–4292 (2012)
14. Auger, F., Hilairet, M., Guerrero, J.M., Monmasson, E., Kowalska, T.O., Katsura, S.: Industrial applications of the kalman filter: a review. IEEE Trans. Ind. Electron. **60**, 5458–5471 (2013)
15. Solanes, J.E., Armesto, L., Tornero, J., Girbes, V.: Improving image-based visual servoing with reference features filtering. In: IEEE International Conference on Robotics and Automation, pp. 3083–3088 (2013)

Author Index

Printed in the United States
By Bookmasters